Inorganic Mass Spectrometry

Inorganic Mass Spectrometry

Principles and Applications

JOHANNA SABINE BECKER

Research Centre Jülich, Germany

BICENTENNIAL 1807 WILEY 2007 BICENTENNIAL

John Wiley & Sons, Ltd

Telephone (+44) 1243 779777

Email (for orders and customer service enquiries): cs-books@wiley.co.uk
Visit our Home Page on www.wiley.com

Other Wiley Editorial Offices

John Wiley & Sons Inc., 111 River Street, Hoboken, NJ 07030, USA

Jossey-Bass, 989 Market Street, San Francisco, CA 94103-1741, USA

Wiley-VCH Verlag GmbH, Boschstr. 12, D-69469 Weinheim, Germany

John Wiley & Sons Australia Ltd, 42 McDougall Street, Milton, Queensland 4064, Australia

John Wiley & Sons (Asia) Pte Ltd, 2 Clementi Loop #02-01, Jin Xing Distripark, Singapore 129809

John Wiley & Sons Ltd, 6045 Freemont Blvd, Mississauga, Ontario L5R 4J3, Canada

Wiley also publishes its books in a variety of electronic formats. Some content that appears in print may not be available in electronic books.

Anniversary Logo Design: Richard J. Pacifico

Library of Congress Cataloging-in-Publication Data

Becker, Johanna Sabine.
Inorganic mass spectrometry : principles and applications / Johanna Sabine Becker.
p. cm.
Includes bibliographical references.
ISBN 978-0-470-01200-0 (cloth : acid-free paper)
1. Mass spectrometry. 2. Chemistry, Inorganic. I. Title.
QD96.M3B38 2007
543′.65—dc22 2007018922

British Library Cataloguing in Publication Data

A catalogue record for this book is available from the British Library

ISBN 978-0-470-01200-0

Typeset in 9/11pt Times by Integra Software Services Pvt. Ltd, Pondicherry, India
Printed and bound in Singapore by Markono Print Media Pte Ltd, Singapore
This book is printed on acid-free paper responsibly manufactured from sustainable forestry in which at least two trees are planted for each one used for paper production.

For my parents

Contents

Foreword

Inorganic mass spectrometry (inorganic MS) has had an important presence in chemical analysis for many years and much has been written, especially since the adoption of inductively coupled plasma mass spectrometry, ICPMS, on a routine basis. Hence the clear need for a comprehensive look at this field in a manner readily understandable with benefit for novice or expert. Dr J. Sabine Becker has risen to this challenge and produced an excellent book that gives a unique perspective through its broad coverage of the field by reviewing the important aspects of it, as we know it today. The reader will find excellent coverage of sample preparation, sample introduction, ion sources, ion separation, ion detection and an extensive illustration of applications. It is further strengthened by the exceptionally high number of references. The writing is clear and concise and even the novice can come to a working level in the subject without having to search external references, or having to wade through heavy mathematical arguments. A very important plus with this offering is that it is a one-author book and offers a coherence of thought and style that no multi-authored work can achieve. Thus as one progresses through Dr Becker's chapters, it is more like moving through a good textbook than a series of review chapters. Beyond major libraries, this offering is well suited for a graduate course in inorganic mass spectrometry.

The introductory chapter is brief but provides an ample introduction to mass spectrometry and leaves one comfortable as he/she moves on to the historical and instrumentation chapters that follow. A few of the basic equations are given as part of the review of basic concepts. In these few pages Dr Becker clearly introduces the concepts of atomic mass units relative to carbon, isotopes and isotope abundance. Figures 1.1 and 1.2 go hand in hand in providing the reader with the three major parts of a mass spectrometer (source, ion separation, detection) and show various alternatives for each of these. The subtle use of color in these and subsequent figures adds an attractive benefit for the reader.

Chapter 1 goes into some nicely done historical detail. This chapter begins in the late 19th century and traces those developments through the years. The numerous figures and pictures show well some of the early ingenuity of those exceptional scientists and Dr Becker is careful to note the importance assigned to mass spectrometry by the scientific community of that time; one measure being the numerous Nobel prizes awarded. The history as she presents it provides an excellent context for the future and current developments that this work highlights. Of particular interest to this reader are the early discussions about Wien, Aston and Thompson.

Chapter 2 highlights ion sources for inorganic mass spectrometry and given its popularity, it is no surprise that the inductively coupled plasma (ICP) receives the most attention. Noteworthy here are Figures 2.1 and 2.2, which provide informative summaries regarding atomization and ionization processes leading to the atomic ion followed by an excellent discussion of the ICP highlighting its operation and discussing solution nebulization for sample introduction. The reader will also find optimization discussion and figures – for example, fraction of ionization in the plasma. After the primary discussion of the ICP as an ion source, Figure 2.12 summarizes the various sample introduction modes associated with the ICP and does so in an easily visualized fashion. Dr Becker continues with discussions on selected sample introduction methods to the ICP including laser ablation, electrothermal vaporization, and hydride generation. This is followed by other sources; namely, spark ion source, laser ion source, glow discharge ion source, thermal surface ion source (TIMS – Figure 2.28 shows elements available), SIMS, and SNMS. Following these, softer ion sources for fragmentation and molecular ion formation; namely, electron ionization, MALDI and ESI are briefly discussed. Table 2.2 provides an excellent comparative summary of the various ion sources.

Chapter 3 continues Dr Becker's excellent presentation and has ion separation as its focus. The chapter begins with well developed text on electric and magnetic field sectors for ion separation and culminates with the combination of both sectors – double focusing ion separation as the most important of the three sector possibilities for high resolution inorganic mass spectrometry. However, the quadrupole mass filter is clearly the most widely used and often sold commercially available ion separation system and therefore, appears prominently in the text. This is followed by the time-of-flight (TOF), which has the ability to take spectra very fast and therefore should be highly amenable to laser ablation, chromatography and other sample introduction methods where a transient sample pulse passes through the ion separator. Two other ion separation methods are discussed in this chapter; ion-trap technology and FTICR (fast Fourier transform ion cyclotron resonance), although these are most widely used for molecular MS. The last part of the chapter reviews the concepts of resolution and abundance sensitivity – important to ICPMS users for any of the three commercially available inorganic MS types.

Chapter 4. Ion detectors are the last of the three major mass spectrometer components. With a particular focus on inorganic MS, a concise and well-written chapter results. Dr Becker begins with a discussion of the classical Faraday cup detector, which in its current configurations are still in use today. The secondary electron multiplier (SEM) is introduced and the explanation is readily understandable to the novice through a carefully designed Figure 4.2. It becomes apparent the electron multiplication is the best choice for low-level signals. She draws attention to the SEM in the dual ion mode for either the analog or pulse counting. For extended dynamic range the combination of Faraday cup and SEM is discussed with high ion current channeled to the Faraday detector and low ion currents to the SEM. The channel electron multiplier (CEM or channeltron) and microchannel plates are the next topics. The CEM is shown in Figure 4.5 in both the analog and pulse counting mode. As a modern alternative to the Faraday cup is the Daly detector (Fig. 4.6). This short chapter finishes with multi-collector devices for true simultaneous multi-element analyses.

When ion sources, ion separation and detection are combined, a mass spectrometer system results. While some choose to build their own, a wide variety of these are commercially available, offering as wide a combination of sources, separation means and detection. Chapter 5 focuses on these systems and is given the general title of "Instrumentation." For inorganic MS, Figure 5.1 and 5.2 illustrate some possible combinations by adding sample introduction to ion sources and separation (the detector is implied). Initially the discussion involves the quadrupole as the ion selector, since it is the most widely used with ICPMS. Collision/reaction cells are also brought into the discussion highlighting the additional power ICPMS has for detection through non-metal and some metal ions. Table 5.1 compares four different ICPMS quadrupole products. The plasma source ion trap is also discussed followed by ion separation using a double focusing sector field instrument with a single detector. Time-of-flight instrumentation is then carefully reviewed and Figure 5.10 illustrates a recent market introduction. Multi-detector (named multicollector) sector field instruments make up the next portion of the chapter and figures of merit for sector field instruments are compared in Table 5.2. Solution sample introduction via nebulization, hydride generation and flow injection are the next topics plus hyphenated methods are touched on here. As an important topic, Laser Ablation ICPMS is included in Chapter 5. At this point Dr Becker departs from ICPMS as the source to a number of other inorganic MS systems, of which Glow Discharge may be the most important today. Overall the discussion provides an excellent overview of the types of instrumentation useful for inorganic mass spectrometry.

Chapter 6 begins the discussion on analytical and practical considerations regarding the use of inorganic MS. Qualitative use of the techniques point out salient issues that need to be considered such as isotope pattern, mass determination and interference problems, which are well illustrated in the accompanying figures. Lengthy discussion is given to quantitative approaches and, as might be expected, internal standards, standard addition and isotope dilution get a lot of attention. Although not in method detail, the analyst can benefit highly from the discussion, which presents good "how to" ideas. Beyond solution sample introduction follows laser ablation for solids. An important focus here is the topic of calibration for quantitative determination by laser ablation. Next the reader will find an often ignored or minimized aspect of the analysis – sample preparation. The chapter is easy reading, but of high importance.

Gaseous and volatile compound analysis is covered in Chapter 7. Since all MS systems ultimately require gas phase ions, such a topic is of high importance in inorganic as well as molecular MS analysis. The chapter is short, yet does cover sample preparation for volatiles as well as some important applications. This chapter is very well referenced.

Chapter 8 addresses one of the most important features of inorganic MS – isotope ratio or isotope abundance measurements. As part of a brief introduction, Figure 8.1 provides a good departure for applications of isotope ratio measurements. The discussion provides insights to the capabilities of inorganic MS for isotope ratio measurements, mass discrimination effects, isotope ratio measurements by various types of MS. Table 8.4 gives some interesting applications of isotope ratio measurements. Multicollector ICPMS is discussed for precise isotope ratios and this section is followed by isotope dilution measurements. The chapter concludes with a focus on long-lived radionuclides, geochemistry and geochronology. As is Dr Becker's pattern, the reference section is very well done.

Chapter 9 is a major offering in fields of inorganic MS. At 205 pages, it details applications of inorganic MS for trace, ultratrace and surface analysis. Indeed, no forward can do justice to this exceptional review of applied inorganic MS, nor with space limitations adequately describe it. Notable however, is the richness of figures and tables drawing together from many resources comparative information along and across the various inorganic MS methods and their unique strengths and weaknesses. In fact, the reader will find 46 tables, 67 figures and 12 reference sections included in Chapter 9. The various MS methods and their numerous applications to a variety of sample types require an exhaustive search and clearly this was done for this book chapter (e.g. Ref. Section I alone has 159 references). Applications include materials science, environmental science and control, biology, bioengineering, medicine, foods, geology and geochemistry, cosmochemistry and space science, long lived radionuclides, forensic applications, and additional further applications. Dr Becker also devotes a section of this chapter to a very important aspect of inorganic MS, polyatomic ion formation. All-in-all this chapter provides either the novice or expert an excellent overview on up-to date applications of these powerful MS techniques.

Chapter 10 briefly addresses future trends as Dr Becker sees them emerging. She looks to higher resolution mass spectrometers for better interference separation, better sensitivity and lower detection limits. With this enhanced analytical power she also predicts increased user friendliness, better isotope ratio measurements, smaller sample sizes, enhanced capability to handle transient signals and further advances in sample introduction.

Ultimately the reader will find this work an excellent addition to their library. As earlier indicated, the text reads well, is crisp and concise and provides excellent up-to-date coverage of inorganic mass spectrometry and the important ancillary topics.

Professor Joseph A. Caruso
2007 Rieveschl Distinguished Research Awardee
Director, Metallomics Center
Department of Chemistry
University of Cincinnati

Preface

For many decades mass spectrometry has occupied an outstanding position among analytical techniques due to its universality and it has covered wide fields of application in atomic physics (determination of exact masses of isotopes, half-lives of radionuclides or of isotope abundances), chemical reaction analysis in ion molecular chemistry, studies of the kinetics of chemical processes and determination of thermodynamic data. It should be noted that no other analytical method has had such a significant influence and huge impact in so many fields of modern science and technology as mass spectrometry. Starting with a historical overview of mass spectrometry, this book presents the fundamentals and instrumentation of the most important inorganic mass spectrometric techniques, describes a wide range of analytical methods and a multitude of applications in materials, biological, geological and environmental research including recent developments in combination with biomolecular mass spectrometry for the special field of proteomics – discussed as metallomics and phosphoproteomics – with fascinating applications in medicine, biology and the life sciences. Of all the inorganic mass spectrometric techniques, inductively coupled plasma mass spectrometry (ICP-MS) has a leading place resulting in an exponential proliferation of exciting research and publications. The field of ultratrace and isotope analysis itself has been undergoing continuous development especially in combination with off-line or on-line trace matrix separation. On-line techniques such as HPLC-(high performance liquid chromatography) or CE (capillary electrophoresis)-ICP-MS have been developed as hyphenated techniques in order to reduce isobaric interferences and matrix effects, to facilitate speciation analysis and to improve their limits of detection of elements.

In the first part of the book, fundamentals of inorganic mass spectrometry (basic principles and developments of ion sources, ion separation systems and ion detectors) and instrumental developments in ICP-MS, LA(laser ablation)-ICP-MS, GDMS (glow discharge mass spectrometry), SIMS (secondary ion mass spectrometry), TIMS (thermal ionization mass spectrometry) and other techniques are described. The second part focuses on a multitude of quite different applications in materials science, environmental science and technology, biology and medicine, bioengineering, food analysis, geology and geochemistry, radioanalytical or forensic applications and cluster research.

This book 'Inorganic Mass Spectrometry: Principles and Applications' will give an insight into the state-of-the-art in mass spectrometry in different challenging tasks and recent developments and will present representative coverage of many fields of different topics. It is concerned with the theoretical and experimental conditions of different types of mass spectrometers and gives an overview of new methodological developments and trends for analytical work and modern applications, especially in survey, trace, ultratrace, surface (micro- and nanolocal analysis, imaging and depth profiling) and isotope analysis.

In the text the recommendations made from the International Union of Pure and Applied Chemistry (IUPAC), the International Union of Pure and Applied Physics (IUPAP) in respect to definitions (terms), symbols, quantities together with the International System of Units (SI) in respect to symbols in physical chemistry are considered.[1]

This book is the result of more than 30 years intensive research, experience and routine work in nearly all the fields of inorganic mass spectrometry described here and provides scientists, researchers, engineers and graduate students in analytical chemistry with a basic knowledge of the many facets and recent trends and applications of this important field of mass spectrometry.

Johanna Sabine Becker

Reference

1. Mills, I., Cvitas, T., Homann, K. *et al., Quantities, Units and Symbols in Physical Chemistry (IUPAC: Green Book)*, Blackwell Science, Oxford (1998).

Acknowledgements

I was first introduced to mass spectrometry after I had completed my PhD in quantum chemistry in 1974 when I was working as a young scientist under the supervision of H. J. Dietze at the Institute of Stable Isotopes (German Academy of Sciences, Leipzig) studying the wide field of stability and instability of isotopes in nature. My first research in mass spectrometry concerned investigations of the double-β decay of ^{96}Zr to ^{96}Mo in Precambrian zircon samples from the Baltic Shield via a precise isotope analysis of molybdenum using thermal ionization mass spectrometry (TIMS). At that time, the precision of mass spectrometers with single ion collection was not sufficient to measure very small isotope variations of molybdenum.

During my research in Leipzig I developed and applied analytical methods for the precise isotope analysis and determination (using the isotope dilution technique) of concentrations of U, Rb, Sr, Nd and Sm for age dating of geological samples from the Ore Mountains in Germany by TIMS. In addition, multi-element analytical techniques for the determination of trace elements on solid samples (geological samples, high-purity materials, ceramics, biological materials) by spark source mass spectrometry (SSMS) and laser ionization mass spectrometry (LIMS) were developed and applied. Furthermore, the design and construction of new types of laser ion sources, basic investigation on the formation of molecular, cluster and atomic ions in laser and spark plasma, and the development of ultrasensitive mass spectrometry (within the framework of cooperation with the Nuclear Research Centre in Dubna, Russia) were steps leading to advances in inorganic mass spectrometry. Mass spectrometric research into laser ablation and plasma deposition processes was also pursued during our work on the laser-induced plasma deposition of thin films, e.g., BN and C layers, W–C multiple layers and thin superconducting films, first in the ultrahigh vacuum ion source of the mass spectrometer and later in the deposition chambers developed for the preparation of superconducting thin layers.

For most of my career I have worked in the Central Division of Analytical Chemistry at the Research Centre Jülich – since 1992 as Head of Mass Spectrometry. In this time I have been concerned with the development and application of inorganic mass spectrometric techniques (such as ICP-MS, LA-ICP-MS, SSMS, LIMS, GDMS, SIMS and SMNS) for trace, ultratrace, isotope and surface analysis in environmental, biological and materials research, in the life sciences, medicine, bioengineering and geological research. Other major research topics are the ultrasensitive determination of long-lived radionuclides in the environment and the development of advanced mass spectrometric techniques for micro- and nanolocal analysis in medicine and the life sciences, e.g., for the quantitative imaging of elements in thin sections of brain tissue (imaging mass spectrometry) and for the analysis of separated metal-containing proteins in 2D gels (metallomics and phosphoproteomics) in combination with biomolecular mass spectrometry. It would not have been possible to explore this huge area of work in inorganic mass spectrometry without fruitful cooperations with the departments at the Research Centre Jülich and also with national and international scientists.

First of all, I would like to express my deepest and warmest thanks to my teacher in the field of mass spectrometry, H. J. Dietze (former head of the Central Department of Analytical Chemistry, Research Centre Juelich). I am greatly indebted to him for stimulating my interest in mass spectrometry, for our very successful scientific cooperation in inorganic mass spectrometry over a period of more than 25 years, and also for his motivation and continuous support in encouraging me to write such a comprehensive book on inorganic mass spectrometry, for countless helpful and fruitful discussions, valuable advice and critical reviews.

I am grateful to all my students, postdoctoral fellows and associates from Germany, USA, France, Israel, Italy, Russia, Belarus, Ukraine, Poland, Egypt and India, who studied at my laboratory and contributed to the experimental development of analytical techniques in inorganic mass spectrometry during my scientific career at the Research Centre Jülich. I am indebted to many scientists worldwide for numerous fruitful discussions accelerating analytical progress in mass spectrometry in the last decade.

I would like to thank J. Carter-Sigglow (Jülich) for a critical revision of the language and J. Cossham as the Publishing Editor from John Wiley & Sons, Ltd for her support and fruitful cooperation.

Last, but not least, I wish to thank my family for their great support, especially my daughter, J. Susanne Becker, for joining me in a new and interesting cooperation on metal-, phosphorus- and selenium-containing proteins (metallo-, phospho- and selenoproteomics) using a combination of LA-ICP-MS and MALDI-MS.

Introduction to Mass Spectrometry

Mass spectrometry is one of the most important analytical techniques used today for the determination of element concentrations especially in the trace and ultratrace range, for surface and isotope analysis, and for the structural analysis of organic and bioorganic compounds, due to its very high sensitivity, low detection limits and the possibility of analyzing very small sample volumes.

Mass spectrometry is based on the physical properties of the atomic nucleus. The atomic nucleus of any chemical element consists of protons and neutrons. In an electrically neutral atom the number of positively charged protons in the nucleus equals the number of negatively charged electrons in the shells. The number of protons ($Z =$ atomic number) determines the chemical properties and the place of the element in the periodic table of the elements. The atomic number Z of a chemical element is given as a subscript preceding the elemental symbol (e.g., ${}_{1}H$, ${}_{6}C$, ${}_{17}Cl$, ${}_{26}Fe$ or ${}_{92}U$). Besides the protons, uncharged neutrons with nearly the same mass in comparison to the protons ($m_n = 1.67493 \times 10^{-27}$ kg versus $m_p = 1.67262 \times 10^{-27}$ kg) stabilize the positive atomic nucleus. In contrast to the mass of the protons and neutrons in the nucleus, the mass of the electrons is relatively small at $m_e = 9.10939 \times 10^{-31}$ kg.

The number of nucleons is equal to the sum of the number of protons ($Z =$ atomic number) and number of neutrons (N) in the nucleus and is defined as the mass number ($A =$ nucleon number)

$$A = Z + N \tag{1}$$

The mass number A of an isotope is given as a superscript preceding the elemental symbol (e.g., ${}^{1}H$, ${}^{12}C$, ${}^{35}Cl$, ${}^{56}Fe$ or ${}^{238}U$).

Isotopes of a chemical element are nuclides with the same number of protons (Z) but a different number of neutrons (N) in the atomic nucleus. Isotopes of a chemical element (e.g., ${}^{1}H$ and ${}^{2}H$ of hydrogen; ${}^{35}Cl$ and ${}^{37}Cl$ of chlorine or ${}^{54}Fe$, ${}^{56}Fe$, ${}^{57}Fe$ and ${}^{58}Fe$ of iron, respectively) have the same number of protons (Z) and possess the same chemical properties, but differ in the number of neutrons (N) and thus in the mass number (A). With increasing Z, the number of neutrons in a stable atomic nucleus is higher than the number of protons. For mono-isotopic elements,

Inorganic Mass Spectrometry: Principles and Applications J. S. Becker

which possess only one stable isotope with an abundance of 100 %, the mass of the isotope is than equal to the atomic weight. Twenty chemical elements in the periodic table are mono-isotopic. Twenty-one chemical elements have two stable isotopes (di-isotopic elements), respectively. Seven relatively light elements consist of three stable isotopes. For the majority of chemical elements more than three stable isotopes (poly-isotopic elements) exist. The isotopic composition of elements is demonstrated in the mass spectrum, which is the 2D representation of measured ion intensities in dependence of the mass-to-charge (m/z) ratio ($z = n \cdot$ e – charge of ion). Each chemical element is characterized by its isotopic pattern in a given mass range (see Section 6.1.1.), which means the number of isotopes at certain mass numbers of a specific element exist and the isotope abundances. For example, rubidium has two isotopes at mass numbers 85 and 87 with the isotope abundances of 72.17 % and 27.83 %. This knowledge is the basis of a qualitative identification of element. If an identification of an element is doubtful (e.g., due to possible isobaric interferences) an exact mass determination of isotopes in the mass spectrum measured at higher mass resolution is required. Therefore in the past, one of the most important tasks of mass spectrometry was the determination of isotopic abundances (the isotopic composition of elements) in nature and later the exact measurement of the atomic mass (mass of atom). The unit of atomic mass u is an abbreviation of 'mass unit' (sometimes amu – atomic mass unit – is used in the literature). The atomic weight of an element (A_r(E) (mean) relative atomic mass) is then the weighted average of all naturally occurring atomic masses of isotopes of this element (E).

By means of mass spectrometry, the mass of atoms and molecules, via mass-separated charged atomic or polyatomic ions, can be determined by measuring the mass-to-charge ratio (m/z), whereby the mass of an atom or a molecule is not measured in g or kg, but in a multiple of the atomic mass constant m_u (atomic mass unit). The atomic mass unit m_u is defined as one-twelfth the mass of a neutral ^{12}C atom, m_a (^{12}C), in its ground state:

$$m_u = 1\ u = m_a(^{12}C)/12 \tag{2}$$

In the older literature before 1961, the mass unit of the oxygen atom ^{16}O (MU) was proposed with:

$$1 MU = m_a(^{16}O)/16 \tag{3}$$

$$\text{whereby} \quad 1\ m_u = 1\ u = 1.000317917\ MU. \tag{4}$$

The atomic mass unit (m_u) is also called the dalton (Da) – in honour of John Dalton. In response to the increase in the use of the name 'dalton' for the unified atomic mass unit among chemists, it was suggested by IUPAC that the unified atomic mass unit (u) be renamed the 'dalton' (Da). The definition of the unit would remain unchanged as one-twelfth the mass of a neutral ^{12}C atom in its ground state. The International Union of Pure and Applied Chemistry (IUPAC) proposed that both units, u and Da, should be allowed in official use.

For the calculation of the mass of an atom, molecule or ion in kg the following equation (Equation 5) for the atomic mass constant m_u can be applied:

$$m_u = 1\ u = 1\ D = 1.6605402 \cdot 10^{-27}\ kg \tag{5}$$

An advantage of the definition of mass unit is that the mass of an atom or molecule can be characterized by a full number – the mass number A, which is equal to the number of nucleons (see Equation 1). The mass of an atom or atomic mass (m_a) can be calculated approximately by the following equation (Equation 6):

$$m_a \approx A \cdot m_u \tag{6}$$

In addition, the elementary charge (e) – charge of the electron or proton – measured in Coulomb (C) is defined as:

$$e = 1.60217733 \cdot 10^{-19} C \tag{7}$$

In mass spectrometry, the mass-to-charge ratio (m/z) is mostly measured as a dimensionless number. Sometimes the unit known as the Thomson [Th], in honour of J. J. Thomson, is applied.

From nuclear physics it is known that the mass of a nucleus is always less than the sum of the masses of its components, the protons and neutrons. This phenomenon – called the mass defect (Δm) – seems to be in conflict with the law of conservation of mass. The mass defect Δm can be calculated by comparing the atomic weight of the nucleus m_k with the sum of the masses of the protons m_p and neutrons m_n:

$$\Delta m = (Z \cdot m_p + N \cdot m_n) - m_k \tag{8}$$

where Z = number of protons and N = number of neutrons.

The mass defect (mass excess) or of a nucleus is equivalent to the binding energy of the nucleons in the nucleus and corresponds to:

$$\Delta m = \Delta E/c^2 \tag{9}$$

according to Einstein's relativity theory, whereby c is the speed of light in a vacuum with $c = 2.99792 \times 10^8\,m\,s^{-1}$. For example, for ^{4}He the mass defect Δm is 0.0304 u. Because 1u = 931.494 MeV c^{-2}, the bond energy of the ^{4}He nucleus is 28.3 MeV. The average bond energy of each of the four nucleons in the nucleus helium is about 7 MeV.

In contrast to isotopes of elements which contain the same number of protons, but a different number of neutrons, isobars possess the same number of nucleons (mass number A) but different numbers of protons (which means they differ in atomic numbers and therefore they are different chemical elements), e.g., ^{40}Ca and ^{40}Ar; ^{58}Fe and ^{58}Ni; ^{92}Zr and ^{92}Mo; ^{142}Ce and ^{142}Nd; ^{204}Hg and ^{204}Pb and others.

The isotopic composition of the elements (including the exact atomic mass and the abundance of the isotopes), the atomic weights of elements, definitions and abbreviations are summarized in Appendix II.

To study the isotopic pattern (isotopic abundances in dependence of m/z ratio) which is different for the chemical elements and to determine the chemical composition of any type of sample (gases, liquids or solids) mass spectrometry is the method of choice. All types of mass spectrometric systems for analysis of inorganic and organic compounds use the same basic principle. A general setup of a mass spectrometer and the principle of operation for the qualitative and quantitative analysis of inorganic or organic compounds are given in Figure 1.1 and Figure 1.2, respectively. The sample to be analyzed (gas, liquid or solid material) is inserted into the ion source. Liquids are introduced, e.g., by nebulization of the solution, and solids by evaporation e.g., by laser ablation, electrothermal evaporization or by using Knudsen effusion. The solid sample is often directly inserted into the ion source (e.g., as electrodes in spark source mass spectrometry (SSMS), as a target in laser ionization mass spectrometry (LIMS) or as cathode material in glow discharge mass spectrometry (GDMS)). In the ion source operating at high vacuum conditions, at low pressure or at atmospheric pressure, the sample material is vaporized, atomized and ionized, whereby mostly positively singly charged ions are used for analytical purposes. The positively charged ions are extracted and accelerated from their original place of generation in the ion source on the entrance slit S of the mass separation system (mass analyzer). Whereas the ion source normally lies at a positive potential, the entrance slit has a potential of zero (see Figure 1.2). After extraction of ions from the ion source to the entrance slit S of the mass spectrometer, the ions then have the potential energy eV, where V is the acceleration voltage and e is the electrical charge of a single charged ion (with $z = 1$), which is the same as the elementary charge of an electron (see Equation 7). This potential energy of the ions is transformed completely into kinetic energy ($\frac{1}{2}\,mv^2$, where $v =$ velocity of ions) after passing through the entrance slit S. If the initial energy of the ions received upon formation is negligibly small in comparison to the accelerating energy of the ions by extraction to the entrance slit, then the following equation is valid due to the principle of the conservation of energy:

$$\mathrm{eV} = \tfrac{1}{2}\,mv^2 \tag{10}$$

This fundamental equation explains that the velocity of heavier ions (v_1 of ions with mass m_1) is lower than of lighter ions (v_2 of ions with mass m_2, with $m_1 > m_2$). Equation (10) is used directly in time resolved measurements, for example in time-of-flight mass spectrometers (ToF-MS). The charged ions of the extracted and accelerated ion beam are separated by their mass-to-charge ratio, m/z, in the mass analyzer. Mass-separated ion beams are subsequently recorded by an ion detection system either as a function of time or simultaneously. Mass spectrometers are utilized for the determination of absolute masses of isotopes, atomic weights, relative abundance of isotopes and for quite different applications in survey, trace, ultratrace and surface analysis as discussed in Chapters 8 and 9.

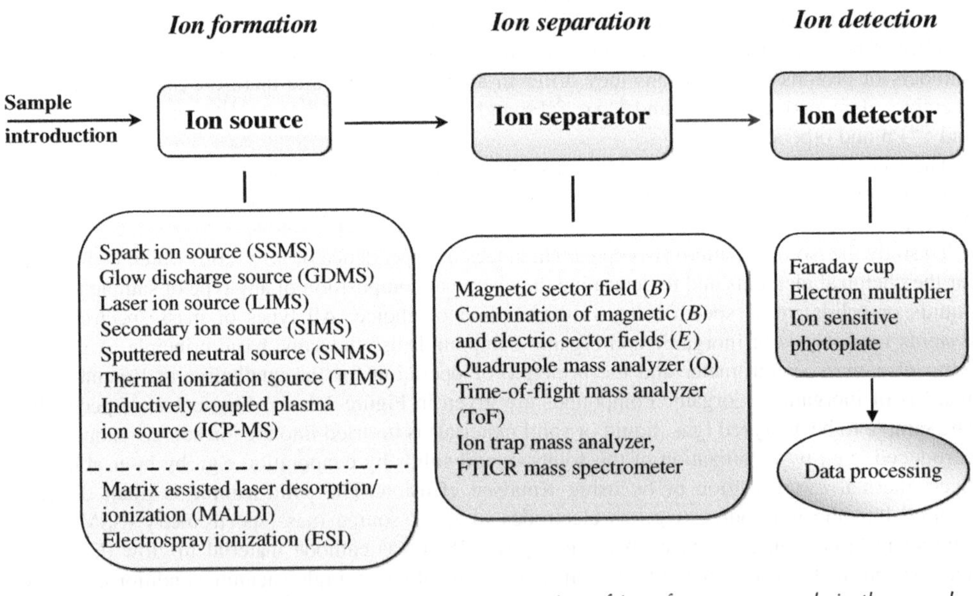

Figure 1.1 Basic diagram of mass spectrometry: generation of ions from compounds in the sample, introduction into the ion source, separation of these ions by their mass-to-charge ratio in the mass separator and detection of ions in the ion detector.

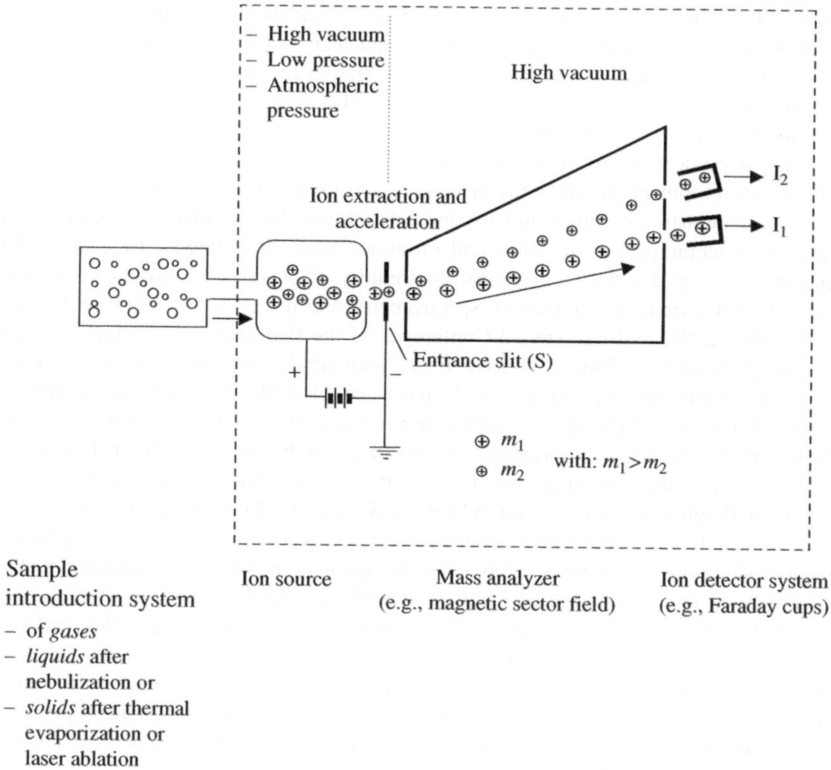

Figure 1.2 *Principle of the operation of a mass spectrometer including sample introduction system, ion source, mass separator (e.g., a magnetic sector field) and ion detector system (e.g., double ion collectors for simultaneous measurements of two separated ion beams).*

Inorganic mass spectrometric techniques, such as ICP-MS, laser ablation ICP-MS (LA-ICP-MS), glow discharge mass spectrometry (GDMS), secondary ion mass spectrometry (SIMS) and accelerator mass spectrometry (AMS), have been established in the last few decades as the most important and sensitive analytical techniques for elemental, surface and isotope analysis at the trace and ultratrace concentration level and for the analysis of species. Furthermore, special mass spectrometric techniques – secondary ion mass spectrometry (SIMS) and sputtered neutral mass spectrometry (SNMS) – have achieved significance as reliable surface analytical techniques with a spatial resolution in the μm and sub-μm range and depth resolution in the nm range, respectively. The capabilities and applications of different inorganic mass spectrometric methods have to be judged with respect to their advantages and drawbacks, also in comparison to non-mass-spectrometric analytical techniques.

In addition, one of the main features of mass spectrometry is, and this is the major advantage in comparison to other atomic and molecular non-mass spectrometric techniques, that it offers the possibility of determining isotope ratios and abundances of isotopes with high precision and accuracy in all types of samples (in solid, liquid and gaseous materials as well). Isotope ratio measurements have applied increasingly for stable isotopes in nature, especially for investigating fine isotope variation in nature or age dating, for process analysis and process control, for determining isotope ratios of radiogenic elements in the nuclear industry, quality assurance of fuel material, for reprocessing plants, nuclear material accounting and radioactive waste control, and for tracer experiments using stable isotopes or long-lived radionuclides in biological or medical studies. On the basis of known isotopic compositions, major, minor, trace and ultratrace element concentrations can be determined in any materials.

Accordingly, developments in all fields of mass spectrometry have been prompted in the past by requests from industry, research and routine laboratories for sensitive and accurate multi-element analytical techniques at the trace and ultratrace level. It is interesting to note that the most important meetings for the international community of plasma and mass spectroscopists – the annual Winter Conference on Plasma Spectrometry alternately in the USA or Europe and two in Asia (2005, 2006) and the Annual Conference of the Federation of Analytical Chemistry and Spectroscopy Societies, FACSS – have been dominated more and more by recent mass spectrometric developments, especially in inductively coupled plasma mass spectrometry (ICP-MS) involving hyphenated techniques resulting in powerful speciation analysis, and nowadays in combination with organic mass spectrometry in protein research, especially for analysis of metal-, selenium- and phosphorus-containing proteins. The International Mass Spectrometry Conference (IMSC, 2006 in Prague) and the annual American Society of Mass Spectrometry Conference (ASMS) have been focused more on organic and especially by bioorganic mass spectrometric developments and application. New developments and applications of surface analytical techniques have been presented at International and European SIMS Conferences and in scientific journals such as *Thin Solid Films, Applied Surface Science, Surface Coating Technology, Chemical Geology* and others. Simultaneously with the increasing installation of ICP-MS instruments worldwide, a rapid increase has also been observed in the number of analytical publications in the *Journal of Analytical Atomic Spectrometry, Analytical Chemistry, International Journal of Mass Spectrometry, Analytical and Bioanalytical Chemistry* (formerly *Fresenius' Journal of Analytical Chemistry), Atomic Spectrometry, Spectrochimica Acta B, Analyst* and others.

Trends in mass spectrometry focus on the improvement of instrumentation, of several techniques in order to minimize sample volume, to improve sensitivity and to reduce detection limits. This is combined with increasing the speed of several analyses, with automation of analytical procedures and subsequently reducing the price of analysis. A minimizing of sample volumes means a reduction of waste volume with the aim of developing 'green chemistry'. Furthermore, new analytical techniques involve a development of quantification procedures to improve the accuracy and precision of analytical data. Special attention in future will be given to the development of hyphenated mass spectrometric techniques for speciation analysis and of surface analytical techniques with improved lateral resolution in the nm scale range.

1

History of Mass Spectrometric Techniques

In its history, mass spectrometry has passed through fascinating scientific epochs of development and supplied important contributions to the world view in quite different fields of science (especially in the life sciences, environmental science and technology, material and nuclear sciences, geosciences, cosmochemistry,[1] planetary and forensic sciences) and in basic studies in chemistry and physics.

Mass spectrometry is more than 100 years old and has yielded basic results and profound insights for the development of atomic physics. The rapid development of nuclear physics, in particular, would be unthinkable without the application of mass spectrometric methods. Mass spectrometry has contributed to conclusive evidence for the hypothesis of the atomic structure of matter. So far mass spectrometry has supplied specific results on the structure of the nucleus of atoms. Nobel prizes have been awarded to a number of scientists (Thomson, Wien, Aston, Paul, Fenn and Tanaka) associated with the birth and development of mass spectrometry, or in which mass spectrometry has aided an important discovery (e.g., for the discovery of fullerenes by Curl, Kroto and Smalley).

Relevant preliminary work for the development of mass spectrography started with Goldstein's discovery of anode rays, which are positive charged gas-phase ions, in gas discharge in the mid-1980s. The experimental arrangement of Goldstein's glow discharge tube as the simplest possible ion source is shown in Figure 1.3 and was firstly published in the '*Sitzungsbericht der Königlich-Preußischen Akademie der Wissenschaften*' in 1886.[2] Goldstein's method for generating positively charged ions in a gas discharge was used in the early mass analysis measurements of Wien,[3] Thomson[4] and Aston.[5] In 1898 Wien analyzed the anode rays by magnetic deflection and found that these rays carried a positive charge. The schematic of a Wien velocity filter with *EB* configuration (E – electrical sector field and B – magnetic sector field) is shown in Figure 1.4. In this experimental arrangement, the deflection of positively charged ions in a magnetic field is compensating by a superimposed deflection in an electrical field. For ions which are transmitted through a Wien

Inorganic Mass Spectrometry: Principles and Applications J. S. Becker

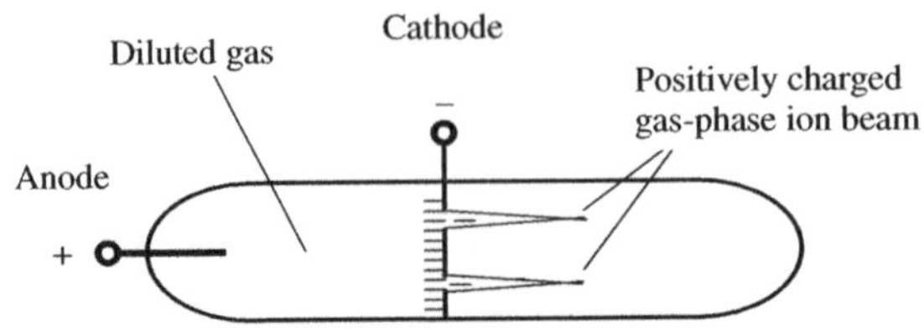

Figure 1.3 *Goldstein's glow discharge tube (1886) for generation of positively charged ions. (C. Brunnée, Int. J. Mass. Spectrom. Ion Proc. 76, 125 (1987). Reproduced by permission of Elsevier.)*

velocity filter it is valid that the electrical force (e E) that acts on an ion is equal the Lorentz force (e v B):

$$e\,E = e\,v\,B \tag{1.1}$$

The velocity of the ion (v) can be described then as the ratio of electrical field force (E) and magnetic field force (B):

$$v = E/B \tag{1.2}$$

$$\text{with} \quad v = \sqrt{2eV/m} \tag{1.3}$$

The Wien velocity filter possesses a high sensitivity and allows analysis over a mass range (using a high accelerating voltage V), but with increasing mass the mass resolution decreases.[3] For his discovery Wien was honoured with the Nobel Prize in 1911.

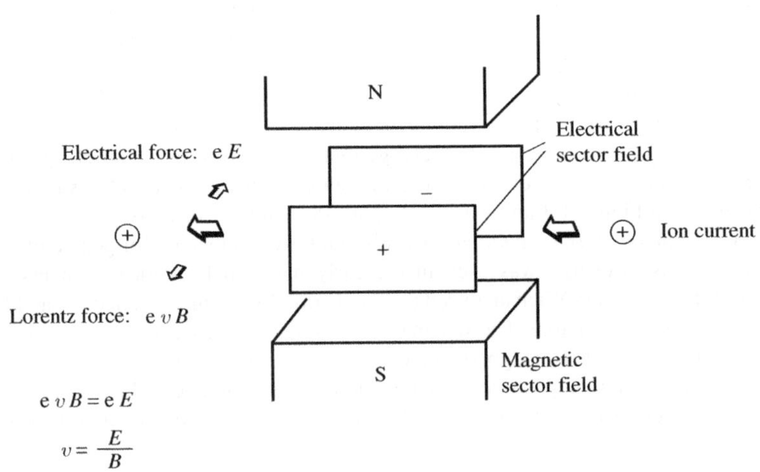

Figure 1.4 *Schematic of a Wien velocity filter with EB configuration: combination of electric (E) and magnetic (B) field (Wien,1898). (C. Brunnée, Int. J. Mass. Spectrom. Ion Proc. 76, 125 (1987). Reproduced by permission of Elsevier.)*

Just 11 years after Goldstein's first publication, Thomson discovered the electron and determined its mass-to-charge (m/z) ratio (in 1897). 'In recognition of the great merits of his theoretical and experimental investigations on the conduction of electricity by gases,' Thomson was honoured in 1906 with the Nobel prize 'for discovery of the electron and determination of its m/z ratio'. Thomson began his investigations of positive rays in 1907 and found that there was a 'different type of positive ions for each element'.[4]

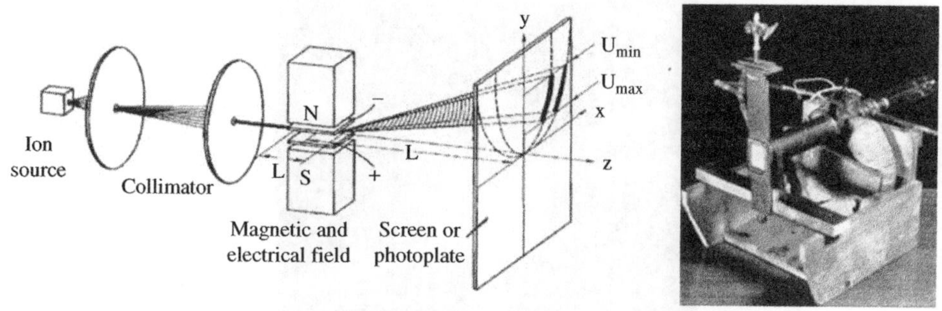

Figure 1.5 *Parabola mass spectrograph constructed by J. J. Thomson (1910) with a discharge tube as ion source, a superimposed electrical field and a magnetic field oriented parallel to it for ion separation, and a photoplate for ion detection. (H. Kienitz (ed.), Massenspektrometrie (1968), Verlag Chemie, Weinheim. Reproduced by permission of Wiley-VCH.)*

The real development of mass spectrometry goes back to 1910 when Thomson[6] found that neon consisted of two different isotopes with masses of 20 and 22 and so he discovered the isotopes of an element with the same chemical behaviour, but different isotope masses. This was the birth of the discovery of isotopes of elements by mass spectroscopy, whereby the history of isotopes is connected with the history of mass spectrometry. However, for mass spectrometry the most important instrumental development was Thomson's first parabola mass spectrograph (1907 and the improved version in 1910, see Figure 1.5). In this parabola mass spectrograph the positively charged ions (positive rays) were formed in a discharge tube. The direction of ion bundle was determined by two apertures. The ions were then deflected by a superimposed electrical field and a magnetic field oriented parallel to it, whereby ions with equal mass, but different energy (and velocity) were detected using an ion-sensitive photoplate in a parabola. In a later work in 1913, Thomson analyzed by means of his parabola mass spectrograph, the line at mass 22 together with the expected more intense line at mass 20 in the mass spectrum of neon. The first mass spectrum that demonstrated two stable isotopes of neon with masses of 20 and 22 and an isotope ratio of about 9 : 1, is shown in Figure 1.6. After the discovery of the neon isotopes by mass spectrography, Thomson tried to separate them by distillation. A few years later Aston[6] found a further very low-abundance neon isotope (of roughly 0.3 %) with a mass of 21 u. With the discovery of isotopes of elements by Thomson in 1910 together with investigations by Mosley[7] on the X-ray line spectra of elements, basic knowledge was obtained about a property of a chemical element which is determined by the atomic number, but not by the mass number. Thomson is accepted as the father of mass spectroscopy today (his fundamental discovery of isotopes by mass spectrometry and the development of first mass spectrograph was, however, not honoured by a Nobel Prize).

In several experiments Thomson also introduced molecules such as $COCl_2$ and hydrocarbons into the discharge ion source of the parabola mass spectrograph and observed the formation of

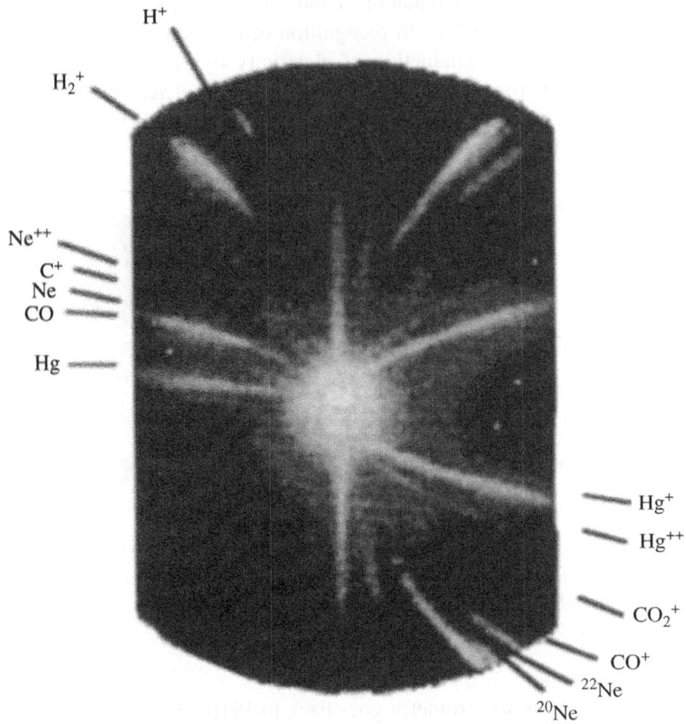

Figure 1.6 *Mass spectrum of neon with masses 20 and 22 u measured by J. J. Thomson (1913) using his parabola mass spectrograph is shown in Figure 1.5. (H. Kienitz (ed.), Massenspektrometrie (1968), Verlag Chemie, Weinheim. Reproduced by permission of Wiley-VCH.)*

many parabolas, revealing the formation of a variety of positively charged fragments (e.g., those corresponding to $COCl_2^+$, CO^+, C^+, O^+, etc.). Because the photoplate, as an ion-sensitive detector, has different sensitivities for the various ions, a quantitation of the relative intensities of the ions was not possible at that time. Especially for mixtures of hydrogen and oxygen, varying in composition over wide ranges, there was little difference in the intensities of the H_2^+ and O_2^+ ions on the photographic plate. Thomson replaced the photographic plate by a Wilson tilted electroscope and Faraday cylinder behind a parabolic slit. By changing the magnetic field and taking 10 s measurements, Thomson demonstrated the first mass spectrum as a plot of the ion current as a function of the mass-to-charge (m/z) ratio. Using the electrical detection system, he was able to achieve more accurate proportions of hydrogen and oxygen content in a given mixture compared to his measurements by photographic ion detection.

Aston, with experience in chemistry as well as physics, began working at the Cavendish Laboratory in 1909 on separating isotopes using mass spectrography. He started to assist Thomson in his study on positive rays. Together they improved on Thomson's original apparatus.

A quite different type of mass spectrometer – the first 180° magnetic sector field mass spectrometer (see Figure 1.7), with directional focusing of ions for isotope analysis, was constructed by Dempster, independently of other instrumental developments in mass spectrometry, in 1918.

Dempster also introduced a monoenergetic source – the electron bombardment ion source – into mass spectrometry[8] and analyzed ions of constant energy, which were extracted from the ion source (A) and were accelerated through a slit into the magnetic sector field. The positive single charged ions were separated in the 180° magnetic sector field (B) according to their mass and could be directed by changing their accelerating voltage or the power of the magnetic sector field on the exit slit. Ions starting with divergence in their direction from the entrance slit of the magnetic sector field were focused on the exit slit. This means that this semicircular uniform magnetic sector field arrangement was the first instrument with directional focusing. The separated ions were collected in a Faraday cup (C) and subsequently measured electrometrically using an electrometer (D). Dempster[8] measured curves for the ion current by gradually changing the magnetic field strength. For selected isotopes maximum ion intensities were observed, whereby the height of a selected maximum was proportional to the abundance of the isotope. With this apparatus (see Figure 1.7) Dempster discovered, in 1920, the isotopes of magnesium and later lithium, potassium and zinc.

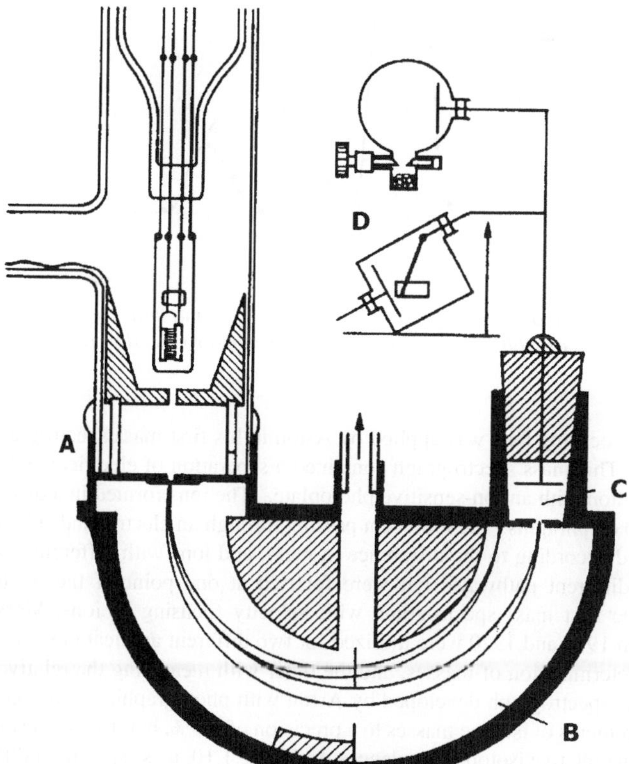

Figure 1.7 *Mass spectrometer from A.J. Dempster (1918). A – ion source; B – electromagnet; C – Faraday cup; D – electrometer. (H. Kienitz (ed.), Massenspektrometrie (1968), Verlag Chemie, Weinheim. Reproduced by permission of Wiley-VCH.)*

The typical mass spectrum of magnesium with three stable isotopes of masses 24 u, 25 u and 26 u using this magnetic sector field mass spectrometer with an electrometer for ion detection is shown in Figure 1.8. Dempster's mass spectrometer was of a simpler design that Aston's mass spectrograph, but could not be used for precise mass measurements. On the other hand, this magnetic sector field mass spectrometer was better suited than Aston's mass spectrograph for measuring the relative abundances of the ionic species and was used for initial investigations of electron impact processes in gases.

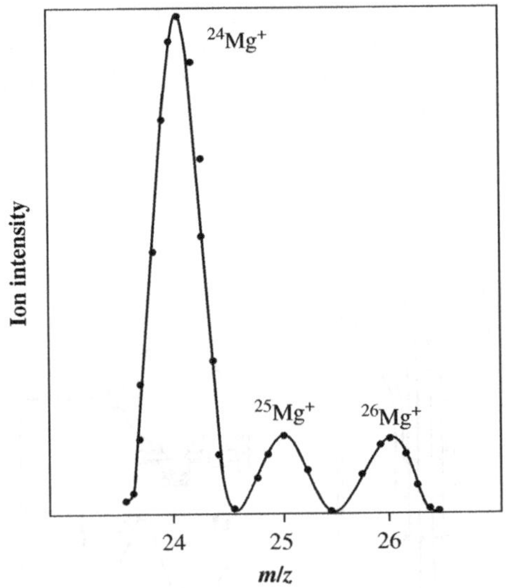

Figure 1.8 *Mass spectrum of magnesium with masses 24u, 25u and 26u (Dempster 1921). (H. Kienitz (ed.), Massenspektrometrie(1968), Verlag Chemie, Weinheim. Reproduced by permission of Wiley-VCH.)*

Another way of focusing ions was applied by Aston in his first mass spectrograph developed in 1919 (Figure 1.9). This mass spectrograph contained a separation of electrical and magnetic fields and used ion detection with an ion-sensitive photoplate.[9] The ions formed in a discharge tube were directed using two diaphragms. The ion beam passing through an electric and subsequent magnetic field was separated according to the difference in mass, and ions with different velocities (kinetic energy) flew on different pathways, but were focused at one point of the photoplate. Aston's instrument was the first mass spectrograph with velocity focusing of ions. Mass spectrographs developed between 1915 and 1920 were ultilized for two different applications: one was concerned with the precise determination of masses, and the other with measuring the relative abundances of isotopes. The mass spectrograph developed by Aston with photographic ion detection was applied mainly for measurements of isotope masses to a precision of 0.1 %, but it was not suited for accurate determination of the relative isotope abundances. In Figure 1.10, mass spectra of different inorganic and organic gases (BF_3, SiF_4, CH_3Br, SO_2 and AsH_3), recorded by Aston in 1920, are presented. Apart from the detection of isotopes of matrix elements (e.g., both boron isotopes at masses 10 u and 11 u or monoisotopic fluorine at mass 19 u), also several impurity elements were detected, for example, C, N, and O at masses 12 u, 14 u and 16 u in the mass spectrum of BF_3, or Hg in the mass spectrum of SO_2 and AsH_3 as contaminant from mercury diffusion pumps applied in the instrument. Furthermore, fragment ions were detected by studying mixtures of light hydrocarbons. Since this time $^{12}C^{16}O^+$ and $^{12}C^{16}O_2{}^+$ ions at masses 28 u and 44 u have been used as reference points for the interpretation of mass spectra and for the calibration of mass scales at light masses. It should be noted that the term 'mass spectrum' was introduced by F. W. Aston in 1920. At this

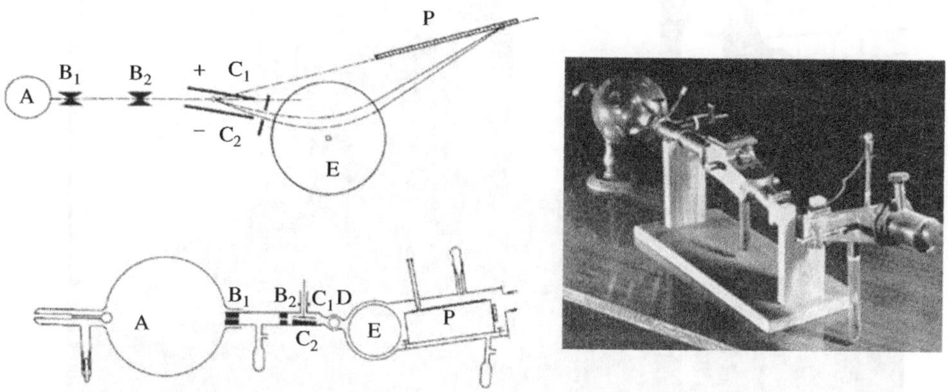

Figure 1.9 *Mass spectrograph from F. W. Aston (1919/20). A – ion source (discharge tube); B_1, B_2 – slits; C_1, C_2 – plates of condensator; D – valve; E – magnet; P – photoplate for ion detection. (H. Kienitz (ed.), Massenspektrometrie (1968), Verlag Chemie, Weinheim. Reproduced by permission of Wiley-VCH.)*

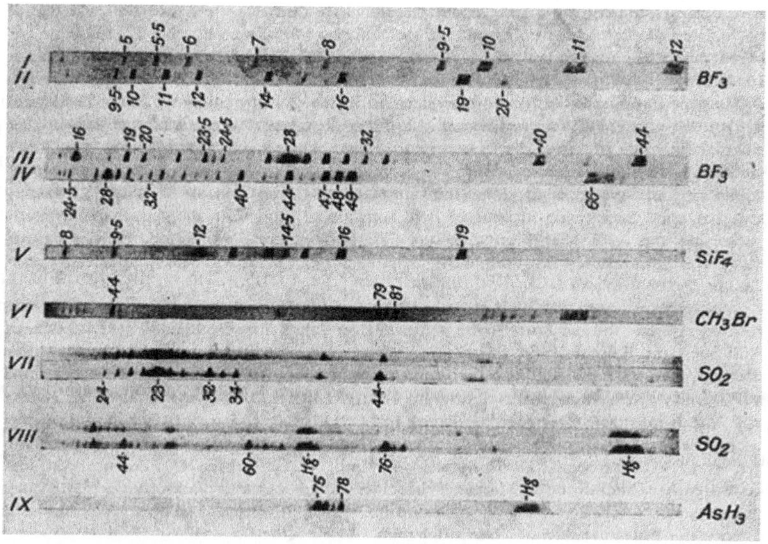

Figure 1.10 *Mass spectra of different gases measured by F. W. Aston in 1920. (H. Kienitz (ed.), Massenspektrometrie (1968), Verlag Chemie, Weinheim. Reproduced by permission of Wiley-VCH.)*

Figure 1.11 *F. W. Aston with second mass spectrograph (1922).*

time mass spectrometers were capable of precise mass determinations and measurement of relative abundances of ions, but could also perform valuable electron impact studies.

In 1921, Aston began to build an improved mass spectrograph. Aston's second mass spectrograph (see Figure 1.11)[10] used curved plates to produce its electric field, so that deflected particles would always be midway between the plates. Aston built the batteries for these plates himself, and developed a large, heavy magnet, capable of producing a powerful magnetic field of 1.6 tesla.

At the beginning of mass spectrometry development, two significant discoveries were made. Firstly, the existence of stable isotopes, which is the basis of mass spectroscopy, was discovered by J. J. Thomson. Secondly, a significant discovery of mass spectrography – made by F. W. Aston – was that the masses of all isotopes are not simple multiples of a fundamental unit, but are characterized by its mass defect. The mass defect ($\Delta m = m_a - Am_u$) (m_a – atomic mass, m_u – atomic mass unit and A – mass number of the isotope) as a function of the atomic number (Z) is presented in Figure 1.12. In 1923 Aston measured mass defects of a multitude of isotopes.[11] By precise determination of the masses of isotopes using mass spectrometry, the experimental evidence for the mass defect of isotopes was obtained and subsequently Einstein's equivalence principle between mass and energy was confirmed. This mass – energy relation postulated by Einstein was first proved experimentally in 1932 by Bainbridge[12] by measuring the masses of the nuclei involved in the following nuclear reaction of hydrogen ^1H with lithium ^7Li:

$$^1H + ^7Li \rightarrow 2\,^4He + Q \qquad (1.4)$$

For this experiment Bainbridge used the high-precision mass spectrograph he had developed by combining a Wien filter with a 180° magnetic analyzer.[13]

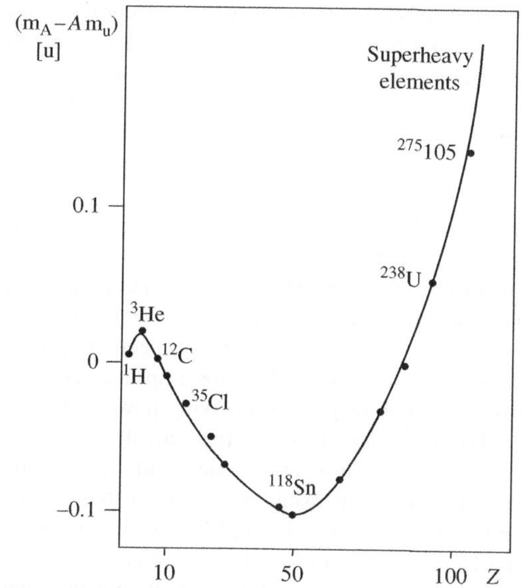

Figure 1.12 Mass defect as a function of atomic number.

With the 180° mass spectrometer with electrometric ion detection developed by Dempster serving for the precise electrometric measurements of isotopic abundances and the mass spectrograph created by Aston with an ion-sensitive photoplate for the exact mass determination of isotopes, a multitude of measurements were carried out for the characterization of naturally occuring stable isotopes of chemical elements. In Figure 1.13, the mass spectra of Cd and Nd measured by Dempster in 1935[14] using a photographic plate for ion detection are shown. Since the beginning of mass spectrography many other stable isotopes have been discovered and studied mass spectrometrically. By 1933 183 isotopes of 66 elements had been discovered. Of the 283 isotopes of 83 elements detected by 1948, 202 isotopes of 71 elements were found by Aston. Therefore Aston is accepted as the actual 'father of the isotopes'. For his discovery, by means of his mass spectrograph, of isotopes in a large number of non-radioactive elements and for his formulation of the 'whole-number rule' Aston was awarded a Nobel Prize in 1922.

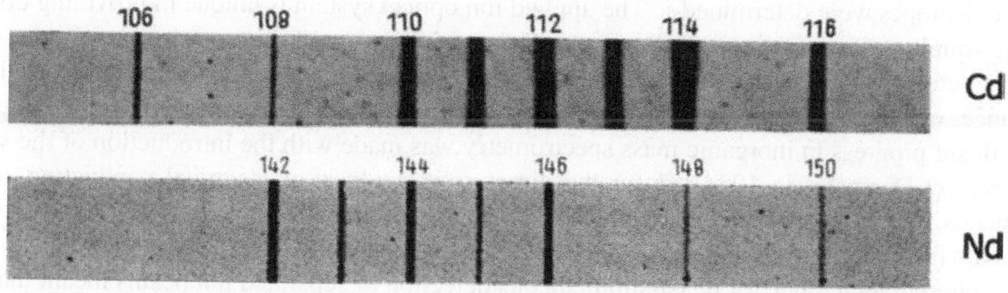

Figure 1.13 Mass spectra of Cd and Nd measured by Dempster (1935). (H. Kienitz (ed.), Massenspektrometrie (1968), Verlag Chemie, Weinheim. Reproduced by permission of Wiley-VCH.)

In 1937, Aston built his third mass spectrograph with an improved magnetic sector field. The slit system for focusing charged ion beams could now be adjusted externally and a more sensitive photographic plate from the Ilford Photographic Company was used. The mass resolution of Aston's final mass spectrograph was about 20 times greater than his first, and this instrument was about 100 times more accurate in the mass determination of separated ions.

Progress in mass spectrometric development was demonstrated, in particular, by a significant improvement in the resolving power of instruments (or mass resolution: $R = m/\Delta m$, if Δm is the smallest mass difference for which two mass lines with masses m and $m + \Delta m$ are resolved) and the sensitivity of the analytical method. Whereas in Thomson's parabolic mass spectrometer the resolving power was only 15, Dempster's 180° magnetic mass spectrometer and Aston's first mass spectrograph had a mass resolution of 100 and 130, respectively. The resolving power was significantly increased to 2000 in Aston's third mass spectrograph. Subsequently, theoretical and practical developments in ion optics enabled further improvements to be made to mass spectrometers and mass spectrographs with respect to resolving power, sensitivity and detection limits of elements. In the 1930s Barber[15] and Stevens[16] theoretically and experimentally studied the effect of the directional focusing of ions in magnetic fields after a deflection of 180°. Hughes and Rojanski[17] demonstrated the directional focusing properties of the electrical field of a cylindrical condenser if the deflection angle is 127°. Herzog and Mattauch[18] generalized the calculations for other deflection angles and Henneberg[19] studied the directional focusing properties of electrical and magnetic fields. The first suggestion of applying directional focusing and velocity focusing simultaneously was published by Dempster and Bartky[20] in 1929. Relevant work on the theory of the ion optical system of mass spectrometry was done in the middle of the thirties and focused on the development and construction of powerful double-focusing mass spectrographs, for example by Mattauch and Herzog[21] in 1934. The practical application of the construction of a double-focusing mass spectrometer was completed in 1936 on the basis of theoretical work on ion optics in instruments with significantly improved mass resolution for the precise determination of isotopes masses by Mattauch and Herzog.[21] This special combination of a 31.8° electrical field and a 90° magnetic field (Mattauch–Herzog geometry) allowed – by the double focusing of ion beams along a 25 cm long photoplate – an improvement of mass resolution of up to 6500. This type of mass spectrometer initiated a new era of so-called high-resolution instruments. Independently of Mattauch and Herzog, Dempster[14] in Chicago, and Bainbridge and Jordan[22] at Harvard University developed instruments with double-focusing properties for ion beams over a small mass range detected on the ion-sensitive photoplate. Whereas in Dempster's mass spectrograph[14] with a 90° deflection condenser and 180° magnetic sector field, a mass resolution of 1000 was observed, Bainbridge and Jordan's apparatus[22] with a 127° deflection condenser followed by a 60° magnetic field allowed a mass resolution of 10 000. In the following years Dempster, Bainbridge, Mattauch and coworkers used double-focusing mass spectrometry to measure a multitude of doublets from which the exact masses of isotopes were determined.[23] The applied ion optical system is unique in providing double focusing simultaneously for ions of all masses. The special feature of this system, namely that on a plane ion detector (e.g., photoplate) the separated ions beams are registered over a relatively large mass range, opened up the application field of multi-element analysis of inorganic solid samples.

Significant progress in inorganic mass spectrometry was made with the introduction of the spark ion source by Dempster in 1934[14,24] for the direct solid analysis of electrical conducting solids (e.g., metals, alloys or graphite). The application of a pulsed radio-frequency (rf) spark source combined with a double-focusing mass spectrograph using Mattauch–Herzog geometry and an ion-sensitive photographic plate for quasi-simultaneous detection of separated ion beams meant that the multi-element trace analysis of nearly all chemical elements even in non-conducting solid samples was possible with detection limits down to the 0.1 $\mu g\ g^{-1}$ concentration range.[25]

Double-focusing mass spectrometers with high mass resolution are rather bulky and expensive instruments so that the development of single magnetic sector field mass spectrometers was of significance in the following years.

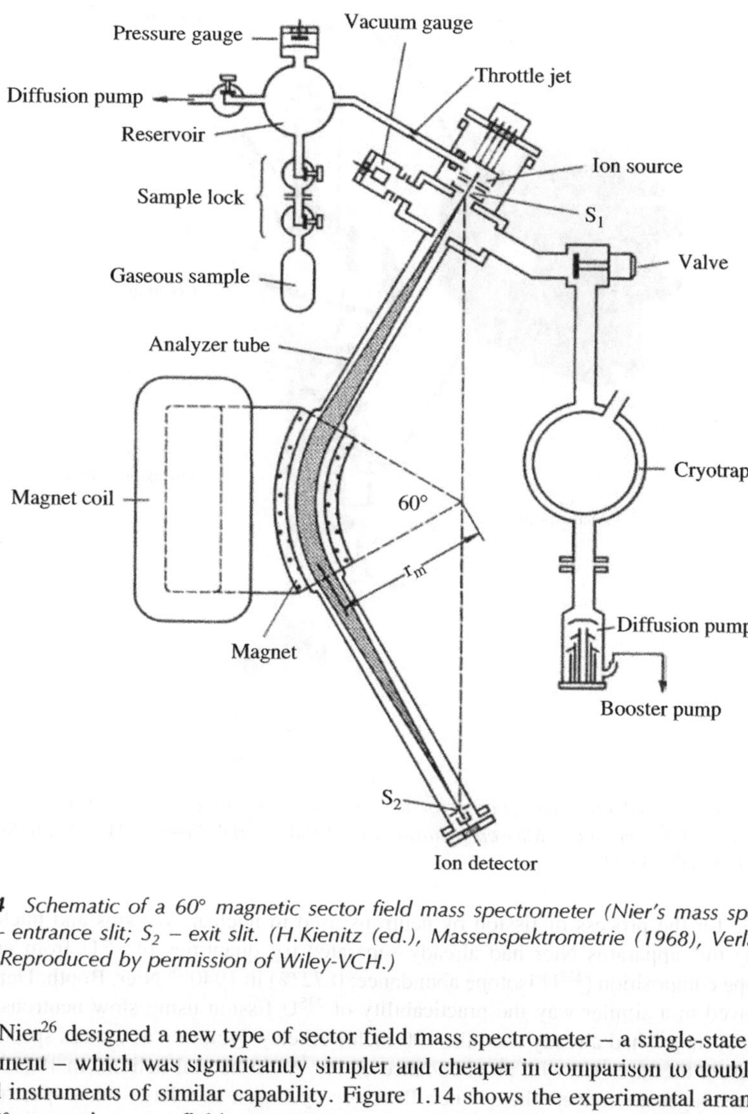

Figure 1.14 *Schematic of a 60° magnetic sector field mass spectrometer (Nier's mass spectrometer, 1940); S_1 – entrance slit; S_2 – exit slit. (H.Kienitz (ed.), Massenspektrometrie (1968), Verlag Chemie, Weinheim. Reproduced by permission of Wiley-VCH.)*

In 1940 Nier[26] designed a new type of sector field mass spectrometer – a single-state 60° sector field instrument – which was significantly simpler and cheaper in comparison to double-focusing sector field instruments of similar capability. Figure 1.14 shows the experimental arrangement of the Nier 60° magnetic sector field mass spectrometer. A schematic of a commercial 60° magnetic sector field mass spectrometer (type CH4, Atlas MAT) for analysis of gases is illustrated in Figure 1.15.

Nier's instrumentation, in comparison to the mass spectrometers existing at that time in 1934, was well developed and was extensively used for a quite different application during the Manhattan Project (1940–1945): for separation of uranium isotopes. It is well known that the ^{235}U isotope

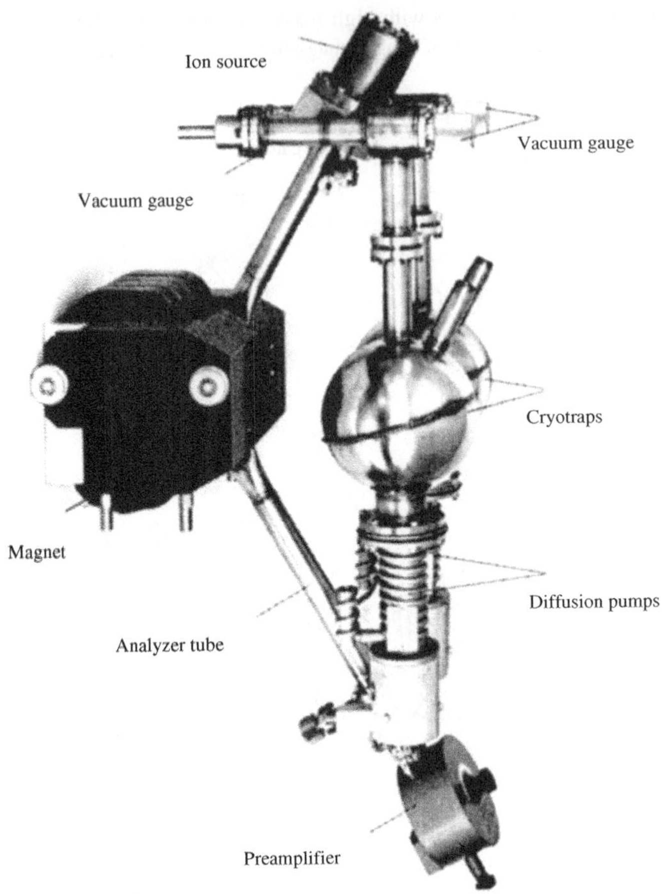

Figure 1.15 Commercial 60° magnetic sector field mass spectrometer (Type CH4, Atlas MAT) for analysis of gases. (H. Kienitz (ed.), Massenspektrometrie (1968), Verlag Chemie, Weinheim. Reproduced by permission of Wiley-VCH.)

is responsible for the process of fission by neutrons used in nuclear weapons and nuclear power plants. Using this apparatus Nier had already separated μg quantities of ^{235}U from uranium of natural isotope composition (^{235}U isotope abundance: 0.72 %) in 1940.[26] Nier, Booth, Dunning, and Grosse[27] proved in a similar way the practicability of ^{235}U fission using slow neutrons. In 1939, Alvarez and Cornog[28] had already used the 100 inch Berkeley cyclotron as a mass spectrograph in order to measure the rare ^{3}He with an abundance of 10^{-7} % in natural helium. In 1941, Lawrence and coworkers began to convert the 37 inch Berkeley cyclotron to a large mass spectrograph for isotope separation of uranium isotopes. By the end of 1941, using this mass spectrograph, uranium was separated, whereby the rare isotope ^{235}U with a natural abundance of 0.72 % was enriched to 3 %. An ion beam current of 5 μA was received at the ion collector for $^{235}U^+$, ten times as large as that used by Nier and coworkers. At the beginning of 1942, 0.2 mg of material isotopically enriched to 30 % ^{235}U was obtained. With some further design changes, ion source modifications, and different collectors, a new unit was installed in the 37 inch magnet in early 1942 (the mass separator was named 'Calutron': *Cal*fornia *U*niversity Cyclo*tron*). The successful application of the Calutron for the isotopic separation of uranium was a precondition for the construction of more efficient electromagnetic separation devices (e.g., at Clinton and Oak Ridge, Tennessee) for the separation of ^{235}U with high enrichment. Solomon[29] noted that in 1945, 3 kg of ^{235}U was produced per day as nuclear material by electromagnetic separation in the USA. After the end of the Second World War the Calutron mass separator was also used for the magnetic separation of other highly enriched stable isotopes which could be applied for a multitude of scientific applications, such as isotope dilution analysis or tracer experiments in biology, medicine, geology, chemistry, material research, forensics and in other fields.

An important application of mass spectrometry in geoscience was the evidence of the ß-decay of ^{87}Rb obtained by Mattauch in 1937 as the fundamental decay process in the well-known Rb–Sr age dating method in geochronology. In continuation of the work by Strassman and Walling[30] in 1938, Ahrens[31] published his results on the development and application of the Rb–Sr geochronological method in 1949, based on the β-decay of ^{87}Rb in ^{87}Sr ($t_{1/2} = 4.88 \times 10^9$ years). The Rb–Sr age dating method was improved by the careful chemical separation of analytes (Rb and Sr) from matrix elements by the ion exchange technique introduced in 1953 by Aldrich.[32] In addition, the development, in particular, of thermal ionization mass spectrometry (TIMS) opened up the possibility of more precise isotope ratio measurements for studying the fine variation of isotope abundances in Nature. Today it is well known that the nuclear decay of naturally occurring radionuclides (e.g., $^{87}Rb \rightarrow ^{87}Sr$; $^{187}Re \rightarrow ^{187}Os$; $^{176}Lu \rightarrow ^{176}Hf$; $^{232}Th \rightarrow ^{208}Pb$; $^{235}U \rightarrow ^{207}Pb$ and $^{40}K \rightarrow ^{40}Ca$) results in a small to large variation of the isotope composition of the elements. These isotopic variations in nature determined by mass spectrometry are the results of naturally occurring decay processes and supply information on the age of minerals and geological formation. With further instrumental developments of sector field mass spectrometers, a major new area of application was improved in the 1950s and 1960s – geochronology for age dating of geological samples and minerals.

Although the physicist Nier was interested mainly in isotope research, including earth and planetary science, he was also aware of the importance of mass spectrometry for chemical analysis of organic compounds, especially for the quick and reliable analysis of complex hydrocarbon mixtures in oil refineries. So a new area of mass spectrometry began in 1942 with the construction of the first commercial mass spectrometer for organic analysis. Financial support for the production of commercial mass spectrometers by the Consolidated Engineering Corporation was indeed provided by the oil industry (Atlantic Refining Corporation). The first mass spectrometers were of the 180° deflection type (Consolidated Engineering Corporation model 21–101 instruments). This was the beginning of the large-scale application of mass spectrometry in analytical chemistry. Due to a series of physical and technical improvements to the vacuum system, ion optics, ion source developments (e.g., of the 'Nier-type' electron impact ion source, which allowed more homogeneous energy distribution of the ions formed) and electronics, single sector field instruments become dominant.[33] These magnetic sector field instruments can reach a mass resolution of a few thousands and were used for separation of ion beams with relatively low ion energy spread (at eV range).

For separation of ion beams with a wide energetic spread (at keV range) double focusing mass spectrometers are necessary. A further technical development to the high-resolution double-focusing mass spectrometer with Mattauch–Herzog geometry was proposed by Ewald[34] in 1946. This instrumentation allowed mass spectrometric measurements with mass resolutions of 30 000–50 000. In the fifties, mass spectrometry was improved by additional theoretical work (investigation of imaging properties for different field arrangements and calculation of imaging errors) by Ewald and Liebl[35] and Hintenberger and König,[36] which resulted in the development of mass spectrometers with high mass resolution. Of special importance, also for inorganic mass spectrometry, and still applied today in modern mass spectrometers, is the double-focusing mass spectrometer constructed by Johnson and Nier in 1952,[37] which employed an electrostatic analyzer symmetrically and a magnetic analyzer asymmetrically and thus allowed a large divergence angle while maintaining second-order directional focusing. Quisenberry *et al.*[38] enlarged this instrument and consequently increased the mass resolution to 75 000, which is of significance for the high-precision mass determination of isotopes. The first commercial double-focusing mass spectrometer was introduced by Kratos in 1957.

Basic investigations in mass spectrometry[39,40] continue to influence instrumental developments. The first application in ion cyclotron resonance mass spectrometry (ICR-MS) was described by Sommer, Thomas and Hipple in 1949.[41] Free radicals were also mass spectrometrically studied by G. C. Eltenton.[42]

Improvements in solid mass spectrometry after the Second World War focused on the development of several types of ion source and mass analyzers. For example, the fundamentals of secondary ion mass spectrometry had already been described by the observation of secondary ions by Thomson in 1910:[5] 'I had the occasion . . . to investigate the secondary Canalstrahlen produced when primary Canalstrahlen strike against a metal plate.' Significantly later in 1949, the first instrumental arrangement for secondary ion mass spectrometric measurements was reported by Herzog and Viehböck.[43] The idea of applying secondary ions for surface analysis was developed in 1950 at the RCA Laboratories, Princeton, where Honig[44] presented results of early SIMS studies.

The evolution in mass spectrometry in the forties and fifties was determined by the requirements of chemical analysis in industry and focused on the development of dynamic mass spectrometers (so-called 'non-magnetic' mass spectrometers). By replacing the heavy magnet, for example, by a quadrupole analyzer, not only was a significant decrease in weight, size and price of the instruments achieved, but faster scanning of electrical fields and shorter analysis time was also possible. Finnigan introduced the first commercial quadrupole mass spectrometer in 1968. Although quadrupole mass spectrometers are not as accurate and precise as double-focusing sector field mass spectrometers they are fast, and today they are the most widely used mass analyzers in organic and inorganic mass spectrometry, for the latter especially by coupling them to an inductively coupled plasma source. Yost and Enke built the first triple quadrupole mass spectrometer as an important tandem mass spectrometric instrument for structure analyses in 1978,[45] and this design was introduced commercially on the analytical market by Finnigan and Sciex in 1982.

The instrumental development of a quadrupole ion trap, which can trap and analyze ions separated by their m/z ratio using a 3D quadrupole radio-frequency electric field, was initiated by Paul and coworkers in the fifties.[46–48] The advantage of Paul's quadrupole ion trap is the reduction of instrumental size at high mass resolution, high sensitivity and the possibility of analyzing compounds at high masses (~100 000 Da), for example proteins and large biomolecules. Furthermore, collision-induced dissociation using a collision gas in the ion trap can solve the interference problems by isobaric ions in mass spectrometry. Finnigan MAT introduced an ion trap system for GC-MS in 1983. 'For the invention of the separated oscillatory field method and its use in the hydrogen maser and other atomic clocks . . . (and) for the development of the ion trap technique' Paul was honoured with the Nobel Prize in 1989.

Another type of dynamic mass spectrometer is the time-of-flight (ToF) analyzer. In 1946, Stephens presented his concept of the linear time-of-flight mass spectrometer (ToF-MS) as the simplest mass separation technique at an American Physical Society meeting in Cambridge, MA.[49] Cameron and Eggers first published the design and showed mass spectra for linear ToF-MS in

1948.[50] At the same time, Goudsmit demonstrated improved mass determination by ToF-MS with an accuracy to the third decimal place for xenon isotopes.[51,52] Progress in the development of the linear ToF by the Bendix Corporation was made by Wiley and McLaren[53] during the fifties. Both designed the first commercial time-of-flight mass spectrometer – the Bendix instrument (Detroit, Michigan) – which was produced in 1958. In a time-of-flight analyzer ions acted upon by the same electrical impulse acquire velocities inversely proportional to their mass, so that they pass the same distance through a field-free space to the detector in a different time. Due to the advantages of the time-of-flight analyzer: unlimited mass range, high ion transmission (there are no slits) very high recording speed (10 000 mass spectra per second) and simplicity of flight tube, this type of dynamic mass spectrometer attained increasing popularity in the following decades. The direct coupling of gas chromatography (GC) and ToF-MS was achieved in the mid-fifties by McLafferty[54] and Gohlke.[55,56] GC-MS became one of the most important hyphenated routine methods and ushered in a new era in bioorganic analysis since biological and medical samples were now also increasingly analyzed with respect to, for example, hormones and steroids.

Time-of-Flight mass spectrometers in combination with a pulsed ion source were used in inorganic mass spectrometry (e.g., ToF-SIMS), but more frequently in organic mass spectrometry. Because of their high sensitivity and ability to detect all separated ion beams ToF analyzers have been used with liquid chromatographic systems for the identification of organic compounds of high molecular weight. In order to improve the mass resolution from several hundred to about 1000, an ion reflector – the so-called reflectron introduced by Mamyrin in 1974[57,58] – was added to the linear ToF. The reflectron compensates for possible differences in the flight times of ions to be analyzed resulting from their kinetic energies.

The development of laser ionization mass spectrometry was started by Honig and Woolston in 1963[59] with studies of laser beam solid surface interaction and ion formation processes. Due to the pulse character of laser-induced ions, ToF analyzers were coupled to laser ion sources in the seventies and produced commercially as LAMMA-500 and later LAMMA-1000 and 2000 (Leybold-Heraeus, Cologne, Germany).

In 1974, Comarisov and Marshall[60] developed Fourier transform ion cyclotron resonance mass spectrometry (FTICR-MS). This technique allows mass spectrometric measurements at ultrahigh mass resolution ($R = 100\,000–1000\,000$), which is higher than that of any other type of mass spectrometer and has the highest mass accuracy at attomole detection limits. FTICR-MS is applied today together with soft ionization techniques, such as nano ESI (electrospray ionization) or MALDI (matrix assisted laser/desorption ionization) sources.

Major progress in inorganic mass spectrometry was made possible by the development of inductively coupled plasma mass spectrometry in 1980 by Houk, Fassel and co-workers[61] and by Date and Gray[62] and especially with the commercial introduction of inductively coupled plasma mass spectrometry in 1983. Since that year, ICP-MS has experienced exponential growth and today plays the dominant role among the inorganic mass spectrometric techniques for trace, ultratrace and isotope analysis.[63,64] The coupling of a laser ablation system to ICP-MS was proposed by Gray[65] in 1985 and has been commercially available on the analytical market since 1990. Nowadays, LA-ICP-MS is the most versatile and sensitive solid mass spectrometric technique in direct trace, ultratrace and surface analysis and is extremely useful for isotope ratio measurements.

Two recently developed mass spectrometric techniques have had a major impact on the analysis of large biomolecules: matrix-assisted laser desorption/ionization mass spectrometry (MALDI-MS) and electrospray ionization mass spectrometry (ESI-MS). MALDI-MS was first introduced by Karas and Hillenkamp[66] and Tanaka *et al.*[67] in 1988 and has experienced an exponential development. It has become a widespread soft ionization technique for bioorganic samples, especially large biomolecules. Fenn and co-workers[68] first published the successful soft ionization technique for large synthetic polymers, and subsequently proteins, using ESI-MS in the late 1980s. An important feature of this technique was the transfer of analytes directly from solution to the gas phase ions with high charge states.[69] Fenn and Tanaka (together with Wüthrich) received the Nobel Prize for chemistry in 2002 in recognition of their contribution to the characterization of biomolecular macromolecules and to mass spectrometry and nuclear resonance spectroscopy (NMR). MALDI and ESI are soft ionization techniques which, combined with low-cost mass analyzers (quadrupole, ion trap or ToF mass spectrometers) or more expensive FTICR-MS, are routinely used today for the analysis of proteins, peptides, synthetic polymers, small oligonucleotides, carbohydrates and lipids. In addition, the multiple-charging phenomenon enables large macromolecules to be detected in mass spectrometers of modest mass-to-charge ratio range. Further solution-based ionization methods, like thermospray and other liquid inlet sources, enables molecules to be introduced directly into a mass spectrometer from chromatographic and electrophoretic separation systems.

Today there are a number of mass spectrometer companies producing a wide variety of different types of quadrupole-based, sector field and ToF instruments for quite different applications in the trace, ultratrace, isotope and surface analysis of inorganic materials and for the structural analysis of organic and bioorganic compounds. Mass spectrometry for the structural analysis of organic compounds including large biomolecules is described elsewhere.

Detailed accounts of the history of mass spectrometry and the detection of isotopes are given by Ewald and Hintenberger,[70] de Bievre,[10] Kienitz,[71] Brunneé,[72,73] Duckworth *et al.*,[74] de Laeter[1] and Budzikiewicz and Grigsby.[75]

References

1. De Laeter, J. R., *Applications of Inorganic Mass Spectrometry*, Wiley-Interscience Series on Mass Spectrometry, John Wiley & Sons, Inc., New York (2001).
2. Goldstein, E., *Berl. Ber.*, **39**, 691 (1886).
3. Wien, W., *Verl. Phys. Ges. Berlin*, **17**, 10 (1898).
4. Thomson, J. J., *Rays and Positive Electricity* Longmans Green and Co., London, (1913).
5. Aston, F. W., *Phil. Mag.*, **39**, 449 (1920).
6. Thomson, J. J., *Phil. Mag.*, **42**, 752 (1910).
7. Moseley, H. G. J., *Phil. Mag.*, **26**, 1024 (1913).
8. Dempster, A. J., *Phys. Rev.*, **11**, 315 (1918).
9. Aston, F. W., *Phil. Mag.*, **38**, 707 (1919).
10. De Bievre, P., *Advances in Mass Spectrometry, Proceedings of the 7th International Mass Spectrometry Conference held at Florence, 30 August to 3 September 1976*, ed. N. R. **7A**, Daly, Heyden & Son Ltd, 395 (1978).
11. Aston, F. W., 2nd edition, Edward Arnold and Co., London (1942).
12. Bainbridge, K. T., *Phys. Rev.*, **39**, 847 (1932); Phys. Rev., **39**, 1021 (1932).
13. Bainbridge, K. T., *Franklin Institute*, **214**, 509 (1933).
14. Dempster, A. J., *Proc. Phil. Soc.*, **75**, 755 (1935).
15. Barber, N. F., *Pr. Leeds Phil. Soc.*, **2**, 427 (1933).
16. Stephens, W. E., *Phys. Rev.*, **45**, 513 (1934).
17. Hughes, A. L. and Rojansji, V., *Phys. Rev.*, **34**, 284 (1929).
18. Herzog, R. and Mattauch, J., *Ann. Phys.* **V**, 358 (1934).
19. Henneberg, W. *Ann. Phys.*, **19**, 335 (1934).
20. Bartky, W., Dempster, A. J., *Phys. Rev.*, **33**, 1019 (1929).
21. Mattauch, J. and Herzog, R., *Z. Phys.*, **89**, 786 (1934).
22. Bainbridge, K. T. and Jordan, E. B., *Phys. Rev.*, **50**, 282 (1936).
23. Mattauch, J. and Flammersfeld, S., *Isotopenbericht*, Z. F. Naturforsch, Sonderheft (1949).

24. Dempster, A. J., *Rev. Sci. Instrum.*, **7**, 46 (1936).
25. Hannay, N. B. and Ahearn, A. J., *Anal. Chem.*, **26**, 1056 (1954).
26. Nier, A. O., *Rev. Sci. Instrum.*, **11**, 212 (1940).
27. Nier, A. O., Booth, E. T., Dunning, J. R. and Grosse, A. V., *Phys. Rev.*, **57**, 546 (1940).
28. Alvarez, L. W. and Cornog, R., *Phys. Rev.*, **56**, 379 (1939).
29. Solomon, A. K., *Warum Atomzertrümmerung*, Bertelsmann Verlag Gütersloh. (1948).
30. Strassman, F. and Walling, E., *Z. Anorg. Allgem. Chem.*, **236**, 78 (1938).
31. Ahrens, H., *Bull. Geol. Soc. Am.*, **60**, 217 (1949).
32. Aldrich, L. T., Doak, J. B. and Davis, G. I., *Am. J. Sci.*, **251**, 377 (1953).
33. Nier, A. O., *Rev. Mod. Phys.*, **18**, 398 (1947).
34. Ewald, H., *Z. Naturforschg.*, **1**, 131 (1946).
35. Ewald, H. and Liebl, H., *Z. Naturforschg.*, **10a**, 842 (1955).
36. Hintenberger, H. and Koenig, L. A., *Z. Naturforschg.*, **12a**, 140 (1957).
37. Johnson, E. G. and Nier, A. O., *Phys. Rev.*, **91**, 10 (1953).
38. Quisenberry, K. S., Scolman, T. T. and Nier, A. O., *Phys. Rev.*, **102**, 1071 (1956).
39. Hipple, J. A. and Stevenson, D. P., *Phys. Rev.*, **63**, 121 (1943).
40. Hipple, J. A. and Condon, E. U., *Phys. Rev.*, **69**, 347 (1946).
41. Sommer, H., Thomas, H. A. and Hipple, J. A., *Phys. Rev.*, **76**, 1877 (1949).
42. Eleton, G. C., *J. Chem. Phys.*, **15**, 455 (1947).
43. Herzog, R. and Viehböck, F. P., *Phys. Rev.*, **76**, 855L (1949).
44. Honig, R. E., *J. Appl. Phys.*, **29**, 549 (1958).
45. Yost, R. A. and Enke, C. G., *J. Am. Chem. Soc.*, **100**, 2274 (1978).
46. Paul, W. and Raether, M., *Z. Physik*, **140**, 262 (1955).
47. Paul, W., Reinhard, H. P. and von Zahn, U., *Z. Phys.*, **152**, 143 (1958).
48. Paul, W. and Steinwedel, H., *Z. Naturforsch.*, **8a**, 448 (1953).
49. Stephens, W., *Phys. Rev.*, **69**, 691 (1946).
50. Cameron, A. E. and Eggers, D. F., *Rev. Sci. Instrum.*, **19**, 605 (1948).
51. Goudsmit, A., *Phys. Rev.*, **74**, 622 (1948).
52. Goudsmit, A., *Phys. Rev.*, **84**, 824 (1951).
53. Wiley, W. L. and McLaren, I. H., *Rev. Sci. Instrum.*, **16**, 1150 (1957).
54. McLafferty, F. W., *Appl. Spectrosc.*, **11**, 148 (1957).
55. Gohlke, R. S., *Anal. Chem.*, **31**, 535 (1959).
56. Gohlke, R. S. and McLafferty, F. W., *Int. J. Mass Spectrom.*, **4**, 367 (1993).
57. Karatev, V. I., Mamyrin, B. A. and Smikk, D. V., *Sov. Phys.-Tech. Phys.*, **16**, 1177 (1972).
58. Mamyrin, B. A., *Int. J. Mass Spectrom.*, **131**, 1 (1994).
59. Honig, R. E. and Woolston, J. R., *Appl. Phys. Lett.*, **2**, 138 (1963).
60. Comisarov, M. B. and Marshall, A. G., *Chem. Phys. Lett.*, **25**, 282 (1974).
61. Houk, R. S., Fassel, V. A., Flesch, G. D., Svec, H. J., Gray, A. L. and Taylor, C. E., *Anal. Chem.*, **52**, 2283 (1980).
62. Date, A. R. and Gray, A. L., *Analyst*, **106**, 1225 (1981).
63. Jarvis, K. E., Gray, A. L. and Houk, R. S., *Handbook of Inductively Coupled Plasma Mass Spectrometry*, Blackie, Glasgow, (1992).
64. Montaser, A. (ed.), *Inductively Coupled Plasma Source Mass Spectrometry*, Wiley-VCH Inc., New York (1998).
65. Gray, A. L., *Analyst*, **110**, 551 (1985).
66. Karas, M. and Hillenkamp, F., *Anal. Chem.*, **60**, 2299 (1988).
67. Tanaka, K., Waki, H., Ido, Y., Akita, S., Yoshida, Y. and Yoshida, R., *Rapid Commun. Mass Spectrom.*, **2**, 151 (1988).
68. Fenn, J. B., Mann, M., Meng, C. K., Wong, S. F. and Whithouse, C. M., *Science*, **246**, 64 (1989).
69. Fenn, J. B., Mann, M., Meng, C. K. and Wong, S. F., *Mass Spectrom. Rev.*, **9**, 37 (1990).
70. Ewald, H., Hintenberger, H., *Methoden and Anwendungen der Massenspektroskopie* Verlag Chemie, Weinheim (1953).
71. Kienitz, H. (ed.), *Massenspektrometrie*, Verlag Chemie, Weinheim (1968).

72. Brunnée, C., *Int. J. Mass Spectrom. Ion Proc.*, **76**, 125 (1987).
73. Brunnée, C. and Voshage, H., *Massenspektrometrie*, Karl Thiemig, München 1 (1964).
74. Duckworth, H. E., Barber, R. C. and Venkatasubramanian, V. S., *Mass Spectroscopy*, 2nd edn, Cambridge University Press, Cambridge, London, New York, New Rochelle, Melbourne, Sydney (1986).
75. Budzikiewicz, H. and Grigsby, R. D., *Mass Spectrom. Rev.*, **25**, 146 (2006).

2

Ion Sources

The ion source is an essential component of all mass spectrometers where the ionization of a gaseous, liquid or solid sample takes place. In inorganic mass spectrometry, several ion sources, based on different evaporation and ionization processes, such as spark ion source, glow discharge ion source, laser ion source (non-resonant and resonant), secondary ion source, sputtered neutral ion source and inductively coupled plasma ion source, have been employed for a multitude of quite different application fields (see Chapter 9).

An important requirement for the mass spectrometric analysis of any sample material (solid, liquid or gaseous) is to produce a constant ion beam of sufficient intensity generated in an appropriate ion source from the sample components. The ions thus formed are then extracted and accelerated in the mass analyzer, separated according to their mass-to-charge ratio and subsequently detected by a sensitive ion detector. The first very simple ion source using glow discharge in a so-called channel ray tube was proposed in 1886 by Goldstein[1] (see Figure 1.3.), who discovered that anode rays consist of positively charged ions. Later, high-vacuum ion sources with electron beam ionization were designed and applied for the analysis of gases. In contrast to gases, the analysis of solid material is more complicated because the solid sample must be evaporated and atomized before ionization. Special ion sources for the analysis of solid samples, e.g., spark, laser and glow discharge ion sources, have been developed. Whereas spark and laser ion sources operate under ultrahigh vacuum conditions, the glow discharge ion source works at a low pressure of a rare gas (e.g., Ar). For the analysis of liquids, the inductively coupled argon plasma (operating at atmospheric pressure) is appropriate for ionizing of analytes in the nebulized solution due to an easy solution introduction system.

The principle processes in the ion source of a mass spectrometer are the evaporation of solid samples or desolvation and vaporization of liquids, atomization of gaseous compounds and ionization of atoms and molecules in order to generate ions which are analyzed mass spectrometrically as summarized in Figure 2.1. In all types of ion sources, during the ionization process singly and multiply charged atomic ions, as well as polyatomic or cluster ions, with quite different ion intensities are formed. In general, in inorganic mass spectrometry the singly charged atomic ions

Inorganic Mass Spectrometry: Principles and Applications J. S. Becker

of the analyte are used for analytical purposes. In the ion sources utilized in inorganic mass spectrometry both positively and negatively charged ions are formed, nevertheless most mass spectrometric techniques only use the positive singly charged atomic ions. In addition, positive singly charged polyatomic and cluster ions formed in ion sources result in disturbing interferences with the atomic ions of analytes in mass spectra (e.g., ArX^+ with X = O, N, C or Ar in ICP-MS and LA-ICP-MS). Several types of polyatomic ions occurring in mass spectra can be generated by ionizing initial or evaporated fragments of compounds directly or by the association of atomic ions with a residual gas atom, with an atom of the matrix elements or the plasma gas in the ion sources of mass spectrometers (see Section 9.11). In a comparison of different inorganic mass spectrometric techniques, the highest polyatomic or cluster ion formation rate was observed during sputter processes in secondary ion mass spectrometry (SIMS). It is interesting to note that caesium polyatomic ions (MCs^+ and MCs_2^+; M – metal or non-metal atom), formed during the bombardment of a solid surface with Cs^+ primary ions in SIMS, are used as analytical ions for surface analysis of samples in routine mode.

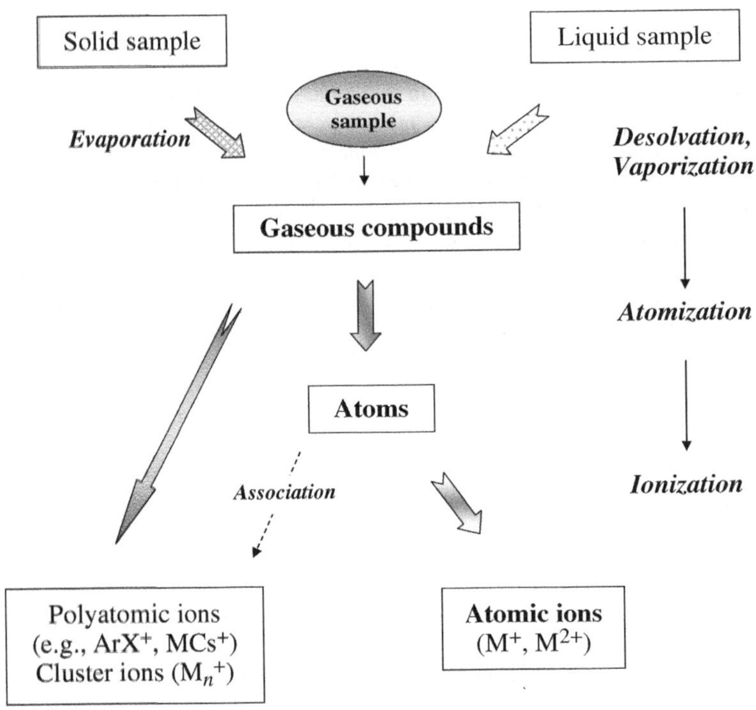

Figure 2.1 *Different processes in an ion source: evaporation of solid samples, desolvation of liquids, atomization of gaseous compounds and ionization of atoms.*

Different mass spectrometric techniques can be classified according to the evaporation and ionization methods applied. Evaporation of solid samples can be performed, for example, by thermal (e.g., on a hot tantalum filament or in a heated graphite furnace) or laser-induced evaporation, and by electron or ion bombardment. Electron ionizaton (EI), ionization during the sputtering process with a primary ion beam, resonant or non-resonant laser ionization or thermal surface ionization can be used as ionization techniques. The following different inorganic mass spectrometric methods can be classified as 'direct ionization' techniques with simultaneous evaporation (atomization) and ionization of evaporated components of samples as so-called 'post-ionization' techniques. In 'post-ionization' techniques, the processes of atomization by evaporation or sputtering of solid samples are separated in time and space from the processes of ionizing the atomic or polyatomic species. In the left part of Figure 2.2, several mass spectrometric techniques with quasi-simultaneous direct evaporation and ionization are summarized, such as inductively coupled plasma ionization, laser ionization, spark ionization and secondary ionization. Post-ionization techniques (see right part of Figure 2.2) are laser ablation combined with an inductively coupled plasma ion source, thermal surface ionization using two filaments (one for evaporation of the sample, the other for ionization of evaporated atoms), sputtered neutral ion source with postionization of sputtered neutrals (e.g., by electron beam, in a plasma gas or by non-resonant or resonant laser irradiation) or glow discharge ionization.

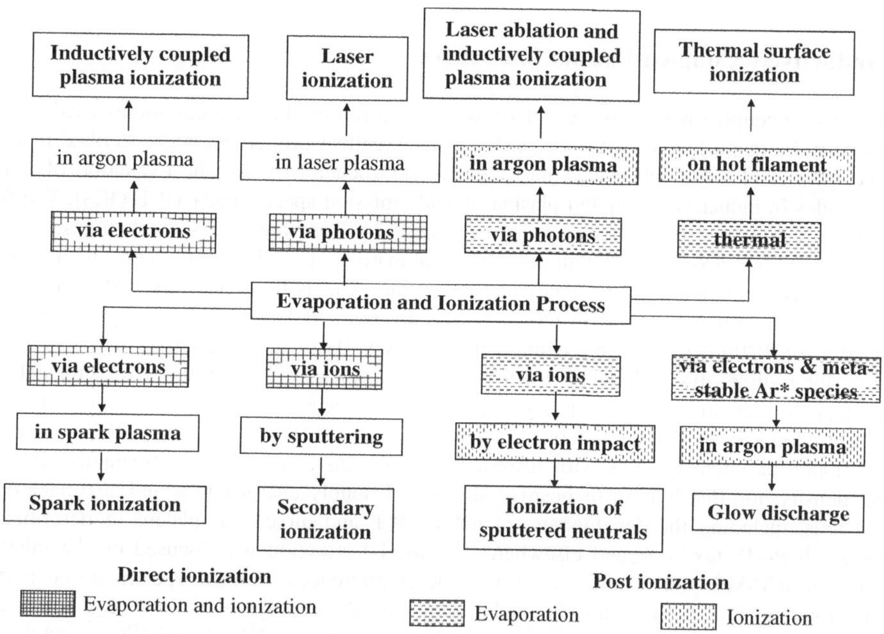

Figure 2.2 *Overview of different types of ions sources in inorganic mass spectrometry and their classification with respect to evaporation and ionization. (J. S. Becker and H. J. Dietze, Int. J. Mass Spectrom. Ion Proc.* ***197****, 1–35. (2000). Reproduced by permission of Elsevier.)*

The introduction of inductively coupled plasma (ICP) in inorganic mass spectrometry means that there is an effective ion source operating at atmospheric pressure. Whereas solid mass spectrometric techniques allow direct analysis of solid samples in ICP-MS, the determination of trace impurities or isotope ratios in solid samples is often carried out after digestion and dissolution of the material. For the determination of trace impurities and isotope ratios in liquids, an additional nebulization of solution before desolvation, atomization and ionization of analytes transported by argon in inductively coupled plasma (ICP), is generally applied. However, in addition to a multitude of advantages (high sensitivity with very low detection limits, easy quantification procedure, excellent routine capability) this technique also has several disadvantages, for example, besides the isobaric interference problems of atomic ions of the analyte and polyatomic ions (which can be often resolved) occasionally time-consuming sample preparation procedures for difficult to dissolve materials, especially if matrix separation is required, loss of volatile analytes, increasing danger of contamination, loss of information on lateral resolution of elements in solid samples after digestion and others can arise. Therefore the coupling of a laser ablation system to an ICP ion source is a useful alternative for a direct solid state mass spectrometric analysis compared to ICP-MS after digestion.

Due to the increasing significance of soft ionization techniques for the analysis of phosphorus-, metal- and metalloid-containing biomolecules, matrix-assisted laser-induced desorption/ionization (MALDI) and electrospray ion sources (ESI) will be discussed briefly at the end of this chapter.

2.1 Inductively Coupled Plasma Ion Source

The inductively coupled plasma source (ICP), which is currently the most commonly used plasma ion source in inorganic mass spectrometry due to its excellent properties, was described in 1964 by Greenfield et al.[2] and introduced into atomic spectrochemistry for the excitation of atoms and molecules in inductively coupled plasma optical emission spectrometry (ICP-OES). The first commercial ICP-OES instrument was constructed in 1974. Compared to established gaseous and solid state mass spectrometric techniques, the combination of an ICP ion source with a mass spectrometer is a relatively young analytical technique. The ICP source was first successfully coupled to a quadrupole mass analyzer by Gray,[3] Houk[4] and Date.[5] Just nine years after the introduction of ICP-OES, the first commercial ICP-MS (Elan 250, Perkin Elmer Sciex) was produced for routine applications. ICP-MS is the most versatile trace, ultratrace elemental and isotope analysis technique available today, and ICP mass spectra are much less complex than ICP optical emission spectra. The development of ICP ion sources was combined with fundamental studies of plasma characteristics with respect to plasma gas and electron temperature, electron number density, ion distribution of positive singly and doubly charged ions and also negatively charged ions, including the development of helium ICP and mixed gas plasma as described in Montaser's ICP-MS textbook and elsewhere.[7–9] Especial attention was focused on the inherent drawback of ICP-MS, the occurrence of isobaric interferences in mass spectra (often named 'isobaric overlaps') of atomic ions of analytes by disturbing polyatomic ions (e.g, $^{56}Fe^+$ and $^{40}Ar^{16}O^+$) or by isobaric singly charged (e.g., $^{40}Ca^+$ and $^{40}Ar^+$ or $^{92}Mo^+$ and $^{92}Zr^+$) and doubly charged atomic ions at the same nominal mass (e.g., $^{28}Si^+$ and $^{56}Fe^{2+}$), respectively. This problem was partly solved by the development of double-focusing sector field ICP-MS in the late 1980s and by the introduction of collision/reaction cell technology in the late 1990s in quadrupole-based ICP-MS (Platform) and magnetic sector field ICP-MS (Isoprobe). Both collision cell ICP mass spectrometers were produced for many years by Micromass Manchester, UK. Furthermore, the formation rate of polyatomic ions can be minimized, for instance by using special sample introduction systems with a desolvator, by hydride generation, by laser ablation and electrothermal evaporation, by operating under cold plasma conditions, by the application of hyphenated techniques (using on-line capillary electrophoresis or liquid chromatography such as HPLC and others), and by the development of special analyte/matrix separation procedures. In this way matrix effects are also minimized.

An inductively coupled plasma source at atmospheric pressure combined with an optical emission or mass spectrometer to form ICP-OES or ICP-MS, respectively, offers the analyst multi-element analysis after a very simple sample preparation and easy solution introduction, mostly for the analysis of liquid samples. Furthermore, a high sample throughput rate of one to a few samples per minute at low detection limits for multi-element determination is possible making ICP-OES advantagoeus in stoichiometric analysis of any materials due to high precision and good accuracy. Today ICP-MS has been widely established as the premier technique in trace element and isotope analysis. In comparison to ICP-OES, ICP-MS offers a larger dynamic range, significantly lower detection limits, the possibility of measuring isotope ratios (including the isotope dilution technique as the definitive method in trace and ultratrace analysis) and thus also broader application fields. The plasma in the atmospheric pressure ICP ion source is formed in a nearly chemically inert environment in a stream of a noble gas. A schematic of an inductively coupled plasma ion source including the quartz plasma torch and induction load coil together with sampler and skimmer as part of the interface region of an ICP mass spectrometer is shown in Figure 2.3.

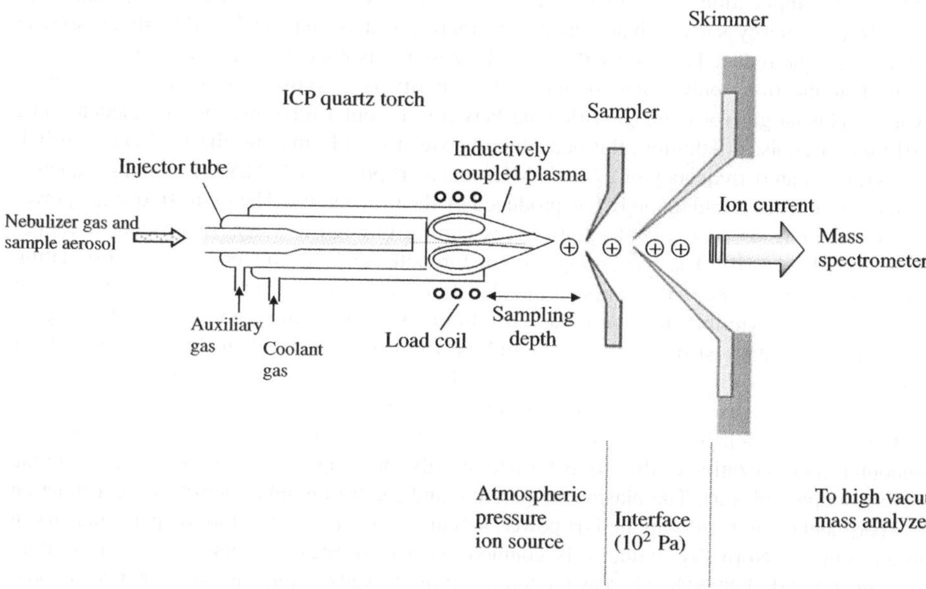

Figure 2.3 *Schematic of inductively coupled plasma source.*

From the sample solution to be analyzed, small droplets are formed by the nebulization of the solution using an appropriate concentric or cross-flow pneumatic nebulizer/spray chamber system. Quite different solution introduction systems have been created for the appropriate generation of an aerosol from a liquid sample and for separation of large size droplets. Such an arrangement provides an efficiency of the analyte introduction in the plasma of 1–3 % only.[6] The rest (97 % to 99 %) goes down in the drain.[7] Beside the conventional Meinhard nebulizer, together with cooled or non-cooled Scott spray chamber or conical spray chamber, several types of micronebulizers together with cyclonic spray chambers are employed for routine measurements in ICP-MS laboratories. The solvent evaporated from each droplet forms a particle which is vaporized into atoms and molecules and is then ionized in the hot inductively coupled plasma (ICP). The fundamental processes of aerosol generation, studies of droplet size distribution and the transport to the ICP with respect to experimental parameters applied are described in the literature.[7–11]

The ICP is sustained in a quartz torch consisting of three concentric tubes with different diameters. Argon, typically used in ICP-MS, flows through the concentric tubes of the ICP torch as shown in Figure 2.3. For several applications of ICP-MS a demountable torch with an inner injector tube of Al_2O_3 is recommended. The sample aerosol is transported along the axis of the torch via the cooler, central channel tube at a nebulizer argon gas flow rate of 0.5–1.2 l min^{-1}. The auxiliary (intermediate gas) argon flows through the middle tube at a gas rate of 0.8–1.2 l min^{-1}. The auxiliary argon gas flow is applied to push the plasma up above the top of the inner tube of the torch to prevent it from overheating. In several applications, other gases such as nitrogen, oxygen or helium are added to argon (mixed gas plasma). Mixed gas plasmas containing only a small amount of foreign gases in a large fraction of argon (e.g., Ar/He, Ar/Xe, Ar/H_2, Ar/O_2, Ar/air, Ar/CH_4 or Ar/N_2) have been used to reduce the formation rate of polyatomic ions, following several isobaric interferences in argon ICP-MS,[7,12] and to enhance the degree of ionization of hard-to-ionize elements.[13,14] By application of helium plasma (which is more expensive) in ICP-MS elements with higher ionization energy such as halogens can be more effective ionized[15] or disturbing isobaric interferences can be reduced e.g., for calcium and potassium isotope ratio measurements.

For most applications only argon of the highest purity is employed as plasma gas in ICP-MS. The Ar plasma gas (or outer gas) flowing between the outer tube and the intermediate tube (10–20 l min^{-1}) cools, in addition, the outer tube between the plasma and the load coil. An ICP source operates at an rf frequency of 27 or 40 MHz and an rf power of 1–2 kW. The high-frequency (or rf – radio frequency) field in the ICP is produced by the rf generator. The radio-frequency power supply (free-running or crystal-controlled) drives current through a water- or air-cooled copper induction coil (load coil). The oscillating current through the load coil produces an oscillating electromagnetic field.[7] The energy input from the rf generator is transferred to the argon gas at atmospheric pressure via this induction load coil. In order to ignite the argon plasma discharge a tesla coil or high-voltage spark is employed and electrons are generated in the argon gas. After the ignition of plasma discharge, the electrons in the plasma are accelerated by the oscillating magnetic field and collide with atoms and molecules in order to ionize the plasma components. At a relatively high plasma temperature very efficient sample volatilization, desolvatization, dissociation of compounds and ionization of the atoms formed, mostly into singly charge ions, take place in the inductively coupled plasma. The plasma temperature and electron number densities are a function of the experimental parameters applied (rf power, nebulizer gas flow rate, solution uptake rate, torch design and others). Normally, inductively coupled plasma operates at a plasma gas temperature of approximately 5000–8000 K whereas the temperature of excited atoms is lower (between about 4000 and 6000 K). The electron temperature lies between 8000 and 10 000 K, the electron density being about $1\text{–}3 \cdot 10^{15}\,cm^{-3}$.[16]

The temperature of the inductively coupled plasma varies with the distance from the load coil and according to the setting of the ICP rf power and nebulizer gas flow rate. A typical profile of the plasma gas temperature along the torch axis as a function of distance from the load coil is shown in Figure 2.4. With increasing distance from the load coil and with a reduction of ICP rf power the gas plasma temperature decreases.

The rf energy is provided to the electrons, so that the rf field accelerates the electrons, which ionize the noble gas (e.g., argon) used for transporting the nebulized aerosols via electron ionization as described in Equation (2.1):

$$Ar + e^- \rightarrow Ar^+ + 2e^- \tag{2.1}$$

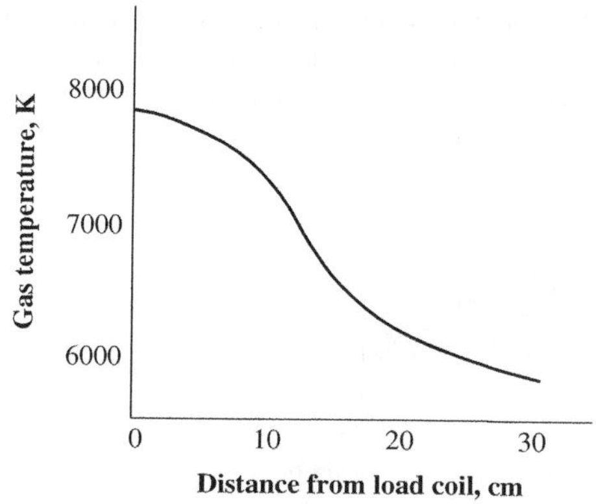

Figure 2.4 *Plasma gas temperature as a function of distance from load coil.*

In order to ionize the Ar atoms the electrons transfer their energy via collision processes to the plasma gas atoms. On the other hand, excited argon atoms are formed and photons emitted by radiative recombination of argon ions with electrons (see Equation 2.2.). The latter lead to a significant increase in the instrumental background in the ICP-MS:

$$Ar^+ + e^- \rightarrow Ar^* + h\nu \qquad (2.2)$$

But bremsstrahlung must also be considered as a part of the background spectrum.[17]
The main ionization processes of an analyte (M) in argon inductively coupled plasma are:

- electron ionization $\quad M + e^- \rightarrow M^+ + 2e^-$ (2.3)
- charge transfer ionization $\quad M + Ar^+ \rightarrow M^+ + Ar$ (2.4)
- Penning ionization $\quad M + Ar^{m*} \rightarrow M^+ + Ar + e^-$ (2.5)

Furthermore, excited metastable species such as Ar^{m*} or M^* are present in the argon plasma of an ICP source due to:

- electron excitation of atoms $\quad M + e^- \rightarrow M^* + e^-$ (2.6)
  of argon $\quad Ar + e^- \rightarrow Ar^* + e^-$ (2.7)

or

- ion–electron radiative recombination $\quad M^+ + e^- \rightarrow M^* + h\nu$ (2.8)

In most application, ICP-MS operates with hot plasma at an rf power of the ICP of about 1200–1300 W (at a plasma temperature of about 8000 K). Under such experimental conditions, the chemical compounds contained in the sample aerosol are decomposed into their atomic constituents. These constituents are ionized to a high degree in the plasma. The ionization efficiency of an ICP source depends on the ionization energy, E_i, (formerly ionization potential) of the element to be analyzed. Elements with an ionization energy of less than about 8 eV are ionized with nearly 100 % yield. With increasing first ionization energy, the ionization efficiency decreases. Figure 2.5 demonstrates the dependence of the degree of ionization α (α = number of single charge atomic ions/number of uncharged atoms) on the first ionization energy, E_i, of chemical elements by inductively coupled plasma ionization. The first and second ionization energies of chemical elements are summarized in the Appendix I. Date and Gray[5] observed a decrease in the ionization yield for elements with increasing ionization energy of 8 eV to 11 eV from 91 % to

12 %. Although for many elements the degree of ionization is about 100 % only 1 in 10^4 to 1 in 10^6 atoms[7] in the original sample are detected due to the extensive loss of ions that occurs during their transport from the plasma via the mass separation system to the ion detector (due to different ion transmission of mass spectrometers).

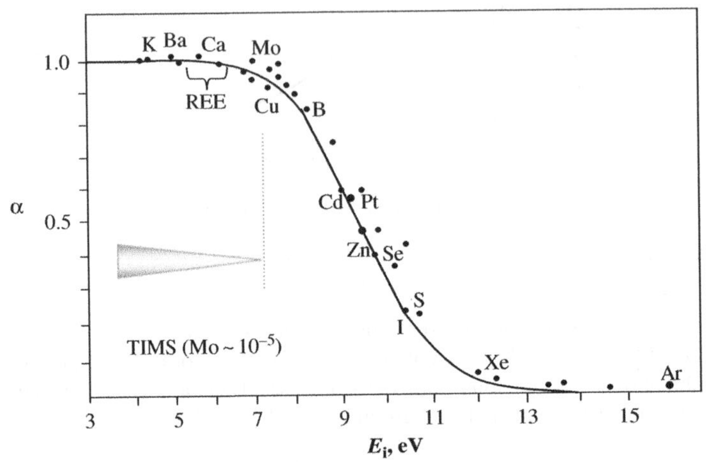

Figure 2.5 *Dependence of ionization degree α on first ionization energy E_i of chemical elements by inductively coupled plasma ionization.*

The ion intensities of analytes, oxide formation rate and intensity of doubly charged ions measured by ICP-MS depend on the experimental parameters used (e.g., nebulizer gas flow rate, rf power of ICP, sample depth, solution uptake rate). In addition, the optimal nebulizer gas flow rate (in respect to maximum analyte ion intensity) for selected rf power is a function of the sampling depth (distance between torch mouth and sampler as part of the interface region, see Figure 2.3). The ion intensity of analyte is a function of the nebulizer gas flow rate. Cs^+ ion intensity has been measured for different sampling depths as a function of nebulizer gas flow rate at an rf power of 1.1 kW by ICP-MS, as shown in Figure 2.6.[18] With increasing sample depth, the optimum nebulizer gas flow rate increases and a decreasing maximum ion intensity is observed.[18] Typical optimization curves, e.g., for $^{138}Ba^+$ and $^{48}Ti^+$, of ion intensity as a function of ICP rf power and nebulizer gas flow rate (using a MicroMist nebulizer with minicyclonic spray chamber), are presented in Figure 2.7 and Figure 2.8, respectively.[19] The maximum ion intensities for $^{138}Ba^+$ and $^{48}Ti^+$ were observed at the rf power of the ICP of > 1200 W and at a nebulizer gas flow rate of 0.86 l min^{-1}. Measurements of stoichiometry and trace impurities in thin barium strontium titanite perovskite layers by ICP-MS were carried out in author's laboratory under these optimized experimental conditions.[19]

The dependence of intensity for $^{88}Sr^+$ ions in comparison to $^{90}Zr^+$ ions on the rf power is illustrated in Figure 2.9. For zirconium, with growing rf power the intensity of the singly charge ion, $^{90}Zr^+$, increases and a maximum intensity is observed at a high rf power of 1300 W (hot plasma). At an operating ICP-MS rf power of ~ 1200–1300 W, the main part of the plasma consists of singly charge atomic ions of the plasma gas (Ar^+), of matrix elements (mostly from water), of analytes and electrons. Besides the positive singly charge atomic ions as the biggest fraction, a small fraction of positive doubly charged ions (≈ 1–3 %) and negatively charged ions

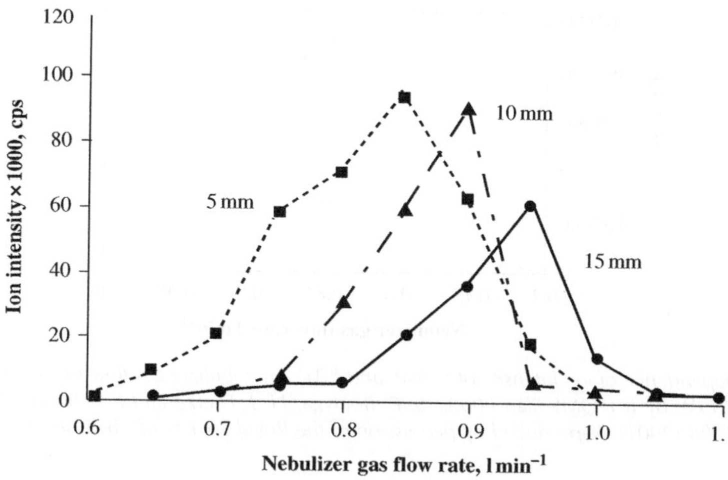

Figure 2.6 *Ion intensity of Cs^+ as a function of nebulizer gas flow rate for different sampling depths at rf power of 1.1 kW measured by ICP-MS. (M. A. Vaughan, G. Horlick and S. H. Tan, J. Anal. At. Spectrom.* ***2****, 765 (1987). Reproduced by permission of the Royal Society of Chemistry).*

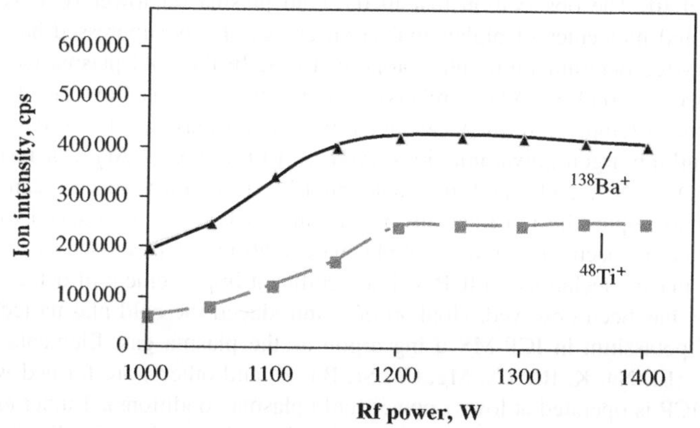

Figure 2.7 *Dependence of ion intensity of $^{138}Ba^+$ and $^{48}Ti^+$ on ICP rf power measured at a nebulizer gas flow rate of 0.86 l min^{-1} by ICP-QMS Elan 6000. (S. F. Boulyga, H. J. Dietze and J. S. Becker, J. Anal. At. Spectrom.* ***16****, 598 (2001). Reproduced by permission of the Royal Society of Chemistry.)*

are formed as well. For analytical purposes in ICP-MS positive singly charged atomic ions of analytes are applied. Doubly charged ions observed with low intensities in ICP mass spectra can cause isobaric interference problems (e.g., $^{186}W^{2+}$ with $m/z = 93$ in the determination of $^{93}Nb^+$ or $^{180}W^{2+}$ with m/z = 90 in the determination of $^{90}Zr^+$ or $^{90}Sr^+$) therefore the experimental parameters of ICP-MS are also optimized with respect to minimum formation of doubly charged ions. In contrast to measurements under hot plasma conditions, for strontium the maximum ion intensity is measured at cold (or cool) plasma conditions (at an rf power of 750 W). The different

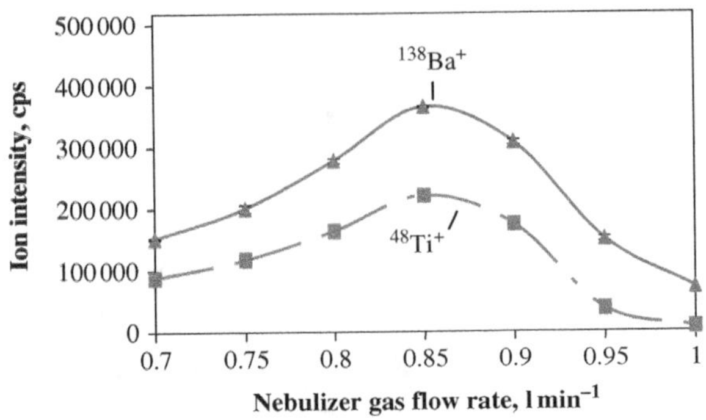

Figure 2.8 Dependence of ion intensity of $^{138}Ba^+$ and $^{48}Ti^+$ on nebulizer gas flow rate measured at an rf power of 1300 W by ICP-QMS Elan 6000. (S. F. Boulyga, H. J. Dietze and J. S. Becker, J. Anal. At. Spectrom. **16**, 598 (2001). Reproduced by permission of the Royal Society of Chemistry.)

behaviour of chemical elements in inductively coupled plasma as a function of the rf power, as demonstrated in Figure 2.9, can be explained by the difference in their first ionization energy, E_i(Sr = 5.7 eV and Zr = 6.8 eV). The varying behaviour for $^{88}Sr^+$ and $^{90}Zr^+$ at differing rf power in ICP-MS has been applied for ultrasensitive ^{90}Sr determination in the presence of zirconium, as proposed in ref.20. The observation that in the cold plasma (at lower rf power) the ionization of elements and molecules of higher ionization energy, E_i, is suppressed has been used for quasi interference-free determination and isotope analysis. In the cold plasma technique at relatively low rf power (~ 600–800 W), a relatively high nebulizer gas flow rate (> 1 l min^{-1}) is applied. At a reduced temperature of the inductively coupled plasma (at low ICP rf power), the formation of disturbing argon polyatomic ions, ArH^+, ArO^+, ArC^+, Ar_2^+, or residual gas ions, C^+, CO^+, O^+, O_2^+, N^+, NO^+, and thus their intensity in the mass spectra, can be suppressed significantly. For example, the ion intensity of Ar^+ and ArH^+ can be reduced by six and four orders of magnitude, respectively, under cool plasma conditions, as discussed by Tanner.[21] As the result of the cold plasma technique in ICP-MS, a significant improvement of detection limits, e.g., for K, Ca and Fe, has been observed. Houk *et al.*[22] introduced the cold plasma technique for the determination of potassium in ICP-MS using argon as the plasma gas. Elements with low first ionization energy (Li, Na, K, Rb, Cs, Mg, Ca, Sr, Ba, Fe and others) are formed with higher ion intensities if the ICP is operated at low rf power (cold plasma conditions). Tanner *et al.*[23] reported on the determination of Fe, Ca, K, Na and Li by ICP-MS with low detection limits in the ng ml^{-1} range and below in aqueous solutions under cold plasma conditions. Under cold plasma conditions a low rf power of ~ 750–850 W was applied during the measurement. This technique allowed, for example an ultrasensitive ^{90}Sr determination due to suppression of the $^{90}Zr^+$ formation which leads to an isobaric interference. The generation of interfering doubly charged tungsten ions at $m/z = 90$ could also be excluded. The main drawback of ICP-MS is the formation of polyatomic ions, which can lead to disturbing isobaric interferences with atomic ions of analyte. Therefore, the rf power of ICP, the nebulizer gas flow rate and the position of the ICP torch were optimized before mass spectrometry measurements with respect to the maximum singly charged atomic ion intensity of analytes, minimum formation rate of doubly charged atomic ions and singly charged polyatomic ions to avoid disturbing isobaric interferences.

The effect of pre-evaporation on ion distributions in ICP-MS was studied by Lui and Beauchemin.[24] The connecting tube (2 or 5 mm i.d., 11 cm long) between the spray chamber

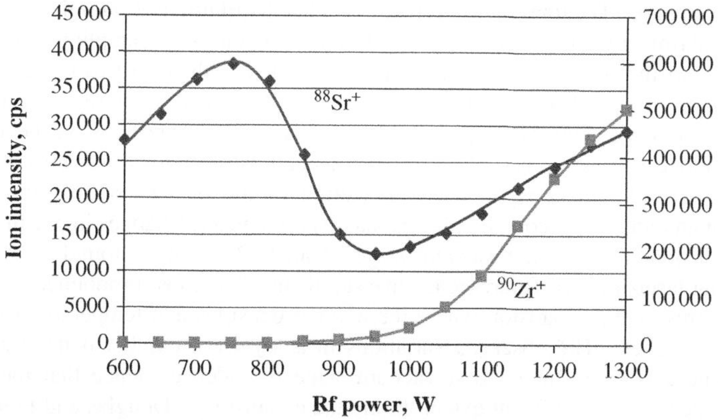

Figure 2.9 *Ion intensities of* $^{88}Sr^+$ *and* $^{90}Zr^+$ *as a function of rf power. (A. P. Vonderheide, M. V. Zoriy, C. Pickhardt, J. Caruso, P. Ostapczuk, R. Hille and J. S. Becker. J. Anal. At. Spectrom.* ***19****, 675 (2004). Reproduced by permission of The Royal Society of Chemistry.)*

and the torch was heated to 400 °C to study the effect of pre-evaporation of the ion distribution in ICP-MS. Axial and radial profiles of 14 analyte ions and 4 polyatomic ions (LaO^+, ArO^+, ArN^+ and $CO_2{}^+$) were measured by ICP-ToF-MS. Upon heating the connection tube, the optimal axial position of all analytes shifted closer to the load coil. An improvement of sensitivity and detection limits, which is commensurate with a reduction of noise resulting from smaller droplets entering the plasma after traversing the pre-evaporation tube, was obtained.[24]

After ion generation in the ICP ion source, the positively charged ions were extracted from the argon plasma (at $\approx$ 100 kPa) via the differentially pumped interface (at a pressure at $\approx$ 130 Pa), using a mechanical pump between the sampler and skimmer cones, into the high vacuum of the mass analyzers. The two-stage differentially pumped interface employed in each ICP mass spectrometer was first developed by Campargue[25] for the molecular beam technique. The problem of the extraction of ions formed in an atmospheric pressure ion source into the vacuum of a mass spectrometer was solved. This was made possible by the insertion of an ion extraction interface with sampling cone and skimmer cone as the boundary to the atmospheric ion source, on the one side, and to ion optics as a part of the high-vacuum mass analyzer on the other side, respectively. The plasma, which expands through the sampling orifice (diameter of orifice = 0.8–1.2 mm), produces a free jet, whereas the centreline flow of the jet passes through the skimmer orifice (diameter of orifice = 0.4–1 mm) in the ion lens system. Smaller sampling orifices (< 0.8 mm) have been applied, but in the case of a high salt content of the analyzed solution clogging effects are possible. By differential pumping of the interface, the atmospheric pressure in the inductively coupled plasma ion source and the high vacuum of the mass analyzer can be matched.[26] To study the process of ion sampling and transport through the vacuum interface that serves as a conduit between atmospheric pressure plasma and the high vacuum mass analyzer, planar laser-induced fluorescence imaging experiments were carried out by Farnsworth's group.[27,28] The authors produced images of the density distribution of barium atoms and ions in the ground state and of exited barium ions in the region between the load coil and the sampling cone of an inductively coupled plasma.[28] The influence of changes of rf power, nebulizer gas flow rate, and in the addition of lithium to the sample (to study matrix effects) on distribution of barium species was investigated. The images reveal that the radial distributions of atomic species across the diameter of the plasma are compressed as the plasma is drawn into the sampling orifice.[28] Changes in the ion flux into the vacuum interface were investigated with regard to operating conditions and sample composition and were demonstrated in radially resolved fluorescence images of the sampling orifice, as were changes in spatial distributions of atoms and ions on the 1 mm scale of the hole. Laser induced fluorescence analyses for plasma diagnostic measurements were employed to study the effects of collisions on the composition of the plasma passing through the first vacuum stage of an inductively coupled plasma mass spectrometer. Changes in plasma composition during the expansion into the first vacuum stage of an ICP-MS were found by Macedone and Farnsworth using laser-induced fluorescence measurements. The most pronounced effects were in the first millimetres of the expansion where the number densities and temperatures of the plasma species were the highest. The observed variations in analyte ion number density along the axis of the supersonic expansion in the first vacuum stage provided evidence that the ion–electron recombination occurs to a significant extent during the expansion.[27] Douglas and French discussed ion–electron recombination in the supersonic expansion downstream from the sample cone, arguing that the slow drop in electron temperature relative to the rate at which the heavy particles cool prevents significant recombination.[29]

The supersonic expansion of the plasma jet from the sample and skimmer cones of the interface in the vacuum of the mass spectrometer is a rapid process where space charge effects occur in the interface region and in the region between the skimmer cone and ion optics. Due to space charge effects, ions with lower mass are deflected more than ions with higher mass, which results in a mass discrimination effect relevant especially in isotope ratio measurements. In addition, a high solvent load enhances capacitive coupling between the load coil and plasma, increasing the strength of the secondary discharge between the plasma and grounded interface. By secondary discharge, the ion kinetic energy spread increases and the ion energy distributions broaden, thus reducing the ion transmission through the mass spectrometer. To increase ion transmission and to avoid secondary discharge, a grounded conductive shield (e.g., of platinum or tantalum) is located between the load coil and ICP torch. Figure 2.10 presents a schematic of the shielded torch used in the double focusing sector field ICP-SFMS Element, Thermo Fisher Scientific. The guard electrode between the load coil and torch is switched on (ground potential) and off (floating potential) electronically. Such a plasma shielded torch can provide an improvement in sensitivity up to a factor of ten with a nebulizer–spray chamber arrangement or using an ultrasonic nebulizer for solution introduction as a result of reducing the ion kinetic energy. Gray[30] proposed plasma shielding with quadrupole-based ICP-MS for measurements under cool plasma conditions and observed a significant reduction in polyatomic ion formation. The characteristics of plasma shielding in comparison to an unshielded torch in double-focusing sector field ICP-SFMS for the ultrasensitive determination of long-lived radionuclides were studied by Becker and Dietze[31] and McLean *et al.*[32] A significant improvement in sensitivity was achieved by using a shielded torch in sector field ICP-MS in comparison to an unshielded torch for selected radionuclides (^{232}Th, ^{237}Np, ^{238}U, ^{239}Pu and ^{241}Am) as illustrated in Figure 2.11.[31] Due to improved sensitivity of ICP-SFMS with shielded torch, the detection limits of these radionuclides could be improved by nearly one order of magnitude in the sub fg ml^{-1} range.[31]

Further designs of ion sources applied in plasma spectroscopy such as electrodeless microwave induced plasmas (MIPs) operating in a noble gas atmosphere at low power (mostly below 200 W) or capacitively coupled microwave plasma using Ar, He or N_2 the as plasma gas (at 400–800 W) were described in detail by Broekaert.[33] Microwave plasmas produced by a magnetron are operated at 1–5 GHz. Their special application fields for selected elements and/or element species are based (due to the low power applied) in atomic emission spectrometry.[33]

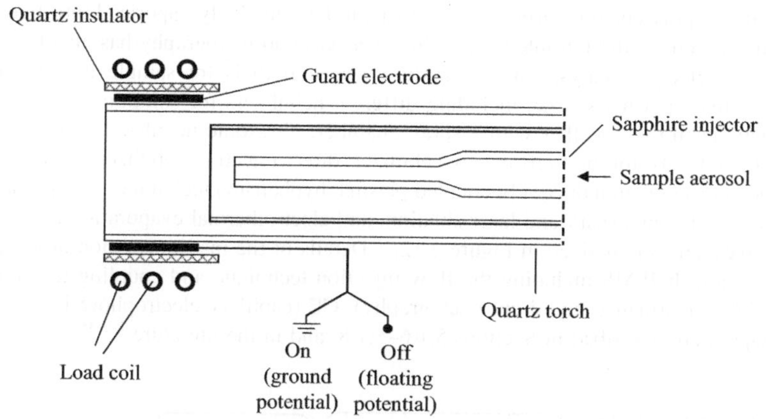

Figure 2.10 *Schematic of shielded torch in ICP-MS (J. S. Becker and H. J. Dietze. Int. J. Mass Spectrom. Ion Proc.* ***197****, 1–35 (2000). Reproduced by permission of Elsevier.)*

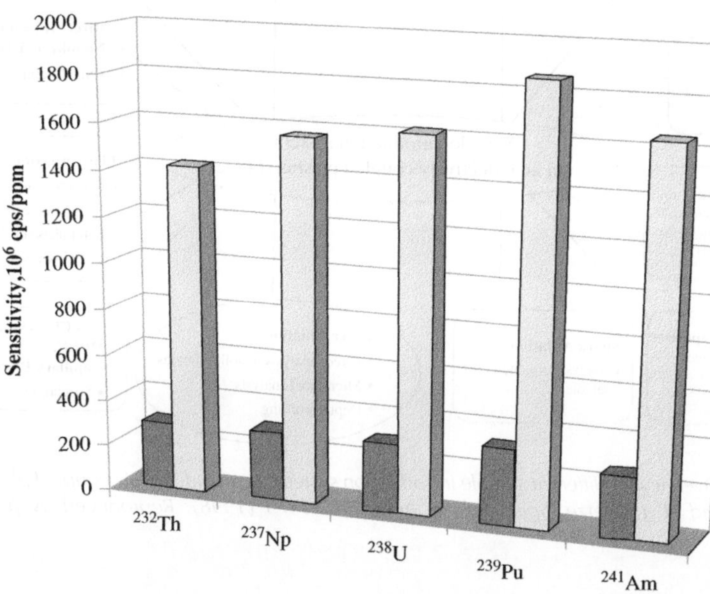

Figure 2.11 *Sensitivity for selected long-lived radionuclides measured by ICP-SFMS (Element, Thermo Fisher Scientific) with (yellow) and without (green) shielded torch (GE – guard electrode). (J. S. Becker and H. J. Dietze. Int. J. Mass Spectrom. Ion Proc.* ***197****, 1–35 (2000). Reproduced by permission of Elsevier.)*

Various efficient devices have been utilized for sample introduction into an inductive plasma source, for example the application of several nebulizers, hyphenated techniques, hydride generation, laser ablation and electrothermal vaporization. The role of the solution introduction system in an inductively coupled plasma source is to convert the liquid sample into a suitable form (e.g., using a nebulizer/spray chamber arrangement) that can be effectively vaporized into free atoms in order to generate ions with high intensities. However, gas chromatography has also been coupled to ICP-MS for selective analysis of gas mixtures. Several tools for sample introduction in an inductively coupled plasma source, including different nebulizers for solution introduction (such as a pneumatic nebulizer together with a spray chamber, ultrasonic nebulizer or microconcentric nebulizer with a desolvator, high-efficiency nebulizer, direct injection nebulizer, the application of hydride generation) into an inductively coupled plasma, hyphenated techniques for speciation analysis, the slurry technique, spark and laser ablation, and electrothermal evaporation for the analysis of solid samples, are summarized in Figure 2.12.[34] Details of the instrumentation of sample introduction systems in ICP-MS including the flow injection technique and coupling techniques such as HPLC (high performance liquid chromatography), CE (capillary electrophoresis) and GC (gas chromatography) are described in Sections 5.1.6–5.1.8. and in the literature.[7–9,35]

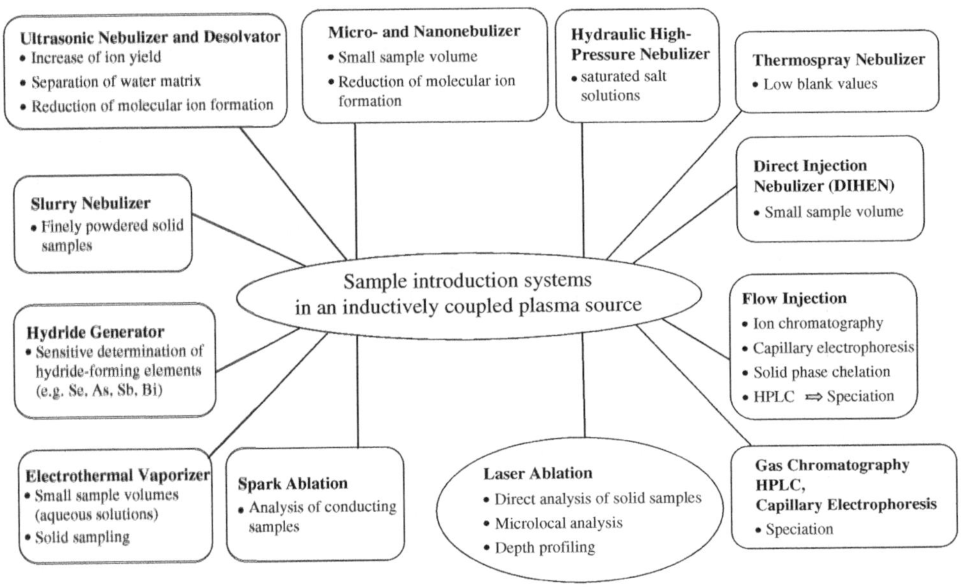

Figure 2.12 *Overview of different sample introduction systems in an inductively coupled plasma source. (J. S. Becker and H. J. Dietze, Spectrochim. Acta **53B**, 1475 (1998). Reproduced by permission of Elsevier.)*

2.1.1 Laser Ablation Coupled to an Inductively Coupled Plasma Ion Source

The most important and frequently used coupling technique with an ICP source in mass spectrometry is a laser ablation device which allows the direct analysis of solid samples without any time-consuming digestion procedure or any restriction of matrix composition.[34,36–43] LA-ICP-MS offers a fast and precise in situ (spatially resolved) measurement of trace elements and isotope ratios with high analytical throughput, minimal sample preparation and at reduced mass spectral interferences.[38,44,45] For bulk analysis on solid samples using an effective laser ablation arrangement together with a sensitive ICP-MS, low detection limits are obtained.[46–48] The growth of the literature on laser ablation in analytical chemistry has been explosive in the last few decades with the use of nanosecond, picosecond and the development of femtosecond lasers for new applications and their important characteristics have been pushed to new limits.[49]

This relatively new solid-state mass spectrometric technique uses the ablation of sample material by a focused laser beam in an argon atmosphere under normal pressure (see Figure 2.13 a). In LA-ICP-MS, different commercial laser ablation systems (e.g., CETAC LSX 200, 500 and 3000, Cetac Technologies, Omaha, NE, USA; Merchantec LUV 266 nm laser microprobe, UP 213, UP 266, UP 266 Macro from New Wave Research, Merchantec, Fremont, CA, USA) or home-made systems[50,51] have been coupled to several types of ICP-MS. The application of excimer lasers with shorter wavelengths in LA-ICP-MS has been proposed, which results in fewer fractionation effects.[43] The size of the laser ablation chamber varies from smaller laser ablation chambers to bigger ones. Normally samples with dimensions of several mm^2 up to 20 mm × 20 mm can be easily measured in commercially available laser ablation cells. For bigger sample sizes (e.g., for 2D gels with a size of 200 mm × 200 mm) or biological tissues, different sections of gel samples or biological tissues have been analyzed one after the other after cutting as described in reference[52]. Large-volume laser ablation chambers (from New Wave Research or CETAC Technologies) are available on the analytical market for the analysis of 1D and 2D gels and have been developed for several applications in different laboratories. To study metal-containing proteins and phosphoproteins on membranes with sizes of 120 mm × 120 mm, a laser ablation cell has been designed and optimized by Jakubowski's group.[53] On the other hand, small laser ablation chambers are advantageous for the sensitive analysis of microsamples by LA-ICP-MS. Such a laser ablation chamber for short wash-out time was designed by Carcia *et al.*[54] A small, transparent laser ablation chamber from PMMA (20 mm in both inner diameter and height,) was constructed and applied in the author's laboratory to analyze single cells and small specimens of biological tissue and doped gels in single-shot measurements by the application of the near field effect in laser ablation ICP-MS to analyze at spatial resolutions in the sub-μm range.[55]

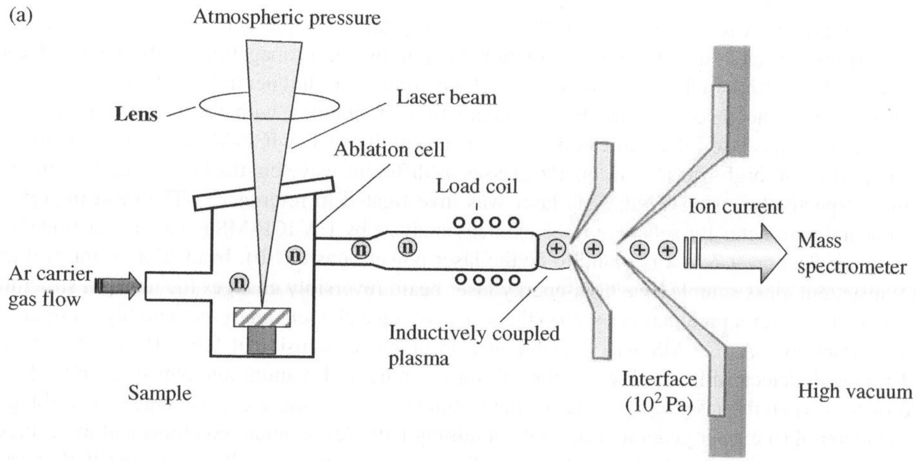

Figure 2.13 *a) Laser ablation coupled to an inductively coupled plasma ion source.*

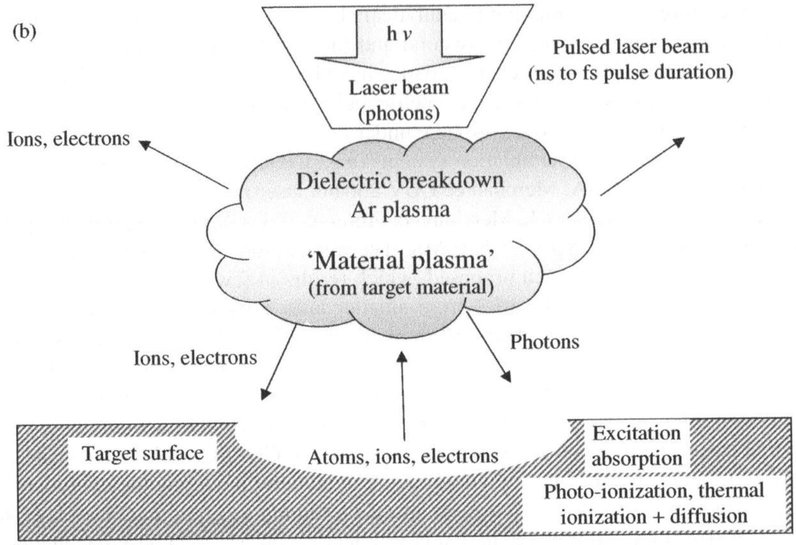

Figure 2.13 *b) Schematic diagram of photon–solid interaction by the laser ablation process.*

Furthermore, to analyze biological tissues of plants or thin sections of human brain tissues or ice cores, cooled laser ablation chambers have been developed and utilized in different application fields.[56,57]

The schematic diagram of photon-solid surface interaction by the laser ablation process is presented in Figure 2.13 b. Laser ablation into an atmosphere of sufficient density produces a laser induced shock wave (thermal expansion of ambient gas). The electrically neutral argon plasma formed with a low degree of ionization also contains unstable, meta-stable excited argon atoms, which do not make any contribution to the ablation process due to their low energy and short lifetime. After the formation of the plasma plume, sample aerosols are produced and released from the laser ablated crater ('ablation pit') along the propagation of the laser-induced shockwave.[58] The 'material plasma' consisting of electrons, ions and neutrals of the target material is formed above the target surface by the interaction of photons with the target material and argon plasma in the laser ablation chamber.[47] The capability of LA-ICP-MS for determination of trace impurities in high transparent quartz glasses at different wavelengths (1064 nm, 532 nm and 266 nm, respectively) using a Nd:YAG laser was investigated in reference[47]. The most important experimental parameter for sensitive trace element analysis by LA-ICP-MS (to avoid fractionation effects – see Section 9.7.2) was found to be the laser power density (Φ). Irradiation of the surface of a transparent glass sample by a high-energy laser beam reversibly changes the material structure (local defects). After a preablation ($\sim$ 50 s) the sensitive trace element analysis of highly transparent glass samples by LA-ICP-MS was possible at a laser power density of $9.6 \times 10^9\,W\,cm^{-2}$ (λ = 1064 nm) with detection limits down to the sub-ng g^{-1} range. Maximum ion intensities of analytes are observed when the laser focus is below the sample surface in these experiments.[47] The ablated sample material (the laser generated aerosol containing particles, evaporated atoms and molecules) is transported under atmospheric conditions with an argon carrier gas flow into the ICP source where the ablated particles are partly atomized and ionized in the argon plasma.[47] Of special interest are the particle formation processes by condensation in the expanded plume and the optimum conditions during the laser ablation process for complete vaporization of particles, the particle size distribution and possible elemental and isotopic fractionation effects. These have been investigated by several groups.[59–64] An explanation of the laser ablation process by mathematical modelling of laser-induced particulate formation is given by Bleiner.[65] Production of laser ablation induced sample aerosols has been visualized using a high-speed camera device equipped with magnification optics (200 times) coupled with a shadow-graph technique by Hirata and Miyazaki.[58] The time resolution of the imaging technique of the formation of laser ablated material was, in these experiments, 1 μs and He was used as the ambient gas. The authors observed a dome-shaped dark area appearing at the ablation pit when an argon fluoride excimer laser ($\lambda = 193\,\text{nm}$) was shot onto the sample surface of a silicon wafer. This dark area reflects changes in refractive index of ambient He probably due to emission of electrons or ions from the ablation pit. The dark area expanded hemispherically from the ablation pit with a velocity of $\sim 1000\,\text{m s}^{-1}$ (for He at 300 K). This was followed by the excitation or ionization of the vaporized sample. The typical speed of particles released from the ablation pit was $\sim 120\,\text{m s}^{-1}$ which was significantly lower than the reported expansion velocity of the plasma plume emitted from the ablation pit ($10^4\,\text{m s}^{-1}$), which is close to the speed of sound ($\sim 1000\,\text{m s}^{-1}$ for He at 300 K).[58] This indicates clearly that the sample particles were rapidly decelerated to the terminal velocity by the drag force of the ambient He. The release angle of sample aerosols from the ablation pit was observed to be very shallow ($< 10°$). This shallow release angle could be due to a down-force induced thermal expansion of the ambient gas just above the ablation pit and laser induced down-force then results in redeposition of sample particles around the laser crater. The redeposition can be minimized by using lower pressure ablation cell (e.g., by evacuation of laser ablation chamber down to 20 kPa).[58]

Studies based on modelling are needed to better define and predict the laser ablation process.[66,67] The modelling of the expansion of the (nanosecond pulsed) laser ablated Cu vapour in He background gas at 1 atm in vacuum was studied by Bogaerts and Chen.[66] Calculation results include the temperature distribution in the target, information about target melting and vaporization, the density profiles of Cu vapour and the background gas in front of the target, as well as the velocity and temperature distributions in the plume. The major influence on the He background gas is the spatial confinement of the vapour plume, the retardation of the expansion velocity, and the lowering of the plume temperature.[66] Fundamental studies of laser ablation using time-resolved shadowgraph and spectroscopic imaging, especially by means of a femtosecond laser, were described by Russo and coworkers.[43] A non-linear laser ablation behaviour (laser induced plasma temperature, electron density and crater depth[68]) as the function of laser power density was observed. With increasing laser power density (from $10^9\,\text{W cm}^{-2}$ up to $10^{11}\,\text{W cm}^{-2}$) the depth of laser crater increased from several hundred nm up to 20 μm. Femtosecond laser pulses are showing promise in achieving matrix independence and elimination of fractionation effects. The plasma induced by a femtosecond laser is cooler and expands more rapidly because it is not influenced by absorption of the laser beam.[43]

The influence of various gas pressure conditions within the laser ablation cell on the particle formation process in laser ablation has also been investigated.[69] In LA-ICP-MS studies at low pressure (down to 2 kPa) a small particle size distribution and a reduction in elemental fractionation effects was obtained. But with decreasing pressure and transport volume of ablated material, a significant decrease in the ion intensities was observed as demonstrated for uranium measurements in the glass SRM NIST 610.[69] However, the laser ablation of solid materials at atmospheric pressure in LA-ICP-MS is advantageous for routine measurements due to lower experimental effort and the possibility of fast sample changing in the ablation chamber. Fractionation effects in LA-ICP-MS have been examined together with a developed approach of solution based calibration by Boulyga *et al.*[70] In this experimental work the authors proposed the insertion of a ultrasonic nebulizer (USN) between the laser ablation chamber and the inductively coupled plasma source of an ICP-MS. Filtering of the ablated aerosol in the wet vapour produced by ultrasonic nebulization resulted in reduction of element fractionation in LA-ICP-MS as demonstrated for the determination of trace impurities in brass and high purity platinum. The mass spectrometric measurements were carried out using a solution based calibration strategy via the USN for quantification of trace element impurities.[70]

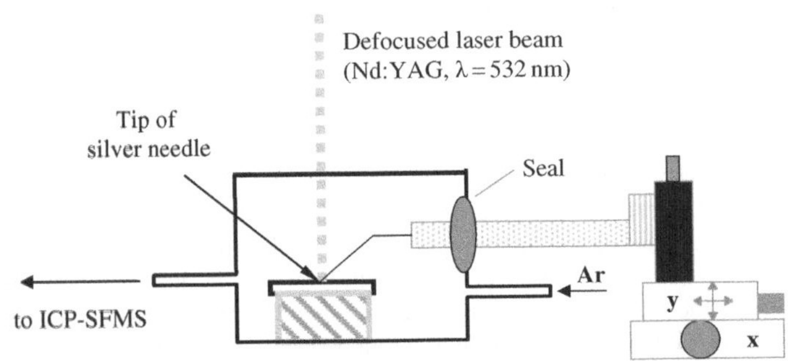

Figure 2.14 *Schematic of near field laser ablation.*

LA-ICP-MS has been established as a promising, powerful and sensitive multi-element analytical technique for the quantitative determination of trace elements in quite different solid materials. At present the lateral resolution of microlocal analysis by LA-ICP-MS is in the low μm range. The microlocal analysis of biological tissues, of single cells or cell organelles requires improved lateral resolution at the nm range and would be possible by application of the near field effect in laser ablation ICP-MS.[55] The experimental setup of near field laser ablation ICP-MS is illustrated in Figure 2.14. The application of the near field effect in LA-ICP-MS was proposed by Becker *et al.* in 2003.[71] The idea of this approach consists in the formation of radiation intensity singularities by means of very small conductive objects immersed into the light radiation field. The resulting electric (linear or non-linear) polarization of such objects causes a considerable enhancement of the radiation intensity in the region comparable with the curvature radius of the object (near field enhancement). Such an object can itself, therefore, be treated as a very small, but intensive source of secondary evanescent near field radiation, which can be much more intensive than the field of the primary beam. If this source is brought so close to the target surface that this secondary radiation cannot diverge to any great degree, it can cause local laser induced surface effects (e.g. local desorption or ablation) limited only by the object size rather than by diffraction effects. The factor of the electrical field enhancement depends, on the one hand, on the form and the material of the field-enhancing object and, on the other hand, on the wavelength and polarization of the light. This factor can theoretically reach 10^4 for a prolate spheroid or a long sharp needle made of silver.[55]

In contrast, the LA-ICP-MS (in comparison to laser ionization mass spectrometry (LIMS) where the ion source operates under high vacuum conditions) at present, in spite of the disadvantage of a higher polyatomic ion formation rate, uses an argon plasma ionization at normal pressure – a promising inorganic mass spectrometric technique for trace, isotope and surface analysis which will be developed for future innovative applications. In the direct analysis of solid samples, compared to ICP-MS measurements of aqueous solutions, a lower polyatomic ion formation rate (especially oxide formation) is observed, but isobaric interferences limit the ultratrace and isotope analysis of analytes in respect to accuracy and precision of analytical data and the limit of detections (LODs) increase. In analogy to ICP-MS, also in LA-ICP-MS many efforts are being made at present in order to solve the interference problems of polyatomic and atomic ions, for example by the application of higher mass resolution in double-focusing sector field mass spectrometers or by the application of a collision cell in ICP-MS in order to dissociate the polyatomic ionic species (see Section 5.1.2.). Another advantage of LA-ICP-MS in comparison to other solid-state mass spectrometric techniques is the better procedure for quantifying analytical data using solution-based calibration as discussed in Section 6.2.6.

2.1.2 Electrothermal Vaporization Coupled to an Inductively Coupled Plasma Ion Source

Sample introduction for analytical atomic spectroscopy remains a significant issue, particularly for the analysis of complex matrices, small sample sizes and those samples in which the analyte is present at ultratrace concentration levels.[72] Electrothermal vaporization (ETV) is applied as an alternative to laser ablation for small sample introduction (e.g., for liquid samples $\sim$10 μl or solid samples $\sim$10 mg) in an inductively coupled plasma source. The liquid sample is placed using a micropipette in a graphite furnace or on a tungsten or tantalum filament surface which is electrothermally heated to different temperatures by passing a high direct current (dc) through it. Using a special heating program the sample (aqueous solution) is dried first at 120 °C. Subsequently a useful trace–matrix separation to improve the detection limits and to reduce matrix effects and inherent interference problems can be performed. The ashing of the sample with organic matrix (biological material) is carried out at 400–1500 °C. Following this the sample is heated step by step for a sequential evaporation of analytes. The analytes are vaporized into a cloud of atoms and molecules at a temperature of 1500–3000 °C. For sample introduction in ICP-MS commercially manufactured or in house electrothermal vaporizers are applied. The electrothermal vaporizer is coupled close to plasma source so that the vaporized sample is transported by a carrier gas flow into the ICP. Direct sample introduction without dilution is thus possible and the contamination danger during sample preparation is minimized. In addition, due to low sample amount and very low solvent content, the formation of disturbing polyatomic ions and subsequent interference problems in ICP-MS are reduced significantly in comparison to continuous solution introduction in to the ICP. This results in an improvement of absolute detection limits of ETV-ICP-MS. In ETV-ICP-MS transient signals have been measured requiring fast electronics or multiple ion collection for precise measurements. In spite of several advantages only a few applications of ETV for sample introduction in ICP-MS are described in the literature. A drawback of this technique is that the analytes may react with the vaporizer material (e.g., graphite) and therefore carbide ions are observed in the mass spectra.

2.1.3 Hydride Generation and Cold Vapour Technique Coupled to an Inductively Coupled Plasma Source

Hydride generation is a widely utilized gas phase sample introduction system in ICP-MS for volatile hydride forming elements such as As, Se, Te, Bi, Sb, Ge, Pb and Sn. The gaseous hydrides (AsH_3, H_2Se, H_2Te, BiH_3, SbH_3, GeH_4, PbH_4, and SnH_4) formed in a continuous flow reaction separation system in the presence of a reducing agent (e.g., $NaBH_4$) in acid media are introduced with a carrier gas into the ICP source of an ICP-MS. The chemical reaction in the hydride generator, the separation of volatile products from liquid phase and the transport of the gaseous species to the excitation or ion source can be performed for liquids in different devices using a continuous flow system or flow injection, e.g. in a U-shaped gas–liquid phase separator or in a spray chamber with a cross flow or MicroMist nebulizer (the chemical reaction occurs inside the spray chamber). As a result of the separation and preconcentration of analyte from the sample matrix (e.g., water) and interfering elements, by the reaction of hydride forming elements with $NaBH_4$ in acid media, an increase in element sensitivity and improvement in detection limits has been observed for different applications with liquids.

In analogy to sample introduction by hydride generation, mercury trace analysis is possible by reducing Hg compounds to the metal using the cold vapour technique or the determination of iodine at the ultratrace level (after oxidation with 70 % perchloric acid of iodide to iodine) via the gas phase.

2.2 Spark Ion Source

Spark Source Mass Spectrometry (SSMS) is one of the oldest mass spectrometric techniques.[73–76] Two kinds of spark ion sources are used for analytical purposes in SSMS: the radiofrequency (rf) spark source and the low voltage direct current (dc) arc ion source. The rf spark ion source was developed by Dempster[77] in 1935 for mass spectrographic trace element analysis in solid samples. A schematic of a spark ion source is shown in Figure 2.15. The ion source consists of a vacuum chamber in which two pin-shaped electrodes of the sample material (length: 10 mm, diameter 1–2 mm) are mounted opposite each other. The electrodes containing the sample material investigated must be electrically conducting. The pin-shaped electrodes are cut from bulk material and the surface should be cleaned (e.g., by etching with high-purity acids). Non-conducting material has to be powdered, mixed with high-purity graphite, silver or gold powder homogenized and pressed to form electrodes. The spark plasma for the formation of ions required for mass spectrometric

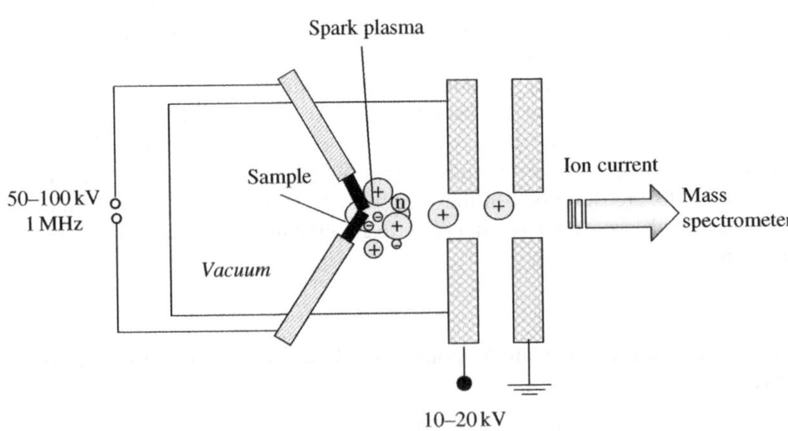

Figure 2.15 Schematic of a spark ion source.

analysis is generated in this spark ion source by applying a pulsed 1 MHz rf voltage of several tens of kilovolts (50–100 kV using a Tesla transformer) across the small gap between the two electrodes. Using *x-y-z* manipulators on both electrode holders the optimized geometric arrangement of electrodes, with respect to maximum total ion current during the measurement, can be achieved. The optimum distance between the electrodes is about 0.1–0.5 mm, whereby the shape of the spark plasma can be observed during sparking with a stereomicroscope. Several SSMS instruments are equipped with an automatic spark gap control device (autospark system) that regulates the optimum distance between the electrodes in respect of maximum total ion current. In most cases, the samples are sparked in an electrode configuration as shown in Figure 2.15. Another experimental arrangement of the ion source is applied to analyze a flat sample using a counter electrode e.g., of silver or tantalum, as demonstrated by Dietze[75] and later by Saprykin *et al.*[78] in so-called gliding SSMS (GSSMS) for the analysis of non-conducting samples (e.g., high purity single crystals). Such an arrangement is useful for the analysis of layers and surfaces and the analytical errors due to inhomogeneities of sample can be reduced. The pulse length and the repetition rate during sparking of the sample are easily optimized by selecting a wide range of exposures. The solid sample material of the electrodes is evaporated, atomized and ionized by electrons in the rf spark plasma, which is ignited when a high voltage of 10^4–10^5 V is applied to the electrodes. The ionization process between two electrodes in the spark ion source can be explained as follows. Electrons are emitted by field emissions (on fine tips, so-called whiskers) at field strengths of 10^6–10^7 V cm^{-1} from a negative electrode (cathode). Due to the influence of the high voltage on the electrodes these electrons were accelerated to the positive electrode (anode). By means of electron bombardment, the anode is heated and material is evaporated. The evaporated material is ionized by accelerated electrons emitted from the cathode. During this process space charging effects between the two electrodes are observed. The reversible discharge will be unstable if the voltage on the electrodes exceeds a critical value and the voltage breaks down to a few volts. The characteristic voltage–current dependence during the vacuum discharge is demonstrated in Figure 2.16. After the reversible breakdown of voltage the discharge of the high frequency spark is extinguished and can continue to exist as a low voltage discharge (arc plasma), when the voltage applied to the electrodes is constant. A characteristic feature of low voltage arc discharge plasma is the formation of multiply charged atomic ions with high ion intensity. The formation of multiply charged ions with high intensity results in a multiplicity of isobaric interferences of singly charged analyte ions with atomic ions of different charges, e.g., for the analysis of silicon in a complex matrix via $^{30}Si^+$, the following multiply charged atomic ions can be observed: $^{60}Ni^{2+}$, $^{90}Zr^{3+}$ and $^{120}Sn^{4+}$. In contrast, in high frequency spark plasma, singly charged atomic ions are formed with the highest ion intensity – this makes the mass spectra easy to interpret. With the high frequency spark the first process of reversible discharge proceeds with a frequency of 1 MHz.

The ions formed in the spark plasma are accelerated with 10–20 kV (acceleration voltage) in the mass spectrometer. The relatively high energy spread of the ions (some keV) generated in the spark plasma requires the application of a double-focusing mass spectrometer (e.g. with Mattauch–Herzog geometry, combination of electrical sector and magnetic sector field), which is advantageous for the sensitive detection of nearly all elements in one mass spectrum by a sensitive photographic plate.

To make the instrument more versatile, some spark source units have been redesigned in the past to make them interchangeable with other types of ion sources (e.g., laser ion source). The insertion of a laser ion source instead of a spark ion source allows the direct analysis of insulators without any treatment, as described by Dietze.[74,75,79]

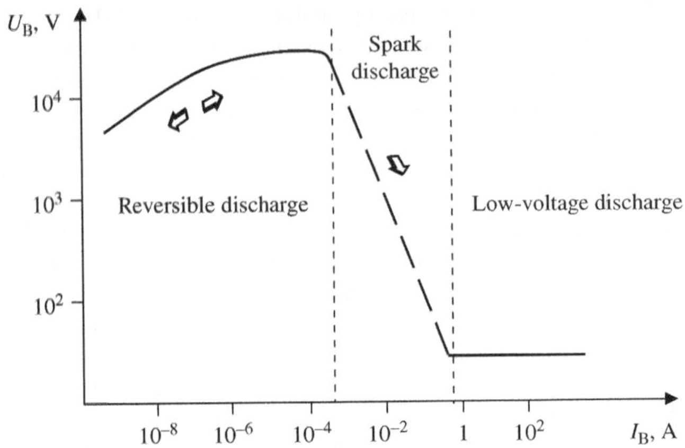

Figure 2.16 *Characteristic current–voltage curve of a spark plasma.*

2.3 Laser Ion Source

2.3.1 Laser Plasma Ionization

After the development of solid-state lasers (ruby and Nd glass lasers) in the early sixties lasers were applied in mass spectrometry,[80] atomic emission spectroscopy (laser induced breakdown spectroscopy – LIBS),[81] microscopy, fluorescence analysis and others. The introduction of the laser in mass spectrometry as an improved sampling device for the trace analysis of bulk samples initiated a breakthrough in the microlocal analysis of inhomogeneous solids, because using a focused laser beam a well-defined volume of solid material can be vaporized and ionized. At present, lasers are widely used in mass spectrometry as a universal tool for the evaporation, atomization and ionization of any material (without any restriction with respect to the solid sample material).[74] Mass spectrometric techniques using a focused laser beam offer an alternative tool to electron and ion beams for microlocal chemical analysis. Laser ionization mass spectrometry (LIMS) is based on the ionization of evaporated and atomized sample material in a laser microplasma under high vacuum conditions. Figure 2.17 illustrates the principles of laser ionization in a high-vacuum laser ion source using the reflection mode. By the interaction of photons with the solid sample surface in the laser ion source at a laser power density of $<10^8$ W cm^{-2}, the sample material is evaporated in the interaction area of photons on the surface, where the gaseous atoms are mainly ionized by electron impact ionization. At this relatively low laser power density fractionation effects occur, that is an enhanced evaporation of elements with low melting point occurs. Furthermore, due in part to the interaction of the evaporated atoms with photons of a certain wavelength, photoionization by resonance ionization can be observed. By the interaction of a focused laser beam at a laser power density of $> 10^8$ W cm^{-2} with the solid surface, the material is evaporated and the gaseous atoms and molecules are ionized in a high temperature plasma. In order to generate laser plasma with a high degree of ionization, a laser power density of about 10^9 W cm^{-2} is required. In this case, pulsed q-switched solid laser systems with small beam divergence are applied, which allow the photons of the laser beams to be focused to a spot diameter down to several μm (e.g., Nd-YAG lasers).

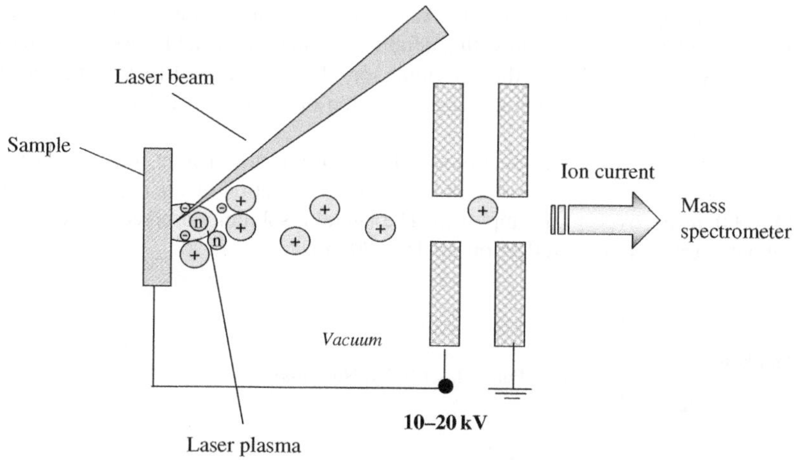

Figure 2.17 Principle of laser ion source.

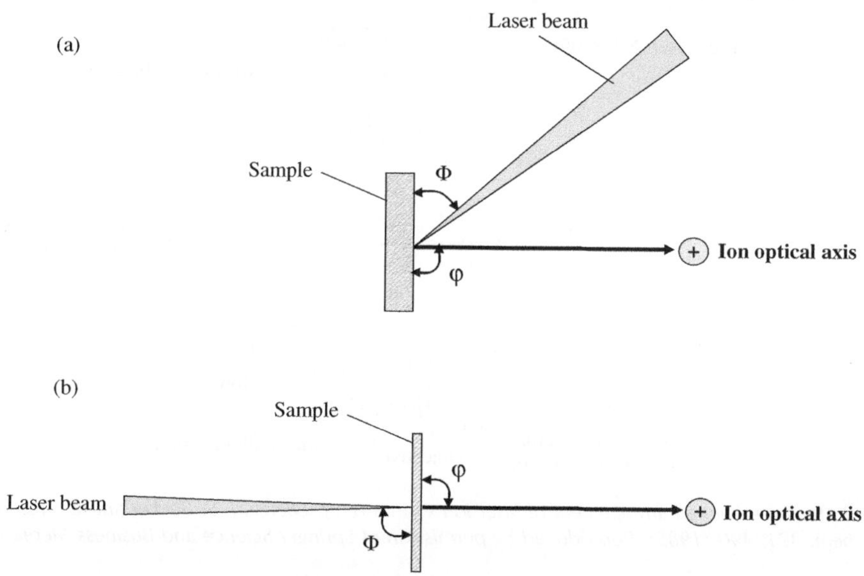

Figure 2.18 Geometric configuration in LIMS: a) reflection mode and b) transmission mode.

In general, two modes are used for the geometric configuration of the laser beam, sample and ion optical axis in the laser ionization source: the reflection mode and the transmission mode (see Figure 2.18 a and b, respectively). The angle (Φ) between the laser beam and the sample surface can be varied in the reflection mode (a) between 45° and 90° and the same range of angle (φ) is used between the ion optical axis and the sample surface. The most frequently used configuration in LIMS is the reflection mode.[79,82,83] The experimental arrangement of a laser ion source in the reflection mode developed by Dietze and Becker is presented in Figure 2.19.[74,79] This laser ion source, with a drift tube in combination with a double-focusing sector field mass spectrometer with Mattauch–Herzog geometry, allowed the determination of a multitude of trace impurities in solid samples to be analyzed with a maximum mass resolution of $m/\Delta m \sim 12\,000$ and detection limits in the low ng g^{-1} range.[84]

In the past, for the analysis of thin sections of tissues or thin transparent foils a laser ionization ion source in the transmission or reflection mode of laser irradiation (using e.g., the LAMMA 500 or LAMMA 1000, respectively) was employed. The lateral resolution observed in the transmission mode was about 1 μm and in the reflection mode 5–20 μm.

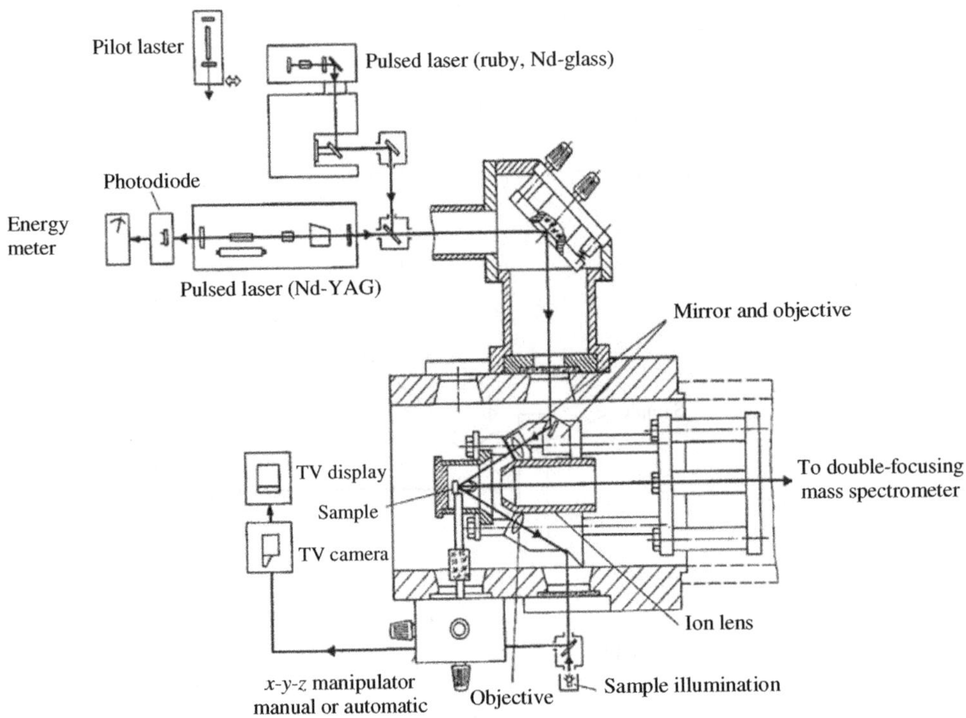

Figure 2.19 *Experimental arrangement of laser ion source (H. J. Dietze, J. S. Becker and Z. Fresenius, Anal. Chem. 321, 490 (1985). Reproduced by permission of Springer Science and Business Media.)*

It is characteristic of such a laser ion source that the experimental conditions for LIMS can be optimized with respect to a stoichiometric evaporation and effective ionization of solid sample material by varying the laser power density Φ as demonstrated in Figure 2.20. Under certain experimental conditions fractionation effects can be avoided. Stoichiometric laser evaporation and ionization of analyzed material is found at a laser power density between 10^9 W cm^{-2} and 10^{10} W cm^{-2}. In this laser power density range, the relative sensitivity coefficients of the chemical elements (RSC = measured element concentration/true element concentration) are nearly one for all the elements in different matrices. With decreasing laser power density ($\Phi < 10^9$ W cm^{-2}) an increasing fractionated evaporation of analytes and an increasing formation of polyatomic ions is observed. In the range of laser power density of 10^7–10^8 W cm^{-2} in weakly ionized plasma, only singly charged atomic ions are formed by thermal surface ionization. In general, the RSCs for low volatile elements are higher than for high melting elements. With a laser power density of about 10^7 W cm^{-2}, only laser induced thermal evaporation of the sample material occurs. The material vaporized during the interaction of the photons of the focused laser beam with the target surface can be post-ionized, e.g., by an electron beam. Whereas for low melting elements with decreasing laser power density Φ an increase in the RSCs occurs. For high melting elements the RSCs decrease with a reduction of laser power density, Φ. In contrast, with increasing laser power density $> 10^{10}$ W cm^{-2}, multiply charged ions with increasing charges and ion intensities are formed so that RSCs are not equal to one (see Figure 2.20). Under these conditions, polyatomic ion formation is minimized due to dissociation processes of these species in hot laser plasma. In addition, the initial energy of the ions increases with growing laser power density. For example, the initial energy of ions formed by the direct interaction of photons with the target surface in vacuum at 10^8 W cm^{-2} up to 5×10^9 W cm^{-2} is thus varied from about 1 eV to 1 keV, respectively. The charge of positive ions and their initial energies as a function of laser power density Φ in the laser plasma is illustrated in Figure 2.21.

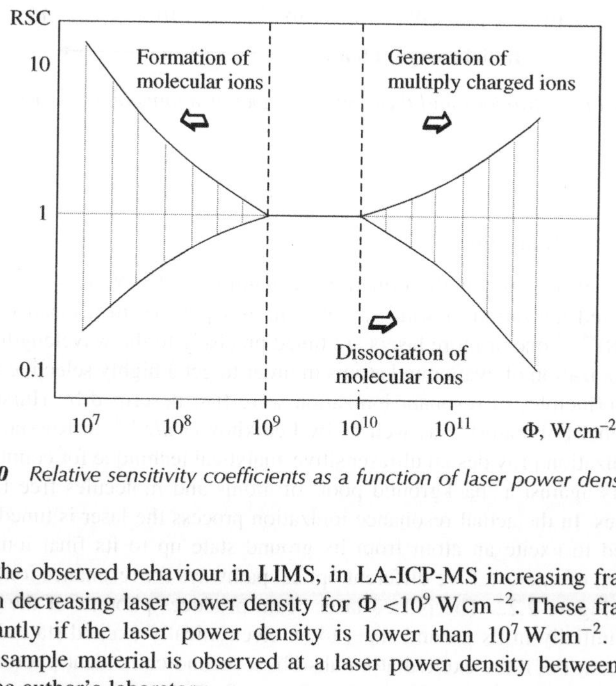

Figure 2.20 *Relative sensitivity coefficients as a function of laser power density in LIMS.*

In analogy to the observed behaviour in LIMS, in LA-ICP-MS increasing fractionation effects are observed with decreasing laser power density for $\Phi < 10^9$ W cm^{-2}. These fractionation effects increase significantly if the laser power density is lower than 10^7 W cm^{-2}. A stoichiometric laser ablation of sample material is observed at a laser power density between 10^9 W cm^{-2} and 10^{10} W cm^{-2} in the author's laboratory.

A high vacuum laser ion source is combined with a dynamic (ToF-MS) or static mass spectrometer (Mattauch–Herzog instrument) for the separation of the ion beams formed and used in LIMS for multi-element major, minor and trace analysis of compact solid samples (bulk analysis).[74]

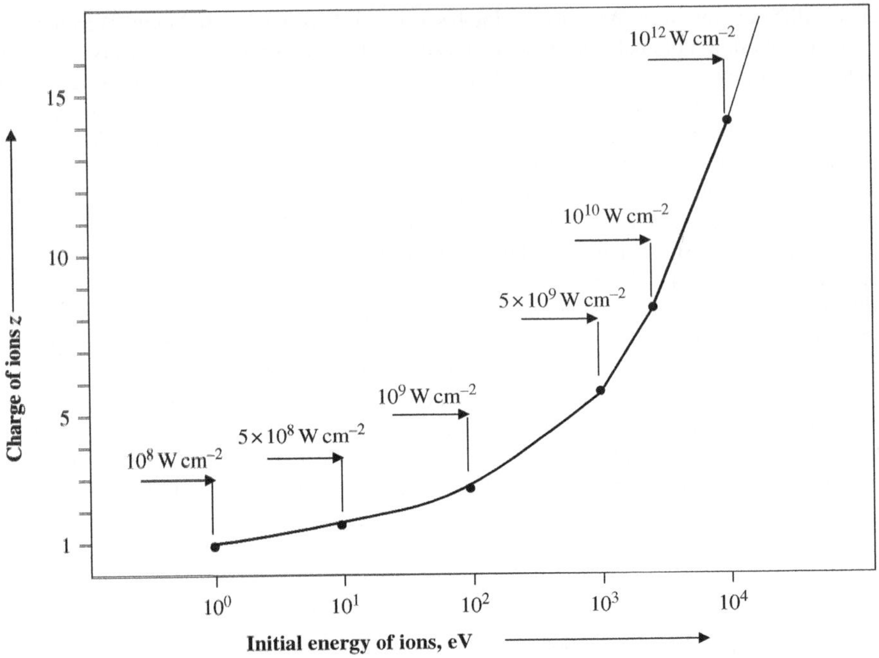

Figure 2.21 Charge of positive ions and their initial energies as a function of laser power density Φ in laser plasma.

2.3.2 Resonant Laser Ionization

Whereas in LIMS only one laser with defined wavelength (e.g., Nd:YAG – 1064 nm) is used for direct vaporization and ionization of solid samples in laser plasma, in resonance ionization mass spectrometry (RIMS)[85–87] one or more lasers are tuned precisely to the wavelength required for the excited states and ionization of evaporated atoms in order to get a highly selective ionization of the analyte. The basic principles of resonant ionization were first described by Hurst and coworkers at Oak Ridge National Laboratory[88] as well as by Letokhov *et al.*[89,90] in Russia. The technology using resonance ionization provides an ultrasensitive analytical technique for counting single atoms of selected elements against a 'background pool' of atoms and molecules free from all possible isobaric interferences. In the actual resonance ionization process the laser is tuned precisely to the wavelength required to excite an atom from its ground state up to its final ionization which is unique to the element of interest. Five different optical pathways for resonance ionization processes are summarized in Figure 2.22. The pulsed laser beam produces photons of the exact energy $h\nu$ to excite an atom, initially in its ground state, to an electronically excited state. In scheme ① an atom absorbs a photon (ω_1), the excited atom absorbs a second coincident photon ω_1 resulting in the ionization of the atom: A [ω_1, $\omega_1\bar{e}$] A^+. In scheme ② the laser is frequency doubled ($2\omega_1$): A [$2\omega_1$, $\omega_1\bar{e}$] A^+. The excitation and thus the resonance ionization is possible using one laser, whereby an atom absorbs a photon ($2\omega_1$) and the excited atom absorbs a second photon (ω_1) in order to ionize the atoms. In schemes ③–⑤ two or more lasers with different wavelengths are required for resonance ionization processes.[88] Stepwise optical excitation and subsequent ionization provides elemental and isotopic selectivity[87] at a high sensitivity with detection limits of up to 10^6 atoms per sample. However, only atoms of the element of interest are selectively ionized by resonance ionization.

Resonance ionization mass spectrometry as a combination of resonance laser ionization with mass spectrometry can be performed on gas atoms only. Therefore, in RIMS of solid samples, before resonance ionization, a neutral gas has to be produced using several methods known from solid state mass spectrometry. During the evaporation of solid material, e.g., by laser evaporation, thermal evaporation or by sputtering with a primary ion beam, the formation of ions should to be avoided. In RIMS, mostly the thermal evaporation of sample from a heated W or Re filament is applied.

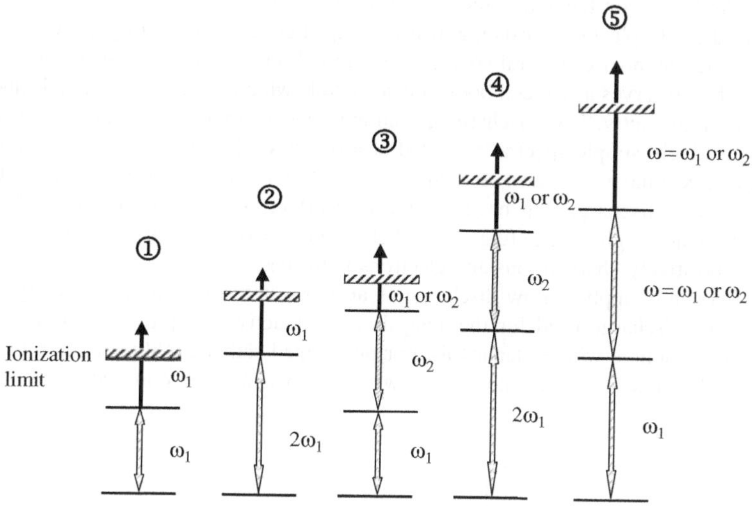

Figure 2.22 *Examples of various pathways available for resonance ionization. (G. S. Hurst et al. Rev. Mod. Phys.* ***51****, 767 (1979). Reproduced by permission of the American Physical Society.)*

State-of-the-art developments in RIMS, especially for the ultratrace determination of long-lived radioisotopes in environmental, biomedical and technical samples, are reported by Wendt *et al.*[91] Obviously, the application of a high vacuum ion source in LIMS and isotope selective RIMS is a great advantage, for example for reducing the formation of disturbing polyatomic ions, although these mass spectrometric trace and isotope analytical techniques are rarely used worldwide because of the expensive experimental arrangement. LIMS has been replaced by LA-ICP-MS. At present, no commercial RIMS instruments are available on the market. RIMS will be important in ultrasensitive trace analysis and especially in isotope ratio measurements of long-lived radionuclides in future, if cheaper tuneable diode lasers are applied.[91]

2.4 Glow Discharge Ion Source

The glow discharge ion source is simple yet versatile, providing information for emission and mass spectrometric methods.[92] A glow discharge ion source combined with a mass spectrometer has supplanted the spark ion source over many decades as the predominant ionization technique of mass spectrometry for determining trace and ultratrace element concentrations in various types of solid samples. Today, glow discharge mass spectrometry is an established analytical technique for analysis of solid samples. Modern approaches to glow discharge in spectrochemical analysis as a more universal method for the direct analysis of solid materials are described in the textbook *Glow Discharge Plasma in Analytical Chemistry* by Marcus and Broekaert.[93] The main preference for GDMS over SSMS is that the atomization of solid material and ionization of neutrals are separated. Furthermore, a more stable ion beam is generated in a glow discharge ion source in comparison to a spark ion source. As a result, in GDMS compared to SSMS, a better precision of analytical data is obtained. In order to use glow discharge for several applications, numerous configurations have been developed and coupled to mass spectrometers as well as optical emission spectrometers.[93]

The glow discharge for the ion source in glow discharge mass spectrometry (GDMS) is a partially ionized gas (a plasma) consisting of nearly equal concentrations of positive and negative charges and a large number of neutral species. The glow discharge cell consists of two electrodes, an anode, which also serves as the cell body, and a cathode which serves as the sample and is made in a pin or planar geometry.[94] As discharge gas an inert gas, commonly argon due to its discharge stability and relatively simple spectra,[92] is used at a pressure of about 10^2 Pa. Other noble gases (He, Ne, Kr and Xe) have been also studied as workable GD plasma gases.[95,96] The plasma gas employed in a glow discharge will have a significant effect on the type of ions produced.[92] By applying a sufficiently high voltage (e.g., 1 kV) between the two electrodes, gas breakdown occurs and plasma of positively charged ions and electrons is formed.

The most frequently applied glow discharge as an ion source in mass spectrometry is a direct current (dc) glow discharge used for the analysis of conductive samples. The principle of a dc glow discharge ion source with a planar (disc) or pin-shaped cathode is illustrated in Figures 2.23 and 2.24. The glow discharge operates as a low energy plasma ion source at a low pressure of

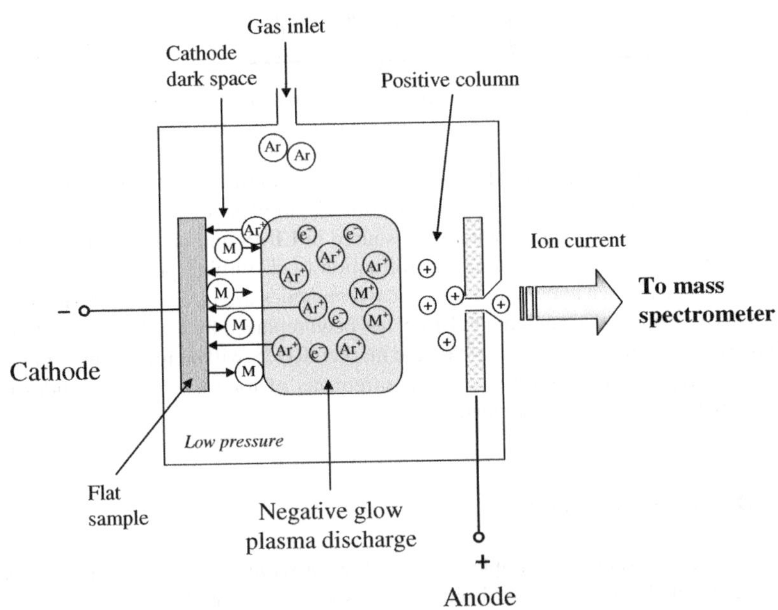

Figure 2.23 *Principle of glow discharge ion source with planar (disc) cathode.*

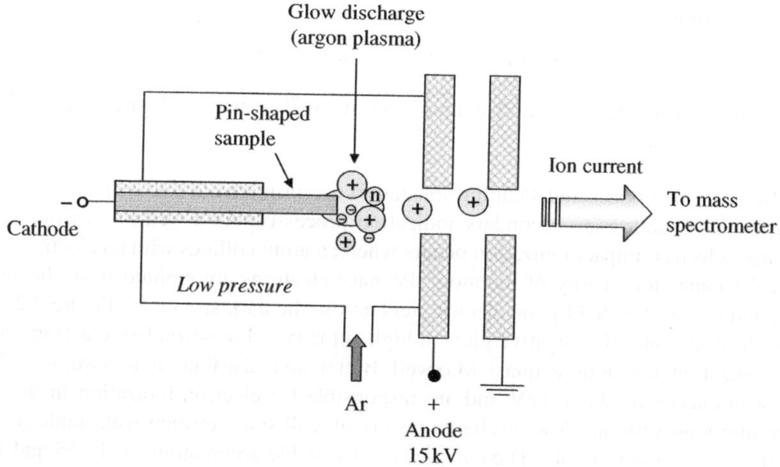

Figure 2.24 *Representation of a glow discharge ion source with pin-shaped cathode.*

$10–10^{-3}$ Pa (argon is generally used as the plasma gas). The Ar^+ ions formed in the glow discharge are accelerated towards the cathode from the sample investigated. The sample material is sputtered at the cathode surface by ion bombardment. Sputtered atoms and molecules are ionized in the glow discharge plasma ('negative glow') by different ionization mechanisms.

The ionization mechanisms[97] taking place in an argon glow discharge are:

- electron impact ionization

$$M + e^- \rightarrow M^+ + 2e^- \tag{2.9}$$

- Penning ionization

$$M + Ar^{m*} \rightarrow M^+ + Ar + e^- \tag{2.10}$$

- charge transfer

$$M + X^+ \rightarrow M^+ + X \tag{2.11}$$

$$MO + X^+ \rightarrow M^+ + O + X \text{ (dissociative)} \tag{2.12}$$

- associative ionization

$$M + Ar^{m*} \rightarrow MAr^+ + e^- \tag{2.13}$$

- photoionization

$$M^* + h\nu \rightarrow M^+ + e^- \tag{2.14}$$

whereby the excited (metastable) atoms M^* are formed by reaction of neutral (M) with excited metastable argon atoms Ar^{m*}:

$$M + Ar^{m*} \rightarrow M^* + Ar \tag{2.15}$$

- cumulative ionization

$$M + e^- \rightarrow M^* + e^- \rightarrow M^+ + 2e^- \tag{2.16}$$

with M – sputtered atom; M* – excited sputtered atom; Ar^{m*} – metastable argon atom; X^+ – ion generated in glow discharge.

Ionization by electron impact and Penning ionization are primary ionization processes, whereas all other types of ionization are secondary ionization processes. As a well-known process in the glow discharge, electron impact ionization occurs when an atom collides with an electron at energy higher than the ionization energy of the atom. Primary electrons are emitted from the surface of the cathode and are accelerated by the electric field across the dark space (see Figure 2.24), where most of the electrons enter the negative glow at high velocity.[93] For secondary electrons generated during the ionization reactions a quasi Maxwell–Boltzmann distribution is assumed. Electrons possess a mean energy of about 4 eV and are responsible for electron ionization in the negative glow.[98] Penning ionization in glow discharge occurs by collision between metastable plasma gas atoms and the atoms of the sample. The energies of metastable argon atoms of 11.55 and 11.72 eV are sufficient to ionize most elements of the periodic table.[98]

An extensive numeric modelling of different collision processes (including the densities and level populations of the plasma species) in direct current (dc) glow discharge – which is still the most widely used source in GDMS – is described by Bogaerts and Gijbels.[94,99,100] Whereas the plasma species (e.g., electrons, gas ions and fast gas atoms) are described by a plasma model, a sputter model is based on the results of the plasma model and deals with the phenomena related to the analyte atoms, i.e., the sputtering process, the thermalization of plasma by elastic collisions with gas atoms and subsequent diffusion of thermalized atoms in the plasma, the ionization of atoms and finally the path of ions in the mass spectrometer.[94] Plasma diagnostics (determination of electron and ion density, electron temperature and plasma potential), in order to check the theoretical results, can be performed by Langmuir probe measurements.[101,102]

As an example, the glow discharge ion source with Grimm-type geometry developed by Jakubowski *et al.*[103] (see Figure 2.25) can be coupled to a commercial mass spectrometer. The positively charged ions formed in the argon plasma of the glow discharge source are extracted and accelerated into a double-focusing sector field mass spectrometer, but also a quadrupole mass analyzer, an ion trap or a ToF mass spectrometer can be employed.

In general, the Grimm-type glow discharge ion source shows stable operation in direct current (dc) mode and pulse mode (at a frequency of 600 Hz and a pulse width of 20 μs) providing either a continuous supply of ions or a time variant ion beam. Hastings and Harrison[92] compared glow discharges coupled to a ToF-MS (LECO Corporation, St. Joseph, MI) using traditional inert gases[104] and alternative reactive gases, where argon was systematically replaced by nitrogen (or oxygen). To decouple the reactive plasma processes from potentially reactive surface reactions, gold as a non-reactive cathode material was employed. With increasing nitrogen and oxygen addition, significant loss of the Ar^+ and ArH^+ ion intensities was observed, but the O_2^+ ion intensity increased. By application of nitrogen as plasma gas only small intensities of N_2^+, N_2H^+, N_3^+ and N_4^+ ions in the dc and pulsed GD mass spectra were measured, with a significantly lower sputter rate in the pulsed mode.[92]

A new dc glow discharge has been designed and interfaced with a ToF-MS by Hieftje and coworkers.[105] This glow discharge ion source is simple to operate, sustainable in a He environment at atmospheric pressure at the 50–100 mA level with a burning voltage of 0.9–1.2 kV. Singly charged atomic ions are observed in GD mass spectra. Because of the high first ionization energy of He (24.6 eV), the presence of He^+ in background mass spectra suggests that a significant amount of energy is available to ionized gaseous analyte species that are free from interfering

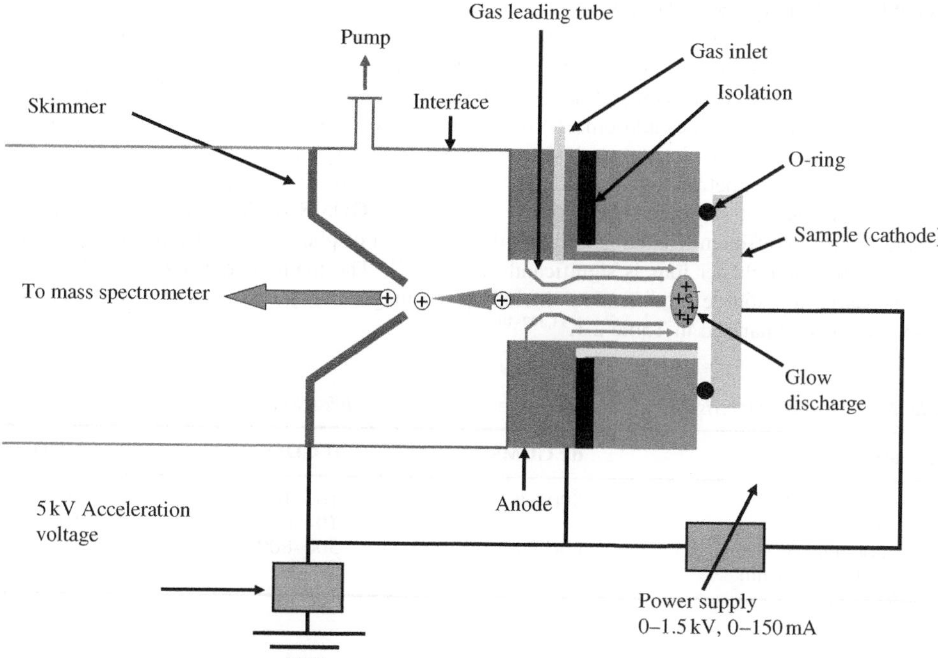

Figure 2.25 *Experimental arrangement of modified Grimm-type dc GD ion source design with flow tube. (Reproduced by permission of N. Jakubowski, Nachrichten aus der Chemie, 2003.)*

background ions. If this system is coupled with hydride generation, a reduction in the continuum background permits excellent limits of detection down to 10 ppt (comparable to those generated by solution-based ICP-ToF-MS) to be obtained.[105]

Whereas a dc glow discharge ion source is suitable for the direct analysis of conducting samples, non-conducting bulk materials or layered systems are not easy to analyze by dc GDMS due to charging-up effects on the surface of the sample cathode so that the positive ions stop bombarding the cathode. In order to increase the application field for dc GDMS, especially for the characterization of non-conductors with respect to trace impurities, further modifications have been introduced, for example by the application of a secondary cathode diaphragm in front of the flat non-conducting sample surface,[106] which produces a thin conducting layer on the sample surface during sputtering with Ar^+ ions of plasma gas (due to redeposition of secondary cathode material). Furthermore, the application of a conducting powdered binder (e.g., of high purity graphite, Ag, Au, Cu, Ga), which is mixed with the non-conducting powdered sample before the pressing of the homogeneous mixture to form an electrode, similar to the technique applied in SSMS for the analysis of non-conducting powders, is generally used in dc GDMS as well. Both techniques for the analysis of non-conductors by dc GDMS have major disadvantages. Firstly, the formation of new disturbing polyatomic ions (e.g., argide or oxide ions etc. of secondary cathode or binder material) is observed in the mass spectra. Secondly, the sample surface is contaminated by secondary cathode material and the detection limits of the elements deteriorate due to dilution of the powdered sample by the conducting binder. The latter process is furthermore related to possible contamination during the preparation of mixed electrodes as well. In order to avoid these difficulties in dc GDMS, radio frequency (rf) glow discharge sources have been developed in several laboratories for direct solid-state analysis, especially of compact insulators (e.g., ceramics, glasses or thick oxide layers on super-resistant alloys) without any restriction.[107–109] Studies of ion intensities and energy distributions of atomic ions of analyte, plasma gas ions, residual gas ions and polyatomic ions (argides, hydrides, oxides and others) formed in the glow discharge have been carried out in order to optimize the trace element analysis in semiconducting and non-conducting materials with the rf GDMS and magnetron-enhanced rf GDMS.[93,110] By utilizing an additional magnetic field located behind the flat sample – magnetically enhanced rf GDMS (rf MGDMS) - the atomization and ionization efficiency is enhanced and the formation of plasma gas ions, residual gas ions and polyatomic and cluster ions is significantly reduced.[111] The main characteristics of dc and rf glow discharge ion source modifications including the magnetically enhanced rf glow discharge ion source are summarized in Table 2.1.[93,107–109,112]

Table 2.1 *Main characteristics of glow discharge ion source modifications.*[108]

Ion Source Type	dc GDMS	rf GDMS	rf MGDMS
Gas (Ar) Pressure, Pa	500–50	100–10	100–1
Power, W	1–5	10–50	10–50
dc Bias Potential, V	800–1200	500–800	$\sim$500
Sputtering Rate, mg/min.	0.05–0.5	0.1–1	0.1–10

DART™ (Direct Analysis in Real Time) as an atmospheric pressure ion source that overcomes many limitations of traditional ion sources for analysis of organic substances was introduced in 2004.[113] DART is based on the interaction of excited gas atoms (helium or nitrogen) with analyte having an ionization energy lower than the energy of the excited gas (e.g., He*, the excited helium state has an energy of 19.8 eV) whereby positively charged ions of analyte are generated by Penning ionization (Equation 2.10) at atmospheric pressure. When helium is used, the dominant positive ion formation mechanism involved is the generation of ionized water clusters followed by proton transfer reactions. Negatively charged ions of analyte are formed via reactions with O_2^- ions which were generated by electron capture of electrons produced by Penning ionization with atmospheric oxygen. Using the DART ion source (coupled to AccuTOF-LC™ from Jeol) simple mass spectra characterized by M^+ and/or $(M+1)^+$ in the positive ion mode and M^- and/or $(M+1)^-$ in the negative ion mode, respectively, have been obtained. Alkali metal cation attachment and doubly charged ions are not observed. The DART ion source combined with a ToF-MS permits easy and fast qualitative and quantitative analysis of gases, liquids and solids (drugs, body fluids or tissues, explosives, chemical weapon agents, environmental compounds, papers, polymers, food, spices, beverages, organic or organometallic compounds and others) in open air without sample preparation.

2.5 Thermal Surface Ionization Source

Thermal ionization mass spectrometry (TIMS) is one of the oldest mass spectrometric techniques, first applied by Dempster in 1918.[114] The thermal emission of positivly charged ions emitted from a salt on a heated surface was first observed by Gehrcke and Reichenheim 12 years before.[115] The thermal surface ionization source is a very simple ion source and operates under high vacuum conditions. TIMS is mostly useful for elements with relatively low ionization energy (E_i) – in general lower than 7 eV (e.g., Rb – 4.18 eV or K – 4.34 eV; the ionization energy of elements together with the relative abundances of isotopes and atomic masses are summarized in Appendix I). In TIMS only a small amount of analyte, about 1 ng or less, is required in the routine mode. In a thermal ionization source, ion formation is based on following principle. If an atom is absorbed on a metal surface and the work function W for electrons leaving the filament metal surface (e.g., Re–4.98 eV; W – 4.58 eV; thoriated tungsten – 2.7 eV) is larger than the first ionization energy (E_i) of the atom, then the atom is ionized due to the Langmuir effect.[116] By heating the metal surface the absorbed ion is evaporated if $W > E_i$, i.e., if the Fermi levels of the atom and of the surface are similar, which is required for an electron exchange between atom to be ionized and the metal of the filament surface. With increasing temperature of the metal surface the probability increases of thermal surface ionization for different elements with low ionization energy. To theoretically describe the process of thermal surface ionization Langmuir and Kingdom[116] modified the Saha equation valid for plasma. The ionization efficiency (α) for a selected analyte in thermal surface ionization can be described by the Saha–Langmuir equation (Equation 2.17.) if a thermodynamic equilibrium is obtained:

$$\alpha = n^+/n_0 = g^+/g_0 \exp\left[(\mathrm{W} - \mathrm{E_i})/\mathrm{kT}\right] \qquad (2.17)$$

where α = ionization yield
n^+ and n_0 = number of emitted ions and neutral atoms, respectively
g^+ and g_0 = statistic weights of ion and atom, respectively
W = work function of the surface
E_i = first ionization energy of the atom to be ionized
k = Boltzmann constant
T = temperature of filament surface in degrees kelvin

For many decades, thermal ionization mass spectrometry was the leading, generally recognized analytical technique for the precise and accurate measurement of isotopic ratios for different elements with a precision of better than 0.01 %.[117] In TIMS, a small volume (1–10 μl) of aqueous sample solution, which contains some ng to μg of the analyte to be analyzed, is deposited on a cleaned filament surface (e.g., high-purity Re) and evaporated to dryness. The experimental arrangement of a single filament thermal ionization source is shown in Figure 2.26a. On the surface of a hot filament (e.g., of rhenium, at a filament temperature of 1000–2500 K) the evaporated sample is ionized. The most frequently applied technique in TIMS works with two filaments arranged opposite each other (see Figure 2.26b). One of the filaments is used for the evaporation of sample by thermal heating and the other for ionization of evaporated atoms and molecules on the hot filament surface. Typical ionization curves in TIMS for several elements are illustrated in Figure 2.27. Using this experimental arrangement the evaporation of sample and ionization of gaseous sample atoms are separated from each other. The advantage of the double filament thermal ionization source compared to a single filament arrangement is better control and optimization of both processes (i.e., evaporation of solid sample and ionization of evaporated atoms, respectively) separately. Furthermore, by increasing the current of the ionization filament stepwise, several analytes with different ionization energies (e.g., Rb and Sr) can be analyzed one after the other. The thermal surface ions formed in a high vacuum ion source are extracted via an ion optical system in a single-focusing mass spectrometer. In order to analyze elements with an ionization energy > 7 eV special 'emitter techniques' have been developed, for example by adding a small quantity of silica gel and phosphoric acid to the sample on the single filament,[118] whereby an enhancement of ionization by at least one order of magnitude is observed. This technique has been applied for U/Th–Pb geochronology and for studies of fine isotope variation in nature and works successfully, especially for lead isotope ratio measurements. Furthermore, in TIMS, under certain experimental conditions, relatively high stability ion currents of polyatomic ions have been observed, which have been used as analyte ions in isotope ratio measurements (e.g., $NaBO_2^+$ and $Li_2BO_2^+$ for boron and lithium, respectively, or AsS^+ for sulphur).[62]

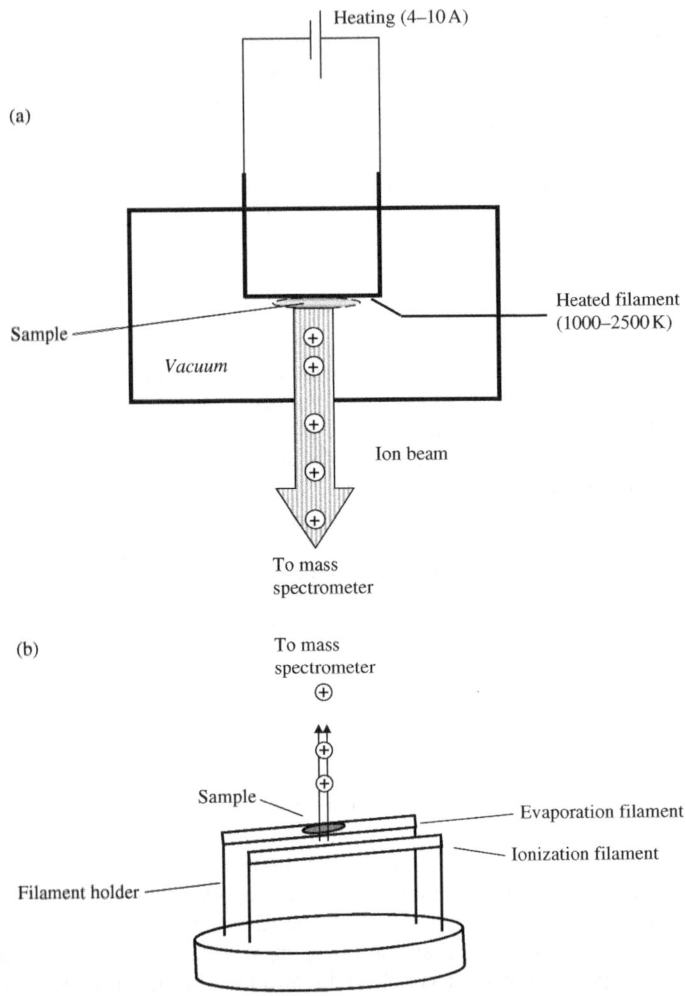

Figure 2.26 *a) Experimental arrangement of a single-filament thermal surface ionization source; b) Schematic diagram of a double-filament arrangement for thermal surface ionization. (H. Kienitz (ed.), Massenspektrometric (1968), Verlag Chemie, Weinheim. Reproduced by permission of Wiley-VCH.)*

A main characteristic of TIMS is that positively or negatively charged ions of the analyte are also formed and used for mass spectrometric analysis. In negative thermal ionization mass spectrometry (NTIMS) elements or molecules with a relatively high electron affinity (E_{ea} > 2 eV) can be

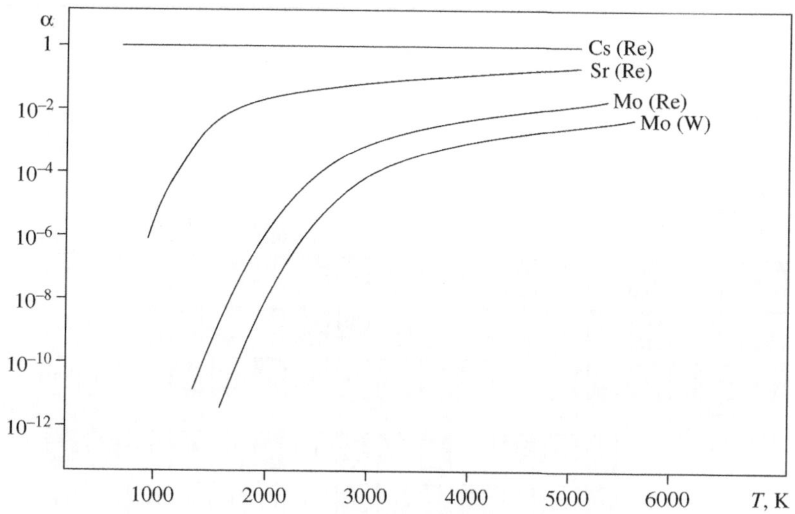

Figure 2.27 *Typical ionization curves in thermal surface ionization.*

investigated. The ionization efficiency for negative ions is a function of the electron affinity of the analyte E_{ea} (e.g., Cl – 3.61 eV, Br – 3.36 eV, CN – 3.17 eV, I – 3.06 eV, Se – 2.12 eV, S – 2.07 eV) and can be described in an equilibrium state by a modified Saha–Langmuir equation (Equation 2.18):

$$\alpha = n^-/n_0 - g^-/g_0 \exp\left[(E_{ea} \quad W)/kT\right] \tag{2.18}$$

where α = ionization yield
n^- and n_0 = number of emitted ions and neutral atoms, respectively
g^- and g_0 = statistic weights of ion and atom, respectively
W = work function of the surface
E_{ea} = electron affinity of the analyzed element
k = Boltzmann constant
T = temperature of filament surface in degrees kelvin.

Besides negative singly charged atomic ions (M^-), polyatomic ions (MO_x^- with $x = 1$–4) are also observed and applied in NTIMS. A further benefit of NTIMS is that mostly high ion intensities were measured and very seldom are isobaric interferences from impurities in mass spectra observed. NTIMS was of interest for Re–Os geochronology, because both elements possess high ionization energy. Heumann *et al.*[119] have demonstrated the possibility of precise isotope ratio measurements for Os via negative ions. Only 1 ng of Os sample is sufficient to produce an ion current of the most abundant OsO_3^- ions of 10^{-11} A. A precision of 0.04–0.01 % for the osmium isotope ratio is observed. In Figure 2.28 the capability of thermal ionization via positively or negatively charged ions as demonstrated by Heumann *et al.*[119] is illustrated.

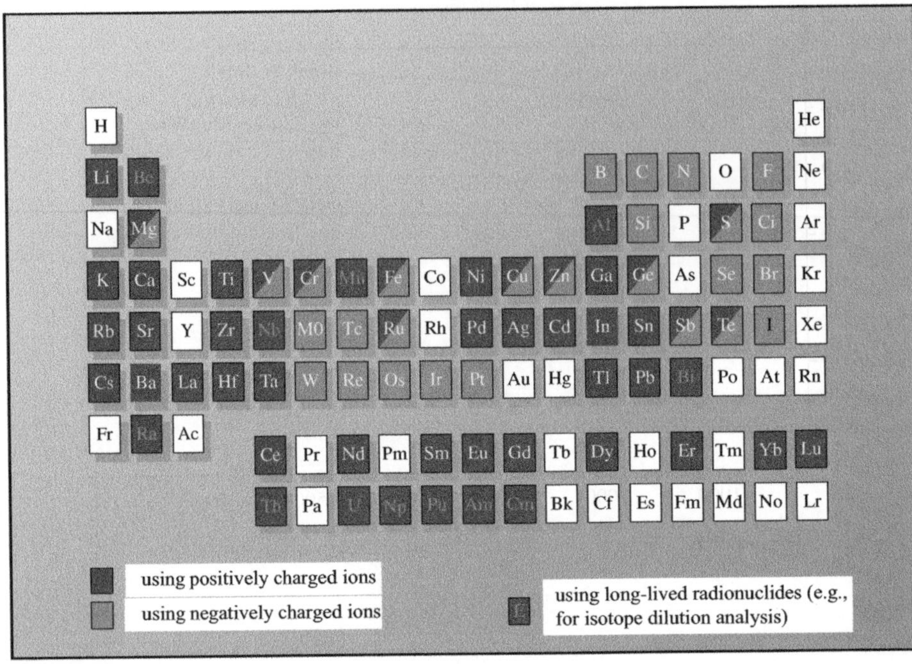

Figure 2.28 *Elements analyzed by thermal surface ionization. (K. Heumann et al. Analyst **120,** 1291 (1996).*

In general, ions generated by thermal ionization possess low initial energies (0.1–0.2 eV), therefore mostly single magnetic sector field mass spectrometers are used for ion separation.

2.6 Ion Sources for Secondary Ion Mass Spectrometry (SIMS) and Sputtered Neutral Mass Spectrometry (SNMS)

Secondary ion mass spectrometry (SIMS) and sputtered neutral mass spectrometry (SNMS) are today the most important mass spectrometric techniques for surface analysis, especially for thin layer analysis, for depth profiling, for the determination of contaminations and element distribution on a solid sample surface.

In SIMS and SNMS, the components of the solid sample surface (or liquid in several special applications) are sputtered during bombardment with a focused primary ion beam (e.g., Ar^+, Cs^+, Ga^+, O^-, O_2^+) of sufficiently high ion energy (0.2–40 keV) in a high vacuum ion source. The choice of primary ion beam depends on the element to be analyzed and the type of information required. The primary ion beams are produced in a primary ion source such as a duoplasmatron ion source, a caesium or a liquid metal ion source. As a function of the ion energy, the implanted primary ions penetrate the solid surface to different depths (1–10 nm, penetration depth) and transfer their kinetic energy as a function of the sample material, their primary ion energy and their mass. A representation of primary ion–solid interaction and the sputtering process is illustrated in Figure 2.29. Part of the transferred energy which is returned to the solid surface via impact cascades induces the sputtering of positively and negatively charged secondary atomic and polyatomic ions and neutrals.[120] A theoretical collision cascade model for the sputtering process is described by Sigmund.[121] Charged and uncharged particles are sputtered and ejected from the target surface. Approximately 1 % of the sputtered material comes off as ions, which can then be analyzed mass spectrometrically. Neutrals are sputtered with the highest intensities.

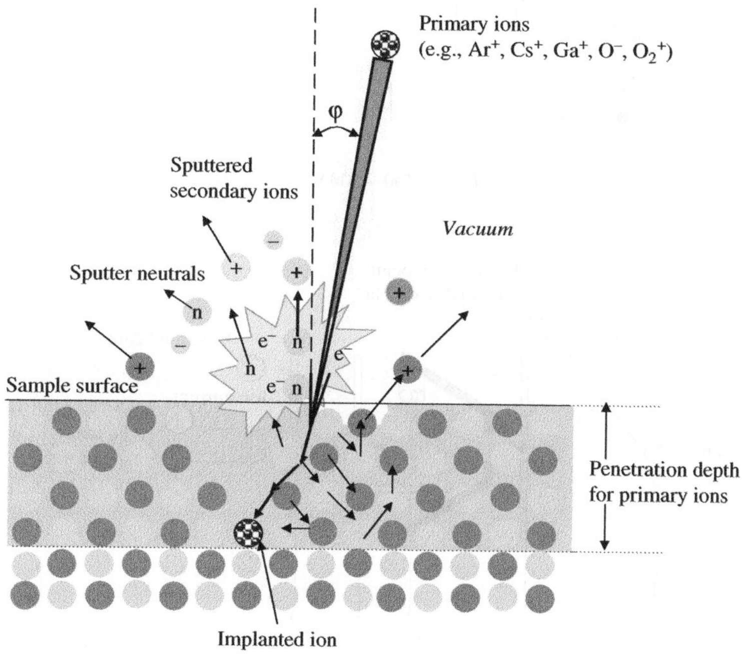

Figure 2.29 *Schematic diagram of the sputter process in a secondary ion source.*

The principles of ion sources which use a primary ion beam for sputtering of solid material on sample surface in a high vacuum ion source of a secondary ion mass spectrometer or a sputtered neutral mass spectrometer are shown in Figure 2.30a and Figure 2.30b, respectively. Whereas in SIMS the positive or negative secondary ions formed after primary ion bombardment are analyzed, in SNMS the secondary sputtered ions are suppressed by a repeller voltage and the sputtered neutrals which are post-ionized either in an argon plasma ('plasma SNMS'[122]), by electron impact ionization ('e-beam SNMS'[123]) or laser post-ionization[124] are used for the surface analysis (for details of the ionization mechanisms see references 122–124).

In general, the sputtered yield of ions from the sample surface and thus the relative sensitivity factors (RSFs) of elements varies for a given primary ion beam according to the experimental parameters used and as a function of the matrix composition by six orders of magnitude. Matrix-dependent RSFs for several types of primary ion beams are used as correction factors for the quantification of analytical data. RSFs (or sputtered ion yield) are dependent on the ionization energy (E_i) and electron affinity (E_{ea}) of elements. A general exponential relationship between positive ion yields and ionization energy and negative ion yields with electron affinity in different

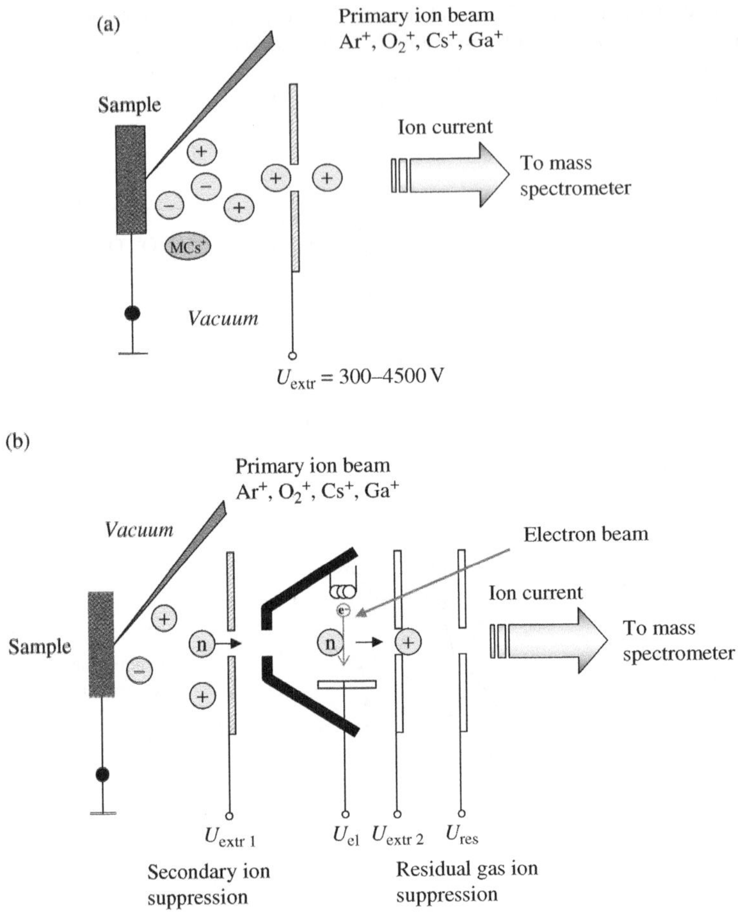

Figure 2.30 *a) Principles of a secondary ion source; b) Schematic diagram of an ion source of a secondary neutral mass spectrometer (SNMS). (J. S. Becker and H. J. Dietze, Int. J. Mass Spectrom. Ion. Proc.* ***197****, 1–35. (2000). Reproduced by permission of Elsevier.)*

matrices (e.g., GaAs, Si, GaP, diamond, Ge, $LiNbO_3$ and InP) has been described by Deline *et al.*[125] The choice of primary ion beam can be made with knowledge of RSFs for several elements in a given matrix. The relative sensitivity factors for elements, for example in a silicon matrix using O_2^+ and Cs^+ primary ions, respectively, are illustrated in Figure 2.31.[126] Positively charged secondary ions are analyzed e.g., by sputtering of silicon surface using an O_2^+ primary ion beam. Whereas halogenides with high ionization energy possess a low element sensitivity, alkali elements with the lowest ionization energy are formed with the highest ion intensities. In contrast, by the use of a Cs^+ primary beam, negatively charged ions are formed and electronegative elements (e.g., H, C, O, N, F, Cl, P, Ge, Se, As, Br, Te, I and Au) are analyzed and the behaviour of element sensitivity is reversed. That means elements with high electronegativity are more sensitively measured than elements with lower ones. The impact on the solid sample by Cs^+ primary ions results in an increase in the Cs concentration on the surface so that the electron leaving potential is decreased and the yield of negatively charged secondary ions is increased. The application of Cs^+ primary ions enhances the ionization yield (sensitivity) of electronegative elements by several orders of magnitude compared to the use of Ar^+ or Ga^+ primary ion beams. In addition, using Cs^+ as a primary beam, MCs^+ polyatomic ions (M – analyte atom of sample investigated) are formed by the reaction of reactive primary ions with analyte during the sputtering process in 'Cs polyatomic ion SIMS'. By the application of MCs^+ and $MCs_2{}^+$ polyatomic ions[127], measured at a mass that is 133 u or 266 u, respectively higher than the mass of analyte (metal or non-metal), it is possible to avoid isobaric interferences that are observed at the mass of the atomic ion M^+.[128] The positively or negatively charged ions formed during sputtering in the secondary ion source were ejected from the target surface are accelerated into the mass analyser.

(a)

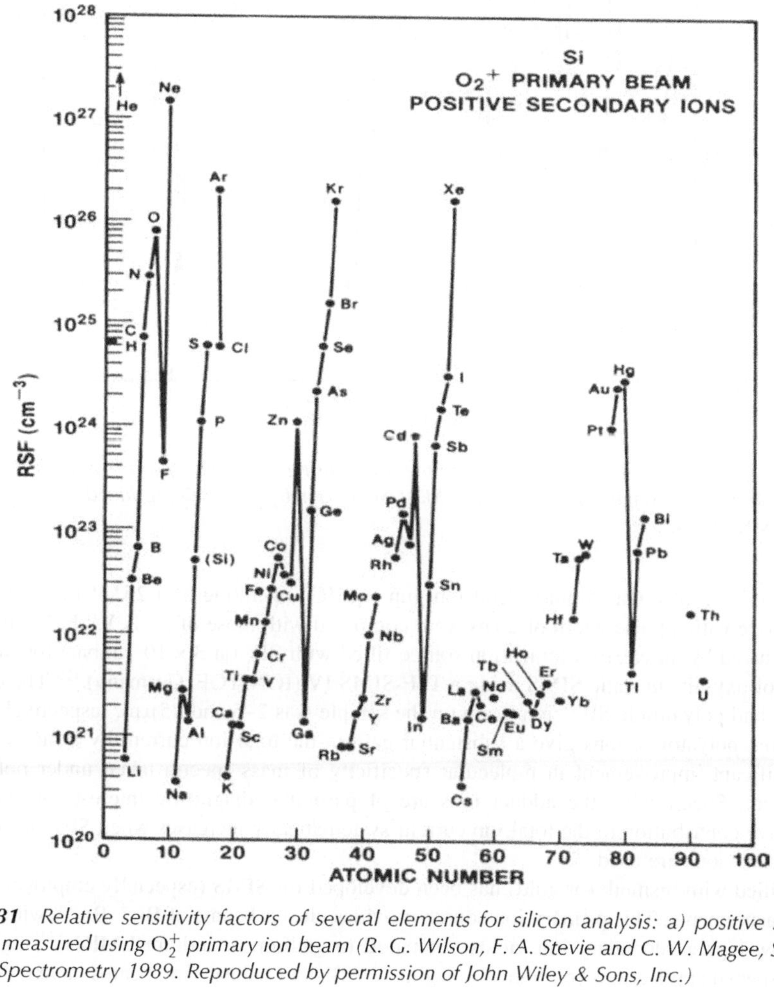

Figure 2.31 *Relative sensitivity factors of several elements for silicon analysis: a) positive secondary ions were measured using* O_2^+ *primary ion beam (R. G. Wilson, F. A. Stevie and C. W. Magee, Secondary Ion Mass Spectrometry 1989. Reproduced by permission of John Wiley & Sons, Inc.)*

(b)

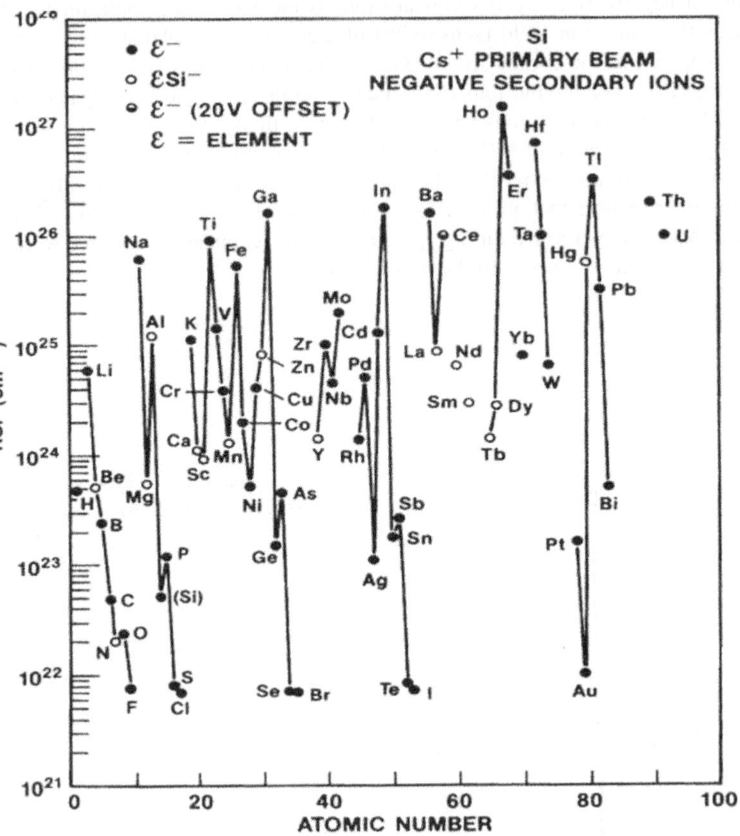

Figure 2.31 b) Negative secondary ions were measured using Cs^+ primary ion beam. (R. G. Wilson, F. A. Stevie and C. W. Magee. Secondary Ion Mass Spectrometry 1989. Reproduced by permission of John Wiley & Sons, Inc.)

The capabilities of a Ga^+ liquid metal ion gun (LMIG) operating at a 25 kV beam voltage in bunched mode with a pulse width of 20 ns were compared with those of a 9 keV SF_5^+ primary ion source produced by an electron ionization source filled with SF_6 (at 3×10^{-6} mbar) for molecular speciation of oxysalts in static SIMS using a ToF-SIMS IV (ION-TOF, Germany).[129] The diameter of the Ga^+ and polyatomic SF_5^+ ion beam on the sample was 2–5 and 25 μm, respectively. It was observed that polyatomic ions give a substantial gain in the total ion current by a factor of 4–10 and a significant improvement in molecular specificity of mass spectra taken under polyatomic bombardment. Specifically, the adduct ions are of particular diagnostic interest for speciation. Their relative contribution to the total ion current systematically increases when SF_5^+ primary ions instead of Ga^+ ions are used.[129]

LMIG filled with bismuth (or gold) has been developed for SIMS (especially employed in ToF-SIMS) measurements. This primary ion source provides cluster beams of Bi_n^+ Bi_n^{2+} with $n = 1$–7 which allow higher secondary ion yields compared to gallium or indium LMIGs[130] and is well suited for special applications (biological imaging).[131] Bismuth cluster ion beams have much better intensities and efficiencies than Au_3^+ cluster primary ion beams and allow better spatial and mass resolution in ToF-SIMS as demonstrated for biological imaging in rat brain sections.[131]

Secondary ion sources (using primary ion beams to sputter solid sample surfaces) are applied not only in SIMS and SNMS, but also for the formation of negative ions in accelerator mass spectrometry (AMS).

2.7 Electron Ionization Source

If electrons of sufficient energy (70 eV) interact with the sample vapour at a pressure of 10^{-5}–10^{-3} Pa positive ions are generated. Electron ionization (EI) in an ultrahigh vacuum ion source is a well known ionization process. Gas compounds, volatile liquids and solids can be ionized in an electron ionization source (formerly electron impact ion source). However, electron ionization also plays a part in all types of plasma ion sources including inductively coupled plasma sources operating at atmospheric pressure. The electron ionization source was introduced by Dempster[114,132] in 1916 as a useful ion source. Electron ionization sources were improved by Smyth,[133] Bleakney[134] and later by Tate[135] and Nier.[136] The electron ionization source is applied for the ion formation of gases or of volatile samples which readily form gases before or during introduction to the mass spectrometer. The gases are introduced directly into the ion source; the basic arrangement is shown in Figure 2.32. The gaseous sample or the vapour of the volatile liquid or evaporated solid (e.g., by Knudsen effusion) is introduced continuously into the collision chamber of the ion source at reduced pressure (1–10 Pa). The pressure in the ion source combined with the high vacuum system of the mass spectrometer is $< 10^{-4}$ Pa. Electrons are emitted from the heated cathode (e.g., filament of tungsten or rhenium, temperature of cathode $T_K \sim 1500$–2000 K) and accelerated to the anode positioned opposite. The accelerating voltage for electrons (V_e) between the cathode and anode is chosen to be between 10 V and 100 V. The electron beam density is about 10 mA mm^{-2}. Gas atoms and molecules are ionized and fragmented by the collision with electrons via the following reactions (For the ionization of an atom or molecule the impacting electron must summon up at least the required energy of the first ionization potential – so-called 'appearance potential'):

$$A + e^- \rightarrow A^+ + 2e^- \quad (2.19)$$

$$AB + e^- \rightarrow AB^+ + 2e^- \quad (2.20)$$

$$AB + e^- \rightarrow A^+ + B + 2e^- \quad (2.21)$$

$$AB + e^- \rightarrow A + B^+ + 2e^- \quad (2.22)$$

Besides the ionization of gases an induced fragmentation of molecules by electron impact (EI fragmentation) also occurs (Equations 2.21 and 2.22). The fragmentation of organic molecules can be used for structural analysis of unknown compounds.

Electron ionization sources produce constant ion beams of about 10^{-8} A with low initial energy spread. The ion current measured depends strongly on the ionization degree of the gas analyzed (type of atoms and molecules). Positive ions and electrons are formed by the interaction of electrons of sufficient energy with gas atoms or molecules. The ion current I^+ is proportional to the pressure (p) of the gaseous sample, to the electron current I_e, the length (l) of the collision chamber and the differential ionization (s) of elements as a function of the ionization energy:

$$I^+ = s \cdot p \cdot l \cdot I_e \quad (2.23)$$

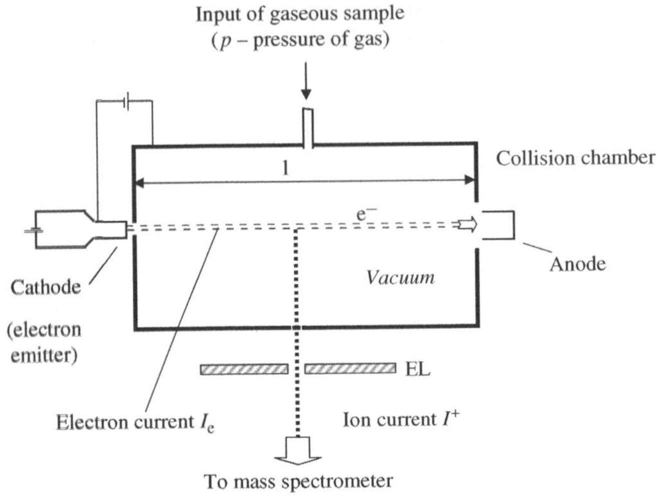

Figure 2.32 *Principles of electron ionization source. Cathode (W, Re): e^- – emission; e^- – energy is a function of accelerating voltage; pressure (p) of gaseous sample < 10 Pa. (H. Kienitz (ed.), Massenspektrometrie (1968), Verlag Chemie, Weinheim. Reproduced by permisssion of Wiley-VCH.)*

The differential ionization (s) of a gas (at a pressure of about 10^2 Pa) is proportional to the number of positive ions and electrons formed via collision with an electron along a path of 1 cm. The parameter s is a function of the energy of the electrons, and the type and density of the gas. In order to obtain a linear dependence of the pressure of the sample in a Knudsen cell and the ion current of the species measured by mass spectrometry, the pressure of the sample is selected so that the mean free path length of the evaporated species is larger than the cell dimensions (2–3 cm). The ionization efficiency curves (i.e., the number of ions produced per cm free path length and per mm Hg pressure in the ionization region of an electron ionization source as a function of electron energy) are illustrated in Figure 2.33. For every type of gas (consisting of atoms or molecules) a specific ionization efficiency curve of the gas can be measured. The minimum energy required for ionization of an atom or molecule is defined as the ionization energy (E_i). Whereas for atoms the first ionization energy varies between 3.89 (for Cs) and 24.59 eV (for He) most molecules have an ionization potential in the 7–15 eV range. With increasing energy of electrons the differential ionization increases for different gases up to a maximum of between 50 and 80 eV and decreases with further increasing electron energy. On the basis of typical ionization efficiency curves, electron ionization sources operate in this electron energy range (mostly at 70 eV) because all the atoms and molecules are ionized to single (positively) charged atomic or polyatomic ions with maximum intensity. At higher electron energy applied in an electron ionization source, the formation of multiply charged ions is observed. Electron ionization mass spectrometry uses the singly charged atomic ions or polyatomic ions generated as analyte ions (such as SiF_3^+, SF_5^+, UF_5^+ or VOF_3^+) for isotope ratio measurements.

The ions formed are extracted by the extraction lens, EL, perpendicular to the electron beam. Due to the low energy spread of ions formed in an electron ionization source (< 1 eV) one single-focusing magnetic sector field mass spectrometer or quadrupole analyzer is sufficient for

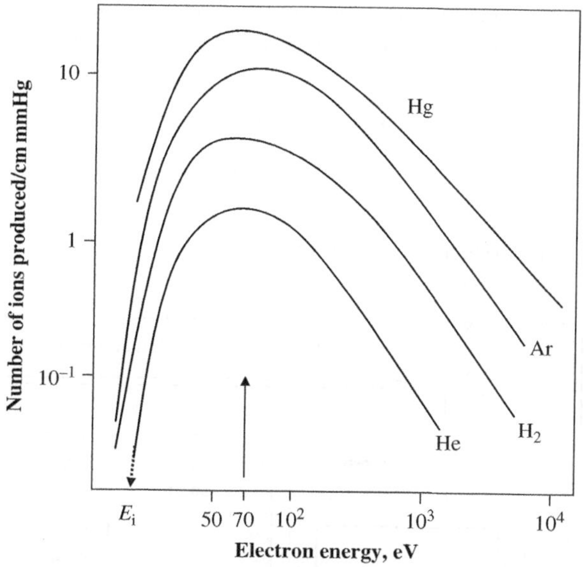

Figure 2.33 *Number of ions produced per cm free path length and per mmHg pressure in the ionization region of an electron ionization source as a function of electron energy. The ionization efficiency curves show a plateau between 50 and 80 eV. (H. Kienitz (ed.), Massenspektrometrie (1968), Verlag Chemie, Weinheim. Reproduced by permission of Wiley-VCH.)*

the separation of the ion beam formed. Because a multitude of fragment ions are formed by collision induced fragmentation, which complicates the mass spectra obtained, electron ionization plays a subordinate role in trace element analysis by inorganic mass spectrometry except in gas chromatography mass spectrometry (GCMS) to analyze volatile metal–organic compounds. Electron ionization is today used as a post-ionization technique, for example in Knudsen mass spectrometry.

Liquids are evaporated prior to introduction into the electron ionization source via a heated sample introduction system. Solid samples are introduced directly into the ion source and are evaporated in it or introduced after evaporation outside, for example in a Knudsen cell coupled to a mass spectrometer or by separate laser evaporation (laser evaporation mass spectrometry). In Figure 2.34 the experimental arrangement of a Knudsen cell together with an electron impact ion source is illustrated. The solid sample is heated by ohmic heating or electron bombardment of the Knudsen cell. Temperatures $< 3000\,°C$ are obtained and can be measured by an optical pyrometer using the window or by thermocouples. An exact temperature determination is the precondition for exact measurement of thermodynamic data. The polyatomic beam generated in the Knudsen cell emerges through a small orifice in the electron ionization source of a mass spectrometer (e.g., single sector field mass spectrometer or ToF analyzer). A movable shutter valve between the Knudsen cell and ion source is used to separate the evaporation chamber and ionization source of the mass spectrometer, for instance in order to measure the instrumental background to correct measured signals by background gases. Using Knudsen effusion mass spectrometry at high temperature, species in complex systems present in the mass spectrum also can be identified and gas phase processes such as fragmentation, gas phase reaction, gas discharge or dissociation of compounds (for determination of dissociation energies) can be studied. For example, the Van't Hoff equation:

$$\delta \ln K/\delta T = \Delta H^0/kT^2 \quad (2.24)$$

enables one to plot ln$K$ versus 1/T and thus determine the reaction enthalpy ΔH^0.

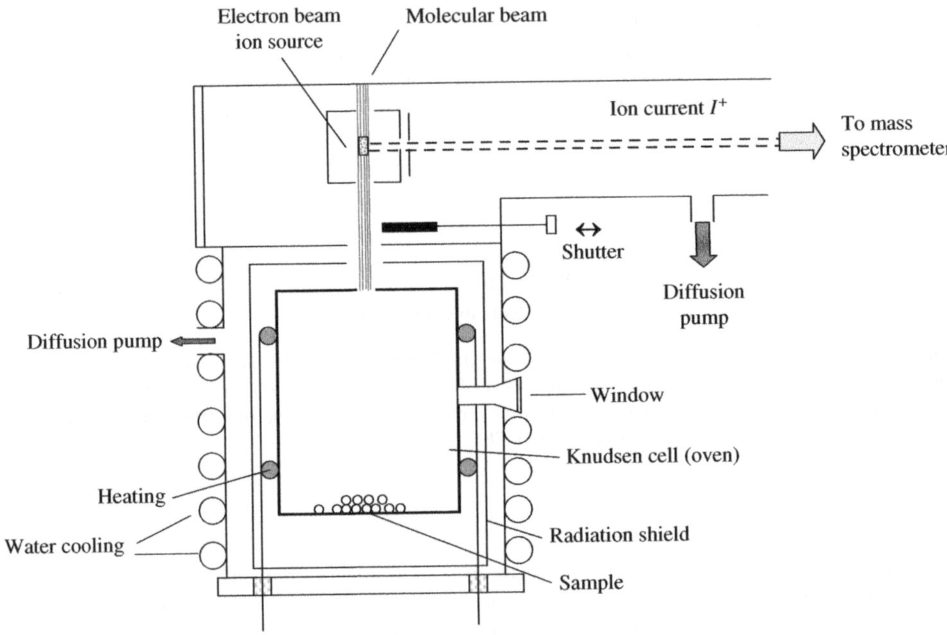

Figure 2.34 Schematic of an electron ionization source with Knudsen cell.

Knudsen effusion mass spectrometry is used to measure the composition of the effusing beam (including the composition of the sample investigated, fragments or newly formed species) and in particular to determine the thermochemical data. The instrumentation is briefly described in Section 5.2.10.[137]

Electron ionization mass spectrometry is used primarily for the analysis of gases (e.g. in dynamic gas source mass spectrometry for precise isotope ratio measurements of, e.g., C, N, O, H and noble gases – see Chapter 7) and for the structural analysis of organic compounds (e.g., in GCMS) as fractionation processes are observed. The ion intensity of the ions formed depends on the energy of electrons, whereby at lower electron energy a decrease in ionization efficiency with less fragmentation and high intensity of polyatomic ions is observed. In addition, low energy electrons are captured by elements and molecules with high electron affinity (e.g., halogens) to produce negative ions which were used as analyte ions in mass spectrometers possessing a negative ion mode. Further ionization techniques such as chemical ionization (CI) using a reagent ion (e.g., CH_4^+) or proton transfer reactions described elsewhere[86] play a subordinate role in inorganic mass spectrometry.

2.8 Matrix Assisted Laser Desorption/Ionization Source

Matrix-assisted laser desorption/ionization mass spectrometry (MALDI-MS) which was first introduced and developed in 1988 by Karras, Hillenkamp[138] and Tanaka[139] is today used as one of the most important gentle ionization techniques for large biomolecules (oligonucleotides, carbohydrates, natural products and lipids), proteins, peptides and polymers. In MALDI-MS the analyte is mixed with a light absorbing matrix solution (e.g., acetonitrile and 0.1 % TFA – trifluoroacidic acid) containing small organic molecules which have a strong absorption of photons at the laser wavelength. The solvent for preparing the matrix solution is removed by drying before mass spectrometric measurement. Commonly used matrices in MALDI-MS are HCCA (α-cyano-4-hydoxycinnamic acid) for the analysis of peptides and glycopeptides or DHB (2,5-dihydroxybenzoic acid) for peptides and small proteins, where the matrix-to-sample ratio is about 5000:1. The analyte molecules are embedded in a surplus of matrix molecules like a solid solution and completely isolated from one other. Careful sample–matrix preparation (mostly after desalting of the sample) is necessary for high quality MALDI mass spectra, and the dried droplet method is frequently used.

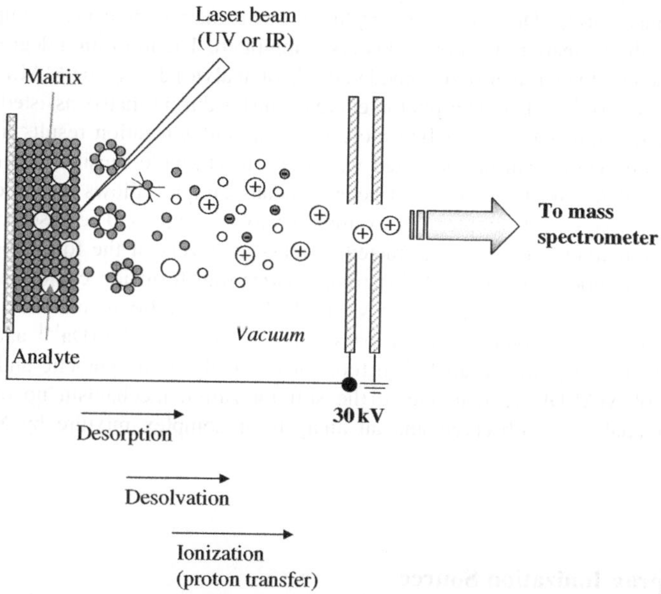

Figure 2.35 *Diagram of MALDI source.*

A schematic of the basic principles of a matrix-assisted laser desorption/ion source is shown in Figure 2.35. By the interaction of a focused laser beam with short pulses and a suitable matrix, the energy of the photons is transferred to the matrix molecules. In MALDI mostly pulsed UV (e.g., nitrogen, $\lambda = 337$ nm, pulse duration: 3–10 ns), but also IR lasers (e.g., Er:YAG, $\lambda = 2.94\,\mu m$ or CO_2, $\lambda = 10.6\,\mu m$ with a higher pulse duration of up to 600 ns) are used. The MALDI mass spectra obtained during soft ionization by UV and IR lasers are identical. The energy density (in MALDI-MS called *fluence*) is about 20–100 mJ cm^{-2}, whereas the laser power density (called *irradiation*) varies between 10^6 and 10^7 W cm^{-2}. In comparison to laser-induced mass spectrometric techniques for the element and isotope analysis of inorganic substances (such as LIMS and LA-ICP-MS), the laser power density in MALDI-MS is 2–3 orders of magnitude lower. In a MALDI source mostly positive singly charged ions or analyte are formed, but also doubly charged ions are observed. As a result of photon interaction with the solid sample/matrix mixture via laser-induced excitation and vaporization the aggregate state of the matrix and analyte molecules is changed from the solid to gaseous phase with expansion of the matrix plume in the high vacuum ion source. During this process of evaporation and expansion, the matrix, which plays a key role in this technique, carries the intact analyte molecules with it. In a second step after desolvation, ionization of analyte molecules take place by gas phase proton transfer reactions and partly by photoionization or ion–molecule reactions. The most probable ionization process in a MALDI source is gas phase proton transfer in the expanded laser induced plume with photoionized matrix molecules. In comparison to laser plasma ionization techniques, which result in fragmentation of organic molecules, the MALDI process is more sensitive. By matrix assisted laser induced desorption and ionization the matrix molecules help to reduce the analyte damage from laser irradiation by absorbing most of the incident photon energy. Whereas the photons of the laser beam heat and decompose the matrix compounds, the complete molecule is gently ionized and extracted in the ion separation system of the mass spectrometer used. Due to a high surplus of small matrix molecules compared to large analyte molecules the formation of sample clusters is inhibited. The ionization degree (the ratio of ions formed divided by the number of desorbed neutrals of analyte) during the MALDI process was found to be less than 10^{-5}.[140] Efficient photon energy transfer during matrix assisted laser induced desorption via matrix molecules to analyte molecules and soft ionization results in a high yield of intact analyte ions which can be measured with pico- to attomole sensitivity. In addition, the capability of MALDI-MS for the analysis of heterogeneous samples allows the structural analysis and identification of biological samples, for example protolytic digests. The desorbed and ionized positively charged analyte ions can be extracted in mass analyzers of the following types: time-of flight (ToF), quadrupole, sector field, ion trap, Fourier transform ion cyclotron resonance or tandem mass spectrometers. The advantage of MALDI-MS is the ability to analyze large molecules (detection of picomoles of proteins) with a mass range of up to 300 000 Da[141] and it is used to an increasing extent as a powerful analytical tool for the analysis of synthetic and biopolymers. Further benefits of MALDI are that due to the soft ionization mechanism no or only a little fragmentation of analytes is observed and an analysis of complex mixture by MALDI-MS is possible.

2.9 Electrospray Ionization Source

The first electrospray ionization (ESI) experiments were carried out by Dole *et al.*[142] in the late sixties. The authors discovered the phenomenon of the generation of multiply charged molecule ions during electrospray ionization. Due to the fundamental work on ESI-MS by Fenn's group in the mid 1980s[143,144] (Fenn was honoured in 2003 with the Nobel Prize[145]) this soft ionization technique in combination with mass spectrometry is today used as one of the commonly employed soft ionization techniques for the investigation of large biomolecules. Electrospray ionization as an atmospheric pressure ion (API) source is routinely used for the analysis of peptides, proteins, carbohydrates, small oligonucleotides, biopolymers and lipids and as the method of choice for coupling techniques such as liquid chromatography together with mass spectrometry. In ESI, the diluted liquid sample is introduced using a syringe pump in a hypodermic needle (nozzle) at a flow rate of 5–20 μl min^{-1}. In the early ESI arrangement, the needle was held at a potential of between 3 and 4 kV relative to the surrounding cylindrical electrode.[146] By the application of a drying (and/or heated) gas (e.g., for an effective solvent evaporation a heated dry nitrogen or air stream is used) the solvent is evaporated from each droplet. As the size of the charged droplet decreases the charge density on the droplet surface increases. Due to Coulomb repulsion between the like charges on the droplet surface, ions are ejected via the so-called 'Taylor cone'. The fine electrosprayed aerosol of charged droplets expands into the aperture of a small (heated transfer) capillary that interfaces the atmospheric pressure ESI source to the first pumping stage (about 10^2 Pa) as proposed by Fenn *et al.*[144] A small amount of electrosprayed ions are extracted via the skimmer orifice into the vacuum of the mass spectrometer (10^{-3}–10^{-4} Pa). The experimental arrangement of an electrospray ion source is illustrated in Figure 2.36. Electrospray using nanoflow rates in nano-electrospray ionization with droplet sizes < 200 nm is an advantage for miniaturized ESI sources and reduces the sample size of the analyte, for instance by applying nano-ESI to a chip. To analyze unstable compounds that are difficult by conventional ESI cold spray ionization can be utilized e.g., to study unstable intermediates, metal complexes, several bio- or supramolecules and others. The application of ESI-MS for the analysis of ionic metal complexes and other inorganic analytes is reviewed by Colton *et al.*[147] and Traeger.[148]

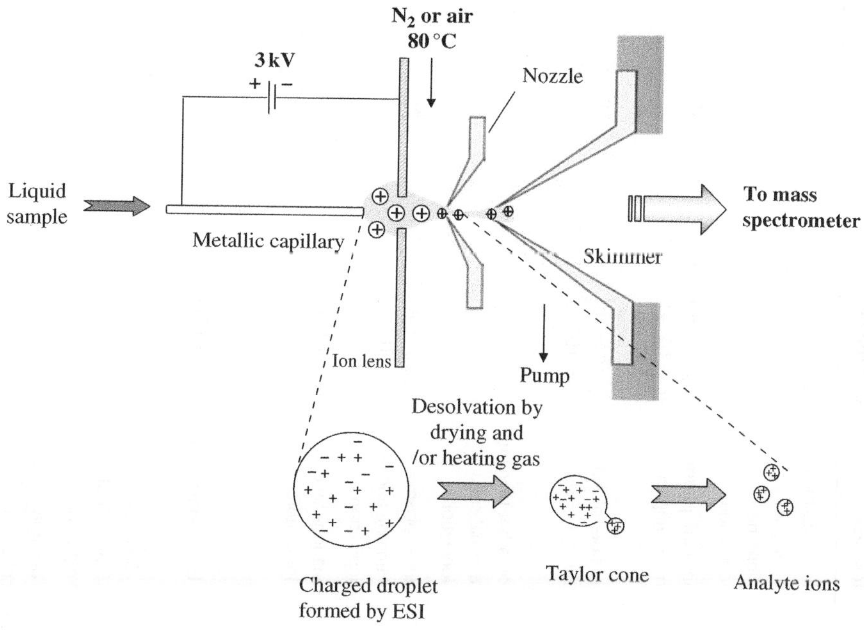

Figure 2.36 *Principles of electrospray ionization.*

Different types of ion sources used in mass spectrometry are compared in Table 2.2. Several plasma ion sources applied in analytical atomic spectrometry were described by Broekaert.[33]

Table 2.2 Comparison of different ion sources used in mass spectrometry.

Ion source	Atomization Ionization	Plasma temperature Electron density	Ions measured	Vacuum condition	Mass analyzer utilized	Application
Inductively coupled plasma ion source	in Ar plasma by e^- impact (Penning ionization)	$T_{plasma} \sim 6000–8000\,K$ $n_e \sim n \times 10^{15}\,cm^{-3}$	M^+	atmospheric pressure	quadrupole, ToF-MS, double-foc. SFMS (reverse Nier–Johnson)	traces, ultratraces, species, isotope ratios
Spark ion source	in spark plasma by e^- impact	$T_{plasma} \sim 50\,000\,K$ $n_e \sim 10^{18}–10^{19}\,cm^{-3}$	M^+	high $< 10^{-4}$ Pa	double-focusing SFMS (Mattauch–Herzog)	traces, high-purity materials, geology
Laser ion source	in laser plasma by e^- impact	$T_{plasma} \sim 10^4 - n \times 10^5\,K$ $n_e \sim 10^{16}–10^{20}\,cm^{-3}$	M^+	high $< 10^{-4}$ Pa	double-focusing SFMS (Mattauch–Herzog)	traces, high-purity ceramics
Thermal surface ionization source	on a hot filament thermal surface ionization		M^+, MX^-	high $< 10^{-4}$ Pa	single-focusing SFMS	isotope ratios isotope dilution geochronology
Glow discharge ion source	in Ar plasma Penning and e^- ionization	$T_{plasma} \sim 3000–6000\,K$ $n_e \sim 10^{15}\,cm^{-3}$	M^+	low pressure	double-focusing SFMS (reverse Nier–Johnson)	traces, high-purity metals
Sputtered ion source	via ions by e^- ionization	$n_e \sim 10^{15}\,cm^{-3}$	M^+; CsM^+, M^-	high $< 10^{-4}$ Pa	double-focusing SFMS, ToF-MS, quadrupole, tandem MS (AMS)	surface analysis material and life sciences, geochronol.
Electron ionization source	by e^- ionization	$n_e \sim 10^{15}\,cm^{-3}$	M^+	high $< 10^{-4}$ Pa	single-focusing MS	stable isotope ratios thermodynamic data
Electrospray ion source	electrospray ioniz., Coulomb repulsion	—	M^{n+}	atmospheric pressure	ToF-MS, FTICR-MS tandem MS	biomolecules
Matrix assisted laser desorption/ionization ion source	desorption by photons proton transfer	—	M^+	$< 10^{-4}$ Pa	ToF-MS, FTICR-MS tandem MS	biomolecules

SFMS – Sector field mass spectrometer

References

1. Goldstein, E., Berl. Ber. **39**, 691 (1886).
2. S. Greenfield, I. J. Jones and C. T. Berry, *Analyst*, **89**, 713 (1964).
3. Gray, A. L. and Date, A. R., *Analyst*, **108**, 1033 (1983).
4. Houk, R. S., Fassel, V. A., Flesch, G. D., Svec, H. J., Gray, A. L. and Taylor, C. E., *Anal. Chem*., **52**, 2283 (1980).
5. Date, A. R. and Gray, A. L., *Analyst*, **106**, 1228 (1981).
6. Browner, R. F. and Boorn, A. W., *Anal. Chem*., **56**, A786 (1984).
7. Olesik, J. W., in *Inorganic Mass Spectrometry* C. M. Barshick, D. C. Duckworth and D. H. Smith (eds.), Marcel Dekker, Inc., New York, Basel (2000).
8. Hill, S. J. (ed.), *Inductively Coupled Plasma Mass Spectrometry*, Sheffield Analytical Chemistry, Sheffield Academic Press, Sheffield (1999).
9. Montaser, A. (ed.), *Inductively Coupled Plasma Source Mass Spectrometry*, Wiley-VCH Inc., New York, (1998).
10. Todoli, J. L. and Mermet, J. M., *Spectrochim. Acta*, **61B**, 239 (2006).
11. Todoli, J. L. and Vanhaecke, F., in *ICP Mass Spectrometry Handbook*, S. M. Nelms (ed.) Blackwell Publishing, Oxford 182 (2005).
12. Montaser, A. and Zhang, H., in *Inductively Coupled Plasma Mass Spectrometry*, A. Montaser, (ed.) Wiley-VCH Inc., New York (1998).
13. Sheppard, B. S., Shen, W. L. and Caruso, J., *J. Am. Soc. Mass Spectrom*., **2**, 355 (1991).
14. Sheppard, B. S., Shen, W. L., Davidson, T. M. and Caruso, J. A., *J. Anal. At. Spectrom*., **5**, 697 (1990).
15. Nam, S. H., Zhang, H., Cai, M., Lim, J. S. and Montaser, A., *Fresenius' J. Anal. Chem*., **355**, 510 (1996).
16. Montaser, A., in *Inductively Coupled Plasma in Analytical Spectrochemistry*, 2nd ed., A. Montaser and D. W. Gologhtly (eds.), Wiley-VCH, New York, (1992).
17. Batal, A., Jarosz, J. and Mermet, J. M., *Spectrochim. Acta*, **36B**, 983 (1981).
18. Vaughan, M. A., Horlick, G. and Tan, S. H., *J. Anal. At. Spectrom*., **40B**, 1347 (1985).
19. Boulyga, S. F., Dietze, H. J. and Becker, J. S., *J. Anal At. Spectrom*., **16**, 598 (2001).
20. Vonderheide, A. P., Zoriy, M. V., Izmer, A. V., Pickhardt, C., Caruso, J. A., Ostapczuk, P., Hille, R. and Becker, J. S., *J. Anal. Atom. Spectrom*., **19**, 675 (2004).
21. Tanner, S. D., *J. Anal. At. Spectrom*., **10**, 905 (1995).
22. Jiang, S.-J., Houk, R. S. and Stevens, M. A., *Anal. Chem*., **60**, 1217 (1988).
23. Tanner, S. D., Paul, M., Beres, S. A. and Denoyer, E. R., *At. Spectrosc*., **16** 16 (1995).
24. Lui, S. and Beauchemin, D., *Spectrochim Acta*, **61B**, 157 (2006).
25. Campargue, R., *J. Phys. Chem*., **88**, 4466 (1984).
26. Brunnée, C., *Int. J. Mass Spectrom. Ion Proc*., **76**, 121 (1987).
27. Macedone, J. H. and Farnsworth, P. B., *Spectrochim. Acta*, **61B**, 1031 (2006).
28. Mills, A. A., Macedone, J. H. and Farnsworth, P. B., *Spectrochim. Acta*, **61B**, 1039 (2006).
29. Douglas, D. J. and French, J. B., *J. Anal. At. Spectrom*., **3**, 743 (1988).
30. Gray, A., *J. Anal. At. Spectrom*., **1**, 403 (1986).
31. Becker, J. S. and Dietze, H. J., *J. Anal. At. Spectrom*., **14**, 1493 (1999).
32. McLean, J. A., Becker, J. S., Boulyga, S. F., Dietze, H. J. and Montaser, A., *Int. J. Mass Spectrom*., **208**, 193 (2001).
33. Broekaert, J. A. C., Analytical Atomic spectrometry with Flames and Plasmas, 2nd edn, Wiley-VCH Verlag GmbH, Weinheim (2005).
34. Becker, J. S. and Dietze, H. J., *Spectrochim. Acta*, **53B**, 1475 (1998).
35. Beauchemin, D., *Anal. Chem*., **76**, 3395 (2004).
36. Becker, J. S., *Spectrochim. Acta*, **57B**, 1805 (2002).
37. Becker, J. S. and Dietze, H. J., *Int. J. Mass Spectrom*., **197**, 1 (2000).
38. Becker, J. S., Pickhardt, C. and Dietze, H. J., *Int. J. Mass Spectr*., **2003**, 283 (2000).
39. Durrant, S. F., *J. Anal. At. Spectrom*., **14**, 1385 (1999).
40. Guenther, D. and Hattendorf, B., *Trends Anal. Chem*., **24**, 2491 (2005).
41. Jochum, K. P., Stoll, B., Herwig, K. *et al. Geoch. Geophys. Geosyst*., **7**, 1 (2006).

42. Mao, S. S., Mao, X. L., Greif, R. and Russo, R. E., *Appl. Phys. Lett.*, **77**, 2464 (2000).
43. Russo, R. E., Mao, X. L., Liu, C. and Gonzalez, J., *J. Anal. At. Spectrom.*, **19**, 1084 (2004).
44. Boulyga, S. F., Desideri, D., Meli, M. A., Testa, C. and Becker, J. S., *Int. J. Mass Spectrom.*, **226**, 329 (2003).
45. Pickhardt, C., Dietze, H. J. and Becker, J. S., *Int. J. Mass Spectrom.*, **242**, 273 (2005).
46. Becker, J. S., Pickhardt, C. and Dietze, H. J., *J. Anal. At. Spectrom.*, **16**, 603 (2001).
47. Becker, J. S. and Tenzler, D., *Fresenius' J. Anal. Chem.*, **370**, 637 (2001).
48. Gastel, M., Becker, J. S., Kueppers, G. and Dietze, H. J., *Spectrochim. Acta*, **52B**, 2051 (1997).
49. Miller, J. C. (ed.) *Laser Ablation: Principles and Applications*, Springer Verlag, Berlin, Heidelberg, New York (1994).
50. Becker, J. S., Breuer, U., Westheide, J., Saprykin, A. I., Holzbrecher, H., Nickel, H. and Dietze, H. J., *Fresenius' J. Anal. Chem.*, **355**, 626 (1996).
51. Westheide, J. T., Becker, J. S., Jaeger, R., Dietze, H. J. and Broekaert, J. A. C., *J. Anal. At. Spectrom.*, **11**, 661 (1996).
52. Becker, J. S., Zoriy, M., Becker, J. Su. *et al.*, *Anal. Chem.*, **77**, 5851 (2005).
53. Feldmann, I., Koehler, C., Roos, P. and Jakubowski, N., *J. Anal. At. Spectrom.*, **21**, 1006 (2006).
54. Carcia, C. C., Lindner, H. and Niemax, K., in *Book of Abstracts of 8th European Workshop on Laser Ablation in Elemental Analysis*, D. Guenther (ed.) Zuerich, (2006).
55. Becker, J. S., Gorbunoff, A., Zoriy, M. V., Izmer, A. V. and Kayser, M., *J. Anal. At. Spectrom.*, **21**, 19 (2006).
56. Kidness, A., Sekaran, N. and Feldmann, J., *J. Clin. Chem.*, **49**, 1916 (2003).
57. Zoriy, M. V., Kayser, M., Izmer, A., Pickhardt, C. and Becker, J. S., *Int. J. Mass Spectrom.*, **242**, 297 (2005).
58. Hirata, T. and Miyazaki, Z., *Anal. Chem.*, **79**, 147 (2007).
59. Figg, D. J., Cross, J. B. and Brink, C., *Appl. Surf. Sci.*, **129**, 287 (1998).
60. Hergenröder, R., *Spectrochim Acta B*, **61**, 284 (2006).
61. Jackson, S. E. and Günther, D., *J. Anal. At. Spectrom.*, **18**, 205 (2003).
62. Koch, J., Feldmann, I., Hattendorf, B. *et al. Spectrochim. Acta*, **57B**, 1057 (2002).
63. Kuhn, H. R. and Guenther, D., *Anal. Chem.*, **75**, 747 (2003).
64. Rodushkin, I., Avelsson, M. D., Malinovsky, D. and Baxter, D. C., *J. Anal. At. Spectrom.*, **17**, 1223 (2002).
65. Bleiner, D., *Spectrochim Acta B*, **60**, 49 (2005).
66. Bogaerts, A. and Chen, Z., *J. Anal. At. Spectrom.*, **19**, 1169 (2004).
67. Bogaerts, A., Chen, Z. Y. and Gijbels, R., *Spectrochim Acta*, **58B**, 1867 (2003).
68. Lui, H. C., Mao, J. L., Yoo, J. H. and Russo, R. E., *Spectrochim. Acta*, **54B**, 1607 (1999).
69. Fliegel, D. and Günther, D., *Spectrochim. Acta*, **61 B**, 841 (2006).
70. Boulyga, S. F., Pickhardt, C. and Becker, J. S., *At. Spectr.*, **25**, 53 (2004).
71. Becker, J. S., Gorbunoff, A., Kayser, M. and Pompe, W., in *Offenlegungsschrift: 30. Juni 2005*, Deutsches Patentamt, München (2003), vol. DE 103 5t4 787 A1.
72. Sturgeon, R. E., Viera, M. A. and Ribero, A. S., in *Second Asia-Pacific Winter Conference on Plasma Spectrochemistry*, ICP Information Newsletter, Inc., Bangkok, Thailand, 46 (2006).
73. Adams, F., Gijbels, R., Van Grieken, R. (eds.), *Chemical Analysis (N.Y.), Vol. 95: Inorganic Mass Spectrometry*, John Wiley & Sons, Inc., New York (1988).
74. Dietze, H. J., *Analytikertaschenbuch*, **10** 249 (1991).
75. Dietze, H. J., *Massenspektroskopische Spurenanalyse*, Akademischer Verlagsgesellschaft Geest & Portig K.-G., Leipzig (1975).
76. Jochum, K. P., *Spectroscopy Europe*, **9**, 22 (1997).
77. Dempster, A. J., *Proc. Am. Phil. Soc.*, **75**, 755 (1935).
78. Saprykin, A. I., Becker, J. S. and Dietze, H. J., *Fresenius' J. Anal. Chem.*, **364**, 763 (1999).
79. Dietze, H. J. and Becker, J. S., *Fresenius Z. Anal. Chem.*, **321**, 490 (1985).
80. Honig, R. E. and Woolston, J. R., *Appl. Phys. Lett.*, **2**, 138 (1963).
81. Rosan, R. C., Healy, M. K. and Mc.Nary, W. F., *Science*, **142**, 236 (1963).
82. Jochum, K. P., Matus, L. and Seufert, H. M., *Fresenius' J. Anal. Chem.*, **331**, 136 (1988).
83. Voigt, H., Heinen, H.-J., Meier, S. and Wechsung, R., *Fresenius' J. Anal. Chem.*, **308**, 195 (1981).

84. Dietze, H. J. and Becker, J. S., *Int. J. Mass Spectrom. Ion Proc.* **82**, R1 (1988).
85. Gruning, C., Huber, G., Klopp, P., Kratz, J. V., Passler, G., Trautmann, N., Waldek, A. and Wendt, K., *Int. J. Mass Spectrom.*, **235**, 171 (2004).
86. Trautmann, N., Passler, G. and Wendt, K. D. A., *Anal. Bioanal. Chem.*, **378**, 348 (2004).
87. Wendt, K., and Trautmann, N., Int. J. Mass Spectrom., **242**, 161 (2005).
88. Hurst, G. S., Payne, M. G., Kramer, S. D. and Young, J. P., *Rev. Mod. Phys.*, **51** 767 (1979).
89. Balikin, V. I., Bekov, G. J., Letokhov, V. S. and Mishin, V. I., *U. Fiz. Nauk*, **132**, 293 (1980).
90. Letokhov, V. S., *U. Fiz. Nauk*, **125**, 57 (1978).
91. Wendt, K., Blaum, K., Bushaw, B. A. *et al.*, *Fresenius' J. Anal. Chem.*, **364**, 471 (1999).
92. Hastings, E. P. and Harrison, W. W., *J. Anal. At. Spectrom.*, **19**, 1268 (2004).
93. Marcus, R. K. and Broekaert, J. A. C., *Glow Discharge Plasmas in Analytical Spectroscopy*, John Wiley & Sons Ltd., Chichester, UK (2003).
94. Gijbels, R., Van Straaten, M. and Bogaerts, A., in *Advanced Mass Spectrometry*, Proc. 13th International Mass Spectrometry Conference, Budapest, I. Cornides (ed.), John Wiley & Sons, Ltd, Chichester (1994).
95. Chapman, B., *Glow Discharge Processes: Sputtering and Plasma Etching*, John Wiley & Sons, Inc., New York (1980).
96. Steers, E. B. M., *J. Anal. At. Spectrom.*, **12**, 1033 (1997).
97. Harrison, W. W., Yang, C. and Oxley, E., in *Glow Discharge Plasma in Analytical Spectroscopy* J. A. C. Broekaert (ed.) John Wiley & Sons Ltd., Chichester, UK, 71 (2003).
98. Anderson, J. M., *J. Appl. Phys.*, **31**, 511 (1960).
99. Bogaerts, A. and Gijbels, R., *Fresenius' J. Anal. Chem.*, **364**, 367 (1999).
100. Bogaerts, A. and Gijbels, R., in *Glow Discharge Plasma in Analytical Spectroscopy*, R. K. Marcus and J. A. C. Broekaert (eds.), John Wiley & Sons Ltd., Chichester, UK, 155 (2003).
101. Fang, D. and Marcus, R. K., *Spectrochim. Acta*, **45B**, 1053 (1990).
102. Fang, D. and Marcus, R. K., *Spectrochim. Acta*, **46B**, 983 (1991).
103. Beyer, C., Feldmann, I., Gilmour, D., Hoffmann, V. and Jakubowski, N., *Spectrochim. Acta*, **57B**, 1521 (2002).
104. Oxley, E., Yang, C., Lui, J. and Harrison, W. W., *Anal. Chem.*, **75**, 6478 (2003).
105. Wetzel, W. C. and Hieftje, G. M., *J. Anal. At. Spectrom.*, **21**, 1367 (2006).
106. Schelles, W., De Gent, S., Maes, K. and Van Grieken, R., *Fresenius' J. Anal. Chem.*, **355**, 858 (1996).
107. Becker, J. S., Saprykin, A. I. and Dietze, H. J., *Int. J. Mass Spectrom. Ion Proc.*, **164**, 81 (1997).
108. Saprykin, A. I., Becker, J. S. and Dietze, H. J., *Fresenius' J. Anal. Chem.*, **359**, 449 (1997).
109. Saprykin, A. I., Melchers, F. G., Becker, J. S. and Dietze, H. J., *Fresenius' J. Anal. Chem.*, **353**, 570 (1995).
110. Jäger, R., *Reports of Research Centre Jülich*, **3447**, 1 (1997).
111. Jäger, R., Saprykin, A. I., Becker, J. S., Dietze, H. J. and Broekaert, J. A. C., *Mikrochim. Acta*, **125**, 41 (1997).
112. Winchester, M. R., Duckworth, D. C. and Marcus, R. K., in Glow Discharge Spectroscopies, R. K. Marcus (ed.), Plenum Press, New York 263 (1993).
113. Cody, R. C., *SpectroscopyEUROPE*, **12**, 12 (2004).
114. Dempster, A. J., *Phys. Rev.*, **11**, 315 (1918).
115. Gehrcke, E. and Reichenheim, O., *Ber. Bunsenges. Phys. Chem.*, **8**, 559 (1906).
116. Langmuir, I. and Kingdom, K. H., *Proc. Roy. Soc. A*, **107**, 61 (1925).
117. Platzner, I. T., *Modern Isotope Ratio Mass Spectrometry*, John Wiley & Sons, Inc., New York (1997).
118. Cameron, A. E., Smith, D. H. and Walker, R. L., *Anal. Chem.*, **41**, 525 (1969).
119. Heumann, K. G., Eisenhut, S., Gallus, S. *et al.*, *Analyst*, **120**, 1291 (1997).
120. Sigmund, P., in *Sputtering by Particle Bombardment I*, R. Behrisch, (ed.) Springer Verlag, Berlin, 18 (1981).
121. Sigmund, P., in *Inelastic Ion-Surface Collision*, N. H. Tolk, J. Tully, C. W. Heiland and C. W. White (eds.) Academic Press, Oxford (1977).
122. Jenett, H., in *Analytikertaschenbuch 16.*, Springer Verlag, Berlin, Heidelberg, 43 (1997).
123. Bock, W., Gnaser, H. and Oechsner, H., in *Proceeding of 10th International Conference on Secondary Ion Mass Spectrometry* B. H. A. Benninghoven, H. W. Werner (ed.) John Wiley & Sons Inc., New York 395 (1997).

124. Benninghoven, A., Janssen, K. T. F., Tümpner, J. and Werner, W. H., *Secondary Ion Mass Spectrometry VIII (Proceedings of the Ninth International Conference on SIMS)*, A. Benninghoven *et al.* (eds.) *John Wiley & Ltd., Sons*, Chichester, 509 (1992).
125. Deline, V. R., Evans, C. A. and Williams, P., *Appl. Phys. Lett.*, **33**, 578 (1978).
126. Wilson, R. G., Stevie, F. A. and Magee, C. W., *Secondary Ion Mass Spectrometry*, John Wiley & Sons, Inc., New York (1989).
127. Gao, Y., *J. Appl. Phys.*, **64**, 3760 (1988).
128. Niehus, E., Grehl, T., Kollmer, F., Moellers, R. and Rading, D., in *5th European Workshop on Secondary ion Mass Spectrometry*, A. Benninghoven *et al.* (eds), 7 Münster (2006).
129. Van Ham, R., Van Vaeck, L., Adams, F. and Adriaens, A., *J. Anal. At. Spectrom.*, **10**, 1088 (2005).
130. Todd, P. J., Schaaf, T., Chaurand, P. and Caprioli, R. M., *J. Mass Spectrom.*, **36**, 355 (2001).
131. Touboul, D., Kollmer, F., Niehus, E., Brunelle, A. and Laprévote, O., *J. Am. Soc. Mass Spectrom.*, **16**, 1608 (2005).
132. Dempster, A. J., *Phil. Mag.*, **31**, 438 (1916).
133. Smyth, H. D., *Proc. Roy. Soc. A*, **104**, 121 (1923).
134. Bleakney, W., *Phys. Rev.*, **34**, 157 (1929).
135. Tate, J. T. and Smith, D. H., *Phys. Rev.*, **46**, 773 (1934).
136. Nier, A. O., *Phys. Rev.*, **18**, 415 (1947).
137. Hilpert, K. and Miller, M., *J. Electrochem. Soc.*, **141**, 2769 (1994).
138. Karas, M. and Hillenkamp, F., *Anal. Chem.*, **60**, 2299 (1988).
139. Tanaka, K., Waki, H., Ido, Y., Akita, S., Yoshida, Y. and Yoshida, R., *Rapid Commun. Mass Spectrom.*, **2**, 151 (1988).
140. Quist, A. P., Huth-Fehre, T., Westman, A. and Sundquist, B. U. R., *Rapid Commun. Mass Spectrom.*, **8**, 149 (1994).
141. Spengler, B. and Cotter, R. J., *Anal. Chem.*, **62**, 793 (1990).
142. Dole, M., Mack, L. L., Hines, R. L., Mocley, R. C., Ferguson, L. D. and Alica, M. B., *J. Chem. Phys.*, **49**, 2240 (1968).
143. Fenn, J. B., Mann, M., Meng, C. K. and Wong, S. F., *Mass Spectrom. Rev.*, **9**, 37 (1990).
144. Fenn, J. B., Mann, M., Meng, C. K., Wong, S. F. and Whithouse, C. M., *Science*, **246**, 64 (1989).
145. Fenn, J. B., *Angew. Chem., Int. Ed.*, **42**, 3871 (2003).
146. Gross, J. H., *Mass Spectrometry*, Springer Verlag Heidelberg, New York, (2004).
147. Colton, R., D'Agostino, A. and Traeger, J. C., *Mass Spectrom. Rev.*, **14**, 79 (1995).
148. Traeger, J. C., *Int. J. Mass Spectrom.*, **200**, 387 (2000).

3

Ion Separation

As an essential part of a mass spectrometer, the ion separation system has the task of separating the fast-flying ions (with different masses m and charges z (with $z = n \cdot e$) formed in an ion source and extracted from this source using an ion optic system) with respect to their different mass-to-charge (m/z) ratios. The separated ion beams are than supplied to the ion detection system for spatial or time resolved ion detection and registration. The mass spectrum is then the 2D representation of ion intensity as a function of the m/z ratio.

Ion separation systems can be classified as 'static ion separation systems' and 'dynamic ion separation systems'.[1,2] The proposed classification is related to the time dependence of the mass spectrometric ion separation systems during the separation process of the ions.

All static separation systems possess a magnetic sector field. *Static ion mass separation devices* are systems with separation fields (magnetic fields, electrical fields and a combination of the two in a mass spectrometer) which possess a high short and long term constancy of fields with respect to the flight time of the ions (apart from the relatively slow time dependent change of these fields for recording a mass spectrum). That means the separation of ion beams is carried out using fields that remain constant in time.

Dynamic ion separation systems are based on another physical principle and use the different flight time of ions with different masses and different velocity (e.g., in ToF mass analyzers). In addition, in dynamic ion separation systems there is a time dependent variation of one or more system parameters, e.g., changing of electrical or/and magnetic field strengths, which means the ion motion during the measurement procedure is crucial for the mass spectrometric analysis.

A critical comparison of several mass separation systems applied in inorganic mass spectrometry was discussed by Brunnée in 1987.[3]

In the following chapter, the focusing and separating effects of different types of ion separation systems on fast flying charged ion beams will be considered.

Inorganic Mass Spectrometry: Principles and Applications J. S. Becker

3.1 Sector Field Analyzer

Sector field analyzers are magnetic or electric sector fields or a combination of the two with ion focusing and separation properties used in static mass spectrometers.

3.1.1 Magnetic Sector Field Analyzer

All known static ion separation systems in mass spectrometry operate with a magnetic sector field for the separation of beams of charged ions of different masses. In general, in a magnetic sector field mass spectrometer a homogeneous magnetic field is used, which is generated by flat pole shoes positioned parallel to each other so that the field strength is the same at every point in the field. The separation of ion beams of charged ions with different masses in a magnetic sector field takes place due to the prism effect and for direction focusing of diverging ions due to the lens effect. A homogeneous magnetic sector field with the angle Φ_m is often used as the simplest and most essential part of a sector field mass spectrometer to separate ion beams with different m/z ratios. Magnetic sector fields with $\Phi_m = 60°$ or $90°$ are preferred for the construction of static mass spectrometers.

In the spectrometer, the trajectory of an ion of mass m that flies into the constant magnetic field at velocity v becomes a circular trajectory with radius r_m. The radius r_m of a moving ion with charge $z = n \cdot e$ and of mass m in a circular trajectory is the result of the equilibrium of the Lorentz force $(n \cdot e \cdot v \cdot B)$ and the centrifugal force $\left(\frac{m \cdot v^2}{r_m}\right)$ in a homogeneous magnetic field with magnetic force B in gauss:

$$n \cdot e \cdot v \cdot B = \frac{m \cdot v^2}{r_m} \tag{3.1}$$

The velocity v of the ion is determined by its kinetic energy $m \cdot v^2/2$, which the ions have acquired during their acceleration at an accelerating voltage V in volts, where the following applies:

$$n \cdot e \cdot V = \frac{m \cdot v^2}{2} \tag{3.2}$$

The radius of trajectory r_m in cm of an ion on its circular path through a homogeneous magnetic sector field with magnetic force B (gauss) and accelerating voltage V (volts) is determined by:

$$r_m = 143.6 \cdot \frac{1}{B}\sqrt{\frac{m \cdot V}{n}}\,[\text{cm}] \tag{3.3}$$

and the velocity of ions v in cm s^{-1} can be calculated by:

$$v = 1.39 \cdot 10^6 \sqrt{\frac{n \cdot V}{m}}\,[\text{cm s}^{-1}] \tag{3.4}$$

In the following considerations, only positive singly charged ions ($z = 1$) will be considered. For example, a singly charged Ar^+ atomic ion with a mass of 40 u possesses, after an acceleration with a voltage V of 5 kV, a velocity v of about $9 \cdot 10^5$ cm s^{-1} and will be deflected in a homogeneous magnetic field B of 6500 gauss to a circular path with a radius r_m of 10 cm.[2]

For a monoenergetic ion beam, the homogeneous magnetic field possesses both focusing and dispersing properties, which means a lens effect for a divergent ion beam with the same mass and a prism effect for ions in the beam with different masses. Both effects are analogous to the effects of an optical lens in light optics.

Lens Effect of Magnetic Sector Field

The focusing properties of a homogeneous magnetic field (lens effect) for the mass separation of ion beams are illustrated in Figure 3.1. A′, A″ and C are the positions of the object, image and centre of curvature, respectively, and l_m' and l_m'' are the linear distances of object and image, respectively, from the boundaries of the magnetic sector field with its angle Φ_m. Ions of mass m and velocity v start from point A′ in the direction of the magnetic field. The middle ion beam of the divergent ion beam with a divergence angle of 2α (α is the half-angular direction spread of the ions) flies perpendicular to the boundary of the magnetic field. After traversing the linear distance l_m' from the ion emitting point A′, the ions enter the magnetic field at the point O′, where the ions are constrained to follow a circular path of radius r_m. When the ions leave the magnetic field perpendicular to the other boundary of the magnetic field at the point O″, the trajectories of the ions are straight again for l_m''. Due to the lens effect of a homogeneous magnetic field with angle Φ_m – similar to an optical cylinder lens used in light optics – the three divergent ion beams starting at A′ with mass m are focused at the image point A″. Therefore, like an optical convex lens, the homogeneous magnetic field possesses direction focusing properties for diverging ion beams (with the same mass and energy) entering this field. That means divergent ions of the same mass and energy are focused so that the image of entrance slit A′ for ions lies at the focal point A″ where the exit slit of the magnetic sector field is installed (see Figure 3.1). Due to the similarity of the lens properties of a homogeneous sector field and a convex lens, all the familiar terms from optics can be applied to a magnetic sector field, such as the distance of focus from the exit boundary of magnet g_m, the focal length f_m and the so-called transverse magnification. In accordance with the known lens equation derived from geometric optics, for a magnetic sector field the following is fulfilled:

$$(l_m' - g_m)\,(l_m'' - g_m) = f_m^{\,2} \tag{3.5}$$

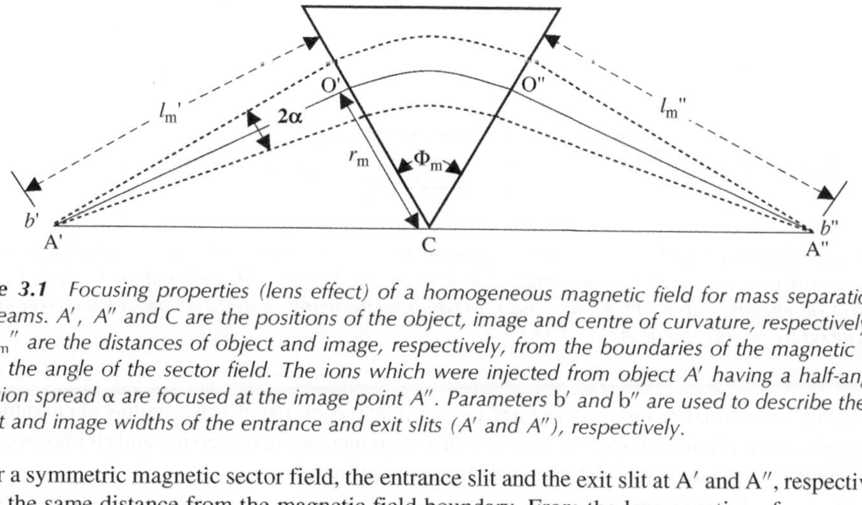

Figure 3.1 Focusing properties (lens effect) of a homogeneous magnetic field for mass separation of ion beams. A′, A″ and C are the positions of the object, image and centre of curvature, respectively, l_m' and l_m'' are the distances of object and image, respectively, from the boundaries of the magnetic field. Φ_m is the angle of the sector field. The ions which were injected from object A′ having a half-angular direction spread α are focused at the image point A″. Parameters b′ and b″ are used to describe the real object and image widths of the entrance and exit slits (A′ and A″), respectively.

For a symmetric magnetic sector field, the entrance slit and the exit slit at A′ and A″, respectively, are at the same distance from the magnetic field boundary. From the lens equation of a symmetric magnetic sector field with $l_m' = l_m''$ Equation (3.6) is valid

$$l_m' = l_m'' = r_m \cdot \mathrm{ctg}\frac{\Phi_m}{2} \tag{3.6}$$

But in practice no 'ion emitting point (A′)' exists, because mass spectrometers use slit systems (entrance and exit slits) of finite width (b' and b'', respectively), and so parameters b' and b'' are introduced to describe the defined object and image widths of the entrance and exit slits (A′ and A″), respectively. The ratio b''/b' is defined as the so-called 'transverse magnification' G_m of a homogeneous magnetic sector field:

$$G_m = b''/b' \tag{3.7}$$

As already discussed, the direction focusing properties of a magnetic field enable diverging ions to enter the field so that the image of the entrance slit for ions lies in the focal plane of the system. The radial (3.2a) and axial (3.2b) focusing properties of a magnetic sector field with oblique ion beam entrance and exit are illustrated in the bottom part of Figure 3.2. In this figure, the ion beam enters the magnet at an oblique angle, enabling radial focusing to be attained by virtue of the fringing field. Ion beams with different radii of trajectories (r_m and $r_m + \Delta r_m$) are focused at A_1'' and A_2'' as imaging points on the imaging curve. The distance of these two imaging points A_1'' and A_2'' on the bent imaging curve is defined as the dispersion distance. In practice, the beam axis dispersion ($D_m = 2\Delta r_m$), as a relevant parameter of a magnetic sector field, is considered as the distance between A_1'' and B_2'' in the plane of the exit slit (S).

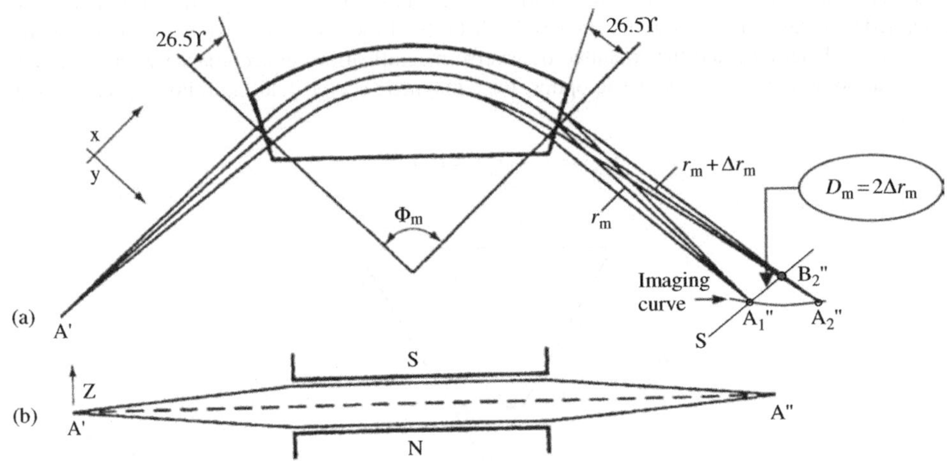

Figure 3.2 *Radial (a) and axial (b) focusing properties of a magnetic sector field with oblique beam entrance and exit.* A_1'' *and* A_2'' *are imaging points of ion beams with different path radii* (r_m *and* $r_m + \Delta r_m$) *on imaging curve. (H. Kienitz (ed.), Massenspektrometrie (1968), Verlag Chemie, Weinheim. Reproduced by permission of Wiley-VCH.)*

By using non-normal beam incidence and curved pole pieces, the effective radius of curvature of the magnetic sector field is extended, which results in an increase in dispersion and an improvement in the abundance sensitivity of the mass spectrometric system.[4]

Prism Effect of Magnetic Sector Field

The prism effect of a homogeneous magnetic sector field is crucial for the separation of different types of ions according to their mass-to-charge ratio. In Figure 3.3 the principle of mass separation of ion beams due to the prism effect of a homogeneous magnetic sector field is illustrated. A monoenergetic ion beam consisting of ions, e.g., of three different masses m_1, m_2 and m_3 (with $m_3 > m_2 > m_1$ of the same ion energy), at point A′, flies perpendicular in the homogeneous magnetic sector field. The ion beams with three different masses are then separated due to the prism effect of the homogeneous magnetic field according to their mass-to-charge ratio so that the ions with lower mass are deflected to a greater extent than the heavier ones. The radius of the ion beams which the ions pass through in the magnetic field can be calculated from Equation (3.3). By differentiating Equation (3.3), the relative change of the deflection radius is obtained for ions of different mass, but of the same energy:

$$\frac{\Delta r_m}{r_m} = \frac{1}{2}\frac{\Delta m}{m} \tag{3.8}$$

For ions of the same mass, but different energy the following is valid:

$$\frac{\Delta r_m}{r_m} = \frac{1}{2}\frac{\Delta V}{V} \tag{3.9}$$

The ion beams of different masses separated according to the m/z ratio emerge from the magnetic field at various angles. Ions with the lowest mass m_1 are focused at point A_1'', ions with medium mass m_2 at point A_2'' and the heaviest ions m_3 are focused at point A_3''. All images of separated ion beams A_1'', A_2'' and A_3'' are on the imaging curve of the magnetic sector field.

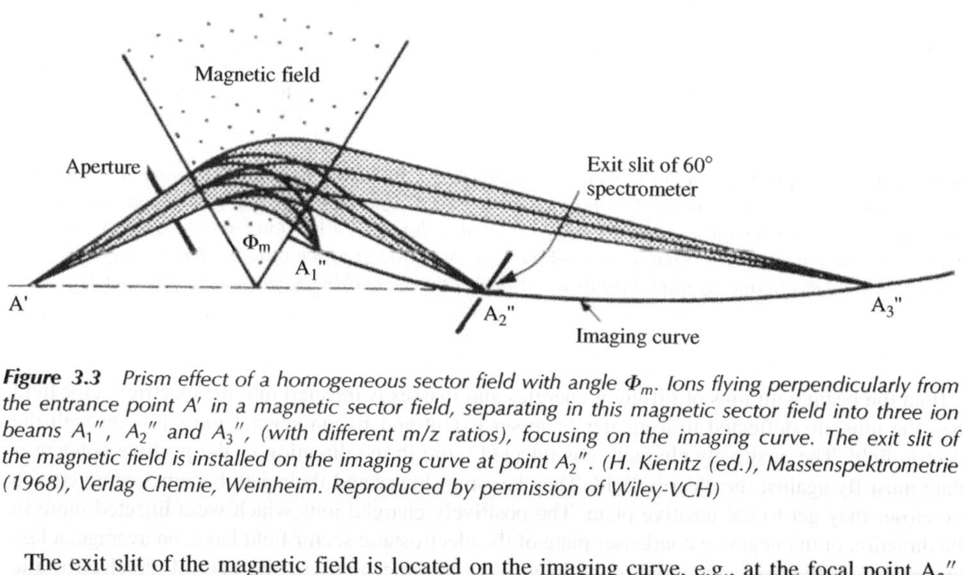

Figure 3.3 *Prism effect of a homogeneous sector field with angle Φ_m. Ions flying perpendicularly from the entrance point A′ in a magnetic sector field, separating in this magnetic sector field into three ion beams A_1'', A_2'' and A_3'', (with different m/z ratios), focusing on the imaging curve. The exit slit of the magnetic field is installed on the imaging curve at point A_2''. (H. Kienitz (ed.), Massenspektrometrie (1968), Verlag Chemie, Weinheim. Reproduced by permission of Wiley-VCH)*

The exit slit of the magnetic field is located on the imaging curve, e.g., at the focal point A_2'' (see Figure 3.3). By changing the magnitude of the magnetic force B all separated ion beams continuously pass the exit slit and can be detected using a suitable ion detection system (see Chapter 4).

3.1.2 Electric Sector Field Analyzer

Besides a magnetic field, an electric sector field also influences the trajectory of the ions. An electric sector field analyzer as a section of the cylindrical condenser consists of two cylindrical condenser plates with opposite potentials, to which a grounded deflection voltage is applied roughly in the middle. A positive voltage is applied to the outer plate of the condenser and a negative voltage to the inner plate, respectively. In mass spectrometry, a section of an electrical cylindrical condenser with radius r_e (the angle of sector field is Φ_e, see Figure 3.4) is used thus generating a radially symmetrical electrostatic field. The positive ions coming from the ion emitting point A′ are injected after passing the distance l_e'' in the cylindrical condenser. The voltage V_0 applied to the condenser is grounded in such a way that the equipotential area, corresponding to the radius r_e, also has zero potential. That means, if the condensor voltage has been correctly chosen, the middle circular trajectory of ions with energy eV is found at radius r_e in the middle between the two condenser plates. The positively charged ions flying through the electric sector field with kinetic energy $\frac{1}{2} m \cdot v^2$ are deflected towards the cylindrical axis of the electric field and leave the electric sector field analyzer.

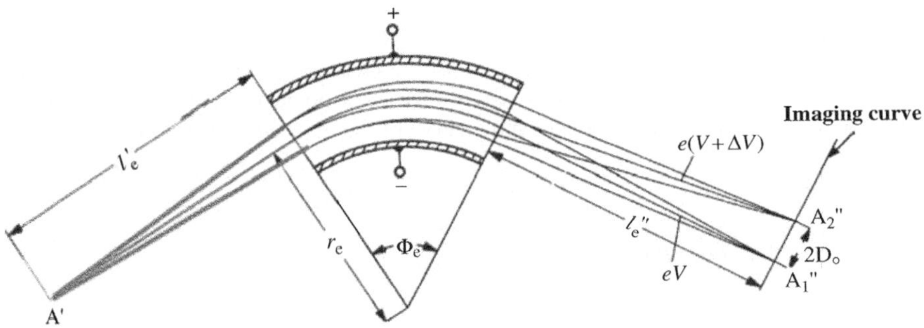

Figure 3.4 Prism and lens effect of an electric sector field with radius r_e and sector angle Φ_e. The ions coming from the ion-emitting point A′ enter the electric sector field after a distance of l_e' and are focused on the imaging curve behind the electric sector field at a distance of l_e''. Ions of energy $e(V+\Delta V)$ and $e(V-\Delta V)$ are focused in two separate images A_1'' and A_2''. (Massenspektrometrie Herausgegeben, Von Kienitz, H., Verlag Chemie, GmbH, Weinheim (1968). Reproduced by permission of Wiley-VCH)

If an ion beam with ions of different energies and masses is injected into the electric sector field then the ions are deflected to a greater or lesser extent and fly on several ion paths through the electric field. The positively charged ions injected more in the direction of the positive condenser plate must fly against the electric field. They become slower and their path becomes more curved the closer they get to the positive plate. The positively charged ions which were injected more in the direction of the negative condenser plate of the electrostatic sector field have, on average, a less curved path in the field compared to the ions flying in the central circular path of the electrostatic sector field. Similar to a magnetic sector field, an electric sector field acts on a divergent ion beam like a convex lens (due to the lens effect) and an ion beam with different kinetic energy acts like a prism (due to the prism effect). Following the parameters of an electric sector field, such as distance of object l_e', distance of image l_e'', focal length f_e and the distance of focus g_e, are combined by the well known lens equation derived from optics:

$$(l_e' - g_e)(l_e'' - g_e) = f_e^2 \tag{3.10}$$

Ions of the same kinetic energy are focused on the imaging curve at points A_1'' and A_2'', respectively (see Figure 3.4). Like a magnetic sector field a so-called 'transverse magnification' of the electric sector field G_e can be defined, because mass spectrometers use slit systems (entrance and exit slits) of finite width (s' and s''), and so parameters s' and s'' are introduced to describe the defined object and image widths of the entrance and exit slits (A′ and A″ in Figure 3.4), respectively. In analogy to Equation (3.7) for a magnetic sector field, the ratio s''/s' is defined as the so-called 'transverse magnification' G_e of a homogeneous electrical sector field:

$$G_e = s''/s' \tag{3.11}$$

As a result of the influence of the electric field on fast moving ion beams, an energy focusing operation (also known as 'velocity focusing') is performed.

For an electric sector field with angle Φ_e it is valid that the electrostatic force (field strength) $(e \cdot E_0)$ equilibrates to the centrifugal force $\frac{m \cdot v^2}{r_e}$ as described in the following equation:

$$e \cdot E_0 = \frac{m \cdot v^2}{r_e} \tag{3.12}$$

where E_0 is the force of the electric field, v is the velocity of ions, e and m are the elementary charge and mass of the ions, respectively and r_e is the radius of the cylindrical condenser (sector field). With the known relation between kinetic and potential energy of an ion $\frac{1}{2} m \cdot v^2 = e \cdot V_0$, the radius of ion trajectory r_e can be calculated:

$$r_e = \frac{2V_0}{E_0} \tag{3.13}$$

The separation of ion beams with different energy V and ΔV in an electric sector field (see Figure 3.4) is applied for their energy focusing only. Monoenergetic ion beams with ions of different masses are not separated in an electrical field. Ions with energy $e \cdot V$ fly on the middle trajectory between the two plates of the electric sector field, whereas ions with higher energy $e(V + \Delta V)$ or lower energy $e(V - \Delta V)$ are less or more deflected. Using this capability, an electric sector field is able to separate ion beams with different kinetic energies.

The combination of an electric sector field with a magnetic sector field to form a double-focusing instrument is advantageous for improving the properties (e.g., the mass resolution) of mass spectrometers.

3.1.3 Combination of Magnetic and Electric Sector Fields – Double-focusing Sector Field Mass Spectrometer

For an adequate separation of ions with two different masses m and Δm by applying a mass separation system with a single homogeneous magnetic sector field there is one prerequisite: namely the similar energy of the ions. This can be done for ions generated, for example, in an electron ionization source or thermal surface ionization source (see Chapter 2). For such ions the energy spread is lower than 1 eV and a single magnetic sector field is sufficient for separation of the ion beam by their m/z ratios. If ions of the same mass possess different kinetic energies then they are not sharply focused on the imaging curve of mass analyzer. Especially, ion sources operating at high voltage (such as a spark ion source) do not produce monoenergetic ion beams and the kinetic energy of the ions is relatively high. If the energy spread of ions formed in the ion source is larger than 10 eV, single-focusing mass separation systems can no longer be employed. The ions formed, e.g., in a spark plasma, can have an energy spread from about one to several keV, therefore this characteristic imposes the necessity of not only directional focusing, but also energy (velocity) focusing. Moreover, ions formed in a glow discharge or in inductively coupled plasma possess, as a function of the experimental parameters used, an energy spread larger than 10 eV. This relatively high energy spread of the ions results in broadening of mass peaks and decreasing mass resolution of the mass spectrometric system. In order to receive sharp images of mass separated ion beams with higher ion energy spread and to obtain a higher mass resolution, magnetic and electric sector fields are combined in powerful double-focusing sector field mass spectrometers. In these instruments both sector fields are so combined, that the energy dispersion of the magnetic sector field is compensated by the energy dispersion of the electric sector field in a suitable way. In addition, by the application of double-focusing mass spectrometry, a higher mass resolution is obtained as required to separate disturbing isobaric interferences of atomic and molecular ions.

By a suitable combination of electric and magnetic fields, realized in double-focusing mass spectrometers, e.g., with Nier–Johnson or Mattauch–Herzog geometry, the energy dispersion of the magnetic field is compensated by the equal but opposite dispersion of the electric field.[1] Ions with a higher energy spread can be detected using these double-focusing sector field mass spectrometers in well mass resolved mass spectra. The general theory of double focusing was developed by Mattauch and Herzog.[5] Different field combinations of electrical and magnetic sector fields for double focusing in mass spectrometry were proposed by Dempster ($\Phi_e = 90°$; $\Phi_m = 180°$),[6] Bainbridge and Jordan ($\Phi_e = 127°$; $\Phi_m = 60°$),[7] Mattauch and Herzog ($\Phi_e = 31.8°$; $\Phi_m = 90°$)[5] and Nier and Johnson ($\Phi_e = 90°$; $\Phi_m = 90°$).[8] The history of double-focusing sector field mass spectrometry is described in Chapter 1. Today, mainly double-focusing mass spectrometers with forward or reverse Nier–Johnson geometry and Mattauch–Herzog geometry are constructed and applied in inorganic mass spectrometry. These double-focusing mass spectrometers will be discussed in the following sections.

Double Focusing for One Mass

Figure 3.5 illustrates the principles of an ion optical system of a double-focusing sector field mass spectrometer with Nier–Johnson geometry. In this configuration, developed by Nier *et al.*[8–10] (see Figure 3.5) consisting of two 90° sector fields: a cylindrical condenser ($\Phi_e = 90°$) and a magnetic sector field ($\Phi_m = 90°$) with deflection of ions in the same direction for direction and energy focusing (so-called 'C geometry'). The image of entrance slit S_1 in the electric sector field becomes symmetrical ($l_e' = l_e''$). The ion beams enter the electric and magnetic sector fields vertically. For double-focusing conditions in mass spectrometers with Nier–Johnson geometry with $\Phi_e = \Phi_m = 90°$, $r_e/r_m = 1.238$ is valid[2] (the radius ratio of electric and magnetic sector field is not freely selectable). By a defined deflection radius of the electric sector field r_e and the magnetic sector field r_m, respectively, energy and also directional focusing of separated ion beams at one position on the imaging curve is achieved. That means that the exit slit of mass spectrometer S_2 (at distance l_m'' from the exit of the magnetic sector field) is installed at this point of the imaging curve for energy focusing and the imaging curve for direction focusing of ions. By varying the magnetic field B ions of various masses can be transferred under double-focusing conditions at defined deflection radius r_m and can be detected one after the other. In the case of double focusing with deflection in the same direction for direction and energy focusing, the velocity dispersion of an electrical field and the backwards calculated velocity dispersion of the magnetic field must be compensated. Following the dispersion coefficients of magnetic and electrical fields K' and K_e'',

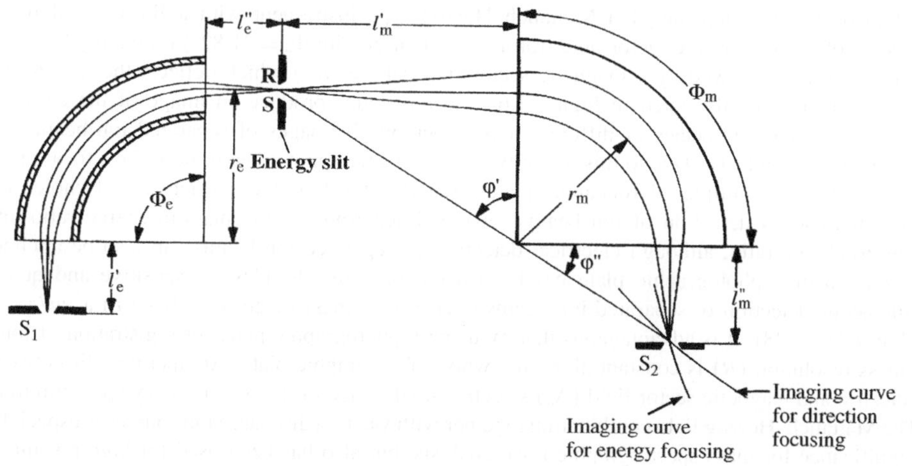

Figure 3.5 Double-focusing sector field mass spectrometer with Nier–Johnson geometry. (H. Kienitz (ed.), Massenspektrometrie (1968), Verlag Chemie, Weinheim. Reproduced by permission of Wiley-VCH)

respectively, mainly functions of the geometry of sector fields (r_e, r_m, Φ_e, Φ_m, l_m' and l_e''), are equal according to Equation (3.14):

$$K' = K_e'' \tag{3.14}$$

That means two ion beams of different energy (eV and $e(V + \Delta V)$, see Figure 3.4) separated in an electrical field that both fly through the energy slit (see Figure 3.5) on its borders (R and S) are focused by directional focusing in the exit slit S_2 of a double focusing mass spectrometer. The general theory of double-focusing mass spectrometry, developed more than 50 years before, is discussed in detail by Ewald and Hintenberger.[11]

A double-focusing mass spectrometer with a Nier–Johnson configuration (with $\Phi_e = \Phi_m = 90°$) permits a large divergence angle and second order directional focusing. An enlarged version of a Nier–Johnson type instrument with a mass resolution of 75 000 has been applied for precise atomic mass measurements.[10] Mass spectrometers with (forward) Nier–Johnson geometry (EB) or reverse Nier–Johnson geometry (BE) are also combined in modern mass spectrometers with different types of ion sources with higher initial energy spreads, e.g., together with a glow discharge ion source, a secondary ion source or an inductively coupled plasma ion source (see Chapter 5). Nier–Johnson type mass spectrometers are applied, not only in inorganic, but also in organic mass spectrometry.

Double Focusing for All Masses Simultaneously

Double focusing is also possible by combination of two sector fields with opposite velocity dispersion, if the deflection of ions in both sector fields is opposite ('S geometry'). In the case of double focusing with deflection of ions in opposite directions the following relation is valid:

$$K' = -K_e'' \tag{3.15}$$

A special experimental arrangement of double-focusing sector field mass spectrometry – the Mattauch–Herzog type mass spectrometer has double-focusing properties for ions of all masses simultaneously. A schematic of a Mattauch–Herzog type instrument with a linear focal plane consisting of an electrostatic sector field (toroidal condenser with $\Phi_e = 31.82°$) for energy focusing of the ion beam, followed by a magnetic sector field ($\Phi_m = 90°$) which deflects the ions in the opposite direction is illustrated in Figure 3.6. The ion optics produce an image of the entrance slit (S_E) for each ion species of different mass whereby all images of separated ion beams are situated on a straight line. Due to this property an ion sensitive photoplate or nowadays focal plane array detectors or multiple microchannel plates (see Chapter 4) is the ion detector of choice for quasi-simultaneous detection of ion beams separated according to the mass-to-charge ratio and energy-to-charge ratio, although electrical detection of separated ion beams can also be applied. The ion sensitive photographic plate has been used for many decades for sensitive and quasi-simultaneous detection of separated ion beams over a wide mass range (e.g., from Li at $m/z = 6$ to U at $m/z = 238$). It is advantageous that by using a photographic plate for registration of ions, the mass resolution (R) is constant along the whole photographic plate. An aperture slit between the electric and magnetic sector field (A_e) selects a small range of the broad energy spectrum ions.

The Mattauch–Herzog field combination together with vacuum discharge ion sources is especially of significance for mass spectrographic trace analysis, but also has been used for high resolution structural analysis of organic substances for a long period.

Double-focusing mass spectrometers with Mattauch–Herzog geometry and high mass resolution have been used for a long time, especially in nuclear physics, for the precise determination of exact

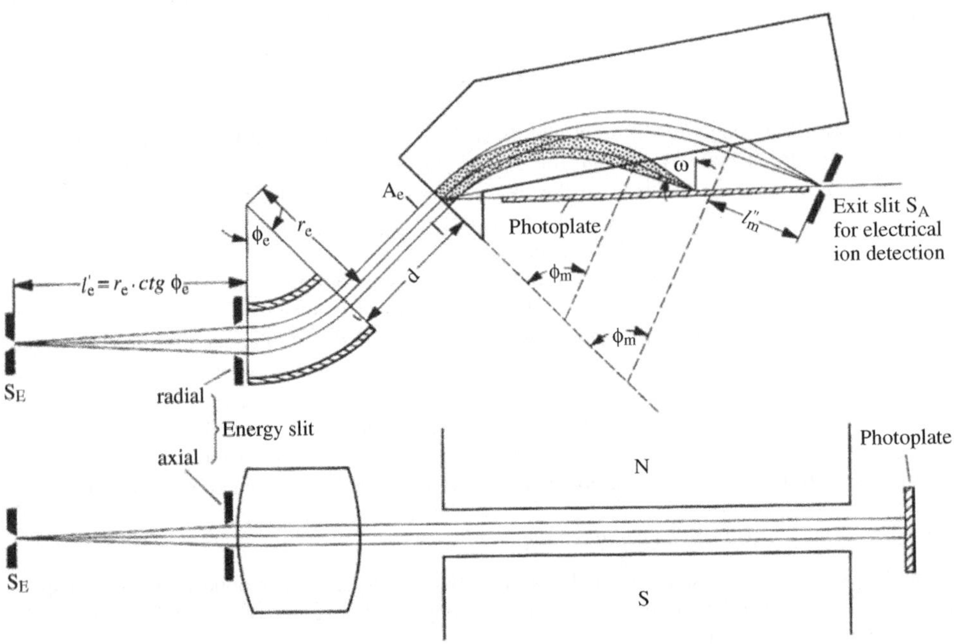

Figure 3.6 *Schematic of a double-focusing sector field mass spectrograph with Mattauch–Herzog geometry with a linear imaging curve (double focusing for ions of all masses simultaneously) modified field combination with $\Phi_m \sim 70°$. In the mass spectrograph, photographic ion detection or focal plane array detectors are used for quasi-simultaneous detection of separated ion beams. (H. Kienitz (ed.), Massenspektrometrie (1968), Verlag Chemie, Weinheim. Reproduced by permission of Wiley-VCH)*

atomic masses. Dietze,[1] Bacon and Ure[12] discuss developments in spark source mass spectrometry using Mattauch–Herzog instruments and describe several applications for the multi-element analysis of trace elements in a variety of different matrices. Hintenberger and König studied double-focusing mass spectrometers with correction of higher order imaging errors.[13]

A mass spectrometer with Mattauch–Herzog geometry was also selected by Nier for space research due to its ability to meet the strict requirements imposed by rockets and satellites without seriously compromising performance and versatility.[14]

Recently, Hieftje *et al.*[15,16] equipped a small double-focusing mass spectrograph built in house with Mattauch–Herzog geometry with several ion sources (such as glow discharge, an inductively coupled plasma ion source or a microwave plasma torch) and a novel array detector for simultaneous ion detection.

Another combination of magnetic and electric sector fields, together with a tandem accelerator, is realized in different types of accelerator mass spectrometers (AMS)[17] applied for carbon-14 dating and extreme ultratrace analysis of long-lived radionuclides at natural isotope abundances (see Chapter 5).

3.2 Dynamic Mass Separation Systems

Dynamic mass separation systems use the fact that ions with different masses (accelerated with the same voltage) possess several velocities and consequently their flight times are different. There are about 50 dynamic separation systems known[2] using several types of ion movements (linear straight ahead, linear periodic or circular periodic as a function of the electric or magnetic sector field applied). The simplest dynamic mass separation system is a linear time-of-flight (ToF) mass analyzer, and a widely applied mass separation system is the quadrupole analyzer.

3.2.1 Quadrupole Mass Analyzer

The quadrupole mass analyzer frequently used in mass spectrometry was introduced by Paul and Steinwedel[18] in 1953 and consists of four hyperbolic or cylindrical rod-shaped electrodes arranged on a circle with the radius r_0. The schematic of a quadrupole mass analyzer with cylindrical or hyperbolic rods is shown in Figures 3.7 and 3.8, respectively. The voltage on these electrodes consists of dc (dc – direct current) and rf (rf – radio frequency) voltages $V \cdot \cos\omega \cdot t$ (with ω as circular frequency), with a superimposed dc voltage U, is applied to the four electrodes. The ions extracted from the ion source are accelerated in an ion beam through the quadrupole mass analyzer in the z direction (see Figure 3.7). The injected ions in the quadrupole on the ion optic axis (in the z direction) oscillate in the resulting quadrupole field of four rod electrodes. The ions moving along the z axis are subjected to the influence of the total electric field resulting from the application of potential to the rods:

$$\Phi_0 = +(U - V \cdot \cos\ \omega \cdot t) \tag{3.16}$$

and

$$-\Phi_0 = -(U - V \cdot \cos\ \omega \cdot t) \tag{3.17}$$

Opposite electrodes of the quadrupole have potentials of the same sign. The electric potential within the rods, $\Phi\ (x, y)$, is decribed by:

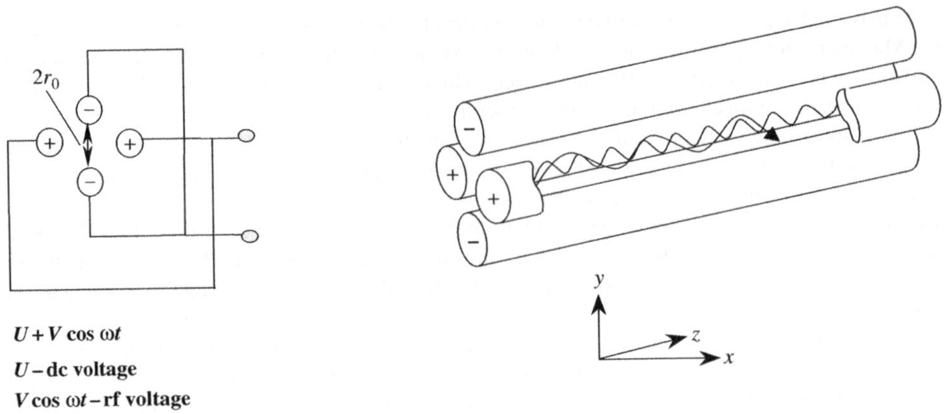

Figure 3.7 Schematic and cross section of a linear quadrupole mass analyzer with cylindrical rod electrodes. (H. Kienitz (ed.), Massenspektrometrie (1968), Verlag Chemie, Weinheim. Reproduced by permission of Wiley-VCH)

$$\Phi\,(x, y) = \frac{(x^2 - y^2)}{r_0^2}\Phi_0 \tag{3.18}$$

where x and y are displacements from the centre line, r_0 is the distance from the centre line to an electrode and Φ_0 is the potential applied to an electrode.[19] For motion of ions in the x direction the force is given by:

$$F_x = -e\frac{2x}{r_0^2}\,\Phi_0 \tag{3.19}$$

The force increases linearly from zero of the centre of the quadrupole. The force in the x direction is independent of the y position, that means the x and y motions are independent and can be considered separately. The ion motion in a quadrupole can be described in the form of the Mathieu equation. Substituting Equation (3.16) in to Equation (3.19) and considering that the acceleration of ions in the x direction is given by $a_x = \mathrm{d}^2x/\mathrm{d}t^2$ then the differential equation of ion motion results in:

$$m\frac{\mathrm{d}^2x}{\mathrm{d}t^2} = -e\frac{2x}{{r_0}^2}(U - V\cos\omega t) \tag{3.20}$$

Collecting terms leads to:

$$\frac{\mathrm{d}^2x}{\mathrm{d}\xi^2} + (a_x - 2q_x\cos\,2\xi)x = 0 \tag{3.21}$$

with $\xi = \omega t/2$ and the Mathieu parameters for x motion (a_x and q_x) are given in Equations (3.22) and (3.23). The oscillations are stable only when the following equations for the ion motion in a quadrupole in the x direction:

$$a_x = \frac{8\,e\cdot U}{m\cdot\omega^2\cdot{r_0}^2} \tag{3.22}$$

and

$$q_x = \frac{4\, e \cdot V}{m \cdot \omega^2 \cdot {r_0}^2} \qquad (3.23)$$

(with r_0 – field radius) are fulfilled. For ion motion in the y direction, the potential has the opposite sign (Equation 3.18). The same Mathieu equation is obtained for the y motion with the substitutions:

$$a_y = -a_x \quad \text{and} \quad q_y = -q_x \qquad (3.24)$$

The ion's motion along the z axis is not affected by the electrical field and its speed is not changed. The solutions to the Mathieu equation are complex and are classified as 'unstable' and 'stable'. An ion trajectory is 'unstable' if the amplitude of oscillation increases exponentially until the ion strikes a rod and is lost (mathematically, x or y exceeds r_0).[19] That means, ions with unstable oscillations in the quadrupole are filtered out. Scanning of masses is carried out by sweeping of the rf voltage, V, whereby the superimposed dc voltage, U, is also swept. By means of these measurements, the U/V ratio is kept constant (slope $0 - a$), as demonstrated in the stability diagram for a linear quadrupole mass analyzer in Figure 3.8. The ions extracted from the ion source with decreasing masses $m_1 > m_2 > m_3 > m_4$ move by scanning voltages U and V on the line through the

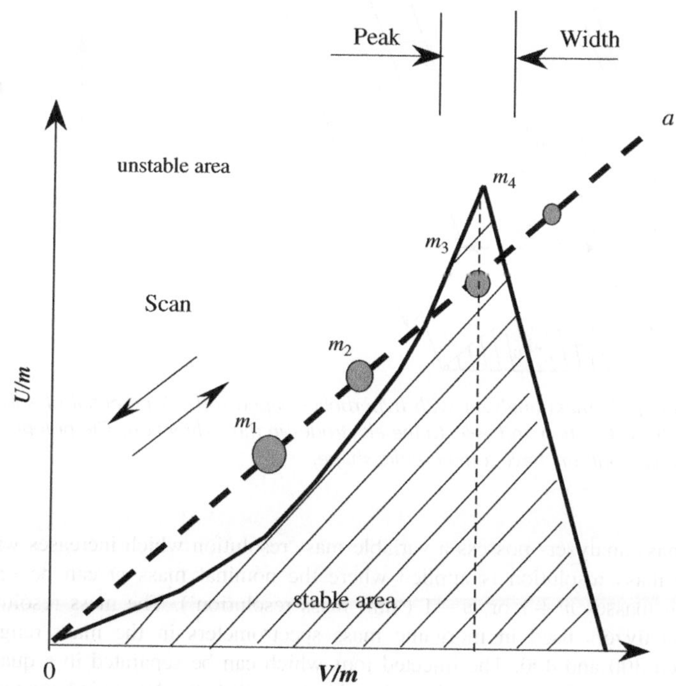

Figure 3.8 *Stability diagram for a linear quadrupole mass analyzer, the first stability region showing a line scan. If the rf and dc voltages applied to the quadrupole are adjusted so that an ion mass* m_3 *is inside the tip of the stability region, then heavier ions of mass* m_2 *and* m_1 *and lighter ions* m_4 *are outside the stability region and are filtered out.*

stable part of the stability diagram of the quadrupole analyzer. As demonstrated in Figure 3.8, the separated ions of mass m_3 in the stable area can be detected using an appropriate ion registration system, whereas ions of other masses, m_1, m_2 and m_4, are filtered out; a quadrupole is therefore often named a 'mass filter'. The precision and accuracy of analytical data is dependent on the resolution of the quadrupole analyzer, as represented by the slope on the line $0 - a$ given by the ratio of the dc voltage U/rf voltage V, which should be constant within 10^{-5}.[3] With a smaller peak width in the stable area the mass resolution increases. Commercial quadrupole systems operate in the 'first' stable region which is shown in Figure 3.8.

Improved separation behaviour is observed in quadrupole analyzers with hyperbolic rods (see Figure 3.9). The quadrupole field is produced by four parallel hyperbolic electrodes, whereby Equations 3.16–3.17 can be applied.

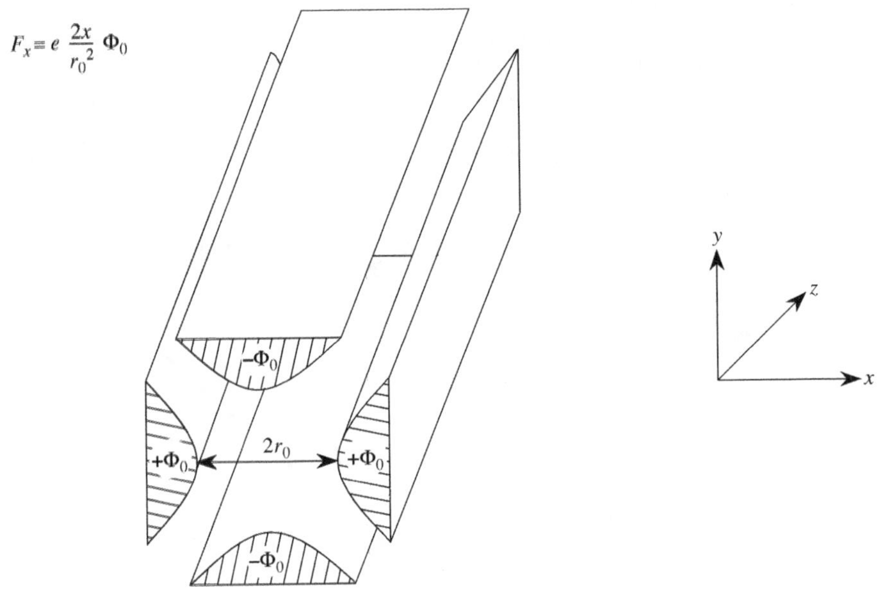

Figure 3.9 Quadrupole mass analyzer with hyperbolic-shaped rods. A potential of $+\phi_0$ is applied to the electrodes in the x direction and $-\phi_0$ to the electrodes in the y direction. The potential at the centre is zero. Equipotential contours have a hyperbolic shape.

Quadrupole mass analyzers possess a variable mass resolution which increases with increasing mass. Mostly a mass resolution is applied where the nominal mass m can be separated from the neighbouring masses $m+1$ or $m-1$ ('unit mass resolution'). The mass resolution ($m/\Delta m$) of quadrupole analyzers used in inorganic mass spectrometers in the mass range < 200 u is normally between 300 and 400. The injected ions which can be separated in a quadrupole mass analyzer possess an ion energy spread in the eV range and can be varied up to 150 eV. The mass resolution of a quadrupole analyzer and abundance sensitivity is improved by analyzing ion beams with lower ion energy, as demonstrated also by the application of a collision cell in ICP-QMS.[20]

Operation of quadrupole mass filters operated in higher zones of stability is described with an emphasis on the potential of elemental mass spectrometry by Du *et al.*[19] These quadrupole mass filters allow a mass resolution in the second stability region of up to 9000 and make it possible to resolve a multitude of isobaric interferences.

Quadrupole analyzers are frequently used as universal mass analyzers in inorganic and organic mass spectrometry as well as having wide fields of application in mass spectrometry due to their high speed, high dynamic range, MS/MS capability, simplicity, easy sample introduction and relatively low cost. In general, a single quadrupole analyzer is used for elemental and isotope analysis in inorganic mass spectrometry (e.g., in ICP-MS) and is sufficient for a multitude of applications in organic mass spectrometry. Quadrupole based mass analyzers have been established in tandem mass spectrometers (e.g., in a triple quadrupole mass spectrometer – QqQ) as a standard tool for liquid chromatography (LC) MS/MS instrumentation or in combination with other mass separation systems (e.g., sector field analyser or time-of-flight instruments). In a triple quadrupole mass spectrometer, the first quadrupole Q_1 is used for mass separation of analytes of interest with mass m_1 which react in the second quadrupole Q_2 (rf only) with a collision gas for collision-induced dissociation (CID) experiments in order to generate metastable ions with different masses from which the third quadrupole selects certain daughter ions of mass m_3. Triple quadrupole mass spectrometers are not used in inorganic mass spectrometry.

3.2.2 Time-of-flight Analyzer

The idea of a time-of-flight (ToF) analyzer for mass separation was first proposed by Stephens in 1946.[21] The first ToF instrument with very low mass resolution was built by Cameron and Eggers in 1948.[22] The first ToF mass spectrometer consisted of a pulsed ion source, a field free drift tube and an oscillograph as the ion detection device. ToF instruments with a mass resolution of practical significance of 50 and 100 were developed by Ionov and Mamyrin[23] and later by Wiley and McLaren,[24] respectively, and form the basis of present day instruments. In time-of-flight (ToF) analyzers pulsed ion beams with different masses (m), start at the same time at the entrance slit of the time-of-flight tube, possess the same kinetic energy, but achieve different velocities (v) according to $E_{kin} = \frac{1}{2} m \cdot v^2$. This means that the ions are extracted simultaneously and accelerated with an acceleration voltage (V) into a field free region with a uniform energy. With respect to the relationship between energy, mass and velocity of ions the flight time of ions is therefore proportional to the square root of the mass. Lighter ions are faster than heavier ones, so that the ions are separated according to their different flight times in the field free flight tube. This can be converted to the mass-to-charge ratio *(m/z)* to obtain a mass spectrum. In Figure 3.10 the schematic of a linear time-of-flight mass spectrometer is shown. Pulsed ion beams consisting of ions with different masses (produced by a pulsed acceleration voltage, e.g., using an ion beam chopper) are accelerated into the field free drift tube at time t_0 and travel at different velocities as a function of their mass. An ion packet near the ion source (1), which starts at time t_0, moves in the direction of the ion detector. Because the velocity of ion beams is a function of their masses the ion packet disperses. First the light ions reach the ion detector at flight time t_l and are detected before the medium ions arrive at flight time t_m (if the charge of all the ions is the same). The last ones to reach the detector are the heaviest ions. So the ion beams are detected at the ion collector one after the other with increasing mass. In a linear flight tube of length d the velocity of ions with the charge $z = n \cdot e$ can be calculated from the well known Equation (3.2). From this equivalence

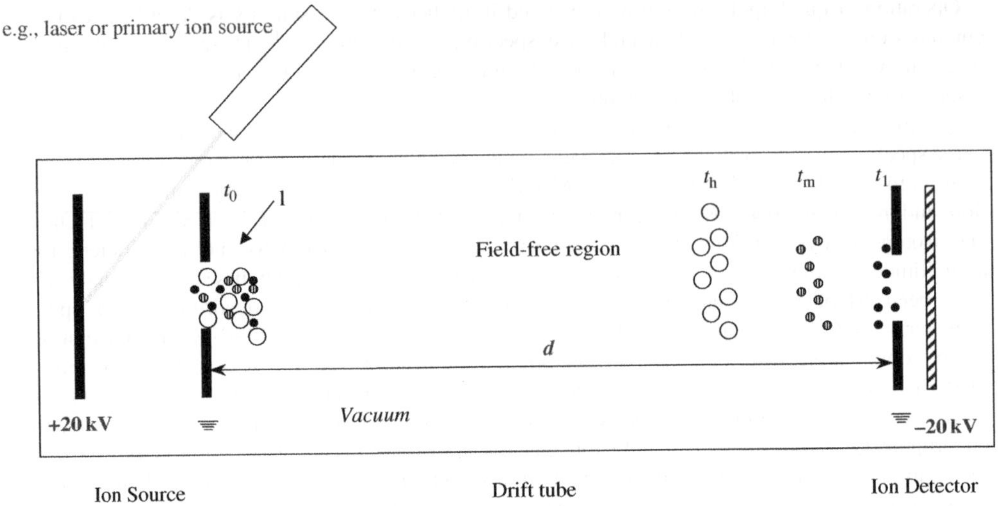

Figure 3.10 Diagram of a linear time-of-flight mass analyzer; separation of ions is shown with three different masses (heavy ions – ○; ions of medium mass – ◍ and light ions – ●). The flight time of heavy ions is longer than that of light ions due to their lower kinetic energy.

relation of kinetic and potential energies, the velocity of ions (v) can be estimated using:

$$v = \sqrt{\frac{2n \cdot e \cdot V}{m}} \quad (3.25)$$

With the length of the field free tube d the flight time of ions t will be defined by:

$$t = \frac{d}{v} = d\sqrt{\frac{m}{2n \cdot e \cdot V}} \quad (3.26)$$

The flight time difference Δt for the detection of two ion beams with different masses m_1 and m_2 is proportional to $d \cdot (\sqrt{m_1} - \sqrt{m_2})$.

The time resolution of ions with different masses and the resulting mass resolution of the ToF mass analyzer are proportional to the length of the drift tube. That means the mass resolution is higher using a longer drift tube compared to a shorter one. On the other hand, the mass resolution is inversely proportional to the accelerating voltage V. Furthermore, in order to increase the mass resolution of ToF mass separation systems, ion sources producing ions with lower energy spread are advantageous. In comparison to more expensive sector field instruments, a simple linear ToF separation system (Figure 3.10) does not have any ion focusing properties. To improve the mass resolution of ToF mass analyzers from a few hundred to several thousand, in 1973 Mamyrin *et al.*[25] introduced the first ToF reflectron to compensate for the ion energy spread. The authors measured a mass resolution of about 5000 in the mass spectrum of rhenium bromide. A schematic of a ToF mass analyzer with reflectron is illustrated in Figure 3.11. This mass spectrometer consists of a pulsed ion source, flight tube with ion reflector and a sensitive ion collector. The reflector for ions is located behind the field free drift region and in front of the ion detector acting as an ion mirror to

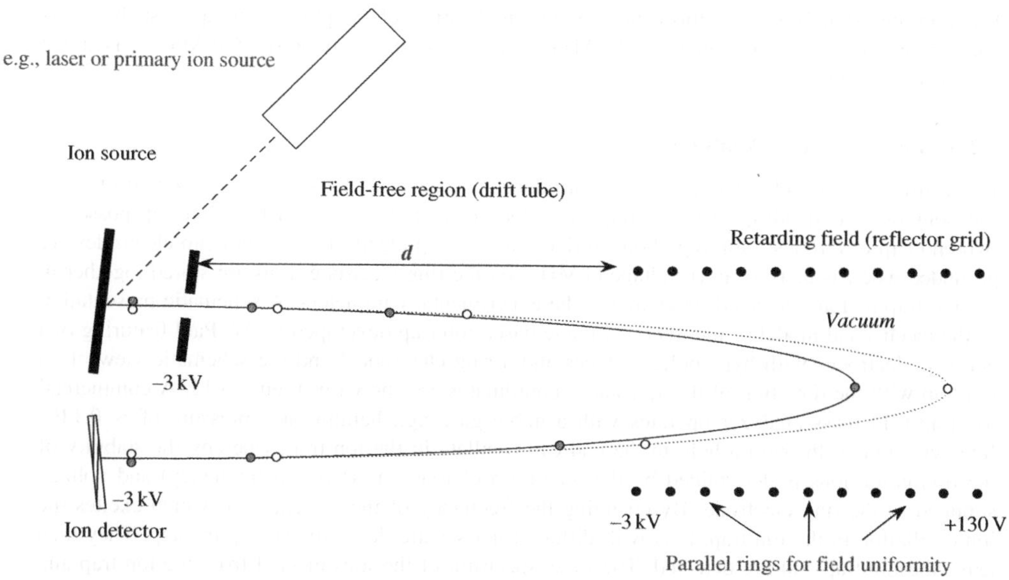

Figure 3.11 *Schematic of a time-of-flight mass analyzer with reflectron proposed by Mamyrin.[11] Ions of higher kinetic energy penetrate more deeply into the reflecting field and are delayed in comparison to ions of lower kinetic energy. The delay compensates for differences in transit time of the field-free regions. Ions produced e.g., by focused laser or primary ion beam.*

focus ions of different ion kinetic energy. The reflection voltage V_r is set at about 1.05–1.10 times the acceleration voltage V, in order to insure that all ions are reflected within the homogeneous portion of the electric field of the device (Figure 3.11).[26] The ions penetrate the electric field of the reflectron until they reach a kinetic energy of zero and are deflected and accelerated in the opposite direction to the ion detector. Ions of higher kinetic energy penetrate more deeply into the reflecting electric field and are delayed in comparison to ions of lower kinetic energy. The delay compensates for differences in transit time of the field free regions due to the initial energy spread of the ions and thus a focusing of ions with the same m/z ratio is realized. As a result, the mass resolution $m/\Delta m$ of the ToF instrument with reflectron is significantly increased compared to simple linear ToF mass analyzers.

A novel design for a gridless two-stage ion mirror in ToF mass spectrometry for space application was developed by Scherer *et al.*[27] The electrostatic fields of the reflectron are generated along two individual adjustable sections of a resistor helix applied to the inner surface of a ceramic cylinder. This design is advantageous for increased homogeneity of electrostatic fields; the usuable inner diameter of ion flight path for a given outer diameter is maximized. An additional electrostatic lens in front of the reflectron allows the geometrical alteration of the shape of the ion beam, and its direction with regard to the ion optical axis. Mass resolution up to 5000 at full width at half maximum (FWHM) has been achieved in triple reflection mode using an additional ion mirror.[27]

The advantage of ToF analyzers is their unlimited mass range for a sensitive analysis of large biomolecules or clusters of up to several millions of daltons and their ability to rapidly obtain mass spectra relevant for the fast analysis of transient signals.[28] The latter property of analyzing fast transient signals is of significance for the application of coupling techniques such as gas chromatography mass spectrometry (GC-MS)[29] or capillary electrophoresis ICP-MS as discussed in Chapters 5 and 9.

3.2.3 Ion Trap Mass Analyzer

In the fifties, Paul and coworkers[18,30,31] introduced the quadrupole ion trap analyzer in order to trap ions in a 3D quadrupole radio frequency electric field. The original Paul ion trap possesses two end cap electrodes with hyperbolic surfaces and a ring electrode. The end cap electrodes are grounded whereas there is an rf voltage (1 MHz) on the ring electrode. Ions are stored together in the ion trap and are detected by changing the experimental parameters. A schematic presentation of the mechanical model of the 3D quadrupole based ion trap developed by W. Paul featuring two end cap electrodes with hyperbolic surfaces and a ring electrode[18] and the schematic view of an ion trap with the direction of the x, y and z coordinates are shown in Figure 3.12. A commercial ion trap (IT) mass analyzer operates with a noble gas, e.g., helium, at a pressure of $\sim$ 0.1 Pa. Ions which enter through a hole in the end cap oscillate in the ion trap, whereby the stability of the oscillating ions is determined by the m/z ratio of ions, the rf (radio frequency) and voltage supplied to the ring electrode. By changing the frequency of the rf generator, which excites the ion oscillation in the ion trap, ions with different masses are destabilized step by step. They then leave the ion trap and are detected. The mass spectrum of the ions ejected from the ion trap and separated according to their m/z ratio is then obtained by scanning the rf voltage (method of operation: mass selective instability mode). Another procedure for ejecting ions from the ion trap is known as the mass selective instability mode whereby a second rf voltage (lower in frequency than that of the ring electrode) is supplied to the end caps. When the rf voltage is scanned on the ring electrode, causing mass selective instability, ions are sequentially brought into resonance with the rf of the end caps. This behaviour results in an increase in translation energy of the trapped ions with defined m/z ratio and the ions are ejected from the ion trap in a well-defined manner.[32] Cooks *et al.*[33] have reported on developments in new geometries of quadrupole ion traps and trap arrays including various linear and rectilinear ion traps, the simplest form being a rectilinear ion trap geometry made from four flat electrodes.

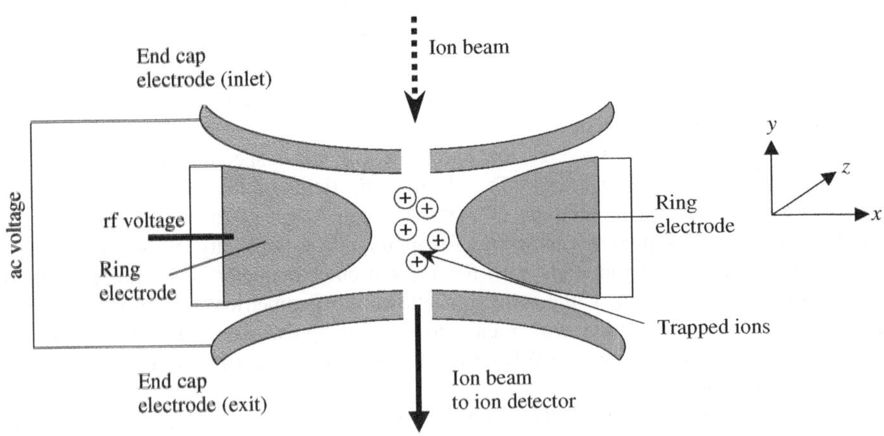

Figure 3.12 Schematic view of an ion trap with the direction of the x, y and z coordinates.

In spite of the significant reduction in instrumental size, in comparison to quadrupole, sector field and Fourier transform mass spectrometers, ion trap mass analyzers also permit mass spectrometric measurements of large organic compounds such as proteins and biomolecules (with masses ~100 000 Da) at high mass resolution and low detection limit (down to 10^{-15} g). In addition, using a suitable collision gas in the ion trap, the isobaric interference problems in mass spectrometry can be solved by collision induced dissociation. The possibilities of applying ion traps are limited by restricted quantification possibilities and low dynamic range. Over the last few decades, quadrupole ion trap mass analyzers (now known as the 3D QIT), together with soft ionization techniques, such as electrospray ionization (ESI) or MALDI, have developed into a sensitive and versatile analytical tool for identifying organic molecules (especially for biomolecular analysis) in order to look at intact ions with masses into the kDa range. The use of low cost quadrupole ion traps as high efficient mass analyzers is well known from their application in tandem mass spectrometers (MS/MS) also coupled together with liquid and gas chromatography (LC-MS and GC-MS, respectively)[34,35] in organic mass spectrometry.

In spite of high mass resolution, ion trap mass analyzers have not been widely applied in inorganic mass spectrometry and no commercial ICP-IT-MS is available on the analytical market. One reason could be the occurrence of space charge effects of the most abundant argon plasma ions in the ion trap analyzer, which results in in-trap collisions and scattering of the analyte ions of interest.[32] As a result, a decrease in the sensitivity of analyte ions has been observed in comparison to ICP-MS with double-focusing sector field geometry. Ion traps have been applied in inorganic mass spectrometry for fundamental studies of ion – molecule reactions, especially for reducing disturbing isobaric interferences (e.g., of $^{90}Sr^+$, $^{90}Y^+$ and $^{90}Zr^+$ for $^{90}Sr^+$ determination or $^{40}Ca^+$ and $^{40}Ar^+$ for Ca isotope ratio measurements)[36] by collision-induced dissociation and chemical reaction of ions with collision gas atoms or molecules, as described by Barshik *et al.*[36] and by Eiden *et al.*[37] Figure 3.13 shows the experimental arrangement of an ion trap combined with an octopole collision cell as a part of a plasma source ion trap (PSIT) mass spectrometer with an inductively coupled plasma or glow discharge ion source described by Koppenaal's group.[36–38] By introducing H_2 into the octopole collision cell behind the skimmer cone (pressure: $\sim 10^{-5}$–10^{-6} Pa), the authors observed a 40-fold reduction in the Ar^+ ions. Furthermore, by the introduction of O_2 ($\sim 10^{-3}$ Pa) into the ion trap, a selective formation of YO^+ and ZrO^+, but not SrO^+ was found, which could be of relevance for a sensitive determination of ^{90}Sr without chemical separation.[36]

An innovative approach for high mass resolution analysis was the introduction of the so-called Orbitrap in 2005 as the second mass analyzer in a hybrid tandem instrument using a linear quadrupole ion trap as the first stage by Thermo Fisher Scientific LTQ™ Orbitrap™. The Orbitrap needs an external storage capable of fast pulse injection of stored ions in this mass analyzer (of the size of a walnut), which does not require the cost expensive and maintenance requiring superconducting magnet of an FTICR-MS.[39,40] Due to the high sensitivity and mass resolution achieved ($m/\Delta m > 100000$) over a wide mass range, the Orbitrap will become of increasing importance in mass spectrometry, e.g., as demonstrated for the analysis of phosphopeptides to study post-translational modifications in nature.[41]

3.2.4 Ion Cyclotron Resonance Mass Analyzer

In a commercially available device based on Penning traps – the Fourier transform ion cyclotron resonance (FTICR) mass spectrometer – the ions are constrained spatially or rather ions can be

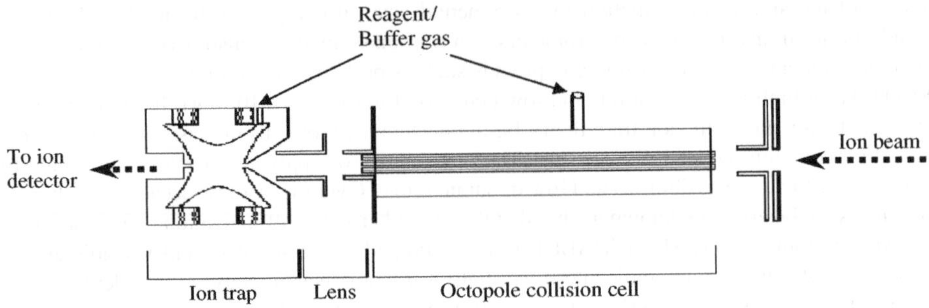

Figure 3.13 *Combination of an ion trap with octopole collision cell in an ICP mass spectrometer. (D. Koppenaal, C. J. Barinaga and M. R. Smith, J. Anal. At. spectrom.* ***9****, 1053 (1994). Reproduced by permission of The Royal Society of Chemistry.)*

trapped on a circular trajectory by a combination of electric and magnetic fields. Fourier transform mass spectrometers were developed from cyclotrons (ion resonance accelerators developed by Lawrence at the University of Berkeley, California in 1930) by Comisarov and Marshall in 1974:[42,43] Both techniques use the fact that ions in a uniform magnetic field move in small circular orbit given by the cyclotron equation:

$$\omega_c = zB/m \tag{3.27}$$

where ω_c = cyclotron frequency, z = charge of ion, m = mass of ion and B = magnetic field strength. In general, the FTICR technique operates in the pulsed mode. Ions which are injected into the Penning trap along the z axis can be trapped on a circular trajectory by a suitable electrostatic field combined with a homogeneous magnetic field B. In an ion cyclotron resonance analyzer, (Figure 3.14) the ion motion is excited by the electric field between two plates 1 and 2 when an rf voltage is applied.[44] Trapping the voltage at plates 3 and 4 keeps the ions in the middle plane. The circular trajectory of the ions is stable if the centripetal (magnetic) force $F = n \cdot e \cdot v \cdot B$ is equal to the centrifugal force:

$$n \cdot e \cdot v \cdot B = \frac{m \cdot v^2}{r} \quad \text{or} \quad n \cdot e \cdot B = \frac{m \cdot v}{r} \tag{3.28}$$

The radius of ion trajectory in the Penning trap is then determined by:

$$r = \frac{m \cdot v}{n \cdot e \cdot B} \tag{3.29}$$

That means the radius of the ion trajectory becomes small if the velocity of the ion is low and the magnetic field B is increased.[44] In order to separate ions with high kinetic energy, strong magnetic fields are required. Therefore in an FTICR mass spectrometer, a superconducting magnet (mostly cooled with liquid nitrogen or helium) with a magnetic field between 3 tesla and 24 tesla

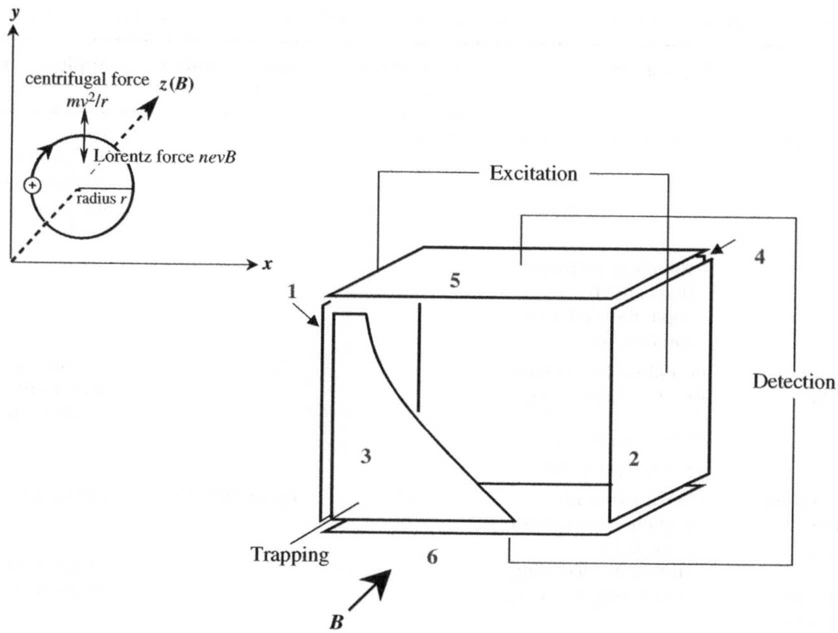

Figure 3.14 *Schematic representation of an ion cyclotron resonance cell (Penning trap) for high resolution mass spectrometry. The magnetic field* B *is oriented and the ions are injected along the* z *axis. The ions are trapped along this axis by a trapping voltage (e.g., 1 V applied to the front and back plates 3 and 4). (Modified from: C. Brunnée, Int. J. Mass Spectrom.* **76**, *125 (1987). Reproduced by permission from Elsevier.)*

is applied. Commercial FTICR-MS (e.g., from Bruker Daltonic) with field strengths up to 15 tesla using a cryocooler refrigeration unit, allow mass spectrometric measurements with a maximum mass resolution up to 10^6 and the highest mass accuracy. Ions of each mass have a characteristic cyclotron frequency (ω_c), see equation 3.27. The motion of positive ions in a uniform magnetic field B is shown in Figure 3.14 (top left). Once the ions are trapped in the analyzer cell (Penning trap), mass analysis is performed by applying a resonant rf electric field across the cell. The ions are accelerated in an rf electric field at their cyclotron resonance frequency. The ions fly on a spiral type path and thus the ions are detected using plates 5 and 6 for ion detection. The motion of the ions in the FTICR trap and the evaluation of transient ion signals using a fast Fourier transform algorithm to obtain high mass resolved spectra of analytes is described in several reviews.[45,46] Extension of dynamic range in FTICR-MS via stored waveform inverse Fourier transform excitation (SWIFT technique) is discussed by Wang *et al.*[47]

Fourier transform ICR mass spectrometers together with any type of ion source, such as nanoESI, MALDI (or also an inductively coupled plasma ion source) permit mass spectrometric measurements to be performed at ultrahigh mass resolution ($R = m/\Delta m = 10^5$–10^6) with a very low detection limit and the highest possible mass accuracy ($\Delta m = 10^{-3}$–10^{-4} Da). In addition, a high mass range is possible and FTICR-MS can be applied for MS/MS experiments.[48] A comparison of different separation systems used in inorganic mass spectrometry is presented in Table 3.1.

Table 3.1 Comparison of different ion separation systems used in inorganic mass spectrometry.

Analyzer	Separation	Mass range (Da)	Mass resolution ($m/\Delta m$)	Application
Magnetic sector field	equilibr: Lorentz force and centrifugal force $n \cdot e \cdot v \cdot B = \frac{mv^2}{2}$ - direction focusing (lens effect) - separation after m/z (prism effect)	1 500	5 000	TIMS, organic ms, double-focusing sector field ms, tandem ms
Electrostatic sector field	equilibr: electrostatic force and centrifugal force $e \cdot E_0 = \frac{mv^2}{2}$ - energy focusing	—	50	double-focusing sector field ms, tandem ms
Double-focusing sector field e.g. Mattauch–Herzog, Nier–Johnson geometry, and reverse geom.	combination of magnetic and electric sector field - direction focusing and energy focusing	~ 350	up to 20 000	organ. ms SSMS, LIMS SIMS, ICP-MS
Quadrupole	Mathieu equation	~ 300	< 400	ICP-MS, SIMS,
Time-of-flight	$E_{kin} = \frac{mv^2}{2}$	< 500 000	up to 10 000	organ. ms, SIMS, ICP-MS

3.3 Mass Resolution and Abundance Sensitivity

Because slits in the range 0.1 to 1 mm width are used in mass spectrometers, the separated ion beams have a defined breath b close to the slits. The capability of the mass spectrometer for separating ion beams with different masses m and $m + \Delta m$ is characterized by the *mass resolution R* (or resolving power) using the following definition:

$$R = m/\Delta m \tag{3.30}$$

The mass resolution $R(m/\Delta m)$ of a mass spectrometer (or part of the mass spectrometer, e.g., of a sector field) gives information on the size of the mass difference Δm of two ion beams with masses m and $m + \Delta m$ in order to separate and detect both ion beams clearly. That means no peak tailing of a neighbouring peak results in an increase of ion intensity at the measured ion intensity of the analyte ion at mass $m + \Delta m$. There are, in general, two definitions of mass resolution (see Figure 3.15). The 10 % valley definition (a) considers two peaks of equal height in a mass spectrum at masses m and $m + \Delta m$ that are separated by a valley which at its lowest point is 10 % of the height of either peak. The mass resolution $m/\Delta m$ should be given for a number of values of m. The peak width definition (b) states that for a single peak of ions at mass m in a mass spectrum, the mass resolution ($m/\Delta m$) is estimated from the width of the peak (Δm, see Figure 3.15) at a height which is a specified fraction (e.g., 5 %) of the maximum peak height. For a

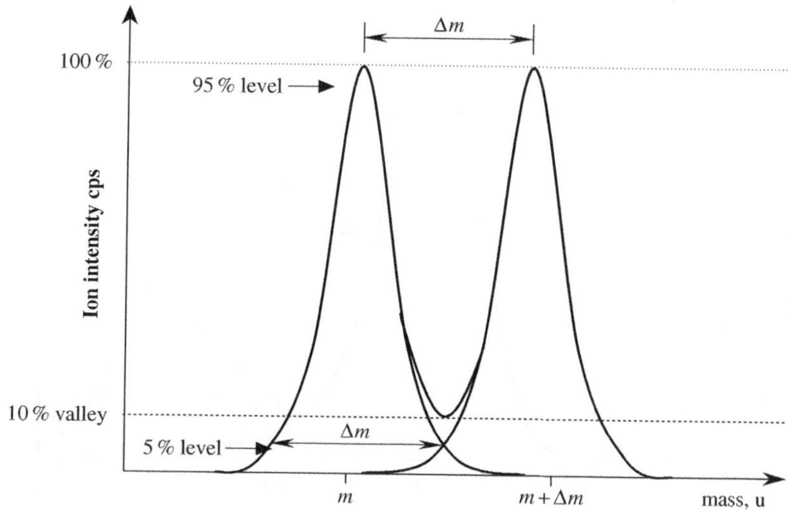

Figure 3.15 *Demonstration of two definitions of mass resolution* ($R = m/\Delta m$)*: 10 % valley definition and peak width definition.*

well separated symmetrical peak recorded with a system which is linear in the range between 5 % and 10 % of the peak, the 5 % peak width definition is equivalent to the 10 % valley definition. A common definition of mass resolution is taken at 50 % of peak height and is called full width at half maximum (FWHM). The required mass resolution ($m/\Delta m$) for mass spectrometric separation of two neighbouring peaks (e.g., of two atomic ions or of an atomic and a molecular ion) can be estimated using the known atomic masses from the IUPAC table value for atomic masses (see Appendix II)[49] for m and Δm whereby the mass of the polyatomic or cluster ions is calculated as the sum of the masses of their atomic constituents. The theoretical mass resolution can be calculated from the geometric parameter for each ion separation system. In order to improve the mass resolution of a mass spectrometric system for solving interference problems, the slit width in the mass spectrometer is minimized. However, an increase in mass resolution results in a loss of ion intensity and sensitivity. The highest mass resolution is possible using Fourier Transform ion cyclotron mass spectrometers (FTICR-MS) with high field strengths as described in Section 3.2.4.

A second important property of mass spectrometric separation systems is the *abundance sensitivity*. Even under sufficient vacuum conditions there is a scattering of ions in the beam by the residual gas. Furthermore, by scattering of particles by the wall of the analyzer tube or by electrostatic repulsion in the ion beam itself or charging effects, so-called 'peak tails' in mass spectra are observed, which result in an increasing energy spread of ions and consequently in a deterioration in the abundance sensitivity. The abundance sensitivity of a mass spectrometer is defined as:

$$\text{Abundance sensitivity} = \frac{\text{Ion intensity at mass } m+1}{\text{Ion intensity at mass } m} \tag{3.31}$$

A schematic illustrating the abundance sensitivity is shown in Figure 3.16. Measurements of the ion intensity (I_B) of the analyte ion of interest B^+ at $m+1$ are too high due to peak tailing (β) of the high intensity (or more abundant) neighbouring peak A^+ at mass m. The abundance sensitivity,

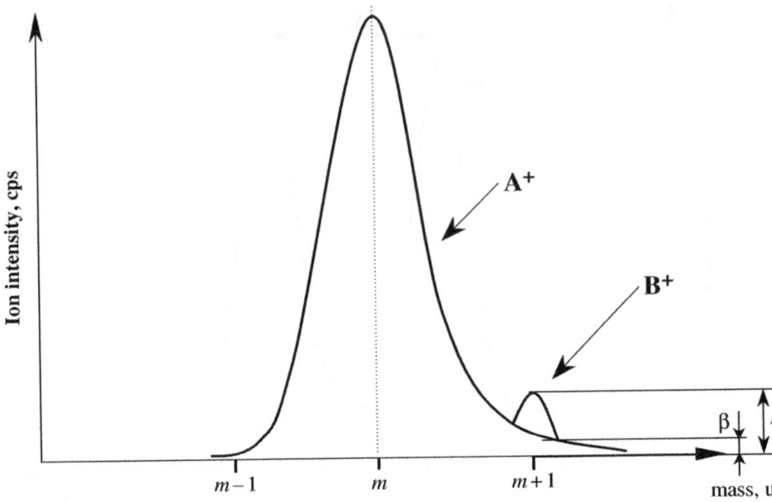

Figure 3.16 *Abundance sensitivity: Influence of peak tailing (β) from the abundant ion peak of mass m on neighbouring ion peak at mass m+1 with low intensity I_B.*

which is best in sector field mass spectrometers with multiple ion collectors (and an additional energy filter to suppress scattered ions), can reach up to 10^{-9}.

The abundance sensitivity can also be considered for ions beams with a mass difference of two mass units. For example, the ultrasensitive measurement of ^{236}U is limited by peak tailing from the major uranium isotope $^{238}U^+$, and ^{90}Sr detection at the ultratrace level is difficult due to peak tailing of the most abundant $^{88}Sr^+$ isotope with a mass difference of two. For $^{238}U^+$ and $^{90}Sr^+$ the abundance sensitivity is defined as:

$$\text{Abundance sensitivity} = \frac{\text{Ion intensity at mass } m+2}{\text{Ion intensity at mass } m} \tag{3.32}$$

In double-focusing sector field ICP-MS spectrometers with single ion detection, the best abundance sensitivity with a mass difference of two is observed at medium mass resolution $m/\Delta m = 4400$ (9×10^{-7}). At low mass resolution ($m/\Delta m = 400$) an abundance sensitivity for ^{90}Sr determination of 2×10^{-5} was measured in the author's laboratory. In quadrupole ICP-MS with and without a hexapole collision cell, an abundance sensitivity of 6×10^{-8} and 6×10^{-7}, respectively, for ^{236}U determination in the presence of abundant $^{238}U^+$ with high ion intensity was determined by Boulyga *et al.*[50] A significant improvement in abundance sensitivity for the determination of the $^{236}U/^{238}U$ isotope ratio has been obtained by multi-collector ICP-MS (Nu Plasma equipped with an ion deceleration filter, so-called 'high abundance sensitivity channel').[51] The application of the ion deceleration system in MC-ICP-MS allows a reduction of peak tailing from $^{238}U^+$ ions at $m/z = 236$ down to 3×10^{-9}, so that $^{236}U/^{238}U$ isotope ratios can be measured in the 10^{-8} to 10^{-9} range. By use of a special ion lens deceleration system in MC-TIMS, an upper limit for $^{236}U/^{238}U$ isotope ratios of 6×10^{10} has been achieved.[52] Accelerator mass spectrometry (AMS) allows uranium isotope ratio measurements with an abundance sensitivity for ^{236}U in the range 10^{-10}–10^{-12}.[53,54]

References

1. Dietze, H. J., *Massenspektroskopische Spurenanalyse*, Akademischer Verlagsgesellschaft Geest & Portig K.-G., Leipzig (1975).
2. Kienitz, H. (ed.), *Massenspektrometrie*, Verlag Chemie, Weinheim, (1968).
3. Brunnée, C., *Int. J. Mass Spectrom. Ion Proc.*, **76**, 125 (1987).
4. De Laeter, J. R., *Application of Inorganic Mass Spectrometry*, Wiley-Interscience Series on Mass Spectrometry, John Wiley & Sons, Inc. New York, (2001).
5. Mattauch, J. and Herzog, R., *Z. Phys.*, **89**, 786 (1934).
6. Dempster, A. J., *Phys. Rev.*, **51**, 67 (1937).
7. Bainbridge, K. T. and Jordan, E. B., *Phys. Rev.*, **50**, 282 (1936).
8. Nier, A. O. and Johnson, W. H., *Phys. Rev.*, **91**, 10 (1953).
9. Nier, A. O. and Roberts, T. R., *Phys. Rev.*, **81**, 507 (1951).
10. Nier, A. O., Roberts, T. R. and Franklin, E. G., *Phys. Rev.*, **75**, 346 (1949).
11. Ewald, H. and Hintenberger, H., *Methoden und Anwendungen der Massenspektroskopie*, Verlag Chemie, Weinheim (1953).
12. Bacon, R. and Ure, A. M., *Analyst*, **109**, 1229 (1984).
13. Hintenberger, H. and Koenig, L. A., *Z. Naturforschg.*, **12a**, 140 (1957).
14. Nier, A. O., *USGS Bulletin*, **1890**, 1 (1989).
15. Barnes, J. H., Schilling, G. D., Sperline, R., Denton, M. B., Young, E. T., Barigina, C. J., Koppenaal, D. W. and Hieftje, G. M., *Anal. Chem.*, **76**, 2631 (2004).
16. Barnes, J. H., Sperline, R., Denton, M. B., Barigina, C. J., Koppenaal, D., Young, E. T., and Hieftje, G. M., *Anal. Chem.*, **74**, 5327 (2002).
17. Kutschera, W., *Int. J. Mass Spectrom.*, **242**, 145 (2005).
18. Paul, W. and Steinwedel, H., *Z. Naturforsch.*, **8a**, 448 (1953).
19. Du, Z., Douglas, D. J. and Konenkov, N., *J. Anal. At. Spectrom.*, **14**, 1111 (1999).
20. Boulyga, S. F. and Becker, J. S., *J. Anal. At. Spectrom.*, **17**, 1202 (2002).
21. Stephens, W., *Phys. Rev.*, **69**, 691 (1946).
22. Cameron, A. E. and Eggers, D. F., *Rev. Sci. Instrum.*, **19**, 605 (1948).
23. Ionov, N. I. and Mamyrin, B. A., *Zh. Tekh. Fiz.*, **23**, 2101 (1953).
24. Wiley, W. L. and McLaren, I. H., *Rev. Sci. Instrum.*, **16**, 1150 (1957).
25. Mamyrin, B. A., Karataev, V. I., Shmikk, D. V. and Zagulin, V. A., *Sov. Phys. JETP*, **37**, 45 (1973).
26. Gross, J. H., *Mass Spectrometry*, Springer Verlag, Heidelberg, New York (2004).
27. Scherer, S., Altwegg, K., Fischer, J. *et al.*, *Int. J. Mass Spectrom.*, **251**, 73 (2006).
28. Mamyrin, B. A., *Int. J. Mass Spectrom.*, **131**, 1 (1994).
29. Gohlke, R. S. and McLafferty, F. W., *Int. J. Mass Spectrom.*, **4**, 367 (1993).
30. Paul, W. and Raether, M., *Z. Physik*, **140**, 262 (1955).
31. Paul, W., Reinhard, H. P. and von Zahn, U., *Z. Phys.*, **152**, 143 (1958).
32. O'Connor, G. and Evans, E. H., in *Inductively Coupled Plasma Mass Spectrometry*. S. J. Hill (ed.), Sheffield Academy Press, 138 (2000)
33. Cooks, G., Ouyang, Z., Gao, L., Fico, M. and Chappell, W., in *17th International Mass Spectrometry Conference, Prague*, Vychodoceska Tiskarna, Sezemic, Czech Republic (2006).
34. Kleintop, B. L., Eades, D. M., Jones, J. A. and Yost, R. A., in *Practical Aspects of Ion Trap Mass Spectrometry*, R. E. March. and Todd J. F. J. (eds.), CRC Press, New York (1995).
35. Yates, N. A., Booth, M. M., Stephenson, J. L., Jr. and Yost, R. A., in *Practical Aspects of Ion Trap Mass Spectrometry*, R. E. March and J. F. J. Todd (eds.), CRC Press, New York (1994).
36. Barshick, C. M., Duckworth, D. C., Smith, D. H. and editors, *Inorganic Mass Spectrometry: Fundamentals and Applications* (In Pract. Spectrosc., 23 (2000)).
37. Eiden, G. C. and Koopenaal, D., *Rapid Commun. Mass Spectrom.*, **11**, 37 (1997).
38. Koppenaal, D. W., Barinaga, C. J. and Smith, M. R., *J. Anal. At. Spectrom.*, **9**, 1950 (1994).
39. Makarov, A., Denisov, E., Lange, O., Horning, S., *J. Amer. Soc. Mass Spectrom.*, **17**, 1758 (2006).
40. Makarov, A. A., *Anal. Chem.*, **72**, 1156 (2000).

41. Muenster, H., Strupat, K., Makarov, A. *et al.* in *17th International Mass Spectrometry Conference*, Vychodoceska Tiskarna, Sezemice, Czech Republic (2006).
42. Comisarov, M. B. and Marshall, A. G., *Chem. Phys. Lett.*, **25**, 282 (1974).
43. Comisarov, M. B. and Marshall, A. G., *Chem. Phys. Lett.*, **26**, 489 (1974).
44. Duckworth, H. E., Barber, R. C. and Venkatasubramanian, V. S. (eds.), *Mass Spectroscopy*, 2nd edn, Cambridge University Press, Cambridge, London, New York, New Rochelle, Melbourne, Sydney (1986).
45. Marshall, A. G. and Grosshans, P. B., *Anal. Chem.*, **63**, 215A (1991).
46. Marshall, A. G., Handrickson, C. L. and Jackson, G. S., *Mass Spectrom. Rev.* **17**, 1 (1998).
47. Wang, T. C. L., Ricca, T. L. and Marshall, A. G., *Anal. Chem.*. **56**, 2935 (1986).
48. White, F. M., Marto, J. A. and Marshall, A. G., *Rapid Commun. Mass Spectrom.*, **10**, 1845 (1996).
49. IUPAC Isotopic Composition of the Elements 1997, *J. Anal. At. Spectrom.*, **14**, 5N (1999).
50. Boulyga, S. F., Matusevich, J. L., Mironov, V. P. *et al.*, *J. Anal. At. Spectrom.*, **17**, 958 (2002).
51. Boulyga, S. F., Kloetzli, U. and Prohaska, T., *J. Anal. At. Spectrom.*, **21**, 1427 (2006).
52. Richter, S., Alonso, A., De Bolle, W., Wellum, R. and Taylor, P. D. P., *Int. J. Mass Spectrom.*, **193**, 9 (1999).
53. Fifield, L. K., *Nucl. Instrum. Methods*, **B172**, 297 (2000).
54. Paul, M., Berkovits, D., Ahmad, I. *et al.*, *Nucl. Instrum. Methods*, **B172**, 688 (2000).

4

Ion Detection Systems

When the separated ion beams leave the mass analyzer system the ions are collected and detected using an appropriate ion detection system inserted in the ultrahigh vacuum of the mass spectrometer. Ion currents at the exit of the mass analyzer are in the range of 10^{-8} to 10^{-19} A. The registration of both high and very small ion currents requires special fast ion detection systems as discussed in this chapter.

4.1 Faraday Cup

In many mass spectrometers for electric ion detection a Faraday cup is generally used today for the direct and accurate measurement of ion currents of separated ion beams, especially if relatively high ion currents are to be measured. As the simplest single ion collector, a Faraday cup located behind the ion exit slit consists of a deep rectanglular bucket (see Figure 4.1a). It collects all the ions which enter it, and guides them to an amplifier. This single ion collector has an electrode with a negative voltage where the positively charged ions deposit their charge. The electrode is located near the entrance of the ion collector. The positive ions collide with the sloping collision surface. The electric current flows away from the electrode surface via the high ohmic resistor by means of the collision of ions on the negative electrode of the Faraday cup; these ions are neutralized on the collision surface. However, secondary electrons are also emitted from the surface, which supply additional positive ions. To suppress the secondary electrons formed escaping from the cup, a secondary electron apperture with a voltage of about −30 V is applied. Consequently, the electrons will be unable to leave the ion collector. The noise inherent in Faraday cup detectors and associated electronic circuits represents a limitation on the sensitivity of the measurements. The dominant source of noise for a Faraday cup is Johnson noise (ΔV):

$$\Delta V = \sqrt{\frac{4k_B RT}{t_m}} \tag{4.1}$$

Inorganic Mass Spectrometry: Principles and Applications J. S. Becker

generated by the high ohmic resistor.[1] The Johnson noise of a Faraday cup is given in Equation (4.1), where k_B is the Boltzmann constant, R is the resistor value in ohm (Ω), T is the temperature in kelvin (K) and t_m is the integration time. When the integration time is increased, a decrease in collector noise (the square root dependence) is observed, which means if the integration time increases by factor of four, then the noise of the Faraday cup decreases by factor of two.[1] Optimum operating conditions are achieved when the current amplifiers are maintained in an evacuated and temperature stabilized housing. The temperature coefficients of high ohmic resistors are in the range of 200 ppm °C^{-1}, therefore if a gain stability better than 2 ppm is required, the temperature of the resistor must be kept stable at better than 0.01 °C.[1]

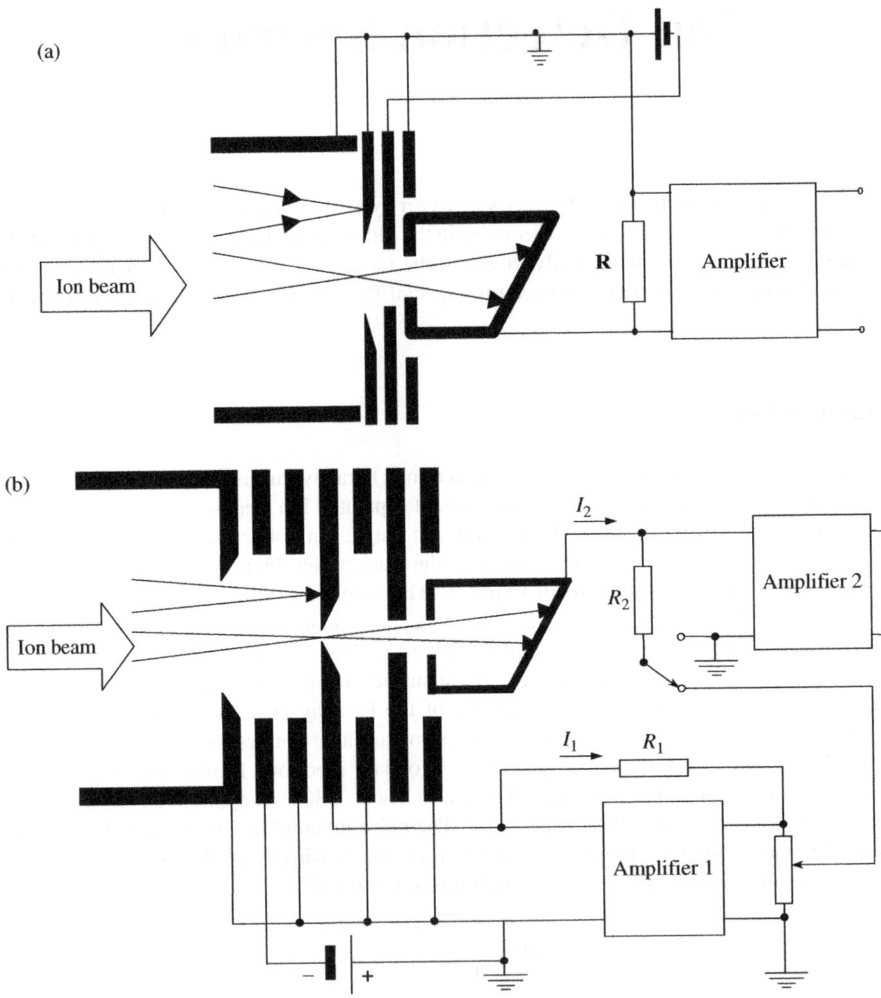

Figure 4.1 Basic principles of a Faraday cup and amplifier circuit: a) incoming charge of ions is converted into a voltage by an operational amplifier with a high ohmic feedback resistor (R) and; b) dual ion detector for direct measurements of two ion currents I_1 and I_2 using two amplifiers with resistors R_1 and R_2.

A dual ion beam collector developed by Nier[2] is illustrated in Figure 4.1 b. Both collectors are connected to two amplifiers for the simultaneous and direct measurement of ion currents in the dual mode. Amplifier 1 works with degeneration whereas amplifier 2 works without. Such a dual ion beam collector is applied, for example, for precise and accurate measurements of isotope ratios, especially of gases in commercial stable isotope ratio mass spectrometers.[2]

4.2 Secondary Electron Multiplier

For a further extension of sensitivity by several orders of magnitude in comparison to less sensitive Faraday cup collectors, a secondary electron multiplier (SEM) is used in mass spectrometers. The secondary electron multiplier is a fast detector of positively charged ions, but also of electrons and photons can be detected. The concept of SEM was originally put forward by Zworykin *et al.* in 1936[3] and developed by Bay[4,5] and Allen.[6] The schematic diagram of a secondary electron multiplier, proposed by Allen in 1950,[7] is shown in Figure 4.2. The secondary electron multiplier is the most common ion detector today with a variety of designs as discussed by Koppenaal *et al.*[8] The principle operating of an SEM is as follows. The positive ions impinge upon the first plate, or conversion dynode, resulting in the formation of secondary electrons. This means that in an SEM the positive ion current is converted into an electron current at the conversion dynode. The ejected secondary electrons are accelerated, focused and impinge on the surface of the second dynode whereby again a multitude of secondary electrons are generated. Repeating this process through 10 to 14 states thus leads to an enormous increase in gain. In order to obtain a high yield of secondary electrons from the implanted ions in the secondary multiplier, the dynodes are connected to successively higher positive potentials. The effectiveness of the remaining stages of the electron multiplier depends upon the geometry of the plates, the interstage voltage, the material and stage of the activation of the plates and the degree of magnetic shielding.[9] A linear focusing design proposed by Allen[6] became the most common arrangement whereby an increasing gain is applied with in stage voltages of up to 300–500 V. The dynode materials include copper – beryllium (2 %) and silver – magnesium alloys (2–4 %), which are activated to obtain thin activated beryllium or magnesium oxide surface layers with a high work function and low dark current. For the detection of low energy ions, a suppressor grid is operated at a negative potential with respect to the first dynode.[10] The SEM operates as an 'open ion detector' inserted into the ultrahigh vacuum of any mass spectrometer either in the current amplification or ion counting mode. The operation mode depends on the magnitude of the high voltage applied to the device. If the SEM operates in a medium amplification range with a gain of 10^3–10^5, then the resulting current at the output can be measured in an analogue mode with conventional amplification techniques. At higher amplification (10^6–10^8), the gain is large enough that individual electron pulses to be readily detected and counted using sensitive pulse counting electronics.[1] Although the sensitivity of mass spectrometers using SEM instead of a Faraday cup can be increased by a factor of 10^5–10^8 and a fast response time is possible, there are several disadvantages, e.g., mass discrimination effects, since ions with lower masses possess a higher velocity than ions with higher masses of the same element, and a deterioration in peak shapes or the dead time of the ion detector. If pulsed ion counting systems are used in mass spectrometers, the dead time of the detector (for counting rates higher than 10^6 cps) is the reason why a lower number of counts are registered than actually occur. Dead time correction of analytical data is required, particularly if extreme isotope ratios are to be measured. In addition, lower precision is observed using electron multipliers compared to Faraday cups due to lower counting statistics.

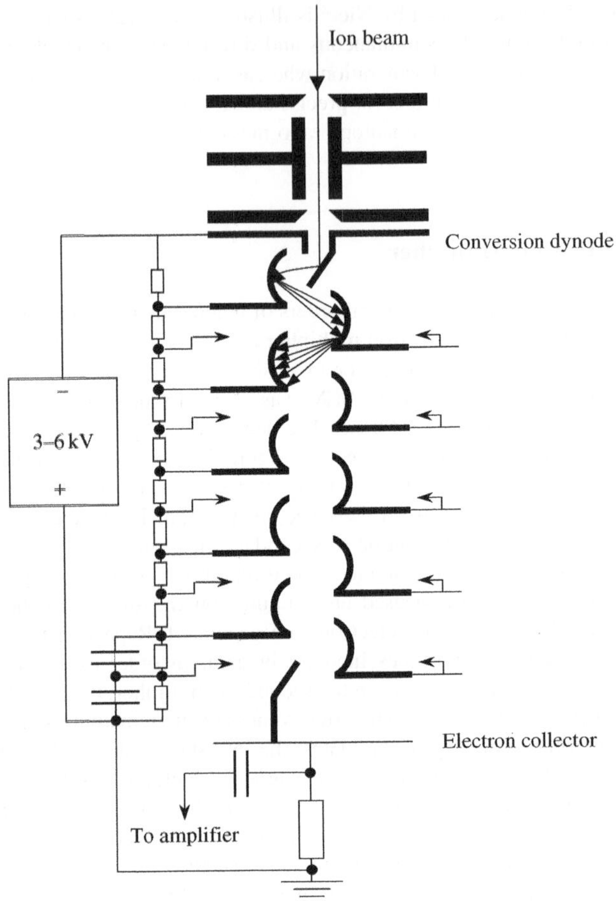

Figure 4.2 *Basic principles of the secondary electron multiplier (SEM).*

It is possible to extend the dynamic range of the mass spectrometer employed up to 10^9, if the dual detector mode of an SEM is applied. A schematic diagram and a photograph of the dual detector mode is shown in Figure 4.3 a and b, respectively. A modern SEM with dual detector mode and high speed amplifier operates in the analogue mode for higher ion currents and in the pulse counting mode for lower ones. Such a dual ion detector and permits mass independent cross calibration and allows mass spectrometric measurements on the dynamic range to be extended up to nine orders of magnitude, that means concentrations from 1 pg ml^{-1} (ppt) up to 1000 μg ml^{-1} (1000 ppm) can be measured. This dual detector mode is utilized e.g., in ICP mass spectrometers.

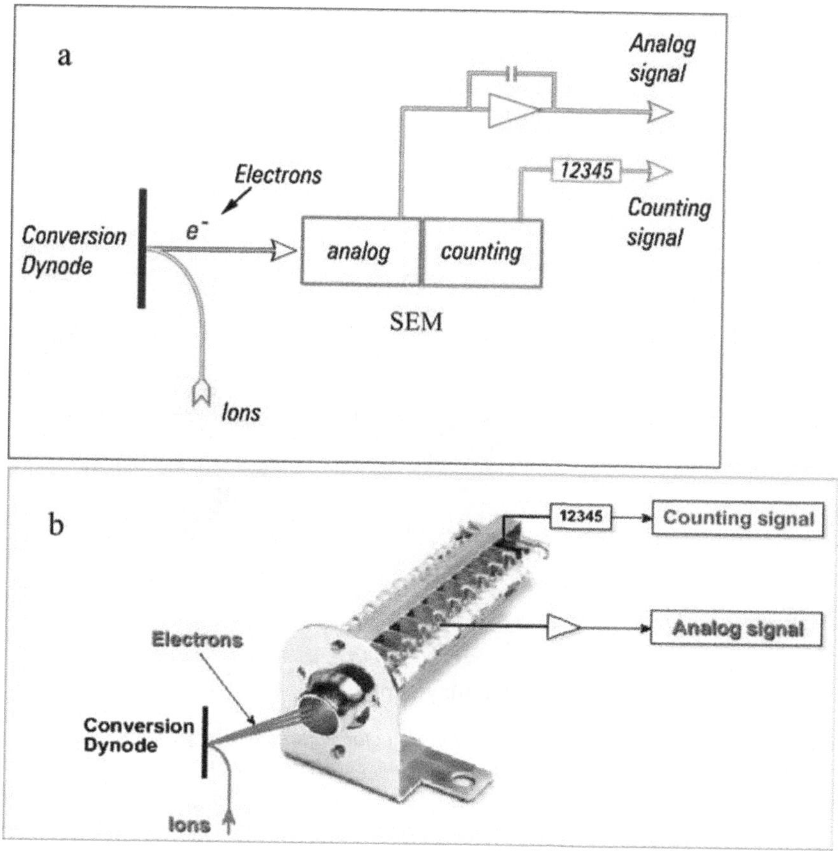

Figure 4.3 Schematic (a) and photograph (b) of dual ion detector mode. (Reproduced by permission of Agilent.)

4.3 Combination of Faraday Cup and Secondary Electron Multiplier

A further possibility of extending the dynamic range is to combine the Faraday cup with a secondary electron multiplier (see Figure 4.4 a). An increase of the dynamic range of the ion detection system of up to 10^{13} is achieved, for example, in the double-focusing sector field mass spectrometer Element (Thermo Fisher Scientific) which is commercially available with an inductively coupled plasma ion source or a glow discharge ion source. In Figure 4.4 b a photograph is shown of the single ion detection system using a dual detector (SEM) and a Faraday cup (together with an electrostatic analyzer, on the right in this photograph) for a significant extension of the dynamic range. This triple mode ion detection system allows matrix elements and trace impurities ($< 0.1\,pg\,ml^{-1}$) to be analyzed in aqueous solution at low mass resolution in one measurement with an automatic cross calibration at minimum integration time (counting mode – 0.1 ms, analogue mode and Faraday cup – 1 ms each). Due to the integral circuit, no delay time is observed with a Faraday cup and automatic switching between the detection mode is performed in < 0.1 ms.

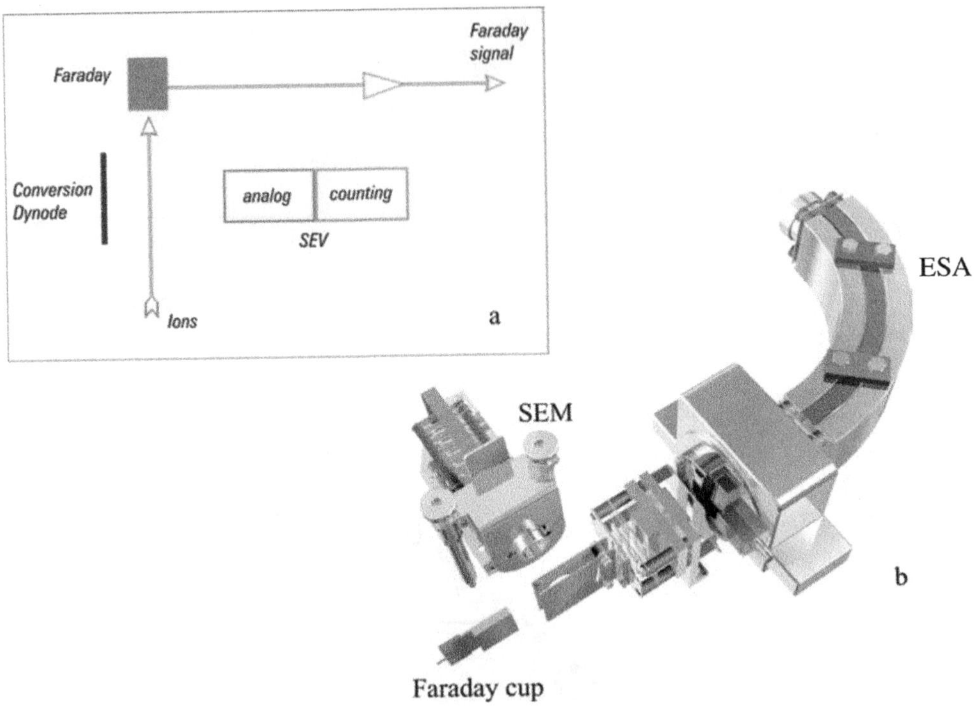

Figure 4.4 Schematic (a) and photograph of ion detector with extended dynamic range (b) in the double-focusing sector field mass spectrometer Element. (Reproduced by permission of Thermo Fisher Scientific, Bremen.)

4.4 Channel Electron Multiplier and Microchannel Plates

Another sensitive ion detector used in mass spectrometers is a channel electron multiplier (CEM, channeltron). The compact version of a channel electron multiplier proposed by Goodrich and Wiley[11] consists of a hollow tube with a semiconducting inner surface and contact leads at the end. A potential difference of about 2 kV between the ends creates a uniform axial field due to current flow on the inner surface. Ions incident on one end generate secondary electrons whereby their transverse velocity causes further impacts on the inner surface whilst the secondary electrons are carried along the tube by the longitudinal field. The amplifier current is collected at the other end of the tube, which can take the form of a spiral.[9] The schematic of a channeltron working in the dual acquisition mode (analogue and pulse-counting mode for an extended dynamic range), is shown in Figure 4.5. The straight focal plane of a double-focusing mass spectrometer with Mattauch – Herzog geometry is ideal for replacing the ion sensitive photographic plate by a multitude of CEMs. Several channeltrons arranged to form a multiple ion collection system was the design proposed by Jochum *et al.*[12] in an old spark source mass spectrometer to replace the photoplate by a more sensitive and precise ion detection system (see Section 4.7.). Microchannel plates (MCP) were developed to reduce the size of the linear channeltron to a few micrometres. In MCPs parallel cylindric channels are drilled with diameters between 4 μm and 25 μm and a hole spacing of about 6 μm to 30 μm. Electron multiplication in the channel is achieved by semiconducting layers on the channel surfaces whereby the plate input side is kept at a negative potential compared to the output side. The amplification of the number of electrons can reach 10^8. Microchannel plate based assemblies for ion detection over a large area are used as the ion detector, for example, in time-of-flight mass spectrometers.

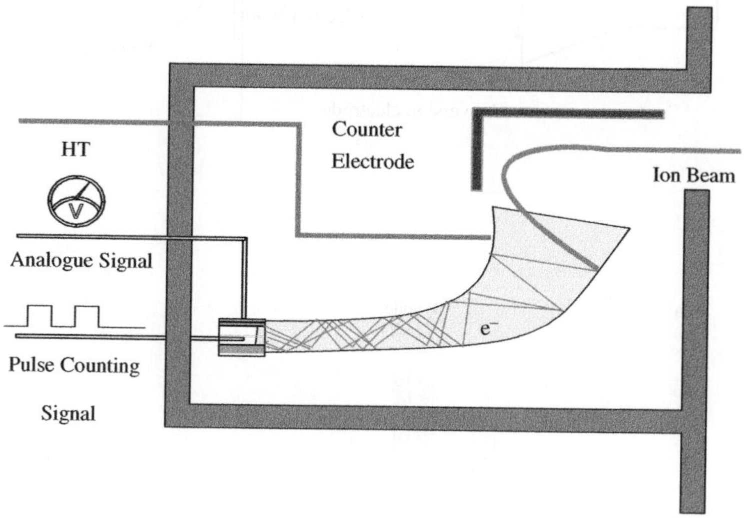

Figure 4.5 *Channel electron multiplier (CEM) working in the analogue and pulse-counting mode. (Reproduced by permission of GV Instruments Ltd.)*

4.5 Daly Detector

A further alternative to the Faraday cup – the Daly detector[13] – is illustrated in Figure 4.6 a. In the Daly detector a conversion dynode, which is at a high negative potential (~ −40 kV), is applied to convert ions into electrons. The Daly detector was developed from an earlier device using a scintillator (e.g., of phosphorus) for the direct detection of positive ions.

The basic operating principles of a Daly ion collector system used in the ICP-MS Platform (VG Instruments) is illustrated in Fig. 4.6 b. The incoming ions are attracted at 15 kV onto the Daly conversion electrode. By ion impact, about 5–6 electrons per impact are produced at the conversion electrode. Consequently, the emitted electrons are repelled onto the scintillator where photons are formed. The photons are multiplied by a photomultiplier, which is situated outside the vacuum, and are registered. The advantage of a modern Daly detector, which has a dynamic range over eight orders of magnitude, is its long lifetime and the fact that neither dead time correction or cross calibration is required.

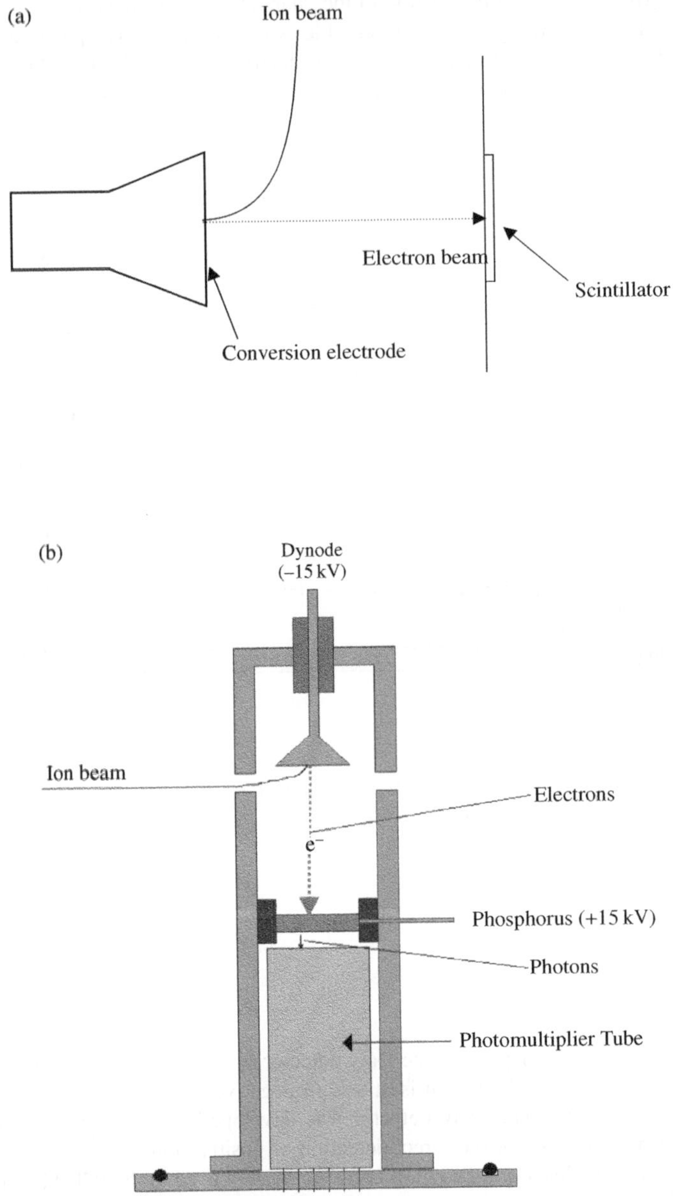

Figure 4.6 *Daly collector system consisting of the conversion electrode and a scintillator (a) together with a photomultiplier (b), this detector is used in the ICP-MS from VG Instruments. (Reproduced by permission of GV Instruments Ltd.)*

4.6 Multiple Ion Collection System

A multiple ion collector device is required for the simultaneous determination of separated ion beams in precise and accurate isotope ratio measurements in order to study, for example, isotope fine variation in Nature or during tracer experiments using enriched stable isotope tracers. In thermal ionization mass spectrometers or in ICP-MS, mostly a system of several Faraday cups (up to 16) and/or ion counters (electron multipliers) is applied. In the photographs in Figures 4.7 and 4.8 examples of multiple ion collector systems are shown from the mass spectrometers MC-ICP-MS

(a)

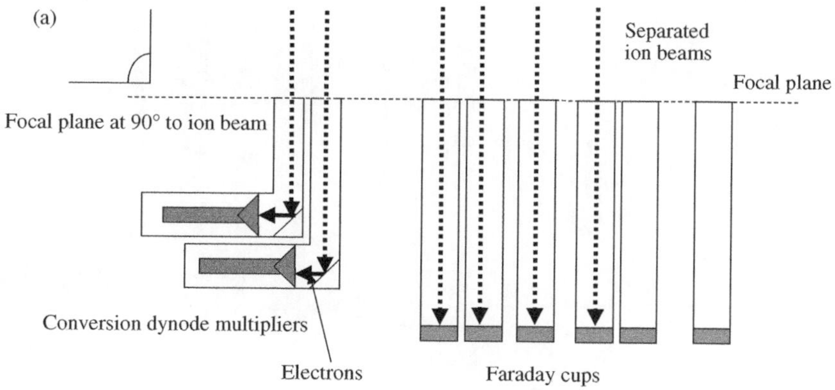

(b)

Figure 4.7 *Schematic (a) and photograph (b) of a multiple ion collector system of MC-ICP-MS IsoProbe (VG Instruments). (Reproduced by permission of Thermo Fisher Scientific, Bremen.)*

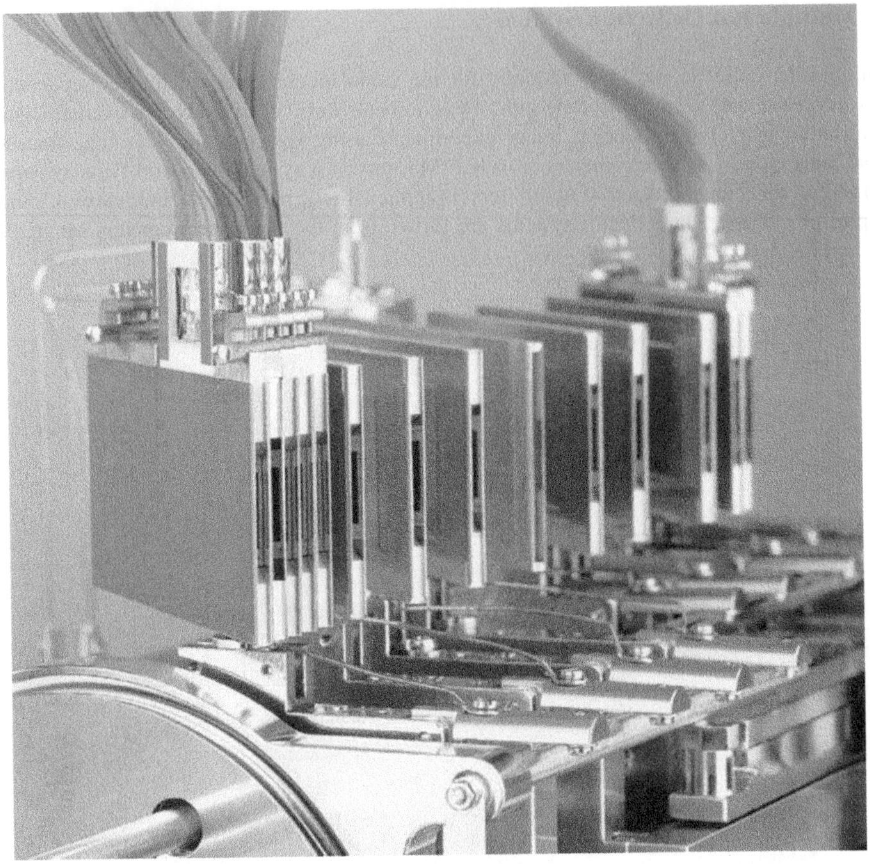

Figure 4.8 *Multiple ion collector array from MC-ICP-MS NEPTUNE and MC-TIMS TRITON (Thermo Fisher Scientific, Bremen). Miniaturized ion counters identical in size to Faraday detectors are mounted in the high mass range to detect, e.g., low-abundance uranium isotopes. (Reproduced by permission of Thermo Fisher Scientific, Bremen.)*

IsoProbe from VG Instruments (Manchester) and MC-ICP-MS Neptune or MC-TIMS Triton both from Thermo Fisher Scientific (Bremen). The Triton and Neptune systems employ miniaturized ion counting devices that are interchangeable with the standard Faraday cup detectors.[1] The multiple collector can be configured with as many as nine Faraday cups and eight miniature ion counters (see Figure 4.8).[1] Multiple ion collectors as a combination of Faraday cups and ion counters are being increasingly applied in mass spectrometry where precise and accurate isotope ratio measurements of very low-abundance isotopes (e.g. ^{234}U, ^{234}U and ^{236}U determination for application in environmental research or nuclear safeguards) in the presence of highly abundant ^{238}U are required or for measurements of transient ion signals of low intensity. In addition, the dynamic range for isotope ratio measurement can be significantly increased.

In 1977, Jochum *et al.*[12,14] developed the multiple ion counting (MC) technique using an old spark source mass spectrometer with 20 separate channeltrons 1.8 mm wide for simultaneous electrical ion detection. The sensitivity was increased by a factor of 20 compared to SSMS with ion detection using a photoplate and the precision of the analytical results was improved.

In a double-focusing mass spectrometer with an in house built Mattauch – Herzog geometry with ICP or glow discharge ion source, Hieftje *et al.*[8,15,16] applied an interesting ion detection system for simultaneous detection of separated ion beams. The focal plane camera (FPC) consisting of an array detector of gold Faraday cups, each coupled to its own integrator, was inserted into a Mattauch – Herzog type instrument (instead of the normally used ion sensitive photoplate, see Section 3.3.1.) in the linear focal plane for simultaneous ion detection over a large m/z range. The focal plane camera (operating at −40 °C, Peltier cooled) possesses 31 active collectors, the collector width and height are 145 μm and 1.6 mm, respectively.[15] The integration time can be varied between 1 ms and 20 s. Limits of detection are comparable to those obtained with a single channel secondary electron multiples (SEM). The iosotope ratio accuracy and precision are better with FPC than when an SEM is employed. The dynamic range has been shown to be linear over seven orders of magnitude.[15] Focal plane detectors (FPD) or array detectors are also applied for detecting ions simultaneously in a small m/z range (e.g., ±2–5 % of the centre mass) in single-focusing sector field mass spectrometers.[17] A comparison of different ion detectors in mass spectrometry is discussed by Koppenaal *et al.*[8]

4.7 Fluorescence Screen and Photographic Ion Detection

Ion detection with photographic plates played an important role in the history and development of mass spectrometry and was used up to the seventies in inorganic mass spectrometry, e.g., in trace analysis, but also in high resolution organic mass spectrometry. The detection of positive ions was made possible by the discovery by Goldstein in 1886[18] that described a fluorescence observed with glass and later with a natural silicate (willemite) during the impact of positively charged ions. Several years later Thomson[19] prepared very uniform fluorescence screens using a fine suspension of willemite in alcohol deposited on glass for ion detection. In 1910, as an alternative to the fluorescent screen, the ion sensitive photographic plate was introducedby Koenigsberger and Kutchewski[20] and independently by Thomson. Thomson was the first to demonstrate that the darkening of photographic plate depends strongly upon the penetrating power of positive ions at given energy and decreases with increasing mass of ions. The energy and the mass of ions are important parameters for the photographic density – ion exposure relationship. In order to reduce the energy loss of positive ions in the gelatine, the thickness of the protected gelatine layer was minimized to gelatine free layers. For photographic ion detection, Schumann plates were found to be the most sensitive and they were modified by Aston in 1925[21] and Bainbridge in 1931.[22] Later, Q plates were made commercially available by Ilford Ltd and photographic plates by AGFA UV2 (until the end of seventies) with a minimum of protective gelatine, but a high silver content for highly sensitive ion detection. Gelatine free plates were prepared by direct evaporation of silver bromide onto the glass plate in vacuum, e.g., by Kodak. The impact of ions on an ion sensitive photoplate transfers their kinetic energy to the silver bromide seeds of the emulsion. After the photoplate has been developed the blackened grains become visible. Several thousand ions are sufficient to produce a visible line. The typical blackening curves (opacity as a function of the log exposure) of the photographic plates are measured by photometry. The isotope ratios and elemental concentrations are deduced from the photographic density (blackness) of the selected isotope lines. The main advantage of a photoplate is that, in the case of Mattauch – Herzog mass spectrometers, a complete mass spectrum over a large mass range is obtained. For example, in a Mattauch – Herzog type instrument the ions from mass 7 u to 236 u are detected quasi-simultaneously with a ratio of the heaviest to lightest mass of ions of 36:1. With a stronger magnetic field B, the mass range can be shifted to higher masses. By using an ion sensitive photoplate as an ion detector, the complete composition of the analyzed sample, which means the matrix composition as well as minor and trace elements, can therefore be readily demonstrated and quantitatively determined. The exposure of photoplates can be extended if the experimental parameters (e.g, the stability of sector fields) are constant and the limits of detection for elements can thus be improved significantly.

The ion sensitive photographic plate for ion detection, which integrates the separated ion beams of all masses, was applied for many decades in mass spectrographs (spark source mass spectrographs or later in laser ionization mass spectrographs with Mattauch – Herzog geometry), e.g., for the exact mass determination of isotopes, for the detection of radionuclides, for a survey analysis of an unknown sample and for the determination of major, minor and trace elements and isotope ratios. For example, the detection of radionuclides (e.g., of ^{87}Rb) and their mass determination was performed by placing, for a considerable period, a second photographic plate directly (layer by layer) on the first ion sensitive photoplate with the complete mass spectrum. In the second photoplate, only the mass spectrum of radionuclides (e.g., of ^{87}Rb due to β^- decay in ^{87}Sr) was detected. By comparing the mass spectra in both plates it was possible to identify the radionuclides, their mass determination and abundances.[23]

In addition, due to the quasi-simultaneous detection of all elements over a large dynamic and mass range the ion sensitive photographic plate was well suited for data storage of all elements (and isotopes) of the samples analyzed. That means that additional information, e.g., the formation rate of cluster ions or of multiply charged species in a sample, could be obtained from the mass spectrum even a considerable time after the measurements. Using mass spectrometers with electrical ion detection, mostly over a small mass range, several isotopes or elements could be analyzed. The disadvantage of photographic ion detection is the time-consuming handling and evaluation of data. Therefore photographic ion detection by ion-sensitive photoplate is no longer of major significance today and has been replaced in modern mass spectrometers by fast, more efficient and sensitive electrical ion detection.

References

1. Wieser, M. E. and Schwieters, J. B., *Int. J. Mass Spectrom.*, **242**, 97 (2005).
2. Nier, A. O., Ney, E. P. and Ingram, M. G., *Rev. Sci. Instr.*, **18**, 294 (1947).
3. Zworykin, V. K., Morton, G. A. and Malter, L., *Proceedings of Institute of Radio Engineers*, **24**, 351 (1936).
4. Bay, Z., *Nature*, **141**, 282 (1938).
5. Bay, Z., *Rev. Sci. Instrum.*, **12**, 127 (1941).
6. Allen, J. A., *Phys. Rev.*, **55**, 966 (1939).
7. Allen, J. A., *Nuclear Energy Series*, NRC-USA (1950).
8. Koppenaal, D., Denton, M. B., Hieftje, G. M. and Barnes, J. H., *Anal. Chem.*, **Nov 1**, 419A (2005).
9. Duckworth, H. E., Barber, R. C. and Venkatasubramanian, V. S., *Mass Spectroscopy* 2nd edn., Cambridge University Press, Cambridge, London, New York, New Rochelle, Melbourne, Sydney (1986).
10. Stickel, R. E., Kellert, F. G., Smith, K. A., Dunning, E. B. and Stebbings, R. F., *Rev. Sci. Instrum.*, **51**, 721 (1980).
11. Goodrich, G. W. and Wiley, C. W., *Rev. Sci. Instrum.*, **33**, 761 (1962).

12. Jochum, K. P., Laue, H.-J., Seufert, H. M. *et al.*, *Fresenius' J. Anal. Chem.*, **359** 385 (1997).
13. Daly, N. R., *Rev. Sci. Instrum.*, **31**, 264 (1960).
14. Jochum, K. P., *Spectroscopy Europa*, **9**, 22 (1997).
15. Barnes, J. H., Schilling, G. D., Sperline, R., Denton, M. B., Young, E. T., Barigina, C. J., Koopenaal, D. W. and Hieftje, G. M., *Anal. Chem.*, **76**, 2531 (2004).
16. Barnes, J. H., Sperline, R., Denton, M. B., Barigina, C. J., Koopenaal, D., Young, E. T. and Hieftje, G. M., *Anal. Chem.*, **74**, 5327 (2002).
17. Birkinshaw, K., *J. Mass Spectrom.*, **32**, 795 (1997).
18. Goldstein, E., *Sitzungsbericht der Königlich Preussischen Akademie der Wissenschaft*, **39**, 691 (1886).
19. Thomson, J. J., *Rays of Positive Electricity* Longmans Green and Co., London (1913).
20. Koenigsberger J. and Kutchewski, J., *Physikalische Zeitschrift*, **11**, 666 (1910).
21. Aston, F. W., *Mass Spectra and Isotopes*, Edward Arnold Co., London (1925).
22. Bainbridge, K. T., *J. Franklin Institute*, **212**, 489 (1931).
23. Brunnée, C. and Voshage, H., *Massenspektrometrie*, Karl Thiemig, München, 1 (1964).

5

Instrumentation

Different mass spectrometric systems are designed to make use of all the ion sources, ion separation and ion detection systems – as described in the previous chapters.

One essential component as an integral part of all mass spectrometers is the vacuum system – not yet discussed – which produces the required vacuum conditions for ion generation (in vacuum and low pressure ion sources), and for the separation and detection of ions. Mass spectrometers operate most efficiently at ultrahigh vacuum conditions. It is essential to maintain a high vacuum in the mass analyzer and the ion detector, including the field free regions of the mass spectrometer in order to minimize the continuous instrumental background, which is visible in the mass spectrum. This means that an ultrahigh vacuum has to be achieved in the main parts of the mass spectrometer in order to avoid disturbing ion signals or increasing background in the mass spectra as a result of the collisions and scattering effects of ions with the residual gas atoms or molecules. The vacuum system of a mass spectrometer consists of two or more pumping stages to generate the high vacuum in the analyzer, ion detector and also in the ion source if it operates at ultrahigh vacuum (such as spark, laser or secondary ion source). A low pressure ion source (such as a glow discharge ion source) normally operates with an additional pumping system arranged directly at the ion source. As one of the pumping stages, rotary vane pumps with a pumping power of $4–16\,m^3\,h^{-1}$ are applied to produce a medium vacuum of several Pa. These mechanical rotary pumps are combined with powerful high vacuum pumps to remove the exhaust from the turbomolecular pumps. The pumping power of the generally water cooled turbomolecular pumps is about $200–6000\,l\,s^{-1}$ to generate an ultrahigh vacuum with a pressure of 10^{-5} Pa and lower. Oil diffusion pumps provide a high pumping speed and were often used in earlier instruments. However, back streamed oil may be deposited on the surfaces of sensitive parts of the mass spectrometer. Oil diffusion pumps and also ion getter pumps installed in old mass spectrometers have been replaced in some laboratories by powerful turbomolecular pumps.[1,2] In order to measure the pressure in mass spectrometers, three types of pressure gauges may be used: thermal conductivity gauges, ionization gauges and diaphragm vacuum gauges, as described elsewhere.[3]

Inorganic Mass Spectrometry: Principles and Applications J. S. Becker

Commercial mass spectrometers with inductively coupled plasma ion sources working at atmospheric pressure of 100 kPa apply the three stage differentially pumped vacuum system to obtain a high vacuum ($<10^{-6}$ Pa) in the mass analyzer and ion detector. In order to use an ICP ion source in mass spectrometry an interface stage is inserted between the atmospheric pressure ion source and the high vacuum mass analyzer. The interface region is pumped by a high speed rotary vane pump if the inductively coupled plasma is in operation. The pressure in the interface region during the measurements is about 133 Pa (1 torr), which is maintained by using the rotary vane pump that is switched off if the ICP-MS is in the 'standby' mode (when the gate valve at the interface is closed). In order to maintain the high vacuum in the mass analyzer, the gate valve in the interface region is under pneumatic switch control and shuts automatically if there is a failure in gas flow or if the plasma is shutdown. The interface stage (see Figure 2.3) is limited by a sampler cone to the ICP source and by a skimmer cone to the ion optics of the mass spectrometer. Sampler and skimmer cones made of nickel or platinum have an orifice with a diameter of about <1 mm for the ion gateway.

The main problem with the quadrupole ICP-MS is the instrumental background due to photons forming in the plasma. To minimize this background, a so-called 'photon stop' is utilized in quadrupole based ICP-MS instruments (e.g., in the Elan mass spectrometers from PerkinElmer). This photon stop is a small metal plate placed in the centre of the ion beam behind the skimmer cone, which prevents photons and neutrals reaching the detector. In other quadrupole ICP-MS devices the instrumental background is reduced by using off axis transfer optics for the ion path in the mass analyzer e.g., as implemented by Agilent in the late 1980s (see Figure 5.1 right column, at the top). The 'off-axis' ion Omega lens arrangement protects the collision cell from neutrals resulting in high ion transmission and low background intensity (5–10 cps). This ion optical system reduces maintenance compared to quadrupole based ICP-MS with a photon stop. An off axis geometry is also realized in the Platform ICP-MS from Micromass, which is not produced any more (see Figure 5.1, right column, in the middle). The off axis hexapole collision cell – inserted between the ion source and the quadrupole analyzer – acts as an ion focusing lens and reduces disturbing isobaric interferences as well.

Due to the non-linear geometry of all sector field instruments (ICP-SFMS see Figure 5.1, right column, at the bottom) the instrumental background is about one order of magnitude better (0.1–0.5 cps) than in quadrupole ICP-MS.

Each mass spectrometer possesses an ion focusing system that uses 'ion lenses' (electrostatic plates) to focus and transfer the charged species extracted efficiently from the ion source at the entrance aperture of the mass analyzer and to separate the ion beam from the neutrals and photons. Neutrals deposited on ion lenses no longer affect ion focusing. Several ion lens designs (mostly cylindrical lenses with different voltages operating as an ion energy filter) have been employed in inorganic mass spectrometry. Detailed discussions of calculated ion trajectories in ICP-QMS from the skimmer tip of an ICP interface stage passing a photon stop through three cylindrical einzel lens elements and a Bessel box are presented in reference [3]. Ions of different masses which possess the same kinetic energy fly on the same trajectory in the electrical field of the ion lens system. However, as discussed in Chapter 2 the ions formed in several types of ion sources have a kinetic energy spread which leads to a defocusing of the ion beam. In addition, a defocusing of the ion beam is observed due to space charge effects during ion extraction.

In analogy to other modern analytical instruments, a computer-based data processing and evaluation system is inserted in all mass spectrometers constructed today. All processes, from sample introduction using an autosampler, e.g., in an ICP-MS, to optimization of experimental parameters in the ion source, ion extraction, separation of ion beams and their registration, the vacuum system and the whole measurement procedure are supported and controlled by a fast and powerful

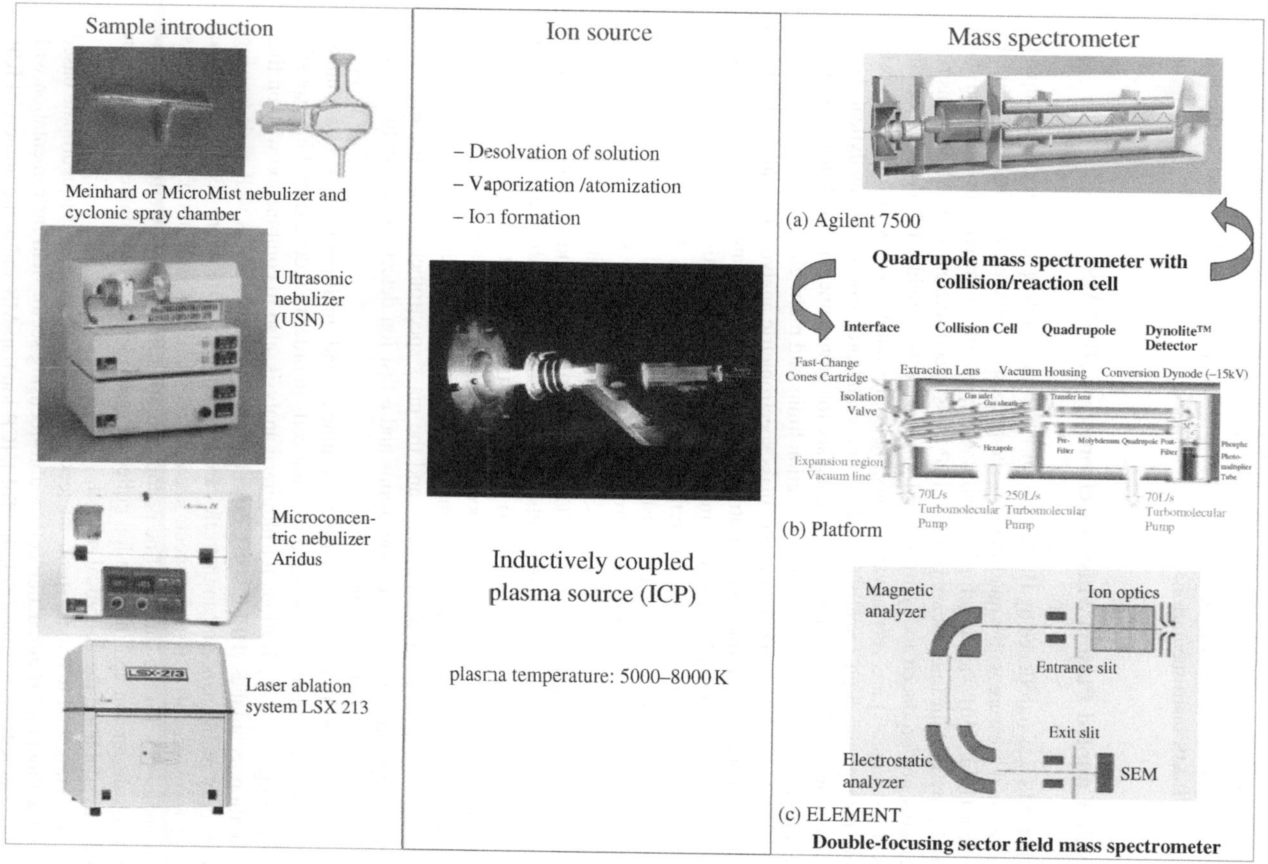

Figure 5.1 *Main parts of an inductively coupled plasma mass spectrometer: sample introduction systems (left column), e.g., Meinhard or MicroMist nebulizer with cyclonic spray chamber, ultrasonic nebulizer, microconcentric nebulizer and laser ablation system (all from CETAC Technologies), ion source (middle column) and several types of mass spectrometers, (a) Agilent 7500 from Agilent, (b) Platform from GV Instruments, or (c) Element from Thermo Fisher Scientific. (Parts of this figure were reproduced with permission from CETAC Technologies, Agilent, GV Instruments and Thermo Fisher Scientific, respectively.)*

computer system. With the aid of a company specific computer program, the acquired mass spectrometric data (ion intensity as a function of time and/or m/z ratio) are converted into mass spectra in order to yield finally element and species concentrations, isotope ratios, depth profiles or images of element and species distribution in investigated samples.

The most frequently used inorganic mass spectrometric technique is ICP-MS, and so the next section focuses on ICP-MS instrumentation.

5.1 Inductively Coupled Plasma Mass Spectrometers (ICP-MS)

As a reminder, ICP-MS was introduced by Houk, *et al.* in 1980[4] by coupling an inductively coupled plasma source at normal pressure with a quadrupole based mass spectrometer. The first commercial ICP-MS instrument (Elan 250) was therefore a quadrupole mass spectrometer introduced on the international market by PerkinElmer, Sciex in 1983. ICP mass spectrometers with different instrumental arrangements are produced today by several companies worldwide. ICP-MS (especially quadrupole based instruments) are very advantageous due to the relatively low instrumental costs and high sample throughput, high sensitivity and accuracy and precision of the analytical data, thus fulfilling requirements for trace and isotope analysis for environmental, nuclear or medical controls and monitoring.[5–10] Due to its excellent detection limits and multi-element capability, ICP-MS is today the most frequently utilized inorganic mass spectrometric analytical technique for fast multi-element determination in the trace and ultratrace concentration range and for precise isotope analysis, mostly in aqueous solutions and solid samples after sample dissolution or directly e.g., by laser ablation for a wide variety of quite different types of sample.[3,11,12] This is demonstrated by the rapid growth of ICP mass spectrometer installations worldwide, and the different types of quadrupole based ICP mass spectrometers (e.g., from Agilent; PerkinElmer Sciex; Thermo Fisher Scientific; Varian and others) or double-focusing sector field instruments with single ion detection (Element, Thermo Fisher Scientific and Nu ATTOM®, Nu Instruments) and multiple ion collectors (Neptune, Thermo Fisher Scientific and Nu Plasma, Nu Instruments). About 20 years after the introduction of ICP-MS, there are more than 5000 ICP-MS instruments installed worldwide.

The main common parts of an ICP mass spectrometer as discussed above are the sample introduction system, the inductively coupled plasma (ICP) ion source for desolvation, atomization and ion formation of introduced sample material, and the mass spectrometer including the mass analyzer system for separation of extracted ion beams and a fast ion detection system to register separated ion beams as illustrated in Figure 5.1.

Instrumental developments (e.g., of sector field instruments with multiple ion collection, introduced in 1992, or the insertion of collision and reaction cells in order to reduce disturbing isobaric interferences), the progress in applications for ultratrace analysis, also in combination with on line hyphenated separation techniques (HPLC, CE), especially routine capability as well as decreasing price and user friendly maintenance mean that sales are increasing by 10 % every year. To improve the analytical performance of ICP mass spectrometers for precise isotope ratio measurements (e.g., for geochronology or for the study of fine isotope variation in nature) powerful instrumentation with high mass dispersion and multiple ion collector systems instead of single ion collection are commercially available on the analytical market.

An overview of commercial ICP mass spectrometers from different companies (quadrupole based ICP-MS with and without collision/reaction cell, double-focusing sector field instrumentation with single and multiple ion collectors, time-of-flight (ToF), ICP-ion trap-MS and non-commercial ICP-Fourier transform ion cyclotron resonance (FTICR) mass spectrometers is given in Figure 5.2. By using ion traps and FTICR mass spectrometers in ICP-MS isobaric interferences of atomic ions (e.g., $^{58}Fe^+$ and $^{58}Ni^+$) requiring high mass resolution can be resolved and an interference free analysis of analyte ions thus become possible.[13] Experimental arrangements of mass analyzers with high mass resolution are illustrated in Figures 3.12–3.14.

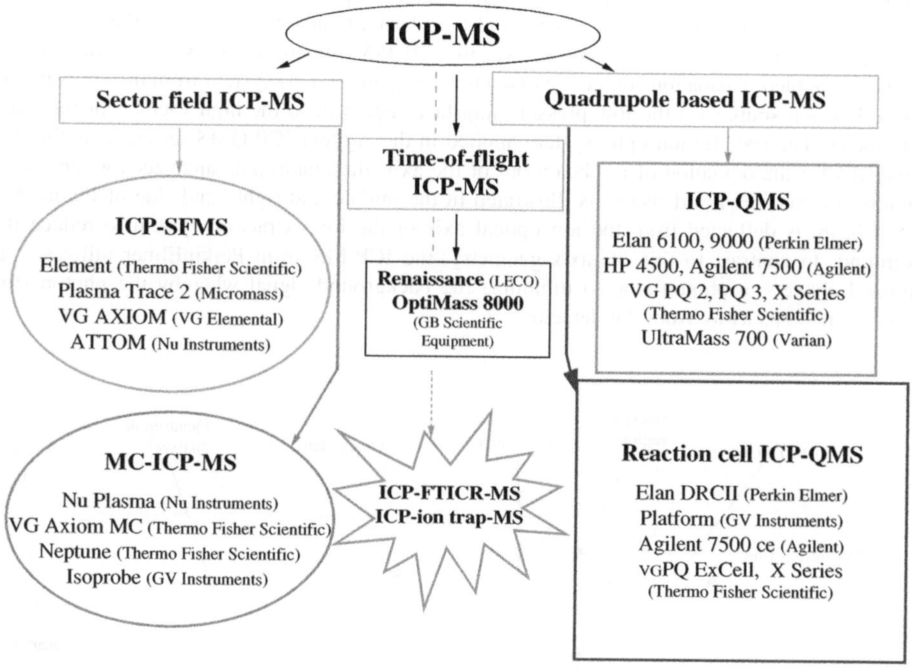

Figure 5.2 *Overview of ICP-MS instrumentation: quadrupole-based ICP-MS with and without collision/reaction cell, sector field ICP-MS with single ion collector (ICP-SFMS) and multiple ion collector ICP-MS (MC-ICP-MS) and time-of-flight ICP-MS.*

In ICP-MS mostly liquids are analyzed. The aqueous solution is nebulized using an effective pneumatic nebulizer (e.g., Meinhard or MicroMist nebulizer with cyclonic spray chamber, ultrasonic nebulizer (USN), microconcentric nebulizer with desolvator (e.g., Aridus or APEX), the latter for the introduction of small volumes of solution)[3] as described in Section 5.1.6.

5.1.1 Quadrupole Based ICP Mass Spectrometers (ICP-QMS)

Several types of quadrupole based ICP-MS with and without a collision/reaction cell (e.g. from various companies such as PerkinElmer Sciex, Norwalk, CT; Agilent Technologies, Tokyo, Japan; Thermo Fisher Scientific, Bremen, Germany; Varian GmbH Mulgrave, Victoria, Australia etc., see Figure 5.2) are available on the analytical market for different applications, especially for multi-element trace analysis, for speciation analysis of selected elements (e.g., Se, Cd, Sb, As) and for isotope ratio measurements. These ICP mass spectrometers differ only marginally in their capabilities (e.g., elemental sensitivity, limits of detection, oxide formation rate, short-term and long-term stability, dynamic range and precision of isotope ratio measurements). A schematic of a quadrupole mass analyzer is shown in Figure 3.7, and the operating mode is described in

Section 3.2.1. The experimental setup of a quadrupole ICP-MS including the solution introduction system, ICP torch, interface, quadrupole mass analyzer, ion detector and pumping system is depicted in Figure 5.3. In this quadrupole ICP-MS from Agilent (Agilent 7500, developed from the former HP 4500 from Hewlett Packard without a collision/reaction cell), the instrumental background is reduced as mentioned above because off axis transfer ion optics are utilized. In the left part of this figure, the solution introduction system and the plasma (shielded) torch with the inductively coupled coil and the rf power supply are shown. In all ICP-MS instruments, the ions formed in the inductively coupled plasma ion source are extracted using an ion lens system from the atmospheric pressure ICP ion source via the low pressure interface region in to the high vacuum of the mass spectrometer. The specific ion optic system applied in the Agilent ICP-QMS series, including the Omega lens for the deviation of ion beam out of the axis, the quadrupole analyzer for separating the ion beams and the ion detector, is illustrated in the middle and right-hand side of Figure 5.3. The ion beam is deflected from the ion optical axis of the ion extraction system to reduce the background. In contrast to this off axis geometry, the ICP-MS from PerkinElmer utilizes – as mentioned above – a photon stop to minimize the background signal whereby the photon stop reflects the photons away from the detector.

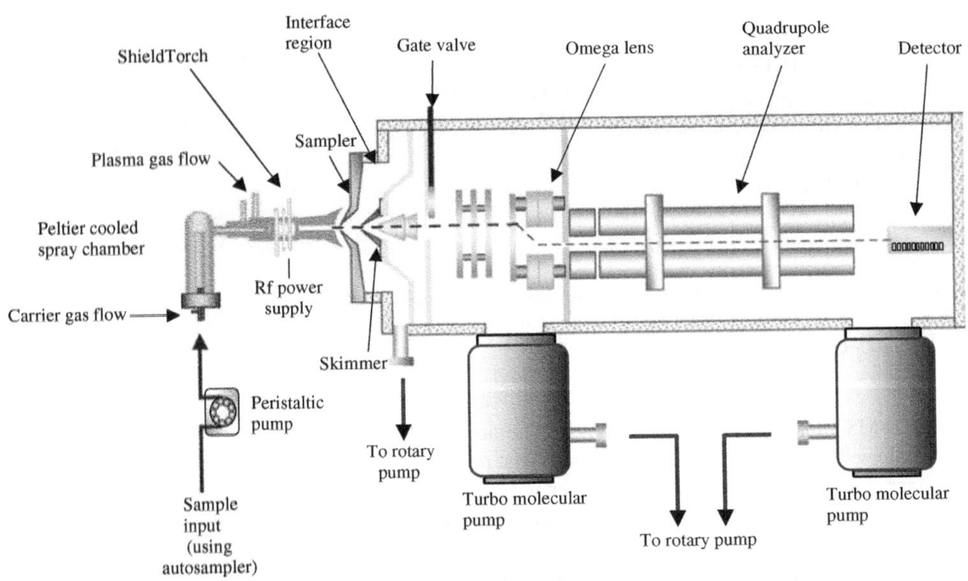

Figure 5.3 *Instrumental outline of a quadrupole-based ICP-MS without collision/reaction cell (Agilent 7500). (Reproduced by permission of Agilent.)*

The mass range of most quadrupole based ICP-MS instruments is limited (1–300 u). By ICP-MS nearly all elements in the mass range of 6 u (for lithium) to 238 u (for uranium) can be measured. Light elements at low concentration levels, such as H, O, N, C and noble gases, cannot be determined because an atmospheric pressure (argon) plasma ion source always contains traces of these light elements in the air and also in aqueous solution acidified with HNO_3.

In several solid samples (e.g., steel containing carbon), at higher concentration of selected light elements (such as C) with a relatively high background the ion signal of analyte can be detected e.g., by direct measurements on solid materials using LA-ICP-MS. For the determination of higher element concentrations of light elements (such as H, O, N and C) alternative element analytical techniques are available, which allow an accurate and precise determination in quite different materials. Because argon is mainly used as the plasma gas, in addition to Ar^+, $Ar_2{}^+$, ArO^+ ArH^+, ArN^+ ions and others mostly of high intensity, all other noble gases (He, Xe and Kr) can also be observed in the ICP mass spectrum as contaminants of argon.

As discussed before, quadrupole based ICP-MS allows multi-element determination at the trace and ultratrace level and/or isotope ratios in aqueous solutions in a few minutes as a routine method with detection limits of elements in the sub pg ml^{-1} range and a precision for determined trace element concentration in the low % range (RSD – relative standard deviation). The precision for isotope ratio measurements varies between 0.1 % and 0.5 % RSD. This isotope ratio precision is sufficient for a multitude of applications, e.g., for evidence of contamination of sample with depleted or enriched uranium in urine (this technique is used in the author's laboratory in a routine mode[14]) or the isotope dilution technique for the quantitative determination of trace element and species concentration after doping the sample with enriched isotope spikes.

An important limiting factor – which reduces the accuracy and precision of trace, ultratrace and isotope analysis – is the occurrence of isobaric interferences of analyte ions with singly or doubly charged atomic ions of other elements or polyatomic ions with the same m/z ratio (see Section 6.1.3). In general, in inorganic mass spectrometry interference free isotopes (if available) are selected for analysis. In quadrupole based ICP-MS without a collision cell, the interference problem is solved by a mathematical correction of possible isobaric interferences (after measurement of ion intensities of disturbing interfering atomic and polyatomic ions) or by the application of the cold plasma technique for selected elements with low ionization potential (such as K, Na, Ca or Sr). In addition, special techniques are applied to reduce interference problems, for example, the use of hydride generation for sample introduction for hydride forming elements such as Se, As, Te, Bi, Sb, Ge, Sn and Pb, utilizing special sample introduction devices (ultrasonic or microconcentric nebulizers with desolvator), separating analytes from matrix off line and applying hyphenated techniques (by on line coupling of HPLC or CE together with ICP-MS).

As an alternative to commercial quadrupole based ICP-MS measurements at low mass resolution ($m/\Delta m \approx 300$), in 1996 Yiang and Douglas[15] proposed a quadrupole ICP-MS system which allows a maximum mass resolution of 9000. At a mass resolution of 5000 the sensitivity was comparable to that of a commercial double-focusing sector field ICP-MS operated at the same mass resolution. Due to the very high continuum instrumental background of about 1000 cps no commercial high resolution quadrupole instrumentation with an inductively coupled plasma source exists and the development of high resolution quadrupole based ICP-MS has ceased.

5.1.2 ICP Mass Spectrometers with Collision or Dynamic Reaction Cell or Collision Reaction Interface

The figures of merit of quadrupole-based ICP-MS, such as the precision of isotope ratio measurements and the detection limits, can be improved significantly, especially for elements which are difficult to determine due to the appearance of isobaric interferences (e.g., by the trace, ultratrace and/or isotope ratio measurements of Ca, Fe, S, As, I or Se).[16–22] The occurrence of interference problem can be minimized by the insertion of a collision/reaction cell in ICP-MS as the result of defined collision induced reactions using selected collision/reaction gases or gas mixtures (such as H_2, He, NH_3, O_2, CH_4 and others). For each analytical problem, which is different, e.g., for U or

K isotope ratio measurements or a multi-element analysis, optimized measurement procedures have to be developed. Several quadrupole based ICP-MS instruments with multipole collision/reaction cells are available on the analytical market, as well as quadrupole, hexapole and octopole cells. Increasing the number of rods in the multipole collision cell allows smaller dimensions and lower gas consumption. An Octopole Reaction Cell (ORC) ICP-MS has been inserted in the Agilent 7500ce/cs series. A collision/reaction cell was introduced into ICP-QMS by Agilent in 2001. A schematic of ICP-ORC-MS Agilent 7500ce from Agilent is depicted in Figure 5.1 a, (top right), the insertion of an on axis high transmission ORC between the ion optics and quadrupole analyzer and a photograph of a small ORC is shown in Figure 5.4 a and b, respectively.

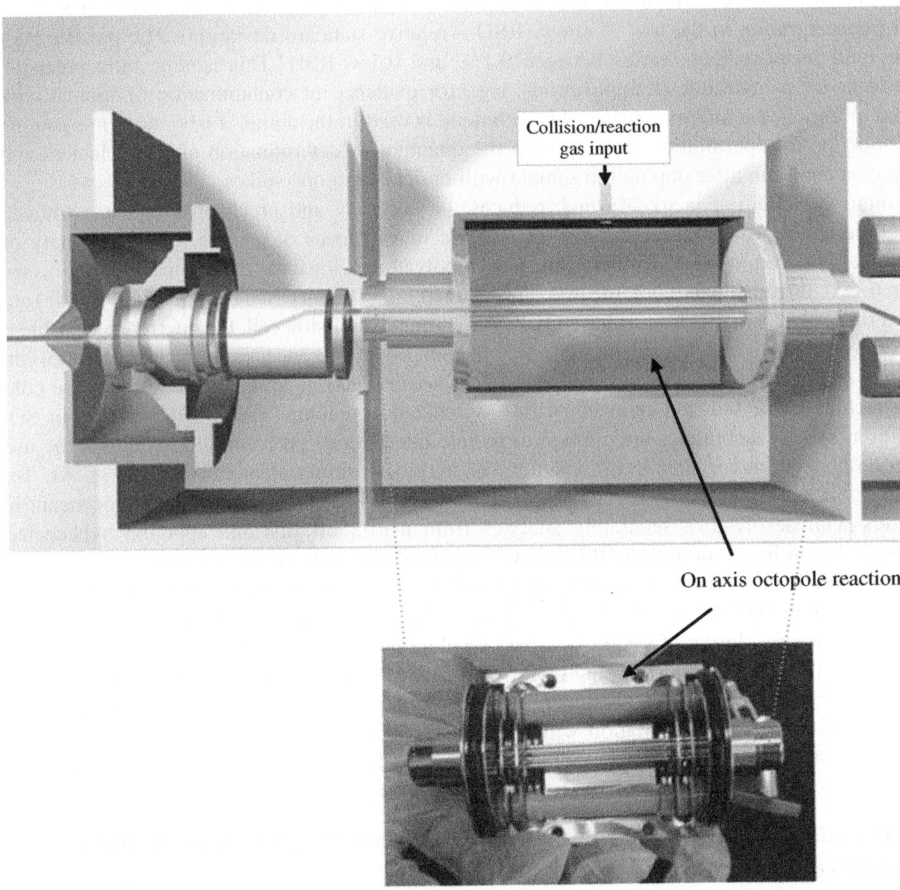

Figure 5.4 *Schematic of inserted on axis octopole reaction cell between the ion lens system and quadrupole analyzer (a) and photograph of octopole reaction cell (b). (Reproduced by permission of Agilent.)*

A quadrupole Dynamic Reaction Cell (DRC) is utilized in the Elan DRC introduced by Perkin Elmer in 1998, see Figure 5.5. The DRC is arranged on axis because a photon stop is inserted

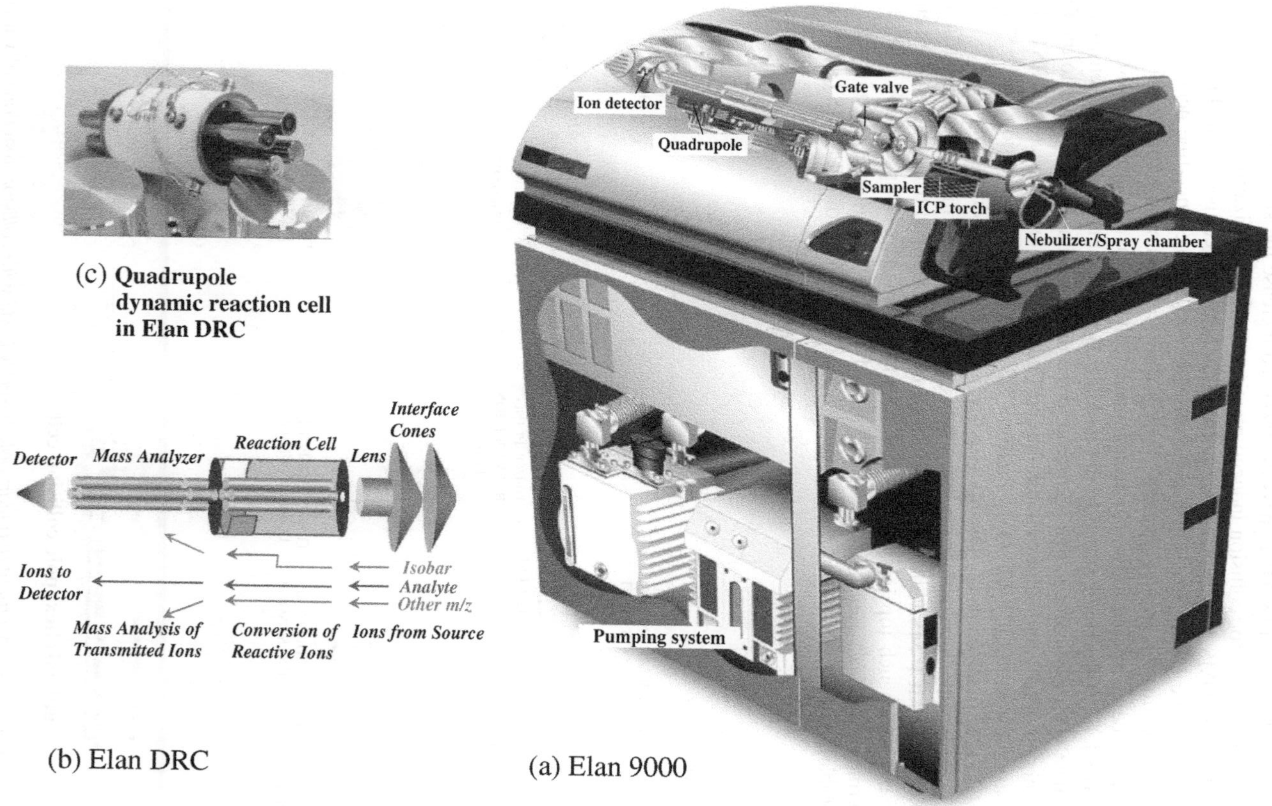

Figure 5.5 *Schematic of ICP-QMS Elan 9000 without collision cell (a); of Elan DRC with quadrupole reaction cell (b), topleft: photograph of DRC (c). (Reproduced by permission of Perkin Elmer, Germany.)*

in all the Elan series of ICP mass spectrometers to prevent the photons and neutrals reaching the detector. An off line hexapole collision cell is arranged in the ICP-CC-QMS, Platform (see Figure 5.1 b, previously created by Micromass in 1997). To increase the signal-to-noise ratio, the linear hexapole collision cell for collision induced reactions, ion thermalization and ion focusing in the Platform ICP-CC-QMS, was replaced by a dual axis hexapole collision cell (Platform XS from GV Instruments, see Figure 5.6[23]) as a more efficient and maintenance-free neutral/photon filter. Using this experimental arrangement, the background signal can be minimized by nearly one order of magnitude compared to the old arrangement with a linear hexapole collision cell as shown in Figure 5.1 b., right-hand column in the middle). However, the analyte ion sensitivities in this arrangement were also reduced and according to the detection limits are similar to the first arrangement or higher. Uranium isotope ratios on the certified isotope reference material NIST SRM U050 were measured with the ICP-CC-QMS Platform XS with improved precision ($^{235}U/^{238}U = 0.052\,784 \pm 0.000\,12$).[23] Unfortunately this mass spectrometer is now produced any more.

In addition to these collision/reaction cell instruments, since 2002 Thermo Fisher Scientific has been selling the XSeries ICP-MS with a hexapole collision cell (developed from ThermoElemental PQExel ICP-MS) as a bench top instrument on the analytical market. A special ion extraction system in Thermo's XSeries^{II}, ion optics together with a hexapole collision cell to minimize the interference problem in ICP-MS, provides the lowest background signals for ICP-QMS (≤ 0.5 cps).

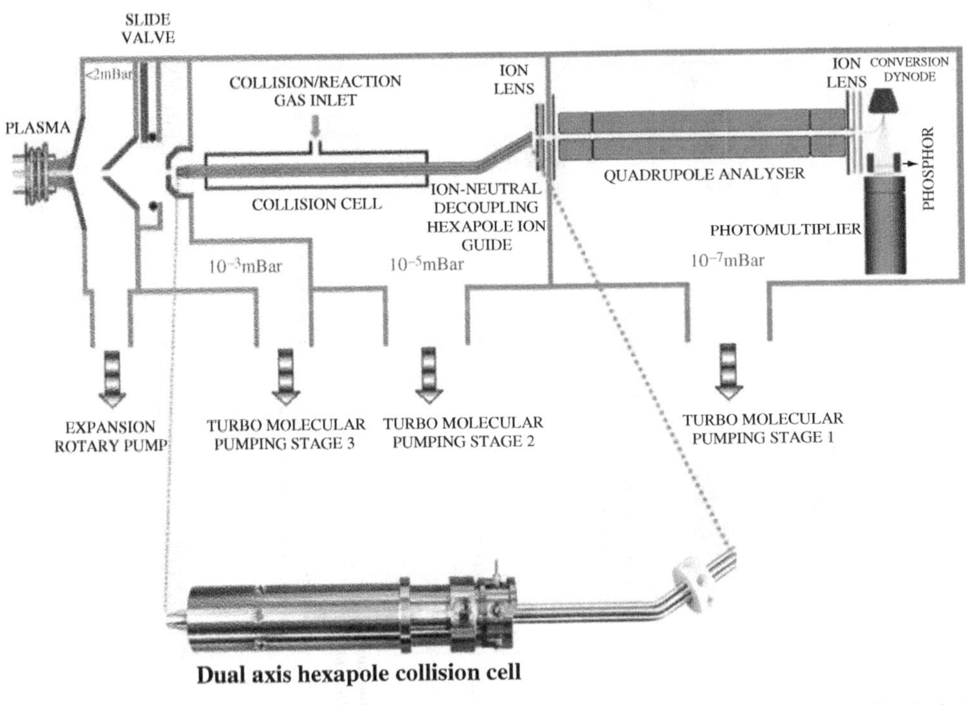

Figure 5.6 *Experimental arrangement of quadrupole based ICP-MS with collision/reaction cell with dual axis hexapole collision cell, Platform XS; photograph of collision cell is shown at bottom (GV Instruments, Manchester) (Reproduced by permission of GV Instruments Ltd.)*[21]

In collision/reaction cell technology noble gases such as He, Xe, Ar, molecular gases (e.g., H_2, O_2, CH_4, NH_3) or gas mixtures (H_2/He) are introduced into a hexapole, octopole gas target or quadrupole dynamic reaction cell of ICP-MS. By collisions of ions formed in the ICP with gas atoms or molecules of the collision gas, the kinetic energy spread of the ions is reduced from several eV to < 0.1 eV, which results in an increased sensitivity to analyte ions due to the higher transmission of ions (observed in the Platform ICP-CC-QMS) and an improved precision of analytical data as studied for isotope ratio measurements of uranium in the author's laboratory.[16] For other ICP-QMS with a multipole gas target, a decrease in sensitivity was observed. In addition, by collision induced reactions of the collision gas with ions of the gas-filled collision/reaction cell, possible isobaric interferences can be minimized. For example, a determination of $^{80}Se^+$ or $^{40}Ca^+$ isotope abundance, which is difficult or impossible by quadrupole based or double-focusing sector field ICP-MS even at high mass resolution, can be advantageously performed by ICP-QMS with a collision/reaction cell. The interfering argon dimer ions ($^{40}Ar_2^+$) or atomic plasma gas ions ($^{40}Ar^+$) are dissociated or neutralized, respectively, in the cell via defined collision induced reactions (using a mixture of He/H_2 or NH_3 as the collision gas).[16] For selenium the signal-to-noise ratio of $^{80}Se^+/^{40}Ar_2^+$ can be significantly improved by more than five orders of magnitude under optimized experimental conditions. In this way, precision for difficult-to-determine isotope ratios in ICP-MS such as $^{82}Se/^{80}Se$ and $^{40}Ca/^{44}Ca$ of 0.20 % and 0.26 %, respectively, was achieved by ICP-CC-QMS (Platform, GV Instruments) in the author's laboratory.[24] An arrangement of an off axis hexapole collision cell similar the quadrupole ICP-MS Platform is realized in the multiple ion collector sector field instrument Isoprobe from the GV Instruments (see Section 5.1.5).

Gas phase ion–molecule chemistry using a collision cell has been widely applied in organic mass spectrometry for controlled collision induced dissociation of organic compounds (CID), e.g., for the analysis of organic compounds and biomolecules.[25] It should be noted that a possible application of the gas target (collision cell) in inorganic mass spectrometry was first proposed in 1983[26] to reduce the background spectrum and improve sensitivity in isotope and trace analysis by exploiting the dissociation of interfering polyatomic ions. At the beginning of the eighties, ICP mass spectrometry and the collision cells in inorganic mass spectrometers as routinely used today were unknown.

However, under certain experimental conditions, not only the dissociation, but also the formation of polyatomic ions with high intensity was observed in the collision cell. For example, $^{31}P^+$ or $^{32}S^+$ determination can be performed using oxide ions with high intensity.[27] Oxide ions of $^{31}P^+$ or $^{32}S^+$ which are interference free at $m/z = 47$ and 48, respectively, are formed with maximum ion intensities in the collision or reaction cell under optimized experimental conditions.[27]

Further applications of ICP-MS with a collision/reaction cell in trace, ultratrace analysis and isotope ratio measurements are described in Chapter 8.[16,17,20,22,27,28]

An innovative approach was developed in the Varian 810 and 820 ICP mass spectrometers. To suppress the background signals, 90° ion optics, utilizing an ion mirror (see Figure 5.7 a) was designed and the ions were selected from neutrals and photons, which do not change their direction passing through an electrical field. Besides minimizing the background signal, the advantage of this setup is also that there is lower contamination at the ion optics, the sampler and the skimmer cones. Due to this special ion optical system, aqueous solutions with high matrix content (up to several hundred $mg\,l^{-1}$) can be analyzed. A further reduction of disturbing interferences is possible by improving the design of the interface region: the Collision Reaction Interface (CRI). Instead of a pressurized multipole (hexapole, quadrupole or octapole collision cell) in the Varian 820 MS, the company proposed introduction of the collision gas (e.g., hydrogen and helium) in front of the mass analyzer in the expanded plasma through the tips of the sampler and/or skimmer cones. In Figure 5.7.b, a schematic of the skimmer cones with the channels for the gas input into the plasma jet is illustrated. In contrast to other ICP-MS with a collision/reaction cell, where the collision induced reactions take place in the extracted ion beam, in the Varian instrument these processes (charge transfer, proton transfer or dissociation of polyatomic ions) occur in the reaction zone at the skimmer orifice (marked in Figure 5.7. b) on the original plasma jet at a significantly higher temperature. The experimental setup of the mass analyzer in the Varian ICP mass spectrometer combining an S-shaped pre-quadrupole and a 3 MHz quadrupole is demonstrated in Figure 5.7 c.

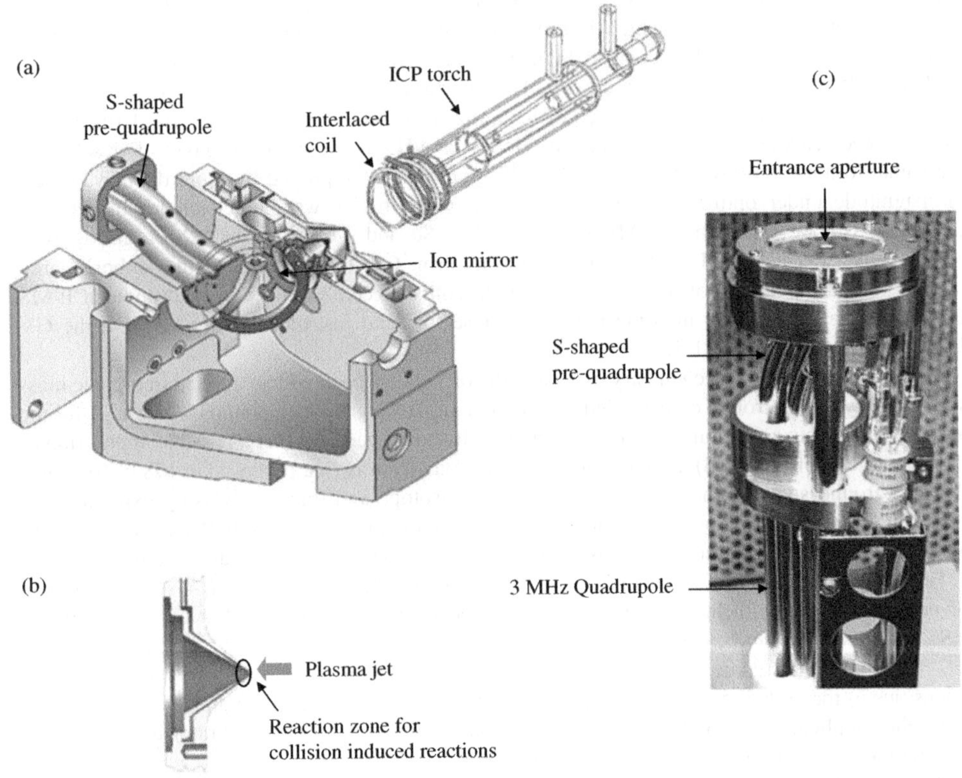

Figure 5.7 *Experimental arrangement a) of 90° reflecting ion optics of Varian ICP-MS; b) cone with channels for collision gas; c) 3 MHz quadrupole analyser with S-shaped pre-quadrupole. (Reproduced by permission of Varian, Inc)*

The figures of merit of several commercial quadrupole based ICP mass spectrometers with and without a collision/reaction cell, produced by different companies, are summarized in Table 5.1.

Collision induced reactions are also applied in ICP-MS with an ion trap for high mass resolution measurements. Two instrument configurations of the plasma source ion trap (PSIT) are shown in Figure 5.8 a and b, respectively. PSIT I utilizes a linear quadrupole for ion transport and mass selection (Figure 5.8 a) while the PSIT II applies direct injection of ions into the trap (without

Table 5.1 Figures of merit for quadrupole based ICP mass spectrometers.

Company	Quadrupole-based ICP-MS			
	Agilent	**Perkin Elmer Sciex**	**Thermo Fisher Scientific**	**Varian**
Without collision cell	**Agilent 7500 a**	**Elan 9000**	**XSeries**	**Varian 810-MS**
Collision/reaction cell	**Agilent** ce[1]/cs[2] octopole cell (ORS)	**Elan DRC II** quadrupole **d**ynamic reaction **c**ell	**XSeries**^{II} hexapole cell (*XS-Minus modus)	**Varian-820-MS** using collision reaction interface
Quadrupole	Hyperbolic, Mo	Binary gold coated ceramic, producing hyperbolic field	Mo rods with prepost filters	Stainless steel rods locked into ceramic mounts, producing hyperbolic field
Ion optics	Omega lens	single cylinder lens	Chicane ion lens	90° reflecting ion mirror
Dynamic range	10^9	10^8	10^8	10^9
Mass range, u	2–260	1–270	2–255	3–256
Reducing background	off axis	photon stop	off axis	90 degree ion optics double off axis quadrupole
Background signal (220 u)	3 cps	2 cps	0.5 cps	< 5 cps
Sensitivity, Mcps/ppm	^{7}Li: 30; ^{89}Y: 80[1]/160[2]	^{24}Mg: 8; ^{115}In: 30	^{7}Li: 6(70*); ^{115}In: 40(120*); ^{238}U: 80(120*)	1000
Limits of detection, ppt	Y: 0.5[1]/0.1[2]	In, U: 0.5	In: 1; Pb: 2; ^{80}Se: 25	< 0.05
Precision (RSD, 20 min)	≤ 3 %	≤ 4 %	≤ 2 %	≤ 3 %
Abundance sensitivity				
Low mass side	$< 5 \times 10^{-7}$	$< 1 \times 10^{-7}$	$< 1 \times 10^{-6}$	$< 1 \times 10^{-7}$
High mass side	$< 1 \times 10^{-7}$	$< 1 \times 10^{-7}$	$< 1 \times 10^{-6}$	$< 1 \times 10^{-6}$
Oxide ions, MO^+/M^+	≤ 1.5 % (≤ 0.8 %, He)	≤ 3 %	≤ 3 % (2 %, Peltier cooled)	< 1 % (CeO^+)
Double charged ions, M^{2+}/M^+	≤ 3 %	≤ 3 %	≤ 3 %	< 3 %

[1] Agilent 7500 ce; [2] Agilent 7500 cs

the linear quadrupole transfer/selection optics, see Figure 5.8 b) as described by Barinaga and co-workers[29] in 1996. Collision induced reactions in a collision/reaction cell to solve the interference problem in ICP-MS were studied by Eiden *et al.*[30,31] Selected collision induced reactions of polyatomic ions (e.g., OH^+, ArC^+, ArN^+, ArO^+, $ArCl^+$, $Ar_2{}^+$, ClO^+, $O_2{}^+$) with He as the reaction gas in the rf quadrupole ion trap, are relevant for precise isotope ratio measurements. For example, the reduction of thorium oxide formation in ICP-MS with a quadrupole ion trap was investigated by Koppenaal and co-workers.[32] The result of all these experiments using a Paul type ion trap in ICP-MS in the mid to late 1990s demonstrated efficient and sensitive detection of atomic ions, although with a slightly reduced dynamic range compared to conventional ICP-QMS. In spite of high mass resolution (e.g., $m/\Delta m \sim 600\,000$ for separation of isobaric $^{129}I^+$ and $^{129}Xe^+$ ions), possible ion trap (and also FTICR) mass spectrometers with an inductively coupled plasma ion source are not yet commercially available due to relatively low sensitivity and

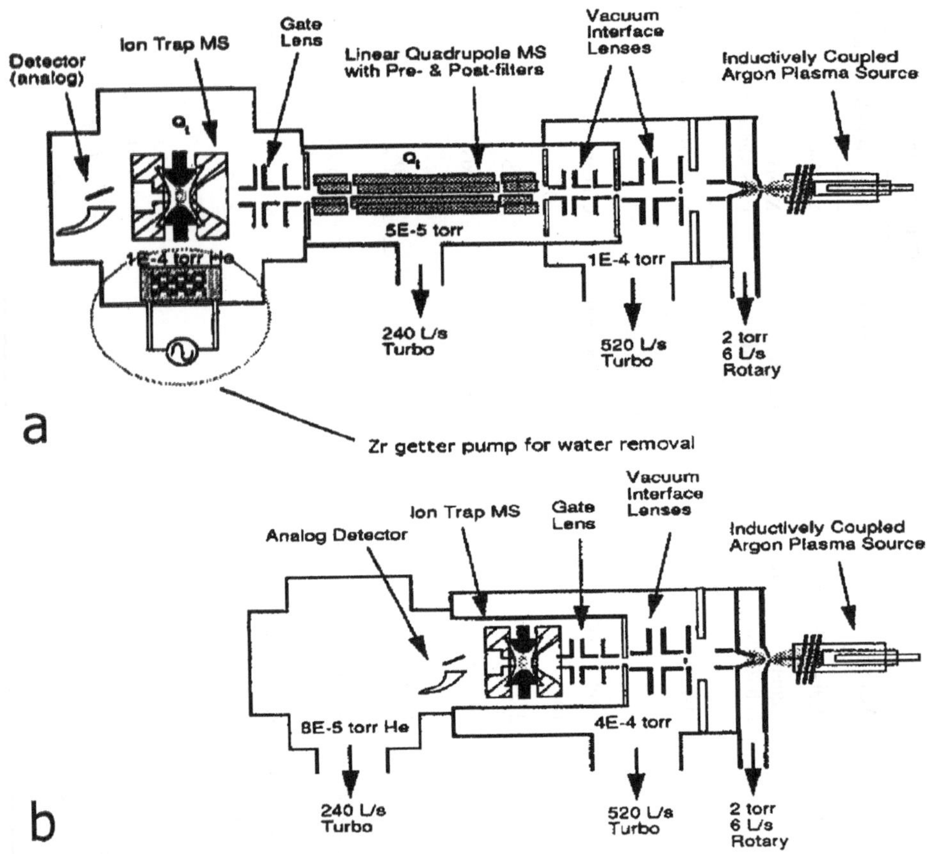

Figure 5.8 *Instrumental configurations of the plasma source ion trap (PSIT): (a) PSIT I utilizes a linear quadrupole for ion transport and mass filtering, (b) PSIT II uses direct injection of ions into the trap (without the linear quadrupole transfer/selection optics). (Barinaga, C. J., Eiden G. C., Alexander, M. L. and Koppenaal, D. W., Fresenius. J. Anal. Chem.,* **355**, *487(1996). Reproduced by permission of Springer Sciences and Business Media.)*

difficulties in quantifying the analytical data. A large Ar^+ intensity relative to the analyte ions in ICP makes ion trap mass spectrometers impractical, because only a limited total number of ions can be stored in the ion trap. In 2006, Koppenaal *et al.*[33] described the coupling of an ICP source to a commercially available high resolution Orbitrap MS system for highly resolved measurements to solve radioanalytical and metallomics problems in future.

Due to the minor importance of these high mass resolution instruments in inorganic mass spectrometry they will not be further discussed in the following chapters.

5.1.3 Double-focusing Sector Field ICP Mass Spectrometers with Single Ion Collector (ICP-SFMS)

A useful possibility for solving the interference problem of atomic ions of the analyte and disturbing polyatomic ions at the same nominal mass is the application of double-focusing sector field ICP-MS (ICP-SFMS) at the required mass resolution (see Section 3.1.3). In 1989, a double-focusing sector field ICP-MS with single ion collector[34,35] (PlasmaTrace from Micromass, Manchester, UK) was developed from an organic mass spectrometer used originally for the structural analysis of complex organic compounds. At present, the Element device with an inductively coupled plasma source, from Thermo Fisher Scientific, formerly Finnigan MAT, Bremen, Germany, is the leading sector field instrument on the market. This type of sector field instrument with an ICP ion source has been accepted today as the most sensitive and powerful analytical tool in inorganic mass spectrometric analysis. Older instrumentations, e.g., the PlasmaTrace, the JMS-Plasma X2 from Jeol, Tokyo, Japan, with reverse Nier–Johnson geometry, are no longer produced due to several disadvantages in hard-and software, difficult maintenance, significantly higher costs compared to quadrupole ICP-MS and other reasons. Double-focusing sector field mass spectrometers are generally designed with reverse Nier–Johnson geometry (see Figure 5.9), whereby a magnetic sector field for directional focusing and an electric sector field for energy focusing of ion beams (BE configuration) are

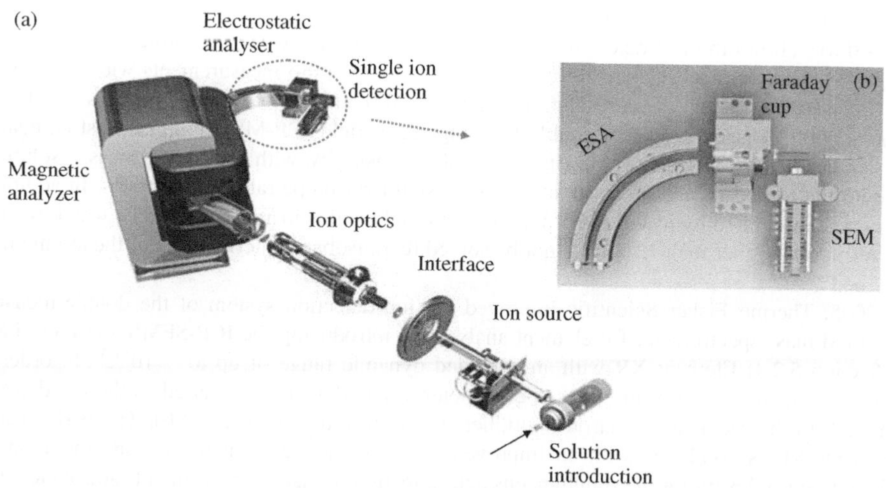

Figure 5.9 Double-focusing sector field ICP-MS (ICP-SFMS, Element XR, Thermo Fisher Scientific, Bremen) with extended dynamic range up to $\sim 10^{13}$ (a), arrangement of ion detector systems is shown at top right, (b). (Reproduced by permission of Thermo Fisher Scientific, Bremen.)[22]

combined. Furthermore, the double-focusing sector field VG Axiom from VG Elemental with Nier–Johnson geometry (EB configuration) is used in several laboratories. Advantageous is the variable slit system so that the mass resolution ($m/\Delta m$ from 300 to $\sim$ 10 000) can be optimized via three computer controlled slit assemblies so that the mass resolution can be optimized for specific applications. In addition, a double-focusing sector field ICP-MS ATTOM is offered from Nu Instruments, Wrexham, UK on the international market.

The analytical performance of selected sector field mass spectrometers with an ICP ion source used today in different laboratories worldwide will be discussed briefly. The performances of double-focusing sector field ICP-MS are similar with respect to the high sensitivity achievable, $\sim 10^9$ cps for 1 μg ml^{-1} of analyte, very low noise (< 0.2 cps) and thus very low detection limits in the sub-fg ml^{-1} range at low mass resolution ($m/\Delta m \approx 300$). Mass spectrometric measurements are carried out at low mass resolution for interference free isotopes. The peak shape (flat top peaks) at low mass resolution allows precise isotope ratio measurements to be made by ICP-SFMS. Furthermore, a high dynamic range is achieved by ICP-SFMS with a single ion collector (using a secondary electron multiplier – SEM) of up to $\sim 10^9$. The difference between the instruments can be found in the slit system used in order to set up the mass resolution of the mass spectrometer. The slit system of the older PlasmaTrace2 with five possible mass resolution settings permitted a variable mass resolution by automatically changing these settings in order to maximize the transmission of ion beams analyzed. In contrast, the ICP-SFMS Element, as the instrument most frequently applied today, possesses a fixed slit system with three mass resolution settings using different slit systems whereby measurements can be carried out at low mass resolution: $m/\Delta m \sim 300$–400, medium mass resolution: $m/\Delta m \sim 3000$–4000 and high mass resolution: $m/\Delta m \sim 10\,000$–12 000. A 'medium' mass resolution ($m/\Delta m \approx 3000$–4000) is required where isobaric interferences of atomic ions of the analyte occur with polyatomic ions at the same nominal mass (e.g., for the separation of $^{56}Fe^+$ and $^{40}Ar^{16}O^+$). The so-called 'high' mass resolution mode in ICP-MS ($m/\Delta m \approx 12\,000$) is employed, for example, for the separation of $^{80}Se^+$ and $^{40}Ar_2{}^+$ interference. The experimental arrangement of the ICP-SFMS Element with solution introduction, ion source, interface, ion optics, magnetic and electrostatic analyzer combined in reverse Nier–Johnson geometry and a single ion detector is illustrated in Figure 5.9.

ICP-SFMS allows the fast determination of a multitude of analytes at ultratrace level (down to sub-fg ml^{-1} range without analyte enrichment[12]) and isotope ratio measurements with a precision of between 0.02 % (at low mass resolution: $m/\Delta m \sim 300$) and 0.2 % (at medium mass resolution: $m/\Delta m \sim 4000$). Disadvantages of double-focusing sector field ICP-MS are higher cost compared to easier quadrupole based ICP-MS and decreasing sensitivity with increasing mass resolution, where no flat top peaks are observed and the precision for isotope ratios deteriorates. In addition, duty cycles are limited by hysteresis of the magnetic field and isobaric interferences which require mass resolution higher than 12 000 cannot be solved (e.g., isobaric interferences of the atomic ions $^{40}Ca^+$ and $^{40}Ar^+$).

In 2005, Thermo Fisher Scientific improved the ion detection system of the double-focusing sector field mass spectrometer for element analysis by introducing the ICP-SFMS and GD-SFMS (see Section 5.2.4) Element XR with an extended dynamic range of up to $\sim 10^{13}$.[36] In order to extend the dynamic range of the mass spectrometer, a Faraday cup is inserted as the ion detector in addition to the secondary electron multiplier (see right top in Figure 5.9 b). The performance of Element XR is combined with an improvement in abundance sensitivity. This instrumental configuration is relevant for the simultaneous determination of matrix, trace and ultratrace elements and is also of importance in isotope analysis for the measurements of extremely small isotope ratios. But it should be considered that a higher matrix input results in contamination problems and memory effects.

5.1.4 Time-of-flight Mass Spectrometers (ToF-MS)

Time-of-flight (ToF) mass spectrometers combine a relatively simple mechanical setup with extremely fast electronic data acquisition. Using ToF mass spectrometers, mass resolutions ($m/\Delta m$) of up to 10 000 and a mass range of up to 500 000 Da together with an ion transmission of more than 10 % and large acceptance volumes is possible. Most ToF mass spectrometers employ ion detectors using multi-channel plates which have a time response < 1 ns and a high sensitivity (single ion signal > 50 mV). The principle of a linear ToF mass spectrometer, which is a very simple design, and one with a reflectron is described in Section 3.2.2 (see also the schematics of both types of ToF mass analyzers in Figures 3.10 and 3.11).

The Bendix Corporation's (USA) time-of-flight mass spectrometer (ToF-MS) with a pulsed electron impact ion source was one of the first commercial non-magnetic sector mass spectrometers on the analytical market as described by Wiley in 1956.[37]

ToF mass spectrometers as dynamic instruments gained popularity with the introduction of matrix assisted laser desorption/ionization (MALDI) and electrospray ionization (ESI) as effective pulsed ion sources for the soft ionization of large biomolecules (up to 10^5 dalton) due to their high ion transmission.[38] ToF mass spectrometers, quadrupole analyzers and/or magnetic sector fields can be combined in tandem mass spectrometers (MS/MS) for the analysis of organic compounds.

In inorganic mass spectrometry, ToF mass analyzers, which are able to measure very fast ion intensities of separated ion beams and thus isotope ratios of elements, tend to be applied in SIMS (e.g., the ToF-SIMS from IonToF Münster, Germany) rather than in ICP-MS. Whereas in SIMS the higher mass range of ToF mass analyzers in comparison to double-focusing sector field instruments is of special interest for the analysis of biomolecules, in ICP-ToF-MS the dominant advantageous property is the fast measurement of transient signals. This benefit is exploited by the application of hyphenated techniques (in combination with capillary electrophoresis – CE or high-performance liquid chromatography – HPLC) for flow injection sample introduction in ICP-ToF-MS for speciation analysis.

Hieftje and co-workers first reported a design for ICP-ToF mass spectrometric systems with both orthogonal acceleration (oa) and axial acceleration (aa) geometry.[39–41] The different experimental setups of ICP-ToF mass spectrometers with orthogonal and axial acceleration are compared in Figures 5.10 and 5.11. Both instruments possess a time-of-flight analyzer with a reflectron to improve the mass resolution.

A ICP-ToF-MS with orthogonal acceleration (oa) geometry has been produced by GBC Scientific Equipment Pty Ltd (Melbourne, Victoria, Australia) since 1993. The OptiMass 9500 ICP-oa-ToF-MS (Figure 5.10) has a 2×0.5 m flight tube fitted with a reflectron ion mirror. Due to the orthogonal configuration with the ion detector, lower background signals are observed compared to an axial arrangement. The analytical performance of the commercial ICP-oa-ToF-MS (detection limits of elements due to a low background at 1 ng l^{-1} range and an abundance sensitivity of 7.4×10^{-5} in the high mass range) has been discussed by Sturgeon *et al.*[42] This instrument is well suited for the acquisition of transient signal information and, although limited to an 8 ms temporal resolution for full mass scan events, the capability of this ICP-ToF-MS is more than sufficient for applications in flow injection, electrothermal vaporization and various chromatographic sources of analyte introduction.[42] Mester *et al.*[43] utilized this equipment for the determination of tributyltin in aqueous solutions by means of ICP-oa-ToF-MS following headspace SPME (solid phase microextraction) sampling and thermal desorption. The precision of isotope ratios (up to 0.04 % RSD), with an HPLC pump to introduce a microlitre volume, in ICP-oa-ToF-MS for the measurement of transient signals was studied by Willie and co-workers.[44]

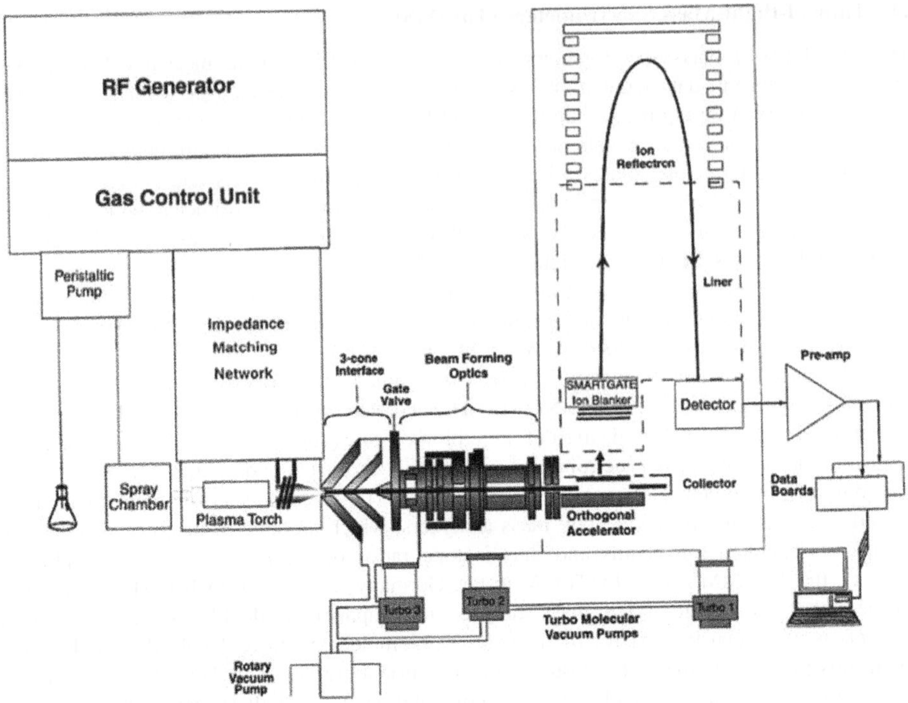

Figure 5.10 Schematic of ICP-ToF-MS OptiMass 8000 (GBC Scientific Equipment). (R. E. Sturgeon et al., J. Anal. At. Spectrom., **15**, 607 (2000). Reproduced by permission of the Royal Society of Chemistry.)

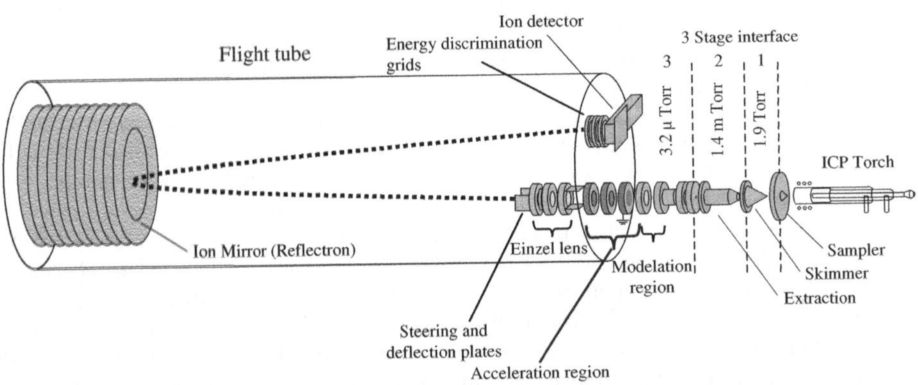

Figure 5.11 Setup of ICP-ToF mass spectrometer Renaissance, LECO (Reproduced by permission of LECO Co.).

The axial time-of-flight ICP-MS was introduced in 1998 by LECO (Renaissance, LECO, St. Joseph, MI) and has been commercially available for several years. A schematic of the ICP-ToF-mass spectrometer is shown in Figure 5.11. The analytical performance of this mass spectrometer, which allows 25 000 spectra to be measured per second and is thus faster than a quadrupole ICP-MS, can be summarized as follows: sensitivity 10^6–10^7 cps ppm^{-1}, detection limits 1–10 ng l^{-1} for all elements, precision 5 %, dynamic range 10^6. Due to their relatively low sensitivity, the ToF mass spectrometers have lower detection limits, about one order of magnitude poorer than in commercial quadrupole-based ICP-MS,[45] so that relatively high analyte concentrations in solution are utilized (from 50–500 μg l^{-1}), e.g., for isotope ratio measurements.[46] The benefits of ICP-ToF-MS for the determination of trace elements Pt, Rh and Pd in body fluids and road dust have been described and the application of flow injection on line pre-concentration reviewed by Adams *et al.*[47,48] The application of different hyphenated techniques in ICP-ToF-MS was reviewed by Leach and co-workers.[49] Because of the sensitivity limitation of ICP-ToF mass spectrometers and the associated higher detection limits (about one order of magnitude poorer than in commercial quadrupole based ICP-MS[45]) they do not open up areas for application where quadrupole based ICP-MS is currently limited. Therefore only a few instruments of this type are installed worldwide. As an inorganic mass spectrometric technique, ICP-ToF-MS is only of minor significance due to its low sensitivity.

5.1.5 Multiple Ion Collector ICP Mass Spectrometers (MC-ICP-MS)

To overcome the limitations of mass spectrometry with a single ion collector (e.g., plasma instabilities – so-called 'flicker noise' – or also in part mass drift effects) multiple ion collector mass spectrometers with an ICP source (MC-ICP-MS) for simultaneous static measurements of mass separated ion currents were designed in the early nineties of the last century. The development of MC-ICP-MS started with the Plasma 54 (VG Elemental) which was developed from the Sector 54 TIMS (VG Elemental). This double-focusing sector field ICP mass spectrometer with forward Nier–Johnson geometry possesses an array of mechanically adjustable Faraday collectors in a block with 9 or 11 (the latter in the University of Michigan instrument) Faraday collectors for precise and simultaneous measurements of high ion currents and a single Daly detector for ion counting of low-abundance isotopes. A schematic of the MC-ICP-MS VG Plasma 54 at the University of Michigan (U-M) is shown in Figure 5.12 a (top left-hand side). The Plasma 54 is equipped with a large energy filter for high abundance sensitivity measurements.[50] The U-M instrument possesses a wide flight tube with an off axis Faraday cup located on the high mass side for measurements of U simultaneously with Pb.[51] Most of the instrument is at ground potential, the ion source and its associated pumping are floated at high potential. The ions formed in the inductively coupled Ar plasma possess a relatively broad energy spread up to ∼ 15 V, which is about two orders of magnitude higher compared to ions formed by thermal surface ionization. The Plasma 54 is constructed from a 30 cm electrostatic analyzer (ESA) used for energy focusing and matching the ion energy dispersion with that of the magnetic sector field (radius of magnet: 54 cm), so that the ions are energy focused in the directional focal plane at the collectors. Effective energy focusing is performed using an additional small ESA with a small voltage offset; the entrance utilizes a deceleration lens for reacceleration of the ions before their detection and registration.[52] The first isotope ratio measurements on aqueous solutions using double-focusing sector field ICP-MS with multiple Faraday detectors were published by Walder *et al.* in 1992[53,54] and measurements on solid samples by LA-MC-ICP-MS were demonstrated one year later by the same group.[55] To achieve isotope ratios with a standard deviation of about 0.01 % small sample sizes of about 10 ng were measured.

At present, the Nu Plasma from Nu Instruments, Wrexham, UK, the Isoprobe from GV Instruments, Manchester, UK, and the Neptune from Thermo Fisher Scientific, Bremen, Germany, are the most precise multiple ion collector ICP mass spectrometers available on the analytical market.

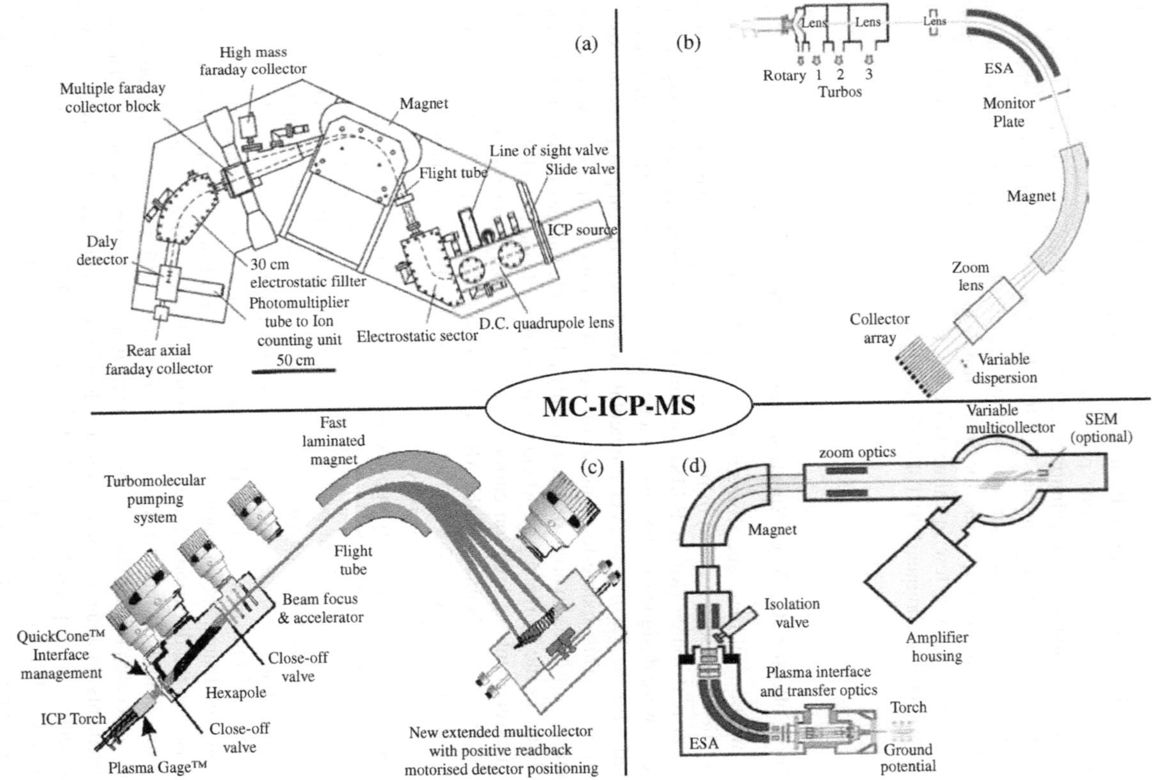

Figure 5.12 *MC-ICP-MS: a) Plasma 54 (with called 'S configuration'), from VG Elemental (M. E. Wieser and J. B. Schwieters, Int. J. Mass Spectrom.,* **242**, *97 (2005). Reproduced by permission of Elsevier), b) Nu Plasma from Nu instruments (Reproduced by permission of NU Instruments Ltd); c) Isoprobe from GV Instruments (Reproduced by permission of GV Instruments Ltd); d) Neptune from Thermo Fisher Scientific (Reproduced by permission of Thermo Fisher Scientific, Bremen); e) Nu 1700 (Reproduced by permission of NU Instruments Ltd).*

(e)

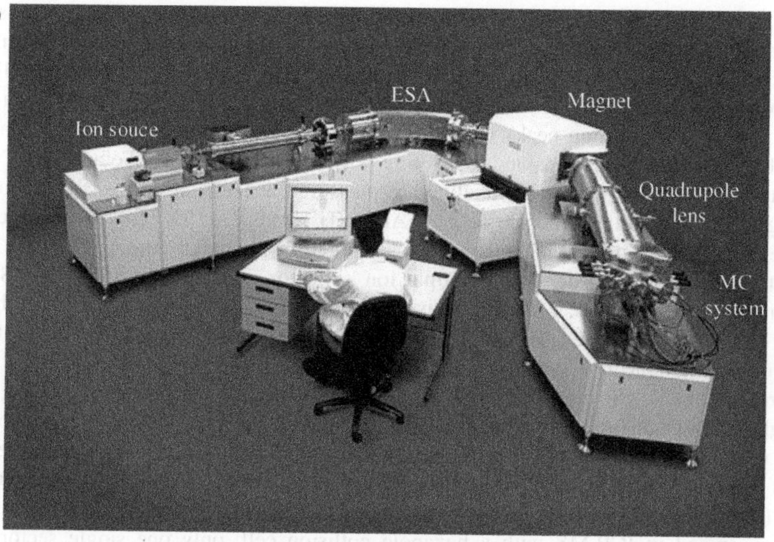

Figure 5.12 *(Continued).*

The schematics of commercially available multiple ion collector ICP mass spectrometers, together with the VG Plasma 54, are shown in Figure 5.12. Three multiple ion collector ICP mass spectrometers, VG Plasma 54 (Figure 5.12 a), Nu Plasma (Figure 5.12 b) and Neptune (Figure 5.12 d), are constructed as double-focusing sector field mass spectrometers combining a magnetic and electric sector field with Nier–Johnson geometry. The Nu Plasma (with Nier–Johnson geometry, called 'standard C configuration') from Nu Instruments shows some general similarities compared to the Plasma 54 (with called 'S configuration'), but it uses fixed multiple ion collectors together with variable dispersion ion optics applied to the ion beams. Ions are extracted with an acceleration voltage of 4 kV behind the skimmer and focused using an ion optical lens system. The instrument uses a 35 cm radius ESA together with a 25 cm radius magnetic sector field. The ion optical system is designed to allow a high ion transmission of the ions formed in the plasma through the instrument. The Nu Plasma is equipped with a variable dispersion zoom optic device to focus the separated ion beams in the fixed multiple Faraday cups of the multiple ion collector systems containing twelve Faraday buckets. In addition, three multipliers for ion counting can be utilized for measuring low abundance isotopes at trace and ultratrace concentration levels. An improved abundance sensitivity, which is required for example for determination of $^{236}U/^{238}U$ isotope ratio is obtained in an MC-ICP-MS (Nu Plasma) equipped with an ion deceleration filter (so-called 'high abundance sensitivity channel').[56] The application of the ion deceleration system in MC-ICP-MS allows a reduction in peak tailing from $^{238}U^+$ ions at $m/z = 236$ down to 3×10^{-9}, so that $^{236}U/^{238}U$ isotope ratios can be measured in the 10^{-8} to 10^{-9} range.

The double-focusing sector field ICP-MS with multiple ion collectors, Neptune (Thermo Fisher Scientific, Bremen, Germany, see Figure 5.12 d) allows precise isotope ratio measurements at a maximum resolving power ($m/\Delta m$, 5 % and 95 %)[57] of 10 000. The practical mass resolution for this configuration is defined by Weyer and Schwieters[57] as resolving power R_{power} by the following equation:

$$R_{power}(5\%,\ 95\%) = m/[m(5\%) - m(95\%)] \tag{5.1}$$

where $m(5\,\%)$ is the mass of peak at 5 % level (see Figure 3.15) and $m(95\,\%)$ is the mass at 95 % peak height. Isotope ratios are measured in the static operation mode, where the magnetic field remains constant. For measurements at high mass resolution using a small entrance slit (behind the ion source and in front of the ion detection system) and a wide exit slit at the detector at high resolving power, wide flat top peaks with two plateaus are obtained. The second plateau appears when both ion beams atomic ions and interfering polyatomic ions (e.g., ^{56}Fe and $^{40}Ar^{16}O^+$) enter the detector slit. The atomic ions of analyte are measured at the first peak plateau shoulder and disturbing interfering polyatomic ions e.g., $^{40}Ar^{16}O^1H^+$ and $^{40}Ar^{16}O^+$ for precise and accurate $^{57}Fe/^{56}Fe$ isotope ratio measurements, are well separated from analyte ions.[57,58]

In contrast to other ICP-MS with multiple ion collector systems, the MC-ICP-MS Isoprobe (GV Instruments, Figure 5.12 c) only consists of a single magnetic sector field together with a hexapole collision cell inserted between the ICP ion source and a single sector magnet field. This innovative development in ICP-MS was created in 1996. Using the hexapole collision cell, as in the quadrupole based ICP mass spectrometer (Platform from GV Instruments, see Figure 5.1 b, centre right) most of disturbing isobaric interference (e.g., $^{40}Ar_2{}^+$ and $^{80}Se^+$) by collision induced reactions with a He/H_2 gas mixture can be significantly suppressed. Furthermore, the kinetic energy spread of ions is reduced from about 20 eV to the < 1 eV range and thus a single focusing mass analyzer (laminated magnet) can be used without need of energy focusing. Due to the relatively low energy spread in ICP-MS with a hexapole collision cell, only one single sector field – a magnet – (see schematic arrangement of MC-ICP-MS Isoprobe on the left-hand side at the bottom in Figure 5.12 c) is sufficient to separate ion beams with different *m/z* ratios for simultaneous multiple detection of separated ion beams. For the determination of isotope ratios, the precision is improved in sector field mass spectrometers compared to ICP-SFMS with a single ion detector, by about one order of magnitude down to 0.001–0.002 % when multiple ion collectors (MC) are applied for the simultaneous detection of ion currents with different *m/z* ratios.[24]

An innovative development and design for the highest precision isotope ratio measurements in double-focusing sector field ICP-MS is represented by the Nu 1700 (the so-called 'Big Mac', see Figure 5.12 e) from Nu Instruments introduced at the Isotope Geochemistry Laboratory, ETH Zurich.[59] This instrument possesses a high mass dispersion and large geometry consisting of a 70° magnet with a 750 mm radius for the average ion path, combined with a 70° electrostatic analyzer (ESA) with a 943 mm ion path radius. The high mass resolution achieved ($m/\Delta m$, up to 10 000) by the MC-ICP-MS can be applied to solve a multitude of inherent isobaric interference problems of polyatomic ions with atomic ions of the analyte. This high resolution MC-ICP-MS possesses 16 Faraday collectors (ten fixed, six mechanically movable ones) and multiple ion counting with a high abundance sensitivity filter. The main fields of application for the expensive MC-ICP-MS Nu 1700 are in geosciences, for studying extremely small isotope variations in nature, for geochronology (age dating of rocks and minerals) and cosmochronology.

An interesting approach is the coupling of an ICP ion source with a double-focusing sector field Mattauch–Herzog type instruments (operating with spark or laser ion source and photoplate ion detection in the past). Advantages compared to Nier–Johnson or reverse Nier–Johnson type instruments employed in sector field ICP-MS, which allows the double-focusing for one mass only (in a small mass range), is that a Mattauch–Herzog mass spectrometer makes double-focusing possible for all masses simultaneously (see Section 3.1.3). Double-focusing ICP-MS with Mattauch–Herzog geometry and multiple channel ion detection (using photodiode array detectors) for a quasi-simultaneous ion detection of transient signals, was developed and described by Cromwell and Arrowsmith[60] and by Hieftje *et al.*[61,62]

The analytical performance of selected commercially available sector field ICP-MS with multiple ion collection compared to double-focusing sector field ICP-MS with a single ion collector are compared in Table 5.2.

Table 5.2 *Analytical Performance for ICP Mass Spectrometers.*

	Sector field ICP-MS with single and multiple ion collectors			
	Element, Thermo Fisher Scientific	**Nu Plasma, Nu Instruments**	**NEPTUNE, Thermo Fisher Scientific**	**Isoprobe, VG Instruments**
Geometry	reverse Nier–Johnson 16 cm laminated magnet, 10.5 cm ESA	Nier–Johnson, radius: 35 cm ESA, 25 cm laminated magnet	Nier–Johnson 81.2 cm dispersion	single 27 cm laminated magnet extended to 54 cm (dispersion) and collision cell
Ion source	on mass potential	high voltage on expansion region	on mass potential	on mass potential
Accelerating voltage	8 kV	6 kV	10 kV	6 kV
Mass resolution ($m/\Delta m$)	300, 4000, 10 000 (10 % valley)	400–2500 (10 % valley) maintaining peak flat	Resolving power (RP): 2000/6000/10 000	400 (10 % valley)
Dispersion		Variable dispersion (max. 17 %)	Variable dispersion (max. 17 %)	17 %, fixed
Dynamic range	$< 10^{13}$	$< 10^{12}$ Faraday cups/ion counter	$< 10^{12}$ Faraday cups/ion counter	$< 10^{12}$ Faraday cups/ ion counter
Mass range, u	7–240	5–250	3–330	4–320
Background signal, 220 u	0.2 cps	0.1 cps	0.2 cps	0.2 cps
Sensitivity, Mcps/ppm	^{115}In :> 1000 ($m/\Delta m = 300$) ^{115}In :> 100 ($m/\Delta m = 4000$) ^{115}In :> 15 ($m/\Delta m = 10000$)	^{115}In :> 1000 ($m/\Delta m = 400$) ^{115}In :> 1000 ($m/\Delta m = 2500$)	^{115}In > 1500 (RP = 2000) ^{115}In > 350 (RP = 6000) ^{115}In > 150 (RP = 10 000)	^{115}In > 1000 ($m/\Delta m = 400$)
Limits of detection, ppt	0.001	0.001	0.001	N/A
Short term stability (RSD)	< 1 % (10 min)	< 1 % (10 min)		

Table 5.2 (*Continued*).

	Sector field ICP-MS with single and multiple ion collectors			
	Element, Thermo Fisher Scientific	**Nu Plasma, Nu Instruments**	**NEPTUNE, Thermo Fisher Scientific**	**ISOPROBE, VG Instruments**
Long term stability (RSD)	< 2 % (1 h)	< 2 % (1 h)		
Isotope ratio precision (RSD)	0.01 % ($m/\Delta m = 300$)	0.002 % (RSD)	0.002 % (RSD)	0.002 % (RSD)
Abundance sensitivity				
Low mass side	$< 1 \times 10^{-7}$	$< 1 \times 10^{-7}$	with RPQ: $< 5 \times 10^{-6}$ ()	$< 1 \times 10^{-7}$
High mass side	$< 1 \times 10^{-7}$	$< 1 \times 10^{-7}$	$< 5 \times 10^{-9}$ ($^{232}Th/^{230}Th$)	$< 1 \times 10^{-7}$
Peak flatness	only at $m/\Delta m = 300$	flat top peak also at high resolution	plateau flatness at high RP	
Detectors	dual mode: SEM and Faraday	12 Faradays + 3 ion counters	9 Faradays + 8 ion counters	9 Faradays + 7 ion counters
Special features	high linear dynamic range	simultaneous isotope analysis zoom lens, fixed collectors pseudo high resolution	simultaneous isotope analysis	simultaneous isotope analysis

5.1.6 Solution Introduction Systems in ICP-MS

The most common aggregate state analyzed in ICP-MS is the liquid state; therefore several devices have been designed for effective nebulization, including aerosol formation, and also for transport (solution introduction) in the inductively coupled plasma source. A stable liquid solution introduction system in the ICP is a relevant precondition for precise and accurate trace and isotope analysis. In ICP-MS, the solution must be converted into an aerosol before atomization of analytes and ionization of atoms and molecules in the ICP, and before their mass spectrometric analysis. Several nebulizers for solution introduction with different solution uptake rates (e.g., Meinhard nebulizer (J. E. Meinhard Associates, Santa Ana, CA) or MicroMist nebulizer (Glass Expansion, Pty., Camberwell, Victoria, Australia) together with an efficient cyclonic spray chamber for filtering large aerosol droplets (Aridus; APEX or ultrasonic nebulizers (USN)) – the latter three with desolvator) have been applied in ICP-MS (see also Figure 5.1, top left). In comparison to very sensitive USN, which nebulizes, in general, relatively high volumes of solution (typically $2\,\mathrm{ml\,min^{-1}}$), microconcentric nebulizers are advantageous due to a significantly lower solution uptake rate. An overview of several sample introduction systems for liquids, solids and gases in ICP-MS, which have been applied, studied and described by several working groups are summarized in Figure 2.12.[3,9,20,63–69] Several application fields of micronebulization in ICP-MS are summarized in Figure 5.13.

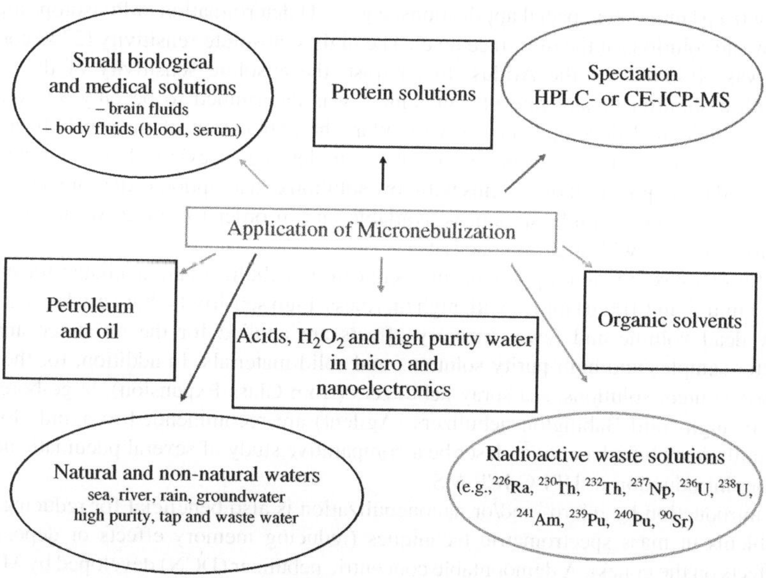

Figure 5.13 *Fields of application of micronebulization in ICP-MS.*

Pneumatic Nebulizers Including Selected Micronebulizers

Pneumatic nebulizers are utilized for the generation of an aerosol with droplet size distribution in the low μm range[3] and consequently for an effective solution introduction into the ICP ion source of the mass spectrometer. The sample solution is therefore transported by a peristaltic pump (for micronebulizers the self-aspiration mode is also applied) and nebulized in a spray chamber. The relatively inexpensive pneumatic nebulizer (Meinhard nebulizer, working with a solution uptake rate of 1 ml min^{-1}) together with an efficient spray chamber arrangement (e.g., a cooled double pass spray chamber to separate the large droplets) has only low analyte transport efficiency (1–20 %). The microflow nebulizer spray chamber arrangement is more effective for smaller solution consumption and enhanced transport efficiency.[3,70] A rapidly growing field of trace element determination involves the analysis of μL or nL solutions, which is of particular concern when only a few μl of a biological, toxic or radioactive solution is available.[12,64,71–73] The development of microanalytical techniques coupled with increasing sensitivity of ICP-MS is especially required for the determination of radionuclides and their isotope ratios in small volumes of radioactive materials in order to reduce the radioactivity handled, to minimize the radioactive waste from an analytical procedure, the irradiation dose to the operator or possible contamination of instruments with radioactive substances.[73,74] A relatively cheap and frequently used concentric micronebulizer, which is robust and works effectively, is the MicroMist microconcentric nebulizer (solution uptake rate, e.g., 0.1 or 0.2 μl min^{-1}). For routine measurements (e.g., determination of uranium and thorium concentration in urine or radioactive waste solutions), the MicroMist nebulizer, together with a small cyclonic spray chamber, yields good precision and accuracy at low % range RSD and rapid wash out of matrix. This high performance nebulizer is followed by the more expensive Aridus microconcentric nebulizer with desolvator (CETAC Technologies, Omaha) with a solution uptake rate of 0.1 μl min^{-1} for special applications, e.g., ^{236}U determination or Pu isotope analysis in radioactive waste solutions at the ultratrace level. The highest absolute sensitivity (2340 counts fg^{-1} for ^{238}U)[75] was observed for the Aridus. In contrast, the absolute sensitivity of the Meinhard nebulizer with a solution uptake rate of 1 ml min^{-1} was determined to be only 97 counts fg^{-1} for ^{238}U. Methodological developments, in particular, the application of micronebulizers for the determination of trace elements in environmental samples were reviewed in reference [76]. The development and the application of systems with low solution consumption is the subject of research for cases where a limited sample amount is available and in order to reduce waste management costs (e.g., for solutions with high radioactivity).

PFA (perfluoroalkoxyethylene) polymer microconcentric nebulizers are available for microflow rates of 20 μl min^{-1} and 100 μl min^{-1} with high inertness, high sensitivity, but very low background signals, low dead volume and fewer memory effects are applied for the ultratrace analysis of semiconductor samples and high purity solutions and solid materials. In addition, for the analysis of high matrix content solutions, sea spray nebulizers (from Glass Expansion), large-bore DIHEN (Meinhard) or high-solid Babington nebulizers (Agilent) are recommended to avoid clogging of nebulizer capillaries. Stefánka *et al.*[77] describe a comparative study of several pneumatic nebulizers for improving introduction in HPLC-ICP-MS.

Solution introduction by micro- and/or nanonebulization is also beneficial for reducing contamination problems in mass spectrometric techniques (reducing memory effects or deposition and clogging effects on the cones). A demountable concentric nebulizer (DCN) developed by Montaser's group[78] was introduced in ICP-MS and the performance and the aerosol diagnostics for solution introduction in ICP-MS were studied. A direct injection high efficiency nebulizer (DIHEN, Meinhard) without a spray chamber was explored for effective solution introduction directly into the ICP of a mass spectrometer.[9,28,64,79] The schematic arrangement together with a demountable ICP torch and torch interface is presented in Figure 5.14 a. The aqueous solutions are introduced into the DIHEN in the continuous flow mode via a syringe pump (e.g., from Havard Apparatus, Inc. Massachusetts). Using the DIHEN, isotope ratio measurements of plutonium and uranium at

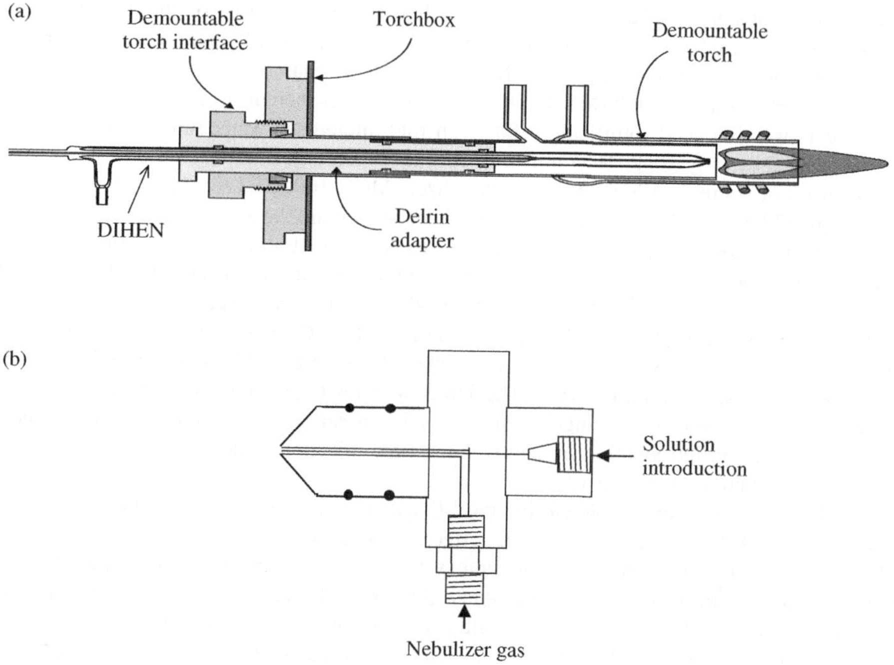

Figure 5.14 *Schematics a) of DIHEN (J. E. Meinhard Associates, Santa Ana, CA) (J. S. Becker, H. J. Dietze, J. A. McLean and A. Montaser, Anal. Chem.,* **71***, 3077 (1999). Produced by permission of American Chemical Society) and b) of microconcentric nebulizer (MCN 100, CETAC Technologies, Ohama. Reproduced by permission of CETAC Technologies)*

ultratrace concentration level have been carried out at the ultratrace level.[9,12,80,81] A disadvantage of DIHEN in ICP-MS is the relatively high cost compared to the MicroMist nebulizer and an increased polyatomic ion formation rate.[82] The sample volume analyzed at 60 μl min^{-1} for the DIHEN still remains relatively high, the nebulization is usually not complete and a portion of the sample remains as radioactive waste, and the oxide formation rate is rather high.[9] With the aid of a demountable DIHEN, improved operational characteristics of micronebulizers have been found and the aerosol properties and analytical analytical performance has been studied, first of all by Montaser's group.[79,83] The same working group discussed the capability of a pneumatic-driven high efficiency nebulizer (HEN) in comparison to an ultrasonic nebulizer in helium ICP-MS.[84] Compared to argon used mostly as the plasma and nebulizer gas in ICP-MS, helium possesses a higher ionization energy (24.6 eV) and its plasma species offer higher energies. In addition He possesses one high abundant isotope (^{4}He – 99.99 %) and thus the formation of argon based polyatomic ions can be avoided. Use of helium instead of argon as the nebulizer gas results in slightly coarser primary aerosol at a solution uptake rate of 100 μl min^{-1} and a gas flow rate of 1 l min^{-1}. In addition a four-fold decrease in analyte transport efficiency through a Scott spray chamber is observed where He is used for nebulization instead of Ar.[84]

The use of a desolvating system improves the sensitivity and can extend the matrix effect with respect to conventional sample introduction apparatus.[85] The current trend is towards total consumption systems. Fundamental studies concerning the analysis of microsamples with common micronebulizer based systems have been reviewed by Todoli and Mermet.[85]

Further developments in solution introduction in ICP-MS have focused on the design of a torch integrated sample introduction system (TISIS) by Cairns *et al.*,[86] in which a microflow concentric nebulizer was coupled to a single pass, low flow volume spray chamber that can be clipped directly onto the base of a standard ICP-MS torch. Using this experimental arrangement, memory and matrix effects were reduced significantly for a ten fold diluted sea-water sample, with the signal being suppressed by only 10–20 % compared with the response in ultrapure water.

A schematic of a simple microconcentric nebulizer (MCN 100, CETAC Technologies, Ohama, NE) is given in Figure 5.14 b. This microconcentric nebulizer and the MicroMist nebulizer together with the Cinnibar spray chamber (from Glass Expansion, Australia) shown in Figure 5.1 (top left) are applied for routine measurements in the author's laboratory, especially if only a small solution volume is available.[70,80] For solution introduction, the free aspiration mode or a nearly pulse free peristaltic pump (Perimax 16, Spetec GmbH, Erding, Germany) are applied. An effective micronebulizer DS-5 (CETAC Technologies, Ohama, NE), similar to MCN shown in Figure 5.11b, was developed by Schaumlöffel *et al.*[87] The DS-5 operates together with a simple single spray chamber with a high precision syringe pump at a constant carrier flow rate of $7\,\mu l\,min^{-1}$ and has been used for solution introduction by nanovolume flow injection in double-focusing sector field ICP-MS as described in Section 5.1.8.

The APEX system (Element Scientific Inc., Omaha) as an improved Aridus nebulizer was introduced for ICP-MS in 2004 for more effective solution introduction at flow rates from $20–400\,\mu l\,min^{-1}$.[88] In this solution introduction system (see Figure 5.15), a microflow PFA nebulizer is combined with a heated cyclonic spray chamber followed by cooling of the nebulized aerosol in a condenser loop and using a multipass condenser cooled by a Peltier element. The APEX solution introduction system results in a significant increase of sensitivity (by a factor of ten in comparison to a standard nebulizer spray chamber arrangement) and a decreasing polyatomic formation rate.[89]

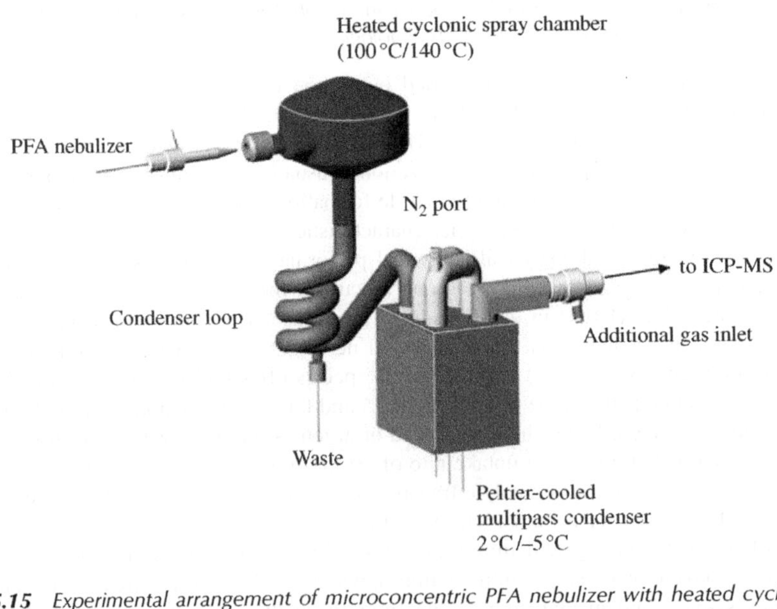

Figure 5.15 *Experimental arrangement of microconcentric PFA nebulizer with heated cyclonic spray chamber and Peltier cooled multipass condenser APEX. (Reproduced by permission of Element Scientific Inc., Omaha).*

A frequently used micronebulizer with heated spray chamber and membrane desolvator is the Aridus from CETAC Technologies, Ohama, NE. The experimental setup of the Aridus II microconcentric nebulizer is shown in Figure 5.16.

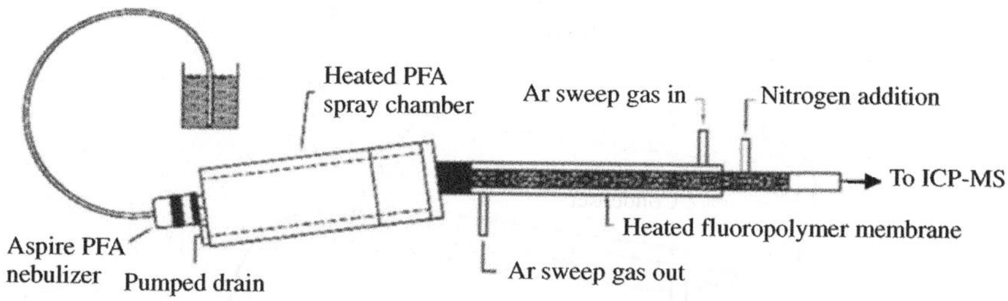

Figure 5.16 *Schematic of microconcentric nebulizer (Aridus II, CETAC Technologies, Ohama). (Reproduced by permission of CETAC Technologies.)*

Ultrasonic Nebulizer

A schematic arrangement and photograph of an ultrasonic nebulizer (USN, CETAC U-6000 AT from CETAC Technologies, Ohama) are shown in Figures 5.17 and 5.1 (bottom left column), respectively. The ultrasonic nebulizer includes a powerful heating and cooling step prior to introduction into the membrane desolvation system. In the membrane desolvator, the water vapour is effectively removed from the aerosol stream and therefore the formation of polyatomic ions such as oxide, hydroxide and hydride ions, which results in serious interference problems in mass spectra, is reduced by one order of magnitude.[12,90] In addition, any possible clogging (a drawback of pneumatic nebulizers) can be avoided because the solution is introduced via the surface of a piezoelectric transducer driven by a generator at a frequency of 200 kHz to 10 MHz.[3] Using USN together with ICP-MS, the element sensitivity can be increased in comparison to pneumatic nebulization,[12,71] which results in a significant improvement in the detection limits for analytes. Increased sensitivity of more than one order of magnitude compared to the MicroMist (MM) or Meinhard nebulizer is observed for the determination of long-lived radionuclides in aqueous solution (e.g., sensitivity for ^{239}Pu: ultrasonic nebulizer (USN) – 22.6×10^9 cps ppm^{-1}, MicroMist nebulizer – 2.9×10^9 cps ppm^{-1}, Meinhard nebulizer – 1.9×10^9 cps ppm^{-1}) as measured by Becker and Dietze.[12] Obviously, the ultrasonic nebulizer is advantageous for very sensitive solution introduction of aqueous solution in ICP-MS, but the consumption of aqueous solution at generally 2 ml min^{-1} is relatively high.[3] Considering this solution consumption using the USN, the sensitivity in cps fg^{-1} is very close to the MicroMist nebulizer, but due to the lower polyatomic ion formation rate, the detection limits of elements using the USN in ICP-MS are improved by about one order of magnitude. A further drawback of the commercial USN from CETAC Technologies is the significantly higher cost in comparison to the pneumatic nebulizer spray chamber arrangement, an inferior precision of 2–3 % (compared e.g., with Meinhard nebulizer of 1 %, RSD) and longer washout times.[3] In conclusion, the USN is a helpful apparatus for solution introduction in routine mode for sensitive trace element analysis in liquids in many laboratories and is applied especially where there is no restriction with respect to sample volume.

Characteristics such as the solution uptake rate, the sensitivity and the limit of detection for uranium determination using several nebulizers (Meinhard, MicroMist nebulizer, Q-DIHEN,

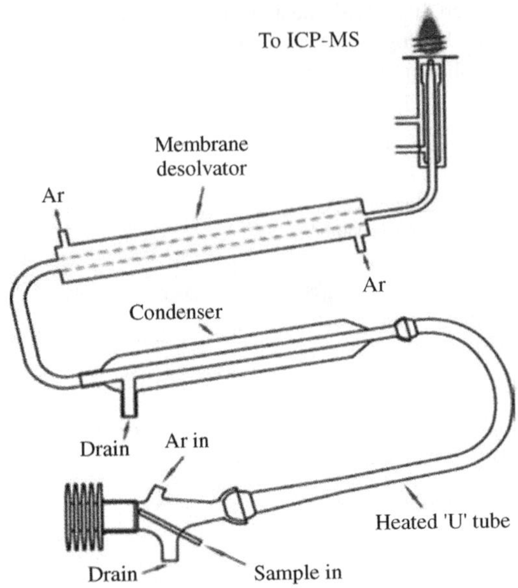

Figure 5.17 Setup of ultrasonic nebulizer with membrane desolvator (USN 6000AT, CETAC Technologies, Ohama). (Reproduced by permission of CETAC Technologies.)

Aridus, USN operated in the continuous mode versus the DS-5 nebulizer used in the flow injection mode) for solution introduction in ICP-SFMS (Element, Thermo Fisher Scientific, Bremen) for uranium determination in aqueous solution[75] are summarized in Table 5.3.

Table 5.3 Characteristics of ICP-SFMS (Element, Thermo Fisher Scientific, Bremen) for uranium determination in aqueous solution using different solution introduction systems.

Nebulizer	Solution uptake rate [ml min^{-1}]	Sensitivity [counts fg^{-1}]	Limit of detection [pg l^{-1}]
Meinhard	1	97	0.9
MicroMist	0.2	570	0.8
Q-DIHEN	0.06	1680	n.d.
Aridus	0.1	2340	0.2
USN[a]	2.0	540	0.07
DS-5	0.007	2418	1600

[a] USN with desolvator; [b] nanovolume flow injection (sample loop: 54 nl)

5.1.7 Hydride Generation and Cold Vapour Technique

Hydride generation for analytical use was introduced at the end of the 1960s using arsine formation (Marshal Reaction) in flame atomic absorption spectrometry (FAAS). A simple experimental setup for a hydride generator is shown in Figure 5.18. Today, hydride generation,[91,92] which is the most widely utilized gas phase sample introduction system in ICP-MS, has been developed into a routine method for the volatile hydride forming elements As, Se, Te, Bi, Sb, Ge, Pb and Sn. The gaseous hydrides (AsH_3, H_2Se, H_2Te, BiH_3, SbH_3, GeH_4, PbH_4, and SnH_4) formed in a continuous flow reaction separation system in the presence of a reducing agent (e.g., $NaBH_4$) in acid media, are introduced into the ICP source of an ICP-MS. The chemical reaction in the hydride generator, the separation of volatile products from the liquid phase and the transport of the gaseous species to the excitation or ion source can be performed for liquids in different devices using a continuous flow system or flow injection, e.g., in a U-shaped gas–liquid phase separator or in a spray chamber with a cross flow or MicroMist nebulizer (the chemical reaction occurs inside the spray chamber). As a result of the separation and pre-concentration of analyte from sample matrix (e.g., aqueous solutions) and interfering elements by the reaction of hydride forming elements with $NaBH_4$ in acid media, an increase in element sensitivity and improvement in detection limits has been observed for different applications on liquids.[92] The most frequently studied elements are As and Se and their species in river, sea, ground, mineral and tap water samples,[93,94] and they are analyzed without any sample pretreatment.

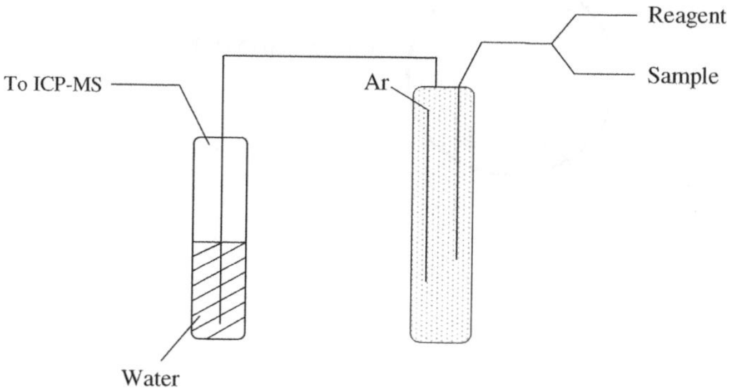

Figure 5.18 *Arrangement of a hydride generator.*

In analogy to sample introduction by hydride generation, it is also possible to perform mercury trace and isotope analysis by reducing Hg compounds to the metal using the cold vapour technique or the determination of iodine at the ultratrace level (after oxidation with 70 % perchloric acid of iodide to iodine) via the gas phase.

5.1.8 Flow Injection and Hyphenated Techniques

A rapidly growing field of trace element determination involves the analysis of microlitre or nanolitre volumes of sample, which is of particular concern when only a sample amount of few milligrams or less is available.[73] Microanalytical techniques coupled with increasingly sensitive ICP-MS, especially flow injection for solution introduction in inductively coupled plasma, offer a unique tool for analyzing very small sample amounts, e.g., for the determination of radionuclides and their isotope ratios in nuclear materials, radioactive waste or clinical samples.[70,74,75,95] Sample introduction into ICP-MS using several types of micronebulizer operating in the continuous mode at low uptake rates (10 to 100 $\mu l\,min^{-1}$), such as the MicroMist (Glass Expansion), microconcentric nebulizer with desolvator (Aridus, CETAC)[96,97] or the direct injection high efficiency nebulizer (DIHEN, Meinhard) can also be used in the flow injection mode.[64] To handle several microlitres of an aqueous solution, a commercial HPLC injection valve has been coupled with a commercial micronebulizer.[80] The small volume of analyte solution is introduced into a continuous flow of 2 % nitric acid according the flow injection principle and transient signals of analytes are measured.[12,66,71,98]

A further miniaturization of solution introduction into ICP-MS allows the analysis of nanolitre sample volumes so that 100 % transport efficiency is achieved and dead volume is minimized using nanovolume (≈ 50 nl) flow injection together with a total consumption nebulizer operating at uptake rates significantly below the $10\,\mu l\,min^{-1}$ level. The microflow nebulizer (DS 5 developed

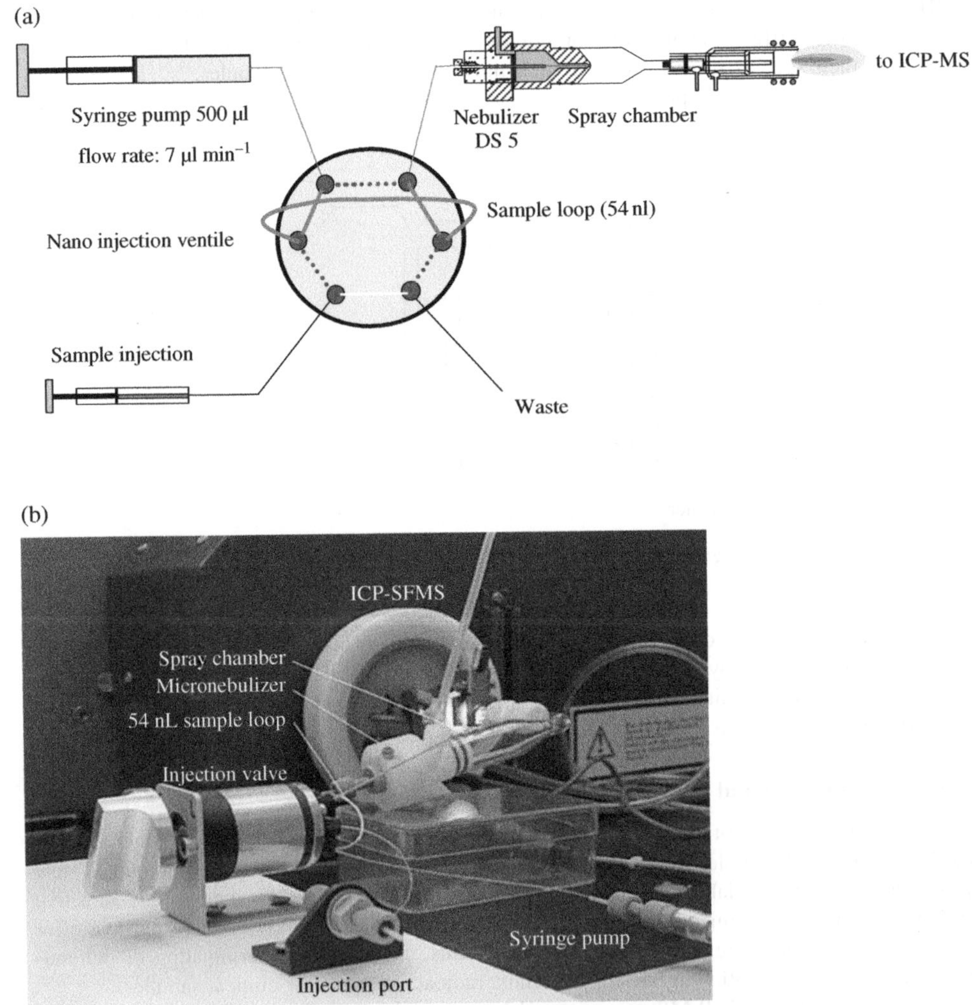

Figure 5.19 Experimental arrangement of nanoflow injection in ICP-MS using the microflow nebulizer DS 5 with spray chamber together with a nanoinjection ventile a) schematic and (b) photograph. (J. S. Becker, Trends in Anal. Chem., **24**, 243 (2005). Reproduced by permission of Elsevier.)

by Schaumloeffel[87]) inserted in the nanoflow injection arrangement is shown in Figure 5.19 a (schematic) and Figure 5.19 b (photograph). The sample solution is injected by means of a nanovolume injector into a continuous flow of carrier liquid at 7 μl min^{-1} prior to ICP-SFMS for the ultratrace determination of uranium and plutonium.

Besides the determination of total elemental concentration in liquids, which is relevant for numerous analytical issues, there is a growing awareness that the chemical form in which trace elements occur in environmental samples or in living organisms should be known. Because the toxicity of elements depends on their chemical form, the excess or the deficiency of some metals (e.g., Cr, Fe, Co, Cu, Zn, Mo, Se, Mn, which are essential elements) is of consequence for living organisms. Speciation analysis has therefore become an important topic in present day analytical research in the life sciences and environmental research.[99–102] The experimental setups of hyphenated ICP-MS techniques for speciation analysis such as CE-ICP-MS[103] (see Figure 5.20) and HPLC-ICP-MS[104] combined with the possibility of on line isotope dilution analysis are illustrated in Figure 5.21. The details of experimental arrangements and a multitude of applications are described in several textbooks, reviews in original papers.[100,101,103,105–107] Trends in speciation

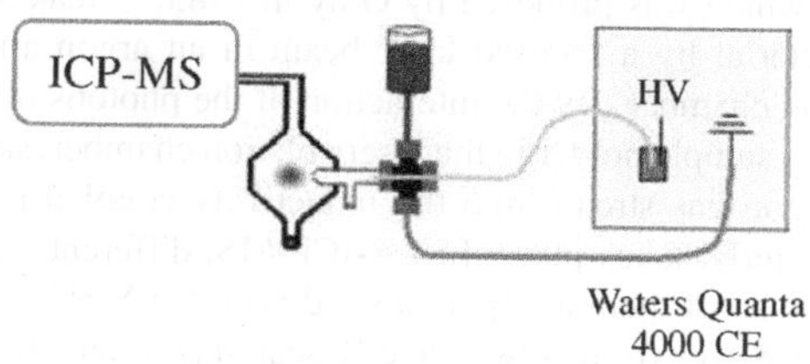

Figure 5.20 *Schematic of hyphenated ICP mass spectrometric techniques by coupling a capillary electrophoresis system (CE) to ICP-MS. (J. S. Becker and H. J. Dietze, Int. J. Mass Spectrom. In Proc.* ***197****, 1 (2000). Reproduced by permission of Elsevier.)*

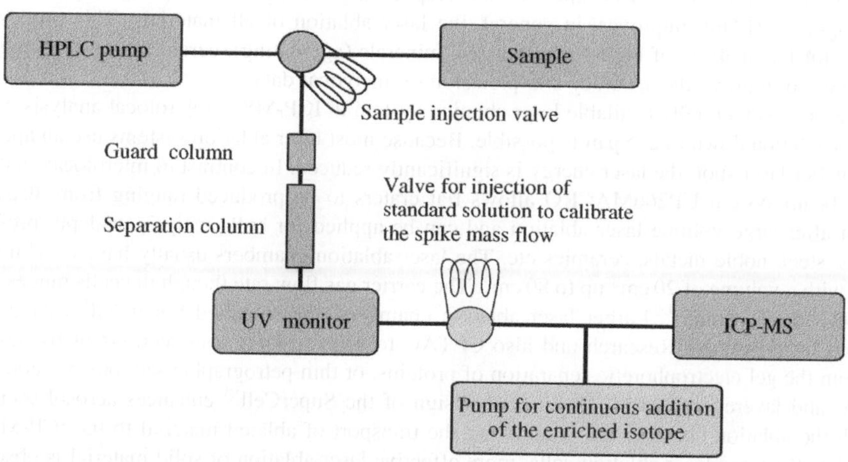

Figure 5.21 *Schematic of hyphenated ICP mass spectrometric techniques: HPLC-ICP-MS for the application of isotope dilution techniques. (K. G. Heumann, L. Rottmann, J. Vogl, J. Anal. At. Spectrom.,* ***12****, 1381 (1994). Reproduced by permission of The Royal Society of Chemistry.)*

using ICP-MS with different on line separation techniques (HPLC, CE, GC, SPME – solid phase microextraction, SFC – supercritical fluid extraction, SEC – size exclusion chromatography and others) are reviewed in.[72,99–101,107–111]

Field flow fractionation (FFF), as a gentle size fractionation coupled to ICP-MS, offers the capability to determine trace metals bound to various size fractions of colloidial and particulate materials.[112] On line coupling of FFF with ICP-MS was first proposed by Beckett in 1991.[113] Separation is achieved by the balance between the field force and macromolecular diffusion in the FFF channel. Depending on the field force used, FFF is classified into different techniques such as sedimentation, gravitational, electrical, thermal and flow FFF.[112]

5.1.9 Laser Ablation Inductively Coupled Plasma Mass Spectrometers (LA-ICP-MS)

LA-ICP mass spectrometry is becoming the method of the choice for trace and isotope analysis of solid samples and is already the most important laser induced technique in inorganic mass spectrometry due to the advantage of direct solid sampling by focused laser irradiation on the sample surface and its ability to provide microscale information.

Laser ablation ICP-MS, which was proposed by Gray in 1985,[114] makes use of the evaporation of mostly solid sample material by a focused laser beam in an argon atmosphere under normal pressure in the laser ablation chamber. By the interaction of the photons of the laser beam with the sample surface mounted on a sample holder in the laser ablation chamber, sample material is ablated and transported with an argon gas stream into the inductively coupled plasma ion source where the ionization of ablated material takes place. In LA-ICP-MS, different commercial laser ablation systems have been used – most of them applying a solid-state Nd:YAG laser – e.g., from CETAC Technologies, Ohama NE, USA or from New Wave Research (former Merchantek), Fremont, CA. For example, the laser ablation systems CETAC LSX 200 and LSX 500 use the 266 nm wavelength with nanosecond pulse, the LSX 213 is optimized for ablation at 213 nm, whereas the LSX 3000 allows measurements to be made at 266, 213 and 193 nm. The laser ablation systems Merchantec LUV 266 nm laser microprobe or UP 213 and UP 266 from New Wave are applied in LA-ICP-MS at a wavelengths of 266 nm and 213 nm, respectively. A higher absorption of photons at a wavelength of 213 nm improves, in general, the laser ablation of all materials. This property is relevant for the analysis of high purity glasses, minerals (age dating), ceramics and high melting metals etc. to improve the accuracy and precision of analytical data.

Using the commercially available laser ablation system in ICP-MS, a microlocal analysis with a spatial resolution down to 2–5 μm is possible. Because most laser ablation systems use an aperture for a smaller laser spot, the laser energy is significantly reduced. In contrast to microlocal analysis, a large beam system UP266MACRO allows flat craters to be produced ranging from 30 μm to 750 μm after large volume laser ablation and can be applied for bulk analysis or depth profiling in glass, steel, noble metals, ceramics etc. The laser ablation chambers usually have a cylindrical design with a volume of 20 cm^3 up to 80 cm^3. The carrier gas flow rate though the cells ranges from 0.5 $l\,min^{-1}$ to 1.5 $l\,min^{-1}$. Larger laser ablation chambers (the so-called SuperCell™) have been designed by NewWave Research and also CETAC to analyze large thin sections of tissues, 2D gels from the gel electrophoretic separation of proteins, or thin petrographic sections of geological samples and layered materials. The special design of the SuperCell™ enhances aerosol transport through the ablation chamber thus improving the transport of ablated material to the ICP-MS. In contrast to the large laser ablation cells, more effective laser ablation of solid material is observed by reducing the cell volume in several cell designs.[115,116]

Furthermore, a few commercial excimer laser ablation systems are available on the analytical market, for the analysis of highest purity glasses, diamonds and highly transparent materials, such as New Wave UP193HE, (Homogenized Excimer) from NewWave or the GeoLasArF from MicroLas, Göttingen, Germany. In addition, by using an excimer laser in LA-ICP-MS, higher energy densities (up to $35\,\mathrm{J\,cm^{-2}}$ compared to $18\,\mathrm{J\,cm^{-2}}$ for the LSX 500 at 266 nm), flat laser craters (~ 10 nm depth) are possible and fewer fractionation effects are observed during laser ablation. In comparison to excimer laser systems, solid laser ablation systems (Nd:YAG lasers) are easier to maintain, cheaper and smaller. Any laser ablation system can be coupled with quadrupole ICP-MS (with and without collision cell),[5,67,117–120] double-focusing sector field ICP-MS[24,121–123] with single and multiple ion collection or ICP-ToF-MS.[124,125] To improve the ablation of solid samples (with respect to sensitivity), femtosecond lasers are employed for various applications. Enhanced sensitivities, and thus increased signal-to-noise ratios and lower detection limits, better stability of ion signals, fewer fractionation effects, i.e., more stoichiometric ablation, are achieved by the laser ablation of solid material with a pulse length in the femtosecond range – working mostly at higher repetition rates (one to several kHz) – compared to nanosecond laser ablation at low repetition rates (10–20 Hz).

The experimental arrangement of LA-ICP-MS using a powerful laser ablation system coupled to a double-focusing sector field ICP-MS (Element, Thermo Fisher Scientific, Bremen) is shown in Figure 5.22. A cooled laser ablation chamber using two Peltier elements in serial connection

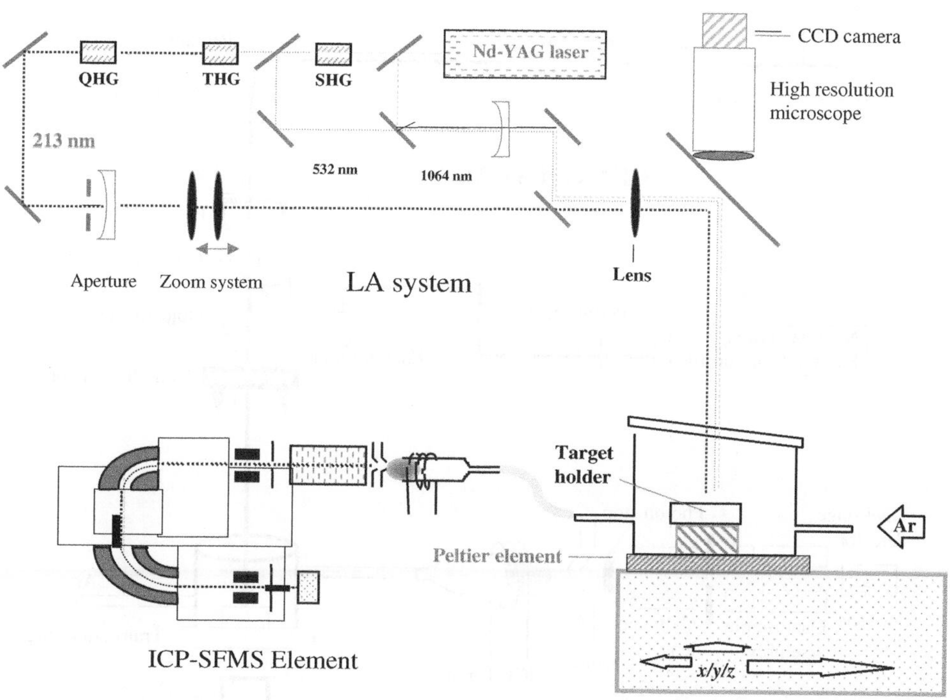

Figure 5.22 Schematic of laser ablation inductively coupled plasma mass spectrometer: a laser ablation system (wavelength 213 nm) is coupled to a double-focusing sector field ICP-MS (Element, Thermo Fisher Scientific, Bremen). The cooled laser ablation chamber using two Peltier elements was developed in the author's laboratory.[126] (M. V. Zoriy, M. Kayser, A. Izmer, C. Pickhardt and J. S. Becker, Int. J. Mass Spectrom., **242***, 297 (2005). Reproduced by permission of Elsevier.)*

under the target holder made of aluminium has been developed in the author's laboratory. This experimental arrangement of LA-ICP-MS with a cooled ablation chamber (see Figure 5.22) is advantageous for quantitative element mapping (element distribution) in thin sections of brain tissue.[126] Pre-investigations indicate that, in comparison to a non-cooled chamber, this setup with a cooled ablation chamber results in an improvement in precision and accuracy (up to one order of magnitude) of uranium isotope ratio measurements on one drop of certified isotope standard reference material deposited on a biological surface.[126] The cooled laser ablation chamber in LA-ICP-MS has been applied for imaging copper, zinc and other elements in thin sections of human brain samples (hippocampus) and of rat brain tumours, as discussed by Becker *et al.*[127,128] Another approach using a cryostatic laser ablation chamber for trace analysis on ice cores is described by Reinhardt *et al.*[129]

A sensitive sector field ICP mass spectrometer or quadrupole based ICP-MS with a laser ablation system has been applied for screening of two dimensional (2D) gels with respect to phosphoproteins and metal containing proteins.[27,121,122,130] The LA-ICP mass spectrometer using a quadrupole based ICP-MS (Elan 9000, Perkin Elmer Sciex) together with a commercial Nd:YAG laser ablation system at a wavelength of 213 nm (e.g., LSX 213, CETAC Technologies Inc., Ohama) is illustrated in Figure 5.23. A similar arrangement has been utilized for numerous applications as described in Chapter 9.[69,117–121,131]

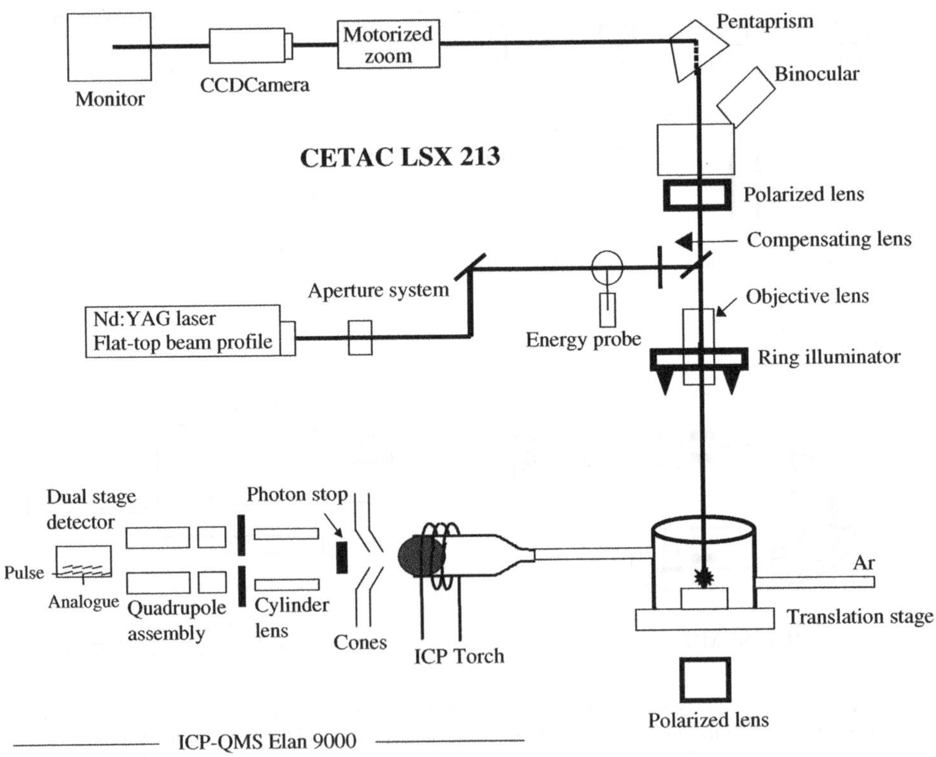

Figure 5.23 *LA-ICP-MS by combining a quadrupole based ICP-MS with a CETAC LSX 213 laser ablation system.*

The application of a laser in mass spectrometers offers the possibility of a direct microlocal analysis with a lateral resolution down to the μm range. The capability of near-field LA-ICP-MS (NF-LA-ICP-MS)[65] created in the author's laboratory provides local information on inhomogeneous element distribution on solid sample surfaces with a spatial resolution in the sub-μm range.

LA-ICP-MS will develop into a dominant solid-state analytical technique in the sensitive bulk analysis of homogeneous non-conducting materials (insulators, ceramics and glasses) in the future. The state-of-the-art, challenges and limitations with respect to isotope ratio measurements by LA-ICP-MS using different instrumentations have been reviewed in reference [131].

5.2 Spark Source Mass Spectrometers (SSMS)

Due to the relatively high initial energy of atomic ions formed in spark plasmas of several keV, spark ion sources are applied together with double-focusing mass spectrometers in a combination of electric sector and magnetic sector fields with Mattauch–Herzog geometry. This experimental arrangement (described in Section 3.1.3.) is advantageous for the sensitive and quasi-simultaneous detection of nearly all elements in one mass spectrum. The experimental arrangement of a double-focusing Mattauch–Herzog instrument with interchangeable spark and laser ion source developed by Dietze and Becker[71,72] is shown in Figure 5.24. This powerful SSMS/LIMS instrument equipped with photographic ion detection has been used for multi-element trace analysis on metals, high-purity materials, geological and biological samples for nearly three decades in the laboratories in Leipzig and Jülich.[132–136] In addition, fundamental studies have been performed on ion formation processes (such as cluster ion formation) in laser and spark plasmas.

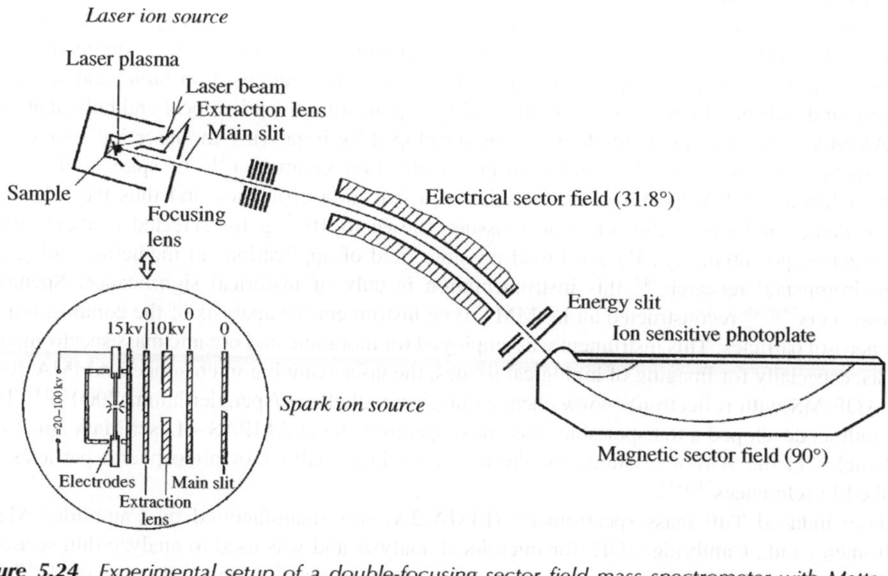

Figure 5.24 *Experimental setup of a double-focusing sector field mass spectrometer with Mattauch–Herzog geometry with interchangeable spark and laser ion source. (J. S. Becker and H. J. Dietze, Int. J. Mass Spectrom. Ion Proc.* ***197****, 1 (2000). Reproduced by permission of Elsevier.)*

A further development in SSMS was achieved by Jochum *et al.*[137,138] in 1997 using the multi-ion counting (MC) technique with 20 separate channeltrons (see Section 4.5). SSMS was for many decades one of the dominant mass spectrometric methods in the trace analysis of compact and powdered high purity metals, semiconductors, ceramics and especially of geological samples. There are only a few laboratories worldwide which still work with old spark source mass spectrometers constructed in the 1960s and 1970s. Due to the high experimental expense and time consuming analysis involved in SSMS and due to the great experimental effort in other mass spectrometric techniques (such as GDMS, ICP-MS and LA-ICP-MS), no mass spectrometers with a spark ion source are currently being constructed and they are not available on the analytical market at present.

5.3 Laser Ionization Mass Spectrometers (LIMS)

LIMS is based on the ionization of evaporated and atomized sample material in laser microplasma under high vacuum conditions (see Figure 2.7). The ions formed in the laser plasma possess a relatively broad kinetic energy spread as a function of the laser power density and laser pulse length applied. Therefore double-focusing mass spectrometers with Mattauch–Herzog geometry are frequently used in LIMS.[136,139] The instrumentation in LIMS was developed by Dietze and co-workers from a double-focusing spark source mass spectrometer by constructing a replaceable ion source as demonstrated in Figure 5.24.[135,136,140,141] By reducing the pulse length and spot size of the laser beam it was possible to improve LIMS with respect to microlocal and trace analysis. LIMS was implemented in the mid-seventies of the last century as a laser microprobe analyzer (LAMMA-500 from the former Leybold-Heraeus, Cologne, Germany) using a time-of-flight (ToF) mass spectrometer with a reflectron allowing high ion transmission and high detection power. In this instrument, a light microscope system was helpful for further focusing of the laser beam down to a diffraction limited spot size of about 0.5 μm (at a wavelength of 266 nm).[142] Because in LAMMA-500 the transmission mode (see Figure 5.25) was utilized, its application was limited to thin microscopic samples, such as histological cross sections, thin biological sections, thin films and foils etc. In order to extend the field of application for microlocal and bulk analysis, the LAMMA-1000 from Leybold-Heraeus was developed by improving the laser ion source and ion extraction directly from the sample surface (reflection geometry).[143] In spite of the most important feature of LAMMA instruments, the high detection efficiency and thus the very low absolute detection limit in thin biological tissues (down to 10^{-20} g for selected elements with low ionization potential, e.g., K) and broad possible field of applications in medicine, biological and environmental research,[144] this instrumentation is only of historical significance. Spengler and coworkers[145,146] reconstructed an LAMMA-type instrument for analysis of the composition of single aerosol particles. This instrument was employed for inorganic and organic mass spectrometric analysis, especially for imaging of biological tissues, the laser scanning microscope LAMMA 2000 (laser TOF-MS with reflectron) (www.chemie.uni-giessen.de/home/spengler/lamma2000/).[147] The same authors developed a transportable laser mass spectrometer (LAMPAS – Laser Mass Analyzer for Particles in the Airborne State) for the direct on line analysis of atmospheric particles as described in references[148,149].

A laser-induced ToF mass spectrometer (LIMA-2A) was manufactured by Cambridge Mass Spectrometry Ltd., Cambridge, UK, for microlocal analysis and was used to analyze thin sections of biological samples in the transmission mode or bulk material in the reflection mode.[150,151] Typical LIMA applications in microelectronics include identification of impurities in dielectrics, microlocal analysis, depth profiling, thick film analysis and investigations on hybrid circuits.

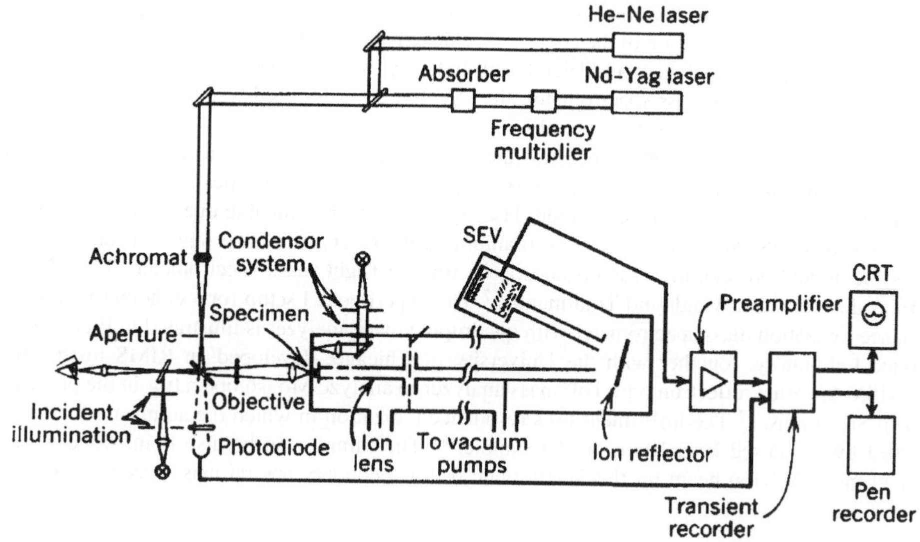

Figure 5.25 Schematic arrangement of the LAMMA-500 with time-of-flight analyzer and reflectron using the transmission mode. (J. S. Becker and H. J. Dietze, Fresenius. J. Anal. Chem. **344**, 69 (1992). Produced by permission of Springer Sciences and Business Media.)

In order to improve the routine capability of laser ionization mass spectrometry, the laser ion source has been equipped with a carousel carrier for a maximum of eight samples. Both ToF-MS with laser ion source (LAMMA and LIMA) were, in the eighties, valuable tools for the qualitative characterization of microsamples and microlocal analysis. Due to serious problems with the quantification of analytical data and high experimental effort, LAMMA and LIMA instruments have not been commercially available for a number of years. Obviously the application of a high vacuum ion source in LIMS is a great advantage, e.g., for reducing the formation of disturbing polyatomic ions, but these mass spectrometric trace analytical techniques are rarely used anywhere in the world due the very expensive experimental arrangement, time consuming evaluation of data (e.g., if ion detection via ion sensitive photoplates is utilized) and quantification problems.

Laser ionization mass spectrometry using double-focusing Mattauch–Herzog type instruments with photoplate ion detection was realized commercially in the Russian EMAL 2 in 1980.[152] The experimental details, features and some applications of this static laser ionization mass spectrometer are discussed by Basova.[153] A pulsed laser ion source was also coupled with a Fourier transform mass spectrometer in 1989 featuring remarkable mass resolution of up to 400 000.[154] Application of a single KrF excimer laser shot at a wavelength of 249 nm on a ceramic sample has allowed the separation of charged $^{88}Sr^{2+}$ ions from single charge atomic ions $^{44}Ca^{+}$.

LIMS has been replaced in recent years by the more powerful and sensitive LA-ICP-MS.

5.4 Resonance Ionization Mass Spectrometers (RIMS)

In a resonance ionization mass spectrometer,[155–158] one or more lasers are applied which are tuned precisely to the wavelength required for optical resonance excitation from the atomic ground state into the excited state up to the final ionization state of evaporated atoms. Consequently, highly sensitive and selective ionization of the analyte is achieved and the problem of isobaric interferences in mass spectrometry, which can be difficult with ICP-MS, SIMS, GDMS or TIMS, is eliminated. In a resonance ionization mass spectrometer, the selective laser induced resonance ionization (see Figure 2.22 in Section 2.3) is usually combined with a quadrupole analyzer, magnetic sector field or ToF mass analyzer in order to measure the isotopic composition of elements.[157,159,160] Due to the pulsed nature of the ionization process, ToF mass analyzers are ideal.[157,161] For selective excitation and ionization of atoms, different laser systems (such as tunable due laser systems) can be applied in RIMS. Solid-state lasers on titanium–sapphire crystals have replaced tuneable dye laser systems and are coupled, for instance, to a time-of-flight mass spectrometer (ToF-MS) in RIMS as described by Wendt and Trautmann.[157] The experimental setup for a coherent multi-step resonance ionization mass spectrometer with quadrupole mass analyzer is illustrated in Figure 5.26. Argonne Laboratory, together with the University of Chicago, developed an RIMS instrument, CHARISMA, with a reflectron type ToF mass analyzer to analyze Mo isotope ratios in the presence of Zr in SiC grains.[162] The instrument uses resonance ionization in which the atoms are thermally desorbed by a focused laser beam on the sample.[163] The sample can be also bombarded with a 5 keV primary Ar^+ ion beam for the SIMS mode or for secondary neutral mass spectrometry.

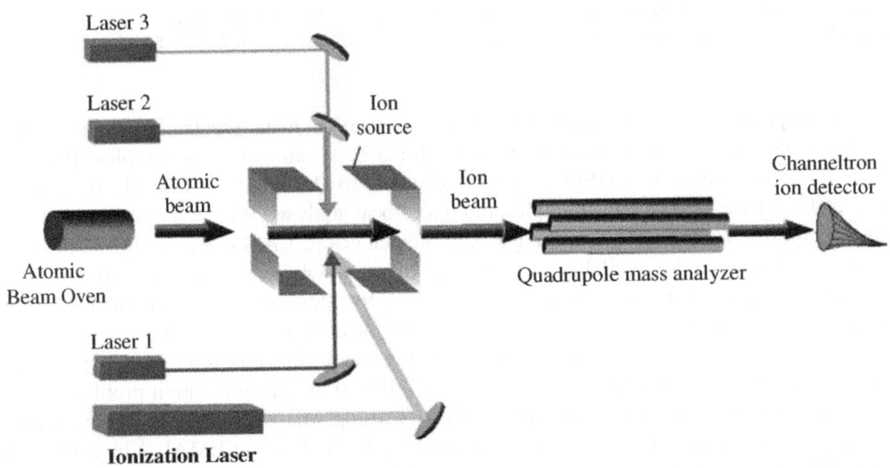

Figure 5.26 Experimental setup of a resonance ionization mass spectrometer (RIMS). (K. Wendt and N. Trautmann, Int. J. Mass Spectrom., **242***, 161 (2005). Reproduced by permission of Elsevier.)*

In spite of the excellent capability and advantages (high selectivity and sensitivity) of RIMS for the ultratrace analysis of isotopes with naturally rare abundance in environmental, geological, medical and nuclear samples, no commercial instrumentation is available to date. In contrast to AMS and RIMS as mono-elemental (element-specific) analytical techniques, ICP-MS and LA-ICP-MS possess, in analogy to GDMS and SIMS, have the ability for multi-element analysis and thus could have the widest fields of application.

5.5 Glow Discharge Mass Spectrometers (GDMS)

At present GDMS[164,165] is one of the most powerful solid-state analytical methods for the direct determination of trace impurities and depth profiling of solids. The positively charged ions formed in the low pressure argon plasma of the glow discharge are extracted and accelerated into the double-focusing sector field mass spectrometer, quadrupole, ion trap or ToF mass analyzer.

One of the commercial glow discharge mass spectrometers with a direct current (dc) GD ion source is the double-focusing sector field instrument with Nier–Johnson geometry, VG-9000 (VG-Elemental, Thermo Instruments), with a maximum mass resolution ($m/\Delta m$) of 10 000. Figure 5.27 shows the schematic setup of the GDMS VG 9000. The capability of the glow discharge mass spectrometer VG 9000, which allows the determination of trace elements in electrically conducting samples, but also in non-conducting materials (e.g., nuclear samples after mixing the sample with a conducting binder or using the secondary cathode technique) with detection limits at the ng g^{-1} concentration range and below and with a reproducibility of about $\pm 10\,\%$ has been demonstrated in different groups.[166–170] This GDMS was available for many decades on the analytical market but is no longer produced.

A GD mass spectrometer from Thermo Fisher Scientific with a direct current (dc) glow discharge ion source based on the mass spectrometric arrangement of the Element (Element GD) has been available on the analytical market since 2005 for sensitive multi-element analysis of trace impurities in conducting samples. The experimental arrangement of the dc GD ion source (Grimm type) and

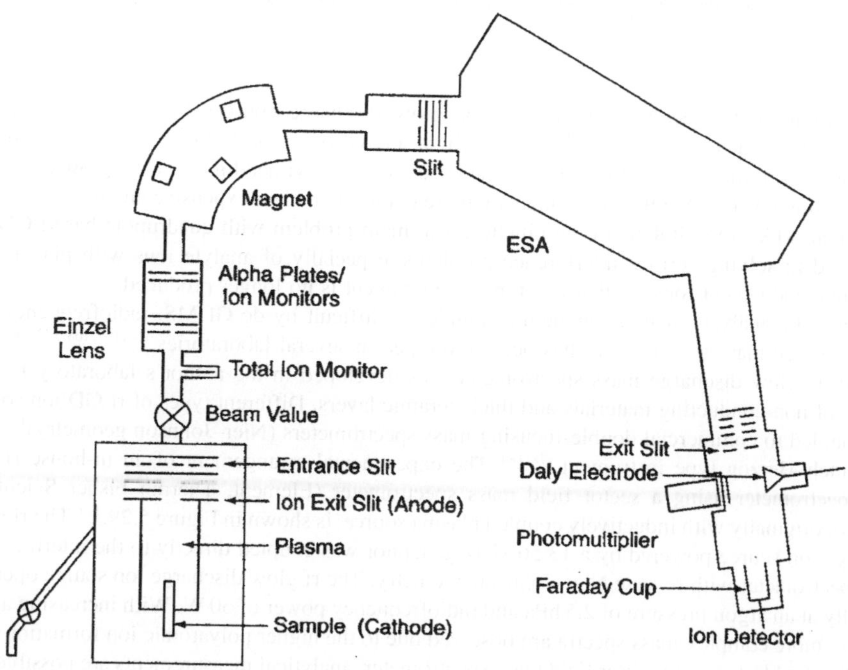

Figure 5.27 *Schematic diagram of direct current (dc) glow discharge mass spectrometer (VG 9000). (Reproduced by permission of GV Instruments Ltd.)*

a photograph is shown in Figure 5.28 a and b, respectively. This GDMS instrumentation enables flat conducting samples to be analyzed, but does not allow any microlocal analysis with a spatial resolution.

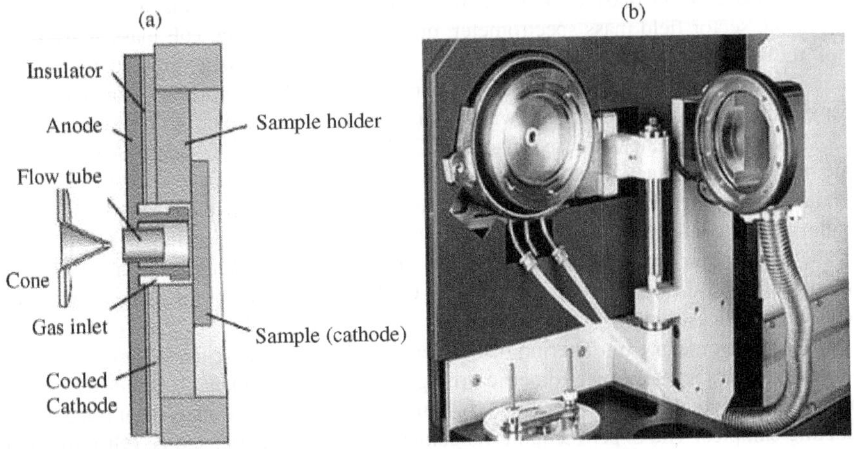

Figure 5.28 *Direct current glow discharge ion source: a) schematic, b) photograph of Element GD (Thermo Fisher Scientific, Bremen, German), Reproduced by permission of Thermo Fisher Scientific, Bremen.)*

In addition to the sector field instruments, glow discharge sources have been coupled to quadrupole mass spectrometers,[171,172] time-of-flight analyzers,[173] ion traps[174,175] or Fourier transform mass spectrometers.[13,176] Quadrupole mass spectrometers with a direct current glow discharge source and interchangeable ICP ion source are more compact and less expensive and were produced by Finnigan MAT (TS-SOLA) in the nineties. The main problem with quadrupole based GDMS is involved in solving serious interference problems, especially of analyte ions with plasma gas polyatomic and cluster ions, so that this type of instrument is no longer produced.

Because the analysis of non-conducting samples is difficult by dc GDMS, radiofrequency (rf) glow discharge mass spectrometry has been developed in several laboratories.[164,167,168,177–183] For example, rf glow discharge mass spectrometry was developed in the author's laboratory for the analysis of non-conducting materials and thick ceramic layers. Different types of rf GD ion source were coupled to commercial double-focusing mass spectrometers (Nier–Johnson geometry)[177] and a Mattauch–Herzog type instrument.[178,184] The experimental arrangement of an in-house rf GD mass spectrometer using a sector field mass spectrometer (Element, Thermo Fisher Scientific, Bremen), originally with inductively coupled plasma source, is shown in Figure 5.29.[177] The rf glow discharge ion source powered by a 13.56 MHz generator was coupled directly to the interface of a mass spectrometer with reverse Nier–Johnson geometry. The rf glow discharge ion source operates optimally at an argon pressure of 2.5 hPa and radiofrequency power of 30 W. With increasing argon pressure, more complex mass spectra are observed due to the higher polyatomic ion formation rate. Using the double-focusing sector field mass spectrometer, analytical measurements are possible at a mass resolution of $m/\Delta m = 300$, 3000 and 10 000. That means a multitude of possible interferences from analyte and polyatomic ions which are observed in rf GDMS can be separated at the required mass resolution. The detection limit of rf GDMS (in the low resolution mode ($m/\Delta m = 300$))

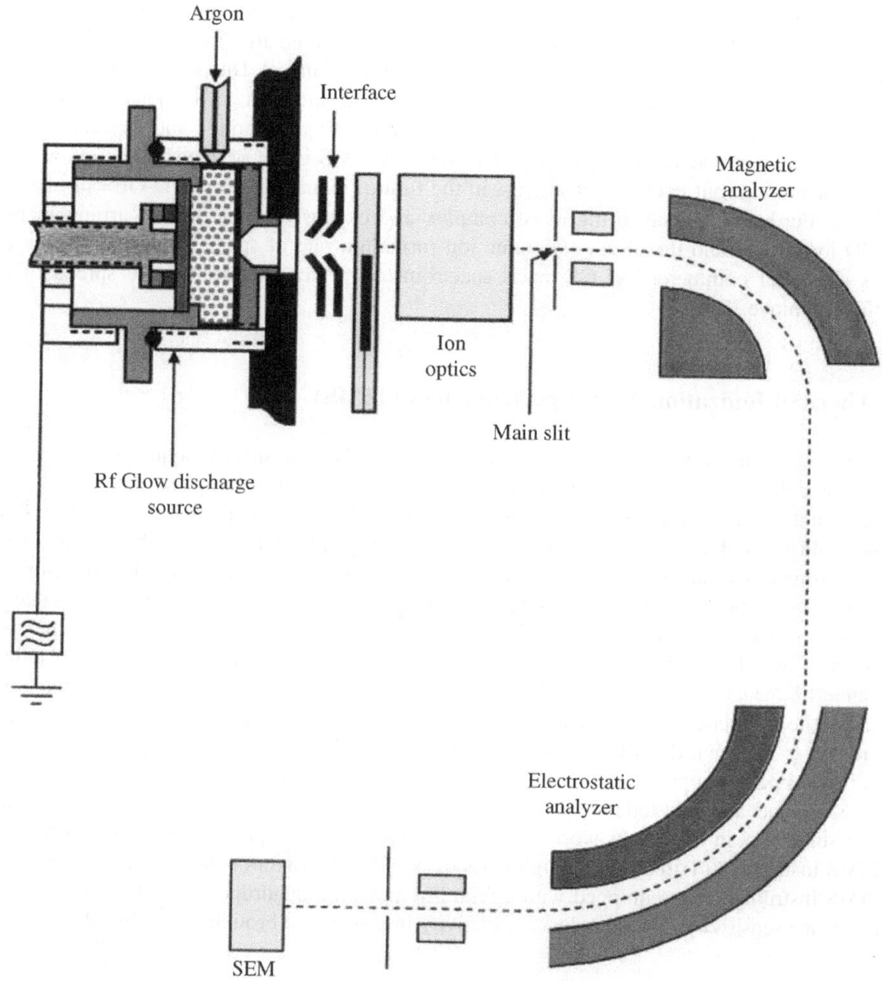

Figure 5.29 *Instrumental setup of a radiofrequency (rf) glow discharge mass spectrometer by combining an ion source developed in house with a commercial double-focusing sector field mass spectrometer with reverse Nier–Johnson geometry (Element, Thermo Fisher Scientific, Bremen, Germany). (J. S. Becker et al., Int J. Mass Spectrom.,* ***164****, 81 (1997). Produced by permission of Elsevier.)*

was determined, e.g., for B in high ohmic GaAs, as $10\,ng\,g^{-1}$. The detection limits in rf GDMS developed at Research Centre Juelich using the Mattauch–Herzog instrument (SM-1B, Varian MAT, Germany) with photoplate and electrical ion detection, which was used for more systematic investigations of fundamental processes in an rf glow discharge, were determined on different non-conducting ceramics to be in the low $\mu g\,g^{-1}$ range. Unfortunately, serious contamination problems were observed in the mass spectrometer with a glow discharge ion source due to the significantly higher ion beam current produced compared to ICP-MS and even to LA-ICP-MS, and consequently strong memory effects occurred. As yet, ultratrace analysis of trace elements (e.g., Ga and As) in aqueous solution by ICP-MS after GDMS of GaAs is only possible after time consuming cleaning procedures (including the ion optical lens system behind the skimmer). Due to the fast development of LA-ICP-MS with significantly higher sensitivity and the fact that lower detection limits for trace elements in the multi-element analysis of non-conducting materials, and especially in the higher mass range, are achievable with LA-ICP-MS, rf GDMS is not of significance in analytical chemistry at present, but could be of interest in the future (in analogy to rf GD-OES in materials science[185]). Furthermore, due to the more complex and expensive experimental arrangement of an rf GD ion source and the high polyatomic ion formation rate of species with the plasma gas, there is a lack of commercial rf GD mass spectrometers which has limited the spread of this powerful technique.

5.6 Thermal Ionization Mass Spectrometers (TIMS)

At present, two magnetic sector field instruments with thermal surface ionization source and multiple ion collectors (Triton, Thermo Fisher Scientific, Bremen, Germany and the IsoProbe-TM, VG Elemental, Manchester, UK) are commercially available on the analytical market. The Triton and the IsoProbe-TM are developments of the former multiple ion collector thermal ionization mass spectrometers Finnigan MAT 262 and the VG Isomass 354, respectively. The development of the ion source optics in the Triton compared to the MC-TIMS MAT 262 to achieve higher ion transmission in the single and double filament technique results in the sensitivity being increased by a factor of two. The thermal ionization source is equipped with a sample turret holding 21 single and double filaments. The ions formed by thermal surface ionization are accelerated in the Triton at 10 kV in the magnetic sector field analyzer. Due to the relatively low initial energy spread of the thermal ions ($\sim 0.5\,eV$), the MC-TIMS instruments only utilize a laminated magnetic sector field for high speed peak jumping and low hysteresis for the mass separation of ion beams. Both MC-TIMS instruments are used today as multiple ion collector mass spectrometers for precise isotope ratio measurements in many labs worldwide. A schematic of the experimental arrangement of the MC-TIMS instrument (Triton) with a high mass dispersion of 818 mm is illustrated in Figure 5.30. MC-TIMS instruments are equipped with a retarding potential quadrupole (RPQ) lens to increase the abundance sensitivity. The abundance sensitivity improvement brought about by RPQ in TIMS

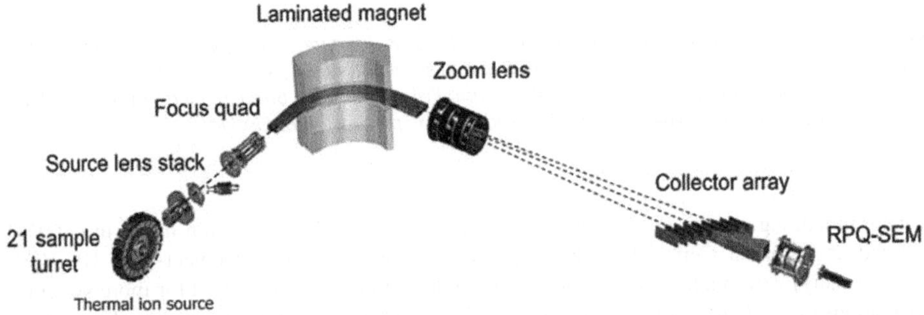

Figure 5.30 Experimental setup of multiple collector thermal ionization mass spectrometer: a) MC-TIMS Triton, Thermo Fisher Scientific, Bremen, Germany. (Reproduced by permission of Thermo Fisher Scientific Bremen.)

(Triton) is roughly a factor of ten better, in comparison to MC-ICP-MS (Neptune), due to the smaller energy spread of the ions formed in thermal surface ionization.[58] By use of a special ion lens deceleration system in MC-TIMS, an upper limit for $^{236}U/^{238}U$ isotope ratios of 6×10^{-10} was found.[186] In contrast, the older VG ISOLAB-120 mass spectrometer with a thermal ionization source, the first electrostatic analyzer (ESA) and a laminated magnet was designed for precise and accurate $^{9}Be/^{10}Be$ isotope ratio measurements. A second 30 cm radius electrostatic analyzer is placed in the front of the entrance to the ion counting detector and a fixed potential difference is applied to the two curved plates of the ESA such that only ions with a specific kinetic energy can emerge.[58] Such a triple sector mass spectrometer is advantageous to reduce the scattered background ions and the second ESA is employed to achieve high abundance sensitivity.

MC-TIMS measurements are relevant for high precision isotope analysis as discussed in Chapter 8.

5.7 Secondary Ion Mass Spectrometers (SIMS) and Sputtered Neutral Mass Spectrometers (SNMS)

A few SIMS and SNMS instruments for surface analysis[187–189] are commercially available on the analytical market. These are SIMS instruments using a double-focusing sector field mass spectrometer (e.g., CAMECA IMS-7f), time-of-flight secondary ion mass spectrometers (ToF-SIMS IV from CAMECA, Cedex, France, or ToF-SIMS 5, the ToF-SIMS 300 from ION-TOF, Münster, Germany and the PHI TRIFT IV from Physical Electronics, USA) and quadrupole based SIMS (SIMS 4550 and 4600 CAMECA, Cedex, France) or the quadrupole based SNMS instruments with SIMS option (INA-X, SPECS GmbH, Berlin, Germany).

The experimental arrangement of a secondary ion mass spectrometer with (forward) Nier–Johnson geometry (CAMECA IMS-7f from CAMECA, France) is shown in Figure 5.31. The CAMECA IMS-7f, developed from the CAMECA series (and older instruments: CAMECA IMS-6f, IMS-5f, IMS-4f and IMS-3f) is equipped with two primary ion sources: the duoplasmatron source producing positively or negatively charged oxygen ions and a caesium primary ion source for the generation of Cs^+ ions. The ion energy of primary ions is mostly chosen as several keV. The secondary ions formed in the ultrahigh vacuum ion during the sputter process are accelerated with up to 10 kV in the sector field mass spectrometer. The CAMECA IMS-7f is able to operate as a secondary ion microscope (to produce images of sputtered secondary ions from the sample surface in the raster area) or in the secondary ion microprobe mode for mass spectrometric bulk, trace and isotope analysis and depth profiling. The raster area is normally not larger than $250 \times 250\,\mu m$. The depth and lateral resolution achievable with the CAMECA IMS-7f is 2–5 nm and 50 nm, respectively. The detection limits obtained in SIMS are a function of the matrix investigated, the mass resolution applied, the isobaric interference of secondary ions, background signals and the primary ions used and varied up to the low $ng\,g^{-1}$ range. A maximum mass resolution $m/\Delta m$ of the double-focusing sector field SIMS of 10 000 can be achieved.

Compared to sector field SIMS the advantage of ToF-SIMS is the one order of magnitude higher transmission of ions (ToF-SIMS: 10^{-1}) and an increased mass range. The basic ToF secondary ion mass spectrometer (ToF-SIMS-IV) developed by ION-TOF, Münster, Germany, is equipped with three primary ion sources for sputtering of sample surface. These are a liquid metal (isotopically enriched ^{69}Ga) primary ion source, a caesium ion source used for sputter etching and analytical purposes, and an electron impact ion source for ionization of Ar, Xe, O_2 or SF_6 used as primary ion beams for sputtering of the solid surface. To compensate for charging effects during the

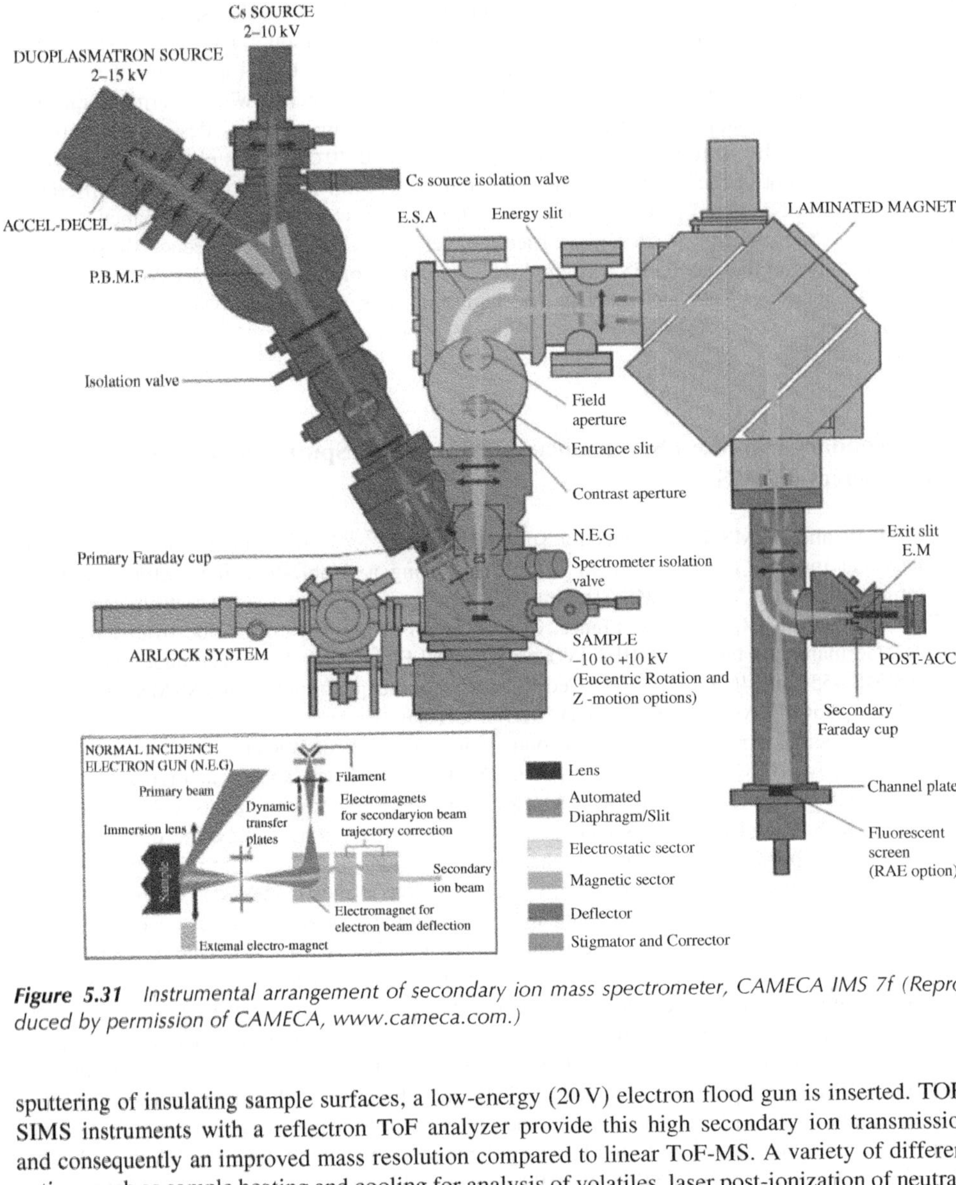

Figure 5.31 *Instrumental arrangement of secondary ion mass spectrometer, CAMECA IMS 7f (Reproduced by permission of CAMECA, www.cameca.com.)*

sputtering of insulating sample surfaces, a low-energy (20 V) electron flood gun is inserted. TOF-SIMS instruments with a reflectron ToF analyzer provide this high secondary ion transmission and consequently an improved mass resolution compared to linear ToF-MS. A variety of different options such as sample heating and cooling for analysis of volatiles, laser post-ionization of neutrals etc. can be applied in ToF-SIMS (ToF-SIMS 5 or 300). The ToF-SIMS 300 (from ION-TOF) allows wafers with a diameter of 300 mm to be analyzed and was designed for application in the semiconductor industry to detect trace metal impurities or doping elements in semiconductors with detection limits up to 10^8 atoms cm^{-2}. In addition to the depth profiling by ToF-SIMS, organic surface contaminants, sub-μm structures and defects can also be analyzed. ToF-SIMS is employed with increasing extent in a multitude of life science studies for imaging of elemental and molecular distribution in thin tissue sections (see Section 9.5.6).

Isotope ratio measurements on solid surfaces using secondary ions formed by the bombardment of the sample surface with primary ions have been performed with the double-focusing SIMS (CAMECA IMS-7f from CAMECA) or ToF-SIMS (e.g., from ION-TOF) instruments, as well. Both types of secondary ion mass spectrometers use a single ion collector (a secondary electron detector for SEM imaging). The double-focusing sector field SIMS and the ToF-SIMS provide detailed elemental and molecular information about surfaces, thin layers and interfaces of the sample, and gives full 3D analytical results. The schematic setup of the ToF-SIMS (ION-TOF) is shown in Figure 5.32. ToF-SIMS has gained widespread application in basic research and high technology industries for providing information on elemental and molecular composition of material surfaces.

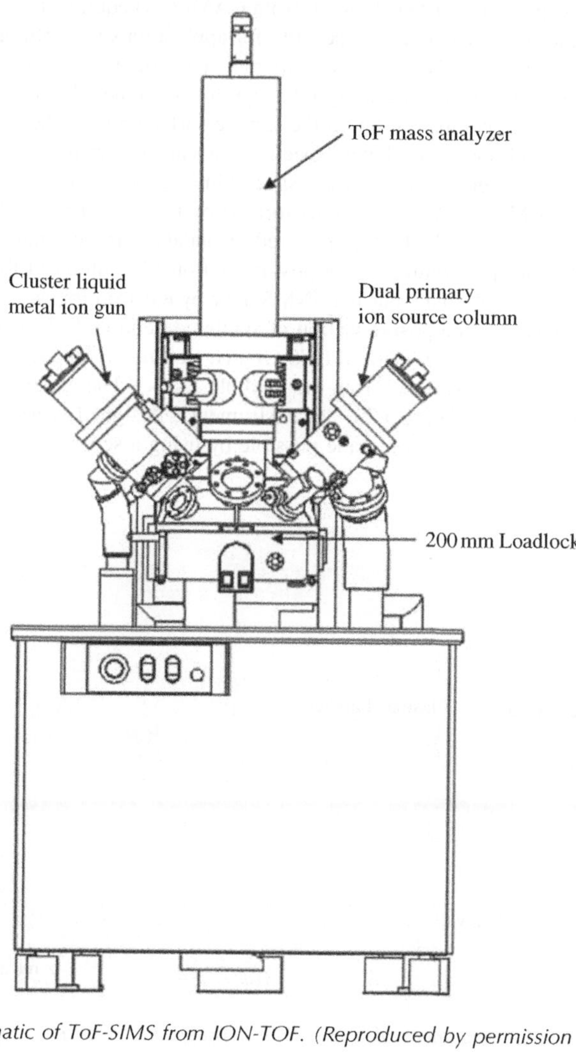

Figure 5.32 Schematic of ToF-SIMS from ION-TOF. (Reproduced by permission of ION-TOF GmbH, Münster, Germany.)

PHI TRIFT IV ToF-SIMS (Physical Electronics, USA) employs three electrostatic analyzers in the ion path to filter the background and metastable secondary ions. Using liquid metal cluster ion guns (such as Au_3^+ ion beam for sputtering of sample surface) increased sensitivity compared to a Ga^+ primary ion beam are obtained (www.phi.com). The application of dual primary ion guns is useful for an effective dual beam depth profiling on multi-layered samples.

A sensitive high resolution ion microprobe (SHRIMP, with a maximum mass resolution of $m/\Delta m$ of 30 000) was designed on the basis of the double-focusing sector field Matsuda type instrument[190] as discussed by Compston in 1996.[191] The first SHRIMP and a modified version for precise isotope ratio measurements were constructed in the 1980s and later produced commercially.[192] The main application of this instrument so far has been *in situ* U–Pb dating of zircon.[192,193]

In addition to double-focusing sector field and ToF-SIMS instruments, a less expensive quadrupole based mass spectrometer (SIMS 4550 from CAMECA) equipped with two primary ion sources (oxygen and caesium) is available especially for applications in depth profiling, including ultradepth resolved elemental analysis in the semiconductor field and materials research and industry. However, insulators can also be analyzed using this instrument by means of a low energy electron flood gun for charge compensation on the sample surface during the sputter process.

Furthermore, several quadrupole based mass spectrometers are known to allow SIMS and SNMS measurements in one instrument (INA-X from Specs GmbH, Berlin and SIMSLAB from VG Scientific). Whereas in SIMS the secondary ions formed by the sputtering of the sample surface with primary ions are analyzed, in SNMS, the sputtered secondary ions are suppressed by a repelling voltage and the neutrals are post-ionized. The post-ionization of neutrals in the SNMS mode is carried out either in an argon plasma ('plasma SNMS'), or by electron impact ionization ('e-beam SNMS') or by using a laser for the postionization of sputtered neutrals.[95,187] The only quadrupole based SNMS instrument – the INA-X from SPECS GmbH, Berlin – commercially available on the surface analytical market was developed from the Leybold INA-3. The experimental arrangement of SNMS INA-X is illustrated in Figure 5.33. This instrument is designed to enable both SNMS and SIMS measurements. SIMS analyses are made possible by using a separate fine-focusing primary

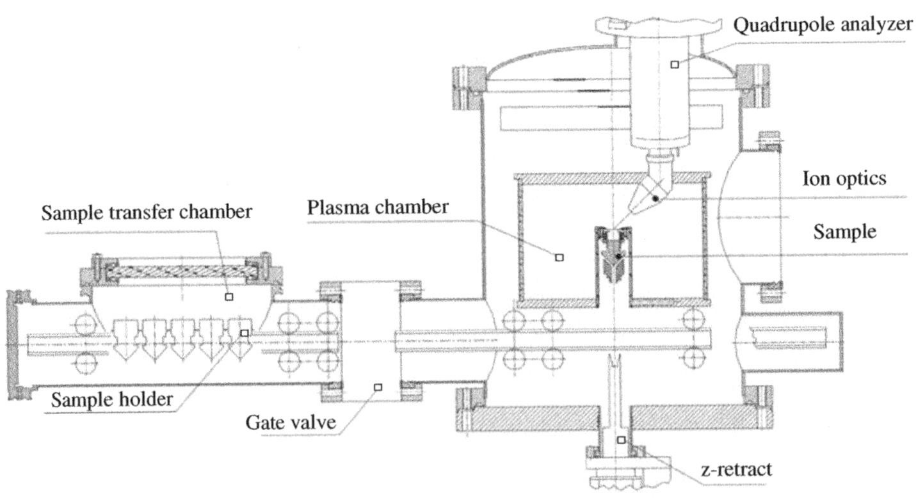

Figure 5.33 *Experimental arrangement of SNMS INA-X. (Reproduced by permission of Prof. Dr. H. Oechsner.)*

ion source IQE 12/38. For special applications, the sensitivity and thus the detection limits of surface analytical measurements by SIMS are improved compared to SNMS and element mapping with lateral resolution down to 100 μm is possible. In addition, *in situ* XPS measurements (XPS – X-ray photon electron spectroscopy) by INA-X are performed without completely switching off of the SNMS plasma. In contrast to mass spectrometric techniques using a primary ion source for sputtering the sample surface, the INA-X applies plasma sputtering using an electron cyclotron resonance plasma and also for post-ionization of sputtered neutrals with the 'electron gas' of this rf plasma. In the direct bombardment mode (DBM) and high frequency mode (HFM) the plasma is applied for sputtering as well as post-ionization. Using the direct bombardment mode, a high depth resolution down to 1 nm is achieved for the analysis of conducting materials. In the high frequency mode, the sample voltage is switched with high frequency to prevent charging effects on the sample surface and consequently insulating materials can be analyzed. The INA-X has so far been a useful analytical tool for characterizing surfaces. SIMS and SNMS are relevant techniques for the surface analysis of H, O, N and C.

Compared to SIMS, in SNMS, the matrix effects are reduced significantly. This characteristic allows easy quantification of analytical data.

In SIMSLAB from VG Scientific, both surface analytical techniques – SIMS and SNMS – have been applied (see Figure 5.34). In this mass spectrometer different types of primary ion sources are available. Ar^+, Cs^+, Ga^+ or O_2^+ primary ions are accelerated in the secondary ion source on the solid sample surface. Similar to the CAMECA IMS-7f, with this experimental arrangement, besides depth profiling, a microlocal analysis can also be performed. The sputtered secondary ions (for SIMS) or the post-ionized sputtered neutrals (for SNMS) – the post-ionization is carried out by an electron beam in an ionizer box (right-hand schematic in Figure 5.34) – are separated

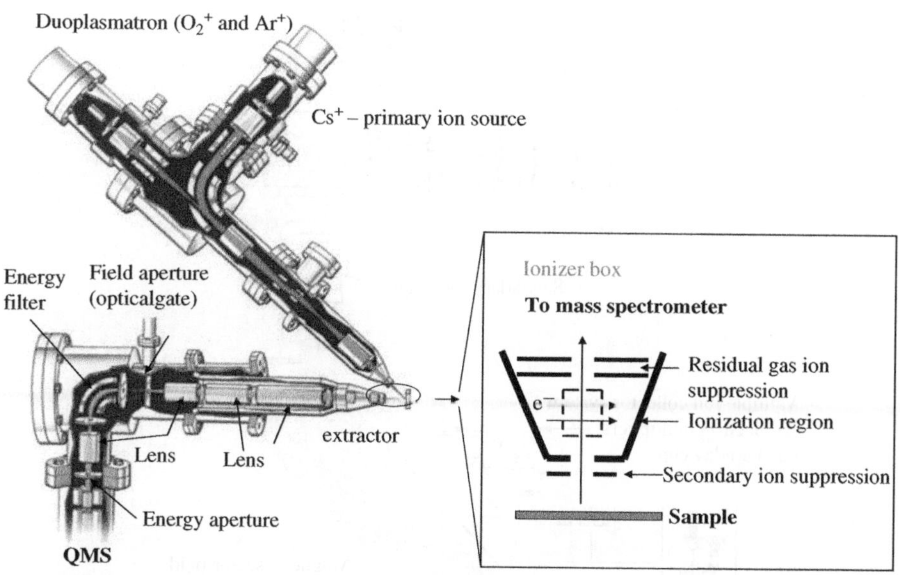

Figure 5.34 *Combination of a secondary ion and sputtered neutral mass spectrometer with a quadrupole analyzer (SIMSLAB, Fisons). (J. S. Becker and H. J. Dietze, Int. J. Mass Spectrom. Ion Proc.* ***197****, 1–35 (2000). Reproduced by permission of Elsevier.)*

after extraction in a quadrupole analyzer and detected with a single ion detection system using a channeltron. Unfortunately, this innovative equipment with quadrupole mass analyzer is no longer produced due to the lower sensitivity of element analysis in the SNMS mode and low mass resolution achievable.

For precise isotope ratio measurements with a spatial resolution in the μm range and below, double-focusing sector field mass spectrometers with a simultaneous multiple ion collector arrangement, SIMS IMS 1270 (maximum mass resolution ($m/\Delta m$) of 40 000) with several electron multipliers and Faraday cups and NanoSIMS 50 for very small samples and imaging (especially for micro- and nanolocal analysis) are available from CAMECA, whereby a spot size of ~ 50 nm using Cs^+ primary ions (and ~ 200 nm using O^- as primary ion beam) can be routinely achieved.[194] The layout of the double-focusing NanoSIMS 50 with a multiple ion collector system is illustrated in Figure 5.35. The samples are mounted in a special sample holder with a diameter of 50 mm and different hole sizes (mostly holes with 10 mm diameter are used). Samples are inserted directly in the holes of sample holder, small particles are pressed into a gold foil or geological samples are embedded in a resin (suitable for high vacuum) and coated with gold in order to avoid charging effects on the sample surface. Small non-conducting samples were analyzed through the holes of a thin metallic top 'grid' or 'plate'. Biological sections are deposited on the polished side of a metallic cylinder or silicon target. In the ion source the typical pressure is in the 10^{-7} to 10^{-8} Pa. The NanoSIMS with (forward) Nier–Johnson geometry uses a laminated magnet with large mass

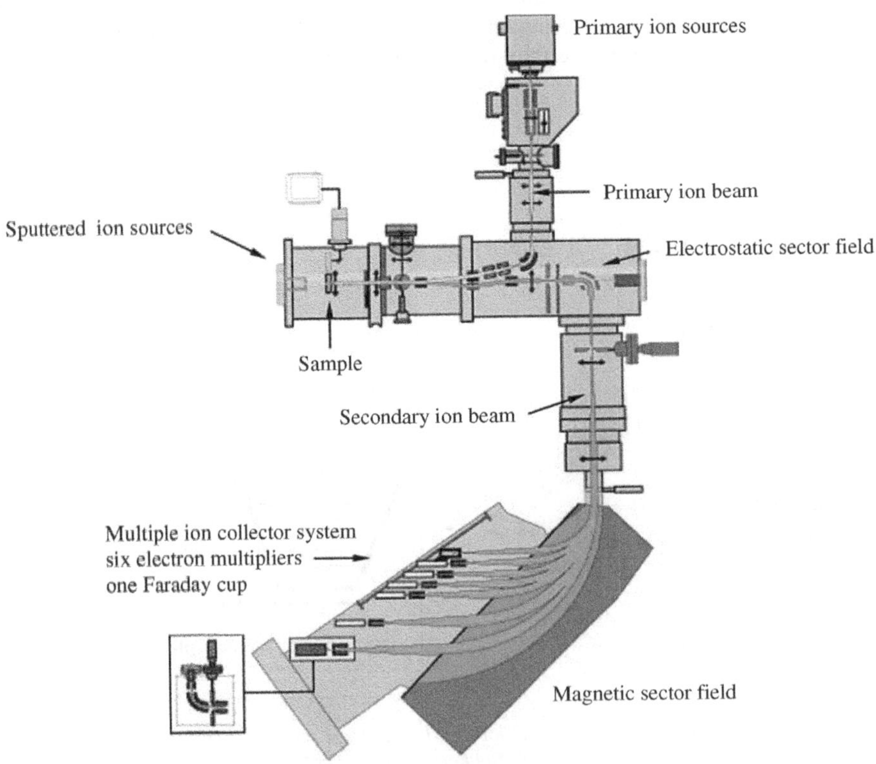

Figure 5.35 Instrumental setup of multiple ion collector NanoSIMS 50 ion microprobe. (Reproduced by permission of CAMECA, www.cameca.com.)

dispersion and is equipped with a duoplasmatron primary ion source (O^-) and caesium primary ion source (Cs^+). Smaller spot size for a given ion beam current and higher efficiency is achieved by coaxial configuration of primary and secondary ion beam in NanoSIMS. Using both reactive primary ions (O^- and Cs^+), secondary ions were generated and analyzed mass spectroscopically in the dynamic mode to obtain ion images. Five movable electron multipliers and one to two Faraday cups are inserted in the NanoSIMS 50 or seven electron multipliers in the NanoSIMS 50L. The mass spectrum is displayed along the focal plane of the double-focusing sector field instrument. In order to solve the interference problems in SIMS, measurements can be performed in the so-called 'high mass resolution mode'. The transmission of ions at a mass resolution ($m/\Delta m$) of 10 000 is about 20 % of the transmission at a mass resolution of 3 500 (which allows a maximum transmission). A normal incidence electron flood gun can be applied for the analysis of insulators, when the gold coating method is not sufficient for high primary ion beams (e.g., using the Cs^+ primary ion gun and analyzing negatively charged secondary ions). The main application of NanoSIMS is in geological science and also in nuclear astrophysics (element analysis and precise isotope ratio measurements at sub μm and nm range spatial resolution) where presolar grains or cometary dust of smaller size will be measured as described in Chapters 8 and 9.

The main features of SIMS and SNMS are compared briefly in Table 5.4.

Table 5.4 *Main features of SIMS and SNMS.*

	SIMS	SNMS
Evaporation	simultaneous evaporation and	using primary ions: Ar^+, Cs^+, O2+, O^-, Ga^+
Ionization	ionization using primary ions: Ar^+, Cs^+, O_2^+, O^-, Ga^+, cluster ions	post-ionization of neutrals: plasma, e-beam, laser (resonant, non-resonant)
Detection limit	0.001–1 $\mu g\,g^{-1}$	10–100 $\mu g\,g^{-1}$
RSC	10^2–10^6	0.3–3
Calibration	using implantation standards (10–20 %, RSD)	
Depth resolution	< 5 nm	< 5 nm
Lateral resolution	0.1–5 μm, 50 nm (NanoSIMS)	< 0.1 μm (Ga^+ source)
Application	Depth profiling, imaging, trace and isotope analysis, micro- and nano-analytics	Depth profiling, imaging
Instruments	CAMECA IMS 7f (CAMECA), ToF-SIMS (ION-TOF), Quad-SIMS 4550 (CAMECA), NanoSIMS 50 (CAMECA)	SIMSLAB (VG Scientific), INA X (Specs GmbH)

5.8 Accelerator Mass Spectrometers (AMS)

The development of accelerator mass spectrometry goes back to the early days of ion accelerators.[195] Accelerator mass spectrometry (AMS) was developed in 1977 by introducing accelerators (cyclotron[196] and tandem accelerator[197]) into mass spectrometry. A schematic diagram of a powerful

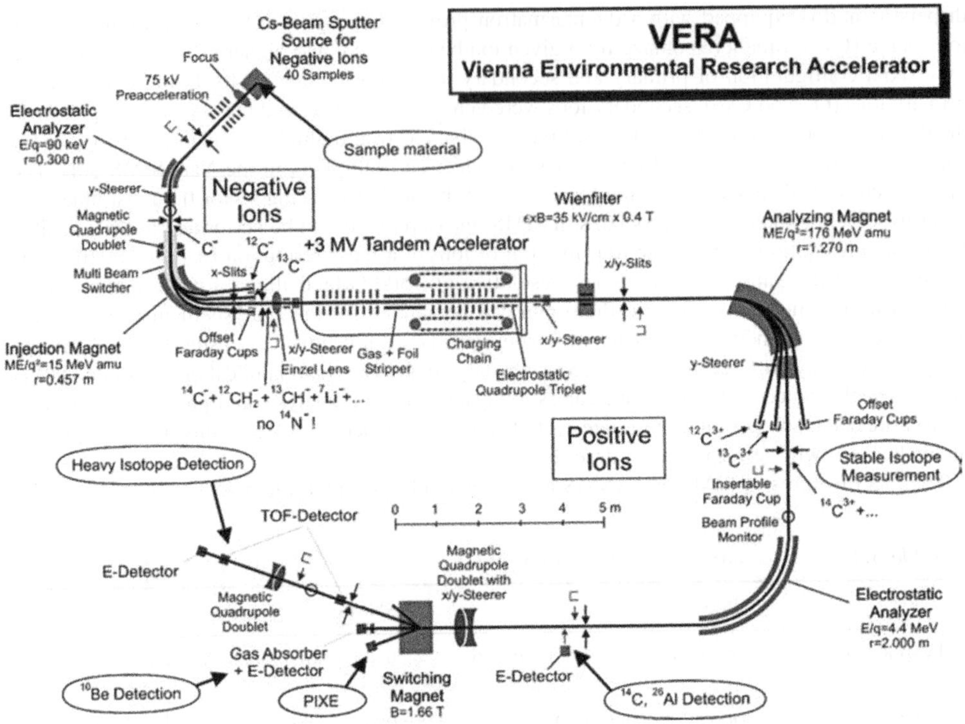

Figure 5.36 Schematic of an accelerator mass spectrometer (VERA–Vienna Environmental Research Accelerator).[195] (Reproduced by permission of Prof. W. Kustschera, VERA laboratory, Faculty of Physics, University of Vienna.)

accelerator mass spectrometer (VERA–Vienna Environmental Research Accelerator) is shown in Figure 5.36.

Measuring extremely low natural isotope abundances down to a level of 10^{-15} (e.g., $^{41}Ca/^{40}Ca$, $^{60}Fe/^{56}Fe$) requires an ultrasensitive method capable of handling a very large dynamic range of isotope analysis.[195] As the most powerful (sensitive and selective) mass spectrometric technique available at present, AMS measures ion currents of abundant stable isotopes (e.g., ^{12}C, ^{13}C) in Faraday cups and the rare radionuclides (e.g., ^{14}C, $t_{1/2} = 5730\,a$) are determined by ion counting in a solid-state or gas detector. ^{14}C dating requires high precision isotope ratio measurements in order to compete with existing beta counting facilities.[195] Most AMS are equipped with a caesium sputter ion source (used in SIMS) for the formation of negative ions, which are accelerated at 20–75 kV in an 'ion preparation stage' (consisting of, e.g., an electrostatic analyzer and magnetic sector field). For the frequently required ^{14}C determination, the mass separated ion beam with $m/z = 14$ ($^{14}C^-$, $^{12}CH_2^-$, $^{13}CH^-$, but no $^{14}N^-$ ions are formed) is then accelerated and decelerated in a 3 MV tandem accelerator with a gas target or thin foil, where disturbing isobaric polyatomic ions are dissociated and negatively charged ions are stripped to positively charged ions. In a final analysis stage, for example, by a magnetic sector field, the interfering and dissociated ions are separated and the analyte ion $^{14}C^{3+}$ is detected ultrasensitively.

Instrumental layouts and developments in AMS are reviewed by Kutschera.[195] Today AMS is the most powerful, sensitive and selective mass spectrometric technique for measuring long-lived radionuclides at the level of natural isotopic abundances (10^{-16} to 10^{-12}). Accelerator mass spectrometry (AMS) allows uranium isotope ratio measurements with an abundance sensitivity for ^{236}U in the range of 10^{-10}–10^{-12}.[198,199]

5.9 Electron Ionization Mass Spectrometers for Stable Isotope Ratio Measurements

Ions formed in an electron ionization mass spectrometer possess a relatively low initial energy spread, therefore single focusing magnetic sector field instruments are sufficient to separate ion beams with different *m/z* ratios. Commercial stable isotope ratio mass spectrometers (SIRMS) allow highly precise and accurate isotope ratio measurements e.g., with the Finnigan MAT 253, of H/D, $^{13}C/^{12}C$, $^{15}N/^{14}N$, $^{18}O/^{16}O$, $^{34}S/^{32}S$ (from SO_2 and SF_6) $^{28}Si/^{29}Si$ as well as Ar, Kr and Xe isotope ratios in gases. The Finnigan MAT 253 SIRMS consists of a dual gas inlet system, an electron impact ion source operating under high vacuum conditions, a single-focusing magnetic sector field and a multiple ion collection system. The formed ions are accelerated with 10 kV in a magnetic sector field (generated by an electromagnet) with a relative large dispersion (180 mm). Precise isotope ratio measurements are carried out by comparison of data on the sample and a standard (using the dual inlet for sample and standard gas introduction in the ion source) by simultaneous determination of ion currents at several *m/z* ratios by means of multiple ion collectors. In the DELTA V Plus (Thermo Fisher Scientific, Bremen) isotope ratio mass spectrometer (with dual gas inlet system) for improvement of transmission the ion optics have been refined and up to ten ion collectors can be accommodated. Isotope ratio measurements can be performed in the mass range up to $m/z = 96$ (compared to $m/z = 1$–150 in the Finnigan MAT 253). The schematic of a stable isotope ratio mass spectrometer for highly precise and accurate isotope analysis on gases DELTA V from Thermo Fisher Scientific is illustrated in Figure 5.37. To convert organic

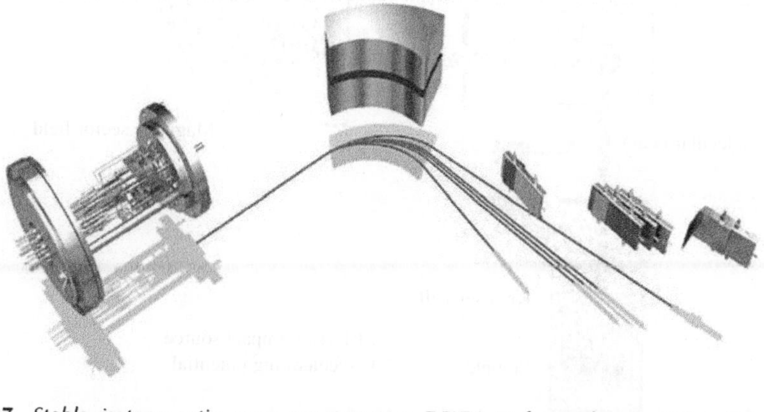

Figure 5.37 Stable isotope ratio mass spectrometer DELTA V from Thermo Fisher Scientific with e⁻ beam ion source and multiple ion collection. (Reproduced by permission of Thermo Fisher Scientific, Bremen Germany.)

or inorganic compounds into a gas which can be analyzed by SIRMS, several sample preparation devices and interfaces are commercially available. Chromatographic systems (GC or HPLC) can also be coupled on line to isotope ratio mass spectrometers to extend the application field.

5.10 Knudsen Effusion Mass Spectrometers

The Knudsen effusion mass spectrometer known as the 'high temperature mass spectrometer' is an analytical instrument for investigating chemical reactions in the condensed and gas phase at high temperatures and for determining thermodynamic data and the gas phase, gas–solid or gas–liquid equilibrium.[200,201] The experimental arrangement of a Knudsen effusion sector field mass spectrometer is illustrated in Figure 5.38.[203] If the Knudsen cell operates at high temperature, the vapour is in equilibrium with the condensed phase (solid sample). The evaporation of a solid sample in the Knudsen cell and the post-ionization of gaseous molecules in an electron beam ion source of, for example, a magnetic sector field mass spectrometer are separated in time and place, which allows the condensed – gas phase system to be studied without disturbance, under equilibrium conditions. The ion beams separated according to the mass-to-charge ratio are detected by means of an electron multiplier and/or a Faraday cup. A Knudsen effusion mass spectrometer with electron impact ion source allows the electron energy to be varied. Consequently, the ion intensities of ionized species separated as a function of electron impact energy are used for the determination of the ionization potential (E_i) of the compound investigated (see Section 2.7).

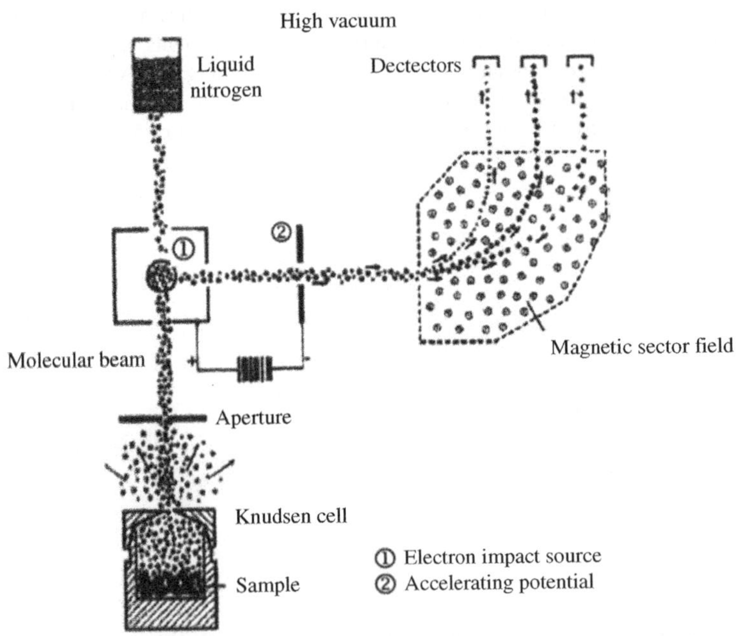

Figure 5.38 Knudsen cell mass spectrometer with e⁻ beam ion source. (K. Hilpert, Fresenius' J. Anal. Chem. **370**, 471 (2001). Reproduced by permission of Springer Sciences and Business Media.)

Knudsen cell mass spectrometers play a role in diagnostics of evaporated samples at high temperature whereby the species present with various *m/z* ratios are monitored continuously. In spite of the importance of Knudsen cell mass spectrometry, no commercial mass spectrometers are now available.

References

1. Busch, K. L., *Spectrosc.*, **16**, 14 (2001).
2. Wutz, M., Adam, H. and Walcher, W., *Theory and Practice of Vacuum Technology*, 5th edn, Vieweg, Braunschweig, Wiesbaden (1992).
3. Montaser, A. E. (ed.), *Inductively Coupled Plasma Source Mass Spectrometry*, Wiley-VCH Publishers, New York (1998).
4. Houk, R. S., Tassel, V. A., Flesch, G. D., Svec, H. J., Gray, A. L. and Taylor, C. E., *Anal. Chem.*, **52**, 2283 (1980).
5. Becker, J. S., Burow, M., Zoriy, M., Pickhardt, C., Ostapczuk, P. and Hille, R., *Atom. Spectr.*, **25**, 197 (2004).
6. Becker, J. S., Zoriy, M., Halicz, L. *et al.*, *J. Anal. At. Spectrom.*, **19**, 1257 (2004).
7. Boulyga, S. F., Testa, C., Desideri, D. and Becker, J. S., *J. Anal. At. Spectrom.*, **16**, 1283 (2001).
8. Epov, V. N., Benkheddsa, K., Cornett, R. J. and Evans, R. D., *J. Anal. At. Spectrom.*, **20**, 424 (2005).
9. McLean, J. A., Becker, J. S., Boulyga, S. F., Dietze, H. J. and Montaser, A., *Int. J. Mass Spectrom.*, **208**, 193 (2001).
10. Tresl, I., De Wannemacker, G., Quetel, C. R. *et al.*, *Environ. Sci. Tech.*, **38**, 581 (2004).
11. Becker, J. S. and Dietze, H.-J., *Int. J. Mass Spectrom.*, **228**, 127 (2003).
12. Becker, J. S. and Dietze, H. J., *J. Anal. At. Spectrom.*, **14**, 1493 (1999).
13. Barshik, C. M., Duckworth, D. C., Smith, D. H. (eds.), *Inorganic Mass Spectrometry: Fundamentals and Applications*. In: *Pract. Spectrosc.*, 23 (2000).
14. Becker, J. S., Burow, M., Boulyga, S. F., Pickhardt, C., Hille, R. and Ostapczuk, P., *Atom. Spectr.*, **23**, 177 (2002).
15. Ying, J.-F. and Douglas, D. J., *Rapid Commun. Mass Spectrom.*, **10**, 649 (1996).
16. Boulyga, S. F. and Becker, J. S., *Fresenius' J. Anal. Chem.*, **370**, 618 (2001).
17. Boulyga, S. F. and Becker, J. S., *J. Anal. At. Spectrom.*, **17**, 1202 (2002).
18. Boulyga, S. F., Becker, J. S., Matusevitch, J. L. and Dietze, H. J., *Int. J. Mass Spectrom.*, **203**, 143 (2000).
19. Boulyga, S. F., Matusevich, J. L., Mironov, V. P., *et al.*, *J. Anal. At. Spectrom.*, **17**, 958 (2002).
20. Izmer, A. V., Boulyga, S. F. and Becker, J. S., *J. Anal. At. Spectrom.*, **18**, 1339 (2003).
21. Izmer, A. V., Zoriy, M. V., Boulyga, S. F. and Becker, J. S., *J. Anal. At. Spectrom.*, **19**, 1278 (2004).
22. Tanner, S. D., Baranov, V. I. and Bandura, D. R., *Spectrochim. Acta*, **57 B**, 1361 (2002).
23. Read, H., Abou Hakra, F., Palacz, Z. and Meffan-Main, S., *European Winter Conference on Plasma Spectrochemistry 2005, Budapest, Hungary*; *(www.gvinstruments.co.uk/icp2.htm).*
24. Becker, J. S., *J. Anal. At. Spectrom.*, **17**, 1172 (2002).
25. Belgacem, O., Bowdler, A., Brookhouse, I., Brancia, F. L. and Raptakis, E., *Rapid Commun. Mass Spectrom.*, **20**, 1653 (2006).
26. Becker, J. S. and Dietze, H. J., *Isotopenpraxis*, **19**, 105 (1983).
27. Boulyga, S. F., Pickhardt, C., Becker, J. Su., Przybylski, M. and Becker, J. S., *Plasma Source Mass Spectrometry: Applications and Emerging Technologies, (8th International Conference on Plasma Source Mass Spectrometry Durham, United Kingdom)*, Royal Society of Chemistry 54 (2003).
28. O'Brien, S. E., Acon, B. W., Boulyga, S. F., Becker, J. S., Dietze, H. J. and Montaser, A., *J. Anal. At. Spectrom.*, **18**, 230 (2003).
29. Barinaga, C. J., Eiden, G. C., Alexander, M. L. and Koppenaal, D. W., *Fresenius' J. Anal. Chem.*, **355**, 487 (1996).
30. Eiden, G. C., Barigina, C. J. and Koppenaal, D. W., *J. Anal. At. Spectrom.*, **14**, 1129 (1999).
31. Eiden, G. C. and Koppenaal, D., *Rapid Commun. Mass Spectrom.*, **11**, 37 (1997).

32. Koppenaal, D. W., Barinaga, C. J. and Smith, M. R., *J. Anal. At. Spectrom.*, **9**, 1953 (1994).
33. Koppenaal, D. W., Barinaga, C. J. and Hager, J. G., in *Book of Abstracts of Second Asia Pacific Winter Conference on Plasma Spectrochemistry*, R. M. Barnes (ed.), ICP Information Newsletters, 142 (2006).
34. Bradshaw, N., Hall, E. F. H. and Sanderson, N. E., *J. Anal. At. Spectrom.*, **4**, 801 (1989).
35. Morita, M., Ito, H., Uchiro, T. and Otsuka, K., *Anal. Sci.*. **5**, 609 (1989).
36. Wills, J., Hamester, M. and Rottmann, L., in *Book of Abstracts of 2005 Asia-Pacific Winter Conference on Plasma Spectrochemistry, Thailand*, R. M. Barnes (ed.), ICP Information Newsletters, 113 (2005).
37. Wiley, C. W., *Science*, **124**, 817 (1956).
38. Karas, M. and Hillenkamp, F., *Anal. Chem.*, **60**, 2299 (1988).
39. Myers, D. P. and Hieftje, G. M., *Microchim. J.*, **48**, 259 (1993).
40. Myers, D. P., Li, G., Yang, P. and Hieftje, G. M., *J. Am. Soc. Mass Spectrom.*, **5**, 1008 (1994).
41. Myers, D. P., Mahoney, P. P. and Hieftje, G. M., *J. Am. Chem. Soc.*, **6**, 400 (1995).
42. Sturgeon, R. E., Lam, J. W. and Saint, A., *J. Anal. At. Spectrom.*, **15**, 607 (2000).
43. Mester, Z., Sturgeon, R. E., Lam, J. W., Maxwell, P. S. and Péter, L., *J. Anal. At. Spectrom.*, **16**, 1313 (2001).
44. Willie, S., Mester, Z. and Sturgeon, R. E., *J. Anal. At. Spectrom.*, **20**, 1358 (2005).
45. Nelms, S. M. (ed.), *ICP Mass Spectrometry*, Blackwell Publishing Ltd, Oxford, (2005).
46. Vanhaecke, F., Moens, L., Dams, R., Allen, L. and Georgitis, S., *Anal. Chem.*, **14**, 1807 (1999).
47. Benkhedda, K., Dimitrova, B., Infante, H. G., Ivanova, E. and Adams, F. C., *J. Anal. At. Spectrom.*, **18**, 1019 (2003).
48. Benkhedda, K., Infante, H. G., Ivanova, E. and Adams, F. C., *Trends in Anal. Chem.*, **21**, 332 (2002).
49. Leach, A. M., McClenathan, D. M. and Hieftje, G. M., in *Handbook of Speciation: Technique and Methodology*, R. Cornelis, J. Caruso, H. Crews, K. Heumann (eds.), John Wiley and Sons, New York (2003) pp. 313.
50. Halliday, A. N., Lee, D.-C., Christensen, J. N. *et al.*, *Geochim. Cosmochim. Acta*, **62**, 919 (1998).
51. Halliday, A. N., Christensen, J. N., Jones, C. E., Walder, A. J. and Freedman, P. A., *Trans. Amer. Geophys. Un.*, **74**, 626 (1993).
52. Halliday, A. N., Christensen, J. N., Lee, D.-C., Hall, C., Luo, X. and Rehkaemper, M., in *Inorganic Mass Spectrometry*, C. M. Barshick, D. C. Duckworth and D. H. Smith (eds.), Marcel Dekker AG, New York, 291 (2000).
53. Walder, A. J. and Freedman, P. A., *J. Anal. Atom. Spectrom.*, **7**, 571 (1992).
54. Walder, A. J. and Furuta, N., *Anal. Sci.*, **9**, 675 (1993).
55. Walder, A. J., Platzner, I. and Freedman, P. A., *J. Anal. Atom. Spectrom.*, **8**, 19 (1993).
56. Boulyga, S. F., Kloetzli, U. and Prohaska, T., *J. Anal. At. Spectrom.*, **21**, 1427 (2006).
57. Weyer, S. and Schwieters, J. B., *Int. J. Mass Spectrom.*, **226**, 355 (2003).
58. Wieser, M. E. and Schwieters, J. B., *Int. J. Mass Spectrom.*, **242**, 97 (2005).
59. Williams, H. M., Peslier, A. H., McCammon, C. *et al.*, *Earth Plan. Sci. Lett.*, **235**, 435 (2005).
60. Cromwell, E. F. and Arrowsmith, P., *J. Am. Soc. Mass Spectrom.*, **7**, 458 (1996).
61. Barnes, J. H., Schilling, G. D., Sperline, R. *et al.*, *Anal. Chem.*, **76**, 2631 (2004).
62. Barnes, J. H., Sperline, R., Denton, M. B. *et al.*, *Anal. Chem.*, **74**, 5327 (2002).
63. Becker, J. S. and Dietze, H. J., *Spectrochim. Acta*, **53B**, 1475 (1998).
64. Becker, J. S., Dietze, H. J., McLean J. A., Montaser, A., *Anal. Chem.*, **71**, 3077 (1999).
65. Becker, J. S., Gorbunoff, A., Zoriy, M. V., Izmer, A. V. and Kayser, M., *J. Anal. At. Spectrom.*, **21**, 19 (2006).
66. Becker, J. S., Soman, R. S., Sutton, K. L., Caruso, J. A. and Dietze, H. J., *J. Anal. At. Spectrom.*, **14**, 933 (1999).
67. Becker, J. S. and Tenzler, D., *Fresenius' J. Anal. Chem.*, **370**, 637 (2001).
68. Boulyga, S. F., Matusevich, J. L., Mironov, V. P. *et al.*, *J. Anal. At. Spectrom.*, **17**, 958 (2002).
69. Pickhardt, C., Becker, J. S. and Dietze, H. J., *Fresenius' J. Anal. Chem.*, **368**, 173 (2000).
70. Gammelgaard, B. and Jensen, B. P., *J. Anal. At. Spectrom.*, **22**, 235 (2007)
71. Becker, J. S., Boulyga, S. F., Pickhardt, C., Becker, J., Buddrus, S. and Przybylski, M., *Anal. Bioanal. Chem.*, **375**, 561 (2003).
72. Schaumloeffel, D., *Anal. Bioanal. Chem.*, **379**, 351 (2004).

73. Becker, J. S., State of the art in determination of long-lived radionuclides by Plasma Mass Spectrochemistry (invited Plenary lecture), in *Book of Abstracts of Second Asia-Pacific Winter Conference on Plasma Spectrochemistry, Bangkok, Thailand.* R. M. Barnes (ed.), ICP Information Newsletters, 162 (2006).
74. Becker, J. S., *Spectrochim. Acta*, **58B**, 1757 (2003).
75. Schaumlöffel, D., Giusti, P., Zoriy, M. V., *et al.*, *J. Anal. At. Spectrom.*, **20**, 17 (2005).
76. Becker, J. S., *Canad. J. Anal. Sci. Spectr.*, **47**, 98 (2002).
77. Stefánka, Z., Koellensperger, G., Stingeder, G. and Hann, S., *J. Anal. At. Spectrom.*, **21**, 86 (2006).
78. O'Brien Murdock, S.-A., Kahen, K., Chirinos, J. R., *et al. J. Anal. At. Spectrom.*, **19**, 666 (2004).
79. Westphal, C. S., Kahen, K., Rutkowski, W. F., Acon, B. W. and Montaser A., *Spectrochim. Acta B*, **59**, 353 (2004).
80. Becker, J. S. and Dietze, H. J., *Fresenius' J. Anal. Chem.*, **364**, 482 (1999).
81. Boulyga, S. F. and Becker, J. S., *Fresenius' J. Anal. Chem.*, **370**, 612 (2001).
82. Becker, J. S. and Dietze, H. J., *Int. J. Mass Spectrom.*, **202**, 69 (2000).
83. Jorabchi, K., Kahen, K., Gray, C. and Montaser, A., Anal. Chem., **77**, 1253 (2005).
84. Jorabchi, K., McCormick, R., Levine, J. A., Liu, H., Nam, S. H. and Montaser, A., *Spectrochim Acta B*, **61**, 945 (2006).
85. Todoli, J. L. and Mermet, J. M., *Spectrochim. Acta B*, **61**, 239 (2006).
86. Cairns, W. R. L., Barbante, C., Capodaglio, G., Cescon, P., Gambaro, A. and Eastgate, A., *J. Anal. At. Spectrom.*, **19**, 286 (2004).
87. Schaumloeffel, D., Ruiz Encinar, J. and Lobinski, R., *Anal. Chem.*, **75**, 6873 (2004).
88. Klemens, P., Feuerbacher, H., Boulyga, S. and Wiederin, D., in *Book of Abstracts of 19th ICP-MS Anwendertreffen, D. Günther (ed.) Zürich* (2004).
89. Becker, J. S., *Trends in Anal. Chem.*, **24**, 243 (2005).
90. Eroglu, A. E., McLeod, C. W., Leonard and McCubbin, K. S. D., *J. Anal. Atom. Spectrom.*, **13**, 875 (1998).
91. Bings, N. H., Stefanka, Z. and Mallada, S. R., *Anal. Chem.*, 203 (2003).
92. Pohl, P., *Trends Anal. Chem.*, **23**, (2004).
93. Gettar, R. T., Garavaglia, R. N., Gautier, E. A. and Batistoni, D. A., *J. Chromatogr.*, **A884**, 211 (2000).
94. Narasaki, H. and Mayumi, K., *Anal. Sci.*, **16**, 65 (2000).
95. Becker, J. S. and Dietze, H. J., *Int. J. Mass Spectrom.*, **197**, 1 (2000).
96. Desideri, D., Meli, M. A., Roselli, C., Testa, C., Boulyga, S. F. and Becker, J. S., *Anal. Bioanal. Chem.*, **374**, 1091 (2002).
97. Zoriy, M., Pickhardt, C., Ostapczuk, P., Hille, R., Becker, J. S., *Int. J. Mass Spectrom.*, **232**, 217 (2004).
98. Schaumloeffel, D., Giusti, P., Zoriy, M., Pickhardt, C., Szpunar, J., Lobinski, R. and Becker, J. S., *J. Anal. At. Spectrom.*, **20**, 17 (2005).
99. Kot, A. and Namiesnik, J., *Trends Anal. Chem.*, **19**, 69 (2000).
100. R. Cornelis, J. C., Crews, H., and Heumann, K. *Handbook of Speciation: Technique and Methodology*, John Wiley & Sons, Inc., New York (2003).
101. Sadi, B. B. M., Vonderheide, A. P., Becker, J. S. and Caruso, J., *Multiple Detection Size-Exclusion Chromatography*, ACS Symposium Series 893, American Chemical Society, Washington, DC, 168 (2004).
102. Szpunar, J. and Lobinski, R., *Hyphenated Techniques in Speciation Analysis.* R. M. Smith (ed.)' RSC Chromatic Monographs, Cambridge (2003).
103. Day, J. A., Caruso, J. A., Becker, J. S. and Dietze, H.-J., *J. Anal. At. Spectrom.*, **15**, 1343 (2000).
104. Heumann, K. G., Rottmann, L. and Vogl, J., *J. Anal. At. Spectrom.*, **12**, 1381 (1994).
105. Kerl, W., Becker, J. S., Dannecker, W. and Dietze, H. J., *Fresenius' J. Anal. Chem.*, **362**, 433 (1998).
106. Lobinski, R., Schaumlöffel, D. and Szpunar, J., *Mass Spectrom. Rev.*, **25**, 255 (2006).
107. Szpunar, J., Lobinski, R. and Prange, A., *Appl. Spectrosc.*, **57**, 102A (2003).
108. Schaumlöffel, D. and Lobinski, R., *Int. J. Mass Spectrom.*, **242**, 217 (2005).
109. Szpunar, J., *Anal. Bioanal. Chem.*, **378**, 54 (2004).
110. Szpunar, J., *Analyst*, **125**, 963 (2000).
111. Szpunar, J., Lobinski, R. and Prange, A., *Appl. Spectrosc.*, **125**, 102A (2000).
112. Siripinyanond, A., Barnes, R. M. and Amarasiriwardena, D. D., *J. Anal. At. Spectrom.*, **17**, 1055 (2002).
113. Beckett, R., *At. Spectr.*, **12**, 228 (1991).
114. Gray, A. L., *Analyst*, **110**, 551 (1985).

115. Becker, J. S., Gorbunoff, A., Zoriy, M. V., Izmer, A. V. and Kayser, M., *J. Anal. At. Spectrom.*, **21**, 19 (2006).
116. Carcia, C. C., Lindner, H. and Niemax, K., in *8th European Workshop on Laser Ablation in Elemental Analysis*, D. Guenther (ed.), Zuerich (2006).
117. Becker, J. Su., Pickhardt, C., Hoffmann, N., Hocker, H. and Becker, J. S., *At. Spectr.*, **23**, 1 (2002).
118. Pickhardt, C. and Becker, J. Su., *Fresenius' J. Anal. Chem.*, **370**, 534 (2001).
119. Pickhardt, C., Brenner, I. B., Becker, J. Su. and Dietze, H. J., *Fresenius' J. Anal. Chem.*, **368**, 79 (2000).
120. Westheide, J. T., Becker, J. Su., Jaeger, R., Dietze, H.-J. and Broekaert, J. A. C., *J. Anal. At. Spectrom.*, **11**, 661 (1996).
121. Becker, J. S., Boulyga, S. F., Becker, J. Su., Pickhardt, C., Damoc, E. and Przybylski, M., *Int. J. Mass Spectrom.*, **228**, 985 (2003).
122. Becker, J. S., Zoriy, M., Pickhardt, C., Przybylski, M. and Becker, J. S., *Int. J. Mass Spectrom.*, **242**, 135 (2005).
123. Boulyga, S. F., Desideri, D., Meli, M. A., Testa, C. and Becker, J. S., *Int. J. Mass Spectrom.*, **226**, 329 (2003).
124. Bings, N. H., *J. Anal. At. Spectrom.*, **17**, 759 (2002).
125. Leach, A. M. and Hieftje, G. M., *J. Anal. At. Spectrom.*, **17**, 852 (2002).
126. Zoriy, M. V., Kayser, M., Izmer, A., Pickhardt, C. and Becker, J. S., *Int. J. Mass Spectrom.*, **242**, 297 (2005).
127. Becker, J. S., Zoriy, M., Dehnhardt, M., Pickhardt, C. and Zilles, K., *J. Anal. At. Spectrom.*, **20**, 912 (2005).
128. Becker, J. S., Zoriy, M. V., Pickhardt, C., Palomero-Gallagher, N. and Zilles, K., *J. Anal. At. Spectrom.*, **77**, 3208 (2005).
129. Reinhardt, H., Kriews, M., Miller, H., *et al.*, *Fresenius' J. Anal. Chem.*, **370**, 629 (2001).
130. Elliott, V. L., McLeod, C. W. and Marshall, P. S., *Anal. Bioanal. Chem.* **383**, 416 (2005).
131. Pickhardt, C., Dietze, H. J. and Becker, J. S., *Int. J. Mass Spectrom.*, **242**, 273 (2005).
132. Becker, J. S. and Dietze, H. J., *Z. Angew. Geol.*, **29**, 136 (1983).
133. Becker, J. S. and Dietze, H. J., *Z. Angew. Geol.*, **32**, 299 (1986).
134. Becker, J. S. and Dietze, H. J., *Fresenius' J. Anal. Chem.*, **344**, 69 (1992).
135. Dietze, H. J., *Massenspektroskopische Spurenanalyse*, Akademischer Verlagsgesellschaft Geest & Portig K.-G., Leipzig, (1975).
136. Dietze, H. J. and Becker, J. S., *Fresenius Z. Anal. Chem.* **302**, 490 (1985).
137. Jochum, K. P., *Spectr. Europa*, **9**, 22 (1997).
138. Jochum, K. P., Laue, H.-J., Seufert, H. M. *et al.*, *Fresenius' J. Anal. Chem.*, **359**, 385 (1997).
139. Conzemius, R. J. and Svec, H. J., *Anal. Chem.*, **50**, 1854 (1978).
140. Dietze, H. J., *Analytikertaschenbuch*, **10**, 249 (1991).
141. Dietze, H. J. and Becker, J. S., in *Laser Ionization Mass Analysis*, A. Vertes, R. Gijbels and F. Adams (eds.) Chemical Analysis Series, John Wiley & Sons, Inc., New York, **124**, 453 (1993).
142. Hillenkamp, F., Unsöld, E., Kaufmann, R. and Nitsche, R., *Appl.Phys.*, **8**, 341 (1975).
143. Schüler, B., Nitsche, R. and Hillenkamp, F., *Scanning Electron Microsc.*, **II**, 597 (1980).
144. Verbueken, A. H., Bruynseels, F. J., Van Grieken, R. and Adams, F., in *Inorganic Mass Spectrometry* F. Adams, R. Gijbels, R. Van Grieken (eds.) *Chemical Analysis*, John Wiley & Sons, Inc., New York, **95**, 173 (1988).
145. Hinz, K. P., Trimborn, A., Weingartner, E., Henning, S., Baltensperger, U. and Spengler, B., *J. Aerosol Science*, **36**, 123 (2005).
146. Vogt, P., Kirchner, R., Scheer, V., Hinz, K. P., Trimborn, A. and Spengler, B., *J. Aerosol Science*, **34**, 319 (2003).
147. Giron-Monzon, L., Manelyte, L., Ahrends, R., Kirch, D., Spengler, B. and Friedhoff, P., *J. Biol. Chem.*, **279**, 49338 (2004).
148. Trimborn, A., Hinz, K. P. and Spengler, B., *Aerosol Sci. Techn.*, **33**, 191 (2000).
149. Trimborn, A., Hinz, K. P. and Spengler, B., *J. Geophys. Res.*, **107**, 8132 (2002).
150. Dingle, T., Griffith, B. W. and Ruckman, J. C., *Vacuum*, **31**, 571 (1981).
151. Utley, A., *Microelectron. Manufact. Testing*, **February** 27, (1990).
152. Gladskoi, V. M. and Belousov, V. I., *Elektron. Ind*, **11**, 95 (1980).

153. Basova, T. A., Boriskin, A. I., Brjuchanov, A. S., Bykovskii, Y. A., Jeremenko, V. M. and Nevolin, V., H., *High Purity Mater. (USSR)*, **3**, 49 (1987).
154. Muller, J. F., Pelletier, M., Krier, G., Weil, D. and Campana, J., in *Microbeam Analysis*, R. P. Russell (ed.) San Fransisco Press, San Fransisco, R. P. Russell (ed.), San Fransisco Press, San Fransisco, 311 (1989)
155. Trautmann, N., Passler, G. and Wendt, K. D. A., *Anal. Bioanal. Chem.*, **378**, 348 (2004).
156. Wendt, K., Blaum, K., Bushaw, B. A. *et al., Fresenius' J. Anal. Chem.*, **364**, 471 (1999).
157. Wendt, K. and Trautmann, N., *Int. J. Mass Spectrom.*, **242**, 161 (2005).
158. Wendt, K., Trautmann, N. and Bushaw, B. A., *Nucl. Instrum. Meth. Phys. Res. B*, **172**, 162 (2000).
159. Moore, L. J., Fassett, J. D. and Travis, J. C., *Anal. Chem.*. **56**, 2770 (1984).
160. Wunderlich, R. K., Hutcheon, I. D., Wasserburg, G. J. and Blake, G. A., *Int. J. Mass Spectrom.*, **115**, 123 (1992).
161. Rimke, H., Herrman, G., Mang, M. *et al.*, *Mikrochim. Acta*, **III**, 223 (1989).
162. Nicolussi, A., Davis, A. M., Pellin, M. J., *et al., Lunar Planet Sci.*, **28**, 1021 (1997).
163. Ma, Z., Thomson, R. N., Lykke, K. R., Pellin, M. J. and Davis, A. M., *Rev. Sci. Instr.*, **66**, 3168 (1995).
164. Marcus, R. K., *Glow Discharge Spectroscopies*, Plenum Press, New York, London, 1 (1993).
165. Harrison, W. W., Hang, W., Yan, X., Ingeneri, K. and Schilling, C., *J. Anal. At. Spectrom.*, **12**, 891 (1997).
166. Betti, M. and Aldave de les Heras, L., *Spectroscopy Europe*, **15**, 15 (2003).
167. Harrison, W. W., Yang, C. and Oxley, E., in *Glow Discharge Plasma in Analytical Spectroscopy*, R. K. Marcus and J. A. C. Broekaert (eds.), Wiley & Sons Ltd., Chichester, UK, 71 (2003).
168. Marcus, R. K. and Broekaert, J. A. C. (eds.), *Glow Discharge Plasmas in Analytical Spectroscopy*, John Wiley & Sons Ltd., Chichester, UK (2003).
169. Milton, D. M. P. and Hutton, R. C., *Spectrochim. Acta*, **24 B**, 1237 (1993).
170. Van Straaten, M., Swenters, K., Gijbels, R., Verlinden, J. and Adriaenssens, E., *J. Anal. At. Spectrom.*, **9**, 1389 (1994).
171. Hutton, R. C. and Raith, A., *J. Anal At. Spectrom.*, **7**, 623 (1992).
172. Jakubowski, N., Stüwer, D. and Tölg, G., *Int. J. Mass Spectrom. Ion Proc.*, **71**, 183 (1986).
173. Myers, D. P., Heintz, M. J., Mahoney, P. P., Li, G. and Hieftje, G. M., *Appl. Spectrosc.*. **48**, 1337 (1994).
174. McLuckey, S. A., Glish, G. L., Duckworth, D. C. and Marcus, R. K., *Anal. Chem.*, **65**, 1606 (1992).
175. McLuckey, S. A., Goeringer, D. E., Asano, K. G., Vaidyanathan, G. and Stephenson, J. L., *Rapid Commun. Mass Spectrom.*, **10**, 287 (1996).
176. Watson, C. H., Barshick, C. M., Wronka, J., Laukien, F. H. and Eylier, J. R., *Anal. Chem.*, **68**, 573 (1996).
177. Becker, J. S., Saprykin, A. I. and Dietze, H. J., *Intern. J. Mass Spectrom. Ion Proc.*, **164**, 81 (1997).
178. Jäger, R., Saprykin, A. I., Becker, J. S., Dietze, H. J. and Broekaert, J. A. C., *Mikrochim. Acta*, **125**, 41 (1997).
179. Jäger, R., Becker, J. S., Dietze, H. J., Broekaert, J. A. C. *Fresenius, J. Anal. Chem.* **358**, 214 (1997).
180. Saprykin, A. I., Becker, J. S. and Dietze, H. J., *J. Anal. At. Spectrom.*, **10**, 897 (1995).
181. Saprykin, A. I., Becker, J. S. and Dietze, H. J., *Fresenius' J. Anal. Chem.*, **359**, 449 (1997).
182. Saprykin, A. I., Becker, J. S., Von der Crone, U. and Dietze, H. J., *Fresenius' J. Anal. Chem.*, **358**, 145 (1997).
183. Winchester, M. R., Duckworth, D. C. and Marcus, R. K., in *Glow Discharge Spectroscopies* R. K. Marcus (ed.) Plenum Press, New York, 263 (1993).
184. Saprykin, A. I., Melchers, F. G., Becker, J. S. and Dietze, H. J., *Fresenius' J. Anal. Chem.*, **353**, 570 (1995).
185. Hoffmann, V., Kasik, M., Robinson, P. K. and Venzago, C., *Anal. Bioanal. Chem.*, **381**, 173 (2005).
186. Richter, S., Alonso, A., De Bolle, W., Wellum, R. and Taylor, P. D. P., *Int. J. Mass Spectrom.*, **193**, 9 (1999).
187. Benninghoven, A., Janssen, K. T. F., Tümpner, J. and Werner, W. H., *Secondary Ion Mass Spectrometry VIII Proceedings of the Ninth International Conference on SIMS*, John Wiley & Sons Ltd, Chichester, UK (1992).
188. Benninghoven, A., Ruedenauer, F. G. and Werner, W. H., *Secondary Ion Mass Spectrometry VI*, A. Benninghoven *et al.* (eds.) John Wiley and Sons Inc., New York (1987) Chapter I.
189. Jenett, H., in *Analytikertaschenbuch 16*, Springer Verlag, Berlin, Heidelberg, 43 (1997).
190. Matsuda, H., *Int. J. Mass Spectrom. Ion Phys.*, **14**, 219 (1974).
191. Compston, W., *J. Roy. Soc. W. Aust.*, **79**, 109 (1996).

192. De Laeter, J. R., *Application of Inorganic Mass Spectrometry*, Wiley-Interscience Series on Mass Spectrometry, John Wiley & Sons, Inc. New York (2001).
193. Ireland, T. R., *Adv. Anal. Geochem.*, **2**, 1 (1995).
194. Stadermann, F. J., Walker, R. M. and Zinner, E., *New Frontiers in Isotope Geoscience*, The University of Melbourne, Victoria, Australia, 163 (2000).
195. Kutschera, W., *Intern. J. Mass Spectr.*, **242**, 145 (2005).
196. Muller, R. A., *Science*, **196**, 489 (1977).
197. Purser, K. H., Liebert, R. B., Litherland, A. E. *et al., Rev. Phys. Appl.*, **12**, 1487 (1977).
198. Fifield, L. K., *Nucl. Instrum. Methods*, **B 172**, 297 (2000).
199. Paul, M., Berkovits, D., Ahmad, I. *et al., Nucl. Instrum. Methods*, **B 172**, 688 (2000).
200. Baba, M. S., in *Introduction to Mass Spectrometry*, S. K. Aggarwal and H. C. Crain (eds.), Indian Society for Mass Spectrometry, Trombay 429 (1997).
201. Hilpert, K. and Miller, M., *J. Electrochem. Soc.*, **141**, 2769 (1994).
202. Hilpert, K., *Fresenius' J. Anal. Chem.* **370**, 471 (2001).

6

Analytical and Practical Considerations

A mass spectrum recorded during measurements provides 2D information on the intensities of ionic species as a function of the 'position' of ion peaks corresponding to the mass of separated ions, whereby the ion intensities of atomic and polyatomic ions and their corresponding positions (mass-to-charge ratio, m/z of ions) are represented on the y- and x-axis, respectively. Such a mass spectrum, which, in contrast to a spectrum obtained by optical emission spectrometry, displays a very simple structure, can be used for both qualitative and quantitative analysis. The occurrence of ion peaks as a function of the mass number together with the knowledge of the isotopic pattern therefore give the analyst qualitative information on the elemental composition of the sample investigated. On the other hand, quantitative data can be obtained from the number of ionized atoms and molecules (ion intensities) measured mass spectrometrically.

6.1 Qualitative Analysis

The first step in the investigation of an unknown sample is to solve the question of the composition of materials with respect to major, minor and trace elements, i.e. the qualitative analysis. The second step is to determine how much the sample contains. Due to the number of relative masses of isotopes appearing, the lines or peaks of ions in the mass spectra are, in general, grouped according to the periodic table. The distances between the lines or peaks of the ions for mass spectrometers with a linear mass scale are nearly constant in the investigated mass range (see in Figure 1.13, for example, the line spectrum of Nd detected using an ion sensitive photoplate).

It is usually simple to find an orientation in the mass spectra. For example, in the mass spectrum of SSMS, LIMS, GDMS or LA-ICP-MS of a graphite or graphite mixture sample, carbon cluster ions (C_n^+ with $n = 1$–24) occur with an exactly constant mass difference of 12 u. As described in

Inorganic Mass Spectrometry: Principles and Applications J. S. Becker

Section 2.2, high purity graphite powder is utilized, for instance, to produce electrically conducting electrodes from non-conducting powdered samples in SSMS or dc GDMS. In addition, over the whole mass range investigated (from 12–240 u), carbon cluster ions are detected in LA-ICP mass spectra of high purity graphite analyzing trace impurities.[1] These carbon cluster ions of the major carbon isotope ^{12}C together with the less abundant ^{13}C isotope (isotopic abundance: 1.1 %) observed in the mass spectra are very helpful for orientation in a spark, glow discharge or laser induced mass spectrum of a carbon containing sample. Due the defined mass of ^{12}C of 12 u, these C_n^+ cluster ions can also be employed for the exact mass calibration of the mass spectrometer in quite different applications of mass spectrometry. The typical abundance distribution of carbon cluster ions as a function of size observed in LIMS, GDMS, SSMS and LA-ICP-MS is discussed in Section 9.10.[2,3] Because most atomic ions (except for some light elements from H to N and the heavy metals thorium and uranium, see Table of Atomic Weights in Appendix II) possess a negative mass defect (see also Figure 1.12) they therefore occur at lower masses than the whole mass number *A*. In a mass spectrum acquired at medium mass resolution, it is easy to distinguish between atomic ions of analyte and carbon cluster ions (e.g., $^{60}Ni^+$ and $^{12}C_5{}^+$) or organic molecular ions, which mostly have a higher mass compared to the carbon clusters due to the mass atomic of hydrogen of 1.007825u.

In Figure 6.1, two parts of the mass spectrum of the glass sample NIST SRM 613 measured by rf GDMS are illustrated.[4] In the mass spectrum of the glass sample, the orientation for a qualitative analysis focuses on the ions from matrix, major and minor elements (Si^+, Al^+, Na^+, Rb^+, Sr^+ or Pb^+) and the plasma gas species, such as $^{40}Ar^{16}O^+$, $^{40}Ar_2{}^+$ or $^{40}Ar^{12}C^+$. The plasma gas ions $^{40}Ar^{2+}$, $^{40}Ar^+$, $^{40}ArH^+$ ions are formed in a glow discharge source with very high intensities and therefore they are skipped in Fig. 6.1a. The qualitative composition of the sample is determined from the position of the peaks in the mass spectrum together with the isotopic pattern (for example, the four lead isotopes $^{204}Pb^+$, $^{206}Pb^+$, $^{207}Pb^+$, $^{208}Pb^+$, which have an isotope abundance of 1.4 %, 24.1 %, 22.1 % and 52.4 %, respectively, and the two Tl isotopes $^{203}Tl^+$ and $^{205}Tl^+$ with isotope abundances of 29.5 % and 70.5 %, see Figure 6.1b). However, besides the atomic ions of analyte polyatomic ions also occur in the mass spectra (e.g.,$^{28}Si^{16}O^+$, $^{28}Si^{40}Ar^+$ or $^{209}Bi^{28}Si^+$ shown in Figure 6.1a and b, respectively), which can disturb the analysis of the sample. As mentioned before, in inorganic mass spectrometry mostly positive singly charged ions are measured. The doubly charged ions of an analyte occurring in mass spectra at the half masses, e.g., $^{63}Cu^{2+}$ and $^{65}Cu^{2+}$, appear in the mass spectra at an *m/z* ratio of 31.5 and 32.5 u, respectively. Doubly charged atomic ions possess the same isotope pattern as singly charged atomic ions and can be used in the same way as singly charged ions for orientation on the mass scale in the mass spectrum (and for quantification). In Figure 6.1 only doubly charged argon ions are observed.

The calibration of the mass scale of a mass spectrum in ICP-MS is easy using a tuning solution containing several analytes over the whole mass range investigated. By combining the knowledge of the isotopic pattern of elements and the position in the mass spectrum (*m/z*), qualitative analysis is performed using all types of mass spectrometers. In modern mass spectrometers, an advanced library and search feature provides access to the information needed for the interpretation of mass spectra in unknown samples.

6.1.1 Isotopic Pattern

Mass spectrometric measurements are based on measuring ion currents of separated ion beams of isotopes. With knowledge of the isotopic composition of the elements investigated (see Table of Isotopic Abundances in Appendix I[5]), a simple identification of chemical elements using singly

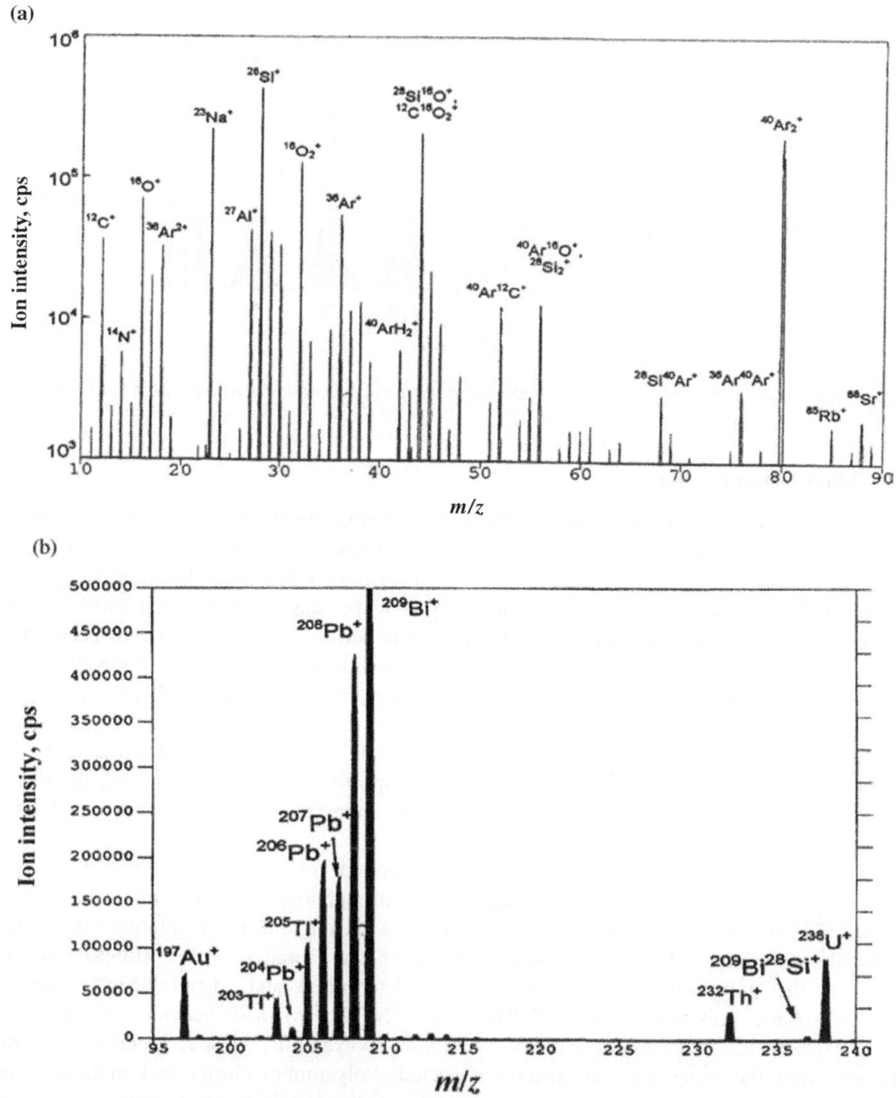

Figure 6.1 Part of a mass spectrum measured by rf GDMS of a glass sample (NIST SRM 613): a) in the mass range m/z 10–90 (analogue mode, highly abundant peaks of $^{40}Ar^{2+}$, $^{40}Ar^{+}$ and $^{40}ArH^{+}$ are skipped) and b) m/z 95–240 (measured in the counting mode). (A. I. Saprykin, J. S. Becker and H. J. Dietze, Fresenius' J. Anal. Chem., **359**, 449(1997), with permission of Springer Science and Business Media.)

charged ions can be performed. Figure 1.8 shows the typical mass spectrum of magnesium with three isotopes, ^{24}Mg, ^{25}Mg and ^{26}Mg, with isotope abundances of 78.99 %, 10.00 % and 11.01 %, respectively, measured by Dempster in 1935.[6] Figure 6.2 illustrates the mass spectrum of hafnium with six isotopes measured by ICP-SFMS.

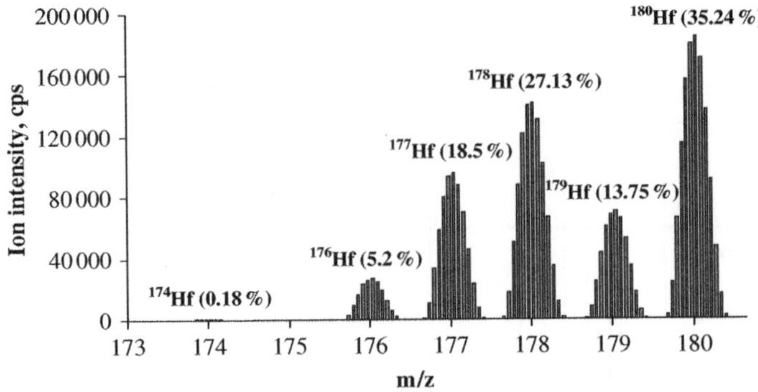

Figure 6.2 *Mass spectrum of hafnium isotopes measured by double-focusing sector field ICP-MS.*

6.1.2 Mass Determination

The mass determination of ionic species (atomic or polyatomic ions) in mass spectrometry is always a comparative measurement, which means the mass of an ionic species is determined with respect to reference masses of elements (or substances) used for mass calibration. The reference mass is thus acquired from the mass unit ($m_u = 1\,u = 1/12$) of the mass of the neutral carbon isotope ^{12}C ($m_u = 1.66 \times 10^{-27}$ kg). A mass calibration is easy to perform in solid-state mass spectrometry if the sample contains carbon (using carbon cluster ions with whole masses, as discussed above). The so-called 'doublet method' was applied formerly, e.g., $^{12}C^+$ ions and doubly charged $^{24}Mg^{2+}$ forming a doublet at the same nominal mass number 12 were considered, where they are slightly displaced with respect to one another. The 'doublet method' is no longer of relevance in modern inorganic mass spectrometry. Orientation in the mass spectra can be carried out via the matrix, minor and trace elements after mass calibration and by comparing the measured isotopic pattern of elements with theoretical values.

In ICP-MS a multi-element tuning solution is applied for the mass calibration of mass spectra. Figure 6.3 shows the mass spectrum for phosphorus determination at $m/z = 31$. In this case, the mass calibration was performed with the aid of a solution of a phosphorus/sulphur mixture. From the known masses of the isotopes of the atomic ions, the masses of the polyatomic ions occurring were determined and identified in accordance with the isobaric polyatomic ions ($^{15}N^{16}O^+$ and $^{14}N^{16}OH^+$). *Mass accuracy* is the deviation of experimental determined mass of an atomic, polyatomic, cluster, molecular or fragment ion from the exact mass of species expected. Polyatomic, cluster and molecular ions were calculated from the atomic masses (see Appendix I), the mass accuracy is usually presented in ppm.

6.1.3 Interference Problems

A general problem in trace, species and isotope analysis is possible isobaric interferences. In all mass spectrometric techniques, the ability to determine several isotopes is limited due to the occurrence of isobaric interferences of atomic ions of analyte with isobaric atomic ions (e.g., $^{92}Zr^+$ and $^{92}Mo^+$) or polyatomic ions (e.g., $^{239}Pu^+$ and $^{238}U^1H^+$) at the same nominal mass. Due to the slight mass difference the separation of interferences of isobaric singly charged atomic ions (e.g., $^{92}Zr^+$ and $^{92}Mo^+$: $m/\Delta m \approx 52\,000$) requires mass spectrometers with high mass resolution such as Fourier

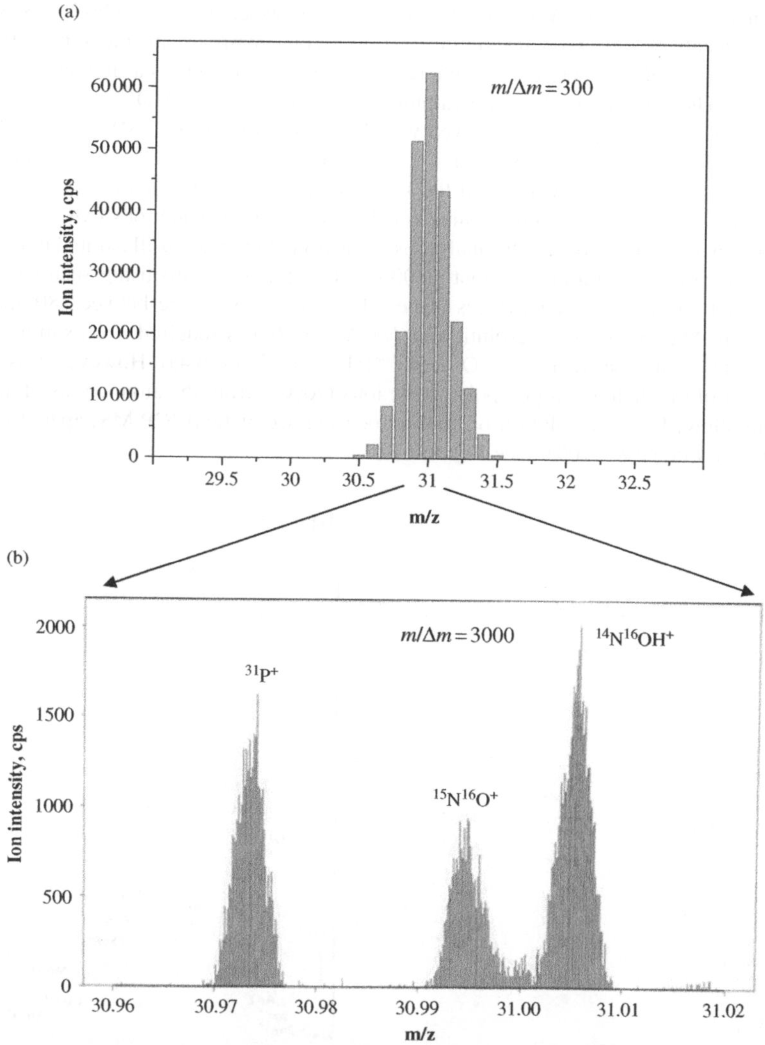

Figure 6.3 *ICP mass spectrum of phosphorus and interfering polyatomic ions measured by ICP-MS at low mass resolution a)* (m/Δm = 300) *and medium mass resolution b)* (m/Δm = 3000).

transform or ion trap mass spectrometers.[7,8] In many cases, interferences of singly charged atomic ions with doubly charged ones (e.g., the separation of $^{92}Mo^+$ and $^{184}W^{2+}$ requires a mass resolution $m/\Delta m \approx 1300$) can be separated by double-focusing sector field mass spectrometry. The apparent interferences of atomic ions of analyte and disturbing polyatomic ions at the same nominal mass in the mass spectra (e.g., $^{41}K^+$ and $^{40}ArH^+$: $m/\Delta m \approx 5000$) are of the greatest importance as demonstrated in numerous studies. The interference problem in inorganic mass spectrometry has been discussed in many papers.[2,9–19] Figure 6.3 shows the mass spectrum of monoisotopic phosphorus ($^{31}P^+$) together with isobaric polyatomic ions $^{15}N^{16}O^+$ and $^{14}N^{16}OH^+$ measured at low and medium mass resolution ($m/\Delta m \approx 300$ (a) and 3000 (b), respectively) by ICP-SFMS. Whereas during measurements at low mass resolution, too high concentrations are measured due to the interference problem, reliable data for phosphorus can be obtained from measurements at medium mass resolution. The formation rate of polyatomic ions such as MO^+, MOH^+, MCl^+, MX^+ or ArX^+ (polyatomic ions of argon are observed with high intensities in ICP-MS and GDMS) with M = matrix or minor element and X = H, O, N, Cl and others. The generation of polyatomic ions in mass spectrometry is dependent on the stability of ionic species and on the experimental parameters used. The mass resolution required for the separation of oxide ions (MO^+) and argide ions (MAr^+) from the atomic ions of the analyte as a function of the mass is illustrated in Figure 6.4. With a maximum mass resolution of about 10 000–12 000 for double-focusing sector field mass spectrometers not all isobaric interferences (especially in the mass range between 80 and 140 u) can be separated. Maximum mass resolution of ($m/\Delta m \approx 10^5$) is required, for example, for the separation of interference between $^{90}Zr^{18}O^+$ and $^{108}Pd^+$ (see Figure 6.4a). However, it is possible to separate most of the diatomic argon polyatomic ions (ArX^+) from the atomic ions of analyte at the maximum allowed mass resolution of double-focusing sector field ICP-MS, apart from some dimers in the mass range 80–100 u (see Figure 6.4b).

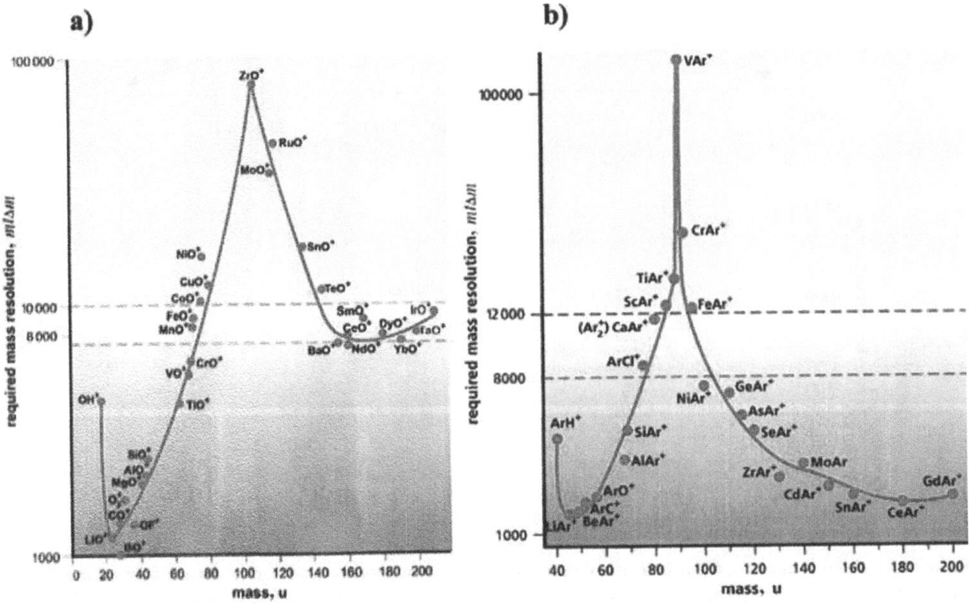

Figure 6.4 Mass resolution required for separation of atomic ions: a) from oxide ions and b) from diatomic argon ions.

Figure 6.5a illustrates part of a real mass spectrum of a radioactive waste solution in the mass range 180–250 u measured by double-focusing sector field ICP-MS (ELEMENT, Thermo Fisher Scientific) at low mass resolution. In this mass spectrum, atomic ions of Re at mass 185 u and 187 u with isotope abundances of 37.4 % and 62.6 %, respectively (Re was used for ^{99}Tc precipitation), ReO^+, ReO_2^+ and ReO_3^+ ions were detected. In particular, the ReO_3^+ and ReO_3H^+ ions appearing at the mass range 233–236 u affected the isotope analysis of uranium due to isobaric interferences.

For the mass spectrometric separation of $^{187}ReO_3^+$ from $^{235}U^+$ ions, a mass resolution ($m/\Delta m$) of 2350 is necessary. In Figure 6.5b part of a mass spectrum in the mass range 234.9–235.1 u obtained at 'medium' mass resolution $m/\Delta m = 3000$ is shown.[20] It was possible to determine the mass of these polyatomic ionic species of Re after spiking the analyte solution with 1 μg l^{-1} ^{235}U.[20]

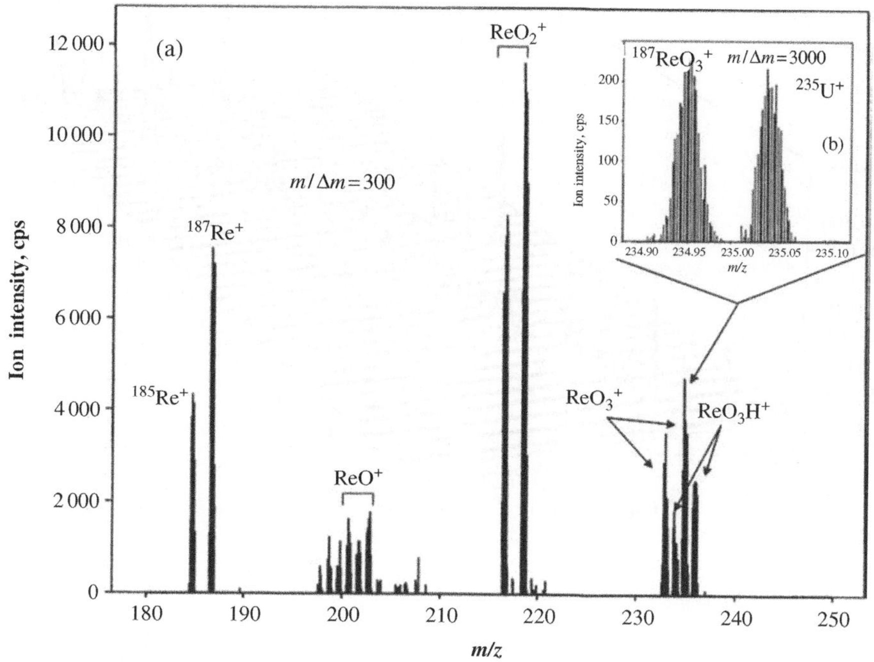

Figure 6.5 *Part of real mass spectrum of a radioactive waste solution in the mass range 180–250 u measured by double-focusing sector field ICP-MS (Element, Thermo) at low mass resolution. Top: part of a mass spectrum in the mass range 234.9–235.1 u obtained at mass resolution* m/Δm = 3000. (J. S. Becker and H. J. Dietze, J. Anal. At. Spectrom., ***14***, 1493(1999). *Reproduced by permission of The Royal Society of Chemistry.*)

Reducing oxide based isobaric interferences in the ICP mass spectrum via gas flow modulation was proposed by Wetzel and Hieftje.[21] After a careful manipulation of the central channel gas flow to impact distinguishable frequency specific behaviour of analyte and oxide ion species and application of a Fourier transform (FT) correction method, contributions from an analyte and oxide species superimposed at a given mass can be mathematically unravelled with a degree of success. Through application of this correction method, a greater than ten-fold error at *m/z* 156 caused by the interference of $^{149}Ce^{16}O^+$ on $^{156}Gd^+$ has been effectively eliminated.[21]

To avoid interferences of isobaric atomic ions of different elements and polyatomic ions at the same nominal mass, off line separation of matrix elements or analyte separation can be applied and/or enrichment and hyphenated techniques such as HPLC- or CE-ICP-MS. Analytical procedures have been developed in the author's laboratory for the determination of spallation nuclides in an irradiated tantalum target using HPLC and CE coupled on line to ICP-MS after dissolution and separation of the tantalum matrix.[12,17] The mass spectra of rare earth elements (REEs)

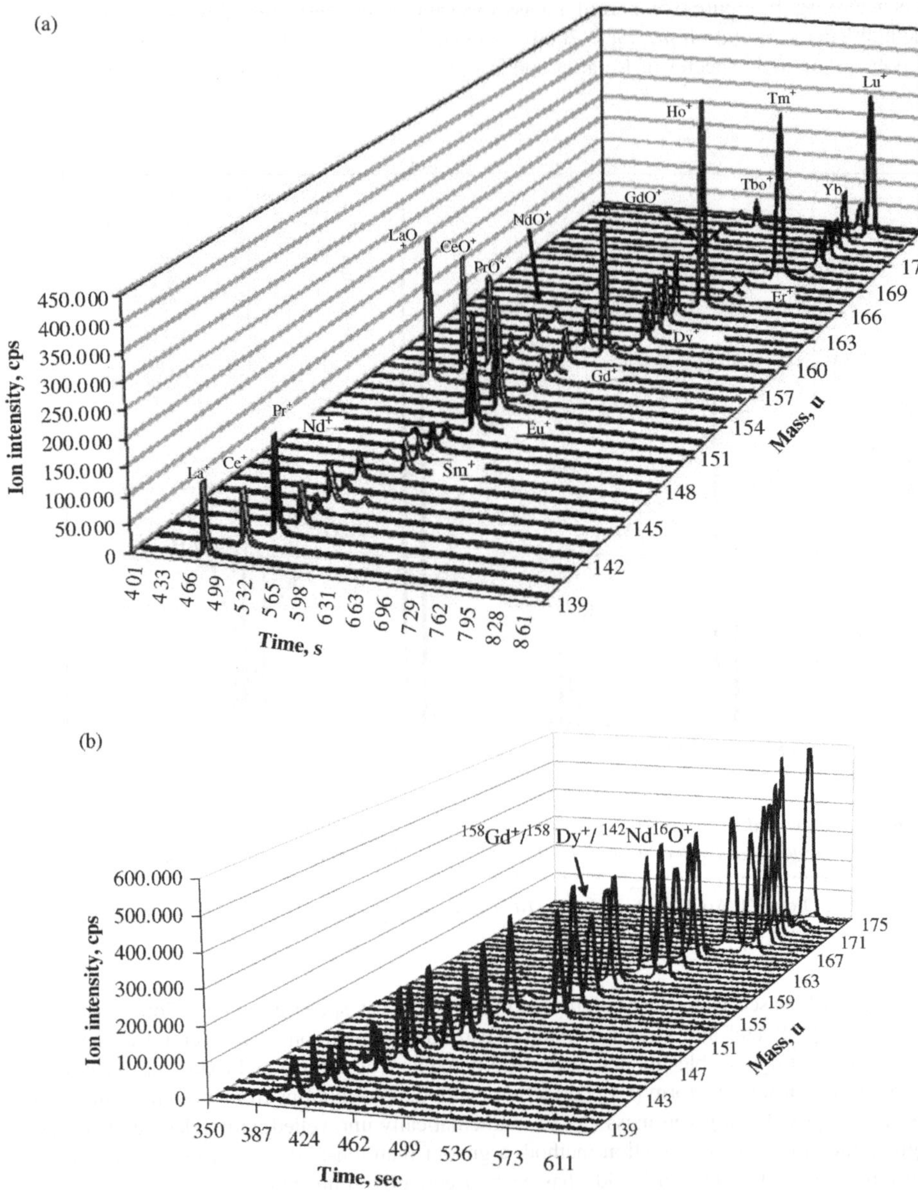

Figure 6.6 *Mass spectrum of rare earth elements (REEs) measured by CE-ICP-MS with on line separation of analytes (50 nl solution volume): a) natural isotopic pattern, concentration of each REE – 800 ng g^{-1}; b) in an irradiated tantalum target (J. S. Becker and H. J. Dietze, Int. J. Mass Spectrom. Ion Proc. 197, 1–35 (2000). Reproduced by permission of the Royal Society of Chemistry.)*

of natural isotopic pattern and those in an irradiated tantalum target separated by online CE-ICP-MS in aqueous solution are compared in Figure 6.6.[12] The nuclide abundances of all REEs in the irradiated tantalum target (Figure 6.6b) differ significantly from the natural isotopic pattern (see Figure 6.6a), and consequently a lot of isobaric interferences of long-lived radionuclides with stable isotopes in the mass spectrum are observed. In the highly radioactive solution isobaric interferences such as $^{158}Gd^+$, $^{158}Dy^+$, $^{158}Tb^+$ and $^{142}Nd^{16}O^+$ are well separated by on line CE-ICP-MS as demonstrated in Figure 6.6b.

Part of a mass spectrum of silicon determination at an *m/z* of 28 in high purity GaAs measured by rf GDMS (Element, Thermo Fisher Scientific with rf glow discharge ion source constructed at our laboratory) is illustrated in Figure 6.7.[22] The interfering polyatomic ions $^{12}C^{16}O^+$ and $^{14}N_2{}^+$ are well separated at a mass resolution $m/\Delta m$ of 3000. In contrast to simple mass spectra in ICP-MS or GDMS, the surface analytical technique SIMS suffers from a huge polyatomic and cluster ion formation rate as the result of the sputtering processes. The serious interference problems in SIMS are demonstrated in Figure 6.8 where polyatomic ion formation by sputtering of borophosphosilicate glass using an $O_2{}^+$ primary ion beam is shown.

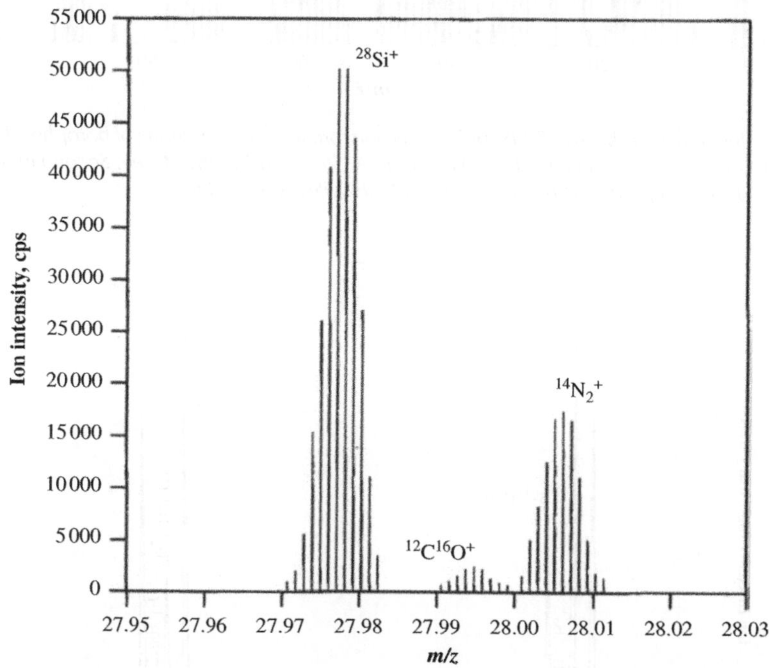

Figure 6.7 *Part of a mass spectrum for silicon determination at m/z = 28 in high purity GaAs measured by rf GDMS (Element, Thermo with rf glow discharge ion source constructed in our lab). The interfering polyatomic ions $^{12}C^{16}O^+$ and $^{14}N_2{}^+$ are well separated at mass resolution m/Δm of 3000. (J. S. Becker et al., Int. J. Mass Spectrom.* ***164****, 81 (1995). Reproduced by permission of Elsevier.)*

The determination of trace impurities in boron nitride of the highest purity for the production of GaAs single crystals is hampered by the formation of disturbing polyatomic and cluster ions.

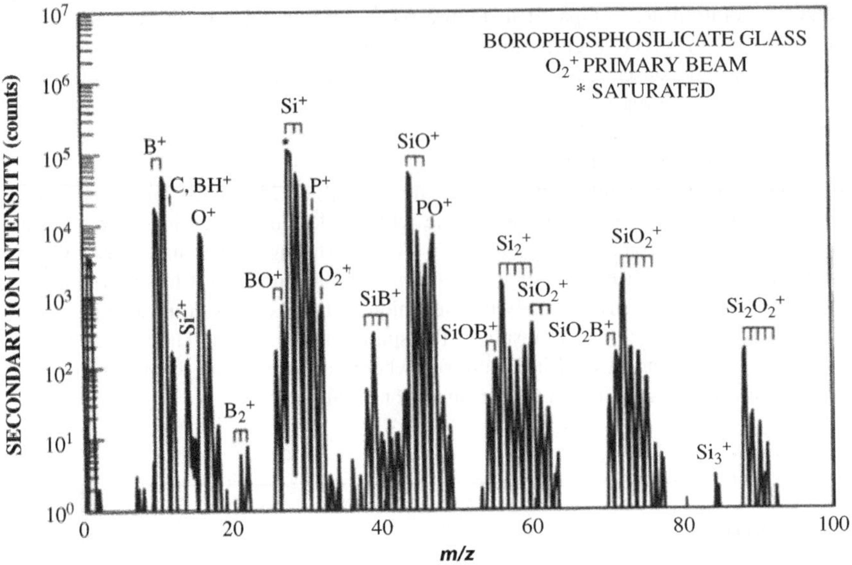

Figure 6.8 Interference problem in SIMS: polyatomic ion formation by sputtering of borophosphosilicate glass using an O_2^+ primary ion beam. (R. G. Wilson, F. A. Stevie and C. W. Magee, Secondary Ion Mass Spectrometry (1989). Reproduced by permission of John Wiley & Sons, Inc.)

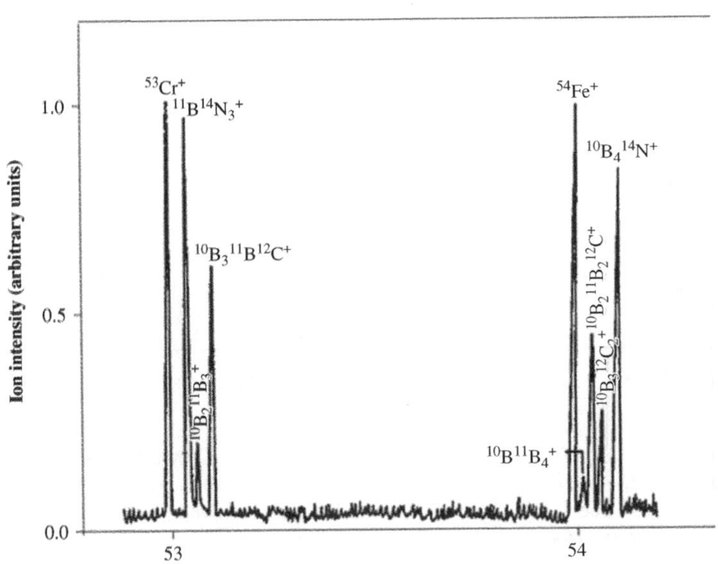

Figure 6.9 Part of laser mass spectrum of boron nitride contaminated with carbon, iron and chromium. (J. S. Becker and H. J. Dietze, Fresenius' J. Anal. Chem., **344**, 69 (1992). Reproduced by permission of Springer Science and Business Media.)

Part of a mass spectrum for the determination of Fe and Cr contamination in boron nitride contaminated with carbon measured by LIMS is shown in Figure 6.9. The analyte ions $^{53}Cr^+$ and $^{54}Fe^+$ due to different masses of isobaric atomic and cluster ions are clearly separated from boron and boron carbide cluster ions as demonstrated in Figure 6.9. Cluster ion formation has been studied by laser ionization mass spectrometry (LIMS) on a boron nitride target.[10]

Ion–molecular reactions are used to resolve isobaric interferences, as discussed, in ICP-MS with a collision/reaction cell or by utilizing ion traps. The mass spectra of Sr, Y and Zr (Fig. 6.10a) without O_2 admitted into the collision cell and (Fig. 6.10b) with $\sim 10^{-3}$ Pa O_2 are different. By introducing oxygen, selective formation of YO^+ and ZrO^+, but not SrO^+, is observed. This behaviour of different oxide formation is relevant for an interference free determination of ^{90}Sr. Ultrahigh mass resolving power ICP mass spectrometry (at $m/\Delta m \approx 260\,000$) selectively removes unwanted ions prior to transfer to the FTICR analyzer cell by gas-phase chemical reactions, e.g., for separation of $^{40}Ca^+$ from $^{40}Ar^+$ obtained with a Fourier transform ion cyclotron resonance (FT-ICR) mass spectrometer equipped with a 3 tesla superconducting magnet.[23]

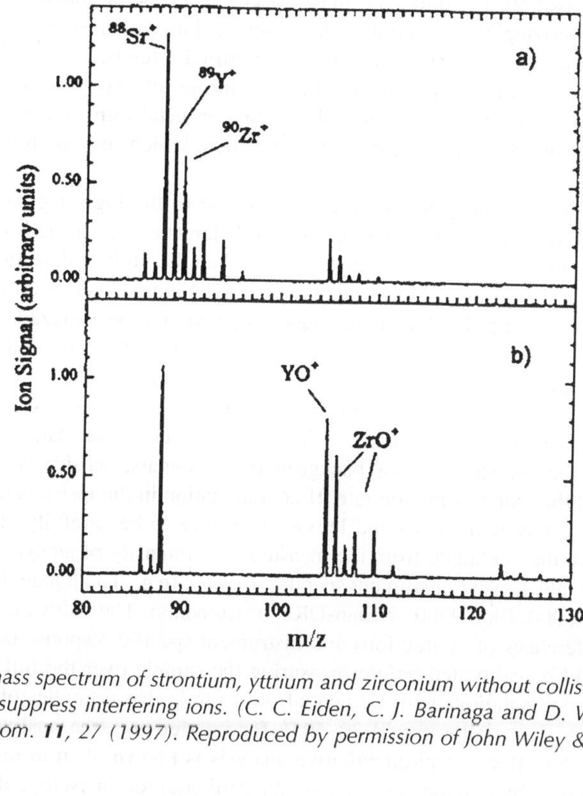

Figure 6.10 *ICP mass spectrum of strontium, yttrium and zirconium without collision gas (a) and using collision gas (b) to suppress interfering ions. (C. C. Eiden, C. J. Barinaga and D. W. Koppenaal, Rapid Comm. Mass Spectrom.* ***11****, 27 (1997). Reproduced by permission of John Wiley & Sons, Ltd.)*

6.2 Quantification Procedures

Quantification of analytical data (ion intensities of analytes) is a fundamental aspect of acquiring the best possible results from mass spectrometric measurements in order to obtain reliable values for concentrations of elements or of element species. The *dynamic range* of a mass spectrometric technique is the total recorded range between the limit of detection (LOD) and the maximum allowed concentration that can be measured (including possible dilution in ICP-MS). Part of the dynamic range is the calibration curve (ion intensity of analyte as a function of analyte concentration). The gradient of a calibration curve ($\Delta I/\Delta$conc) indicates the *sensitivity* of the experimental arrangement. Several calibration strategies applied in inorganic mass spectrometry, such as external calibration, standard addition or isotope dilution techniques, will be discussed in the following section.

6.2.1 Semi-quantitative Analysis

A semi-quantitative measurement is often used for a quick analysis of an unknown sample with respect to a rough determination of the matrix and major and trace element concentrations, and can be performed as the first step before a quantitative analysis of selected elements or element species by inorganic mass spectrometry. A semi-quantitative analysis gives the analyst a quick overview of the elemental composition. In solid-state mass spectrometry, semi-quantitative measurements are helpful for characterizing the material under investigation with respect to minor and trace impurities in a given matrix, especially if no suitable certified reference material (CRM) is available for the quantification procedure. In addition, by means of inorganic mass spectrometry, e.g., LA-ICP-MS, fingerprinting of a small sample of glass or metal splinters, projectiles, car paints etc. can provide information on the origin of the specimen, which may be relevant for forensic investigations.

ICP-MS offers excellent semi-quantitative capability due to the high degree of ionization for most elements. In ICP-MS, one or more internal standard elements are selected which are assumed to be absent in the sample (or the concentration is lower than the limit of detection, such as Sc, Y, In, Ge, Rh, Te, Eu, Tb, Ho and Re). These internal standard elements with well-known concentrations (e.g., at $10\,\mu g\,l^{-1}$) are added to the diluted aqueous solution to be analyzed. Using the internal standard elements, possible instabilities (of quadrupole or sector field mass analyzers resulting in instrumental drifts, plasma flickering, variation in transport of solution or ion suppression due matrix effects etc.) can be identified and corrected. A daily performance test is recommended to check the capability of ICP-MS after optimization of the experimental parameters. This performance test includes the measurement of background ion signals, sensitivity and precision for selected elements and the oxide formation rate. If contamination in the instrument is observed then the solution introduction system, cones and lens system have to be carefully cleaned. The semi-quantitative results are then obtained from the measured ion intensity of analytes by a comparison of data from response tables using instrument specific software (e.g., TotalQuant in the PerkinElmer Software in the Elan 6000, Elan 9000 or Elan-DRC instruments). The software includes a correction of isobaric interferences of atomic ions and instrument specific response tables for elements. The instrument should be calibrated before measuring the sample over the full mass range using multi-element calibration solutions with elements from Li to U. Uncertainties of semi-quantitative analysis in ICP-MS are generally around 30–50 % for interference free analytes. If the concentration range of analytes after a semi-quantitative analysis is known, then in the following step a quantitative measurement by external calibration, standard addition or isotope dilution techniques is recommended.

A semi-quantitative multi-element analysis – limited in accuracy and precision – is performed in solid-state mass spectrometry if the concentration of at least one element – used as an internal standard element – is known. For example, if trace impurities have to be analyzed in high purity copper metal, then Cu with a known concentration of 100 % is chosen as the internal standard element. Concentrations of impurities are calculated with respect to the known matrix concentration using the following analysis equation:

$$c(x) = \frac{c(v) \cdot I(x) \cdot A(v)}{I(v) \cdot A(x)} \tag{6.1}$$

where x is the element to be determined, v is the internal standard element, c is the element concentration, I the ion intensity measured and A is the isotopic abundance.

Trace element concentrations are obtained in the semi-quantitative mode by ICP-MS, LA-ICP-MS, SSMS or SNMS with an error factor of about 0.3–3 for most elements. The results of semi-quantitative trace analysis, e.g., for high purity materials, are sometimes sufficient to estimate the purity of the matrix investigated.

In contrast, due to significantly large matrix effects SIMS does not allow semi-quantitative analysis of solid samples. SIMS only enables a qualitative analysis of sample composition – but as a function of the matrix investigated and the primary ion beam parameters chosen it does offer high detection power for trace elements. Due to inherently high polyatomic ion formation during the sputtering process of the solid sample in SIMS, even a qualitative trace analysis is sometimes difficult especially in the case of analyzing a complex matrix composition of the investigated material. A semi-quantitative measurement in SIMS is possible by means of the well-established method of 'Cs molecular ion SIMS', that is the application of diatomic MCs^+ ions (M – metal or non-metal atom), which are formed by bombardment of a solid surface with Cs^+ primary ions in SIMS. Caesium polyatomic ions[24] are very suitable compounds because caesium is mono-isotopic, and reactive and stable polyatomic ions exist for the elements. By the application of MCs^+ polyatomic ions – measured at a mass which is 133 u higher than the mass of M^+ – it is possible to avoid interferences which are observed at the mass of analyte ion M^+. Advantageous for analytical purposes is the fact that the formation of these MCs^+ polyatomic ions in SIMS is less dependent on matrix[24] although the formation of MCs^+ polyatomic ions is strong dependently on the Cs surface concentration. If MCs^+ ions or partly $MCs_2{}^+$ polyatomic ions are used in depth profiling or trace analysis, the intensity of the polyatomic ions is normalized with respect to Cs^+ or $Cs_2{}^+$ ions, respectively, to consider the different sputter yield and possible variation during the sputtering process. This technique can also be applied for quantitative bulk analysis and depth profiling if relative sensitivity coefficients are experimentally determined on reference materials.

6.2.2 One Point Calibration in Solid-state Mass Spectrometry Using a Certified Reference Material

Certified reference materials (CRMs) are mainly applied to validate the analytical procedure developed for routine analysis in order to determine the accuracy of analytical data, the recovery for selected elements, the uncertainty of trace element determination and the detection limits. Otherwise, in solid-state mass spectrometric techniques, such as SSMS, LA-ICP-MS, GDMS, SNMS or SIMS, one point calibration using CRMs has been established as an important calibration strategy to obtain reliable analytical data. The one point calibration is performed using the experimentally determined relative sensitivity coefficients (RSCs) on a suitable CRM with a similar trace/matrix composition. An RSC of a chemical element is defined as the ratio of the measured element concentration (experimentally determined) divided by the certified element concentration (accepted or recommended value of element concentration) in a given matrix.

$$\text{RSC} = \frac{\text{Experimentally measured element concentration}}{\text{Certified element concentration}} \tag{6.2}$$

Because the RSCs of chemical elements depend on the analytical procedure applied, the RSCs are measured by inorganic mass spectrometric techniques using CRMs – if available with the same matrix or a similar matrix composition compared to the sample investigated – under identical experimental conditions. Consequently, the influence of matrix composition on the analyte ion formation (matrix effect) is considered. In general, the RSCs of elements vary for the different solid-state mass spectrometric methods mentioned above (except SIMS) depending on the experimental parameters used and the matrix composition of sample by about one order of magnitude (SSMS, LA-ICP-MS, SNMS and GDMS: roughly from 0.2 to 3). Figure 6.11 summarizes the RSCs measured by LA-ICP-MS for three geological reference materials, granite NIM-G, lujavrite NIM-L and syenite NIM-S, from the National Institute for Metallurgy, South Africa, which have been used for quantitative trace analysis of geological samples or zeolites.[25] To overcome differences due to crystallization and to transfer inhomogeneous samples into homogenous targets, fused lithium borate targets (by fusion with a lithium borate mixture (90 % $Li_2B_4O_7$, 10 % $LiBO_2$)) of geological samples and certified references materials were prepared. This special technique of sample preparation known from X-ray analysis enables improved matrix matching of samples and CRMs. Figure 6.12 compares the RSCs of two certified reference materials often used in the geological sciences (NIST SRM 612 glass standard and BRC 2G basaltic glass prepared from BCR-2 reference material) measured by LA-ICP-MS. In this measurement, the internal standard element was strontium. As also observed in other experiments[25,26,27] and already mentioned above, the RSCs varied for both reference materials between 0.2 and 3. Semi-quantitative determination of trace elements without CRM is therefore possible by LA-ICP-MS with an error factor in this range. Figure 6.13 compares the LA-ICP-MS results for trace elements in the basalt glass (KL2-G from the Kilauea volcano on Hawaii) using different glass reference materials (NIST SRM 612 and BCR-2G) with the recommended value.[28] For most elements (except Zn and Sn) adequate agreement of analytical results was found, showing that the use of BCR-G2 as reference material lead to results which are in better agreement with the reference values[28] .

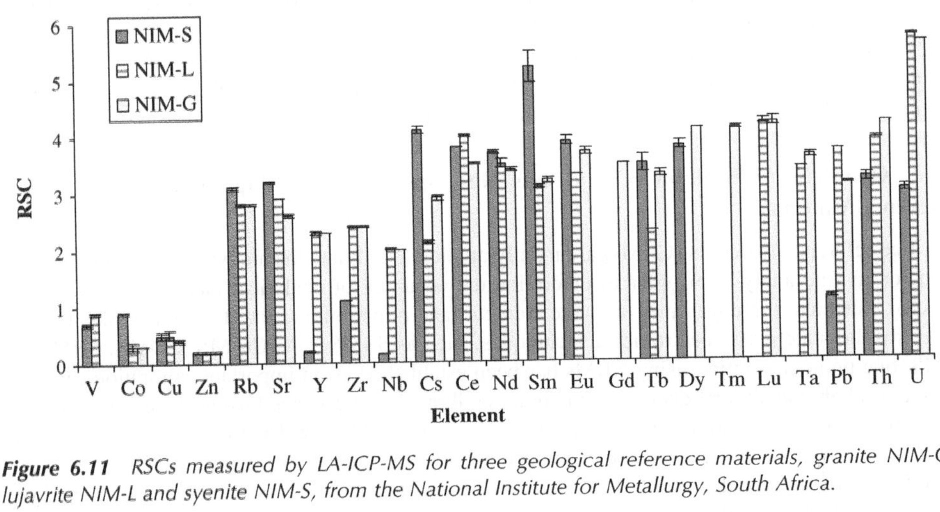

Figure 6.11 *RSCs measured by LA-ICP-MS for three geological reference materials, granite NIM-G, lujavrite NIM-L and syenite NIM-S, from the National Institute for Metallurgy, South Africa.*

A comparison of sensitivities for several elements in LA-ICP-MS and ICP-MS – relevant for solution-based calibration in LA-ICP-MS described below – are illustrated in Figure 6.14.

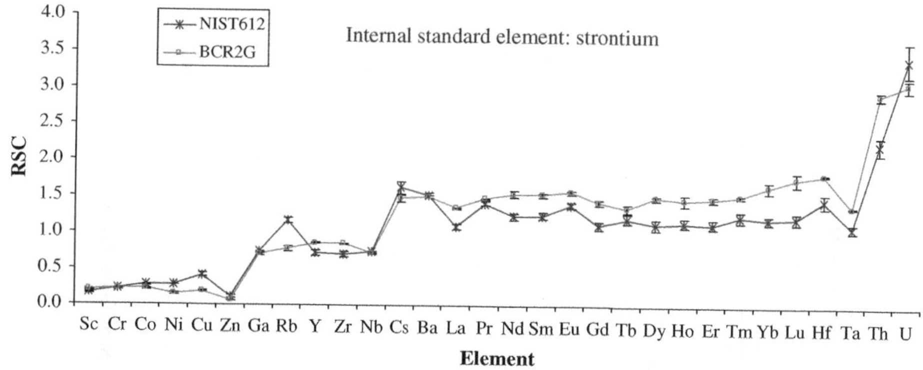

Figure 6.12 *RSCs of two standard reference materials (NIST SRM 612 glass standard and BRC 2G basaltic glass prepared from BCR-2 reference material) measured by LA-ICP-MS.[30] (J. S. Becker et al., Mikrochim. Acta* **135***, 71 (2000). Reproduced by permission of Springer Science and Business Media.)*

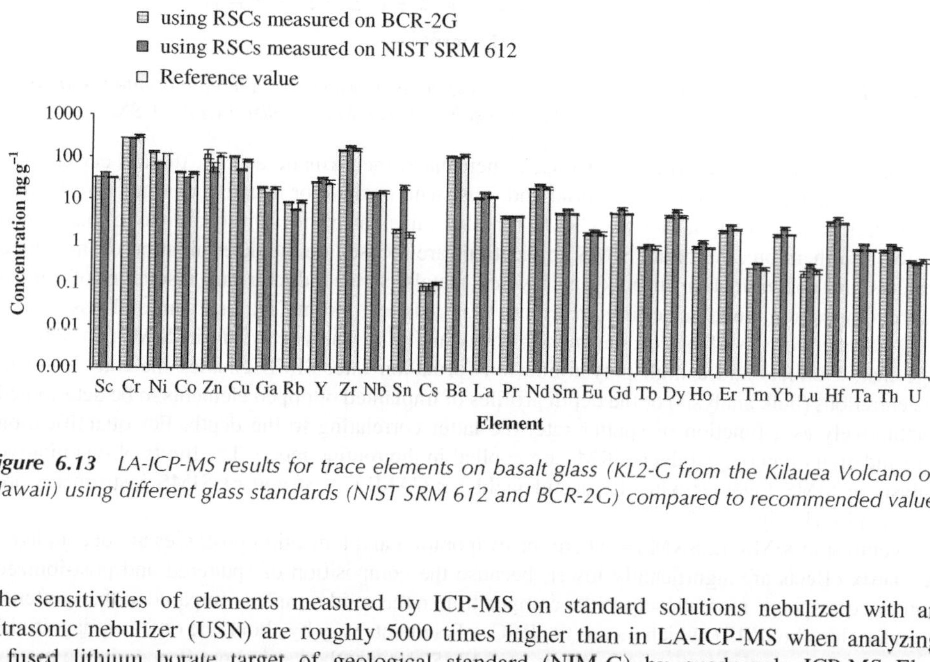

Figure 6.13 *LA-ICP-MS results for trace elements on basalt glass (KL2-G from the Kilauea Volcano on Hawaii) using different glass standards (NIST SRM 612 and BCR-2G) compared to recommended value.*

The sensitivities of elements measured by ICP-MS on standard solutions nebulized with an ultrasonic nebulizer (USN) are roughly 5000 times higher than in LA-ICP-MS when analyzing a fused lithium borate target of geological standard (NIM-G) by quadrupole ICP-MS Elan 6000.[29] The highest element sensitivity (see also Figure 6.12) is observed for the heavy radionuclide elements Th and U. In general, a similar dependence is found for RSCs from masses under optimized experimental conditions in LA-ICP-MS and ICP-MS using the ICP-QMS Elan 6000.

In contrast to all other mass spectrometric techniques, in secondary ion mass spectrometry (SIMS) the relative sensitivity coefficients of elements show a very strong variation, from the matrix and experimental conditions used, of up to six orders of magnitude. The RSCs vary considerably in

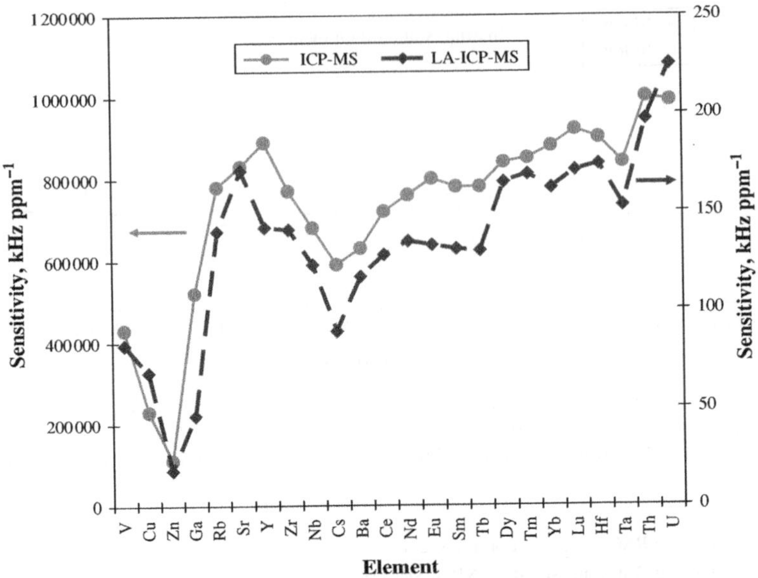

Figure 6.14 Comparison of sensitivities for several elements measured on a lithium borate fused target of geostandard NIM-G by LA-ICP-MS and standard solutions ICP-MS nebulized with a USN.

SIMS as a function of the primary ions used, their mass, their kinetic energy, the ion current, the angle of incidence of the ion, the material and ions (with positive or negative charge) investigated etc. Therefore if there is a lack of CRMs for a certain matrix no semi-quantitative analysis is possible. Furthermore, important SIMS applications are focused on the determination of implanted elements and contamination in bulk and layered metals, semiconductor and insulators of simple matrix composition. Suitable certified reference materials (e.g., from Charles Evans and Associates, USA) are therefore available on the market for quantifying analytical data in SIMS. In spite of large matrix effects and a high polyatomic ion formation rate, SIMS enables the trace element concentrations (bulk analysis) or the depth profiles of implanted or doped elements to be determined quantitatively as a function of sputter rate, the latter correlating to the depth. For quantification and validation of analytical data, CRMs are applied in the routine mode. The limits of quantitative analysis in SIMS together with the useful yield for CAMECA 3f and 6f SIMS instruments are discussed by Hervig *et al.*[31]

In contrast to SIMS, in SNMS – where the evaporation and ionization processes are decoupled – the matrix effects are significantly lower, because the composition of sputtered and post-ionized neutrals corresponds more closely to the composition in the solid sample (compared to the sputtered secondary ions in SIMS), which means the RSCs of elements vary by about one order of magnitude. Consequently, a semi-quantitative analysis by SNMS can also be carried out if no suitable matrix matched CRM is available. This is relevant for thin film analysis, especially for the determination of elemental concentration profiles in depth, for studying the stoichiometric composition of thin films and interdiffusion effects.

For the largest field of application for LA-ICP-MS – geological research – many different geological and glass standard reference materials are available. Using standard reference materials in LA-ICP-MS, analytical results for trace analysis in homogeneous samples can be obtained with an accuracy better than 10 %, and a precision of 2–5 % is also possible.[25] If no suitable CRM of similar matrix composition is available only a qualitative or semi-quantitative analysis can be performed, or synthetic laboratory standards have to be prepared for calibration purposes (see Section 6.5).

6.2.3 Quantification of Analytical Data via Calibration Curves in Mass Spectrometry Using Certified Reference Materials or Defined Standard Solutions

An easy calibration strategy is possible in ICP-MS (in analogy to optical emission spectroscopy with an inductively coupled plasma source, ICP-OES) because aqueous standard solutions with well known analyte concentrations can be measured in a short time with good precision. Normally, internal standardization is applied in this calibration procedure, where an internal standard element of the same concentration is added to the standard solutions, the samples and the blank solution. The analytical procedure can then be optimized using the internal standard element. The internal standard element is commonly applied in ICP-MS and LA-ICP-MS to account for plasma instabilities, changes in sample transport, short and long term drifts of separation fields of the mass analyzer and other aspects which would lead to errors during mass spectrometric measurements.

As a calibration procedure in ICP-MS via calibration curves, *external calibration* is usually applied whereby the blank solution is measured followed by a set of standard solutions with different analyte concentrations (at least three, and it is better to analyze more standard solutions in the same concentration range compared to the sample). After the mass spectrometric measurements of standard solutions, the calibration curve is created as a plot of ion intensities of analyte measured as a function of its concentration, and the linear regression line and the regression coefficient are calculated. As an example of an external calibration, the calibration curve of $^{239}Pu^+$ measured by ICP-SFMS with a shielded torch in the pg l^{-1} range is illustrated in Figure 6.15. A regression

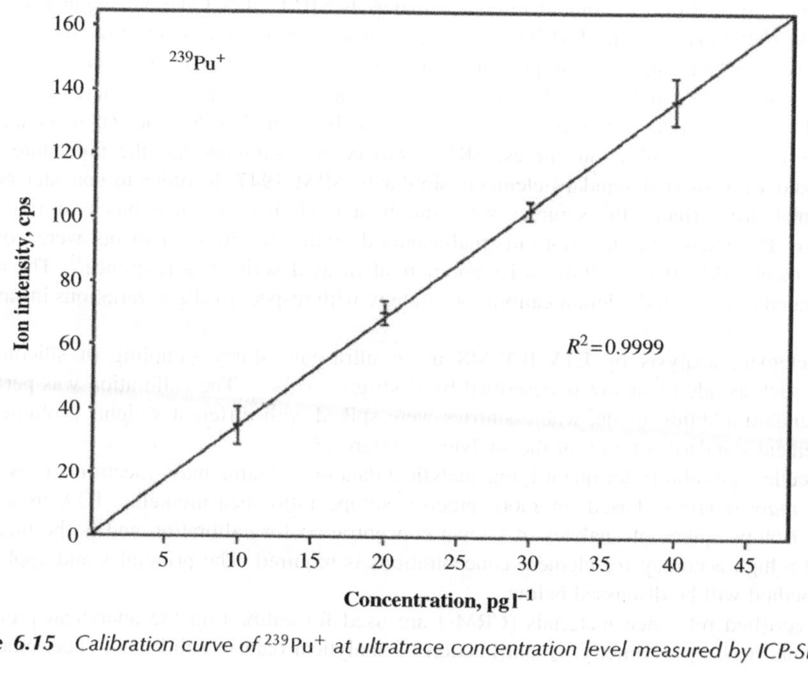

Figure 6.15 *Calibration curve of $^{239}Pu^+$ at ultratrace concentration level measured by ICP-SFMS.*

coefficient (R^2 of 0.9999) was calculated. This calibration curve was applied for quantitative plutonium determination in aqueous solutions at the ultratrace level. If a sample with unknown element concentrations is analyzed, the element concentration in the samples is calculated using the measured calibration curve considering possible interferences, the blank value and the dilution factor. Since pure standard solutions in high purity aqueous solutions (acidified with 2 % HNO_3) are generally applied for the creation of the calibration curve, no matrix effects or variation in element sensitivities and possible drifts were considered. If solutions with high matrix content have to be analyzed, then matrix matching of analyte standard solutions is recommended so that matrix elements are added to the standard solutions in a similar concentration range to the sample solution. First, the matrix matched standard solutions are analyzed and after that the sample solutions as described above using the external calibration mode with respect to trace element contents. The determination of element concentration is performed using measured calibration curves. Variation in sensitivity can be checked using the internal standard element and by repetition of measurements of standard and sample solutions.

External calibration is the most frequently used calibration technique in ICP-MS in the routine mode.

Another calibration technique – *standard addition* – minimizes matrix effects because analytes with well defined increasing concentrations are added to a set of sample solutions to be analyzed. The measured calibration curve in the standard addition mode plots the measured ion intensities of analytes versus the concentration added to the sample solution. The concentration of analytes in the undoped sample is then determined by extrapolation of the calibration curve with the x-axis. Matrix matching is subsequently performed and the matrix effects (signal depression or interference problems) are considered. An example of the standard addition technique is described in Section 6.2.6 using solution based calibration in LA-ICP-MS.

Christopher *et al.*[32] describe a multi-element standard addition method using collision cell ICP-MS for the certification of standard reference materials SRM 1947 Lake Michigan Fish Tissue using SRM 1946 Lake Superior Fish Tissue as the control sample. Prior to digestion, the SRM 1947 samples were spiked with a 1 g aliquot of a multi-element internal standard solution containing Nb, Y, Ru, Rh, In, Pr and Au, and each analytical samples was spiked with increasing aliquots of a multi-element spike solution containing As, Cu, Fe, Mn, Rb, Se and Zn to create a set of standard addition calibration curves. SRM 1946 control samples and the procedure blanks were spiked with internal standard elements similar to SRM 1947. In order to consider possible instrumental drift effects, the samples were run in a random order in a block of five repeat sequences. The measured analyte-to-internal-standards ratios for most elements were corrected for instrumental drift using a third order polynomial fit as described in reference[32]. The authors demonstrated a robust and tolerant calibration strategy with respect to slight variations in sampling a matrix.

Multi-element analysis by ETV-ICP-MS using ultrasonic slurry sampling of silicon based minerals such as talc or quartz is described by Rodriguez *et al.*[33] The calibration was performed in the standard addition mode, where slurries were spiked with different weights (volumes) of a multi-element standard solution of the analytes of interest.[33]

An excellent possibility for quantifying analytical data in inorganic mass spectrometry is *isotope dilution analysis* (IDA) based on more precise isotope ratio measurements. IDA uses highly enriched isotope spikes of analytes of known concentration for calibration and is the method of choice if a high accuracy for element concentrations is required. The principles and applications of this method will be discussed below.

Solid certified reference materials (CRMs) are used for calibrating the analytical procedures in inorganic mass spectrometry. Quantification of analytical data in solid mass spectrometry via calibration curves (external calibration or standard addition) is performed in an analogous way to the analysis of liquids. For the analysis of solid samples by LA-ICP-MS, GDMS, SIMS or SNMS via calibration curves, a set of certified standard reference materials (at least three) of similar matrix composition with different analyte concentrations is necessary. Such certified reference materials are available for trace analysis in glass matrices using the set of NIST certified reference materials of the same matrix composition, SRMs 610–617, Trace Elements in Glass. For example, NIST 610, 612 and 614 is a soda lime silicate glass doped with 61 trace elements nominally at 500, 50 and 1 $\mu g\,g^{-1}$, respectively. The NIST SRM glass series is used extensively in geoanalysis as microanalytical reference materials (as primary and secondary standards).[31,34,35] These samples were originally certified as bulk reference materials, and the homogeneity of elements is based on test of about 0.5 g of sample material. However, if smaller sample amounts are investigated by microlocal analysis (such as LA-ICP-MS), inhomogeneous trace element distribution may be observed. In addition, a multitude of certified reference materials with different matrix compositions are on offer for analytical quality assurance for trace analysis of geological, biological or medical materials, metals, ceramics and waters (e.g., from NIST – National Institute of Standards and Technology, Gaithersburg, USA (formerly NBS); from the NRCC – National Research Council Canada, Toronto, Canada; from BAM – Bundesanstalt für Materialforschung, Berlin, Germany; or from IRMM – Institute for Reference Materials and Measurements, Geel, Belgium, where BCR® from IRMM is a registered trademark of the European Commission). In addition to reference materials for trace element analysis, NIST and IRMM also supply isotope standard reference materials. The certificates carry a certified value with its uncertainty, which is related either to SI units or an internationally accepted reference.

Trends in certified reference materials for the speciation of trace elements are discussed by Cornelis *et al.*[36,37]

Relevant for analysis of geological samples a geochemical database for reference materials and isotopic standards, *GeoReM* (http://georem.mpch-mainz.gwdg.de), has been developed by Jochum *et al.*[34] This database is helpful for reference materials and isotopic standards of geochemical, mineralogical and environmental interest, such as rock powders, synthetic and natural glasses, as well as mineral, isotopic, river water and seawater reference materials. The database contains more than 900 geological reference samples include rock powders originating from the USGS (United States Geological Survey) or GSJ (Geological Survey of Japan), synthetic and natural reference glasses originating from NIST, MPI-DING,[38,39] as well as minerals, isotope standards (e.g., NIST SRM 981 for lead isotope analysis), river or sea water reference materials. In addition, published analytical data and compilation values (major and trace element concentrations, radiogenic and stable isotope ratios) are summarized.

As an example of a possible calibration strategy in solid-state mass spectrometry, Figure 6.16 presents the calibration curves of U and Th measured on fused lithium borate targets of geological reference materials NIM-N, NIM-G and NIM-L. These calibration curves can be applied for U and Th determination in geological samples and also in zeolites. Applying external calibration using these calibration curves, the Th and U concentrations have been determined in zeolites at the $\mu g\,g^{-1}$ level.[25] The sample was prepared in order to achieve ideal matrix matching by fusion with a lithium borate/metaborate mixture. The results of quantification by external calibration and RSCs as correction factors using NIM-L as standard reference material are in good agreement with the data obtained from neutron activation analysis (NAA) for the zeolite sample TSM 180 in Figure 6.16.[25] Suitable matrix matched CRMs are not usually available for calibration procedures with solid-state mass spectrometry, therefore synthetic laboratory standards (prepared in-house) can be used (see Section 6.2.5).

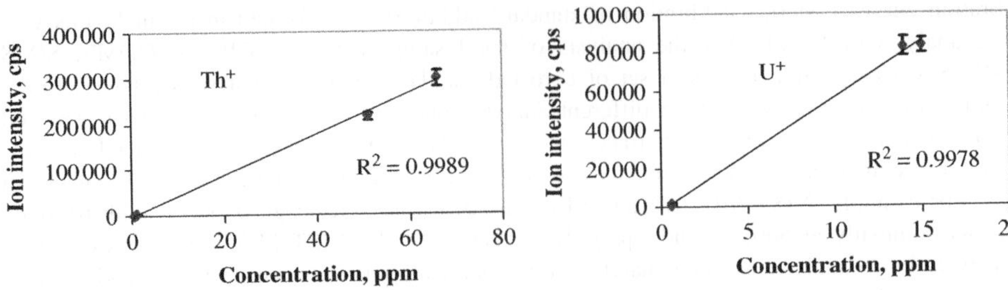

Element	LA-ICP-MS (RSC using NIM-L)	LA-ICP-MS (calibration curve)	NAA
Th	15±0.4	14±0.4	15±1.5
U	8±0.1	8±0.1	9±0.9

Figure 6.16 *Calibration curves of U and Th measured on fused lithium borate targets of geological reference materials NIM-N, NIM-G and NIM-L by LA-ICP-MS.*

6.2.4 Isotope Dilution Technique

An advantageous calibration strategy based on isotope ratio measurements (see Chapter 8) using mass spectrometry is the well established isotope dilution analysis (IDA). IDA is one of the most powerful and accurate analytical methods for determining quantitative data (concentrations of elements or element species) in the sample investigated. Isotope dilution techniques were developed more than 50 years ago and applied for major and trace element determination in TIMS by Ingram.[40] Today IDA can be used with every kind of mass spectrometer in different analytical fields where reliable results are required for the determination of element concentrations in bulk, on the surface or in speciation analysis. Isotope dilution mass spectrometry involves the measurement of isotopes of the same element; thus eliminating differences in chemical behaviour. During the entire analytical procedure including sample preparation, analyte separation and enrichment, a quantitative recovery of analyte is not necessary once the spike and sample have been equilibrated. This characteristic makes the isotope dilution technique more stable with respect to errors during all the chemical processing steps in comparison to other element analytical techniques. The isotope dilution technique is used routinely several laboratories, especially in trace and ultratrace analysis and for the accurate determination of the different species of an element.[23,41–49] In IDA one or two highly enriched isotope tracers or 'spikes' (the latter is the so-called double spike technique) of the element to be determined with well known concentrations are added to the sample (mixed and well homogenized with the solid sample or aqueous solution). A schematic of the isotope dilution technique is shown in Figure 6.17. The determination of the trace element concentration is performed by measuring changed isotope ratios in the sample–spike mixture (*X*) compared to those in the sample (*S*) and highly enriched isotope tracer (*T*) using Equation (6.3) for the isotope dilution:

$$Q_s = Q_T \cdot (T - X)/(X - S) \cdot m_s/m_T \tag{6.3}$$

where Q_s is the element concentration in sample; Q_T is the element concentration in high-enriched tracer, T is the isotope ratio of two selected isotopes in highly enriched tracer; S is the isotope ratio of these two selected isotopes in the sample; X is the measured isotope ratio of the two selected isotopes in the mixture; m_S or m_T is the atomic mass of the element in nature and of

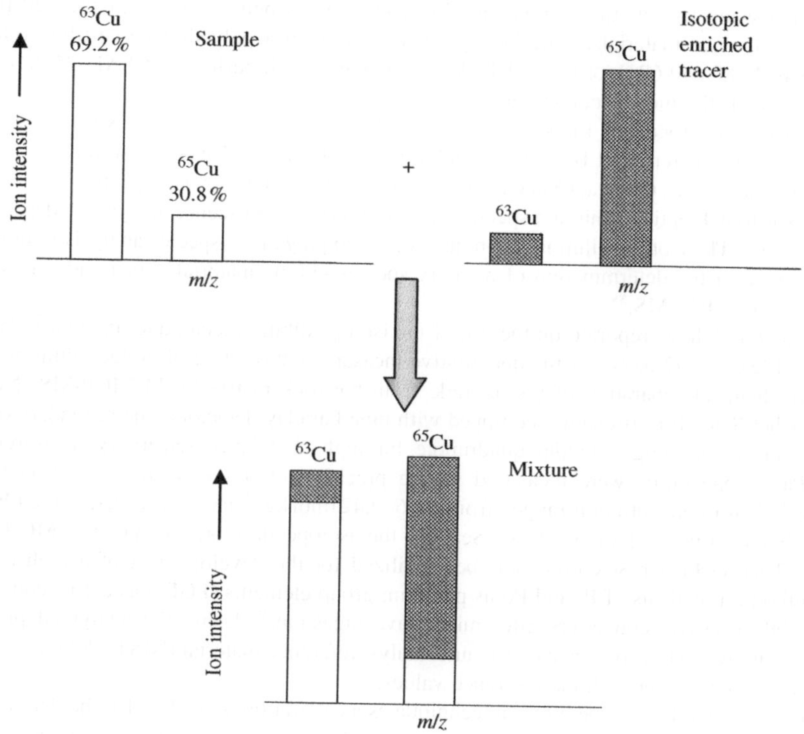

Figure 6.17 *Principles of the isotope dilution technique.*

the isotopically enriched element, respectively. IDA is applicable to all elements with at least two stable isotopes or long-lived radionuclides (e.g., use of ^{129}I radionuclide for iodine determination).

The fundamentals and several applications of isotope dilution mass spectrometry requiring accurate isotope ratio measurements are reviewed by Heumann.[50,51] Today isotope dilution mass spectrometry (IDMS) is recognized as a primary measurement method, by means of which accurate results with sufficiently small uncertainties can be achieved and therefore it has been used in certifying the composition of reference materials. A requirement of isotope dilution analysis in mass spectrometry is to achieve equilibration of spike and sample so that very careful sample preparation steps, especially in solid mass spectrometry, are necessary when a homogeneous sample spike mixture is to be prepared.

For application in the isotope dilution technique and for tracer experiments using isotope enriched spikes, Merck (Darmstadt, Germany) have launched the production of new sets of isotopically enriched materials, such as ^{53}Cr or ^{111}Cd enriched spike calibration solutions, which were characterized by the Institute for Reference Materials and Measurements (IRMM, Geel, Belgium).[52]

The advantage of IDMS in comparison to other quantification strategies (external calibration or standard addition) is that analyte recovery does not have to be quantitative, providing that isotopic equilibrium has been achieved between all of the analyte and the added spike material. High accuracy analysis by IDMS is now well established.[53] In the following paragraphs selected examples of IDMS use will be discussed briefly.

To an increasing extent, the isotope dilution technique is being applied in the certification of standard reference materials, e.g., for Hg determination in a polyethylene certified reference material (CRM 680 and 681), or Cd and Tl determination in high purity Zn (CRM 325/2R), where ID-ICP-MS yields the most accurate data.[45]

The certification procedure for seven trace metals (Ba, Ca, Li, Mg, Mn, Na and Sr) in the certified reference material FEBS-1 (National Research Council Canada, Institute for National Measurement Standards, Ottawa, Canada) based on fish otolith matrix by isotope dilution – ICP-MS in comparison to ICP optical emission spectrometry and X-ray fluorescence analysis, is described by Sturgeon *et al.*[48] The isotope dilution technique is also employed for species analysis in biological systems,[46] e.g., for the determination of mercury species in tuna material,[54] or in aquatic systems using cold vapour ICP-MS.[55]

Regelous *et al.*[56] have reported on the use of the isotope dilution technique (using a ^{233}Pa spike with a half-life of 26.97 days) for the quantitative measurement of 20 fg of protactinium in silicate rocks after chemical separation of the actinide from the rock matrix by MC-ICP-MS (Neptune, Thermo Fisher Scientific, Bremen – equipped with nine Faraday detectors, one secondary electron multiplier and a retarding potential quadrupole for high abundance sensitivity measurements). ^{231}Pa/^{233}Pa isotope ratios were measured with a precision of 0.5 %, with a detection limit of 200 ag ml^{-1}. ^{226}Ra concentration ranges from 0.36–0.42 fmol kg^{-1} have been determined by MC-ICP-MS (Nu Instruments) in the Ross Sea via the isotope dilution technique.[57] MC-ICP-MS (Neptune, Thermo Fisher Scientific) has been utilized for the developement of a high accuracy IDMS method for analysis of Pt and Pd as platinum group elements (PGEs) as active components in automobile catalytic converters after microwave digestion.[58] Using the analytical procedure developed, MC-ICP-MS provides data for autocatalyst reference material (NIST SRM 2556) which show excellent agreement with the reference values.

Viera *et al.*[59] used the isotope dilution technique as a calibration procedure for the determination of As, Ge, Hg, Pb, Se and Sn in coal slurries using chemical vapour generation combined with electrothermal vaporization ICP-MS (CVG-ETV-ICP-MS).

An on line dialysis flow system coupled to ICP-MS (VG Plasma Quad PQII, Thermo Elemental, Winsford, Chaeshire, UK) has been employed to determine trace elements in serum samples by IDMS. Isotope dilution was performed on samples incubated with enriched ^{65}Cu, ^{66}Zn, ^{77}Se and ^{206}Pb for 24 h at 36 °C prior to dialysis to quantify total element concentrations.[60]

In past years, on line chromatographic coupling techniques such as HPLC and CE coupled to ICP-MS with the isotope dilution technique have been used for element quantification in speciation analysis. An interesting application of the isotope dilution technique in medical research was proposed recently by Prange and co-workers,[61,62] who added highly enriched ^{34}S, ^{65}Cu, ^{68}Zn and ^{116}Cd spikes to the interface of the CE-ICP-MS system. The authors separated isoforms of metallothionein (e.g., of rabbit liver) by capillary electrophoresis and quantified S, Cd, Cu and Zn concentrations in isoforms by ICP-SFMS using the isotope dilution technique.[61] A new selenized yeast reference material (SELM-1) for methionine, selenomethionine (SeMet) and total selenium content has also certified by an intercomparison exercise.[63]

The triple spike isotope dilution approach using GC-ICP-MS for monitoring the degradation and solublization of butyltin compounds during *in vitro* gastrointestinal digestion was described by Sanz-Medel's working group.[33] In this study, monobutyltin (MBT), dibutyltin (DBT) and tributyltin were spiked with the enriched tin isotopes of masss 119, 118 and 117 u, respectively. This triple spiked butyltin solution was added to the sample containing natural butyltin species to quantify these species by measuring the isotope ratios.[33]

A method for the determination of selenomethionine in yeast by a combination of high performance liquid chromatography (HPLC) and dried droplet laser ablation ICP-MS using a species specific isotope dilution technique with the aid of a ^{74}Se enriched SeMet spike, was proposed by Yang *et al.*[64] Several reverse isotope dilution samples were prepared to quantify the concentration of the ^{74}Se-enriched SeMet spike. Dried microdroplets of each chromatographic fractionation were ablated from the polystyrene substrate and quantified for SeMet.

A further application of the isotope dilution technique is the determination of selenomethionine in human blood serum by capillary HPLC-ICP(ORC) MS.

The isotope dilution technique in LA-ICP-MS[65] has been applied for the direct analysis of sulfur species in gas oil, diesel or heating fuel as described by Boulyga *et al.*[66] In addition, these authors have developed an analytical method for the direct determination of halogens (Cl, Br and I) in powdered geological and environmental samples using isotope dilution laser ablation ICP-MS. The detection limits of LA-ICP-IDMS are in the low and sub-$\mu g\,g^{-1}$ range for halogens.[67]

6.2.5 Quantification in Solid-state Mass Spectrometry Using Synthetic Laboratory Standards

The quantification of analytical results in solid-state mass spectrometry proves to be difficult, especially if no suitable certified reference material (with the same matrix) is available.[68] Matrix matched laboratory standards are therefore prepared in many laboratories. The preparation of synthetic laboratory standards has been demonstrated, for instance, for trace element determination in ceramic components of the solid oxide fuel cell,[69] in graphite,[1] for the determination of long-lived radionuclides in concrete material[26] or in brain tissues.[70] Synthetic (matrix matched) solid standard samples have been prepared from the compounds of the matrix elements doped with trace elements in given concentrations.[26,69] For example, a GaAs crystal as a laboratory standard for trace element analysis in mass spectrometry doped with known concentrations of different elements in the $\mu g\,g^{-1}$ range, has been prepared by the vertical Bridgman technique as described in reference[71]. This multi-element GaAs crystal has been used for the determination of RSCs of doping elements in the given matrix by LA-ICP-MS, SSMS, rf-GDMS and SIMS.[71] Table 6.1 summarizes the RSCs in SSMS, rf GDMS and LA-ICP-MS in a synthetic GaAs laboratory standard and the results of determination of dopants by different analytical methods, such as SIMS and ICP-MS and ICP-OES (optical emission spectrometry with an inductively coupled plasma source),

Table 6.1 *Results of determination of dopants by different analytical methods and RSCs in SSMS, rf GDMS and LA-ICP-MS in a synthetic GaAs laboratory standard (concentration in $\mu g\,g^{-1}$).*

Dopant	SIMS	ICP-MS	ICP-OES	SSMS	RSC	rf GDMS	RSC	LA-ICP-MS	RSC
Zn	1208 ± 90	827 ± 22	910 ± 50	870[b]	1.0	870[b]	1.0	870[b]	1.0
B	17.8 ± 1.2	19.5 ± 0.7	18 ± 6	8.7 ± 1.5	0.5	8.2 ± 2.9	0.4	8 ± 1	0.4
Ge	—	20.5 ± 0.6	< 40	11 ± 4	0.5	—	—	36 ± 1	1.8
Sn	13.5 ± 2.0[a]	6.0 ± 0.2	< 40	4.2 ± 1.2	0.7	< 10	—	23 ± 1	3.8
Sb	—	49 ± 1	45 ± 12	14 ± 4	0.3	—	—	132 ± 4	2.8
P	—	328 ± 30	—	1290 ± 260	3.9	850 ± 100	2.6	—	—
S	450 ± 80	316 ± 20	390 ± 100	720 ± 140	1.9	475 ± 62	0.8	74 ± 1	0.2
Se	400 ± 75	395 ± 12	420 ± 60	315 ± 48	0.8	120 ± 22	0.3	48 ± 1	0.1
Te	113 ± 27	97 ± 3	110 ± 30	108 ± 15	1.0	43 ± 14	0.4	62 ± 1	0.6

[a] Possible inhomogeneity
[b] Internal standard element

both of the latter techniques were utilized after dissolution of GaAs sample and dilution (the concentrations of trace elements in the solid GaAs laboratory standard are given in $\mu g\,g^{-1}$). The relative standard deviation (RSD) of measured trace element concentrations varies in the low percent range by up to 10–20 %.

The calibration technique employing synthetic laboratory standards can be used for trace analysis as well as for compact and finely powdered solid samples or soft tissues. Pickhardt *et al.*[1] describe the preparation and analysis of a synthetic graphite standard where the doped concentration of 14 trace elements of $0.5\,\mu g\,g^{-1}$ was determined with an accuracy of 1 to 7 % and a relative standard deviation (RSD) of 2 to 13 %. Improvement of detection limits by a factor of ten has been achieved for the analysis of high purity graphite with LA-ICP-MS under wet plasma conditions due to lower background signal and increased element sensitivity. In addition to one point calibration of relative sensitivity coefficients, solution based calibration has been applied to high purity graphite as an alternative, quick and reliable calibration procedure, by Pickhardt *et al.*,[1] which has led to analytical results for trace element determination with an accuracy similar to that obtained by the use of synthetic laboratory standards for the quantification procedure.

Quantitative trace element analysis of diamond by LA-ICP-MS using different synthetic multi-element carbon based standards (e.g., cellulose pellets) is discussed by Rege *et al.*[72], whereby ^{13}C was used for internal standardization. Concentrations of 41 elements were determined in two fibrous diamonds from Jwaneng Botswana (JWA 110 and 115) by relative sensitivity coefficients measured using the synthetic cellulose standard. The analytical data were verified by means of instrumental neutron activation analysis (INAA) and proton induced X-ray emission (PIXE).[72]

Trace element determination in ceramic perovskite layers ($La_{0.65}Sr_{0.3}MnO_3$) for the solid oxide fuel cell has been carried out by rf GDMS and LA-ICP-MS using synthetic laboratory standards which were synthesized from high purity compounds doped with trace elements in the $20–500\,\mu g\,g^{-1}$ range.[69,73] Quantification via calibration curves was compared with one point calibration using relative sensitivity coefficients (RSCs) which varied in rf GDMS between 0.4 and 2. The relative standard deviations in the determination of trace elements in perovskite layers were better than 15 %. The disadvantage of the proposed preparation of synthetic laboratory standards and their characterization in connection with the quantification procedure is the time-consuming step of solid mass spectrometry.

For the calibration of analytical data on thin sections of brain tissue measured by LA-ICP-MS, matrix matched laboratory standards with well defined element concentrations have been prepared. The procedure of preparing matrix matched synthetic laboratory standards for the quantitative imaging of essential elements such as Cu and Zn or toxic elements (Th and U) in thin sections of brain tissue (hippocampus) is summarized in Figure 6.18. Three synthetic standard solutions containing the elements of interest (Cu, Zn, U and Th) were prepared in defined concentrations. Three slices of the same brain tissue (each about 0.65 g) were spiked with selected standard solutions. The final concentrations in brain tissue were 10, 5 and $1\,\mu g\,g^{-1}$ of Cu and Zn and 1, 0.05 and $0.01\,\mu g\,g^{-1}$ of Th and U. The fourth slice was not spiked and was used for blank correction. All tissue brain samples were carefully homogenized and centrifuged at 5000 rpm for 5 min. After that, samples were frozen at a temperature of $-50\,°C$. Frozen matrix matched synthetic laboratory standards of human brain tissue were cut into sections $20\,\mu m$ in thickness and placed onto a glass substrate. The laboratory standards and thin sections of brain tissue prepared on one glass substrate were then measured by LA-ICP-MS under the same experimental conditions in the same run.[70]

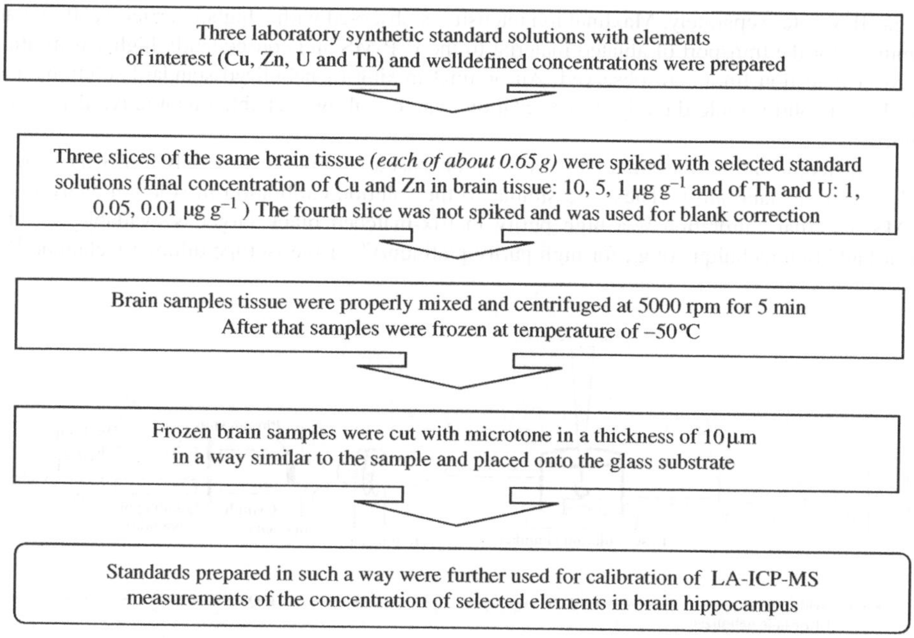

Figure 6.18 *Schematic for preparation of synthetic laboratory standards for analysis of thin sections of brain tissue (hippocampus). (J. S. Becker et al., Anal. Chem. 77, 3208 (2005). Reprinted by permission of American Chemical Society.)*

6.2.6 Solution Based Calibration in LA-ICP-MS

Calibration and quantification procedures are easier in LA-ICP-MS compared to other solid-state mass spectrometric techniques because the laser ablation and the ICP ion source operate at normal pressure and the laser ablation of solid samples and ionization of analytes are separated in space and time. Therefore the advantage of solution calibration in ICP-MS can be applied in this solid-state analytical technique. The introduction of solution based calibration, which is only possible in LA-ICP-MS, was an innovative step in the development of this sensitive mass spectrometric technique. A number of different calibration approaches using aqueous standard solutions in the 'dual gas flow technique' have been discussed by various authors.[74,75] In the 'dual gas flow injection technique', the nebulized standard solution and the laser ablated sample material are mixed in the *y*-piece and the two gas flows from the nebulizer (e.g. ultrasonic nebulizer) and laser ablation chamber are added. Using solution based calibration with the addition of a standard solution, Leach *et al.*[75] determined minor elements in steel reference materials with a relative accuracy of a few %. In comparison to the so-called 'dual gas flow technique' proposed in the literature, where the argon flow rates through the nebulizer and ablation cell add up to 1 l min^{-1} (e.g. 0.45 l min^{-1} and 0.6 l min^{-1}, respectively in reference[75]), the 'mono gas flow technique' was analogously developed in the author's laboratory.[29] In this experimental arrangement, the ultrasonic or microconcentric nebulizer with desolvator (USN 200 or Aridus from Cetac Technologies, Omaha, NE, USA) was coupled on line to the laser ablation chamber, as demonstrated in Figure 6.19. The main advantage of this arrangement compared to the dual gas flow technique is a better optimizing capability for both gas flow rates separately. Maximal ion intensity is observed with a higher carrier gas flow rate (1 l min^{-1}) for the transport of ablated material to the ICP-MS and consequently high sensitivities and lower detection limits are observed. An optimal mixing of nebulized standard solutions and laser ablated solid sample directly in the ablation chamber allows reliable quantitative data to be obtained.

Three different calibration strategies for solution based calibration in LA-ICP-MS have been developed in our laboratory. These are similar to the solution calibration in solution analysis by ICP-MS: external calibration if a high purity matrix-matched blank target is available,[29,71] the standard addition technique (e.g., for high purity platinum)[76] or the isotope dilution technique.[43]

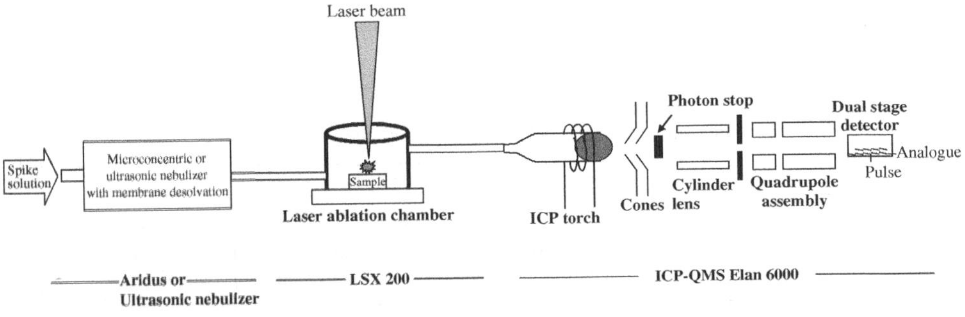

*Figure 6.19 Schematic setup of LA-ICP-MS with solution based calibration using microconcentric or ultrasonic nebulizer with membrane desolvator. (J. S. Becker et al., Int. J. Mass Spectrom. **237**, 13 (2004), reprinted by permission of Elsevier.)*

External Calibration Technique for Solution Based Calibration in LA-ICP-MS

External calibration by nebulization of standard solutions in LA-ICP-MS has been developed for high purity graphite[1] and for geological samples.[29] For the measurement procedure on inhomogeneous geological samples, homogeneous lithium borate fused targets are prepared by powdering, homogenizing and fusing with a lithium borate mixture in a muffle furnace at 1050 °C.[29] In order to achieve matrix matching, the standard solutions are nebulized with an ultrasonic nebulizer (USN) during solution calibration and simultaneously a blank target (e.g. a high purity lithium borate target for geological samples or a high purity graphite target) is ablated with the focused laser beam. The homogeneous geological (lithium borate fused) or graphite targets are measured using the same experimental arrangement, where 2 % nitride acid is simultaneously nebulized with the USN. The analytical results of LA-ICP-MS, e.g., on various geological samples, are in good agreement with the reference values and the results of other trace analytical methods. The calibration curve of barium for the trace analysis of fused lithium borate targets of geological glasses by external calibration in solution calibration with LA-ICP-MS is illustrated in Figure 6.20. The regression coefficient was determined as 0.9998. The analytical results of trace element determination by LA-ICP-MS on a geological sample of andesite StHs6/80 measured by LA-ICP-MS using RSCs and calibration curves by solution-based calibration are compared in Figure 6.21. In general, more accurate data were obtained by utilizing calibration curves compared to one point calibration by means of RSCs.[29] In this experiment a similar experimental setup was employed to that proposed by Pickhardt *et al.*[29] (coupling of an ultrasonic nebulizer with a laser ablation chamber for solution based calibration).

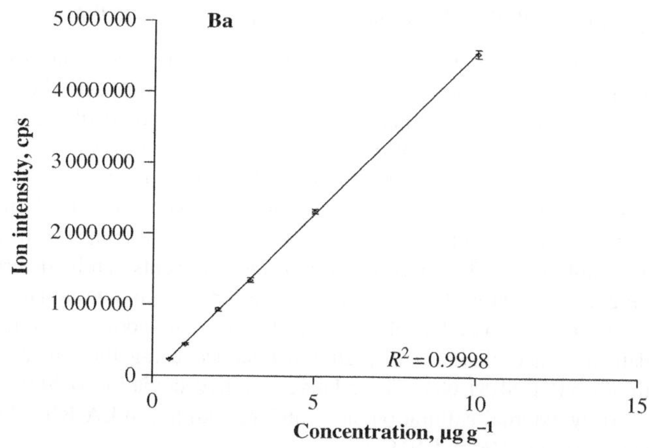

Figure 6.20 Calibration curve of barium measured by solution based calibration using LA-ICP-MS.

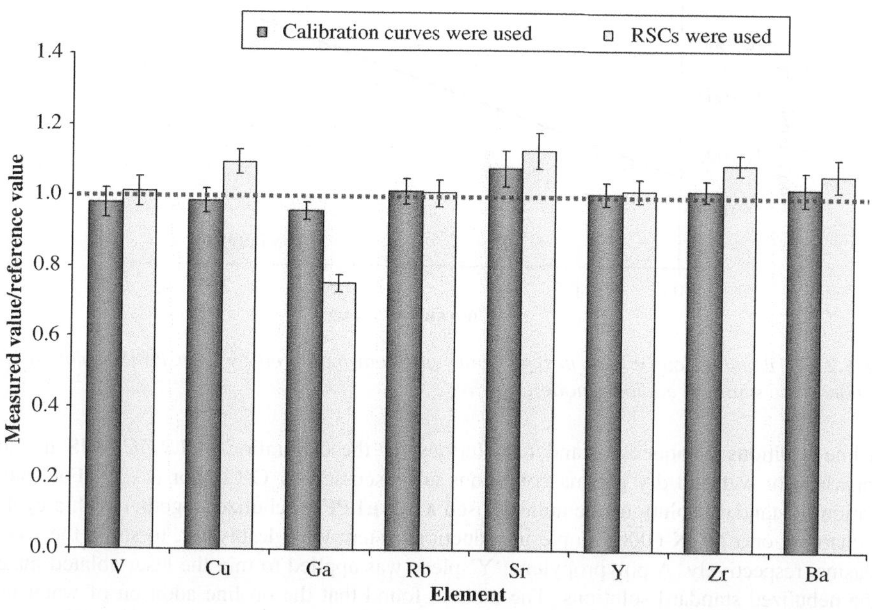

Figure 6.21 Trace element measured by LA-ICP-MS on a geological sample andesite StHs6/80 using RSCs and calibration curves.

Bings[77] quantified 11 trace metals in lubricating oil with the aid of solution based calibration in LA-ICP-MS using a ToF mass spectrometer. It has been shown that good analytical performance in terms of precision (6 % on average) and low detection limits in the ng g^{-1} range can be obtained with this easy and fast quantification procedure in LA-ICP-MS.[77]

Standard Addition Technique With Solution Based Calibration in LA-ICP-MS

The standard addition mode was developed for the multi-element trace analysis of high purity metals, semiconductors and insulators (especially for compact ceramics) where neither matrix matched standard reference materials nor high purity blank targets are available. For example, the quantification of results for ultratrace analysis on high purity platinum was carried out using the standard addition mode (well known in ICP-MS and ICP-OES), which means the multi-element standard solutions (with different concentrations of analytes) were nebulized successively e.g., with a USN. Simultaneously, the high purity platinum sample was ablated with a focused laser beam during solution calibration. The concentration of trace elements can be determined from the calibration curves. Figure 6.22 shows the calibration curve for iridium determination in high purity platinum measured by solution based calibration in the standard addition mode. The accuracy of the analytical data for trace elements in platinum obtained using this quantification strategy was tested on platinum NIST SRM 681. The relative standard deviation (RSD) of trace element determination ($N = 6$) by external calibration and standard addition in LA-ICP-MS using solution based calibration was between 2 % and 10 %.[1,29]

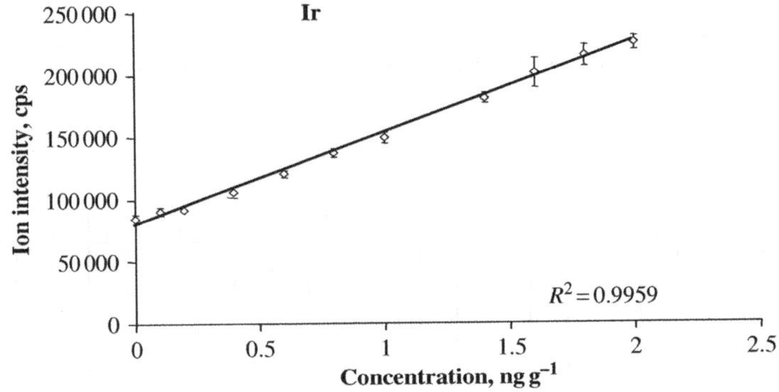

Figure 6.22 *Calibration curve of Ir in high purity platinum measured by LA-ICP-MS (solution based calibration in the standard addition mode).*

On line additions of aqueous standard solutions for the calibration of LA-ICP-MS including a comparison of wet and dry plasma conditions are discussed by O'Connor *et al.*[78] For solution calibration of standard solutions the authors used a 100 μl PFA nebulizer together with a cyclonic spray chamber or a MCN-6000 sample introduction system with desolvator, to study the wet and dry plasma, respectively. A polypropylene 'Y' piece was applied to mix the laser ablated material and the nebulized standard solutions. The authors found that the on line addition of water is the preferred mode of operation for quantification by LA-ICP-MS, i.e., wet plasma is more stable (improved standard deviation of sensitivity ratios).

A special setup for solution based calibration using a micronebulizer (DS-5, CETAC, Omaha, NE) inserted in LA-ICP-MS as shown in Figure 6.23 was proposed by Becker *et al.*[70] The arrangement was developed for the analysis of thin sections of brain tissue. During the laser ablation of thin samples, defined standard solutions with increasing analyte concentration were nebulized (standard addition mode). The calibration curves (standard addition approach) for uranium and thorium determination in thin sections of brain tissue are illustrated in Figure 6.24. The regression coefficients for both calibration curves are better than 0.999. Figure 6.25 demonstrates

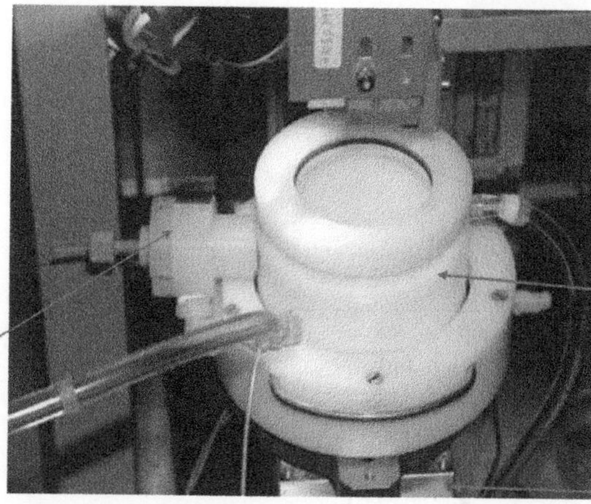

Figure 6.23 *Experimental arrangement of solution based calibration using a micronebulizer (DS-5, CETAC) inserted in a cooled laser ablation chamber for imaging of thin sections of brain tissue. (J. S. Becker et al., Anal. Chem.* ***77****, 3208 (2005). Reproduced by permission of American Chemical Society.)*

the $^{232}Th^+/^{238}U^+$ intensity ratio measured during the standard addition mode for different standard solutions utilizing micronebulization in LA-ICP-SFMS. The $^{232}Th^+/^{238}U^+$ intensity ratio is nearly constant for all measured concentrations of standard solution. To obtain quantitative data in solution based calibration the different sensitivities of ICP-MS and LA-ICP-MS must be considered. The correction factor:

$$C_{\text{LA-ICP-MS}} = \frac{\text{Concentration internal standard element determined by solution based calibration}}{\text{True concentration of internal standard element in the sample}} \quad (6.4)$$

was determined by the use of an internal standard element. In thin sections of brain samples either U or Th was selected, because both elements had been found to be homogeneously distributed in the brain samples investigated. The concentration of the internal standard element can be determined using an alternative analytical method (e.g., by ICP-MS after digestion).

On line Isotope Dilution Analysis in LA-ICP-MS

A suitable calibration strategy has been demonstrated for the determination of selected elements, i.e., Pb, Ag and Tl in a small amount of platinum nanoclusters. In on line isotope dilution analysis with LA-ICP-MS, an isotope enriched tracer solution was nebulized by the Aridus microconcentric nebulizer with desolvator. After about 180 seconds, laser ablation of the investigated platinum nanocluster sample was started and after an additional time of 180 seconds 2 % nitric acid nebulized instead of the tracer solution. The isotope ratios of lead ($^{204}Pb/^{206}Pb$), silver ($^{107}Ag/^{109}Ag$) and thallium ($^{203}Tl/^{205}Tl$) – with $^{63}Cu/^{65}Cu$ for internal standardization – were monitored during the whole experiment. The measuring time for one experiment was <10 min.

The fabrication of regular arrays of metallic nanoparticles by molecular templating is of great interest in order to prepare nanometre structures for future use in nanoelectronics, optical and chemical devices.[43] A sensitive, rapid and powerful direct analytical method is required for the quantitative analysis of high purity platinum or palladium nanoclusters produced by biomolecular

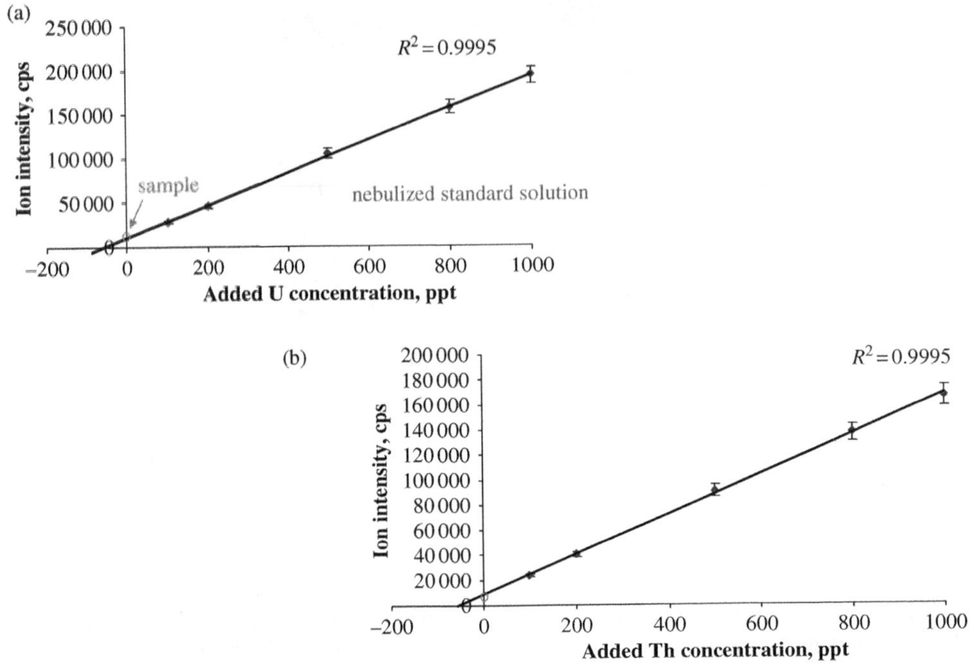

Figure 6.24 *Calibration curves for uranium (a) and thorium (b) determination in brain samples via the standard addition mode using micronebulization in LA-ICP-MS. (J. S. Becker et al., Anal. Chem.* ***77****, 3208 (2005). Reproduced by permission of American Chemical Society.)*

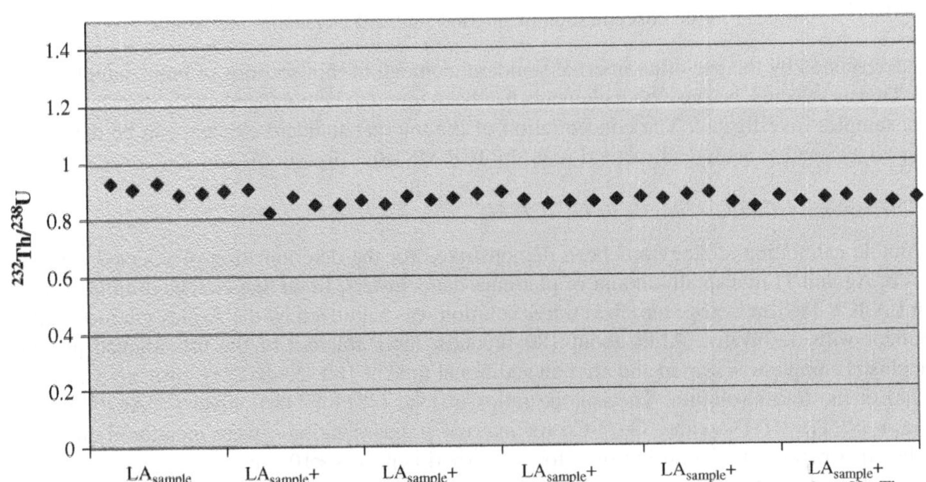

Figure 6.25 *$^{232}Th^+/^{238}U^+$ intensity ratio measured via standard addition mode using micronebulization in LA-ICP-MS. (J. S. Becker et al., Anal. Chem.* ***77****, 3208 (2005). Reproduced by permission of American Chemical Society.)*

templating. Due to possible contamination during the digestion of small amounts of sample (several mg) LA-ICP-MS without sample preparation is advantageous so that semi-quantitative data on the composition of trace impurities can be obtained in a short time. Semi-quantitative data (with an error factor of 2–3) are sometimes sufficient for the characterization of material purity. It is interesting to note that superconducting properties have been observed for these platinum nanoclusters so that with increasing purity the superconducting properties are improved.

The method of choice for accurate analytical data in LA-ICP-MS is the isotope dilution technique, which allows absolute concentrations to be measured. This has been demonstrated by the determination of Cu on small amounts of platinum nanoclusters. Figure 6.26 summarizes the isotope ratios measured during this procedure as demonstrated for ($^{63}Cu/^{65}Cu$). Starting with the analysis of the highly enriched isotope spike solution of Cu (containing 89.2 % ^{65}Cu versus 30.8 % ^{65}Cu in the IUPAC table), the isotope ratio of the spike solution was measured using the Aridus microconcentric nebulizer with membrane desolvator. After 160 s the laser was switched on and a change of isotope ratio was observed. Highly enriched spike solution with a trace element concentration in the lower $ng\,ml^{-1}$ concentration range was nebulized, desolvated and the dry aerosol was transported with argon into the laser ablation chamber where it was mixed with the ablated material and an equilibrium was achieved. The precision of the isotope ratio measured in the mixture was about 6.8 %. After a further 140 s only 2 % HNO_3 was nebulized and a Cu isotope ratio was measured in the sample which agrees well with the IUPAC table value. The precision of these isotope ratio measurements by LA-ICP-MS was 3.4 %. The platinum SRM NIST 681 was analyzed in order to verify this analytical technique. The different element sensitivities in LA-ICP-MS and ICP-MS must be considered in all solution based calibration strategies. These differences in element sensitivities were corrected using a defined correction factor via an internal standard element. Pb was determined in NIST 681 at $13.1 \pm 1.1\,\mu g\,g^{-1}$ (certified value: $12 \pm 2\,\mu g\,g^{-1}$) by means of LA-ICP-MS using solution calibration with the isotope dilution technique.

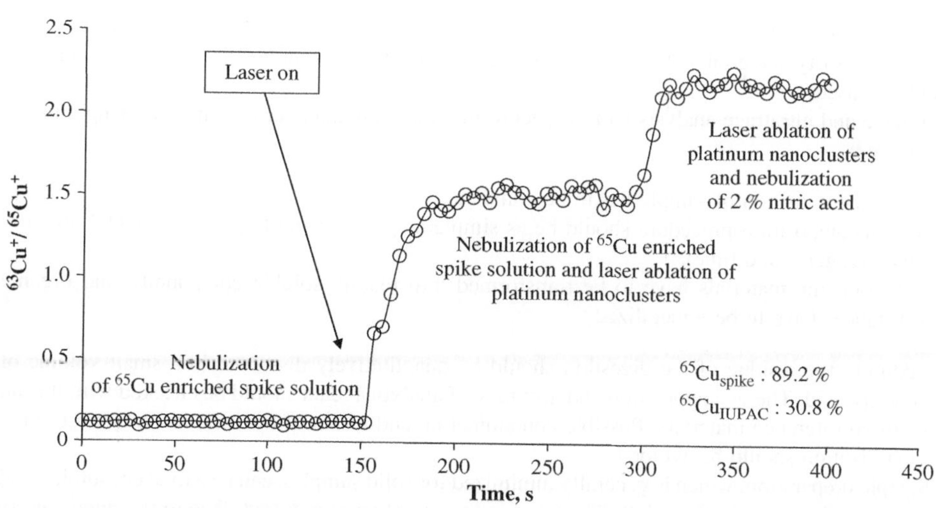

Figure 6.26 *On line isotope dilution in LA-ICP-MS for determination of Cu in platinum nanoclusters. (J. S. Becker et al., Int. J. Mass spectrom.* ***237****, 13(2004). Reproduced by permission from Elsevier.)*

On line isotope dilution in LA-ICP-MS using a microflow nebulizer (DS-5) inserted in the laser ablation chamber (experimental setup is shown in Figure 6.23) for the analysis of uranium in NIST glass standard reference material (SRM) 612 and NIST apple leaves SRM 1515 is described by Pickhardt *et al.*[79] In the analysis of SRMs, firstly the $^{235}U/^{238}U$ isotope ratio in the standard was measured and 2 % HNO_3 was nebulized, secondly this isotope ratio was determined in the mixture of ablated sample and nebulized isotope enriched tracer, and thirdly only the isotope enriched tracer solution was nebulized and analyzed (in analogy to the experiment shown in Figure 6.26). A low solution uptake rate of 7 $\mu l\,min^{-1}$ for the DS-5 reduced the amount of isotope enriched tracer solution needed for isotope dilution analysis.

Further details of different strategies in solution calibration are described in the literature.[1,29,76,79] Precise and accurate measurements of isotope ratios, which is one of the major advantages of mass spectrometric techniques, are a requirement for the application of isotope dilution techniques in trace analysis, which is also the main goal of the application of isotope dilution in solution based calibration in LA-ICP-MS.

6.3 Sample Preparation and Pretreatment

The purpose of sample preparation and pretreatment is to transform the sample into an appropriate form for mass spectrometric measurements. Depending on the analytical task, sample preparation includes several steps such as cutting, cleaning of surfaces, milling, homogenizing or digestion and dilution to turn a solid material into the right size and form (e.g., to prepare electrodes for SSMS or targets for LA-ICP-MS) or to prepare aqueous solutions for ICP-MS. Contamination during sample preparation is still the biggest source of incorrect analytical data in trace analysis. Therefore, sample preparation and also the sample collection steps require very careful sample handling under clean working conditions (dust should be excluded by using clean benches and clean rooms) in order to avoid possible contamination or loss of analytes. For trace element analysis (to determine trace impurities at the $pg\,g^{-1}$ to $fg\,g^{-1}$ level and below) the use of high purity reagents and water, and clean bottles and tools are necessary. Clean vessels made of an inert high purity material with a small volume and a minimal amount of high purity reagents should be used.

In trace and ultratrace analysis with respect to minimal systematic errors, the following must be considered:

- The decomposition of sample has to be complete.
- The decomposition procedure should be as simple as possible (no large experimental arrangement, reagents and time).
- All inorganic materials have to be transformed into readily soluble compounds, and organic substances have to be mineralized.

In addition, the residue of the digestion should be quantitatively dissolved in a small volume of high purity acid. The decomposition and accuracy of analytical data should be checked with the aid of certified reference materials. Possible contamination and losses of trace elements by absorption or volatilization should be avoided.

Sample preparation, which is generally minimized for solid samples, can be a time consuming and labour intensive process, especially if trace matrix separation is required. Due to the huge variety of different matrices, of more or less complex matrix composition, many different decomposition principles have been developed.

6.3.1 Sample Preparation for Analysis of Solids

Sample preparation for solid-state mass spectrometry is required if an inhomogeneous powdered (or compact) sample has to be investigated by bulk analysis with respect to trace impurities. Therefore the first step is milling followed by homogenizing. After that, for SSMS and GDMS the powdered sample is mixed with binder (e.g. high-purity graphite, silver) and electrodes or targets are pressed as described by Dietze and Becker.[80]

No sample preparation is necessary for homogeneous materials in solid-state mass spectrometry. For trace analysis on compact solid samples only a special cleaning of the surface is performed before measurements are made. No sample preparation is also required for microlocal analysis of inhomogeneous materials or depth profiling for layered systems by LA-ICP-MS, SIMS, SNMS or GDMS. Several mass spectrometric techniques require the sample or sample surface to be electrically conducting. In such cases, the surface is coated by a thin metal layer, e.g., in SIMS, to avoid charging effects, or a secondary cathode is installed in direct current GDMS.

6.3.2 Sample Preparation for ICP-MS

An advantage of ICP-MS compared to all other atomic mass spectrometric techniques including TIMS is that usually only simple sample preparation (e.g., by microwave induced digestion of solid samples) is necessary. Sample preparation steps for ICP-MS analyses are similar to those of ICP-OES. Concentrated solutions are analyzed after dilution with high purity water only. In order to correct mass drifts of the instrument, an internal standard element like In or Ir with known concentration (e.g., $10\,\mu g\,l^{-1}$) is added. The solution is then acidified with HNO_3 to stabilize the metal ions in aqueous solution.

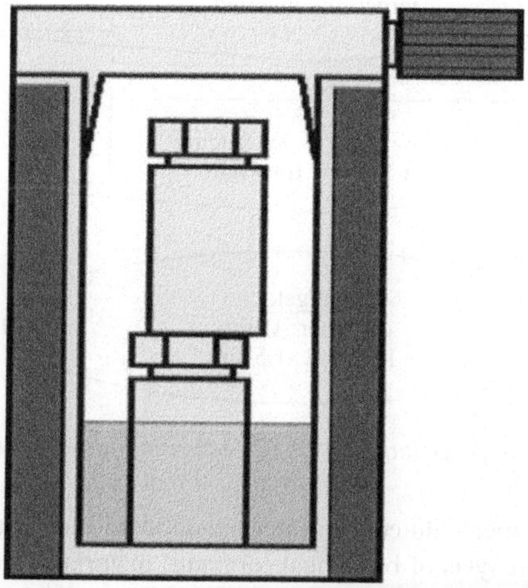

Figure 6.27 *Two small PFTE crucibles in a vessel used for microwave induced digestion of small sample amounts.*

Solid samples are digested with suprapure sub-boiled HNO_3 to avoid contamination in an acid mixture using HF, HCl or $HClO_4$ (e.g., in an open vessel directly or in a closed vessel by microwave induced digestion). The sample digestion of difficult-to-dissolve sample material in closed vessels, in a microwave oven carried out under temperature and pressure control, is very effective, mostly complete, less time consuming and has the advantage of avoiding possible contamination and the lost of highly volatile analytes (such as As, Se, Hg and others). For microwave induced digestion of small amount materials (e.g., 1–100 mg of biological or medical samples) two small crucibles from PFTE are inserted a vessel as illustrated in Figure 6.27. The microwave induced digestion of the sample is than carried out in 2 ml HNO_3 (subb. suprapure), 0.5 ml H_2O_2 (suprapure), 0.2 ml HF (suprapure) at an elaborated digestion program as described (e.g., 10 min at 150 W, 10 min at 0 W and 10 min at 300 W using a CEM Mars 5 microwave oven).

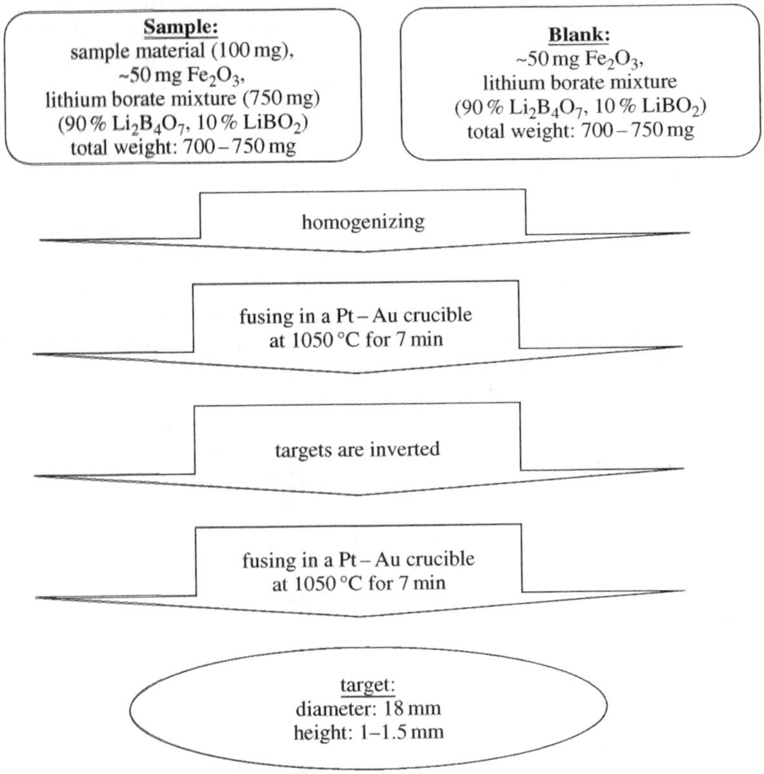

Figure 6.28 *Sample preparation in LA-ICP-MS using fused lithium borate target.*

The digested samples are then diluted, the internal standard element is added and the solution has to be acidified. For the analysis of biological (organic) materials (including urine) a mixture of HNO_3/H_2O_2 has been proposed for a quantitative sample digestion.[81] Soils, ceramics and geological samples are dissolved with a mixture of HNO_3/HF or else, after fusion digestion with lithium metaborate in a platinum crucible, deposited with gold at 1000 °C in a furnace. The schematic of the sample preparation of fused lithium borate mixture is summarized in Figure 6.28. After fusion of the sample with a lithium borate/metaborate mixture it is then dissolved with diluted HNO_3, the fusion product is diluted with high purity water and analyzed by ICP-MS. However, such sample treatment may involve the introduction of systemic errors in the determination of trace element concentrations and isotope ratios also using the isotope dilution technique. For example, in the case of Pu isotope analysis, plutonium oxide (PuO_2) formed at high temperature is insoluble.[82]

Sample preparation also includes certain matrix separation steps (e.g., by ion exchange, extraction chromatography and others), analyte enrichment (by co-precipitation and also ion exchange, e.g., Pu separation is performed using TEVA resin[83]) and preconcentration steps (e.g., by evaporation of solvent) off line or on line by high performance liquid chromatography (HPLC) or capillary electrophoresis (CE) in order to improve the detection limits and avoid disturbing interference problems in ICP-MS.

6.3.3 Trace Matrix Separation and Preconcentration Steps

Chemical separation of matrix and preconcentration of analytes is used to avoid matrix effects, clogging effects on the cones and disturbing interferences of analyte ions with polyatomic ions of matrix elements and plasma gases. A trace/matrix separation method is required for ultratrace analysis.

The ultratrace yield must be 100 % (except when applying stable isotopes in the isotope dilution technique and radioactive tracers for process control), the procedure should be simple and only a small amount of high purity reagents should be required. To avoid contamination, high purity closed micro containers and vessels of quartz, PFA or PFTE must be used. The mostly frequently applied trace matrix separation technique is liquid–liquid extraction with various organic solvents (e.g., MIBK, HDEHP) and separation of analyte by ion exchange (e.g., for Tc or Th separation). Electrolysis[44] and separation by volatilization, e.g. for trace analysis in high purity GaAs,[84] are recommended for special analytical tasks. Matrix effects and contamination of sample introduction system will be minimized after analyte separation. This analytical strategy is advantageous in combination with analysis of small sample sizes.[85]

For some applications, precipitation and co-precipitation, which is often incomplete, is utilized. As an example, for trace matrix separation the procedure for plutonium determination by ICP-SFMS in urine after trace matrix separation is summarized in Figure 6.29.[86,87] The limit of detection for ^{239}Pu ultratrace determination in one litre of urine based on enrichment factor of 100 using the DIHEN in ICP-SFMS was $1.02 \cdot 10^{-18}$ g ml^{-1}.[86]

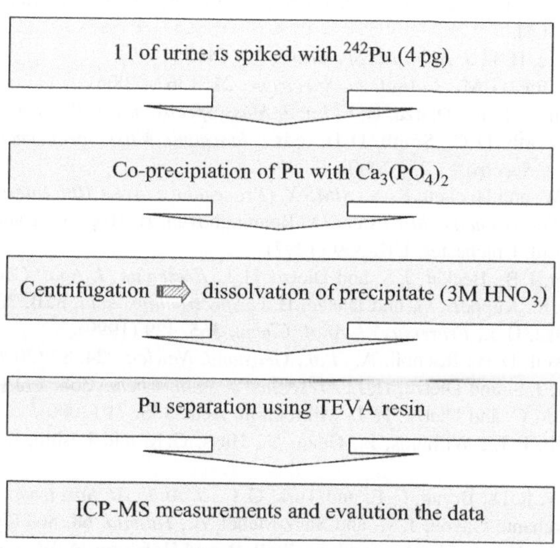

Figure 6.29 *Procedure for Pu determination in urine after trace matrix separation. (M. Zoriy, C. Pickhardt, P. Ostapczuk, R. Hille, J. S. Becker, Int. J. Mass Spectrom* **232***, 217 (2004). Reproduced by permission of Elsevier.)*

The atmosphere of the laboratory is loaded with particulates from different sources (floor, walls, furniture, equipment, clothes, persons etc.). Especially elements that are highly abundant in the Earth's crust (Si, Al, Fe, Ca, Na, K, Mn, Ti) and all elements of anthropogenic pollution (Mg, Cu, Cd, Pb, Ni, Co, Zn and Mn) are always present and can cause serious contamination.

References

1. Pickhardt, C. and Becker, J. S., *Fresenius' J. Anal. Chem.*, **370**, 534 (2001).
2. Becker, J. S. and Dietze, H. J., *Fresenius' J. Anal. Chem.*, **359**, 338 (1997).
3. Becker, J. S. and Dietze, H. J., *J. Anal. At. Spectrom.*, **10**, 637 (1995).
4. Saprykin, A. I., Becker, J. S. and Dietze, H. J., *Fresenius' J. Anal. Chem.*, **359**, 449 (1997).
5. Böhlke, J. K., De Leater, J. R., De Bievre, P. *et al.*, *J. Phys., Chem. Ref. Data*, **34**, 57 (2005).
6. Dempster, A. J., *Proc. Phil. Soc.*, **75**, 755 (1935).
7. Barinaga, C. J., Eiden, G. C., Alexander, M. L. and Koppenaal, D. W., *Fresenius' J. Anal. Chem.*, **355**, 487 (1996).
8. Kluge, H.-J., Bollen, G., Carlberg, C., Moore, R. B. and Quint, Q. W., *Adv. Mass Spectrom.*, **13**, 420 (1995).
9. Becker, J. S. and Dietze, H. J., *Int. J. Mass Spectrom.*, **202**, 69 (2000).
10. Becker, J. S. and Dietze, H. J., *Int. J. Mass Spectrom.*, **73**, 157 (1986).
11. Boulyga, S. F., Dietze, H. J. and Becker, J. S., *Mikrochim. Acta*, **137**, 93 (2001).
12. Day, J. A., Caruso, J. A., Becker, J. S. and Dietze, H. J., *Int. J. Mass Spectrom.*, **15**, 1343 (2000).
13. Hattendorf, B. and Guenther, D., *Anal. Chem.*, **73**, 5494 (2001).
14. Houk, R. S., Fassel, V. A., Flesch, G. D., Svec, H. J., Gray, A. L. and Taylor, C. E., *Anal. Chem.*, **52**, 2283 (1980).
15. Izmer, A. V., Boulyga, S. F. and Becker, J. S., *J. Anal. At. Spectrom.*, **18**, 1339 (2003).
16. Kaschner, R., Becker, J. S. and Seifert, G., *Int. J. Mass Spectrom.*, **176**, 103 (1998).
17. Kerl, W., Becker, J. S., Dannecker, W. and Dietze, H. J., *Fresenius' J. Anal. Chem.*, **362**, 433 (1998).
18. O'Brien, S. E., Acon, B. W., Boulyga, S. F., Becker, J. S., Dietze, H. J. and Montaser, A., *J. Anal. At. Spectrom.*, **18**, 230 (2003).
19. Zoriy, M. V., Halicz, L., Ketterer, M. E., Pickhardt, C., Ostapczuk, P. and Becker, J. S., *J. Anal. At. Spectrom.*, **19**, 362 (2004).
20. Becker, J. S. and Dietze, H. J., *J. Anal. At. Spectrom.*, **14**, 1493 (1999).
21. Wetzel, W. C. and Hieftje, G. M., *J. Anal. At. Spectrom.*, **21**, 1367 (2006).
22. Becker, J. S., Saprykin, A. I. and Dietze, H. J., *Int. J. Mass Spectrom. Ion Processes*, **164**, 81 (1995).
23. Barshik, C. M., Duckworth, D. C., Smith, D. H. (eds.), *Inorganic Mass Spectrometry: Fundamentals and Applications*. In: *Pract. Spectrosc.*, 23 (2000).
24. Gao, Y., Erickson, J. W. and Hockett, R. A., *SIMS X, (Proceedings of the 10th International Conference on Secondary Ion Mass Spectrometry, Muenster)*, A. Benninghoven, B. Hagenhoff and H. W. Werner (eds.), John Wiley & Sons, Ltd, Chichester, UK, 339 (1997).
25. Pickhardt, C., Brenner, I. B., Becker, J. S. and Dietze, H. J., *Fresenius' J. Anal. Chem.*, **368**, 79 (2000).
26. Gastel, M., Becker, J. S., Kuppers, G. and Dietze, H. J., *Spectrochim. Acta*, **52B**, 2051 (1997).
27. Becker, J. S. and Dietze, H. J., *Fresenius' J. Anal. Chem.*, **365**, 429 (1999).
28. Jochum, K. P., Dingwell, D. B., Rocholl, A. *et al.*, *Geostand. Newlett.*, **24**, 87 (2000).
29. Pickhardt, C., Becker, J. S. and Dietze, H. J., *Fresenius' J. Anal. Chem.*, **368**, 173 (2000).
30. Becker, J. S., Pickhardt, C. and Dietze, H. J., Mikrochim. Acta, **135**, 71 (2000).
31. Hervig, R. L., Mazdab, F. K., Williams, P., Guan, Y., Huss, G. R. and Leshin, L. A., *Chem. Geol.*, **227**, 83 (2006).
32. Christopher, S. J., Day, R. D., Bryan, C. E. and Turk, G. C., *J, Anal. At. Spectrom.*, **20**, 1035 (2005).
33. Rodriguez, P. F., Marchante-Gayón, J. M. and Sanz-Medel, A., *Talanta*, **68**, 869 (2006).
34. Jochum, K. P., Nohl, U., Herwig, K., Lammel, E., Stoll, B. and Hofmann, A. W., *Geostand. Geoanal. Res.*, **29**, 333 (2005).
35. Pearce, N. J. G., Perkins, W. T., Westgate, J. A. *et al.*, *Geostand. Newslett.*, **21**, 115 (1997).

36. Cornelis, R., Caruso, J., Crews, H. and K. G. Heumann (eds.), *Handbook of Speciation: Technique and Methodology*, vol. 1, John Wiley & Sons, Inc., New York (2003).
37. Cornelis, R., Crews, H., Donard, O. F., Ebdon, L. and Quevauviller, P., *Fresenius' J. Anal. Chem.*, **370**, 120 (2001).
38. Jochum, K. P., Pfänder, J., Woodhead, J. D. *et al., Geochem. Geophys. Geosyst.*, **6**, 1 (2005).
39. Jochum, K. P., Stoll, B., Herwig, K. *et al., Geochem. Geophys. Geosyst.*, **7**, 1 (2006).
40. Ingram, M. G. and Hayden, R. J., *Handbook on Mass Spectrometry*, Nuclear Series, NRC-USA (1954).
41. Willbold M., Jochum K. P., Raczek I., Amini M., Stoll B. and Hoffmann, A. W., *Anal. Bioanal. Chem.*, **377**, 117 (2003).
42. Willbold M. and Jochum, K. P., *Geostand. Geoanal. Res.*, **29**, 63 (2005).
43. Becker, J. S., Pickhardt, C. and Pompe, W., *Int. J. Mass Spectrom.*, **237**, 13 (2004).
44. Boulyga, S. F., Desideri, D., Meli, M. A., Testa, C. and Becker, J. S., *Int. J. Mass Spectrom.*, **226**, 329 (2003).
45. Klingbeil, P., Vogl, J., Pritzkow, W., Riebe, G. and Müller, J., *Anal. Chem.*, **73**, 1881 (2001).
46. Schaumlöffel, D. and Lobinski, R., *Int. J. Mass Spectrom.*, **242**, 217 (2005).
47. Stoll, B. and Jochum, K. P., *Fresenius' J. Anal. Chem.*, **364**, 380 (1999).
48. Sturgeon, R. E., Willie, S. N., Yang, L. *et al., J. Anal. At. Spectrom.*, **20**, 1067 (2005).
49. Szpunar, J. and Lobinski, R., *Hyphenated Techniques in Speciation Analysis*, R. M. Smith (ed.), RSC Chromatic Monographs, Cambridge (2003).
50. Heumann, K. G., in *Inorganic Mass Spectrometry*, R. G. F. Adams and R. van Griecken, (eds.), John Wiley & Sons, Inc., New York, 301 (1988).
51. Heumann, K. G., *Int. J. Mass Spectrom.*, **118/119**, 575 (1992).
52. Tresl, I. and Quétel, C. R., *J. Am. Soc. Mass Spectrom.*, **16**, 708 (2005).
53. Papadakis, P., Taylor, P. D. P. and De Bievre, P., *J. Anal. At. Spectrom.*, **12**, 791 (1997).
54. Schaumlöffel, D., Ruiz Encinar, J. and Lobinski, R., *Anal. Chem.*, **75**, 6873 (2004).
55. Stuerup, S., Chen, C., Jukosky, J. and Folt, C., *Int. J. Mass Spectrom.*, **242**, 225 (2005).
56. Regelous, M., Turner, S. P., Elliot, T. R., Rasotami, K. and Hawkesworth, J., *Anal. Chem.*, **76**, 3584 (2004).
57. Forster, D. A., Staubwasser, M. and Henderson, G. M., *Marine Chem.*, **87**, 59 (2004).
58. Simpson, L. A., Hearn, R. and Catterick, T., *J. Anal. At. Spectrom.*, **19**, 1244 (2004).
59. Viera, M. A., Ribero, A. S. and Curtius, A. J., *Microchim. J.*, **82**, 127 (2006).
60. Giné, M. F., Bellato, A. C. S. and Menegàrio, A. A., *J. Anal. At. Spectrom.*, **19**, 1252 (2004).
61. Prange, A., Schaumloeffel, D., Brätter, P., Richarz, A.-N. and Wolf, C., *Fresenius' J. Anal. Chem.*, **371**, 764 (2001).
62. Schaumlöffel, D., Prange, A., Marx, G., Heumann, K. G. and Brätter, P., *Anal. Bioanal. Chem.*, **372**, 155 (2002).
63. Mester, Z., Willie, S. N., Yang, L. *et al., Anal. Bioanal. Chem.*, **385**, 168 (2006).
64. Yang, L., Sturgeon, R. E. and Mester, Z., *J, Anal. At. Spectrom.*, **20**, 431 (2005).
65. Encinar, J. R., Schaumloeffel, D., Ogra, Y. and Lobinski, R., *Anal. Chem.*, **76**, 6635 (2004).
66. Boulyga, S. F., Heilmann, J. and Heumann, K. G., *Anal. Bioanal. Chem.*, **382**, 1808 (2005).
67. Boulyga, S. and Heumann, K. G., *Int. J. Mass Spectrom.*, **242**, 291 (2005).
68. Becker, J. S., *Spectrochim. Acta B*, **57**, 1805 (2002).
69. Westheide, J. T., Becker, J. S., Jaeger, R., Dietze, H. J. and Broekaert, J. A. C., *J. Anal. At. Spectrom.*, **11**, 661 (1996).
70. Becker, J. S., Zoriy, M. V., Pickhardt, C., Palomero-Gallagher, N. and Zilles, K., *Anal. Chem.*, **77**, 5851 (2005).
71. Becker, J. S., Soman, R. S., Sutton, K. L., Caruso, J. A. and Dietze, H. J., *J. Anal. At. Spectrom.*, **14**, 933 (1999).
72. Rege, S., Jackson, S., Griffin, W. L., Davies, R. M., Pearson, N. J. and O'Reilly, Y., *J, Anal. At. Spectrom.*, **20**, 601 (2005).
73. Jaeger, R., Saprykin, A. I., Becker, J. S., Dietze, H. J. and Broekaert, J. A. C., *Mikrochim. Acta*, **125**, 41 (1997).
74. Günther, D., Cousin, H., Magyar, B. and Leopold, I., *J. Anal. At. Spectrom.*, **12** 165 (1997).
75. Leach, J. J., Allen, L. A., Aeschliman, D. B. and Houk, R. S., *Anal. Chem.*, **71**, 440 (1999).
76. Becker, J. S., Pickhardt, C. and Dietze, H. J., *J. Anal. At. Spectrom.*, **16**, 603 (2001).

77. Bings, N. H., *J, Anal. At. Spectrom.*, **17**, 759 (2002).
78. O'Connor, C., Sharp, B. L. and Evans, P., *J. Anal. At. Spectrom.*, **21**, 556 (2006).
79. Pickhardt, C., Izmer, A., Zoriy, M. V., Schaumlöffel, D. and Becker, J. S., *Int. J. Mass Spectrom.*, **248**, 136 (2006).
80. Dietze, H. J. and Becker, J. S., *in Laser Ionization Mass Analysis*, A. Vertes, R. Gijbels and F. Adams (eds.) Chemical Analysis Series, John Wiley & Sons, Inc., New York, **124**, 453 (1993).
81. Becker, J. S., Burow, M., Boulyga, S. F., Pickhardt, C., Hille, R. and Ostapczuk, P., *At. Spectr.*, **23**, 177 (2002).
82. Boulyga, S. F. and Becker, J. S., *J. Anal. At. Spectrom.*, **17**, 1143 (2002).
83. Becker, J. S., Zoriy, M., Halicz, L., Teplyakov, N., Müller, C., Segal, I., Pickhardt, C. and Platzner, I. T., *J. Anal. At. Spectrom.*, **19**, (2004).
84. Becker, J. S., Soman, R. S., Becker, T., Panday, V. K. and Dietze, H. J., *J. Anal. At. Spectrom.*, **13**, 983 (1998).
85. Schaumlöffel, D., Giusti, P., Zoriy, M., Pickhardt, C., Szpunar, J., Lobinski, R. and Becker, J. S., *J. Anal. At. Spectrom.*, **20**, 17 (2005).
86. Zoriy, M. V., Pickhardt, C., Ostapczuk, P., Hille, R., Becker, J. S., *Int. J. Mass Spectrom.* **232**, 217 (2004).
87. Becker, J. S., *Trends Anal. Chem.* **24**, 243 (2005).

7

Mass Spectrometric Techniques for Analysis of Gaseous Materials and Volatile Compounds

A wide field of application for mass spectrometry is the analysis of gases and volatile compounds. The analysis of complex gaseous mixtures mostly requires the combination of separation and mass spectrometric techniques, for example by gas chromatography mass spectrometry (GC-MS), which is applied as a routine method in many laboratories today.[1] Not only thermally stable and volatile species, but also those species which can be converted into the gaseous state, are preferentially separated by gas chromatography and detected by mass spectrometry, e.g., by using ICP-MS for element speciation. GC-ICP-MS has the advantage of enhanced sensitivity (with nearly 100 % transport efficiency of analytes) in comparison to liquid chromatography (LC) combined with ICP-MS since samples are introduced into the plasma in a gaseous form. Atomization and ionization efficiencies of analytes in GC-ICP-MS are higher since no energy from the plasma is required for desolvation and volatilization.[2]

This chapter focuses on selected developments and applications of mass spectrometry for the analysis of gases and volatile compounds relevant in inorganic mass spectrometry. A brief introduction to isotope ratio measurements by gas source mass spectrometry is presented in Section 8.3.

7.1 Sampling and Sample Preparation of Gases and Volatile Compounds

The determination of volatile elemental species in biological or environmental samples, such as body fluids, tissues, soils, plants or water, generally requires a careful preconcentration and clean-up procedure in order to separate the analytes from matrix material. Several existing sample preparation procedures and applied measurement techniques (especially GC-ICP-MS in combination with

Inorganic Mass Spectrometry: Principles and Applications J. S. Becker

hydride generation, to form highly volatile covalent compounds from As, Sb, Hg, Sn, Pb, Bi, Cd, Se, Te and Ge) are reviewed by Alonco and Encinar.[3] In most cases, hydride generation by the addition of a defined volume of sodium borohydride to the sample solution is coupled to a cryogenic trapping step in a chromatographic column immersed in liquid nitrogen.[3] The hydrides thus formed are transported from the sample solution to the GC column using helium as carrier gas. Any excess hydrogen formed is separated in front of the ICP torch with the help of a three-way valve connected after the GC column during the preconcentration step.[3,4] Packed column gas chromatography, which was applied as the first chromatographic technique for elemental speciation, does not provide sufficiently high resolution for the analysis of complex environmental and biological materials so that capillary columns with improved resolution are advantageous. In addition, by using GC with capillary columns coupled to a powerful ICP-MS, a significant reduction in carrier gas flow rate of about one order of magnitude (to 1–4 ml min^{-1}) is observed together with an increased sensitivity due to a lower dilution factor in the mobile phase. Only small sizes (1–2 μl) of sample are injected into the column because one limit of GC with a capillary column is the relatively small loading capacity. To combine GC with ICP-MS, several heated interfaces have been developed to avoid analyte condensation, e.g., using a heated stainless steel transfer line connected to the ICP torch through a six valve inside the GC oven,[5] a quartz transfer line heated to 240 °C,[6] a stainless steel tube heated to 270 °C insulated with glass fibre tape (deactivated fused silica tube inside a water cooled torch).[7] Furthermore, a transfer capillary consisting of a heated steel tube is inserted into the torch injector and a Tl solution is introduced into the ICP torch using a T-piece connection with a Meinhard nebulizer and cyclonic spray chamber.[8]

Photochemical alkylation to generate volatile species by UV irradiation of aqueous solutions containing low carboxyl acids, such as formic, acetic, propionic and butyric acids, was proposed by Sturgeon *et al.*[9,10] For isotope ratio measurements on organic compounds, CO_2, N_2 and H_2O are generated by oxidative combustion $[C, H, N, O] + O_2 \rightarrow CO_2 + N_2 + H_2O$ during the reaction at around 850 °C in a sealed quartz or Pyrex tube with CuO as the oxygen source. Alternative reduction reagents are Cr, Zn, Mn or pure C. Following combustion, the sealed tubes are broken one by one on a vacuum line, the product gases are released and cryogenically separated and collected. While preparing samples for nitrogen isotopic analysis, CaO can be added to the quartz tube as a CO_2 absorber.[11] Before mass spectrometric analysis, H_2O is removed from the gas mixture and can be used for D/H determination by conversion to H_2. Such hydrogen gas preparation requires a reducing agent like uranium at a temperature of 800 °C.[12] Stable isotope ratio mass spectrometers with an electron impact ion source (see Section 5.9) for precise and accurate measurements of light stable isotopes of hydrogen, carbon, oxygen and nitrogen of gaseous compounds usually have a dual inlet system for measuring (and comparing) sample and standard under the same experimental conditions. Details of sample preparation and inlet systems for gas isotope mass spectrometers comprising valves, pipes, capillaries, connectors and gauges are described by Ghosh and Brandt.[11] In modern instruments, interfacing stable isotope ratio mass spectrometry (IRMS) with gas chromatography (GC) has been applied for compound specific isotope analysis. Online pyrolysis IRMS techniques have been developed for $\delta^{18}O$ measurements in milligram quantities of cellulose, carbohydrates and aromatic compounds in the presence of glassy carbon at temperatures in excess of 1200 °C.[13]

7.2 Applications of Inorganic Mass Spectrometry for Analysis of Gases and Volatile Compounds

Numerous applications concern the mass spectrometric analysis of quite different gases and highly volatile compounds. For example, the determination of trace metal impurities in ethylene gas for real time continuous online process control by ICP-MS has been applied in the polymer industry by Klinkenberg *et al.*[14] Trace impurities in high purity gases used, for example, in semiconductor manufacturing, can be determined by the impinger sampling method or after particle filter sampling by ICP-MS.[15] Whereas the impinger method is applied for the determination of metal impurities in CO gas,[16] for the impurity analysis of silane the particles are trapped on the surface of a Teflon membrane filter of 0.1 μm pore size.[17] Electronic grade gases used in semiconductor manufacturing processes have also been introduced directly into ICP-MS for the determination of trace impurities at the ultratrace level.[18] For the qualitative and semiquantitative determination of selected elements (As, Cd, Hg and Pb) in cigarette smoke, the gas is introduced directly into the ICP-MS[19,20] (see also Section 9.3.2.)

GCMS has been used for determination of organochloride pesticides and polyaromatic hydrocarbons (PAH) in environmental samples (contaminated water, soil or plant samples)[21,22] in the routine mode. GCMS and LC-MS/MS have been applied for a quasi-simultaneous determination of about 400 pesticide residues in grain after solvent extraction.[23] Of special interest for inorganic mass spectrometry such as GC-ICP-MS, are volatile organic compounds containing so-called 'heteroelements' such as P, S and halogens. Organophosphorus pesticides (OPPs), which are widely used in agriculture for crop protection, have been analyzed by GC-ICP-MS in river water after solid phase microextraction (SPME) as proposed by Sanz-Medel's group.[24] Phosphorus and other heteroelements (S, Cl, and Br) present in selected pesticides have been detected in the ng l^{-1} range and below. In the same group, SPMW-enantioselective gas chromatography with classical electron capture detection (ECD) and ICP-MS (via the measurement of $^{31}P^+$) is employed for the chiral speciation of the pesticide ruelene in environmental samples.[25]

A sensitive and highly selective methodology for the determination of a wide range of pesticides based on the quasi-simultaneous element specific determination of P, S, Cl, Br and I by gas chromatography hyphenated with octopole reaction cell ICP-MS, has been developed in Prange's working group.[26] This method was applied to the screening of different fruit extracts. Detection limits down to the ng l^{-1} level were obtained for the different pesticides analyzed. Peaks were identified by comparing the retention times obtained and their chemical composition.[26]

Phosphoric acid triesters in human blood plasma have been determined by GC-ICP-MS after SPME,[27] after their separation from blood plasma, with detection limits of 50 ng l^{-1} for tripropyl phosphate, 17 ng l^{-1} for tributylphosphate and 24 ng l^{-1} for triphenylphosphate.

The first coupling of a capillary gas chromatography to a sector field ICP-MS for the speciation of organometallic compounds present in a synthetic sample was described by de Smaele *et al.*[28] Transient ion signals of $^{120}Sn^+$, $^{202}Hg^+$ and $^{208}Pb^+$ were measured using $^{126}Xe^+$ as an internal standard (to correct possible drifts of the magnetic field and plasma instabilities).

The generation and identification of volatile arsenic species by photochemical alkylation during UV irradiation of aqueous solutions was studied in Sturgeon's working group.[9,10] Volatile arsenic species were identified by ICP-ToF-MS in combination with GCMS. It was found that As(III) formed arsine in the presence of formic acid, trimethylarsine in acetic acid, triethylarsine in propionic acid and triproplylarsine in butyric acid.[9] For example, a methylation efficiency of about 75 % was observed at an As(III) concentration of 50 μg l^{-1} in 3.1 M acetic acid. Sturgeon's group identified 13 organoarsenic species including several new ones. The same working group analyzed dibutyl tin (DBT) and tributyl tin (TBT) in sediments by species specific isotope dilution plasma source mass spectrometry using GC for sample introduction and analyte separation.[29] A comparison was made between sector field ICP-MS (GC-ICP-SFMS) and quadrupole-based ICP-MS (GC-ICP-QMS), where a greater than two-fold improvement was observed in the precision of the determined $^{120}Sn/^{117}Sn$ isotope ratio in butyl tin species in certified standard reference materials (marine sediment, PACS-2, NRCC, Ottawa, Canada) using GC-ICP-SFMS with detection limits of 0.4 and 0.3 ng g^{-1} for TBT and DBT, respectively, in 0.5 g of sample material.[29]

In order to characterize volatile selenium and mercury species in plant tissues or environmental samples, GC-ICP-MS after solid phase microextraction is the method of choice. Different analytical procedures have been created in several working groups.[30] GC-ICP-MS has been employed for the analysis of selenoaminoacids in food supplements.[31,32] Non-volatile selenium species are often analyzed using GC-ICP-MS after appropriate derivatization of the analytes, as discussed by B'Hymer and Caruso.[33] However, the Se–Hg antagonism in Se- and Hg-containing compounds in Indian mustard, *Brassica juncea*, a selenium-accumulating plant, was investigated and it was found that selenium/mercury-containing compounds of high molecular weight may be protein associated. Apart from the volatile species, no volatile species containing Se and Hg could be found.

GC-ICP-MS has been employed to analyze airborne particles with respect to selected species (e.g., of Hg and Pb).[34,35] Speciation analysis with GC-ICP-MS has been employed for the detection of organometallic compounds in tobacco smoke using a smoke machine.[36] Toxic metals (e.g., Pb, Cd, As, Sn) in tobacco smoke present a significant health risk (carcinogens, nephrotoxins, hepatotoxins and neurotoxins). Chromatograms of derivatized samples measured by GC-ICP-MS demonstrated the presence of an unknown organolead and organotin (butyl tin) in the smoke samples investigated.[36]

The methylmercury species (MeHg) known as a toxic compound has been increasingly studied in biological and environmental samples[37,38] by gas chromatography ICP-MS, and the isotope dilution method has been applied for quantitative MeHg determination.[37] Methylated species of Sb, As and Sn (mono-, di- and trimethlylated species) have been investigated along the profile depth in an urban soil sample from the Ruhr basin by Duester *et al.*[39] The species analysis was performed by pH gradient hydride generation purge and trap gas chromatography combined with GC-ICP-MS. Monomethylated Sb and As species as the dominant species were detected at the ng g^{-1} level and below. The soil profiles measured by GC-ICP-MS showed a decrease of species content with decreasing depth.[39]

As an inorganic mass spectrometric technique for the analysis of gases, GC-ICP-MS differs from other atomic spectrometric techniques (e.g., GC-ICP-AES or AAS) due to its capability to provide information on isotopic composition. This outstanding feature of mass spectrometry can be utilized in the isotope dilution technique as a quantification strategy in order to improve the accuracy of analytical data and to study the sources of elemental species. The analytical performance of capillary GC-ICP-MS has been evaluated using a time-of-flight mass spectrometer for the analysis of volatile metalloid compounds of tin and antimony.[40] Isotope ratios ($^{121}Sb/^{123}Sb$ and $^{118}Sn/^{120}Sn$) of volatile Sb and Sn species (100 pg SbH_3, $MeSbH_2$, SnH_4 and $BuSnH_3$) were measured with a precision of 0.72–0.81 %. By increasing the amount of species up to approximately 500 pg it is possible to achieve a precision for tin isotope ratio measurements of 0.5 %. This precision for isotope ratios using a GC-ICP-ToF mass spectrometer is not sufficient to study very small isotope fractionation within the biogeochemical cycles. Multiple collector ICP-MS (MC-ICP-MS) is therefore required, which allows more precise isotope ratio measurements. The identification and quantification of alkylated tin compounds such as dimethyldiethyl tin, trimethylethyl tin and propyltrimethyl tin in European municipal waste deposits by GC-ICP-MS was carried out by Feldmann's working group.[41]

A gas chromatograph coupled to a MC-ICP-MS for the precise determination of isotope ratios as part of the speciation application has been described, for example, for the elements S, Pb, Hg and Sb.[2] Transient signals of sulfur isotope ratios ($^{32}S/^{34}S$) have been measured in an isotopic gas standard (PIGS 2010, IRMM) to determine SF_6 using GC-MC-ICP-MS (Isoprobe, Micromass, UK) with a hexapole collision cell.[42] For data evaluation of chromatographic peaks, peak integration limits were defined by the determination of a 'uniform isotope ratio zone' inside the transient signal, which resulted in a precision of 0.03–0.05 % RSD on $^{32}S/^{34}S$ depending on the concentration injected.[42] GC-MC-ICP-MS has been employed to study $PbEt_4$ species (injected in organic solvent or derivatized from isotopic lead isotope standard reference material (NIST SRM 981)) via $^{207}Pb/^{206}Pb$ isotope ratios by Krupp *et al.*[8,43] The authors compared the detection limit and precision achieved with lead isotope ratios for a transient signal of tetraethyl lead ($PbEt_4$) species using two MC-ICP-MS instruments (Axiom and Isoprobe) in comparison to quadrupole ICP-MS; all three instruments were coupled to GC. Absolute detection limits of 3 fg and 60 fg and a precision of 0.015 % RSD and 0.2 % RSD for GC-MC-ICP-MS and GC-ICP-QMS, respectively, were obtained. The same group measured isotope ratios of selected volatile methylated and ethylated mercury (MeHgEt and $HgEt_2$) or lead ($PbEt_4$) species with capillary GC-MC-ICP-MS (Axiom).[44] Transient signals of two volatile mercury species, MeHgEt and $HgEt_2$, are illustrated in Figure 7.1. A thallium solution ($10\,\mu g\,l^{-1}$) was introduced via a T-piece connecting the GC and spray chamber to the plasma torch and measured as a continuous ion intensity as a function of time (see Figure 7.1). Tl was used for internal standardization and mass bias correction via the $^{205}Tl/^{203}Tl$ isotope ratio. The $^{200}Hg/^{198}Hg$ isotope ratios were measured via the transient signals of both mercury species with a precision of about 0.05 % (RSD).[44]

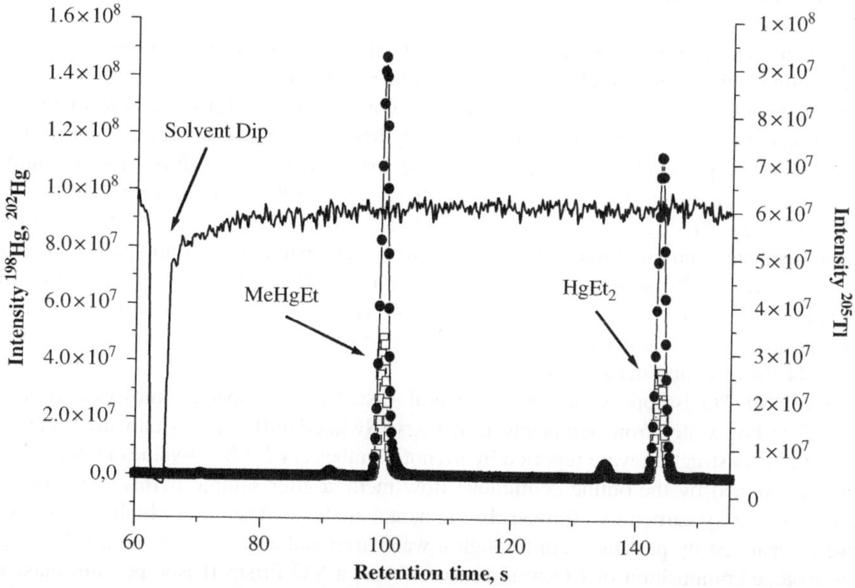

Figure 7.1 *Isotope ratios on transient signals with capillary GC-MC-ICP-MS. Chromatogram of mercury species obtained with GC coupled to Axiom MC-ICP-MS. Open squares:* ^{198}Hg *trace, filled circles:* ^{202}Hg *trace, black line:* ^{205}Tl *trace continuously introduced as liquid solution. (E. Krupp and O. Donard, Int. J. Mass Spectrom.* ***242****, 233 (2005). Reproduced by permission of Elsevier.)*

Cryotrapping cryofocusing capillary GC coupled to the MC-ICP-MS (Isoprobe) has been employed for isotope ratio measurements on trimethyl antimony (Me_3Sb) produced by anaerobic digestion of sewage sludge in a laboratory fermenter.[45] For two different fermentation experiments a significant enrichment of ^{123}Sb by GC-MC-ICP-MS was obtained in the range of 10‰–19‰($\delta^{123}Sb$).

A combined use of single and multiple isotope enriched spike solution for the detection of butyl tin compounds in sediments by GC-ICP-MS has been evaluated in order to validate an analytical procedure proposed for routine determination.[46] A mixture of mono-, di- and tributyl tin (MBT, DBT and TBT, respectively) enriched in ^{119}Sn and a triple spike solution containing each butyl tin species enriched with a different tin isotopes (^{118}Sn and ^{117}Sn) were employed to develop the analytical technique. This demonstrated the advantage provided by the combined use of single and multiple spike solutions to validate a methodology for the determination of butyl tin content by GC-ICP-MS after a quantitative solid–liquid extraction procedure for species in unknown sediment.[46]

Prange and Jantzen[6] analyzed 25 organometallic species (methyl, butyl, or phenyl) of tin, mercury and lead by GC-ICP-MS. The species were separated using a capillary column 30 m in length with a GC coupled to an ICP-MS. The detection limits of tin, lead and mercury organometallic species were determined as 50, 100 and 120 fg, respectively. Organotin species in environmental samples (wet sediments) were analyzed using sodium tetraethylborate for direct ethylation.[6]

Donard and co-workers reported on the application of double-focusing sector field GC-ICP-MS with a single ion collector and multi-collector arrangement for trace metal speciation.[2]

7.3 Stable Isotope Ratio Measurements of Gases and Volatile Compounds

One important area of stable isotope ratio measurements (SIRMS) is the analysis of gases using offline or online techniques (see also the discussion in Section 8.3.).

For example, little is known about the isotopic composition of formaldehyde in the atmosphere. Formaldehyde is a chemical intermediate in hydrocarbon oxidation. The carbon ($\delta^{13}C$) and hydrogen (δD) isotopic composition of atmospheric formaldehyde is analyzed using continuous flow gas chromatography isotope ratio mass spectrometry.[47] Isotope ratios were measured using GC-IRMS (Finnigan MAT 253 stable isotope ratio mass spectrometer, single-sector field with electron impact ion source and multiple ion collection) with a precision of ± 1.1 and $50\%(1\sigma)$ for $\delta^{13}C$ and δD, respectively. The accuracy of the online continuous flow isotope technique was verified by calibrating three aliquots of the gas phase standard via the offline dual inlet IRMS technique. The concentration of formaldehyde in ambient air was determined on IRMS major ion peak areas (i.e., mass 44 for $\delta^{13}C$ and mass 2 for δD).[47]

Studies of $^{18}O/^{16}O$ isotope variations in several vascular plant species, mosses and environmental surface bog water from temperate peat bogs (Switzerland) used as climatic archives for paleoclimatic reconstruction were reported by Ménot-Combes *et al.*[48] $\delta^{18}O$-values in organic material were determined by the online continuous flow method after sample pyrolysis at 1080 °C in the presence of glassy carbon in a Carlo-Erba elemental analyzer. The gases obtained (CO, N_2 and H_2O) were separated by passing them through a water trap and a GC column in a helium carrier gas. The isotope composition of CO was measured with a VG Prism II isotope ratio mass spectrometer relative to the VSMOW isotope standard. The overall analytical uncertainty is 0.08‰ for water $\delta^{18}O$-values.[48]

Stable isotope ratio measurements of 'life science elements', such as carbon, hydrogen and nitrogen, vary slightly, but significantly in major compartments of the Earth.[11] Owing mainly to anthropogenic activities (e.g., burning fossil fuel) the $^{13}C/^{12}C$ isotope ratio of CO_2 in the atmosphere has changed over the past 200 years from 0.0111073 to 0.0110906 (about 1.5 parts per thousand). In between interglacial warm periods and glacial maxima, the $^{18}O/^{16}O$ ratio of precipitation in Greenland has changed by as much as 5 parts per thousand (0.001935–0.001925). These small changes in isotope ratios measured by high precision stable isotope ratio mass spectrometry (using single-focusing magnetic sector field mass spectrometers with a highly efficient electron impact ion source and a multi-collector assembly) reflect natural fractionation processes that have left their signature in natural archives and consequently enable global climatic change to be studied. Such natural archives provide information about how different components of the environment are coupled and have interacted in the past. For example, air inclusions in firn and ice provide an important natural archive for reconstructing historical trace gas concentrations as well as isotope records.[11] In addition, isotope ratio measurements in tree rings can be used for climatic information for the past 10 000 years whereas records in ice cores can go back in time for as much as 400 000 years. The isotopic composition of paleo-water and paleo-atmospheric CO_2 can be obtained directly from ice core samples and trapped inclusions within ice cores. Sedimentary deposits and corals provide an indirect approach to the past, mainly providing information on climate evolution on a geological timescale. The isotopic signature is difficult to extract from these sources, e.g., oxygen isotopes from carbonates must be measured from selected species (foraminifera) and the fossils must have remained immune to isotopic fractionation over time. Ghosh and Brand reviewed stable isotope ratio measurements in global climate change research.[11] Stable isotopes and biosphere–atmosphere interactions including factors affecting the oxygen isotope ratios of plant organic materials, stable isotopic composition of soil organic matter, the study of global change and Earth system science and ecosystem scale processes have been reviewed by Flanagan *et al.*[49]

References

1. Gohlke, R. S. and McLafferty, F. W., *Int. J. Mass Spectrom.*, **4**, 367 (1993).
2. Moldovan, M., Krupp, E. M., Holliday, A. E. and Donard, O. F. X., *J. Anal. At. Spectrom.*, **19**, 815 (2004).
3. Alonso, J. I. G. and Encinar, J. R., in *Handbook of Element Speciation, Techniques and Methodologies*, Vol. I, R. Cornelis, J. Caruso, H. Crews, K. Heumann (ed), John Wiley & Sons, Ltd, Chichester 163, (2003).
4. Donard, O. F., Rapsomanikis, S. and Weber, J. H., *Anal. Chem.*, **58**, 772 (1986).
5. Gallus, S. M. and Heumann, K. G., *J. Anal. At. Spectrom.*, **11**, 887 (1996).
6. Prange, A. and Jantzen, E., *J. Anal. At. Spectrom.*, **10**, 105 (1995).
7. Evans, E. H. and Caruso, J. A., *J. Anal. At. Spectrom.*, **8**, 427 (1993).
8. Krupp, E. M., Pécheyran, C., Pinaly, H. *et al.*, *Spectrochim. Acta*, **56B**, 1233 (2001).
9. Guo, M. X., Sturgeon, R. E., Mester, Z. and Gardner, G. J., *J. Anal. At. Spectrom.*, **20**, 702 (2005).
10. McSheehy, S., Guo, M. X., Sturgeon, R. E. and Mester, Z., *J. Anal. At. Spectrom.*, **20**, 709 (2005).
11. Ghosh, P. and Brand, W. A., *Int. J. Mass Spectrom.*, **228**, 1 (2003).
12. Bigeleisen, J., Perlmann, M. J. and Prosser, H., *Anal. Chem.*, **24**, 1356 (1952).
13. Ferretti, D. F., Lowe, D. C., Martin, R. J. and Brailsford, G. W., *J. Geophys. Res. Atmos.*, **105**, 6709 (2000).
14. Klinkenberg, H., Van Born, W. and Kip, B. J., *Spectrochim. Acta*, **52**, 1695 (1997).
15. Collard, J. M. and Kishi, Y., in *ICP Mass Spectrometry Handbook*, S. M. Nelms (ed.), Blackwell Publishing Ltd, Oxford 462, (2005).
16. Anderson, P. C., Cooper, G. and Houlding, V. H., *Semicond. Int.*, **21**, 127 (1998).
17. Suzuki, I. and Tarutani, K., *Anal. Sci.*, **13**, 833 (1997).
18. Hutton, R. C., Bridenne, M., Coffre, E., Marot, Y. and Simondet, F., *J. Anal. At. Spectrom.*, **5**, 463 (1990).
19. Chang, M. J., Naworal, J. D. and Connell, C. T., *J. Anal. At. Spectrom.*, **21**, 574 (2006).
20. Chang, M. J., Naworal, J. D., Walker, K. and Connell, C. T., *Spectrochim. Acta*, **2003 B**, 1979 (2003).
21. Douglas, G. S., Burns, W. A., Bence, A. E., Page, D. S. and Boehm, P., *Environ. Sci. Techn.*, **38**, 1958 (2004).
22. Gianguzza, A. and Orecchio, S., *Polycyclic Arom. Comp.*, **26**, 37 (2006).
23. Pang, G. F., Lui, Y. M., Fan, C.-L. *et al.*, *Anal. Bioanal. Chem.*, **384**, 1366 (2006).
24. Figalgo-Used, N., Montes-Bayón, M., Blanco-González, E. and Sanz-Medel, A., *J. Anal. At. Spectrom.*, **20**, 876 (2005).

25. Fidalgo-Used, N., Montes-Bayón, M., Blanco-González, E. and Sanz-Medel, A., *J. Anal. At. Spectrom.*, **21**, 876 (2006).
26. Proefrock, D., Leonard, P., Wilbur, S. and Prange, A., *J. Anal. At. Spectrom.*, **19**, 623 (2004).
27. Shah, M., Meja, J., Cabovska, B. and Caruso, J. A., *J. Chromatogr. A*, **1103**, 329 (2006).
28. De Smaele, T., Moens, L. and Dams, R., in *Plasma Source Chemistry – Developments and Applications*, Royal Society of Chemistry, Cambridge 109, (1997).
29. Yang, L., Mester, Z. and Sturgeon, R. E., *J. Anal. At. Spectrom.*, **18**, 1365 (2003).
30. Mounicou, S., Shah, M., Meija, J., Caruso, J. A., Vonderheide, A. P. and Shann, J., *J. Anal. At. Spectrom.*, **21**, 404 (2006).
31. Péréz Méndez, S., Blanco González, E. and Sanz-Medel, A., *J. Anal. At. Spectrom.*, **15**, 1109 (2000).
32. Péréz Méndez, S., Montes Bayón, M., Blanco González, E. and Sanz-Medel, A., *J. Anal. At. Spectrom.*, **14**, 1333 (1999).
33. B'Hymer, C. and Caruso, J. A., *J. Chrom. A*, **1114**, 1 (2006).
34. Leal-Granadillo, I. A., Alonso, A. I. G. and Sanz-Medel, A., *Anal. Chem.*, **423**, 21 (2000).
35. Pecheytan, B., Lalere, B. and Donard, O. F. X., *Environ. Sci. Techn.*, **34**, 27 (2000).
36. Elobeid, M., Chai, Y., Clarke, D., Hannigan, R. and Russ, J., in *Plasma Source Mass Spectrometry*, G. Holland and D. R. Bandura (eds.), The Royal Society of Chemistry, Cambridge, 80 (2005).
37. Hintelmann, H. and Nguyen, H. T., *Anal. Bioanal. Chem.*, **381**, 360 (2005).
38. Luque de Castro, M. D. and Jiménez-Carmona, M. M., *Trends in Anal. Chem.*, **17**, 441 (1998).
39. Duester, L., Diaz-Bone, R. A., Koester, J. and Hirner, A. V., *J. Environ. Monit.*, **7**, 1186 (2005).
40. Haas, K., Feldmann, J., Wennrich, R. and Stärk, H.-J., *Fresenius' J. Anal. Chem.*, **370**, 587 (2001).
41. Mitra, S. K., Jiang, K., Haas, K. and Feldmann, J., *J. Environ. Monit.*, **7**, 1066 (2005).
42. Krupp, E. M., Pécheyran, C., Meffan-Main, S. and Donard, O. F. X., *Anal. Bioanal. Chem.*, **378**, 250 (2004).
43. Krupp, E. M., Pécheyran, C., Meffan-Main, S. and Donard, O. F. X., *Fresenius' J. Anal. Chem.*, **370**, 573 (2001).
44. Krupp, E. M. and Donard, O. F. X., *Int. J. Mass Spectrom.*, **242**, 233 (2005).
45. Wehmeier, S., Ellam, R. and Feldmann, J., *J. Anal. At. Spectrom.*, **18**, 1001 (2003).
46. Rodriguez, P. F., Alonso, J. I. G. and Sanz-Medel, A., *J. Anal At. Spectrom.*, **2005**, 1076 (2005).
47. Rice, A. L. and Quay, P. D., *Anal. Chem.*, **78**, 6320 (2006).
48. Ménot-Combes, G., Burns, S. J. and Leuenberger, M., *Earth Plan. Sci. Lett.*, **202**, 419 (2002).
49. Flanagan, L. B., Ehleringer, J. R. and Pataki, D. E., *Stable Isotopes and Biosphere-Atmosphere Interactions: Processes and Biological Controls*, Elsevier Academic Press, Amsterdam, Boston, Heidelberg (2005).

8

Isotope Ratio Measurements and their Application

An outstanding feature of inorganic mass spectrometry is its determination of precise and accurate isotopic abundances and isotope ratios. Isotopes of the same element (of the same number of protons or atomic number of element, Z) are, by definition, nuclides with different mass m and mass number A ($A = Z + N$) due to the different number of neutrons (N) in the nucleus. Isotope analyses are of special interest for characterizing the composition of samples with respect to stable and unstable isotopes in quite different concentration ranges – from the analysis of matrix elements down to the trace and ultratrace concentration level.[1–9] Of 1700 isotopes, nearly 16 % (264 isotopes) are stable. The chemical elements Tc, Pm, Th, U and the transuranic elements do not possess stable isotopes.

Measurement of the isotope abundances of a chemical element is based on the fact that the sum of all abundances of isotopes with the same Z is 100 %. For example, copper possesses two stable isotopes with $m/z = 63$ and 65. If the isotope ratio ^{63}Cu/^{65}Cu has been determined, e.g., by mass spectrometry, then the isotope abundance of ^{65}Cu (χ) can be obtained by:

$$\chi(^{65}Cu) = 100/(1 + {}^{63}\mathrm{Cu}/{}^{65}\mathrm{Cu}) \tag{8.1}$$

From the measured ^{63}Cu/^{65}Cu isotope ratio the isotope abundances of ^{63}Cu and ^{65}Cu are then calculated in a natural sample as roughly 69.2 % and 30.8 %, respectively. Small deviations from the IUPAC table value[10] could be evidence of fine isotope variation in nature.

ICP-MS, LA-ICP-MS and GDMS have the advantage that atomic ions are formed with the highest intensities so that these are monitored and it is therefore easy to calculate the isotope abundances of an element. In TIMS, atomic ions are preferentially monitored, but several elements form polyatomic ions with higher intensity (e.g., oxide ions). Some of these have three or four oxygen atoms, thereby complicating the calculations and increasing the abundance uncertainties owing to the amplification of oxygen uncertainties.[7] Details on the calculation of

Inorganic Mass Spectrometry: Principles and Applications J. S. Becker

isotope abundances of chemical elements if polyatomic ions occur are described in the textbook 'Modern Isotope Ratio Mass Spectrometry' by Platzner.[7]

Furthermore, isotope analysis is relevant for determining the atomic weight ($A_r(E)$) of elements. The $A_r(E)$ is the average of all masses of all naturally occurring stable isotopes (taking into account the abundances of isotopes) of a chemical element (see Appendix I[10]). By consideration of the masses of isotopes (m_i) and the known relative abundances of all stable isotopes (χ_i) with $i = 1$ to n of a selected chemical element, the average atomic weight ($A_r(E)$) of this element can be calculated:

$$A_r(E) = \sum_{i=1}^{i=n} (m_i \cdot \chi_i) / \sum_{i=1}^{i=n} \chi_i \quad (8.2)$$

Because the sum of the relative abundances of all stable isotope of an element is 100 % Equation (8.2) can be written as:

$$A_r(E) = \sum_{i=1}^{i=n} (m_i \cdot \chi_i) / 100\,\% \quad (8.3)$$

The atomic weights of the chemical elements are summarized in Appendix II.

For example, the isotopic composition and the atomic weight of neodymium,[11] dysprosium[12] and erbium[13] have been determined using synthetic mixtures prepared gravimetrically from highly enriched isotopes of neodymium in the form of oxides of well defined purity by TIMS. No natural isotopic variation was found in terrestrial neodymium, dysprosium or erbium samples. These isotopic compositions of Dy[12] and Er[13] measured by TIMS are accepted as the best measurements from a single terrestrial source as noted in the table of isotopic composition of elements, 2001.[10]

In general, the abundances of stable isotopes of a chemical element are not constant, larger isotope variations are expected for lighter elements compared to heavier ones. Possible sources of isotope variation in nature are observed as the result of isotope fractionation effects (e.g., the fractionation of oxygen isotopes by photosynthesis or respiration[14]), isotope exchange reactions, geological processes during formation and transport of minerals, wash out processes, weathering or cosmogenic effects. Furthermore, the radiogenic decay of unstable isotopes, e.g., the β^- decay of ^{87}Re with a half-life of 8×10^{10} a and consequently the ingrowth of the (daughter) nuclide ^{87}Sr, yielded changes in isotope abundances in nature. In addition, isotope variations, especially also as a result of anthropogenic contamination (e.g., due to nuclear fallout from nuclear weapons test or nuclear power plants[15,16]) and others, result in small changes of isotope abundances.

Variation of an isotope ratio is usually presented as δ value of an isotope expressed as per mil [‰]:

$$\delta = (R_{sample}/R_{standard} - 1) \times 10^3 \quad (8.4)$$

where R represents the measured isotope ratio of the low abundance isotope to the main isotope (e.g., $^2D/^1H$, $^{13}C/^{12}C$, $^{18}O/^{16}O$ or $^{34}S/^{32}S$).

One problem is the availability of reliable reference materials for isotope ratio measurements. Certified isotope standard reference materials are available from NIST (National Institute of Standard Reference Materials, http://www.nist.gov/srd/, formerly NBS – National Bureau of Standards) and IRMM (Institute for Reference Materials and Measurements, http://www.irmm.jrc.de). There are isotope standard reference materials for light elements (H, Li, B, C, N, O), Mg (magnesium metal isotopic standard: NIST SRM 980), and for moderately heavy elements such as K (potassium chloride isotopic standard: NIST SRM 985), Cr (chromium metal isotopic standard: NIST

SRM 979), Ni (nickel metal isotopic standard: NIST SRM 986), Rb (rubidium chloride isotopic standard: NIST SRM 984) and Sr (strontium carbonate isotopic standard: NIST SRM 987). In addition, isotope reference materials are available for heavy elements such as Tl (thallium metal isotopic standard NIST SRM 997), Pb (NIST lead standard reference materials SRM 981–983) or U (uranium oxide NIST isotope standard U 005, U020, U350, U500 or U930) and others. The most important isotope standard reference materials applied in inorganic mass spectrometry are summarized in the table in Appendix V.[17]

In 2005, De Laeter[18] discussed the role of isotope reference materials for the analysis of 'non-traditional' stable isotopes. At present, no isotopically certified reference materials exist for a large number of elements, including Cu, Zn, Mo and Cd, and it is important that this situation be rectified as soon as practicable.[18] Before the isotopically certified reference materials become available for selected elements, suitable reference materials can be created as a standard if sufficient and reliable isotope data have been obtained by interlaboratory comparisons.[18] For example, the $^{176}Hf/^{177}Hf$ isotope ratio was measured using hafnium oxide from Johnson Matthey Chemicals, JMC–475, for hafnium isotope ratio measurements with different multi-collector mass spectrometers (ICP-MS and TIMS) as summarized in Table 8.1.[7,19–25] However, no isotope SRM is certified for the element Mo either. Mo isotope analysis is relevant, for example, for studying the isotope fractionation of molybdenum during chemical processes or the isotope variation of molybdenum in nature as the result of the predicted double β decay of ^{96}Zr or ^{94}Zr.[18,26–28] The spectroscopically pure sample from Johnson Mattey Specpure is proposed as a laboratory standard reference material if sufficient and reliable isotope data are collected via an interlaboratory comparison.

Table 8.1 *$^{176}Hf/^{177}Hf$ isotope ratio measured by MC-ICP-MS and MC-TIMS.*

	$^{176}Hf/^{177}Hf$	RSD (ppm)	Ref.
Mean JMC-475	0.282 161	13 (2σ)	Halliday *et al.*[21]
MC-ICP-MS (Plasma 54)	0.282 149	14 (2σ)	Halliday *et al.*[22]
MC-ICP-MS (Plasma 54)	0.282 155	14 (2σ)	Blichert-Toft *et al.*[20]
MC-ICP-MS (Nu Instrument)	0.282 189	12 (2σ)	Kleinhanns *et al.*[23]
MC-TIMS	0.282 155	14 (2σ)	Nowell *et al.*[24]
MC-ICP-MS (NEPTUNE)	0.282 158	11 (1σ)	Schwieters *et al.*[25]

J. S. Becker, *J. Anal. At. Spectrom.*, **17**, 1172 (2002). Reproduced by permission of the Royal Society of Chemistry

Precise and accurate isotope analyses by mass spectrometry have attained growing importance in the last few years due to instrumental improvements with respect to sensitivity, detection limits, precision and accuracy.[1] As mentioned before, because the isotope abundances of several elements are not constant and vary as a result of nuclear, biological, chemical, geochemical and physical processes, isotope ratio measurements are required for different research and application fields. Isotope ratio measurements are therefore necessary for elements with two or more isotopes for investigating isotope variation in nature,[5,7,29–34] in the geosciences (geochemistry and geochronology for age dating of rocks),[26,34–36] for determining the origin of foodstuffs, plants or minerals, for characterizing nuclear samples and for radioactive waste control.[1,37] Furthermore, isotope analysis has been performed for environmental monitoring,[38–40] for characterizing medical samples (blood, urine, faeces, hair and tissue analysis), also for the health control of exposed persons,[41–44] in the human body[45] and many other applications. For elements which possess only one stable isotope (stable monoisotopic elements, e.g., iodine), but at least one long-lived radionuclide (e.g., ^{129}I), isotope ratio measurements (e.g., $^{129}I/^{127}I$) are of interest for two applications: firstly to study contamination of this long-lived radionuclide in the environment[46,47] and secondly to utilize the long-lived radionuclide for quantification purposes by applying the isotope dilution technique.[46,48] In addition, enriched stable isotopes or radionuclides have been used by means of isotope ratio measurements for tracer experiments.[49–55] Selected application fields for isotope ratio measurements are summarized in Figure 8.1.

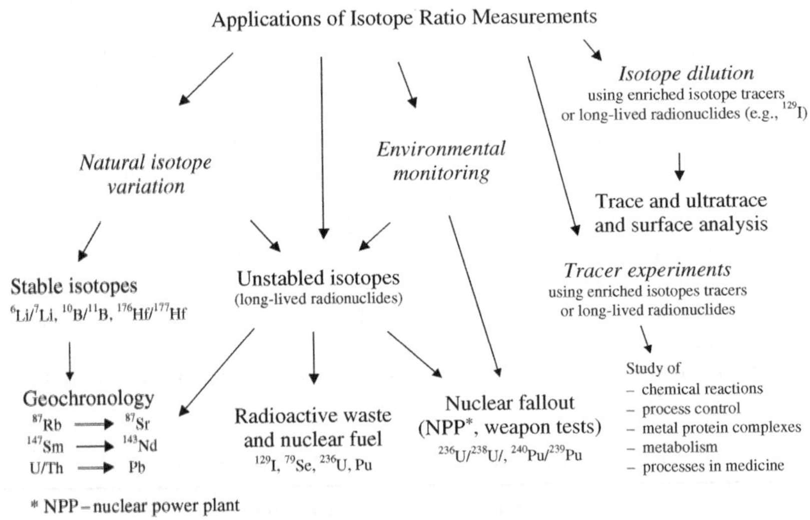

Figure 8.1 *Fields of application for isotope ratio measurements. (Modified from J. S. Becker, J. Anal. At. Spectrom.,* ***17****, 1172 (2002). Reproduced by permission of The Royal Society of Chemistry.)*

The improvement of mass spectrometers for isotope ratio measurements with respect to more efficient and powerful ionization techniques, improved ion separation systems and especially the development of sensitive ion detectors using single ion and multiple ion collectors with increased abundance sensitivity has been accelerated by the necessity of obtaining the most precise and accurate isotope ratio data from the smallest possible amount of sample. Progress has been achieved in the refinement of the variable multiple collector system in MC-ICP-MS, improvements in current amplifier technology to achieve ultimate precision and accuracy, increased abundance sensitivity, improved ion optics using zoom optics, enlarged geometry of mass spectrometers and increased signal-to-noise ratios.

8.1 Capability of Inorganic Mass Spectrometry in Isotope Ratio Measurements

For measurements of isotope ratios or isotope abundances, any of the mass spectrometers discussed in the previous chapters, such as SSMS, LIMS, GDMS[56] and LA-ICP-MS,[6] are of benefit for the direct isotope analysis of solid samples. SSMS and LIMS are rarely applied in isotope analysis due to their relatively low precision. Several applications of the isotope dilution technique as a calibration strategy in SSMS, mostly on geological samples, are known.[57–59] GDMS has been mostly applied in multi-element trace analysis and depth profiling and plays only a minor role in isotope ratio measurements. Betti and co-workers have developed analytical techniques for isotope analysis in samples of concern for nuclear applications.[56,60] In addition, as a powerful surface analytical technique (e.g., for depth profiling), SIMS is well suited for precise isotope ratio measurements on solid samples.[56,61] Due to its spatial resolving power in the μm range and below, as an *in situ* technique (in the ion microprobe mode) SIMS provides the possibility of measuring isotope ratios directly within single mineral grains without any sample preparation.[56] An excellent review on SIMS measurements of stable isotopes to study fractionation effects and isotope variation especially of light elements, is given by Ireland.[61] The precision of several mass spectrometric techniques for isotope ratio measurements using single and multiple ion collection is compared in Table 8.2.

Table 8.2 *Precision of mass spectrometric methods.*

Direct methods			Post-ionization methods		
	Single ion collector	***Multiple ion collector***		***Single ion collector***	***Multiple ion collector***
SSMS	3–5 %	≈ 1 %	TIMS	0.01–1 %	0.001–0.01 %
LIMS	3–5 %	—	GDMS	0.03–1 %	—
SIMS	0.01–1%	0.002–0.1%	SNMS	0.02–1 %	—
ICP-MS	0.02–0.2%	0.001–0.01%	LA-ICP-MS	0.08–1 %	0.006–0.02 %
			RIMS	5 %	—

J. S. Becker, *J. Anal. At. Spectrom.*, **17**, 1172 (2002). Reproduced by permission of the Royal Society of Chemistry

Other mass spectrometric techniques such as RIMS[62–65] and AMS[66–69] possess high isotope selectivity for extreme ultratrace and isotope analysis of, in particular, radiotoxic isotopes (^{14}C, ^{41}Ca, ^{90}Sr, ^{99}Tc, ^{210}Pb, ^{236}U and plutonium isotopes) in the environment, in cosmochemistry, radiodating, nutrition and biomedical research. RIMS has become as an ultrasensitive and selective analytical technique for the determination of extremely low isotope abundances.[8,63,65,70] In spite of the excellent analytical features of RIMS (detection limit for isotopes: 10^6 atoms per sample) and exciting applications for the determination of extremely low abundances and isotope ratios of long-lived radionuclides, such as all plutonium isotopes (including ^{238}Pu), ^{236}U or ^{41}Ca, but no commercial instrument is available on the analytical market.

As one of the most sensitive and selective mass spectrometric techniques, AMS is utilized for measuring long-lived radionuclides at the level of natural isotopic abundances (10^{-16} to 10^{-12}).[5,67,69,71] For example, for isotope ratio measurements of $^{41}Ca/^{40}Ca$ at the natural level of 10^{-14}–10^{-15} AMS is the method of choice. The minimum detectable isotopic ratio for $^{129}I/^{127}I$ of about $< 10^{-12}$ can only be achieved by AMS.[72] The largest application field for AMS is ^{14}C measurements for age dating in archaeology, geology, geophysics and other sciences.[5] However, RIMS and AMS involve complicated and expensive experimental equipment and the analysis, including sample preparation, is time consuming.

Being a valuable isotope analytical technique in routine work for high precision isotope ratio measurements, TIMS is applied in many laboratories worldwide for isotope ratio measurements especially for elements with ionization potentials $< 7\,eV$,[7] such as alkali and earth alkali elements, rare earth elements (REE), uranium and plutonium. It is advantageous that the interference problem occurs relatively seldom in TIMS, especially if the negative thermal ionization technique for elements and molecules with electron affinities $> 2\,eV$ (Ir, W, Os, Re, Pt, Cl and Br) is applied. TIMS with multiple ion collectors achieves a precision of up to 0.001 % thus permitting the study of fine isotope variation in nature and age dating. For example, small isotope fractionations of cadmium in lunar material[73] studied by TIMS (double spike technique) are a result of the redeposition of Cd from the Cd rich vapour cloud associated with volcanism. The volatile nature of Cd, which is the heaviest element to show isotope fractionation in nature, is of importance for explaining the isotope variation.[73] Double spike calibrated TIMS is applied for measurements of mass-dependent fractionation in Se, Fe and Cr.[74] A precision of $\pm 0.2‰$ on $^{80}Se/^{76}Se$, $^{56}Fe/^{54}Fe$ and $^{53}Cr/^{52}Cr$ isotope ratios was achieved with small sample sizes.

For many decades, TIMS was the isotope analytical technique of choice, but due to instrumental developments in ICP-MS, especially with multiple ion collectors (MC-ICP-SFMS), and the advantages of ICP-MS in comparison to TIMS (e.g., higher element sensitivities, faster isotope ratio measurements, comparable precision and accuracy, practically no restriction on the ionization potential of chemical elements, time independent mass fractionation and the possibility of additional multi-element analysis at trace and ultratrace level and fewer, less time-consuming sample preparation steps[75]), TIMS will be replaced in future by powerful ICP-MS to an ever greater extent.

Of all the different mass spectrometric techniques for isotope analysis (such as ICP-MS, LA-ICP-MS, TIMS, GDMS, AMS, SIMS, RIMS and isotope ratio mass spectrometry of gases), the greatest proportion of published papers today concern ICP-MS with single and multiple ion collection.[19] Due to its benefits, ICP-MS has now become a widely accepted method for isotope analysis and allows isotope ratios to be measured in a short time with good accuracy and precision.[9,19,75–78] As discussed above, as a powerful and universal tool, ICP-MS has opened up new applications for isotope ratio measurements of elements with a high first ionization potential, which are difficult to analyze with TIMS (such as Mo, Hf, Fe). Of all the heavy metals studied, uranium was favoured by ICP-MS and LA-ICP-MS.

In comparison to quadrupole ICP-MS (ICP-QMS), double-focusing sector field ICP-MS (ICP-SFMS) with single ion detection offers a remarkable improvement in sensitivity, detection limits and precision at low mass resolution. Whereas in ICP-SFMS a lot of disturbing isobaric interferences of analyte and polyatomic ions can be separated at the required mass resolution, several polyatomic ions disturbing the isotope analysis (e.g., $^{40}Ar^{16}O^+$ for ^{56}Fe and $^{129}Xe^+$ for ^{129}I determination) are removed in a collision cell by selected ion–molecule reactions with a collision/reaction gas such as H_2, He, NH_3 or O_2. Using ICP-SFMS with a single ion detector, isotope ratio measurements at low mass resolution with a precision of up to 0.02 % are possible whereas at medium mass resolution the precision is one order of magnitude poorer due to lower sensitivity. The precision can be improved by one order of magnitude via the simultaneous detection of mass separated ion currents of isotopes using multiple ion collectors in ICP-SFMS (MC-ICP-SFMS).

Because ICP-MS with different instrumentations and sample introduction systems (besides solution nebulization, also laser ablation or hyphenated methods, such as HPLC, CE, SPME) is today the most frequently used analytical technique for precise and accurate isotope ratio measurements, the following section will mainly focus on this form of mass spectrometry with an inductively coupled plasma source.

8.2 Limits for Precision and Accuracy of Isotope Ratio Measurements and How to Solve the Problems

Several effects, like instrumental mass bias, isobaric interferences, instrumental background, contamination of the solution introduction system and the sampler and skimmer cone and lens system, mass scale drift effects, plasma instabilities, drifts of ion intensities and matrix effects, limit isotope ratio measurements with respect to precision and accuracy of the analytical data.[19,79] Detection limits and precision for isotope analysis are strongly dependent on the mass spectrometer used (quadrupole based with and without collision cell, sector field with single ion collector or with multiple ion collectors), matrix composition and sample preparation steps (separation and enrichment factors).

The mass discrimination effect plays an important role in all mass spectrometric techniques. In general, the mass discrimination decreases with increasing mass as demonstrated for ICP-MS in Figure 8.2. In ICP-MS, the mass discrimination is a result of the space charge effects of the expanding ion beams. Time independent discrimination of ions with different mass occurs if the ions – formed in the inductively coupled plasma – leave the skimmer cone. Due to the Coulomb repulsion of positively charged ions, a loss of transmission through the ion optical lens system is observed. During ion extraction the light ions are deflected to a greater extent than the heavy ones, resulting in a measured isotope ratio of lighter to heavier isotopes which is smaller than the true value. Mass discrimination effects (mass bias) – depending on the experimental parameters used – decrease in ICP-MS with increasing atomic weight of the elements and are corrected by measuring the mass discrimination factor using a suitable isotope standard reference material as described in reference[19]. Isotope ratio determination for light elements (e.g., $^{6}Li/^{7}Li$ or $^{10}B/^{11}B$) is difficult as mentioned above, due to the relatively large mass bias effect. Possible drifts of mass bias during the measurements resulting in systematic errors in measured isotope ratios are corrected if the samples and standard are analyzed alternately. In a thermal surface ionization source, preferential evaporation of lighter isotopes and subsequent mass discrimination effects are found. Whereas in ICP-MS the mass discrimination effect is time independent, in TIMS time dependent (but smaller) mass discrimination effects are observed.[7,19] A comparison of different mass discrimination effects in TIMS and ICP-MS is given in Table 8.3.

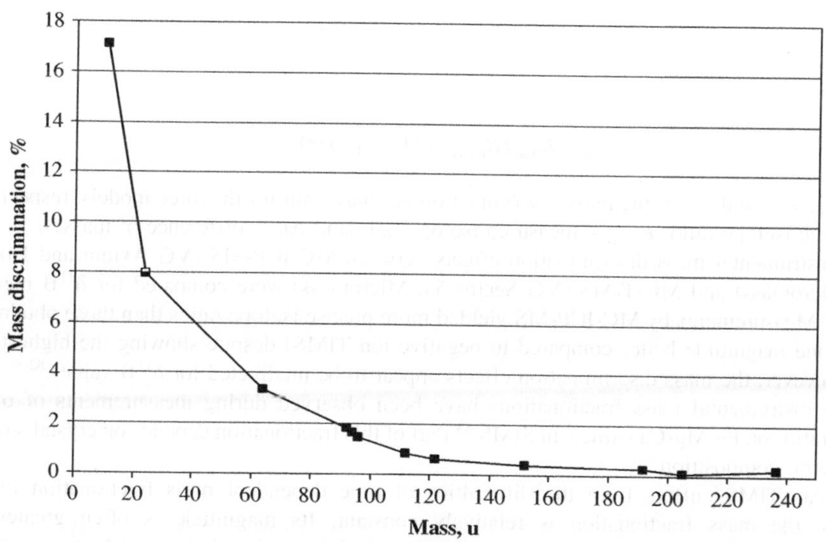

Figure 8.2 *Mass discrimination effect in ICP-MS. (J. S. Becker, J. Anal. At. Spectrom.,* ***17****, 1172 (2002). Reproduced by permission of The Royal Society of Chemistry.)*

Table 8.3 Comparison of mass discrimination in ICP-MS and TIMS.

	TIMS	ICP-MS
Origin	Preferential evaporation of the lighter isotope from the sample filament	Preferential radial deflection of lighter ions from the center of ion beam (Coulomb repulsion due to space charge effects of charged ions → loss in ion transmission)
Observation	$(R_{L/H})$ meas > $(R_{L/H})$ true	$(R_{L/H})$ meas < $(R_{L/H})$ true
Correction method	① Calibration with isotope reference standard (for stable, radioactive, radiogenic and elements with 2 isotopes e.g, U, Pb, Cu) ② Internal normalization, applicable for elements with 3 and more isotopes, where at least one isotope ratio is well established and invariant in nature e.g., Sr, Hf, Nd	① and ② as for TIMS ③ External spiking by a mass non-interfering element with a well established (invariant in nature) isotope ratio e.g., ^{203}Tl, ^{205}Tl for Pb or ^{69}Ga, ^{71}Ga for Ge isotope analysis
Mass discrimination	TIME DEPENDENT	TIME INDEPENDENT

J. S. Becker, *J. Anal. At. Spectrom.*, **17**, 1172 (2002). Reproduced by permission of the Royal Society of Chemistry

Several approaches are available for the correction of mass discrimination:[80,81]
Power law:

$$R_{true}/R_{meas} = (1+\varepsilon_{pow})^{\Delta m} \tag{8.5}$$

Exponential law:

$$R_{true}/R_{meas} = \exp(\varepsilon_{exp}\Delta m) \tag{8.6}$$

Linear law:

$$R_{true}/R_{meas} = (1+\varepsilon_{lin}\Delta m) \tag{8.7}$$

with ε_{pow}, ε_{exp} and ε_{lin} as the mass discrimination per mass unit for the three models, respectively; R_{true} – true isotope ratio; R_{meas} – measured isotope ratio and Δm – difference of masses.

The instrumental mass discrimination effects between MC-ICP-MS (VG Axiom and Isoprobe from Micromass) and MC-TIMS (VG Sector 54, Micromass) were compared for δ^{11}B measurements.[82] Measurements by MC-ICP-MS yielded more precise isotope ratios than those obtained by TIMS (one magnitude better compared to negative ion TIMS) despite showing the highest mass bias. However, the mass discrimination effects appear to be unaffected for δ^{11}B values.[82]

Large instrumental mass fractionations have been observed during measurements of oxygen isotope ratios on Fe-Mg-Ca garnets in SIMS.[83] Part of this fractionation depends on crystal structure and mineral composition.

Whereas TIMS suffers from the difficulties of time dependent mass fractionation in MC-ICP-MS, the mass fractionation is relatively constant. Its magnitude is often greater than that of TIMS and matrix effects are possible[84]. It is known that isotopic reference materials used for the calibration of isotope ratio measurements are available only as a pure metal or compound. Possible matrix effects during isotope ratio measurements, especially in solid-state mass spectrometry, should be investigated and considered. In ICP-MS or TIMS, matrix effects are minimized if the analyte is carefully separated from the matrix element. Matrix effects are of significance, for example, in boron isotope ratio measurements of sub-ng sized samples of foraminiferal-hosted boron,[85] which has been performed by total evaporation negative ion thermal ionization mass spectrometry (TE-NTIMS) using the Triton mass spectrometer with multiple ion collection, (Thermo Fisher Scientific, Bremen). For these small sample sizes, blank contamination is possible as an additional serious problem. This problem can be overcome provided strict protocols are followed, ultraclean reagents are used and filaments are loaded in a stream of pure nitrogen gas. With TE-NTIMS a precision and accuracy of $\pm 0.7‰$ (2 σ) has been achieved, and in standard addition experiments on NIST SRM 951 boric acid standard, the $^{11}B/^{10}B$ isotope ratios have been measured within the error of certified value using both seawater and carbonate matrices.[85]

Another limit source of uncertainty in isotope ratio measurements by mass spectrometry is the dead time of the ion detector for counting rates higher than 10^6 cps, because a lower number of counts are usually registered than actually occur. Dead time correction of the detector is required if extreme isotope ratios are measured by channel electron multipliers and pulsed counting systems.[86]

Furthermore, high isotope abundance sensitivity is desirable for the measurement of low abundance isotopes in the presence of neighbouring high abundant isotopes.[87] The contribution of the neighbouring peak to the intensity of a measured isotope – the abundance sensitivity – influences the accuracy of isotope abundance measurements.[88] For example, the determination of ^{90}Sr ultratraces in the presence of a high strontium concentration of natural isotope abundance is not possible at low abundance sensitivity of the mass spectrometer due to peak tailing of $^{88}Sr^+$ on $m/z = 90$.[89,90] In the determination of ^{90}Sr an improvement of abundance sensitivity in sector field ICP-MS (Element) from 2×10^{-5} to 9×10^{-7} has been observed by changing the mass resolution from $m/\Delta m = 300$ to 4000. In comparison using helium as the collision gas in quadrupole based ICP-MS the abundance sensitivity for $^{236}U/^{238}U$ determination has been determined as 6.3×10^{-8}.[87]

In addition, the occurrence of isobaric interferences of analyte ions with isobaric polyatomic ions can hamper the accuracy and precision of isotope ratio measurements (see also Section 6.1.3). The main factors affecting the accurate and precise determination, for example, of ^{236}U using ICP-MS, are the isobaric interference of the $^{235}UH^+$ molecular ion on $^{236}U^+$ analyte ions, and the presence of $^{238}U^+$ and $^{235}U^+$ peak tails. Isobaric interferences of atomic ions can be separated with a Fourier transform ion cyclotron resonance mass spectrometer (ICP-FTICR-MS).[91] However, this type of instrumentation is of minor significance in isotope ratio measurements, because no commercial instrument is currently available. Isobaric interference problems of analyte ions and polyatomic ions in ICP-MS and LA-ICP-MS have been solved as discussed above, by the application of double-focusing sector field ICP-MS at the required mass resolution, by the application of ICP-MS with a collision/reaction cell, by the use of special nebulizers with desolvator, or by the application of hyphenated techniques such as ion chromatography, HPLC and CE with ICP-MS.[92–96] If the analyte occurs at extremely low concentrations, matrix separation and enrichment of the analyte has been proposed by several authors.[39,89,97,98] Trace/matrix separation, performed offline or online, is being used more and more frequently for isotope ratio measurements in order to avoid possible isobaric interferences, matrix effects and to reduce the detection limits.

Isotope abundances which are free from all sources of bias are defined as 'absolute' isotope abundances. The absolute isotope composition of elements can be measured by MC-TIMS and MC-ICP-MS via gravimetric synthetic mixtures or standard solutions from highly enriched isotopes, as demonstrated for neodymium,[11] erbium[13] and samarium,[11,13,99] respectively.

8.3 Isotope Ratio Measurements by Gas Source Mass Spectrometry

The majority of elements analyzed with respect to isotope ratios in order to study isotope variations in nature are light gaseous elements such as H, O and N and gas forming elements, e.g., C and S. Gas source mass spectrometry (GSMS) with electron impact (EI) ion source produces nearly mono-energetic ions (similar to TIMS) and is an excellent tool for the high precision isotope analysis of light elements such as H, C, N and O, but also for S or Si.[7,100,101] Precise and accurate measurements of isotope ratios have been carried out by gas source mass spectrometers with multiple ion collectors by a sample/standard comparison and the δ values of isotope ratios were determined (see Equation 8.4). Electron impact ionization combined with mass spectrometry has been applied for elements which readily form gaseous compounds (e.g., CO_2 or SO_2) for the isotope analysis of carbon and sulfur, respectively).

Stable isotope ratio measurements of $^{2}H/^{1}H$, $^{13}C/^{12}C$, $^{15}N/^{14}N$, $^{18}O/^{16}O$ and $^{34}S/^{32}S$ in a broad variety of chemical compounds in gaseous, liquid and solid form are relevant for applications in life sciences, environmental research and geology.[7] For example, GSMS was utilized to study isotope effects in planetary atmospheres and the early Solar System.[102] On the basis of precise isotope ratio measurements by GSMS, mass independent isotopic compositions have been found in O_3, CO_2, N_2O, and CO in the Earth's atmosphere and in carbonate from a Martian meteorite, which suggests a role for mass independent processes in the atmosphere of Mars. The observed mass independent isotopic compositions for meteoritic oxygen and sulfur may derive from chemical processes in the presolar nebula, and their distributions could provide insight into early Solar System evolution. In addition, photochemical mass independent variation of sulfur isotopes in achondritic meteorites has been found[103] and oxygen isotope variation during the thermal decomposition of carbonates were observed.[104] In further applications of stable isotope ratio measurements of light elements by GSMS the existence of an ^{16}O rich gaseous reservoir in the solar nebula[105] and isotope effects in ozone formation have been studied.[106] Analytical techniques including extraction, oxidative pyrolysis or grinding in vacuum for stable isotope analysis by GSMS have been utilized to characterize fluid and gaseous inclusions.[107] Ghosh and Brandt have reviewed applications of stable isotope ratio measurements in global climate change research.[32]

To analyze the isotope ratios of gas species, gas chromatography (GC) has been coupled to an isotope ratio mass spectrometer (GC-IRMS) as demonstrated for the ^{13}C and ^{14}N analysis of volatile organic compounds and related gases.[108] GC was also coupled to two different MC-ICP-MS for isotope ratio measurements of short transient signals on lead species to study fractionation effects.[109] For the biomethylation of antimony during sewage sludge fermentation isotope enriched antimonite [^{123}Sb (V)] tracer and cryotrapping GC-ICP-MS have been applied.[110] Different volatile antimony species (monomethyl-, dimethyl- and trimethylantimony) have been quantified via the measured $^{123}Sb/^{121}Sb$ isotope ratios. In addition, tracer experiments using ^{77}Se and subsequent isotope ratio measurements by GC-ICP-MS have been performed to determine dimethylselenide in human breath after ingestion of ^{77}Se enriched selenite.[111] The isotope dilution technique using single and multiple enriched spike solutions combined with GC-ICP-MS was applied to quantify volatile butyltin compounds in Sanz-Medel's working group.[112]

Further applications of mass spectrometry for the analysis of gases are discussed in Chapter 7.

8.4 Isotope Ratio Measurements by Quadrupole based ICP-MS

Numerous applications of quadrupole ICP-MS (without a collision cell) for isotope ratio measurements can be found in quite different fields. For example, magnesium isotope ratios on

NIST SRM 980 and in ^{26}Mg spiked plant nutrient solutions for tracer experiments in biological research have been performed using the Elan 6000 (Perkin Elmer).[113] The accuracy of the measured isotope standard solution–NIST SRM 980 and enrichments with ^{26}Mg spike – was 0.1–0.3 % ($^{26}Mg/^{24}Mg_{measured}$ 0.13958 ± 0.00036 versus $^{26}Mg/^{24}Mg_{certifed}$ 0.13932 ± 0.00026). In long term measurements on isotope standard solutions the relative standard deviation (RSD) varied from 0.2–0.04 % for the $^{26}Mg/^{24}Mg$ isotope ratio. The nutrient solutions investigated contained high amounts of calcium and potassium in the $mg\,l^{-1}$ range. The Mg concentration in the ^{26}Mg spiked nutrient solution was determined by the reverse isotope dilution technique, and the results are in agreement with those of inductively coupled plasma optical emission spectrometry (ICP-OES). The mass spectrometric technique for Mg isotope ratio measurements is routinely applied in author's laboratory as part of tracer experiments using enriched stable Mg isotopes to study the transport phenomena of nutrients in plants. To classify wine according to its geographic origin, boron isotope ratios have been measured by quadrupole ICP-MS (Elan 5000, Perkin Elmer).[114]

The introduction of the collision cell in ICP-MS represents progress in ICP-MS instrumentation which is relevant for improving the precise determination of isotope ratios in comparison to commercial quadrupole ICP-MS without a collision cell. The first investigation of the basic principles and application of the collision cell to improve sensitivity in inorganic ultrasensitive mass spectrometry was published 24 years ago by Becker and Dietze.[115]

By the collision of ions with the collision gas (e.g., He and/or H_2, NH_3) in the gas target, the polyatomic ions are dissociated and the atomic argon ions formed from plasma gas in ICP are neutralized. Originally the collision cell was introduced in ICP-MS in order to dissociate disturbing argon based molecular ions (ArX^+, X = O, N, C, H or Ar) and to neutralize the plasma gas ions (Ar^+).[116,117] Reaction chemistry and collision processes in the gas cell for resolving isobaric interferences in ICP-MS are discussed in references.[116,118] The effect of reducing formation of interfering ions by up to eight orders of magnitude was observed using the Elan 6100 DRC (Perkin Elmer)[116,117] and by up to four orders of magnitude using the Platform (Micromass).[119,120] Due to the removal of isobaric interferences by collision and gas phase reactions it is possible to perform isotope analysis of Ca, Fe and Se (which are disturbed in inductively coupled Ar plasma mass spectrometry by Ar^+, ArO^+ and $Ar_2{}^+$, respectively). It should be noted that new interferences with molecular ions have also been observed using the collision cell in the Platform instrument. The effect of adventitious water in a hexapole collision cell using a VG PQ ExCell ICP-MS to investigate the origins of polyatomic ions has described by Dexter *et al.*[121] An additional effect of the collision of atoms (e.g., He) or molecules (H_2, NH_3, CH_4) in the collision cell with the analyte ions is the loss of ion kinetic energy, which results in a reduction of the energy spread ('cooling') of the ions from several eV to < 0.1 eV, in improved ion transmission, sensitivity of elements and precision in the determination of isotope ratios.

ICP-MS with a hexapole collision cell is advantageous for isotope ratio measurements of difficult to analyze elements.[119] Due to possible interferences with Ar^+, ArO^+ and dimer argon molecular ions $Ar_2{}^+$ the measurements of the isotope ratios $^{40}Ca/^{44}Ca$, $^{56}Fe/^{57}Fe$ and $^{78}Se/^{80}Se$ by ICP-MS are extremely difficult or completely impossible. Whereas for these isotope ratios precisions between 0.2 and 0.3 % were observed using ICP-MS with hexapole collision cell, a precision of 0.07 % was measured for a $10\,\mu g\,l^{-1}$ uranium isotope reference solution ($^{235}U/^{238}U \approx 1$) in the author's laboratory.[16,122,123]

Moens *et al.*[124] demonstrated that ion–molecule reactions in a dynamic reaction cell (DRC) of ICP-QMS allow the $^{87}Rb^+/^{87}Sr^+$ isobaric interference to be eliminated. A special ion–molecular reaction in a collision cell was studied in order to avoid this isobaric interference of atomic ions by the Elan 6100 DRC (Perkin Elmer). The authors analyzed a mixed Sr, Rb solution. After introducing methylfluoride into the collision cell Sr formed selective fluoride polyatomic ions, but Rb did not. An interference free Sr isotope analysis is thus possible via its fluoride ions in the presence of Rb without chemical separation. This technique is applied for Rb–Sr age dating of geological samples, the precision of the age results thus obtained is of the same order as the precision of the TIMS results.

8.5 Isotope Ratio Measurements by Laser Ablation ICP-MS

For direct isotope ratio measurements on solid samples, laser ablation ICP-MS (LA-ICP-MS) is applied to an increasing extent in geology, geochronology, biology and environmental research.[6]

An analytical procedure has been developed for the determination of the $^{235}U/^{238}U$ isotope ratio (and U and Th concentration) in urine at the trace and ultratrace level by LA-ICP-MS. After drying of a homogeneous sample, the urine was analyzed directly by LA-ICP-MS. In order to study the performance of the analytical methods developed, matrix matched synthetic laboratory standards doped with ^{230}Th (IRMM 60), uranium with natural isotope composition ($^{235}U/^{238}U = 0.00725$) and uranium isotope standard reference material (NIST U-930) at low $ng\,l^{-1}$ levels were prepared. The recovery for thorium and uranium concentration measured on the synthetic urine laboratory standard by LA-ICP-MS varied between 91 % and 104 %. The precision and accuracy of the analytical methods was determined for uranium concentration on urine laboratory standards (at uranium concentration $0.1\,ng\,ml^{-1}$) as 7 % and < 1 %, respectively. The analytical technique developed was applied for $^{235}U/^{238}U$ isotopic ratio measurements in urine samples, in order to detect any possible incorporation of depleted or enriched uranium in exposed persons. This technique could be of relevance in forensic science, if only a small amount of dried urine sample is available for investigation.[41]

An analytical procedure has been proposed for precise uranium isotope ratio measurements in a thin uranium layer on a biological surface by LA-ICP-MS using a cooled laser ablation chamber.[125] One drop of uranium isotope standard reference materials NIST, 350, NIST 930, of our isotopic laboratory standard CCLU 500 (20 µl, U concentration $200\,ng\,l^{-1}$) and of uranium with natural isotopic pattern were deposited on the leaf surface and analyzed by LA-ICP-MS at well defined laser crater diameters of 10, 15, 25 and 50 µm. A precision for measurements of isotope ratios in the range of 2.1–1.0 % for $^{235}U/^{238}U$ in selected isotope standards was observed whereby the precision and the accuracy of isotope ratios compared to the non-cooled laser ablation chamber was improved.[125]

Isotope ratio measurements were carried out by LA-ICP-MS on protein spots of a brain sample with Alzheimer's disease in two dimensional (2D) gels after tracer experiments with highly enriched ^{65}Cu and ^{54}Fe isotope tracers.[49] In Figure 8.3 part of the 2D gel and the transient signals for $^{65}Cu^+$ and $^{63}Cu^+$ are shown. Isotope ratio measurements on protein spots did not reveal any changes of isotope abundances of ^{65}Cu, whereas an enrichment of ^{54}Fe was found in several protein spots in the 2D gel. The $^{54}Fe/^{56}Fe$ isotope ratios measured in selected protein spots of an Alzheimer protein sample after tracer experiments are summarized in Figure 8.4.[49]

Selected applications in isotope ratio measurements by LA-ICP-MS using different equipment are summarized in Table 8.4.[50,126–136]

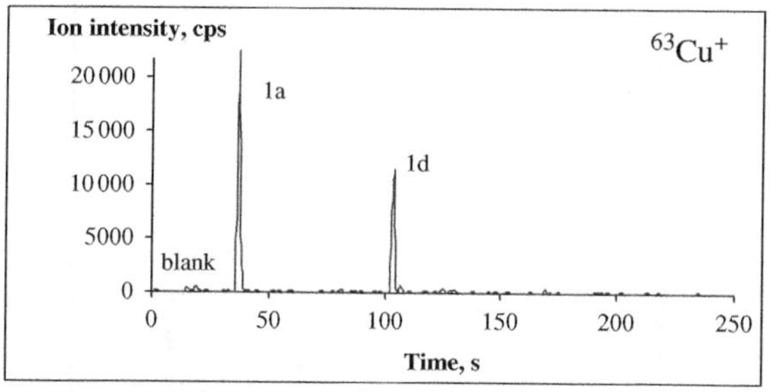

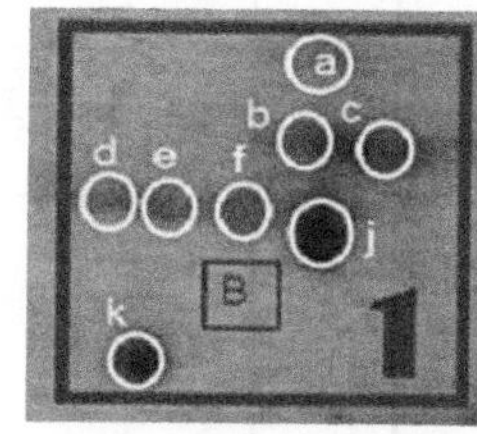

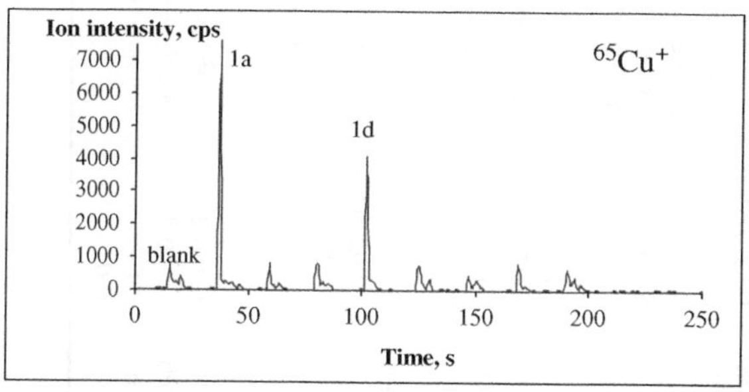

	$^{65}Cu/^{63}Cu$
protein 1a	0.414 ± 0.038
protein 1d	0.413 ± 0.054
table value	0.446
enriched tracer	26.78

Figure 8.3 *Transient signals of ^{63}Cu and ^{65}Cu in protein spots of Alzheimer's brain samples measured by LA-ICP-MS (isotopic-enriched ^{65}Cu spike solution was doped to the gel after 2D gel electrophoresis). (J. Su. Becker et al., Int. J. Mass Spectrom.,* **242**, *135 (2005). Reproduced by permission of The Royal Society of Chemistry.)*

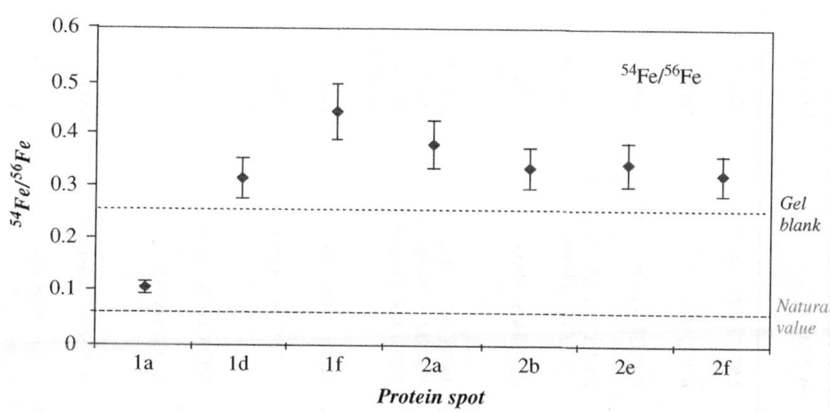

Figure 8.4 *$^{54}Fe/^{56}Fe$ isotope ratios in selected protein spots of Alzheimer's brain samples measured by LA-ICP- MS (isotopic-enriched ^{54}Fe spike solution was doped to the gel after 2D gel electrophoresis). (J. Su. Becker et al., Int. J. Mass Spectrom.,* **242**, *135 (2005). Reproduced with permission of Elsevier.)*

Table 8.4 *Selected applications in isotope ratio measurements by LA-ICP-MS.*

Sample	ICP-MS	Laser system/ Wavelength	Measured isotope ratio	RSD (%)	Application	Reference
Mantle sulfides	*MC-ICP-MS (Nu Plasma)*	*Merchantek LUV/266 nm*	$^{187}Os/^{188}Os$ $^{187}Re/^{188}Os$	*0.05–1*	*geochronology*	*Pearson et al.*[131]
SRM glass, minerals	*ICP-QMS (X-7)*	*Surelite I Continuum/266 nm*	$^{206}Pb/^{207}Pb$ $^{208}Pb/^{207}Pb$	*0.2–0.4* *0.3–0.4*	*geology*	*Crowe et al.*[126]
Pb buttons	*ICP-QMS (Elan 6100)*	*GeoLas ArF/193 nm*	$^{106}Pd/^{108}Pd$ $^{195}Pt/^{194}Pt$	*2.0* *2.9*	*geology*	*Vanhaecke et al.*[135]
Zircon grain	*LA-MC-ICP-MS (Neptune)*	*NewWave*	^{207}Pb, ^{206}Pb	*0.22*	*geology*	*Bouman et al.*[152]
Molybdenite	*ICP-SFMS (Element)*	*UV laser probe/266 nm*	$^{187}M/^{185}Re$ $M = (Re + Os)$	*0.2*	*geochronology*	*Malinovsky et al.*[130]
Zircon, opal	*ICP-SFMS (Element)*	*Brilliant Quantel/266 and 213 nm*	$^{206}Pb/^{238}U$ $^{207}Pb/^{206}Pb$ $^{208}Pb/^{232}Th$	*0.099–0.021* *0.65–1.98* *0.41–18.3*	*geochronology*	*Tieplo*[134]
Glasses	*MC-ICP-MS (Axiom)*	*LSX 200 266 nm*	$^{11}B/^{10}B$	*0.1*	*geology*	*Roux et al.*[129]
Human bones	*ICP-SFMS (Element)*	*Ablascope/ 213 nm*	$^{87}Sr/^{86}Sr$	*0.1–0.2*	*biology and life science*	*Prohaska et al.*[132]
Flower leaves	*ICP-SFMS (Element)*	*Ablascope/ 213 nm (cooled LA ablation chamber)*	$^{235}U/^{238}U$ $^{234}U/^{238}U$ $^{236}U/^{238}U$	*0.4–1.3* *1.6–2.0* *1.0–2.1*	*biology and life science*	*Zoriy et al.*[125]
NIST SRM 611, Zircon, opal	*MC-ICP-MS (Plasma 54)*	*LSX 100 213 nm*	$^{234}U/^{238}U$ $^{230}Th/^{238}U$	*0.02–0.04* *0.05*	*geochronology*	*Stirling et al.*[133]
Protein spots in 2D gels	*ICP-SFMS (Element)*	*Ablascope/ 213 nm (cooled LA chamber)*	$^{67}Zn/^{64}Zn$ $^{65}Cu/^{63}Cu$ $^{54}Fe/^{56}Fe$	*5* *5* *5*	*metallomics (tracer experiments)*	*Becker et al.*[49]
Geological and biological samples	*MC-ICP-MS (Axiom)*	*LSX 200 & New Wave LUV 266 nm/213 nm*	$^{87}Sr/^{86}Sr$ $^{84}Sr/^{86}Sr$	*0.002–0.09* *0.01–0.8*	*geochronology*	*Waight et al.*[136]
Geological and environ. samples	*ICP-SFMS (Element)*	*LINA-Spark/1064 nm*	$^{37}Cl/^{35}Cl$ $^{81}Br/^{79}Br$ $^{129}I/^{127}I$	*3* *3* *3*	*environmental, geology*	*Boulyga & Heumann*[50]
Geological glasses	*ICP-SFMS (Element)*	*New Wave UP/213 nm*	$^{208}Pb/^{206}Pb$ $^{207}Pb/^{206}Pb$	*0.1–0.2* *0.1–0.2*	*earth science, geochemical research*	*Jochum et al.*[128]
NIST Cu SRM 976 *NIST Fe SRM 665*	*MC-ICP-MS (Nu Plasma 500)*	*GeoLas 200CQ/ 193 nm*	$^{65}Cu/^{63}Cu$ $^{56}Fe/^{54}Fe$	*0.04* *0.05*	*fundamental*	*Hirata et al.*[127]

8.6 Multiple Ion Collector Mass Spectrometry for High Precision Isotope Ratio Measurements

Accurate and precise isotope analysis were performed for the study of fine isotope variation in nature for especially light elements such as H, C, O, Li, B, S and Si. Isotope variation has also been described for Ca, Fe, Cu, Sr, Nd, Hf, Os, Pb, U due to nuclear, physical or chemical processes. Whereas gas source mass spectrometric techniques are employed to study isotope variations of the light elements such as hydrogen, carbon and oxygen, multiple ion collector mass spectrometers (ICP-MS, TIMS and SIMS) are today the most precise techniques for isotope ratio measurements of all the other heavier elements.[11,51,137–141] Developments of multiple ion collector mass spectrometry (TIMS and MC-ICP-MS) for isotope ratio measurements and applications are reviewed by Wieser and Schwieters.[9] The long term performance characteristics of MC-ICP-MS (Neptune, Thermo Fisher scientific) for Sr, Nd and Hf isotope ratio measurements are examined by Nowell *et al.*[142] In terms of long term reproducibility and accuracy, this MC-ICP-MS is comparable to that routinely produced by MC-TIMS (TRITON, Thermo Fisher Scientific). For isotope ratio measurements by MC-ICP-MS of a 200 ng ml^{-1} solution of the isotope standard reference material NIST SRM 987 by MC-ICP-MS, and applying a mathematic correction for Kr and Rb interferences as described in reference 142, the reproducibility for $^{87}Sr/^{86}Sr$ ranges between 6 and 36 ppm (2SD) and the long term reproducibility is 38 ppm (2SD). By MC-TIMS (TRITON, Thermo Fisher Scientific) a reproducibility of 6 to 8 ppm (2SD) for $^{87}Sr/^{86}Sr$ isotope ratio measurements was obtained.[142]

Studies of iron isotopic composition in human blood and dietary iron sources revealing that lighter iron isotopes are enriched along the food chain and the potential iron isotope effects in biomedical research have been investigated by Walczyk and von Blanckenburg.[45] The authors present a model that appears to suggest that preferential absorption of lighter isotopes, rather than the release of heavier iron isotopes, exerts the main influence on the iron isotope composition of blood in healthy adults. Consequently, the human body is enriched with lighter iron isotopes, as is observed in animals in general. Possible mechanisms, including the reduction of ferric iron prior to absorption and/or kinetic effects within the enterocyte during uptake, distribution and/or release of absorbed iron, are discussed.[45]

Most applications of isotope ratio measurements by MC-ICP-MS focus on the study of isotope variation in Nature and on geological applications, especially in geochronology. The physical methods of geochronology (age dating of rocks) as an essential field in the earth sciences, which would not have been possible without mass spectrometry, are based on precise and accurate isotope ratio measurements, i.e., age dating using the following radioactive decay: $^{87}Rb \rightarrow {}^{87}Sr$, $^{147}Sm \rightarrow {}^{143}Nd$, $^{238}U \rightarrow {}^{206}Pb$, $^{235}U \rightarrow {}^{207}Pb$, $^{232}Th \rightarrow {}^{208}Pb$, $^{187}Re \rightarrow {}^{187}Os$, $^{176}Lu \rightarrow {}^{176}Hf$). Textbooks on isotope ratio measurements, including several applications in age dating by MC-ICP-MS, have been published by Platzner[7] and De Laeter.[26] In addition, several applications will be discussed in this chapter.

For example, accurate and reproducible U and Th isotope ratios for the geochronology of carbonates were obtained when analyzing small sample sizes (0.05–0.34 ng of Th measured by MC-ICP-MS in comparison to 0.3–1.4 ng Th for TIMS to obtain comparable precision).[140] An analytical technique for the measurement of ^{226}Ra–^{230}Th–^{238}U disequilibria in volcanic rocks is MC-ICP-MS. MC-ICP-MS, which is much less time-consuming than TIMS, also yields a better precision (2–3 times) for $^{232}Th/^{230}Th$ isotope ratios, was proposed by Pietruska *et al.*[143]

The results of uranium isotope ratio measurements by MC-TIMS (TRITON Thermo Fisher Scientific) on a single uranium oxide reference particle (10 μm) are illustrated in Figure 8.5.[9,144] The high precision of multiple ion counters and high detection power in the MC-TIMS are

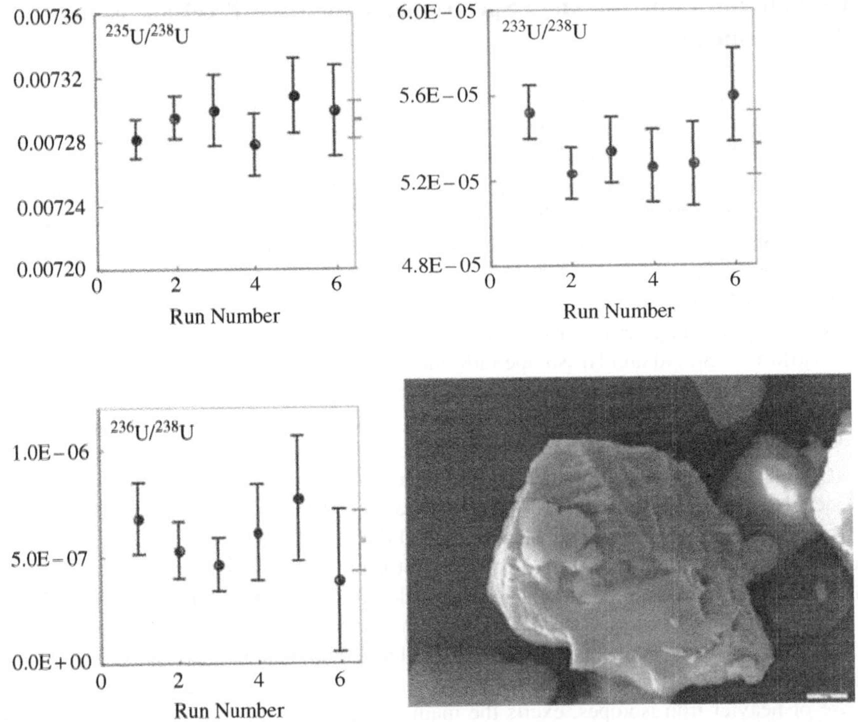

Figure 8.5 Precision of uranium isotope ratios measured by MC-TIMS (Triton, Thermo Fisher Scientific, Bremen). (J. S. Schwieters et al., Geochim. Cosmochim. Acta, **68**, *A60 (2004). Reproduced by permission of Elsevier.)*

advantageous for studying variations of uranium isotopes ^{235}U and ^{236}U in the micron size UO_2 particle.

An analytical technique for high precision Hg isotope analysis in environmental samples using gold trap MC-ICP-MS has been proposed by Hintelmann *et al.*[145] Environmental samples are measured after pyrolyzation at high temperature whereby Hg is released and collected in a gold trap. The authors found significant Hg isotope variation for several samples with elevated MeHg contents (whereby the light isotopes are enriched relative to the sample containing predominantly inorganic Hg).

Molybdenum isotope ratio measurements by MC-ICP-MS (Plasma 54) have been carried out using Zr or Ru 'elemental spikes' to study the mass discrimination during the whole analytical procedure including sample preparation.[146] A 'laboratory fractionation' of Mo isotopes of about 0.15 % is observed during ion exchange by offline Mo separation. Using this analytical technique, possible natural isotope variation of Mo can be determined with a precision of ± 0.02 %.

Further applications of LA-MC-ICP-MS using the Plasma 54 by Zr isotope ratio measurements in zircon and baddeleyite samples are described by Hirata.[147] The isotope ratios $^{92}Zr/^{90}Zr = 0.333\,94 \pm 0.000\,02$ and $^{96}Zr/^{90}Zr = 0.054\,63 \pm 0.000\,01$ have been determined on a Merck reagent with a precision of 0.01–0.02 % and 0.03–0.04 %, respectively. Neither isotopic variation in the

$^{92}Zr/^{94}Zr$ isotope ratio (due to the radiogenic contribution of ^{92}Nb) nor isotopic heterogeneity was found for any of the investigated zircon and baddeleyite samples.

The origin and migration of the Alpine Iceman has been studied using MC-ICP-MS Nu 1700 by comparing the isotopic composition of radiogenic $^{206}Pb/^{204}$ Pb and $^{87}Sr/^{86}Sr$ in teeth, bones and soils.[148] A variation of the $^{87}Sr/^{86}Sr$ isotope ratio from 0.717 to 0.719 was observed and the biological age of the Iceman determined via isotope ratio measurements in various tissues. The age of the Alpine Iceman, who lived 5200 years ago, was found to be 46 years.

Relatively large thallium isotope variations (in the order of 5‰) in a number of iron meteorites in the early solar system have been found by Halliday's working group. In addition, evidence was obtained for the former existence of long-lived ^{205}Pb with a half-life of 15 Myr, which is thought to have been present in the early solar system. Tl isotope fractionation also occurs during weathering.[149]

Re-assessment of silicon isotope reference materials (IRMM-018 and NBS 28 isotope standard) has been performed using both high resolution multiple ion collector ICP-MS and gas mass spectrometry.[101,139] Valkiers[101] calibrated the isotope ratio measurement using synthetically prepared ratios of the three stable silicon isotopes (^{28}Si, ^{29}Si and ^{30}Si) in a series of gas mass spectrometric measurements (Finnigan MAT 271) via SiF_3^+ ions of several isotopes. The high precision high resolution multicollector ICP-MS measurements of silicon on IRMM-018 and NBS 28 by Reynolds *et al.*[139] yielded a discrepancy for Si isotope composition compared with the data from Valkiers.[101] The MC-ICP-MS results conform to the certified values for NIST standard SRM 990 that was used to determine the atomic weight of Si, with a $^{30}Si/^{29}Si$ ratio that is over 6‰ lower for the same atomic weight. In order to avoid problems with future assessments of stable Si isotope variations, the NBS 28 silica sand standard (IRM 8546) should remain the zero point. The authors recommend an inter-laboratory calibration of NBS 28 and other references materials to solve the observed discrepancies and establish a reliable scale for reporting Si isotopes.

Van den Boorn *et al.*[150] have reported on the determination of silicon isotope ratio measurements in silicate materials by MC-ICP-MS (at a mass resolution of 2500 to resolve isobaric interferences) after sodium hydroxide sample digestion and purification of silicon. $\delta^{29}Si$ and $\delta^{30}Si$ have been determined for several silicon isotope standard reference materials. A precision for $\delta^{30}Si$ of 0.18–0.41‰ was achieved. Precise and accurate measurements of isotope ratios on transient signals by HPLC-MC-ICP-MS for nuclear application was performed by Günther-Leopold *et al.*[151]

Selected applications of isotope analysis on stable isotopes by MC-ICP-MS compared to ICP-SFMS are summarized in Table 8.5.[29,45,55,109,152–158]

8.7 Applications of Isotope Dilution Mass Spectrometry

The principle of the isotope dilution analysis (IDA) is described in Section 6.4. Due to its advantages as a definitive and accurate analytical method for the determination of element concentration via isotope ratio measurements, IDA is being increasingly applied in mass spectrometry, especially in ICP-MS and LA-ICP-MS as one of the most frequently used techniques. For example, the isotope dilution technique is employed in species analysis in biological systems,[54] e.g., for the determination of mercury species in tuna material,[55,159] or in aquatic systems.[55] Further applications of the isotope dilution technique are the determination of selenomethionine in human blood serum by capillary HPLC-ICP (ORC) MS[160] or sulfur speciation in gas oil, diesel or heating fuel by LA-ICP-MS.[161] Evans and co-workers have reported on the high accuracy analysis of sulfur in diesel fuel by IDA.[162] ICP-SFMS[163] has been employed for Si species analysis in biological or clinical samples and

Table 8.5 Application of stable isotope ratio measurements.

Samples	Method	Stable isotopes	Precision	References
Human blood liver, muscle	MC-ICP-MS (Neptune) separation by anion exchange chromatography	^{54}Fe, ^{56}Fe, ^{57}Fe	^{56}Fe–0.0049 % ^{57}Fe–0.0071 %	Walczyk *et al.*[45]
Silicates	MC-ICP-MS (Neptune) separation by anion exchange chromatography	^{54}Fe, ^{56}Fe, ^{57}Fe	^{56}Fe–0.0049 % ^{57}Fe–0.0071 %	Schoenberg *et al.*[156]
Aquatic systems (MeHg) (isotope dilution analysis)	ICP-SFMS (Element) cold vapor technique	^{199}Hg, ^{200}Hg, ^{201}Hg ^{202}Hg,	$^{199}Hg/^{202}Hg$, $^{200}Hg/^{202}Hg$ $^{201}Hg/^{202}Hg$ 0.6–1.4 %; LOD: 0.2 $ng\,l^{-1}$	Stürup *et al.*[55]
Organic solutions ($PbEt_4$ MeHgEt, $HgEt_2$)	GC-MC-ICP-MS (Axiom)	^{198}Hg, ^{199}Hg, ^{200}Hg, ^{201}Hg, ^{202}Hg, ^{204}Pb, ^{206}Pb, ^{207}Pb ^{208}Pb	$^{201}Hg/^{198}Hg$–0.077 % $^{207}Pb/^{206}Hg$–0.5 %	Krupp and Donard[109]
Soils (depth profiling)	ICP-SFMS (Element) separation by ion exchange	^{87}Sr, ^{86}Sr, ^{207}Pb, ^{206}Pb	$^{87}Sr/^{86}Sr$–0.08 % $^{207}Pb/^{206}Pb$–0.07 %	Prohaska *et al.*[155]
Silicates	MC-ICP-MS (Neptune) separation by ion exchange	^{87}Sr, ^{86}Sr	$^{87}Sr/^{86}Sr$–0.0015 %	Balcaen *et al.*[151]
CuS	MC-ICP-MS (Nu Plasma)	^{65}Cu, ^{63}Cu	$^{65}Cu/^{63}Cu$–0.01 %	Ehrlich *et al.*[29]

GC-ICP-QMS for mercury species in aquatic systems[164] or in sea sediments[165] using the isotope dilution technique as employed in different working groups. The isotope dilution technique – using a ^{233}Pa spike with a half-life of 26.97 days – was utilized for the quantitative measurement of 20 fg of protactinium in silicate rocks after chemical separation of the actinide from rock matrix by MC-ICP-MS (Neptune, Thermo Fisher Scientific, Bremen – equipped with nine Faraday detectors, one secondary electron multiplier and a retarding potential quadrupole for high abundance sensitivity measurements).[166] $^{231}Pa/^{233}Pa$ isotope ratios were determined with a precision of 0.5 %, and the detection limit was 200 $ag\,ml^{-1}$. Concentrations of ^{226}Ra from 0.36–0.42 $fmol\,kg^{-1}$ in the Ross Sea were determined by MC-ICP-MS (Nu Instruments) via the IDA.[167]

Furthermore, online chromatographic coupling techniques such as HPLC and CE coupled to ICP-MS with the isotope dilution technique have been used for element quantification in speciation analysis. Species analysis by ICP-SFMS in medical research has been performed by Prange *et al.*[168,169] on separated isoforms of metallothionein (e.g., of rabbit liver) by capillary electrophoresis (CE). For quantitative determination of S, Cd, Cu and Zn in isoforms, highly enriched ^{34}S, ^{65}Cu, ^{68}Zn and ^{116}Cd spikes are added to the analyte solution in the interface region of CE.

An analytical method for the direct determination of halogens (Cl, Br and I) in powdered geological and environmental samples using isotope dilution laser ablation ICP-MS (LA-ICP-IDMS using a LINA-Spark™ atomizer coupled to ICP-SFMS Element 2) with detection limits in the low and sub-$\mu g\,g^{-1}$ range for halogens has been developed by Boulyga and Heumann.[50] This isotope dilution technique applied to halogen determination in sediment and rock reference materials opens up the future possibility of time effective and accurate halogen determination at the trace concentration level in many matrices.

To an increasing extent, the isotope dilution technique is being applied in the certification of standard reference materials, e.g., Hg determination in a polyethylene certified reference material (CRM 680 and 681), or Cd and Tl determination in high-purity Zn (CRM 325/2R) where ID-ICP-MS yielded the most accurate data.[170]

8.8 Isotope Ratio Measurement of Long-Lived Radionuclides

Whereas in the past the determination of long-lived radionuclides was dominated by conventional radioanalytical techniques such as α spectrometry, for a few years now ICP-MS has been increasingly used for isotope analysis.[1] Of special interest is the isotope analysis of natural radionuclides (^{238}U, ^{235}U and ^{234}U) and especially of artificial radionuclides (^{236}U, ^{239}Pu, ^{240}Pu, ^{241}Am, ^{129}I, ^{90}Sr and others) in environmental samples for evidence of natural isotope variation or environmental contamination in order to evaluate the sources.[47,171,172] Mass spectrometric measurements have been performed to provide evidence of several long-lived radionuclides (e.g., of ^{236}U or ^{240}Pu/^{239}Pu) as nuclear fallout from nuclear power plants (NPP) or nuclear weapons tests (environmental monitoring). For example, Boulyga *et al.*[173] describe the determination of the ^{236}U isotope to monitor the spent uranium from nuclear fallout using inductively coupled plasma mass spectrometry collected in the vicinity of the Chernobyl nuclear power plant (NPP). Sector field ICP-MS (ICP-SFMS) and quadrupole based ICP-MS with and without a hexapole collision cell (ICP-CC-MS) have been used for uranium isotope analysis. In addition, a multiple ion collector ICP-MS (MC-ICP-MS) has been employed for high precision isotope ratio measurements. The detection limit for ^{236}U was in the fg g^{-1} range and the abundance ratio sensitivity for ^{236}U/^{238}U was 3×10^{-7}, 6×10^{-7}, 6×10^{-7} and 5×10^{-6} in ICP-CC-MS, quadrupole ICP-MS, MC-ICP-MS and ICP-SFMS at mass resolution $m/\Delta m = 300$, respectively. Interlaboratory comparison yielded good accuracy (0.4 % to 1.6 %) of ^{236}U/^{238}U isotope ratios ranging from 1.5×10^{-3} to 3.2×10^{-4} measured in samples containing 100 ng uranium. The ^{236}U/^{238}U isotope ratios and spent uranium fraction has been determined in soil samples from Chernobyl using single-detector ICP-MS and MC-ICP-MS. In comparison to the ^{235}U/^{238}U isotope ratio, the ^{236}U/^{238}U isotope ratio provided a more sensitive and accurate determination of the proportion of spent uranium from the Chernobyl NPP in the spent/natural uranium mixture in soil samples down to 0.1 %. The concentration of the spent Chernobyl uranium in the upper 0–10 cm soil layers in areas investigated in the vicinity of Chernobyl NPP amounts from 2.4×10^{-9} g g^{-1} to 8.1×10^{-7} g g^{-1} depending mainly on the distance from the Chernobyl reactor. The distribution of ^{235}U/^{238}U and ^{236}U/^{238}U isotope ratios as a function of soil depth in the vicinity of Chernobyl and the variation of these uranium isotope ratios in all investigated soil samples are summarized in Tables 8.6 and 8.7.

Table 8.6 *Distribution of ^{235}U/^{238}U and ^{236}U/^{238}U isotope ratios over the soil depth in the Chernobyl vicinity.*

Depth	^{235}U/^{238}U	RSD (%)	^{236}U/^{238}U	RSD (%)
0–5 cm	0.010 06	2.9	0.000 959	7.2
5–10 cm	0.008 69	3.6	0.000 545	8.6
10–15 cm	0.007 76	3.8	0.000 229	9.6
15–20 cm	0.007 25	11	0.000 035	32

Table 8.7 Variation of uranium isotope ratios in contaminated soil samples in the vicinity of Chernobyl.

Isotope ratio	Measured value	IUPAC (Nature)
$^{234}U/^{238}U$	0.000 068–0.000 117	0.000 06
$^{235}U/^{238}U$	0.007 39–0.010 06	0.007 25
$^{236}U/^{238}U$	0.000 058–0.000 972	$< 10^{-10}$

An analytical technique for determining ^{236}U (or ^{239}Pu) at an ultratrace level in environmental samples has been developed in the author's laboratory.[87] In order to reduce $^{235}UH^+$ formation, D_2O (heavy water) was used as a solvent for the dissolution and dilution of uranium samples. Abundance sensitivity was improved by the use of medium mass resolution ($m/\Delta m = 4450$) in comparison to low mass resolution in ICP-SFMS. Part of a mass spectrum from $m/z = 238$–241 measured by ICP-SFMS in Figure 8.6 demonstrates the low background at $m/z = 239$ if a natural uranium sample is diluted with D_2O. For solution introduction, the performances of several different sample introduction systems were compared. The performances of several nebulizers (Meinhard, Aridus and ultrasonic nebulizer) in ICP-SFMS were also studied. It was found that for all nebulization systems, a reduction in UH^+/U^+ is observed for D_2O as compared to H_2O as solvent. Optimum results were obtained in ICP-SFMS for a desolvating microconcentric nebulizer system (Aridus) with a minimum hydride formation rate of 9×10^{-7} and a limit for $^{236}U/^{238}U$ isotopic ratio measurements of 3–5×10^{-7}. A comparing application of three commercially available solution introduction systems in sector field ICP-MS devices yielded a good agreement of results between single collector and MC-ICP-MS. A method relatively free from interferences for plutonium isotope ratio measurements in bioassay samples by TIMS is described by Elliot *et al.*[174]

As is well known, ammunition containing depleted uranium (DU) was used by NATO, for example, in the former Yugoslavia. To evaluate the origin of DU (enrichment process of natural uranium or reprocessing of exhausted nuclear fuel) it is necessary to directly detect the presence

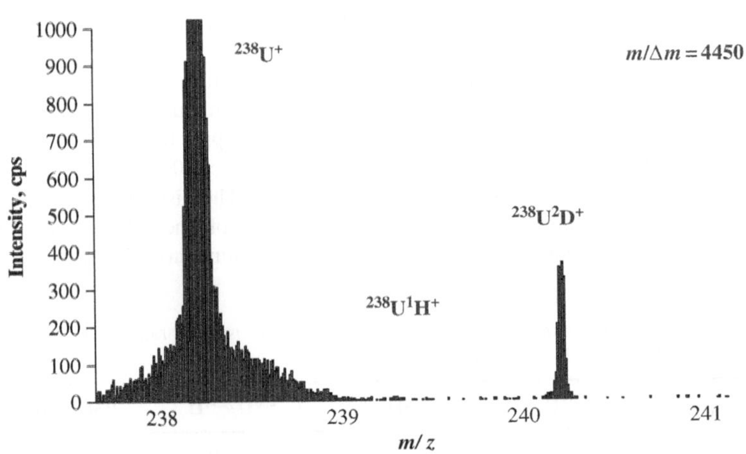

Figure 8.6 ICP-SFMS spectrum of $^{238}UH^+$ measured at mass resolution of 4450 on natural uranium diluted with D_2O. (M. Zoriy, L. Halicz, M. Ketterer, P. Ostapczuk, R. Hille, and J. S. Becker, J. Anal. At. Spectrom., **19**, 675 (2004). Reproduced by permission of the Royal Society of Chemistry.)

of activation products (^{236}U, ^{239}Pu, ^{240}Pu, ^{241}Am and ^{237}Np) in the ammunition. After a selective separation of ultratraces of transuranium elements from the uranium matrix, the actinides are analyzed by ICP-MS compared to α-spectrometry. ^{242}Pu and ^{243}Am are added to calculate the chemical yield. Plutonium is separated from uranium by extraction chromatography with tri-*n*-octylammine (TNOA) with a decontamination factor higher than 10^6. After elution, plutonium is determined by ICP-MS (^{239}Pu and ^{240}Pu) and α-spectrometry ($^{239+240}Pu$) after electroplating. The $^{240}Pu/^{239}Pu$ isotope ratio was measured in a penetrator sample of 0.12 ± 0.04 at a concentration level between 7×10^{-12} g g^{-1} and 2×10^{-11} g g^{-1}. In general, ICP-MS and α-spectrometry results were in good agreement.

The $^{234}U/^{238}U$ isotope ratio has been widely used as a tracer for geochemical processes in underground aquifers. The determination of $^{234}U/^{238}U$ isotope ratios in environmental waters has been performed with ICP-QMS (HP 4500) equipped with a high efficiency nebulizer and membrane desolvator.[175] Uranium was stripped according to the radiochemical separation method from a stainless steel planchet used for the radiometric counting of alpha particles and measured by ICP-QMS leading to results in very good agreement with those of α-spectrometry. The measurement precision ($^{234}U/^{238}U$) was 3–5 % for bottled mineral water (uranium concentration 1 μg l^{-1}). The problem of disturbing interferences from the formation of platinum argides (platinum appears to be an impurity present in stainless steel) was studied.[175] Small isotope variations of $^{234}U/^{238}U$ were detected as the result of alpha emission of ^{238}U to ^{234}Th, which decays to ^{234}U via chemical activity (recoil effect) following beta decay through ^{234}Pa. The natural isotope variations of $^{234}U/^{238}U$ in water sampled from Israel were found between to be 7.28×10^{-5} to 10.9×10^{-5} compared to the IUPAC table value[176] of 5.54×10^{-5}.[177]

Of special interest is the determination of ^{90}Sr, which appears as a radionuclide in the decay series of nuclear fission and can therefore be found in nuclear waste or be released by nuclear accidents. Current methods for the detection of this radionuclide are time consuming and may be prone to a large variety of interferents. An analytical technique has been proposed for the determination of ^{90}Sr ($t_{1/2} = 29.1$ a) in the presence of zirconium in urine at ultratrace level in the author's laboratory.[89] The cold plasma technique and measurement at medium mass resolution were investigated to remove this, as well as other contributions to the background at $m/z = 90$. The performance of ICP-SFMS was investigated, which allows ^{90}Sr to be measured with a detection limit of 3 pg l^{-1} for water samples. The method was applied for the analysis of ^{90}Sr extracted from urine using a crown ether extraction resin and concentrated (enrichment factor: 200). High levels of natural strontium in the separated fraction (of about 1 μg ml^{-1}) mean higher detection limits (80 pg l^{-1}) due to $^{88}Sr^+$ at $m/z = 90$ and the relatively low abundance sensitivity of ICP-SFMS at medium mass resolution of 6×10^{-7}. This detection limit in separated fractions corresponds to a detection limit of 0.4 pg l^{-1} in the original urine sample. The recovery of ^{90}Sr determined by the analytical method in spiked urine samples was in the range of 82–86 %.[89]

ICP-MS with a collision cell is well suited to the sensitive determination of the long-lived and rare ^{129}I and ^{79}Se radionuclides. The long-lived radionuclide ^{129}I($t_{1/2} \approx 1.57 \times 10^7$ a) is mainly released into the environment from anthropogenic sources (such as nuclear fallout from nuclear weapons tests, nuclear accidents and by emissions from nuclear fuel reprocessing plants) and is of interest for environmental monitoring, for thyroid dose reconstruction in contaminated regions after the accident at Chernobyl NPP and for radioactive waste control. The determination of the $^{129}I/^{127}I$ ratio in environmental samples requires an abundance sensitivity down to 10^{-10}–10^{-11}. The problems involved in ^{129}I determination by ICP-MS are the high background caused by ^{129}Xe impurities in the Ar plasma and the insufficient abundance ratio sensitivity of $^{129}I/^{127}I$. Using a mixture of He and H_2 as the collision/reaction gases in ICP-MS with a hexapole collision cell, an efficient reduction of the disturbing background intensity of $^{129}Xe^+$ ions is observed (see Figure 8.7.)[19] and consequently the detection limit for ^{129}I in comparison to sector field ICP-SFMS (without collision

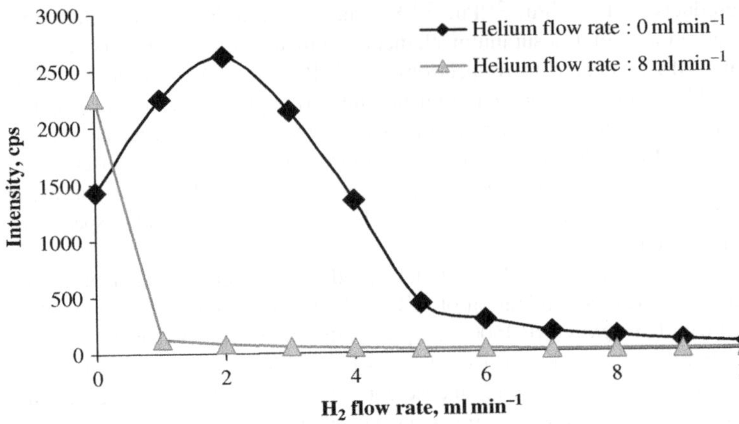

Figure 8.7 *Background intensity in ICP-CC-QMS with hexapole collision cell at m/z=129 as a function of collision gas flow rates. (J. S. Becker, J. Anal. At. Spectrom.,* ***17****, 1172 (2002). Reproduced by permission of The Royal Society of Chemistry.)*

cell) is decreased by nearly two orders of magnitude (from $100\,\mathrm{ng\,l^{-1}}$ to $3\,\mathrm{ng\,l^{-1}}$). In the experiment with ICP-SFMS, a special solution introduction device for introducing volatile iodine via the gas phase into the ICP was developed. In this device, the sample solution containing iodide was mixed with 70 % perchloric acid for oxidation of analyte using a Y-shaped glass tube. The mixture was introduced into a glass chamber made in the author's laboratory via a Meinhard nebulizer, whereby the volatile iodine is separated from the solution and introduced into the ICP-MS. The apparatus coupled to a sector field ICP-SFMS was utilized for iodine determination in biological reference materials (e.g., Apple or Citrus Leaves NIST SRM 1515 and NIST SRM 1572, respectively) by the isotope dilution technique. The biological reference materials were analyzed after microwave assisted digestion thus avoiding any possible loss of volatile iodine. For the determination of $^{129}\mathrm{I}/^{127}\mathrm{I}$ isotope ratios at the ultratrace concentration level in contaminated soil sediments, Izmer *et al.*[46,47] have proposed a hot extraction procedure (T ~ 1000 °C) for the volatile analyte with oxygen as the carrier gas. The sample introduction device is coupled to an ICP-MS with collision cell (ICP-CC-QMS, Platform, Micromass) and the isotope analysis of iodine is performed without any additional sample preparation. A cooling finger (see Figure 8.8) is inserted for intermediate

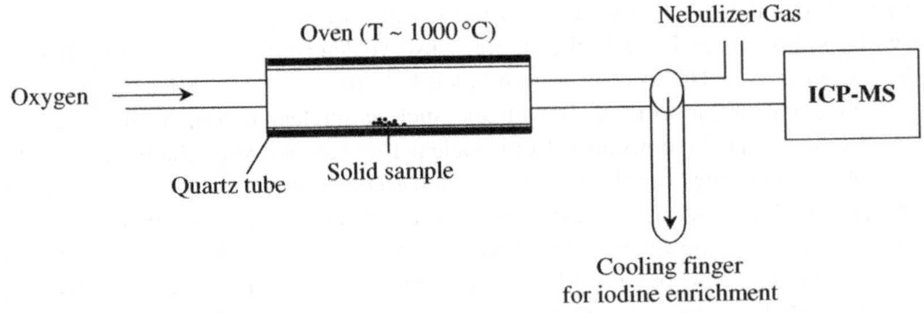

Figure 8.8 *Device with cooling finger for introduction of iodine by hot extraction into the ICP-CC-QMS via the gas phase.*

iodine enrichment. A mixture of O_2 and He as collision gases is introduced in the hexapole collision cell to reduce the xenon background. Subsequently, detection limits for ^{129}I determination in Ocean Sediment Environmental Radioactive Standard (SRM 4357) of 0.4 pg g^{-1} (without cooling finger: 30 pg g^{-1}) have been achieved and the $^{129}I/^{127}I$ isotope ratio measured at the 5×10^{-7} level. In comparison to these results, the $^{129}I/^{127}I$ isotope ratios between 0.0075 and 0.45×10^{-10} were obtained by ultrasensitive AMS.[72]

The determination of the long-lived radionuclide $^{79}Se(t_{1/2} \approx 65\,000\,a)$ by ICP-MS is restricted by several isobaric interferences with atomic and polyatomic ions such as $^{79}Br^+$, $^{39}K^{40}Ar^+$, $^{63}Cu^{16}O^+$, $^{158}Gd^{2+}$ and $^{158}Dy^{2+}$. These interferences can be minimized using a hydride generator as the solution introduction device. The disturbing argon hydride formation $^{38}Ar^{40}Ar^1H^+$ is additionally suppressed using a H_2/He mixture as the reaction/collision gas in the hexapole collision cell of ICP-CC-QMS. Whereas the detection limit of ^{79}Se determination in ICP-SFMS is 100 pg ml^{-1} by applying hydride generation,[178] in ICP-CC-QMS the detection limit can be reduced to 5 pg ml^{-1} with additional suppression of $^{38}Ar^{40}Ar^1H^+$ formation. However, it should be noted that new interferences may appear in ICP-MS with a collision cell when using hydrogen as the collision gas. Therefore all possible new interferences must be studied and considered carefully. At a very low concentration of the long-lived radionuclides investigated, as expected, in environmental samples, new interferences are observed under certain experimental conditions such as hydride ions ($^{78}Se^1H^+$, $^{127}IH_2^+$, $^{235}UH^+$ or $^{238}UH^+$) at masses 79 u, 129 u, 236 u and 239 u for the determination of ^{79}Se, ^{129}I, ^{236}U and ^{239}Pu at ultratrace level. Therefore the application of a solution introduction system with desolvator (e.g. the microconcentric nebulizer Aridus from CETAC) is advantageous for the ultrasensitive determination of ^{236}U and ^{239}Pu, and an alternative collision gas (for example oxygen) is helpful for ^{129}I determination. The hydride formation of uranium using different solution introduction systems has been studied in the author's laboratory as a function of experimental parameters and the use of heavy water as solvent investigated.[15,87] The detection limits of the selected long-lived radionuclides ^{93}Mo, ^{99}Tc, ^{107}Pd, ^{126}Sn, ^{151}Sm, ^{236}U and ^{239}Pu measured by ICP-CC-QMS with a hexapole collision cell varied between 0.003 and 0.3 pg ml^{-1}.[19] Possible interferences were eliminated using a hydride generator, aerosol desolvation or using a chemical reaction with the collision gas. Another possibility for solving the interference problem for ^{79}Se determination is the application of ETV-ICP-MS after a single chemical separation step by ion exchange to extract selenium from the highly radioactive fission product solution (10^{10} Bq l^{-1}) produced by reprocessing of nuclear fuel.[179]

Ruthenium isotope analysis by ICP-QMS with collision/reaction cell (ELAN DRC II, Perkin Elmer Sciex) has been performed for the quantitative determination of ruthenium fission products (^{101}Ru, ^{102}Ru and ^{104}Ru) in contaminated groundwater samples and water extracts of vadose zone sediments from the Hanford Site.[180] Samples were analyzed directly by ICP-MS after a single pass through a 1 ml bed volume of Dowex AG 50W-X8 100–200 mesh hydronium based cation exchange resin. At a total Ru concentration of 0.1 ng ml^{-1} or higher, a relative standard deviation of triplicate replicates for the determination of ruthenium isotopes using a low flow Meinhard type nebulizer and long acquisition times (150 000 ms) was achieved at less than 0.5 %. Possible isobaric interferents (^{96}Mo and ^{96}Zr) in the environmental isotopic fractionation of ruthenium were assumed to be negligible, and therefore the results of ruthenium isotope ratio measurements ($^{101}Ru/^{104}Ru$ and $^{102}Ru/^{104}Ru$) reflect the source of variable contributions from ^{239}Pu and ^{235}U fission in groundwater and sediment extracts.[180]

Selected applications of isotope ratio measurements for plutonium by ICP-MS and LA-ICP-MS are summarized in Table 8.8.[97,155,171,181–183]

Table 8.8 *Applications of ICP-MS and LA-ICP-MS for plutonium isotope ratio measurements.*

Samples	Method	Radionuclides	Detection limits/Isotope ratios measured	References
Soils	ICP-SFMS (VG Axiom) after separation	^{239}Pu, ^{240}Pu, ^{241}Pu	$^{240}Pu/^{239}Pu$ ~ 0.18–0.35 $^{241}Pu/^{239}Pu$ ~ 0.003–0.004	Ketterer *et al.*[97]
Soils and sediments	ICP-SFMS (Element)	^{239}Pu, ^{240}Pu	$^{240}Pu/^{239}Pu$ ~ 0.18–0.2	Nygren *et al.*[155]
Soils (Chernobyl)	LA-ICP-SFMS (Element) (LINA-Spark)	^{239}Pu, ^{240}Pu	$0.25\,pg\,g^{-1}$ $^{240}Pu/^{239}Pu = 0.4$	Boulyga *et al.*[182]
Moss, plant, soil, sediment	ICP-SFMS (VG Axiom) after separation	^{239}Pu, ^{240}Pu,	$^{240}Pu/^{239}Pu$ ~ 0.19–0.26	Agarande *et al.*[179]
Lake and river water	ICP-SFMS (Element) MC-ICP-MS (NuPlasma) after separation	^{239}Pu, ^{240}Pu (in 100 l sea water)	$\sim 0.1\,fg\,l^{-1}$ (ELEMENT) $\sim 0.01\,fg\,l^{-1}$ (NuPlasma) $^{240}Pu/^{239}Pu \sim 0.17$	Becker, Halicz *et al.*[171]
Urine	ICP-SFMS (Element) after separation (1 l urine)	^{239}Pu, ^{240}Pu	$1 \times 10^{-18}\,g\,ml^{-1}$ $^{240}Pu/^{239}Pu \sim 0.14$ (synth. sample)	Zoriy *et al.*[183]

8.9 Applications of Isotope Ratio Measurements in Geochemistry and Geochronology

Thermal ionization and gas source mass spectrometry (TIMS and GSMS, respectively) have been used for many decades and were in the past the dominant techniques for isotope ratio measurements on stable isotopes in geochemistry and geochronology. With the introduction of high precision and accurate multiple ion collector mass spectrometers, and especially of MC-ICP-MS, isotope analysis is now being used to an increasing extent in geochemistry to study isotope variation in nature. ICP-MS is now an established technique for the determination of isotope ratios and abundances as described in previous reviews.[2,18,19] Isotope ratios are essentially unaffected by crystal fractionation, therefore they have been utilized for petrogenic, tectonic and related studies.[26] To study the isotope variation of material from the earth's mantle or the geochemistry of oceanic volcanic material (e.g., ocean basalts and midocean basalts – MORBs), materials are selected which are less contaminated with minerals of the lithosphere. TIMS has successfully been applied for lithium isotope ratio measurements for studying fine variation in nature (especially in geological materials) as described by Chan.[184] The range of lithium isotope compositions in terrestrial geological samples varies up to 60–70‰ (δ^7Li) whereas ocean water is uniform in its lithium isotopic composition with $\delta^7Li = 31.6 \pm 1.6$‰.

The isotope variation of boron isotopic composition in aqueous fluids in environments on the Earth's surface has been reviewed by You.[185] The boron isotope composition ($\delta^{11}B$) is expressed as a per mil (‰) deviation from the boron isotope standard NIST SRM 951:

$$\delta^{11}B = \{[(^{11}B/^{10}B)_{sample}/(^{11}B/^{10}B)_{SRM951}] - 1\} \times 10^3 \tag{8.8}$$

A multitude of analytical techniques have been developed by TIMS using several positively charged polyatomic ions with alkali elements (such as $Na_2BO_2^+$, $Li_2BO_2^+$, $Rb_2BO_2^+$ or $Cs_2BO_2^+$) or negatively charged oxide ions (BO_2^-).[185] For a few years now, sector field ICP-MS with single collectors, but especially with multiple ion collectors has been the technique of choice for precise boron isotope analysis. For example, boron isotopes ($\delta^{11}B$) and boron concentration in a 138 cm long sediment profile from a small town in Finland has been measured[186] by collision cell ICP-MS (Elan 6100 DRC) after leaching by aqua regia. A boron rich layer in the sediment profile with a depletion of ^{11}B was found as evidence of a historic fire in the early nineteenth century. Boron fractionation between brachiopod calcite and seawater has been examined by $^{11}B/^{10}B$ isotope ratio measurements using MC-ICP-MS (Plasma 54).[187]

Boron isotope ratios have been studied in a geothermal system from New Zealand (Ngawha) by MC-ICP-MS (Axiom with eight movable Faraday collectors, from Thermo Electron).[188] The $\delta^{11}B$ values range between −3.1 ‰ and −3.9 ‰, which does not indicate any marine input into the system. A direct determination of boron isotopes ($\delta^{11}B$) on natural and synthetic glass samples at < 1‰ precision at the ng level has been proposed using LA-ICP-MS with multiple electron multipliers.[129]

Lead isotope ratio measurements have been performed in geological glasses by Jochum *et al.* using LA-ICP-SFMS.[128,189] The authors observed an improved precision for $^{208}Pb/^{206}Pb$ and $^{207}Pb/^{206}Pb$ isotope ratios using a 193 nm Nd-YAG laser compared to a laser with a longer wavelength (213 nm) and changed pulse length (5 ns compared to 2.8 ns). The reason for this improvement can be explained by a change in the size of the ablated particles and the fact that localized heating of the sample is reduced.[189]

The growth of Earth like planets has been studied by Halliday by mixing new material, volatile accreted material lost into space and compositional change using several isotope systems like W, Sr and Pb.[137] The detected Tl isotope variation on iron meteorites due to ^{205}Pb β-decay to ^{205}Tl[138] has allowed new conclusions to be drawn on the formation of elements in the solar system. The tungsten isotopic composition of 35 iron meteorites was studied by MC-ICP-MS. Variations in the $^{182}W/^{184}W$ isotopic ratio reflect either time intervals of metal silicate differentiation or result from the burnout of tungsten isotopes caused by prolonged exposure to galactic cosmic rays.[190]

An important application of isotope ratio mass spectrometry is geochronology, which is based on the decay of a radioactive nuclide (e.g., ^{87}Rb, ^{187}Re or ^{238}U) to a stable isotope. The total number of daughter atoms D_t at time t is then equal to the sum of the daughter atoms formed due to the radioactive decay and the number of primordial daughter atoms (D_0) at time $t = 0$:

$$D_t = D_0 + n(e^{\lambda t} - 1) \tag{8.9}$$

with λ as the disintegration constant of the radioactive nuclide, whereby the half-life of parent nuclide $t_{1/2} = \ln 2/\lambda$.

Geochronological measurements (isochrone methodology) are based on the radioactive decay of the parent nuclide to the daughter nuclide using the fundamental Equation (8.8) for calculating the ages of minerals.

In the rubidium–strontium age dating method, radioactive ^{87}Rb isotope with a natural isotope abundance of 27.85 % and a half-life of 4.8×10^{10} a is fundamental to the β decay to the isobar ^{87}Sr. The equation for the Rb–Sr method can be derived from Equation (8.9):

$$(^{87}Sr/^{86}Sr)_t = (^{87}Sr/^{86}Sr)_0 + (^{87}Rb/^{86}Sr)_t(e^{\lambda t} - 1) \tag{8.10}$$

Whereas the abundance of ^{87}Sr in rubidium rich rocks changes over time due to the radioactive β decay of ^{87}Rb as a function of the primordial rubidium concentration and the age of the mineral, the abundance of the stable ^{86}Sr isotope and consequently the $^{86}Sr/^{88}Sr$ ratio is constant in nature. The constant $^{86}Sr/^{88}Sr$ isotope ratio is often used for internal standardization during strontium isotope ratio measurements of $^{87}Sr/^{88}Sr$. In the rubidium–strontium age dating method, the isotope ratios $(^{87}Sr/^{86}Sr)_t$ and $(^{87}Rb/^{86}Sr)_t$ are measured mass spectrometrically (by TIMS or ICP-MS) and the primordial strontium ratio $(^{87}Sr/^{86}Sr)_0$ at $t = 0$ and the age t of the rock can be derived from the isochrone (graph of measured $^{87}Sr/^{86}Sr$ isotope ratios (represented on the ordinate) as a function of the $^{87}Rb/^{86}Sr$ ratio (on the abscissa) in several minerals with different primordial Rb concentrations). The age of the minerals will be determined from the slope of the isochrone $(e^{\lambda t} - 1)$, and the primordial isotope ratio $(^{87}Sr/^{86}Sr)_0$ from the point of intersection with the ordinate (see Figure 8.9).

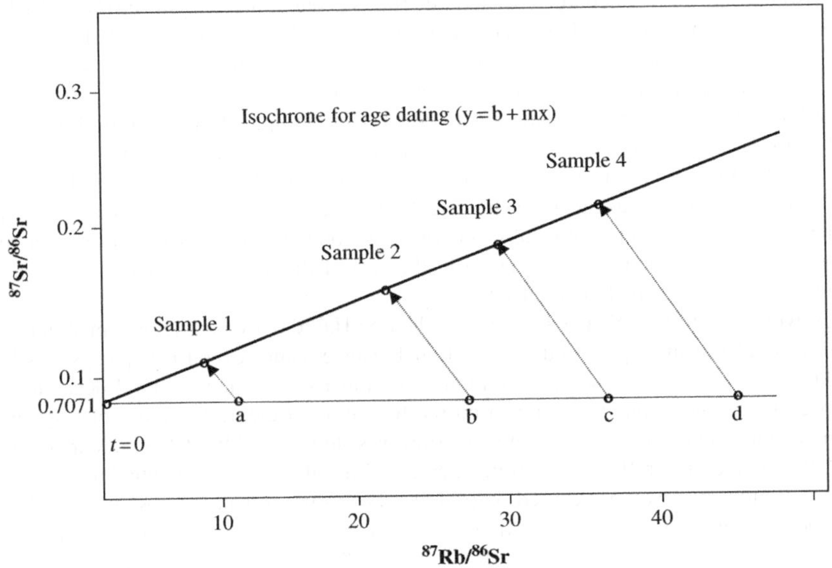

Figure 8.9 *Schematic of isochrone for Rb–Sr age dating; with* $b = (^{87}Sr/^{86}Sr)_0$ *and* $m = (e^{\lambda t} - 1)$

There are a multitude of papers concerning the age dating of rocks by mass spectrometry on the basis of different geochronological systems with well defined constants of radioactive decay. They include such systems as Rb–Sr, K–Ar, K–Ca, Re–Os, Nd–Sm, U–Pb, Th–Pb, Pb–Pb or Lu–Hf, which are discussed in detail in the literature.[26] Therefore only a few relevant examples were briefly sketched here.

A routine method for $^{87}Sr/^{86}Sr$ isotope ratio measurements by ICP-QMS on geological samples for age dating after digestion and trace matrix separation is described in reference[36]. For the determination of zircon ages, TIMS and LA-ICP-MS are utilized via the Rb–Sr and U–Pb age dating techniques, respectively.[191] The Rb–Sr isochrone of granite, constructed by biotite, feldspar, and apatite in the Diabie orogen, yields an age of 118 ± 3 Ma, while the zircon U–Pb dating for granite gave an age of 127 ± 3 Ma. Whereas the zircon age is interpreted as a timing of magma crystallization, the young Rb–Sr isochrone ages record the timing of Sr diffusion closure during the slow cooling.

For an accurate *in situ* $^{234}U/^{238}U$ and $^{230}Th/^{232}Th$ isotope microanalysis of silicate glasses and iron oxides, an excimer laser ablation system (He1Ex, Lambda Physik Compex 110i ArF;

wavelength – 193 nm, pulse length – 25 ns) has been coupled to an MC-ICP-MS (Neptune, Thermo Fisher Scientific).[192] The MC-ICP-MS was equipped with an RPQ filter to improve the abundance sensitivity. The low abundant isotopes ^{234}U and ^{230}Th required measurements using the SEM, whereas ^{238}U, ^{235}U and ^{232}Th can be measured using Faraday cups. A precision was achieved for the measured $^{234}U/^{238}U$ and $^{230}Th/^{232}Th$ isotope ratios of 5 ‰ and 8 ‰ at a spatial resolution of 90 μm in the samples investigated containing $100\,\mu g\,g^{-1}$ of analyte. This corresponds to an uncertainty of ±1800 a for a 100 000-year-old sample and ±6500 a for a 300 000-year-old sample. Matrix matched reference materials are needed for accurate and precise U-series chronologies by LA-MC-ICP-MS, even under conditions where instrumental U/Th fractionation effects have been minimized. The analysis of uranium – hematite standard (UHS) induced a U/Th fractionation that is slightly different (~ 3 %) from aluminosilicate glass reference materials.[192]

Molybdenite samples from Cu–Au–Mo and magnetite deposits in the Yangtze River valley (China) have been examined by Re–Os age dating using ICP-MS and NTIMS. The results of the two mass spectrometric techniques are almost identical. The Re–Os ages of 16 molybdenite samples range from 134.7 ± 2.3 to 143.7 ± 1.6 Ma.[193] Europium isotope variation in Allende CAIs with a deficit for ^{153}Eu of up to 1‰ have been detected by MC-ICP-MS.[194] The lack of Sm isotope fractionation in the same samples makes fractionation by kinetic processes an unlikely case of the ^{153}Eu deficit. The authors interpreted that based on the low ionization potential of Eu, in which electromagnetic separation of ionized gas preferentially depletes the nebular gas in a heavy isotope.

The beta decay of ^{176}Lu to ^{176}Hf with a half-life of $\sim 3.7 \times 10^{10}$ a makes this system a suitable geochronometer and an isotope tracer to study geological systems and dynamic processes in the earth and other planets (e.g., controlled crust–mantle evolution).[195] The determination of the ^{176}Lu decay constant by calibration against the U–Pb decay schemes yields a mean value of $1.865 \pm 0.01510^{-11}\,a^{-1}$. Lu and Hf isotope ratios and concentration have been measured via isotope dilution analysis by MC-ICP-MS (Isoprobe, Micromass), while for the U and Pb isotope ratio measurements MC-TIMS (VG Sector 54, Micromass) was applied. MC-ICP-MS allows high precision analysis of small sample sizes (as low as 5 ng Hf) that are not possible in TIMS based age comparison studies. The existence of strongly unradiogenic hafnium in Early Archaean and Hadean zircons implies that enriched crustal reservoirs existed on Earth 4.3 billion years ago and persisted for 200 million years or more.[195] Using a method for Lu and Hf isotope analysis from silicate rocks and mineral samples involving MC-ICP-MS (Isoprobe, Micromass), after a careful two-stage trace matrix separation procedure by cation exchange (using TODGA resin from Eichrom) including the separation of rare earth elements and Zr, $^{176}Hf/^{177}Hf$ and $^{176}Lu/^{177}Hf$ isotope ratios on rock standards (BHVO-1 and BCR-2) were obtained that are in agreement with published values.[196]

LA-ICP-MS using a multi-collector instrument for Pb–Pb dating via $^{206}Pb/^{204}Pb$ and $^{207}Pb/^{204}Pb$ isotope ratio measurements of natural rutile crystals has been utilized to study geological temperature–time histories.[197] The significance of Devonian-Carboniferous igneous activity in Tasmania on the basis of U–Pb dating has been measured by sensitive high resolution ion microprobe (SHRIMP).[198]

High precision determination of $^{87}Rb/^{85}Rb$ isotope ratios by MC-ICP-MS (Isoprobe, Micromass)[199] is advantageous in comparison to MC-TIMS (VG Sector 54, Micromass) due to an improved precision, and fraction capability problems can be easily considered using admixed Zr for mass bias correction via the $^{92}Zr/^{90}Zr$ isotope ratio. Rb isotopic composition in geological materials (basalt, granite and greywacke) and synthetic samples (in house mica standard, NBS 607 K-fsp standard) have been examined with a reproducibility of $^{87}Rb/^{85}Rb$ measurements of 0.02–0.05 % (2 σ). No anomalous isotope fractionation was observed for the samples investigated.

Selected applications of isotope analysis in geology and geochronology by LA-ICP-MS are summarized in Table 8.8.

References

1. Becker, J. S., *Int. J. Mass Spectrom.*, **242**, 183 (2005).
2. Becker, J. S., *J. Anal. At. Spectrom.*, **20**, 1173 (2005).
3. Heumann, K. G., *Int. J. Mass Spectrom.*, **118/119**, 575 (1992).
4. Heumann, K. G., Eisenhut, S., Gallus, S. *et al.*, *Analyst*, **120**, 1291 (1996).
5. Kutschera, W., *Int. J. Mass Spectrom.*, **242**, 145 (2005).
6. Pickhardt, C., Dietze, H.-J. and Becker, J. S., *Int. J. Mass Spectrom.*, **242**, 273 (2005).
7. Platzner, I. T., *Modern Isotope Ratio Mass Spectrometry*, John Wiley & Sons, Chichester 145 (1997).
8. Wendt, K. and Trautmann, N., *Int. J. Mass Spectrom.*, **242**, 161 (2005).
9. Wieser, M. E. and Schwieters, J. B., *Int. J. Mass Spectrom.*, **242**, 97 (2005).
10. Böhlke, J. K., De Leater, J. R., De Bievre, P. *et al.*, *J. Phys., Chem. Ref. Data*, **34**, 57 (2005).
11. Zhao, M., Zhou, T., Wang, J., Lu, H., Fang, X., Guo, C., Li, O. and Li, C., *Rapid Commun. Mass Spectrom.*, **19**, 2743 (2005).
12. Chang, T. L., Li, W. J., Zhao, M. T., Wang, J. and Qian, Q. Y., *Int. J. Mass Spectrom.*, **207**, 13 (2001).
13. Chang, T. L., Zhao, M. T., Li, W. J., Wang, J. and Qian, Q. Y., *Int. J. Mass Spectrom.*, **177**, 131 (1998).
14. Budzikiewicz, H. and Grigsby, R. D., *Mass Spectrom. Rev.*, **25**, 146 (2006).
15. Boulyga, S. F. and Becker, J. S., *Fresenius' J. Anal. Chem.*, **370**, 612 (2001).
16. Boulyga, S. F., Becker, J. S., Matusevitch, J. L. and Dietze, H. J., *Int. J. Mass Spectrom.*, **203**, 143 (2000).
17. Taylor, P. D. P., De Bievre, P. and Valkiers, S., in *Handbook of Stable Isotope Analytical Techniques 1*, P. A. Groot (ed.), Elsevier, Amsterdam 905 (2004).
18. De Laeter, J. R., *Geostand. Geoanal. Res.*, **29**, 53 (2005).
19. Becker, J. S., *J. Anal. At. Spectrom.*, **17**, 1172 (2002).
20. Blichert-Toft, J., Chauvel, C. and Albarède, F., *Contrib. Mineral. Petrol.*, **127**, 248 (1997).
21. Halliday, A. N., Christensen, J. N., Lee, D.-C., Hall, C., Luo, X. and Rehkaemper, M., in *Inorganic Mass Spectrometry*, C. M. Barshick, D. C. Duckworth and D. H. Smith (eds.), Marcel Dekker AG, New York 291 (2000).
22. Halliday, A. N., Lee, D.-C., Christensen, J. N. *et al.*, *Geochim. Cosmochim. Acta*, **62**, 919 (1998).
23. Kleinhanns, L. C., Kreissig, K., Kamber, B. S., Meisel, T., Nägler, T. F. and Kramers, J. S., *Anal. Chem.*, **74**, 67
24. Nowell, G. M., Kempton, P. D., Noble, S. R. *et al.*, *Chem. Geol.*, **149**, 211 (1998).
25. Schwieters, J., Jung, G., Pesch, R. and Hamester, M., *Application Report 1*, ThermoFinnigan (2001).
26. De Laeter, J. R., *Application of Inorganic Mass Spectrometry*, Wiley-Interscience Series on Mass Spectrometry, John Wiley & Sons, Inc. New York (2001).
27. Wetzel, K., Becker, J. S. and Dietze, H. J., *Isotopenpraxis*, **19**, 221 (1983).
28. Wieser, M. E. and De Laeter, J. R., *Phys. Rev.*, **C 64**, 024307 (2001).
29. Ehrlich, S., Butler, I., Halicz, L., Rickard, D., Oldroyd, A. and Matthews, A., *Chem. Geol.*, **209**, 259 (2004).
30. Gagnevin, D., Daly, J. S., Waight, T. E., Morgan, D. and Poll, G., *Geochim. Cosmochim. Acta*, **69**, 1899 (2005).
31. Galy, A., Yoffe, O., Janney, P. E. *et al.*, *J. Anal. At. Spectrom.*, **18**, 1352 (2003).
32. Ghosh, P. and Brandt, W. A., *Int. J. Mass Spectrom.*, **228**, 1 (2003).
33. Jochum, K. P., Stoll, B., Herwig, K., Amini, M., Abouchami, W. and Hofmann, A. W., *Int. J. Mass Spectrom.*, **242**, 281 (2005).
34. Sime, N. G., De La Rocha, C. L. and Galy, A., *Earth Planet. Sci. Lett.*, **232**, 51 (2005).
35. Yuan, H. L., Gao, S., Liu, X. M., Li, H. M., Guenther, D. and Wu, F. Y., *Geostand. Geoanal. Res.*, **28**, 353 (2004).

36. Zoriy, M. V., Rashad, A., Pickhardt, C., Mohsen, H. T., Förstel, H., Helal, A. I., Zahran, N. F., Becker, J. S., *At. Spectr.*, **24**, 195 (2003).
37. Günther-Leopold, I., Kobler Waldis, J., Wernli, B. and Kopajtic, Z., *Int. J. Mass Spectrom.*, **242**, 197 (2005).
38. Boulyga, S. F., Desideri, D., Meli, M. A., Testa, C. and Becker, J. S., *Int. J. Mass Spectrom.*, **226**, 329 (2003).
39. Boulyga, S. F., Zoriy, M., Ketterer, M. E. and Becker, J. S., *J. Envir. Mon.*, **5**, 661 (2003).
40. Zengh, J. and Yamada, M., *Sci. Tot. Environ.*, **340**, 199 (2005).
41. Becker, J. S., Burow, M., Zoriy, M., Pickhardt, C., Ostapczuk, P. and Hille, R., *At. Spectr.*, **25**, 197 (2004).
42. Ejnik, J., Todorov, T., Mullick, F., Squibb, K., McDiarmid, M. and Centeno, J., *Anal. Bioanal. Chem.*, **382** 73 (2005).
43. Epov, V. N., Benkheddsa, K., Cornett, R. J. and Evans, R. D., *J. Anal. At. Spectrom.*, **20**, 424 (2005).
44. Moheghehghi, A. H., Shanks, S. T., Zigmond, J. A., Simmons, G. L. and Ward, S. L. A., *J. Radioanal. Nucl. Chem.*, **263**, 189 (2005).
45. Walczyk, T. and von Blanckenburg, F., *Int. J. Mass Spectrom.*, **242**, 117 (2005).
46. Izmer, A. V., Boulyga, S. F. and Becker, J. S., *J. Anal. At. Spectrom.*, **18**, 1339 (2003).
47. Izmer, A. V., Zoriy, M. V., Boulyga, S. F. and Becker, J. S., *J. Anal. At. Spectrom.*, **19** 1278 (2004).
48. Kerl, W., Becker, J. S., Dietze, H.-J. and Dannecker, W., *J. Anal. At. Spectrom.*, **11**, 723 (1996).
49. Becker, J. Su., Zoriy, M., Pickhardt, C., Przybylski, M. and Becker, J. S., *Int. J. Mass Spectrom.*, **242**, 135 (2005).
50. Boulyga, S. and Heumann, K. G., *Int. J. Mass Spectrom.*, **242**, 291 (2005).
51. Kent, A. J. R., Jacobsen, B., Peate, D. W., Waight, T. E. and Baker, J. A., *Geostand. Geoanal. Res.*, **28**, 417 (2004).
52. Pritzkow, W., Vogl, J., Koeppen, R. and Ostermann, M., *Int. J. Mass Spectrom.*, **242**, 309 (2005).
53. Rodriguez-Gonzalez, P., Marchante-Gayon, J. M., Alonso, J. I. G. and Sanz-Medel, A., *Spectrochim. Acta B*, **60**, 151 (2005).
54. Schaumlöffel, D. and Lobinski, R., *Int. J. Mass Spectrom.*, **242**, 217 (2005).
55. Stürup, S., Chen, C., Jukosky, J. and Folt, C., *Int. J. Mass Spectrom.*, **242**, 225 (2005).
56. Betti, M., *Int. J. Mass Spectrom.*, **242**, 169 (2005).
57. Dietze, H. J., *Analytikertaschenbuch*, **10** 249 (1991).
58. Dietze, H. J., *Massenspektroskopische Spurenanalyse*, Akademischer Verlagsgesellschaft Geest & Portig K.-G., Leipzig (1975).
59. Jochum, K. P., Seufert, M. and Best, S., *Fresenius Z. Anal. Chem.*, **309**, 308 (1981).
60. Betti, M., Rasmussen, G. and Koch, L., *Fresenius' J. Anal. Chem.*, **355**, 808 (1996).
61. Ireland, T. R., in *Handbook of Stable Isotope Analytical Techniques, 1*, P. A. d. Groot (ed.), Elsevier, Amsterdam 652 (2004).
62. Erdmann, N., Herrmann, G., Huber, G. *et al.*, *Fresenius' J. Anal. Chem.*, **359**, 378 (1997).
63. Gruning, C., Huber, G., Klopp, P. *et al.*, *Int. J. Mass Spectrom.*, **235**, 171 (2004).
64. Nunnemann, M., Erdmann, N., Hasse, H. U. *et al.*, *J. Alloys Comp.*, **271**, 45 (1998).
65. Trautmann, N., Passler, G. and Wendt, K. D. A., *Anal. Bioanal. Chem.*, **378**, 348 (2004).
66. Oughton, D. H., Skipperud, L., Fifield, L. K., Cresswell, R. G., Salbu, B. and Day, P., *Appl. Radiat. Isotopes*, **61**, 249 (2004).
67. Steier, P., Golser, R., Kutschera, W., Priller, A., Vockenhuber, C. and Liechtenstein, V., *Nucl. Instrum. Meth. Phys. Res. B*, **188**, 283 (2002).
68. Tuniz, C., Bird, J. R., Fink, D. and Herzog, G. F., *Accelerator Mass Spectrometry*, CRC Press, Boca Raton, Boston, London, New York, Washington (1998).
69. Vockenhuber, C., Ahmad, I., Golser, R. *et al.* Steier, P. and Winkler, S., *Int. J. Mass Spectrom.*, **223–224**, 713 (2003).
70. Trautmann, N., Passler, G. and Wendt, K. D. A., *Anal. Bioanal. Chem.*, **378**, 348 (2004).
71. Purser, K. H., Liebert, R. B., Litherland, A. E., *Rev. Phys. Appl.*, **12**, 1487 (1977).
72. Szidat, S., Schmidt, A., Handl, J. *et al.*, *Kerntechnik*, **65**, 160 (2000).
73. Schediwy, S., Rosman, K. J. R. and De Laeter, J. R., *Earth Planet. Sci. Lett.*, **243**, 326 (2006).

74. Johnson, T. M. and Bullen, T. B., in *Handbook of Stable Isotope Analytical Techniques 1*, P. A. d. Groot (ed.), Elsevier, Amsterdam 622 (2004).
75. Becker, J. S., *Int. J. Mass Spectrom.*, **242**, 183 (2005).
76. Becker, J. S., *Spectrochimica Acta, Part B: At. Spectr.*, **58B**, 1757 (2003).
77. Becker, J. S., Soman, R. S., Sutton, K. L., Caruso, J. A. and Dietze, H. J., *J. Anal. At. Spectrom.*, **14**, 933 (1999).
78. Desideri, D., Meli, M. A., Roselli, C., Testa, C., Boulyga, S. F. and Becker, J. S., *Anal. Bioanal. Chem.*, **374**, 1091 (2002).
79. Appelblad, P. K., Rodushkin, I. and Baxter, D. C., *Anal. Chem.*, **73**, 2911 (2001).
80. Quétel, C. R., Vogl, J., Prokaska, T. *et al. Fresenius' J. Anal. Chem.*, **368**, 148 (2000).
81. Russ, G. P. and Bazan, J. M., *Spectrochim. Acta*, **42 B**, 49 (1987).
82. Aggarwal, J. K., Mezger, K., Pernicka, E. and Meixner, A., *Int. J. Mass Spectrom.*, **232**, 259 (2004).
83. Vielzeuf, D., Champenois, M., Valley, J. W., Brunet, F. and Devidal, J. L., *Chem. Geol.*, **223**, 208 (2005).
84. Albarède, F. and Beard, B., *Mineral. Geochem.*, **55**, 113 (2004).
85. Forster, G. L., Ni, Y., Haley, B. and Elliott, T., *Chem. Geol.*, **230**, 161 (2006).
86. Ramebäck, H., Berglund, M., Vendelbo, D., Wellum, R. and Taylor, P. D. P., *J. Anal. At. Spectrom.*, **16**, 127 (2001).
87. Zoriy, M. V., Halicz, L., Ketterer, M. E., Pickhardt, C., Ostapczuk, P. and Becker, J. S., *J. Anal. At. Spectrom.*, **19**, 362 (2004).
88. Boulyga, S. F. and Becker, J. S., *J. Anal. At. Spectr.*, **17**, 1202 (2002).
89. Vonderheide, A. P., Zoriy, M. V., Izmer, A. V. *et al.*, *J. Anal. At. Spectrom.*, **19**, 675 (2004).
90. Zoriy, M. V., Ostapczuk, P., Halicz, L., Hille, R. and Becker, J. S., **242**, 203 (2005).
91. Barshik, C. M., Duckworth, D. C. and Smith, D. H., *Inorganic Mass Spectrometry*, Marcel Dekker Inc., New York (2000).
92. Barrero Moreno, J. M., Betti, M. and Garcio Alonso, J. I., *J. Anal. At. Spectrom.*, **12**, 355 (1997).
93. Day, J. A., Caruso, J. A., Becker, J. S. and Dietze, H.-J., *J. Anal. At. Spectrom.*, **15**, 1343 (2000).
94. Kerl, W., Becker, J. S., Dannecker, W. and Dietze, H. J., *Fresenius' J. Anal. Chem.*, **362**, 433 (1998).
95. Kuczewski, B., Marquart, C. M., Seibert, A., Geckeis, H., Kratz, J. V. and Trautmann, N., *Anal. Chem.*, **75**, 6769 (2004).
96. Perna, L., Jernström, J., Aldave de Heras, L., de Pablo, J. and Betti, M., *Anal. Chem.*, **75**, 2292 (2003).
97. Ketterer, M. E., Hafer, K. M., Link, C. L., Kolwaite, D., Wilson, J. and Mietelski, J. W., *J. Anal. At. Spectrom.*,, **19**, 241 (2004).
98. Ketterer, M. E., Jordan, J. A., Szechenyi, S. C., Hudson, D. D. and Layman, R. R., *J. Anal. At. Spectrom.*, **15**, 1569 (2000).
99. Zhou, T., Zhao, M., Li, J. *et al.*, *Fenxi Huaxue*, **33**, 1252 (2005).
100. Valkiers, S., Ruβe, K., Taylor, P., Ding, T. and Inkret, M., *Int. J. Mass Spectrom.*, **242**, 319 (2005).
101. Valkiers, S., Ruβe, K., Taylor, P. D. P., Ding, T. and Inkret, M., *Int. J. Mass Spectrom.*, **242**, 319 (2005).
102. Thiemens, M. H., *Science*, **283**, 341 (1999).
103. Rai, V. K., Jackson, T. L. and Thiemens, M. H., *Science*, **309**, 1962 (2005).
104. Miller, M. F., Franchi, I. A., Thiemens, M. H. *et al.*, *PNAS*, **99**, 10988 (2002).
105. Krot, A. N., McKeegan, K. D., Leshin, L. A., MacPherson, G. J. and Scott, E. R. D., *Science*, **285**, 1051 (2002).
106. Gao, Y. Q. and Marcus, R. A., *Science*, **293**, 259 (2001).
107. Dallay, L., Lucchini, R. and Sharp, Z. D., in *Handbook of Stable Isotope Analytical Techniques*, Vol.1, P. A. d. Groot (ed.), Elsevier, Amsterdam 62 (2004).
108. Meier-Augstein, W., in *Handbook of Stable Isotope Analytical Techniques*, Vol.1, P. A. d. Groot (ed.), Elsevier, Amsterdam 153 (2004).
109. Krupp, E. M. and Donard, O. F. X., *Int. J. Mass Spectrom.*, **242**, 233 (2005).
110. Wehmeier, S. and Feldmann, J., *J. Environ. Monit.*, **7**, 1194 (2005).
111. Kremer, D., Ilgen, G. and Feldmann, I., *Anal. Bioanal. Chem.*, **383**, 509 (2005).
112. Rodriguez, P. F., Alonso, J. I. G. and Sanz-Medel, A., *J. Anal At. Spectrom.*, **2005**, 1076 (2005).
113. Dombovari, J., Becker, J. S. and Dietze, H. J., *Int. J. Mass Spectrom.*, **202**, 231 (2000).
114. Coetzee, P. P. and Vanhaecke, F., *Anal. Bioanal. Chem.*, **383**, 977 (2005).
115. Becker, J. S. and Dietze, H. J., *Isotopenpraxis*, **19**, 105 (1983).

116. Bandura, D. R., Baranov, V. I. and Tanner, S. D., *J. Anal. At. Spectrom.*, **15**, 921 (2001).
117. Tanner, S. D., Baranov, V. I. and Bandura, D. R., *Spectrochim. Acta*, **57 B**, 1361 (2002).
118. Bandura, D. R., Baranov, V. I. and Tanner, S. D., *J. Am. Soc. Mass Spectrom.*, **13**, 1176 (2002).
119. Boulyga, S. F. and Becker, J. S., *Fresenius' J. Anal. Chem.*, **370**, 618 (2001).
120. Boulyga, S. F., Dietze, H.-J. and Becker, J. S., *Mikrochim. Acta*, **137**, 93 (2001).
121. Dexter, M. A., Appelbad, P. K., Ingle, C. P., Batey, J. H., Reid, H. J. and Sharp, B. L., *J. Anal. At. Spectrom.*, **17**, 183 (2002).
122. Becker, J. S. and Dietze, H.-J., *Fresenius' J. Anal. Chem.*, **368**, 23 (2000).
123. Becker, J. S. and Dietze, H.-J., *J. Anal. At. Spectrom.*, **14**, 1493 (1999).
124. Moens, L., Vanhaecke, F., Bandura, D. R., Baranov, V. I. and Tanner, S. D., *J. Anal. At. Spectrom.*, **16**, 991 (2001).
125. Zoriy, M. V., Kayser, M., Izmer, A., Pickhardt, C. and Becker, J. S., *Int. J. Mass Spectrom.*, **242**, 297 (2005).
126. Crowe, S. A., Fryer. B. J., Samson, I. M. and Gagnon, J. E., *J. Anal. At. Spectrom.*, **18**, 1331 (2003).
127. Hirata, T., Hayano, Y. and Ohno, T., *J. Anal. At. Spectrom.*,, **18**, 1283 (2003).
128. Jochum, K. P., Stoll B., Herwig, K., Amini M., Abouchami, W. and Hofmann, A. W., *Int. J. Mass Spectrom.*, **242**, 281 (2005).
129. Le Roux, P. J., Shirey, S. B., Benton, L., Hauri, E. H. and Mock, T. D., *Chem. Geol.*, **203**, 123 (2004).
130. Malinovsky, D., Rodushkin, I., Axelsson, M. D. and Baxter, D. C., *J. Geochem. Expl.*, **81**, 71 (2004).
131. Pearson, N. J., Alard, O., Griffin, W. L., Jackson, S. E. and O'Reilly, Y., *Geochim. Cosmochim. Acta*, **66**, 1037 (2002).
132. Prohaska, T., Latkoczy, C., Schultheis, G., Teschler-Nicola, M. and Stingeder, G., *J. Anal. At. Spectrom.*, **17**, 887 (2002).
133. Stiring, C. H., Lee, D.-C., Christensen, J. N. and Halliday, A. N., *Geochim. Cosmochim. Acta*, **64**, 3737 (2000).
134. Tieplo, M., *Chem. Geol.*, **199**, 159 (2003).
135. Vanhaecke, F., Resano, M., Garcia-Ruiz, E., Balcaen, L., Koch, K. R. and McIntosch, K., *J. Anal. At. Spectrom.*, **19**, 632 (2004).
136. Waight, T., Baker, J. and Peate, D., *Int. J. Mass Spectrom.*, **221**, 229 (2002).
137. Halliday, A. N., *Nature*, **427**, 505 (2004).
138. Rehkämper, M. and Halliday, A. N., *19. ICP-MS* Anwendertreffen, Zurich (2004).
139. Reynolds, B. C., Georg, R. B., Oberli, F., Wiechert, U. and Halliday, A. N., *J. Anal. At. Spectrom.*, **21** 266 (2006).
140. Seth, B., Thirwall, M. F., Houghton, S. L. and Craig, C. A., *J. Anal. At. Spectrom.*, **18**, 1323 (2003).
141. Williams, H. M., Peslier, A. H., McCammon, C., Halliday, A. N., Levasseur, S., Teutsch, N. and Burg, J.-P., *Earth Planet. Sci. Lett.*, **235**, 435 (2005).
142. Nowell, G. M., Pearson, D. G., Ottley, C. J., Schwieters, J. and Dowall, D. P., in *Plasma Source Mass Spectrometry*, G. Holland and S. D. Tanner (eds.), The Royal Society of Chemistry, Cambridge 307 (2003).
143. Pietruska, A. J., Carlson, R. W. and Hauri, E. H., *Chem. Geol.*, **188**, 171 (2002).
144. Schwieters, J. B., Bouman, C., Tuttas, D. and Wieser, M. E., *Geochim. Cosmochim. Acta*, **68**, A 60 (2004).
145. Xie, Q., Lu, S., Evans, D., Dillon, P. and Hintelmann, H., *J. Anal. At. Spectrom.*, **20**, 515 (2005).
146. Anbar, D., Knab, K. A. and Barling, J., *Anal. Chem.*, **73**, 1425 (2001).
147. Hirata, T., *Chem. Geol.*, **176**, 323 (2001).
148. Müller, W., Fricke, H., Halliday, A. N., Culloch, M. T. and Wartho, J.-A., *Sci. Tot. Environ.*, **303**, 862 (2003).
149. Nielsen, S. G., Rehkämper, M., Porcelli, D. *et al.*, *Geochim. Cosmochim. Acta*, **69**, 2007 (2005).
150. van den Boorn, S. H. J. M., Vroon, P. Z., van Belle, C. C., van der Wagt, B., Schwieters, J. B. and van Bergen, M. J., *J. Anal. At. Spectrom.*, **21**, 734 (2006).
151. Günther-Leopold, I., Wernli, B., Kopajtic, Z. and Günther, D., *Anal. Bioanal. Chem.*, **378**, 241 (2004).
152. Balcaen, L. I. L., De Schrijver, I., Moens, L. and Vanhaeke, F., *Int. J. Mass Spectrom.*, **242**, 251 (2005).
153. Bouman, C., Cocheri, A., Robert, M., Schwieters, J. B. and Wieser, M. E., *Geochom Cosmochim. Acta (Suppl. 1)*, **67**, A 44 (2003).
154. Evans, P., Elahi, S., Lee, K. and Fairman, B., *J. Environ. Monit.*, **5**, 175 (2003).
155. Nygren, I., Rodushkin, I., Nilsson, I. and Baxter, C., *J. Anal. At. Spectrom.*, **18**, 1426 (2003).

156. Prohaska, T., Wenzel, W. W. and Stingeder, G., *Int. J. Mass Spectrom.*, **242**, 243 (2005).
157. Schöberg, R. and von Blanckenburg, F., *Int. J. Mass Spectrom.*, **242**, 257 (2005).
158. Zoriy, M., Varga, Z., Pickhardt, C. *et al. J. Environ. Monit.*, **7**, 514 (2005).
159. Snell, J. P. and Quetel, C. R., *J. Anal. At. Spectrom.*, **20**, 447 (2005).
160. Encinar, J. R. Schaumlöffel, D., Yasumitsu, O. and Lobinski, R., *Anal. Chem.*, **76** 6635 (2004).
161. Boulyga, S. F., Heilmann, J. and Heumann, K. G., *Anal. Bioanal. Chem.* **382**, 1808 (2005).
162. Evans, P., Wolff-Briche, C. and Fairman, B., *J. Anal. At. Spectrom.*, **16**, 964 (2001).
163. Klemens, P. and Heumann, K. G., *Fresenius' J. Anal. Chem.*, **371**, 758 (2001).
164. Demuth, N. and Heumann, K. G., *Anal. Chem.*, **73**, 4020 (2001).
165. Lambertsson, L., Lundberg, E., Nilsson, M. and Frech, W., *J. Anal. At. Spectrom.*, **16**, 1296 (2001).
166. Regelous, M., Turner, S. P., Elliot, T. R., Rasotami, K. and Hawkesworth, J., *Anal. Chem.*, **76**, 3584 (2004).
167. Forster, D. A., Staubwasser, M. and Henderson, G. M., *Marine Chem.*, **87**, 59 (2004).
168. Prange, A., Schaumlöffel, D., Braetter, P., Richarz, A.-N. and Wolf, C., *Fresenius' J. Anal. Chem.*, **371**, 764 (2001).
169. Schaumlöffel, D., Prange, A., Marx, G., Heumann, K. G. and Brätter, P., *Anal. Bioanal. Chem.*, **372**, 155 (2002).
170. Vogl, J., *J. Anal. At. Spectrom.*, **22**, 475 (2007).
171. Becker, J. S., Zoriy, M., Halicz, L. *et al.*, *J. Anal. At. Spectrom.*, **19**, 1257 (2004).
172. Zoriy, M., Ostapczuk, P., Halicz, L., Hille, R. and Becker, J. S., *Int. J. Mass Spectrom.*, **242**, 203 (2005).
173. Boulyga, S. F., Matusevich, J. L., Mironov, V. P., Kudrjashov, V. P., Halicz, L., Segal, I., McLean, J. A., Monster, A., Becker, J. S. *J. Anal. At. Spectrom.*, **17**, 958 (2002).
174. Elliot, N. L., Bickel, G. A., Linauskas, S. H. and Paterson, L. M., *J. Radioanal. Nucl. Chem.* **267**, 637 (2006).
175. Mas, J. L., Ma, R., McLeod, C. W., Gonzalez-Labajo, J., Cox, A. and Watson, P., *Anal. Bioanal. Chem.*, **386**, 152 (2006).
176. IUPAC Isotopic Composition of the Elements 1997, *J. Anal. At. Spectrom.*, **14**, 5N (1999).
177. Halicz, L., Becker, J. S., Pickhardt, C. *et al.*, *Int. J. Mass Spectrom.*, **249–250**, 296 (2006).
178. Hoppstock, K., Becker, J. S. and Dietze, H. J., *At. Spectr.*, **18**, 180 (1997).
179. Compte, J., Bienvenu, P., Brochard, E., Fernandez, J. M. and Andreoletti, G., *J. Anal. At. Spectrom.*, **18**, 702 (2003).
180. Brown, C. F., Dresel, P. E., Geiszler, K. N. and Farmer, O. T., *J. Anal. At. Spectrom.*, **21**, 955 (2006).
181. Agarande, M., Benzoubir, S., Neiva-Marquesw, A. M. and Bouisset, P., *J. Environ. Monit.*, **72**, 175 (2004).
182. Boulyga, S. F., Tibi, M. and Heumann, K. G., *Anal. Bioanal. Chem.*, **342**, 3478 (2004).
183. Zoriy, M., Pickhardt, C., Ostapczuk, P., Hille, R. and Becker, J. S., *Int. J. Mass Spectrom.*, **232**, 217 (2004).
184. Chan, L.-H., in *Handbook of Stable Isotope Analytical Techniques*, vol. 1, P. A. e. De Groot (ed.), Elsevier, Amsterdam 122 (2004).
185. You, C.-F., in *Handbook of Stable Isotope Analytical Techniques*, vol. 1, Elsevier, Amsterdam 142 (2004).
186. Peltola, P. and Åström, M., *Appl. Geochem.*, **21**, 941 (2006).
187. Lécuyer, C., Grandjean, P., Reynard, B., Albarède, F. and Telouk, P., *Chem. Geol.*, **186**, 45 (2002).
188. Aggarwal, J. K., Sheppard, D., Mezger, K. and Pernicka, E., *Chem. Geol.*, **199**, 331 (2003).
189. Jochum, K. P., Stoll, B., Herwig, K. and Willbold, M., *J. Anal. At. Spectrom.*, **21**, 666 (2006).
190. Markowski, A., Quitté, G., Halliday, A. N. and Kleine, T., *Earth Planet. Sci. Lett.*, **242**, 1 (2006).
191. Xie, Z., Zheng, Y. F., Zhoa, Z. F. *et al., Chem. Geol.*, **231**, 214 (2006).
192. Bernal, J. P., Eggins, S. M. and McCulloch, M. T., *J. Anal. At. Spectrom.*, **20**, 1240 (2005).
193. Mao, J., Wang, Y., Lehmann, B. *et al., Ore Geology Rev.*, **29**, 307 (2006).
194. Moynier, F., Bouvier, A., Blichert-Toft, J., Telouk, P., Gasperini, D. and Albarède, F., *Geochim. Cosmochim. Acta*, **70**, 4287 (2006).
195. Scherer, E. E., Münker, C. and Mezger, K., *Science*, **293**, 683 (2001).
196. Connelly, J. N., Ulfbeck, D. G., Thrane, K., Bizzaro, M. and Housh, T., *Chem. Geol.*, **233**, 126 (2006).
197. Vry, J. K. and Baker, J. A., *Geochim. Cosmochim. Acta*, **70**, 1807 (2006).
198. Black, L. P., McClenaghan, M. P., Korsch, R. J., Everard, J. L. and Foudouzlis, C., *Austral. J. Earth Sciences*, **52**, 807 (2006).
199. Nebel, O., Mezger, K., Scherer, E. E. and Muenker, C., *Int. J. Mass Spectrom.*, **246**, 10 (2005).

9

Fields of Application in Trace, Ultratrace and Surface Analysis

Inorganic mass spectrometry has become established as a powerful and sensitive bulk, surface and isotope analytical technique and can be universally employed in a wide field of quite different applications. Selected relevant fields of application in inorganic mass spectrometry summarized in Figure 9.1 have been extended to include the trace and ultratrace analysis of environmental, biological, medical and geological materials, water samples, and also of technical products such as metals, alloys, ceramics, glasses, polymers etc.[1] In addition, the analysis of element species can be carried out by (mostly hyphenated) mass spectrometric techniques and/or by a combination of inorganic mass spectrometry (ICP-MS) with biomolecular mass spectrometry (MALDI/ESI-MS). Several mass spectrometric techniques enable surface analysis for depth profiling of thick and thin film and multi-layer systems, for analyzing doping elements, for the determination of lateral element distribution of any surface (imaging) and in bulk, for microlocal analysis and characterizing surface contamination. As a main feature of inorganic mass spectrometry, as discussed before, precise and accurate isotope ratio measurements are additionally performed for studies of isotope variation in nature, age dating, for quantification purposes (by means of the isotope dilution technique), in tracer experiments using isotopically enriched tracers and determination of long and medium lived radionuclides (see Chapter 8).

Selected figures of merit of the most important inorganic mass spectrometric methods are compared in Table 9.1 for the determination of trace impurities in solid materials directly and after digestion. This table gives an overview of the variation of relative sensitivity coefficients (RSC, see Chapter 6), the possibilities for calibration procedures, accuracy and precision, and the advantages and limitations of the analytical technique. In Figure 9.2 the power of detection and measuring concentration range of inorganic state mass spectrometric techniques are compared. The

Inorganic Mass Spectrometry: Principles and Applications J. S. Becker

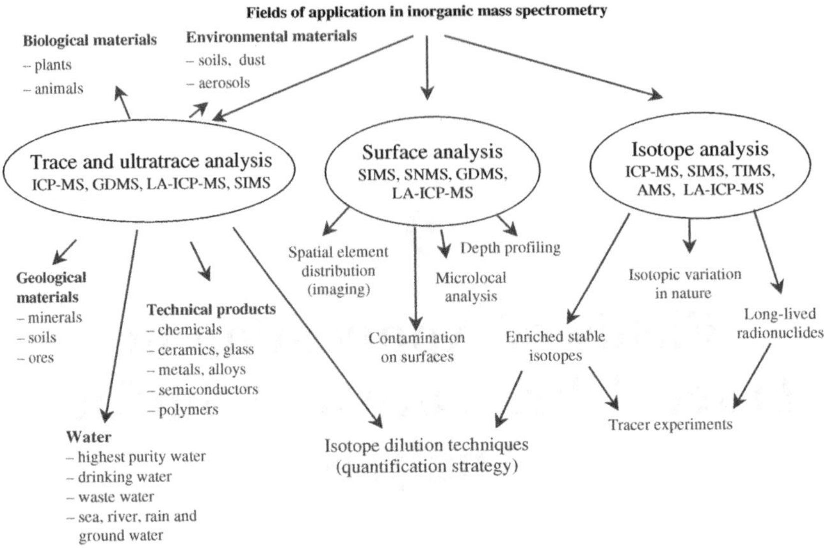

Figure 9.1 *Fields of application in inorganic mass spectrometry. (J. S. Becker, and H. J Dietze, Int. J. Mass Spectrom., **197**, 1 (2000) Reproduced by permission of Elsevier.)*

lowest detection limits are observed with ICP-MS on high purity aqueous solutions. The detection limits for ICP-MS on solid samples are some orders of magnitude higher considering the dilution factor of the digested or dissolved solid sample, the blank values of the chemical used, the instrumental blank and possible interferences with atomic and polyatomic ions.

9.1 Materials Science

Many important processes such as corrosion, catalysis or adhesion effects on surfaces depend on the elemental composition of solid materials, especially of the sample surface or of interface region. Several inorganic mass spectrometric techniques such as ICP-MS, LA-ICP-MS, GDMS, SIMS and partly SNMS are widely used as sensitive analytical techniques with multi-element capability for the characterization of solid samples and surfaces in materials science, e.g., for the determination of trace and ultratrace elements in metals and alloys, semiconductors, insulators (glasses, ceramics) or in layered systems.[1–3] The fields of application for trace and ultratrace analysis by inorganic mass spectrometry in materials science and microelectronics are summarized in Figure 9.3.[2] The detection range and units applicable in the determination of concentration and the common abbreviations for the high-purity of materials are summarized in Tables 9.2 and 9.3, respectively. Bulk analysis is performed directly by solid-state mass spectrometric techniques such as LA-ICP-MS, GDMS or SIMS with only minimal sample preparation or after digestion and dilution of the digested sample by ICP-MS. Detection limits of elements may be as low as the $ng\,g^{-1}$ concentration range and below depending on the elements studied, the matrix composition and the analytical method chosen. For the analysis of thin and thick layers and layered systems, several surface analytical techniques are available, such as SIMS, GDMS, SNMS and LA-ICP-MS with quite different depth resolutions. Whereas dynamic SIMS and SNMS are suitable for

Table 9.1 Comparison of inorganic mass spectrometic methods for determination of trace elements in solid materials.

Method	Detection limit [$\mu g\, g^{-1}$]	RSC	Calibration	Accuracy/ Precision	Advantage	Limits
GDMS	0.0005–0.1	0.2–5	SRM[a]	++/++	highly sensitive	insulators
SIMS	0.001–0.1	10^3–10^6	ion implanted standards	+/+	surface analysis	huge matrix effect, interferences
SNMS	10–100	0.3–3	ion implanted standards	+/+	surface analysis	poor detection power
SSMS	0.001–0.1	0.3–3	SRM	+/+	high sensitive at high $m/\Delta m$	expensive, time consuming
LIMS	0.001–0.1	0.8–2	SRM	+/+	for insulators highly sensitive	expensive, time consuming
ICP-MS	0.000 001–0.001	—	solution calibration	++/++ IDA[b]: +++/++	fast, cheap highly sensitive	interferences, contamination danger
LA-ICP-MS	0.000 01–0.001	0.3–3	SRM, online solution based calibration	+/+	fast, highly sensitive, insulators	interferences
ID-TIMS	0.000 005–0.1	—	by isotope dilution	+++/++	excellent quantification possibility, highest precision	no multi-element capability, time consuming sample preparation

[a] SRM – standard reference material; [b] IDA – isotope dilution analysis.
(J.S. Becker and H.J. Dietze, *Int. J. Mass Spectrom.*, **228**, 127 (2003). Reproduced by permission of Elsevier.)

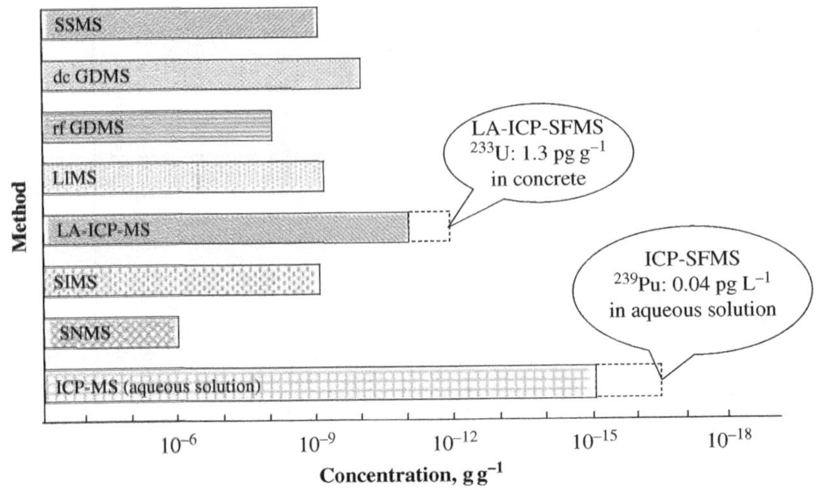

Figure 9.2 *Power of detecting and measuring concentration range of inorganic mass spectrometric methods. (J. S. Becker and H. J. Dietze, Int. J. Mass Spectrom.,* **228***, 127 (2000). Reproduced by permission of Elsevier.)*

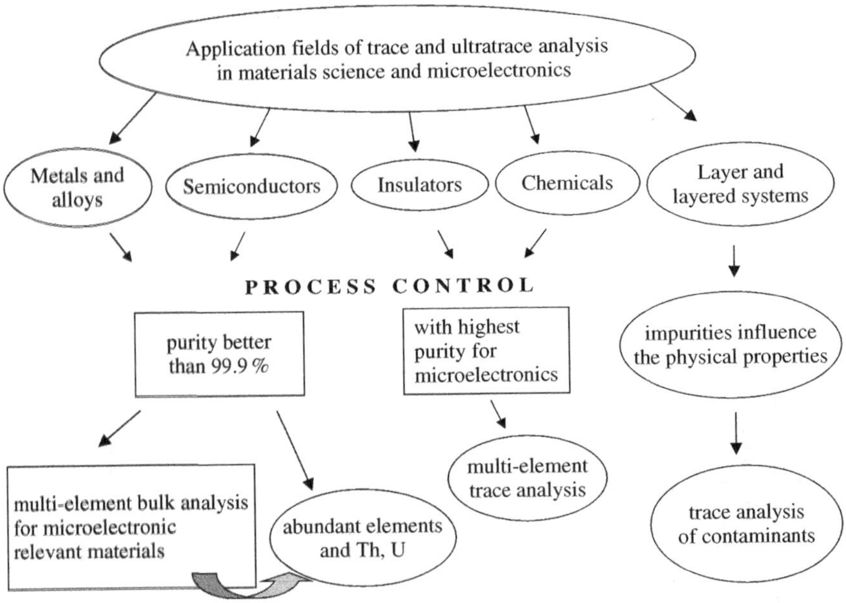

Figure 9.3 *Application fields for trace and ultratrace analysis in materials science and Microelectronics. (J. S. Becker and H. J. Dietze, Int. J. Mass Spectrom., Ion Proc.* **197***, 1–35 (2000). Reproduced by permission of Elsevier.)*

Table 9.2 *Detection range and units in determination of concentration.*

Weight [g]	Concentration	Unit	^{238}U atoms ml^{-1}		
10^{-3} (milli)	mg g^{-1}	0.1%	2.53×10^{18}		
10^{-6} (micro)	μg g^{-1}	ppm	2.53×10^{15}	Traces	↓
10^{-9} (nano)	ng g^{-1}	ppb	2.53×10^{12}		
10^{-12} (pico)	pg g^{-1}	ppt	2.53×10^{9}	Ultratraces	↓
10^{-15} (femto)	fg g^{-1}	ppq	2.53×10^{6}		
10^{-18} (atto)	ag g^{-1}	part per quintillion	2530		
10^{-21} (zepto)	zg g^{-1}	part per sextillion	2.53		
10^{-24} (yocto)	yg g^{-1}	part per septillion	0.00253		

(J. S. Becker and H. J. Dietze, *Int. J. Mass Spectrom.*, **228**, 127 (2003). Reproduced by permission of Elsevier.)

Table 9.3 *Common abbreviations for high purity materials.*

Purity	Abbreviation	Maximum metallic impurities
99.9 %	39 or 3N	1000 μg g^{-1}
99.99 %	49 or 4N	100 μg g^{-1}
99.999 %	59 or 5N	10 μg g^{-1}
99.9996 %	59.6 or 5N6	4 μg g^{-1}
99.9999 %	69 or 6N	1 μg g^{-1}
99.999 99 %	79 or 7N	100 ng g^{-1}

(J. S. Becker and H. J. Dietze, *Int. J. Mass Spectrom.*, **228**, 127 (2003). Reproduced by permission of Elsevier.)

elemental depth profiling of thin layers and layered systems, it is also possible to analyze thick layers with GDMS and LA-ICP-MS due to the higher sputter or laser ablation rate, respectively. Using these techniques for the depth profiling of thick layers or layered systems, in general the depth resolution is significantly diminished due to the higher sputter rate. Because SIMS and SNMS possess difficulties in analyzing thick layers (> several μm) directly, the analysis of depth profiles are mostly measured on cross sections of samples or by pre-sputtering with a second ion gun before measurement. The disadvantages of SIMS, GDMS and SNMS are the relatively expensive instrumentation compared to LA-ICP-MS.

The most frequently applied analytical methods used for characterizing bulk and layered systems (wafers and layers for microelectronics; see the example in the schematic on the right-hand side) are summarized in Figure 9.4. Besides mass spectrometric techniques there are a multitude of alternative powerful analytical techniques for characterizing such multi-layered systems. The analytical methods used for determining trace and ultratrace elements in, for example, high purity materials for microelectronic applications include AAS (atomic absorption spectrometry), XRF (X-ray fluorescence analysis), ICP-OES (optical emission spectroscopy with inductively coupled plasma), NAA (neutron activation analysis) and others. For the characterization of layered systems or for the determination of surface contamination, XPS (X-ray photon electron spectroscopy), SEM-EDX (secondary electron microscopy combined with energy disperse X-ray analysis) and

GD-OES (glow discharge optical emission spectrometry) are applied. AES (auger electron spectroscopy), AFM (atomic force microscopy) and TRXF (transmission reflection X-ray fluorescence analysis) have been successfully used, especially in the semiconductor industry and in materials research.

Of all the analytical techniques inorganic mass spectrometry has occupied a favoured place for the characterization of initial materials including such multi-layered systems due to its specific properties, such as high sensitivity, low detection limit, high dynamic range and the capability to determine isotope ratios.

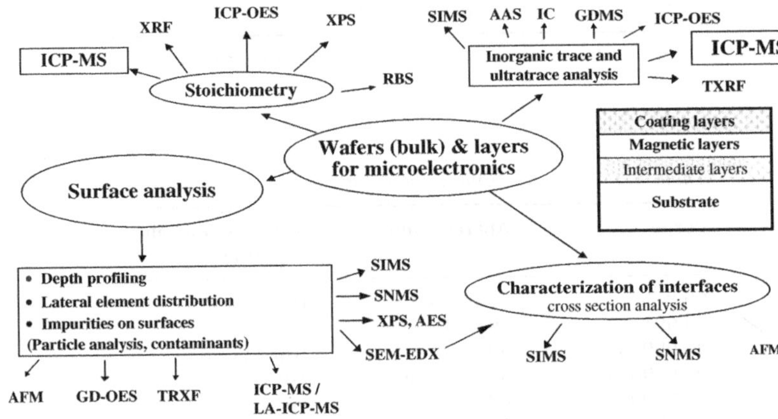

Figure 9.4 *Analytical methods for bulk and layered systems and applications. (J. S. Becker and H. J. Dietze, Int. J. Mass Spectrom.,* ***197****, 1 (2000). Reproduced by permission of Elsevier.)*

9.1.1 Trace and Ultratrace (Bulk) Analysis of Metals and Alloys

A typical field of application for inorganic mass spectrometry in materials research, the semiconductor industry and metallurgy is the analysis of contaminants at trace concentration levels in the purity control of electrically conducting metallic samples including refractory metals and high temperature alloys (e.g. Al, Sc, Cu, Cr, Ni, Mo, Au, Pt, Ta, W, Ti and Ti-Al-, Fe-, Co-, Cr-, Pt- or Ni-based alloys, Ti-W alloys and others). For many decades, due to the high sensitivity and low detection limits, solid-state mass spectrometric methods such as GDMS, SSMS or LA-ICP-MS were employed in addition to non-mass spectrometric techniques (e.g., XRF, GD-OES or SEM-EDX and atomic spectrometry) for the multi-element analysis of trace elements in metals and alloys (e.g., in process control or for corrosion studies) with a minimum of sample preparation. In the past, SSMS was the method of choice and was often used for the multi-element trace element analysis of high purity metals, which can be analyzed directly using two electrodes (with a diameter of $\sim$ 1–3 mm and a length of $\sim$ 10 mm) for compact conducting sample materials. Today there are only a few laboratories worldwide that still work with the old spark source mass spectrometers constructed in the 1960s and 1970s. SSMS was applied in the author's laboratory until the end of nineties for purity determination of solid samples of high purity metals (Al, Cu, Sc, Fe, Ag) including refractory metals (W, Mo, Nb, Ta) or noble metals that are insoluble in acid, for difficult mixtures (e.g., for ultratrace analysis of Ru and Pt), or for purity control of graphite and several alloys (zircaloy, high-temperature Cr–Ni-based alloys or steel). The disadvantage of SSMS is the time consuming evaluation of analytical data using ion sensitive photoplates and the quantification problems if no suitable matrix matched standard reference material is available. An example of SSMS application for high purity indium by a Japanese semiconductor company (DOWA Mining Co., Ltd)[4] is given in Table 9.4. Relatively low detection limits for light elements, such as Al, Si, P, S, K and Ca in the low $ng\,g^{-1}$ range have been observed, whereby the detection limits were mainly determined by blank values (instrumental background). With the detection limits for light difficult to analyze elements in the low $ng\,g^{-1}$ range, SSMS is one of the trace analytical techniques which are able to determine the purity of 6N–7N material (purity 99.9999 % and 99.999 99 %, respectively). The sum impurity of high purity indium given in Table 9.4 is $< 0.65\,\mu g\,g^{-1}$, which means the purity of the metal with respect to the analyzed metals is 99.999 935 %. The absolute detection limits of SSMS with photographic ion detection are in the 0.01 to 1 ng range. In addition, the mass spectra of SSMS are very simple due to the low polyatomic ion formation rate.

Table 9.4 *Result of trace analysis of high purity indium and zinc measured by spark source mass spectrometry (SSMS) and glow discharge mass spectrometry (GDMS), respectively.*

High purity indium (SSMS)			**High purity zinc (GDMS)**		
		Concentration, $\mu g\,g^{-1}$			
Al < 0.005	Mn < 0.01	Cd < 0.07	Na < 0.01	Ca < 0.05	Cd < 0.05
Si < 0.005	Fe < 0.01	In < 0.02	Al < 0.01	Cr < 0.03	Sb < 0.01
P < 0.005	Ni < 0.01	Sn < 0.06	Si < 0.01	Fe < 0.01	Sn < 0.01
S < 0.005	Cu < 0.01	Sb < 0.04	S < 0.03	Ni < 0.01	Tl < 0.01
Cl < 0.01	Zn < 0.02	Te < 0.07	Cl < 0.03	Cu < 0.05	Pb < 0.01
K < 0.01	Ge < 0.04	Au < 0.04	K < 0.01	Ag < 0.01	Bi < 0.01
Ca < 0.01	As < 0.02	Ti < 0.08			
Cr < 0.01	Se < 0.03	Bi < 0.06			

DOWA Mining Co., Ltd. Semiconductor Materials Group, Tokyo, Shape of sample, shot, ingot, disk, rod for MBE.
(J. S. Becker and H. J. Dietze, *Int. J. Mass Spectrom.*, **228**, 127 (2003). Reproduced by permission of Elsevier.)

A typical application field for laser ionization mass spectrometry (LIMS) in multi-element analysis is trace analysis in the purity control of metallic samples. Metallic specimens with a high boiling point, especially, have been found to be ideal for analysis by LIMS.[5] Commercialization of laser ionization ToF-MS (especially the LAMMA and LIMA instruments) and the development of LIMS instrumentation from old spark source mass spectrometers (SSMS) with Mattauch–Herzog geometry by replacing the ion sources made them fit for use in materials research and metallurgy in the eighties, e.g., in the analysis of trace elements including light difficult to determine non-metals (B, C, N, O) in alloys (such as steel,[6–8] brass) and segregates formed during brazing and in corrosion studies.[9,10]

The application of SIMS, SNMS, SSMS and GDMS in quantitative trace analysis for conducting bulk material is restricted to matrices where standard reference materials (SRMs) are available. For quantification purposes, the well characterized multi-element SRMs (e.g., from NIST) are useful. In Table 9.5 the results of the analysis by SNMS and the RSCs (relative sensitivity coefficients) for different elements in a low alloy steel standard (NBS 467) are compared with those of SSMS. Both solid-state mass spectrometric techniques with high vacuum ion sources allow the determination of light non-metals such as C, N, O and P in steel, and the RSCs for the elements measured vary from 0.5 to 3 (except C). RSCs are applied as a correction factor in the analytical method used to obtain quantitative data. Since no suitable certified standard reference materials for trace elements are available for most of the metals and alloys investigated by mass spectrometry, this means that no relative sensitivity coefficients can be determined in different matrices and consequently the quantification of analytical data in SIMS, SNMS, GDMS and SSMS is difficult. In contrast, the quantification procedure is easy in ICP-MS after digestion or dissolution and dilution of solid samples, and matrix effects can be avoided after matrix separation, as demonstrated for the trace impurity determination in scandium, tantalum and zircaloy.[11–13] However, compared to ICP-MS, after digestion of solid samples the danger of contamination in solid mass spectrometry is significantly lower. Due to the expensive apparatus and time consuming analysis, for direct analysis on metals and alloys, SSMS has been replaced in the past few decades by the more powerful and efficient GDMS, which uses a fast electrical ion detection and yields more precise results in trace element determination.

Table 9.5 *Comparison of trace element analysis on low alloy steel standard (NBS 467) measured by SNMS and SSMS (concentration in $\mu g\, g^{-1}$).*

Element	Certified value	SSMS		SNMS	
		Concentration	RSC	Concentration	RSC
B	2	2	1	—	—
C	110	542	4.9	572	5.2
N	40	23	0.6	69	1.7
O	40	129	3.2	106	2.7
Al	1600	1170	0.7	4288	2.7
P	330	261	0.8	230	0.7
Ti	2660	2223	0.9	3401	1.3
V	410	320	0.8	491	1.2
Cr	360	285	0.8	534	1.5
Mn	2750	3893	1.4	2776	1.0
Co	740	751	1.0	698	0.9
Ni	880	1230	1.4	867	1.0
Zr	944	940	1.0	922	1.0

(J. S. Becker and H. J. Dietze, *Spectrochim. Acta*, 53B, 1475 (1998). Reproduced by permission of Elsevier.)

The difficulties of different solid-state mass spectrometric techniques (GDMS, SSMS and SIMS) in the ultratrace analysis of hard to dissolve refractory materials (e.g., high purity molybdenum and tungsten) were demonstrated by Grasserbauer in 1988.[14] Whereas for Fe and Ni in molybdenum, the data measured by three different solid mass spectrometric techniques are in sufficient agreement, for some elements such as Ca, K, Mg, Si, As and Cr the measured concentrations vary by one order of magnitude and more. The quantitative multi-element bulk analysis of refractory metal molybdenum using SIMS with detection limits from $10^{-7}\, g\, g^{-1}$ to $10^{-12}\, g\, g^{-1}$ has been reported by Virag *et al.*[15] The direct element analysis of refractory metals, their carbides and oxides, has been performed by SSMS and GDMS in comparison to X-ray fluorescence analysis.[16] Disadvantages and limitations are observed for wet chemical techniques for the trace element analysis of refractory metals and their compounds due to the danger of contamination and if the sample is difficult to dissolve.

Today, as a direct solid-state analytical technique, dc GDMS is more frequently applied for multi-element determination of trace contaminants, mostly of high purity metallic bulk samples (or of alloys) especially for process control in industrial laboratories. The capability of GDMS in comparison to GD-OES (glow discharge optical emission spectrometry) is demonstrated in a round robin test for trace and ultratrace analysis on pure copper materials.[17] All mass spectrometric laboratories in this round robin test used the GDMS VG 9000 as the instrument, but for several elements (e.g., Ag and Fe) a huge variation of measured concentrations was found in different laboratories.[17] Since 2005 the GD mass spectrometer (Thermo Fisher Scientific) has been applied for multi-element trace analysis in conducting materials with detection limits in the $ng\,g^{-1}$ range and below as demonstrated, for instance, for Th determination at the ppt concentration level in high purity copper matrix.[18] Matschat *et al.*[19] employed GDMS (Element GD, Thermo Fisher Scientific) for multi-element trace analysis of ultrahigh purity copper and iron. A high dynamic range up to 12 orders of magnitude was obtained due to the combination of a secondary electron multiplier with a Faraday cup (see Section 4.3) in this double-focusing sector field instrument with extended dynamic range (Element GD). Thus the matrix element and trace impurities can be measured with a single analysis. Two series of powder samples were prepared for each of the copper and iron matrices. The powders were quantitatively doped with a solution of graduated and defined concentrations for 40 and 20 analytes, respectively. Stable pressed homogenized pellets doped by solutions of analytes with defined analyte concentrations were applied for the quantification procedure via an external calibration procedure. The comparison of certified values of different certified reference materials (CRMs) with the measured values based on calibration with the pressed powder samples led to deviations less than 30 % for most of the investigated analytes in the CRMs.

Wet chemical analytical sample preparation techniques combined with a sensitive mass spectrometry technique such as ICP-MS are employed for multi-element determination of trace element concentrations in metals and alloys after their dissolution. The easy calibration procedure is advantageous, especially if no matrix matched standard reference materials are available for quantification in solid-state mass spectrometry, and by trace/matrix separation the detection limits are improved. For example, an analytical method for trace element determination of rare earth elements and other impurities in high purity scandium (purity: 4N) by ICP-MS was proposed by Panday *et al.*[13] Bis(2-ethyl hexyl)-orthophosphoric acid in toluene was used to extract scandium selectively from an aqueous phase. The recovery for liquid – liquid extraction of most elements (except Zr (2 %), Er (83 %), Tm (55 %), Yb (43 %), Lu (32 %), Hf (8 %), Th (4 %) and U (1 %)) was quantitative. Nearly 30 elements at various concentrations were measured by quadrupole based ICP-MS (Elan 5000) via external calibration and isotope dilution techniques in comparison to neutron activation analysis and SSMS. Using the matrix removal and isotope dilution technique with ICP-MS, about 20 elements were quantified and the lowest detection limits, at 2–10 $ng\,g^{-1}$, were measured.[13] Results obtained for various elements after matrix removal and without removal of the scandium matrix (only dilution after dissolution of high purity scandium) yielded good agreement of analytical data for most elements. The relative standard deviations (RSDs) of trace element determination in high purity metals using the analytical techniques developed varied between 5–10 %. Furthermore, trace analysis of osmium metal in the form of a highly enriched stable ^{187}Os isotope formed by radioactive β-decay of ^{187}Re with a half-life of 4.23×10^{10} a in nature has been carried out by isotope dilution ICP-MS.[20] This resulted in the determination of 21 trace elements in the concentration range 1 $\mu g\,g^{-1}$ (Zr) to 181 $\mu g\,g^{-1}$ (Cr). ICP-OES and total reflection X-ray fluorescence analysis (TRFA) as non-mass spectrometric techniques were applied to validate the analytical data measured by ICP-MS.[20]

Trace and ultratrace impurities (Ti, Zn, Ga, Nb, Sn, Sb, Te, several lanthanides, Ta, Ir, Pt, Pb, Bi and U) in the $\mu g\,g^{-1}$ range and below in wet digested steel samples (with aqua regia in a microwave oven) have been determined by ICP-ToFMS.[21] For Ca determination in steel, an analytical procedure was introduced with microwave assisted digestion and matrix separation by flow injection ICP-MS to solve the interference problem ($^{12}C^{16}O_2^+$ and $^{28}Si^{16}O^+$ on analyte ion $^{44}Ca^+$) after treatment with H_2SO_4 and HF,[22] and a detection limit of 0.6 $\mu g\,g^{-1}$ was obtained. The determination of trace and ultratrace impurities in high purity (4N) copper samples, after digestion and dilution, by ICP-SFMS using continuous nebulization, flow injection (to reduce the deposition of Cu at the cones) and laser ablation for direct sample introduction, was validated with the results of electrothermal AAS and ICP-OES.[23] Reliable results were obtained by ICP-MS with limits of detection of trace elements at ng g^{-1} and below.

For a couple of years now, as a solid-state mass spectrometric technique, LA-ICP-MS is the method of choice for precise trace and ultratrace analysis of metals and alloys due to the possibility of easy calibration of the analytical data. LA-ICP-MS using solution based calibration has been applied for a direct trace analysis to study impurities in platinum nanoclusters for future application in nanotechnology.[24,25] In order to develop an appropriate analytical mass spectrometric technique to obtain reliable data, standard reference material has been employed. Table 9.6. compares the detection limits of trace element determination on high-purity platinum reference material (NIST SRM 681) in LA-ICP-QMS using solution-based calibration[24] and GDMS[26]. For selected trace elements such as the noble metals (Ag, Rh, Ir and Pd) lower detection limits were obtained in LA-ICP-MS using a quadrupole mass spectrometer compared to GDMS. By applying sector field ICP-MS, a significant improvement in detection limits was found as demonstrated for Ir and Pb in platinum matrix.

Table 9.6 *Detection limits of trace element determination on a high purity platinum sample (NIST SRM 681) in LA-ICP-QMS (Elan 6000) using solution based calibration[24] and GDMS[26].*

Element	**Detection limit (μg g^{-1})**	
	LA-ICP-MS	**GDMS**
Cu	0.04	0.009
Rh	0.003	0.006
Ag	0.002	0.01
Pd	0.003	0.005
Hf	0.006	n.d.
Ir	0.001 (0.000 04*)	0.03
Pb	0.003 (0.0009*)	1
U	0.003	n.d.
RSD (%)	5	5–10
Accuracy (%)	< 8	< 15
Analysis Time (h)	≈ 1	≈ 2

[a] LA-ICP-SFMS (ELEMENT, Thermo Fisher Scientific) n.d. – not determined.
(J. S. Becker and H. J. Dietze, *Int. J. Mass Spectrom.*, **228**, 127 (2003). Reproduced by permission of Elsevier.)

LA-ICP-MS has been utilized for the determination of the noble metal concentrations in high purity gold in the author's laboratory. The results of trace analysis on gold powder (Lot # 034535 from the High Purity Standards Company) are summarized in Table 9.7. The solution based calibration strategy in LA-ICP-MS was applied to quantify the measured ion intensities. Whereas only three element concentrations are given by the High Purity Standards Company, measured by arc emission spectrometry, LA-ICP-MS with multi-element capability allows a more precise and accurate analysis of noble metals in a gold sample.

An analytical method has been developed that allows easy standardless semiquantitative measurements of elemental composition from single shot analysis by LA-ICP-ToFMS to identify alloys of different elemental composition.[27,28] This method directly compares adjusted ion signals of elements with the total mass spectrometric ion signals to produce information on the elemental composition of alloy. A dynamic range of three orders of magnitude for element concentration from 0.1 to 100 % has been achieved for single laser shot measurements. This technique is largely 'immune' to matrix effects in laser ablation ICP-MS.

Table 9.7 *Determination of noble metals in high-purity gold (99.999 %) Lot # 034535 by LA-ICP-MS.*

Element	LA-ICP-MS[a]	Certified value[b] (High Purity Standards)
	Concentration ($\mu g\,g^{-1}$)	
Ru	0.34 ± 0.03	—
Rh	0.19 ± 0.02	—
Pd	1.2 ± 0.1	3
Ag	5.5 ± 0.6	6
Re	< 0.5	—
Ir	< 1	—
Pt	0.40 ± 0.03	1

[a] *LA-ICP-MS with solution calibration (MCN Aridus, CETAC).*
[b] *Arc emission spectroscopy.*

An interesting approach is the application of multiple ion collector mass spectrometry (MC-TIMS and MC-ICP-MS) for the determination of Cd and Tl in high purity zinc metal after trace matrix separation by the certification of reference materials from the Bureau Communie de Référence (BCR).[29] Accurate and precise element concentrations in high purity zinc metal have been obtained with both mass spectrometric techniques via precise isotope ratio measurements using the isotope dilution strategy the analytical data.

Trace impurities in noble metal nanoclusters, used for the fabrication of highly oriented arrays on crystalline bacterial surface layers on a substrate for future nanoelectronic applications, can influence the material properties.[25] Reliable and sensitive analytical methods are required for fast multi-element determination of trace contaminants in small amounts of high purity platinum or palladium nanoclusters, because the physical, electrical and chemical properties of nanoelectronic arrays (thin layered systems or bulk) can be influenced by impurities due to contamination during device production[25] The results of impurities in platinum or palladium nanoclusters measured directly by LA-ICP-MS are compared in Figure 9.5. As a quantification procedure, the isotope dilution technique in solution based calibration was developed as discussed in Chapter 6.

In the past, SSMS was employed for trace element analysis in high purity graphite. Especially difficult to determine elements can be measured without problems (at mass resolution $m/\Delta m \approx 10\,000$). In Table 9.8, the results of trace element determination on high purity reactor graphite measured by double-focusing sector field SSMS with Mattauch–Herzog geometry and LA-ICP-MS (using a quadrupole instrument) are compared.[30] In-house graphite laboratory standards doped with 15 analytes at defined concentrations were prepared for quantification purposes. The detection limits in LA-ICP-QMS, determined mostly in the sub-ppm range, can be improved by about one order of magnitude using a more sensitive double-focusing sector field mass spectrometer.

Further applications for the determination of trace impurities in high purity metals and alloys by SSMS, GDMS,[31,32] TIMS,[33,34] ICP-MS/LA-ICP-MS[11–13,20,23,30,35–43] [11]ETV-ICP-MS[44] are summarized in Table 9.9.

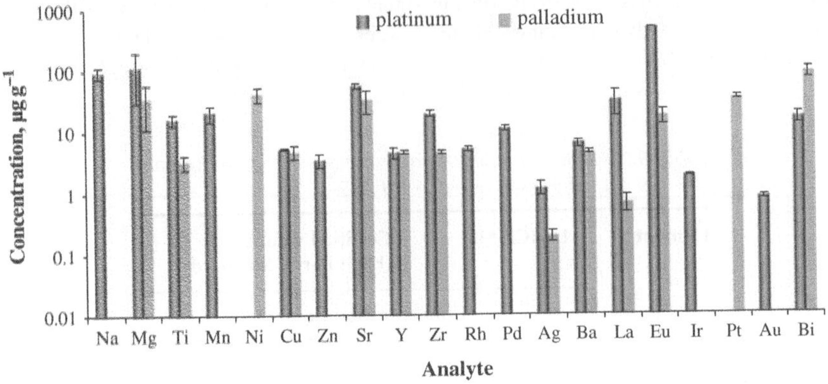

Figure 9.5 *Impurities in platinum and palladium nanoclusters measured by quadrupole-based ICP-MS and double-focusing sector field ICP-MS.*

Table 9.8 *Results of trace analysis in high purity reactor graphite measured by LA-ICP-MS and SSMS.*

Element	LA-ICP-MS concentration ($\mu g\, g^{-1}$)	SSMS[a]
Li	0.7± 0.2	< 1
Mg	< 0.2	< 1.3
Ti	< 0.12	0.20 ± 0.04
Cr	< 2.5	0.13 ± 0.03
Co	< 1	0.10 ± 0.02
Mo	< 0.3	0.04 ± 0.01
Ba	2.2± 0.6	4.6 ± 0.9
Ce	0.3± 0.1	0.41 ± 0.05
Nd	< 0.05	< 0.03
Sm	< 0.03	< 0.02
U	< 0.01	< 0.008

[a] SSMS – spark source mass spectrometer: Mattauch–Herzog geometry ($m/\Delta m \sim 10\,000$).

Table 9.9 *Selected applications of inorganic mass spectrometry in trace analysis of high purity conducting materials.*

Samples	Equipment	Analyzed elements	Limits of detection	References
Tungsten Molybdenum	SIMS, SSMS, GDMS	20–30 elements	low $ng\, g^{-1}$ range	Grasserbauer[14]
Aluminium Gallium	GDMS SSMS	17 elements	0.0003–0.008 $\mu g\, g^{-1}$	Mykytiuk[31]
Aluminium based alloys	GDMS	33 elements	0.2 $ng\, g^{-1}$ (U, Th)	Venzago and Weigert[32]

Platinum powder	GDMS (VG 9000)	Mg, Al, Ca, Fe, Cu, Zn, Ru, Rh, Pd, Ag,Te, Ir, Pb, Au	3–30 ng g^{-1}	von Straaten *et al.*[26]
Titanum	TIMS (THQ) isotope dilution	Cu, Pb, Cd, Ni, Fe U, Th	0.07 ng g^{-1} (U) –4 ng g^{-1} (Ni)	Beer and Heumann[33]
Cobalt	TIMS (THQ) isotope dilution	Fe, Zn, Tl, Cd, Th, U	0.007 ng g^{-1} (U) –1 ng g^{-1} (Tl)	Beer and Heumann[34]
Antimony	ICP-QMS (Plasma Quad3)	Se, Pd, Ag, In, Ta, W, Pt, Au, Hg, Tl	0.3 ng g^{-1} (Cd) 3.9 ng g^{-1} (Pb)	Balarama Krishna *et al.*[35]
Au, Re Pd, Pt	LA-ICP-QMS (Elan 6000)	56 elements, solution based calibration	0.6 ng g^{-1} (Ho) 1.2 µg g^{-1} (Se)	Boulyga *et al.*[36]
Scandium	SSMS, ICP-QMS (Elan 5000)	43 elements ICP-MS	SSMS: 0.05 µg g^{-1} (Tb); ICP-QMS: 0.002 µg g^{-1} (Tb)	Panday *et al.*[13]
Gold	ICP-QMS (Elan 6000)	As (flow injection, hydride generation)	55 ng g^{-1}	Becotte-Haigh *et al.*[37]
Indium	ICP-QMS (Plasma Quad)	38 elements	0.03 µg g^{-1} (U)	Grazhulene *et al.*[38]
Chromium	ICP-QMS (Elan 6100 DRC)	Nb, Zr	2 ng g^{-1} (Nb) 5 ng g^{-1} (Zr)	Hattendorf *et al.*[39]
Tantalum	ICP-QMS (Seiko SPQ 9000)	B P	90 ng g^{-1} 40 ng g^{-1}	Kozono *et al.*[40] Kozono *et al.*[41]
Tantalum	ICP-QMS (Elan 5000)	25 elements	10 ng g^{-1} (Ce)	Panday *et al.*[12]
Copper	ICP-SFMS (Element)	9 elements, 49 LODs	0.1 ng g^{-1} (Ir) –6 µg g^{-1} (B)	Pattberg and Matschat[23]
Plutonium Beryllium	ICP-QMS (VG Plasma QuadPQ2) SSMS "MS 702"	52 elements	for Pu matrix: 4 ng g^{-1} (Cu) 20 ng g^{-1} (Ca)	Pilon *et al.*[42]
Platinum	LA-ICP-QMS (Elan 6000)	Co, Ni, Cu, Zr, Ru, Rh, U Ag, Pd, In, Sn, Hf, Ir, Pb	1 ng g^{-1} (Ir) –200 ng g^{-1} (Sn)	Becker *et al.*[24]
Graphite	LA-ICP-QMS (Elan 6000)	Li, Sc, Ti, Cr, Mn, Co, Zn, Sr Mo, Sn, Ba, Ce, Hf, Pt, Au	5 ng g^{-1} (Yb, Pr) –50 ng g^{-1} (Ce)	Pickhardt and Becker[30]
Osmium	ICP-QMS (Elan 5000)	Na, Mg, Al, Ti, Cr, Ni, Zn, Sr, Zr, B, Mn, Fe, Co, Cu	ng g^{-1} range	Becker and Dietze[20]
Silver alloy	LA-ICP-MS ICP-MS	Zn, Cd, Sn, Sb, Au, Pb, Bi	sub-µg g^{-1}–2 µg g^{-1}	Devos *et al.*[43]
Zircaloy	ICP-QMS (Elan 5000)	34 elements isotope dilution	0.05 µg g^{-1} (Rb, Sb)	Panday *et al.*[11]
Ni alloy	ETV-ICP-QMS	Pb, Bi, Te	4 pg g^{-1}	Hinds *et al.*[44]

9.1.2 Semiconductors

For the production of semiconductor devices, extreme purity of all materials inserted and applied during handling is required. The smallest impurities on semiconductor surfaces, interfaces or in bulk can result in disturbances in function of microelectronic chips. The electrical properties of semiconductors such as silicon, gallium arsenide, germanium, indium antimonide and others are dependent on the trace and ultratrace element concentrations. Besides metal and non-metal impurities the concentration of dopants (e.g., as a function of depth) also has to be determined. With increasing complexity of systems and miniaturization of integrated circuits, requirements have been tightened for powerful and sensitive trace impurity analytical techniques. Besides non-mass spectrometric techniques several mass spectrometric techniques have also been utilized for the analysis of high purity solid semiconducting materials. The rapid development of microelectronics technology has brought with it improvements in the analytical investigations of semiconductor materials by inorganic mass spectrometry, from SSMS through LIMS to GDMS and LA-ICP-MS as direct mass spectrometric techniques and ICP-MS after dissolution of semiconducting samples for trace impurity analysis and for determining the concentration of dopants. Reviews on the development and application of LA-ICP-MS, ICP-MS, LIMS and GDMS in materials research and for trace element analysis in semiconductors are presented in references.[1,9,45]

As an old mass spectrometric technique, SSMS mostly had an antiquated ion detection system using an ion sensitive photographic plate and was utilized for the ultratrace analysis of high purity graphite or GaAs.[30,46,47] One advantage of SSMS (and also LIMS) with an ultra-high vacuum ion source (compared to LA-ICP-MS), is the ability to measure light, difficult to determine non-metals such as C, B, N and O with detection limits down to the low $ng\,g^{-1}$ range.[47–49] In addition, silicon concentration in silicon doped GaAs single crystals for optical devices prepared by the vertical gradient freeze (VGF) technique was measured by SSMS between 5.5×10^{17} and 3.1×10^{18} atoms cm^{-3}.[49] One serious problem of SSMS is the quantification of analytical data if no matrix matched standard reference material exists. Consequently, no relative sensitivity coefficients of elements (RSCs) can be determined; that means that for all elements the RSCs are assumed to be equal to one and therefore only semiquantitative data in SSMS are obtained. For several application fields such as the characterization of so-called 'impurity profiles' in microelectronic materials semiquantitative data with additional information of limits of detection of elements are sufficient enough.

Of all the mass spectrometric techniques, SIMS is most frequently used for the characterization of semiconducting materials due to the excellent detection limits of trace impurities in bulk analysis, determination of doping elements and the capability to perform depth profiling and microlocal analysis. The trend in SIMS measurements is to analyze smaller and smaller sample sizes and thinly layered systems on microelectronic devices. For example, SIMS has been utilized for the quantitative determination of Ge in a SiGe quantum well ($Si_{1-x}Ge_x$ layers and interfaces), and the results compared to low energy RBS.[50] The thickness of the analyzed quantum well was about 12 nm and it was situated at a depth of about 60 nm below the surface. The SIMS measurements were performed using an oxygen primary ion beam in an area of $350\,\mu m \times 350\,\mu m$. The results obtained using two analytical techniques agree well.

Trace impurities including B, N, F, Mg, P, Cl, Cr, Fe, Ni, Co, Cu, Zn, Ge, As, Se, Mo, Sn and Sb in semiconductor materials (Si, SiGe, $CoSi_2$, GaAs and GaN) can also be measured by TEAMS (trace element accelerator mass spectrometry), which utilizes a secondary ion source in an accelerator mass spectrometer.[51] TEAMS is an interesting complementary technique to SIMS and can eliminate interferences that may not be resolvable in SIMS.

Radiofrequency glow discharge mass spectrometry (rf GDMS) for characterization of semiconductors has been applied for the bulk analysis of trace elements in semiconducting silicon carbide and gallium arsenide in the author's laboratory.[52,53] An interesting finding is that single positively charged ions of sample material formed in the glow discharge have about 10 eV higher average energy than the ions of the discharge and residual gas ions.[54,55] Therefore an effective separation of analyte ions from discharge and residual gas ions due to the energy difference was achieved by adjusting the ion transfer optics (width and position of energy slit) in a double-focusing sector field mass spectrometer with Mattauch–Herzog geometry.[56] Limits of detection (LOD) in the sub-$\mu g\,g^{-1}$ range were obtained for Li, Be, B, Na, Mg, Al, P, Cl, Sc, V, Cr, Mn and Co. For all other elements the LODs achieved were between $1\,\mu g\,g^{-1}$ and $100\,\mu g\,g^{-1}$ (e.g., Si, S, Fe, Zr, Hf, Ta, W, Re, Os, Ir, Pt, Au, Hg Tl, Pb, Bi and REE).[53] The analytical performance of rf glow discharge mass spectrometry was also investigated for the trace analysis of semi-insulating gallium arsenide crystals using a double-focusing sector field mass spectrometer with reverse Nier–Johnson geometry.[52] For quantification purposes a synthetic GaAs laboratory standard crystal with a defined element concentration of dopants (Zn, B, Si, Sn, S, Se and Te) in the $\mu g\,g^{-1}$ range was prepared by the liquid encapsulation vertical Bridgman technique. A selected piece of synthetic laboratory standard was analyzed directly by SIMS, SSMS, GDMS and LA-ICP-MS. For the quantification of SIMS measurements, single element ion implantation GaAs certified reference standards were utilized. After dissolution of the GaAs sample in high purity HNO_3–H_2O_2 and dilution, the concentration of doped elements in the laboratory standard were measured by ICP-OES and ICP-MS. By using the results (average concentration) of SIMS, ICP-MS and ICP-OES for the selected piece of synthetic GaAs, laboratory standard relative sensitivity factors (RSC) were measured for the solid-state mass spectrometric techniques SSMS, GDMS and LA-ICP-MS.[46] The experimentally determined RSCs were employed to correct the measured element concentrations in an unknown GaAs sample. In addition, in order to improve the detection limits in ICP-MS, a trace matrix separation by selective volatilization of GaAs matrix by converting the matrix elements into their chlorides in a stream of premixed argon and chlorine $(5+1)$ at relative low temperature ($\sim 280\,°C$) was applied.[52,57] Owing to the high volatilities of the chlorides of Ga and As, it is possible to separate the matrix elements, leaving behind those impurities which do not readily form volatile chlorides. This selective volatilization of matrix elements allows a multi-element trace analysis free from interferences with detection limits in the low $ng\,g^{-1}$ range, which are about one order of magnitude better than without matrix separation. Furthermore, a better quantification of analytical data is possible for the multi-element determination of trace impurities in high purity GaAs (99.9995 %) by ICP-MS after matrix separation, compared to GDMS in the low ppb-at concentration range. In Table 9.10 the results of the determination of trace elements in high purity GaAs by direct dc GDMS measurement and ICP-MS after matrix separation are compared. Relatively worse detection limits in ICP-MS (e.g., for Al) could be explained by possible contamination during the time-consuming sample preparation described in reference.[46]

Similar to the analytical procedure for trace analysis in high purity GaAs wafers after matrix separation, discussed previously,[52] the volatilization of Ga and As has been performed via their chlorides in a stream of aqua regia vapours (at $210\,°C$) using nitrogen as the carrier gas for trace/matrix separation.[58] The recoveries of Cr, Mn, Fe, Ni, Co, Cu, Zn, Ag, Cd, Ba and Pb determined after a nearly quantitative volatilization of matrix elements (99.8 %) were found to be between 94 and 101 % (except for Ag and Cr with 80 %). The concentrations of impurities measured by ICP-DRC-MS (Elan 6100 DRC, PerkinElmer Sciex) after matrix separation were compared with ICP-SFMS (Element 2, Thermo Fisher Scientific) and total reflection X-ray fluorescence analysis (TXRF Phillips). The limits of detection obtained for trace elements in GaAs were in the low $ng\,g^{-1}$ range and below.[58]

An efficient analytical procedure for the trace impurity analysis of Mg, K, Cr, Mn, Fe, Ni, Co, Cu, Zn, Ag, Cd, Ba and Pb in high purity silicon powder after a very simple matrix separation via microwave assisted volatilization of silicon as silicon fluorides, was proposed by Ueng *et al.*[59] The authors observed a nearly 99 % volatilization of the matrix element silicon from the

Table 9.10 *Results of trace analysis on high purity GaAs measured by GDMS and ICP-MS (concentration in ppb-at)[46].*

Element	GDMS	ICP-MS
Mg	< 0.5	< 0.6
Al	(6)	< 9
Ca	< 5	< 3
V	< 0.2	< 2
Cr	3	< 5
Mn	< 0.3	< 0.5
Fe	< 0.2	< 1
Ni	< 0.7	< 1
Zn	< 1	< 2
Sn	< 1	< 0.3

(J. S. Becker and H. J. Dietze, *Int. J. Mass Spectrom. Ion Proc.* **197**, 1 (2000). Reproduced by permission from Elsevier.)

powdered material using a HF/HNO_3 (5:1) vapour produced *in situ* in the microwave vessel of a microwave digester (at 600 W and a reaction time of 40 min). The recoveries of trace elements analyzed by ICP-DRC-MS (Elan DRC, PerkinElmer Sciex) in comparison to double-focusing sector field ICP-MS (Element 2, Thermo Fisher Scientific) varied in the range 83–103 % for silicon powder. To reduce the interference problem on $^{39}K^+$, $^{52}Cr^+$ and $^{56}Fe^+$ in ICP-MS with a collision/reaction cell, the DRC mode using NH_3 as the reaction gas was applied. Detection limits for metal impurities were determined for most elements in the sub-ng g^{-1} range.[59] The same working group proposed closed vessel microwave assisted volatilization of Ge and Se matrix via their chlorides in the presence of a vapour of aqua regia in the microwave vessel and determined 12 trace elements by ICP-QMS in analogy to silicon trace analysis with limits of detection in the low ng g^{-1} range.[60]

Recently, ICP-QMS (Elan 6000) was employed for determination of trace impurities of Cu, Zn, Na, Si, Cr, Mo, Rb, Pb, Ca, Ti, Fe and Cd in the caesium iodide single crystals in the author's laboratory. After cleaning of surfaces, a small amount of high purity CsI single crystal (100 mg) was dissolved with high purity water (Milli-Q Plus), the solution was acidified with sub-boiled HNO_3 and analyzed without any further additional sample preparation steps. As a result, in two different samples, impurities of Zn, Cu, Pb and Rb at low and sub-μg g^{-1} ranges were found. The analytical data were quantified via an external calibration procedure. Limits of detection in ICP-QMS were obtained in the ng g^{-1} range. Additional depth profile measurements by LA-ICP-MS on CsI single crystals directly resulted in different distributions of Cu, Zn and Pb as a function of depth for two single crystals from different producers. In both crystals an inhomogeneous distribution of Zn was found. Whereas for one crystal an enrichment of Pb on the sample surface was observed, in the other crystal a depletion was found. Pb, but also In, Sn, Tl, Eu and other elements, was employed as the doping element in CsI crystals to improve the luminescence properties of CsI single crystals. Quantification of analytical data measured by LA-ICP-MS can be performed by solution based calibration (e.g., by standard addition technique to consider matrix effects and using Cs as internal standard element measured using doubly charged ions). In spite of easier quantification procedures in ICP-MS, direct solid-state mass spectrometric techniques possess relevant advantages compared to ICP-MS with respect to sample preparation (no, or only a few, sample preparation steps are required) and consequently the contamination danger is significantly lower if solid specimens are analyzed without chemical sample preparation. With respect to the time consuming sample preparation steps for ICP-MS measurements (especially if a suitable trace matrix separation procedure is utilized), solid-state mass spectrometric measurements are preferred. A comparison of the strengths and limitations of ICP-MS for the analysis of dissolved solid samples versus LA-ICP-MS as an established solid-state mass spectrometric technique in bulk analysis is given in Table 9.11.

Table 9.11 *Comparison of ICP-MS versus LA-ICP-MS for trace (bulk) analysis of high purity solid samples.*

	ICP-MS	**LA-ICP-MS**
Sample preparation	expensive if matrix separation required	small
Quantification possibility	excellent	SRM required, solution based calibration possible
Precision	isotope dilution: 1–5 %	for homogeneous samples: mostly 5–10 %
Accuracy	good	good, if SRM available
Contamination danger	enormous	low
Inhomogeneous sample	no problems	difficult
Time consuming step	sample preparation	quantification procedure

(J. S. Becker, *Spectrochim. Acta*, 57B, 1805 (2002). Reproduced by permission of Elsevier.)

Selected applications of inorganic mass spectrometry for the multi-element trace analysis of semiconducting materials are summarized in Table 9.12.

Table 9.12 *Selected application of inorganic mass spectrometry in trace analysis of high purity semi-conducting materials.*

Samples	**Equipment**	**Analyzed elements**	**Limits of detection (measured conc.)**	**References**
Gallium arsenide	GDMS SSMS	23 elements	GDMS: 0.4 $ng\,g^{-1}$ (Mn) SSMS: 0.8 $ng\,g^{-1}$ (Mg)	Mykytiuk *et al.*[31]
Gallium arsenide	SIMS, SSMS, rf GDMS, ICP-MS matrix separation	Zn, B, Si, Ge, Sn, Sb P, S, Se, Te	ICP-MS: low $ng\,g^{-1}$ rf GDMS: 10 $\mu g\,g^{-1}$ (Sn)	Becker *et al.*[46]
Gallium arsenide	ICP-DRC-MS ICP-SFMS, TXRF matrix separation	Cr, Mn, Fe, Ni, Co, Cu, Zn, Ag, Cd, Ba, Pb	sub-$ng\,g^{-1}$ 0.8 $ng\,g^{-1}$ (Cr, Ba) 0.03 $ng\,g^{-1}$ (Co)	Sahayam *et al.*[58]
Silicon (SiO_2)	ICP-IDMS, INAA RBS	implanted Sb	(areal density Sb 4.8×10^{16} atoms cm^{-2})	Pritzkow[146]
Silicon powder	ICP-DRC-MS matrix separation	Mg, K, Cr, Fe, Ni, Co, Cu, Zn, Ag, Cd, Ba, Pb	low–sub-$ng\,g^{-1}$ 0.1 $ng\,g^{-1}$ (Co) 12 $ng\,g^{-1}$ (Fe)	Ueng[59]
Germanium Selenium	ICP-DRC-MS matrix separation	Mg, Cr, Mn, Fe, Ni, Co, Cu, Zn, Ag, Cd, Ba, Pb	low $ng\,g^{-1}$ Ge matrix: 0.4 $ng\,g^{-1}$ (Cd)	Ueng[60]

9.1.3 Ceramics, Glasses, Polymers and Other Non-conductors

Ceramics are widely used as materials largely resistant to high temperature, oxidation and corrosion processes in the engineering and electronics industry. Among such ceramics are inorganic sintered compounds that include nitrides, carbides, borides, silicates and oxides. Trace impurities are of special interest in ceramic materials utilized in the microelectronics industry as substrate material for thin film preparation or as crucible materials for the growing of single crystals (e.g., GaAs growing in a p-BN crucible by the liquid encapsulated Czochalski process). Trace element analysis in ceramics, glasses and others makes great demands on the analysis procedure owing to the physical and chemical properties involved. Analytical techniques that do not require any sample preparation are advantageous for difficult to digest ceramics and glasses.

As a sensitive analytical technique for the determination of trace impurities, LA-ICP-MS is well suited for the analysis of glasses[61] or ceramics because no conducting sample material is required. For example, on thin alkali-resistant glass filaments, used as reinforcing material in the construction industry, trace analysis has been carried out directly on very thin solid strands (without any sample preparation steps) by LA-ICP-MS by fixing a bundle of thin glass fibres (with a filament diameter of about 10–20 μm) onto the special thin tape of a target holder.[62] The fibres were ablated in the ablation chamber using a Nd-YAG laser at a wavelength of 266 nm, pulse duration of 5 ns, repetition frequency of 20 Hz, energy per pulse of 5 mJ and laser power density of 1.4×10^8 W cm^{-2}. The multi-element trace analytical method – developed for high-purity glass fibres – was applied to determine the chemical composition of thin alkali resistant glass and basalt fibres with finishing additives used in fine concrete for the building industry. The analytical results were quantified using standard reference materials (SRMs) of glass matrix such as the NIST 612 glass SRM and geological reference glasses (KL-2G and ML3B-G). The results of the determination of major, minor and trace element concentrations in glass fibres measured by LA-ICP-MS using quadrupole based and sector field ICP-MS are in good agreement, as shown in Fig. 9.6. By optimizing the analytical techniques it was possible to achieve detection limits in the sub-ng g^{-1} range.[62] The developed LA-ICP-MS method can be applied for quality control of trace impurities in high purity optical glass fibres. Such high purity glass fibres with a covering layer to reduce signal loss are applied for transmitting information via light pulses. The application of a laser with a shorter wavelength would be advantageous for such an analytical task (e.g., Nd-YAG laser at 213 nm) for improved ablation of transparent material. The advantage of LA-ICP-MS as a fast and direct analytical technique is that no time-consuming sample preparation is required. Problems could occur in the analysis of trace impurities in transparent samples. To optimize the analytical procedure for determining trace impurities in transparent quartz glasses, studies of LA-ICP-MS at different wavelengths of the Nd-YAG laser have been performed in the author's laboratory.[61] At the start of the interaction of the sample surface with the laser radiation at a wavelength of 1064 nm, the high purity SiO_2 quartz glass is almost completely transparent. However, the high energy radiation of the laser beam on the quartz surface gives rise to defects in the glass surface (change in structure of the glass matrix) whereby these defects act as absorption centres and thus encourage sample ablation. Calibration of LA-ICP-MS measurements was carried out via calibration curves (external calibration) recorded by the analysis of certified NIST quartz glass reference materials for trace analysis. The detection limits of elements are in the low ng g^{-1} range and below.[61] Furthermore, a benefit of LA-ICP-MS compared to other solid-state mass spectrometric techniques is the simple calibration procedure using solution based calibration, even if no standard reference materials are available, as discussed in Chapter 6. An innovative approach for quantitative trace element analysis by online isotope dilution analysis (IDA) in LA-ICP-MS, using a microflow nebulizer inserted in the laser ablation chamber, was utilized for uranium determination in NIST glass SRM 612.[63] With the online isotope dilution technique in LA-ICP-MS[25,63] three measurement steps are performed: (a) laser ablation of the sample and nebulization of 2 % HNO_3 to determine the natural isotope ratio of analyte, (b) nebulization of the enriched isotope tracer and laser ablation of the sample to measure the isotope ratio in the mixture, and (c) nebulization of the enriched isotope tracer only, to measure its isotope ratio. Following the formula of the isotope dilution technique, element concentrations can be determined using an internal standard element. The application of online LA-ID-ICP-MS yielded a concentration of $36.2 \pm 1.1\,\mu g\,g^{-1}$ for uranium in NIST glass SRM 612, which agrees with the certified value.[63]

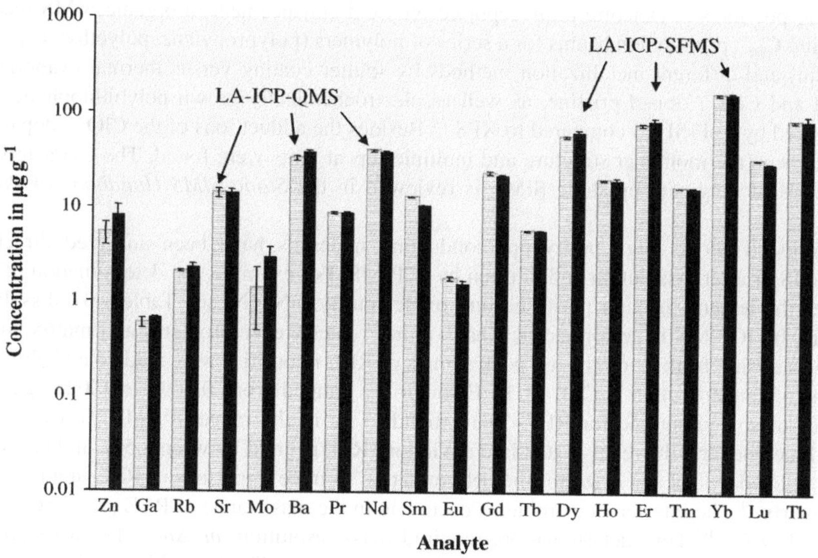

Figure 9.6 *Element concentrations in glass fibres measured by LA-ICP-MS (left column: Elan 6000, Perkin Elmer SCIEX) in comparison to LA-ICP-SFMS (right column: Element, Thermo Fisher Scientific). (J. Su. Becker, C. Pickhardt, N. Hoffmann, H. Hocker and J. S. Becker, At. Spectr.,* ***23****, 1 (2002). Reproduced by permission of the Perkin-Elmer Corporation.)*

LA-ICP-MS using lasers at different wavelengths (266 nm and 193 nm)[64] has been applied for the analysis of trace impurities of synthetic high purity CaF_2 crystals. Especially the determination of long-lived radionuclides in microelectronically relevant high purity materials is attaining increasing importance because ultratraces of α-emitting natural long-lived radionuclides (U and Th) disturb the electronic properties of electronic devices. The analysis of uranium and thorium traces in (Ba, Sr)TiO_3 ferroelectric materials by ICP-MS has been reported by Fukuda and Sayama.[65]

The direct determination of eight trace impurities (Na, Mg, Ca, V, Mn, Fe, Ni and Ga) in Al_2O_3 ceramic based powders by ICP-MS using electrothermal evaporation (ETV) with slurry sampling has been studied by Wende and Broekaert.[66] The authors investigated the capability of several palladium group modifiers. Optimum results were obtained with the $PdCl_2$ modifier when 500 ng Pd was added to 100 μg sample weight of Al_2O_3 injected into the ETV. Detection limits achieved by ETV-ICP-MS varied between 70 ng g^{-1} and 1 μg g^{-1}, and the RSDs range from 5–10 %. Certified NIST SRM 699 was analyzed in order to validate the analytical method developed.

Other studies concerned trace element analysis and investigation of trace impurities and heterogeneities in polymers.[67,68] Mass spectrometric techniques for polymer characterization have been required to monitor the different steps of the manufacturing processes and to determine the concentration of impurities. LA-ICP-MS for direct analysis is a challenging technique for heterogeneity studies and the determination of trace impurities in polymers.[67,68] In addition, the capability to generate chemical information on the organic and inorganic surface components makes SIMS extremely well suited for controlling the doping steps which are crucial in the preparation of conducting polymers.[69] Delcorte *et al.* reported on yield enhancement in organic SIMS measurements using C_{60}^+ primary ion beams for a series of polymers (polypropylene, polyethylene, leucine enkaphalin) and different metallization methods by sputter coating versus thermal evaporation.[70] Undoped and ClO_4^- doped pristine, as well as electrochemically grown polybithiophene (PBT) was analyzed by ToF-SIMS compared to XPS.[71] Besides the adduct ions of the ClO_4^- dopant, low fragments from the monomer structure and multiple repeat units were found. The characterization and analysis of polymers by static SIMS is reviewed in the *Static SIMS Handbook of Polymer Analysis*.[72]

As discussed above, high purity non-conducting materials have been analyzed directly by LA-ICP-MS or after dissolution and dilution by ICP-MS. For example, the determination of traces of rare earth elements in high purity erbium oxide (purity: 5N–6N, see Table 9.3), dissolved in nitric acid, by ICP-MS is performed by Zhank *et al.*[73] Interference problems and matrix suppression effects arising from Er on the ion peak signals of REE impurities were studied and eliminated. The limit of quantification (LOQ) of REE impurities ranged from 0.0090 to 0.025 μg g^{-1}, the recoveries of spiked sample for REEs were found to be in the range 90.3–107 % by using the proposed method, and the relative standard deviation (RSD) varied between 2.5 % and 6.7 %. ICP-MS was utilized by Takaku *et al.* for the determination of trace impurities of REE in a high purity Gd_2O_3 matrix.[74] The interference problems occurring in the mass range of REE (e.g. $^{152}Gd^{16}O^{1}H^+$ at $m/z = 169$ for $^{169}Tm^+$ determination, required mass resolution, $m/\Delta m = 14500$) were overcome by using doubly charged ions (REE^{2+}) as analyte ions. The intensities of doubly charged atomic ions compared to the singly charged ones (REE^{2+}/REE^+) varied between 0.25 % for Lu and 11.7 % for Ce. The limit of detection of REE using doubly charged atomic ions was found to be in the sub-pg g^{-1} range.

In the past, non-conducting samples (e.g., ceramics, glasses and other non-conductors) have been analyzed by SSMS and GDMS after powdering, mixing with a conducting powder (e.g. high purity graphite or silver), homogenizing and pressing of two electrodes or one pin or flat target for dc GDMS. The results of trace element analysis on boron nitride using SSMS are compared to those of LIMS in reference.[75] SSMS for non-conducting single crystals (e.g., CaF_2) by so-called gliding SSMS (GSSMS) using a special experimental arrangements with a tantalum counter electrode has been developed by Saprykin *et al.*[76] Due to the direct analysis of small single crystals, contamination that may occur in ICP-MS during dissolution and matrix separation can be avoided. SSMS as one of the oldest mass spectrometric techniques dominated the trace analysis of high purity ceramic and geological samples about 20–30 years ago. However, in recent years SSMS has been replaced by GDMS and LA-ICP-MS for direct analysis of trace impurities on solid samples.

GDMS has become recognized as one of the most powerful techniques for trace, surface and isotope analysis on solids.[77,78] Using the commercial devices dc GDMS VG 9000 (VG Elemental) or GDMS (from Thermo Fisher Scientific), this method is restricted to the determination of trace elements in electrically conducting or semiconducting materials with detection limits in the $ng\,g^{-1}$ range and below. The main drawback of direct current (dc) GDMS is, however, analyzing non-conducting samples.[79] Besides mixing non-conducting powdered samples with a high purity metal or high purity graphite powder, the implementation of a secondary cathode technique in dc GDMS has been developed to investigate non-conducting refractory insulators. Using dc GDMS with a secondary cathode on the sample surface to make the sample surface electrically conducting, it is possible to determine 57 trace elements in ZrO_2.[79] The detection limits of trace elements in the ZrO_2 matrix are about $100\,ng\,g^{-1}$. The reproducibility is found to be better than 10 % RSD, and the accuracy of the 'standardless' raw results is within a factor of 2–3 of the known concentration. The quantification of analytical data is performed by one point calibration using relative sensitivity coefficients (RSCs).

The more elegant possibility of analyzing insulating materials directly whereby no charging up effects occurring on the sample surface, is the radiofrequency (rf) powered GDMS developed by Marcus[78,80] and in the author's laboratory.[55–53,81] Recent trends in glow discharge spectroscopy are discussed by Jakubowski et al.[82] In our laboratory a home-made rf glow discharge ion source was coupled with a double-focusing mass spectrometer with Mattauch–Herzog geometry (in the original mode used with a low arc discharge ion source for trace analysis of metals and alloys) and with a double-focusing sector field mass spectrometer with reverse Nier–Johnson geometry (originally with an ICP ion source). The characterization and optimization of an rf glow discharge ion source interfaced directly to the Mattauch-Herzog instrument SM 1B without modification of the ion extraction and acceleration device of this double-focusing sector field mass spectrometer is described in reference.[55] The determination of trace elements by rf GDMS compared to other analytical techniques was performed, for example, in silicon carbide[53] and $La_{0.65}Sr_{0.3}MnO_3$[83] ceramics and thick layered non-conducting systems.[84] In general, higher detection limits (in the $\mu g\,g^{-1}$ range) in rf GDMS compared to dc GDMS were observed. A comparison of the main characteristics of GDMS modifications for different applications is given in Table 9.13. The direct trace element analysis of glasses by magnetically enhanced rf GDMS with a commercial double-focusing mass spectrometer (originally with an ICP source) by utilizing an additional ring-shaped magnet located

Table 9.13 *Main characteristics of GDMS modifications for different applications.*

Parameters \ Ion source type	dc GDMS	rf GDMS
Gas (Ar) pressure, Pa	500–50	100–10
Power, W	1–5	10–50
Direct current bias potential, V	800–1200	500–800
Sputtering rate, mg min^{-1}	0.05–0.5	0.1–1
Total ion current, A	10^{-11}–10^{-9}	10^{-10}–10^{-8}
Detection limits:		
Conductors	0.1–100 $ng\,g^{-1}$	1–100 $ng\,g^{-1}$
Insulators	(using secondary cathode)	1–10 $\mu g\,g^{-1}$
Quantification	using SRMs and lab standards	using SRMs and lab standards
Applications	high purity metals, depth profiling	non-conducting materials

Table 9.14 Selected applications of inorganic mass spectrometry in trace analysis of high purity non-conducting materials.

Samples	Equipment	Analyzed Elements	Limits of detection	References
ZrO_2	GDMS (VG 9000)	57 elements	0.0006–185 $\mu g\,g^{-1}$	Schelles and van Grieken[79]
BN, BC_4	LIMS	F, Na, Mg, Al, P, S, Cl, K, Ca	low $ng\,g^{-1}$ range	Becker and Dietze[75]
Si_3N_4, WC	SSMS	Ti, Fe, Co, Zn, Cu		
Silicon carbide	rf GDMS LIMS	B, Al, Ti, V, Cr, Cu, Zr	rf GDMS and LIMS: 10 $\mu g\,g^{-1}$ (Cr)	Jaeger *et al.*[53]
Al_2O_3	ETV-ICP-QMS (Elan 5000)	Ca, Fe, Ga, Mg, Mn, Na, Ni, V	0.07 $\mu g\,g^{-1}$ (Ga) –1.1 $\mu g\,g^{-1}$ (Na)	Wende and Broekaert[66]
Glass fibres	LA-ICP-SFMS Element	Sb, Cs, REE, Hf, Ta, Pb, Th, U	0.03 $ng\,g^{-1}$ (U, Th)	Becker *et al.*[62]
SiO_2	LA-ICP-QMS (Elan 6000)	Li, B, Na, Mg, Al, K, Ca, Ti, V, Cr, Fe, Mn, Cu, Zr, Ba, Pb	0.02 $ng\,g^{-1}$ (Cu) –20 $ng\,g^{-1}$ (Al)	Becker and Tenzler[61]
TiO_2	TIMS (THQ) IDA*	Cu, Pb, Tl, Cd, Ni, Cr, Fe, U after matrix separation	1.8 $ng\,g^{-1}$ (Tl) 2.5 $\mu g\,g^{-1}$ (Fe)	Beer and Heumann[86]
SiC, ZrO_2, Al_2O_3 matrix separation	ICP-QMS (VG Plasma Quad PQ2)	Li, B, Na, Mg, Al, Ca, Sc, Ti, V, Cr, Mn, Fe, Ni, Co, Cu, Zn, Ga, Sr, Y, Zr, Nb	5 $ng\,g^{-1}$ (Co) 4.7 $\mu g\,g^{-1}$ (Na)	Kohl *et al.*[88]
Al_2O_3	ICP-SFMS (Element)	V, Cr, Mn, Fe, Ga, Co, Ni, Cu, Zn	0.4 $ng\,g^{-1}$–1400 $ng\,g^{-1}$	Jakubowski *et al.*[87]
La_2O_3	HPLC-ICP-MS	Ce, Pr, Nd, Sm	0.11 $\mu g\,g^{-1}$ (Sm) 0.7 $\mu g\,g^{-1}$ (Ce)	Quin .[89] *et al*
$La_{0.65}Sr_{0.3}MnO_3$	LA-ICP-MS, ICP-MS, SSMS	20 elements	low $\mu g\,g^{-1}$ range and below	Becker *et al.*[54]
$La_{0.65}Sr_{0.3}MnO_3$	rf GDMS	B, Mg, Al, Ni, Co, Zn	10 $\mu g\,g^{-1}$	Jäger *et al.*[83]
$La_{0.65}Sr_{0.3}MnO_3$ layers Ni-cermet, YSZ	LA-ICP-MS, SSMS, ICP-MS (Elan 5000) rf GDMS, SIMS	B, Mg, Al, P, Si, Ti, Cr, Co, Ni, Cu, Zn, Zr, Ba, Ce, Pb	low $\mu g\,g^{-1}$ range	Becker *et al.*[81]

*IDA - isotope dilution analysis

behind the flat sample in a glow discharge ion source resulted in an enhanced sputtering and ionization efficiency of glass samples.[85] The main limitation of rf GDMS is the relatively high formation rate of diatomic ions, like MO^+, MN^+ or MAr^+ (M – matrix element). Interferences between these ionic species and the analyte ions can lead to serious problems in quantification.[56] However, another remaining problem is the relatively low accuracy if no suitable matrix matched standard reference materials are used.[85]

Several applications of inorganic mass spectrometry (TIMS,[86] GDMS,[53] LIMS, ICP-MS,[66,81,87–89] LA-ICP-MS[61,62,81,90] and ETV-ICP-MS[66]) in the trace analysis of high purity non-conducting materials are collected in Table 9.14.

9.1.4 Thin and Thick Film Analysis

The surface of a solid sample interacts with its environment and can be changed, for instance by oxidation or due to corrosion, but surface changes can occur due to ion implantation, deposition of thick or thin films or epitaxially grown layers.[91] There has been a tremendous growth in the application of surface analytical methods in the last decades. Powerful surface analysis procedures are required for the characterization of surface changes, of contamination of sample surfaces, characterization of layers and layered systems, grain boundaries, interfaces and diffusion processes, but also for process control and optimization of several film preparation procedures.

SIMS and SNMS are versatile analytical techniques for the compositional characterization of solid surfaces and interfaces in materials research.[92–94] As one of the most important applications, both surface analytical techniques allow depth profile analysis (concentration profile as a function of the depth analyzed) to be performed in materials science and the semiconductor industry with excellent depth resolution in the low nm range. For depth profiling in materials science, *dynamic* SIMS and SNMS using high primary ion beam doses are applied. Both techniques permit the analysis of light elements such as H, O, C and N, which are difficult to measure with other analytical techniques.

In blue light emitting diodes, AlGaN/GaN multiple quantum wells (MQWs) are the layers that determine the efficiency of electron–hole recombination.[95] High quality layers with dislocation density in GaN on a sapphire substrate and low piezoelectric fields have been grown using the facet controlled epitaxial lateral overgrowth technique. A ToF-SIMS IV with a $^{69}Ga^+$ primary ion source (25 keV) as the analysis beam and a 1 keV Ar^+ as sputtering ion beam was applied for depth profiling and analysis of 3D distribution of Al (image size 50 μm) in these structures to study the Al incorporation in $Al_xGa_{1-x}N$/GaN multiple quantum wells deposited on GaN with (0001) and (1122) facets. The lateral profiles of the Al concentration were integrated over eight periods through the depth, and the depth profiles reveal ten periods of the MQWs on both facets of the sample. It was demonstrated that the Al incorporation is dependent on GaN facet orientations and Al concentration.[95] When GaN related materials are used for blue optoelectronic devices, impurities which accumulate along dislocations in the material may be responsible for leak currents which limit the device performance and lifetime. The dependence of impurity incorporation, especially of C, O, Al and Si, on the polar direction of GaN growth was investigated by SIMS using a Cs^+ primary ion beam.[96] Furthermore, Ga films were deposited under conditions used for growing device quality materials on a sapphire substrate while controlling their polar direction. Depth profiles of Ga and impurities of 1 μm GaN film were measured by SIMS using an O_2^+ primary ion beam. It was found that the polarity of the GaN film influences the incorporation of impurities.[96]

SIMS has become a powerful tool in the measurement of ultrashallow implants, both for the dose and junction depth. SIMS using a 1 keV primary oxygen ion beam was applied for the analysis of implanted and annealed wafers for sub-100 nm technology as described by Ehrke *et al.*[97] Measurements of junction depth and total dose were carried out on an Atomika quadrupole based SIMS. The quantification of Ge and B doped in SiGe in routine mode by SIMS utilizing a beam of low energy oxygen ions is reported in reference.[98] The main advantages of using an oxygen primary ion beam compared to Cs^+ primary ions are the better depth resolution and easier quantification.[98] There are a number of instrumental limitations in SIMS which are inherent in the sputtering process itself for the characterization of ultrashallow implants or thin layers, e.g., the presence of a transient region, a huge matrix effect with changing element sensitivities, crater wall effects and crater bottom roughening. These effects make it difficult to obtain quantitative depth profiles. Under low energy sputtering conditions the transient region can be reduced and the surface topography could become more pronounced. An analytical technique for the accurate depth profiling of low energy As and P implants (at implant energy of 200 keV and 2 keV) in silicon wafers by SIMS measurements (CAMECA IMS WF) is discussed in reference.[99] Monitoring negatively charged analyte ions As^- and P^- using O_2^+ flooding (at 300 eV impact energy of primary ions, oxygen pressure was a few 10^{-7} mbar) in combination with a low energy Cs^+ primary ion beam (at 150–500 eV impact energy) for sputtering of sample material, resulted in an improvement of element sensitivity (more than one order of magnitude) and removed the ion yield variation at the native oxide/silicon interface.[99]

Quantitative depth profiling using polyatomic MCs^+ and MCs_2^+ ions instead of atomic ions $M^{\pm}$ ions is well established in surface analysis using SIMS. The MCs^+ technique, which reduces matrix effects significantly, was proposed by Gao in 1988.[100] The formation of MCs^+ and MCs_2^+ has been explained by the recombination of sputtered neutral atoms (M) with Cs^+ ions.[101]

The ToF-SIMS dual beam mode represents an interesting approach for depth profiling where the sputtering and analysis process is completely decoupled. The sample surface is sputtered using e.g., the caesium primary ion beam so that the caesium concentration of the sample surface and the depth resolution is determined by the energy and angle of the primary ion beam. At lower beam energies, the Cs concentration of the surface increases, leading to a lower working function. Low depth resolution Cs–Xe cosputtering has been introduced in order to optimize the optimum MCs_x^+ yield, for example in ultrashallow depth profiling. If the sputtering conditions, and hence the caesium surface concentrations, are kept constant by using atomic primary ion beams (Ga^+ and Bi^+) as well as cluster primary ions like Bi_3^+ and C_{60}^+, increased MCs^+ ion yields (and consequently depth profiling sensitivity) by up to a factor of 1000 is observed if cluster ions are employed instead of Ga^+.[101] Investigations of the ion yields of MCs_2^+ polyatomic ions for the qualitative detection of electronegative elements such as N, F, S and Cl in SIMS (a quantitative evaluation of the data obtained is difficult). The application of MCs_2^+ in depth profiling on a Ta/Si multi-layer system are described by Bock *et al.*[102]

Depth scale calibration of an SIMS depth profile requires the determination of the sputter rate used for the analysis from the crater depth measurement. An analytical technique for depth scale calibration of SIMS depth profiles via an online crater depth measurement was developed by De Chambost and co-workers.[103] The authors proposed an *in situ* crater depth measurement system based on a heterodyne laser interferometer mounted onto the CAMECA IMS Wf instrument. It was demonstrated that crater depths can be measured from the nm to μm range with accuracy better than 5 % in different matrices whereas the reproducibility was determined as 1 %.[103] SIMS depth profiling of CdTe based solar cells (with the CdTe/CdS/TCO structure) is utilized for growing studies of several matrix elements and impurities (Br, F, Na, Si, Sn, In, O, Cl, S and Te) on sapphire substrates.[104] The SnO_2 layer was found to play an important role in preventing the diffusion of indium from the indium containing TCO layer.

The quantitative determination of Ge in SiGe quantum well structures ($Si_{1-x}Ge_x$ layers and interfaces)[50] by SIMS is compared to low energy RBS. The thickness of the analyzed quantum well was about 12 nm and it was situated at a depth of about 60 nm below the surface. The SIMS measurements were performed using an oxygen primary ion beam over an area of $350 \times 350\ \mu m^2$.

The results obtained using the two analytical techniques agree well. Depth profiling of BN delta-doped silicon samples grown by magnetron sputtering was performed using SIMS with low energy O_2^+ primary ions at a high depth resolution in the 1 nm range.[105] An international standardization of the depth profiling of dopants in silicon by SIMS has been promoted by the International Standard Organization (ISO)/Technical Commission.

The capability of SNMS as a powerful surface analytical technique is demonstrated in the following three examples. Figure 9.7 presents a depth profile obtained by SNMS on part of a thin-film layered structure on a glass substrate of a solar cell based on amorphous silicon. Amorphous silicon thin film solar cells consist of several layers (thickness: 5–500 nm, see schematic of a solar cell structure at the top of Figure 9.7). Inside the (intrinsic) i-layer (a-Si:H) pairs of electrons and holes are generated by the photoelectric effect. In order to improve the depth resolution, layered model systems are analyzed by SNMS and SIMS to study interface effects.[106] In spite of the complicated thin layered structure of amorphous silicon layered systems doped with H, C and H, C, H, and B, respectively, for thicknesses in the 5–500 nm range (see Figure 9.7), SNMS allows quantitative depth profiling, thus achieving a depth resolution of a few nm.[106,107] The relative sensitivity factors of the elements are determined using model layers in order to obtain quantitative depth profiles with SNMS and SIMS. SIMS is utilized even when the limits of detection reached by SNMS are not sufficient. Examples of SIMS measurements are the depth profiles of ^1H, ^2D in ZnO layers and of P near the n/i interface.[106] Depth profiles of solar control coating on glass (SnO_x/Ni-Cr/Ag/SnO_x/Ni-Cr/Ag/SnO_x/glass) have been performed by combining SNMS and SIMS (using MCs^+ ions as analyte ions) with Auger electron spectroscopy (AES) and electron microprobe analysis (EPM). The Ni/Cr ratio in the nm thin layers was quantified, and the Cr fraction amounted to less than one monolayer.[108]

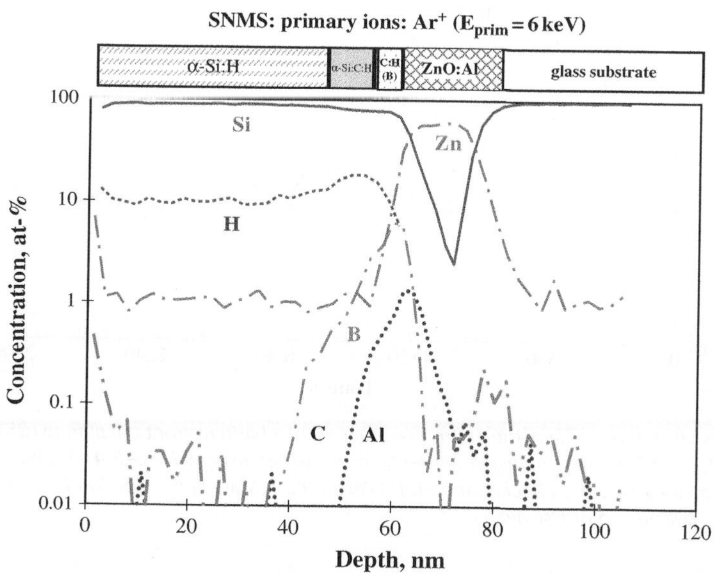

Figure 9.7 Depth profile of part of a solar cell system. (J. S. Becker and H. J. Dietze, Int. J. Mass Spectrom. Ion Proc. **197**, 1(2000). Reproduced by permission of Elsevier.)

Depth profiling by electron gas SNMS using an INA-X instrument on conducting and insulating films (e.g., 80 nm Ge layer on silicon or $Ti/TiSi_2/SiC$ layer structure) and 6 μm thick steel coatings on the nano- and micrometre scale were performed by Jorzick *et al.*[109] Variable sputtering rates can be adjusted ranging from some nm min^{-1} up to 1–2 μm min^{-1} to analyze thin and thick layers and layered systems with high depth resolution.[109] Elements which are difficult to analyze, such as C or O, become quantitatively detectable down to several 10 μg g^{-1}.[110]

Quantitative characterization of plasma–polymer films, especially of ultrathin fluorinated carbon plasma polymer films, has been performed by ToF-SIMS to study changes in the surface composition and molecular distribution. CF_x films on silicon and polyethylene terephthalate (PET) substrates were exposed to a pulsed Ar/CHF_3 plasma by varying the deposition time from 10–90 s.[111–113] The results show differences in film growth and CF_x cross linking for the silicon and PET substrates.[111]

The SNMS depth profile (ion intensity as a function of sputter time) for the matrix elements of a $Ba_{0.7}Sr_{0.3}TiO_3$ layer on a silicon substrate with $Pt/TiO_2/SiO_2$ buffer layers is illustrated in Figure 9.8. Inhomogeneity of the perovskite layer was detected especially for Sr. Furthermore, an interdiffusion of matrix elements of the $Ba_{0.7}Sr_{0.3}TiO_3$ layer and of the Pt barrier layer was observed.

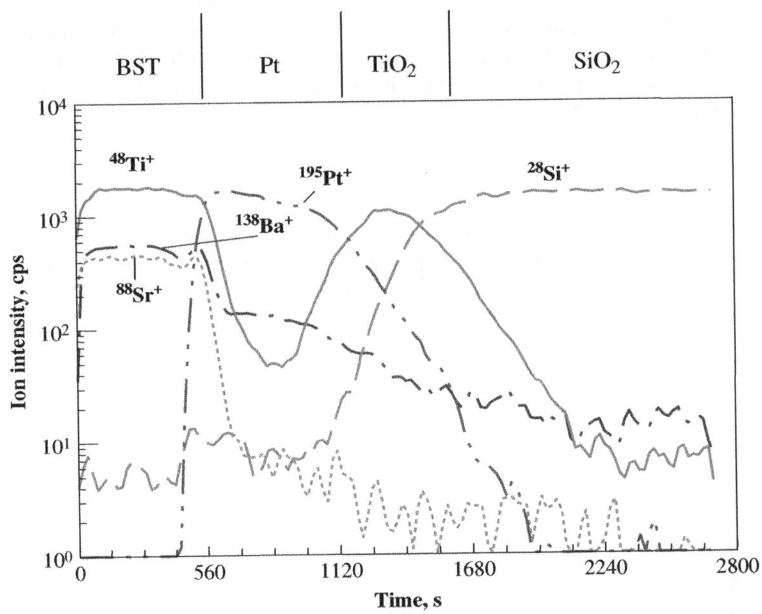

Figure 9.8 *SNMS depth profile of the matrix components of a barium strontium titanite (BST) perovskite layer on Si with a $Pt/TiO_2/SiO_2$ diffusion barrier layer measured using SIMSLAB 410 (FISONS Scientific, 10 keV, Ar^+ primary ions). (J. S. Becker and H. J. Dietze, Int. J. Mass Spectrom. Ion Proc.* ***197****, 1(2000). Reproduced by permission of Elsevier.)*

The application of SNMS with an extremely high depth resolution of a W/Si multi-layer system is illustrated in Figure 9.9. The multi-layer system consists of 69 double layers. The thickness of a double layer is 3.5 nm.[114] Practically no degradation of depth resolution was observed as a function of sputter time (i.e. with sputter removal of the sample material).

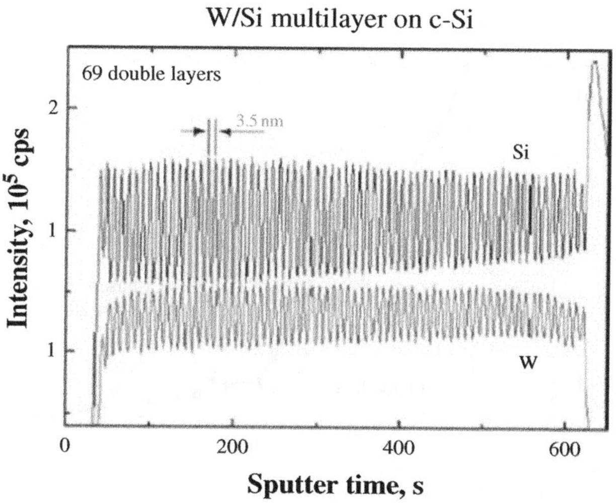

Figure 9.9 *SNMS depth profile of a W/Si multilayer, the thickness of a double layer is 3.5 nm. (Reproduced with permission of Prof. Dr. H. Oechsner, Kaiserslautern, Germany.)*

Trace element accelerator mass spectrometry (TEAMS) can also be applied for depth profiling as demonstrated for H, C, O, Al and Si implants (and impurities) in semiconductor GaN substrate.[115]

Furthermore, the capability of alternative mass spectrometric surface analytical techniques (e.g. GDMS), which, in contrast to SIMS, do not possess the analytical ability to determine lateral element distribution, enables the direct depth and trace analysis of thick layered systems. Due to lower matrix effects compared to SIMS, in GDMS semiquantitative depth profiles of matrix, minor and trace elements are obtained. GDMS can also be used for the quantitative determination of element concentrations as a function of sputtered depth. A depth profile of a 5 μm thick non-conducting chromium oxide layer on a chromium based alloy measured by rf GDMS is shown in Figure 9.10. In this figure, the CrO^+/Cr^+ ion intensity ratio is measured as a function of sputter time.[84] A sputter time of 60 min correlated to a depth of 8 μm. Furthermore, a Mo/W multi-layer system with four molybdenum layers (thickness of a single layer was 60 nm) and four tungsten layers (thickness of a single layer was 100 nm) has been studied by rf GDMS.[56] In the depth profile obtained, the eight single layers are clearly resolved whereby a depth resolution ranging from 15 nm at a depth of 100 nm to 35 nm at a depth of 600 nm is typically reached. The depth resolution is mainly limited by the uneven shape of the sputter crater observed in GDMS.

A depth profile analysis of trace and matrix elements (B, Na, Ni, Fe, Mg, V, Al and C) in a 26 μm Si layer on a SiC substrate measured by GDMS, yielded impurity profiles, for example, with constant Ni contamination in the Si layer and enrichment at the interface layer.[45] However, with respect to depth profiling of thin layers using dc GDMS with a depth resolution between 50 and 500 nm, this technique plays a subordinate role compared to the commercially available and cheaper GD-OES (glow discharge optical emission spectrometry).

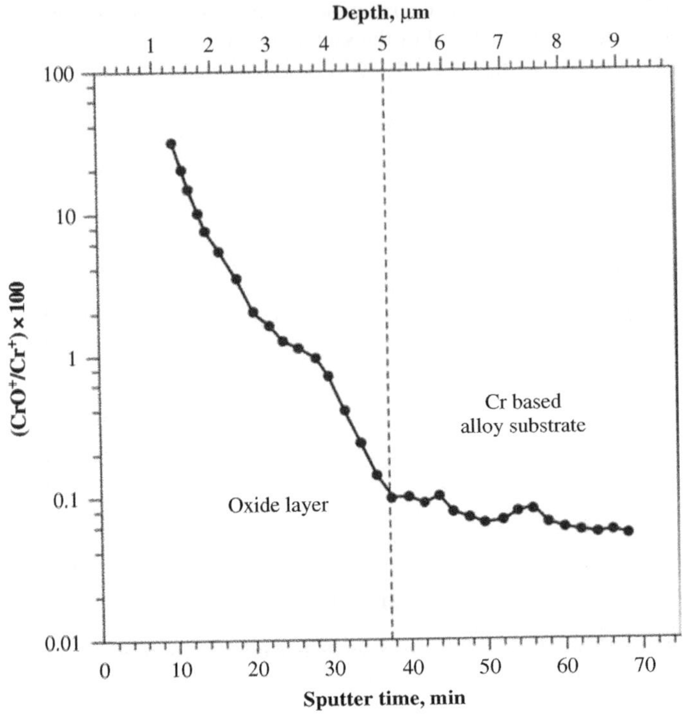

Figure 9.10 *Depth profile for a non-conducting chromium oxide layer on chromium based alloy measured by rf GDMS. (A. I. Saprykin, J. S. Becker et al., Fresenius' J. Anal. Chem.,* ***358****, 145 (1997). Reproduced by permission of Springer Science and Business Media.)*

A quite different application of GDMS is the measurement of hydrogen and deuterium concentration, including depth profile analysis, e.g. in a gold electroplated layer on a CuSn substrate as described in reference.[116] The relative sensitivity coefficient of hydrogen was evaluated by measurements of titanium standard reference material.

As a ceramic multi-layer device, the high temperature solid oxide fuel cell (SOFC) converts chemical energy into electrical energy very efficiently. For trace analysis in the thick ceramic layers of SOFC (anode: Ni-YSZ cermet, YSZ – yttrium stabilized cermet, electrolyte: YSZ, and cathode: $La_xSr_yMnO_3$) sensitive solid-state mass spectrometric techniques, such as LA-ICP-MS and rf GDMS, have been developed in the author's laboratory.[81] For example, inhomogeneous yttrium stabilized zirconium oxide layers for solid oxide fuel cells were analyzed directly with respect to trace impurities in the $\mu g\,g^{-1}$ concentration range using LA-ICP-MS.[117] Thick perovskite layers ($La_{0.65}Sr_{0.3}MnO_3$) of SOFC were characterized with respect to several trace impurities by both rf GDMS[83] and LA-ICP-MS.[117] For quantification purposes, synthetic matrix matched laboratory standards of high purity powders of the basic material ($La_{0.65}Sr_{0.3}MnO_3$) with trace elements in the 20–500 $\mu g\,g^{-1}$ range were prepared and analyzed. It was found that changes of discharge parameters (rf power, pressure of discharge gas (Ar) in the rf glow discharge) have no significant influence on element sensitivity. The detection limits were found to be in the 10 $\mu g\,g^{-1}$

range.[83] Thick ceramic layers and non-conducting bulk materials for solid oxide fuel cells (anodic Ni cermet, YSZ and $La_{0.65}Sr_{0.3}MnO_3$) were characterized by LA-ICP-MS in comparison to ICP-MS after selective digestion, rf GDMS and SSMS, respectively.[81] Mass spectrometric studies serve to check all the processes – characterization of the purity of the initial compounds, possible contaminations with impurities during production of layers, decomposition and sintering of layers during electrochemical processes at high temperatures in SOFC and after use.

Depth profiles of matrix elements on Mn- and Co-perovskite layers of fuel cathodes have been measured by LA-ICP-MS in comparison to other well established surface analytical techniques (e.g., SEM-EDX).[118] On perovskite layers at a spatial resolution of 100 μm a depth resolution of 100–200 nm was obtained by LA-ICP-MS. The advantages of LA-ICP-MS in comparison to other surface analytical techniques (such as XPS, AES, SIMS, SNMS, GD-OES, GDMS and SEM-EDX) are the speed, flexibility and relatively low detection limits with an easy calibration procedure. In addition, thick oxide layers can be analyzed directly and no charging effects are observed in the analysis of non-conducting thick layers.

In many technical applications, metallic, alloyed or ceramic construction materials are subject to corrosive, in particular to oxidizing, environments at high temperatures, which results in a deterioration of their physical and mechanical properties. To reduce the corrosion effects on materials, the surfaces are covered with temperature resistant coatings, which are protective layers several hundred μm in thickness. The most commonly used system for high temperature Ni-based alloys is overlay coatings of the MCrAlY type (with $M = Ni$ or Co). The development and application of coating systems which guarantee reliable component protection during long term service is a crucial requirement for this material type in industrial gas turbines. Interdiffusion of matrix elements between coating and substrate (base material) after oxidation in air at a temperature of 980 °C for several thousand hours affects the coating life due to deterioration of the mechanical properties or/and oxidation performance. LA-ICP-MS is the method of choice for studying cross sections of NiCrAlY based coatings on high temperature alloys to investigate diffusion effects at the interface.[119] The capability of LA-ICP-MS as a microanalytical method has been tested by a comparison using the line scan and single point modes at laser energies of 2 mJ and 4 mJ. An alloy certified reference material BAM-328-1 (Bundesanstalt für Materialprüfung (Federal Institute for Materials Research and Testing), Berlin, Germany) with a similar matrix composition was employed for quantification purposes. The relative sensitivity coefficients (RSCs) of the elements varied between 0.2 and 2. In addition, quantification via calibration curves and solution based calibration was considered. The measured LA-ICP-MS profiles are comparable with those of SEM-EDX analysis. Mass spectrometric measurements of lateral element distribution in NiCrAlY-based coatings on high temperature alloys revealed, for example, an increasing loss of aluminium due to diffusion from the coating to the alloy after oxidation in air under various experimental conditions (see Figure 9.11 a and b for Al and Co, respectively).[119]

Depth profiling analysis of high-tech layered industrial materials (such as thermographic materials, printing plates, CsBr:Eu phosphorus screens and SiC) by LA-ICP-MS is described by Balcaen *et al.*[120] The thicknesses of the layers varied between 1 μm and 200 μm. For the experiments, a 193 ArF excimer based laser ablation system (GeoLas 200M, MicroLas, Germany) was equipped with quadrupole ICP-MS with a dynamic reaction cell (LA-ICP-DRC-MS) to suppress possible isobaric interferences.[120]

The application of LA-ICP-ToF-MS and LA-ICP-QMS for depth profiling of various titanium based coatings on steel and tungsten carbide and Ti based single layers is discussed in references[121,122], respectively. Thickness determination was performed by LA-ICP-QMS with 5 % RSD, a laser ablation rate of $\leq$ 100 nm per laser shot (Nd-YAG laser at wavelength 266 nm using a laser energy of 1.5 mJ at 120 μm laser spot size), and a depth resolution of 2.5 μm was observed.[122]

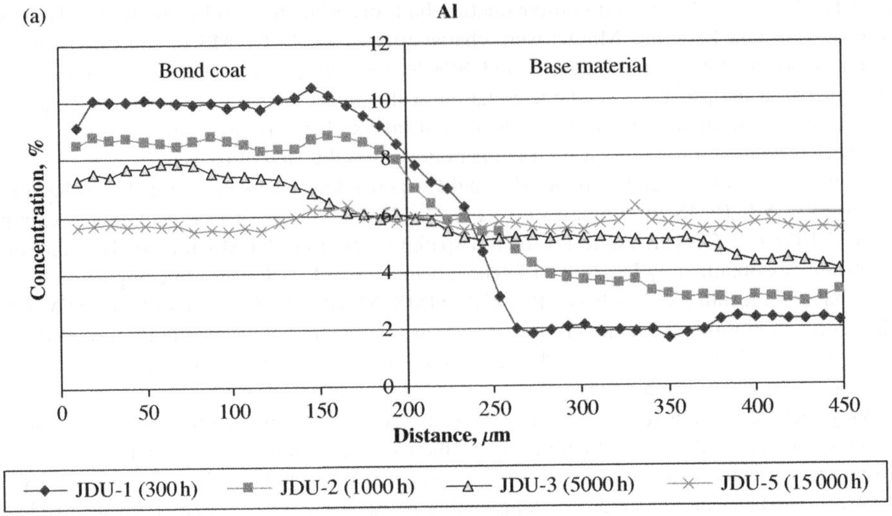

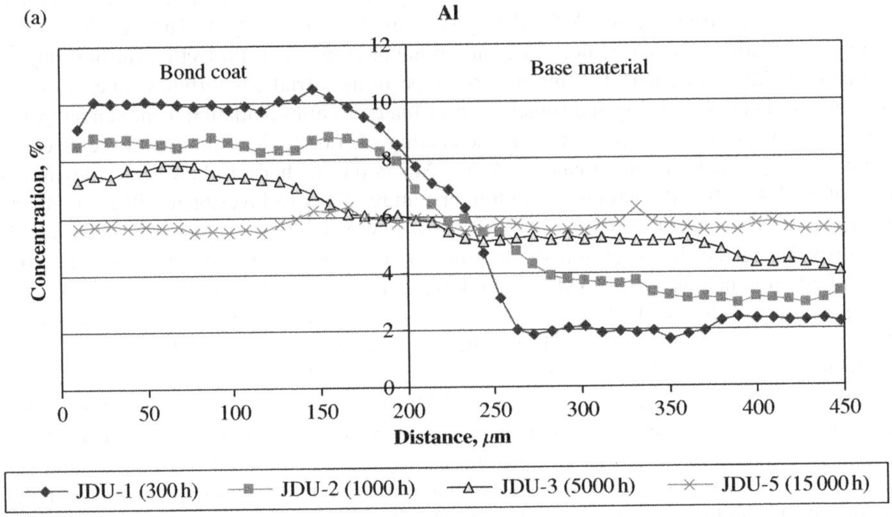

Figure 9.11 *Distribution of Al (a) and Co (b) in NiAlCrY based alloy and coating after oxidation in air at 980 °C and different annealing times measured by LA-ICP-MS. (A. Izmer, M. Zoriy, C. Pickhardt, W. J. Quadakkers, V. Shemet, V. Singheiser and J. S. Becker, J. Anal. At. Spectrom.,* ***20****, 918 (2005). Reproduced by permission of The Royal Society of Chemistry.)*

With LA-ICP-ToF-MS, using the 193 nm ArF laser (laser energy 100 mJ at 120 μm laser beam diameter), a depth resolution of 200 nm per laser shot was measured.[121] LA-ICP-MS was utilized for depth profiling of copper coatings on steel with certified copper coating thicknesses from about 6 to 200 μm.[123]

Nanostructured thin inorganic–organic hybrid layers (HfO_2–SiO_2 and ZrO_2–SiO_2) prepared by a modified sol-gel technique were analyzed using three surface analytical approaches (SIMS, LA-ICP-MS and XPS) to characterize the lateral and depth distribution of several matrix elements and species.[124] By applying a combination of analytical techniques it was possible to observe the homogeneous in depth nanocomposition of the thin film.

Barrier layers for Cu metallization in surface acoustic wave (SAW) devices, which are increasingly used in the information technique and telecommunications industry, have been investigated by SIMS depth profiling in comparison to AES (Auger Electron Spectrometry).[125] Development trends in SAW devices focus on smaller structures, higher input power or higher frequency. Two Cu metallization systems (of 150 nm thickness) on a $LiNbO_3$ substrate with different barrier layers: (a) Ta (20 nm) / Cu (150 nm) / Ti (30 nm) layer system deposited by electron beam evaporation, and (b) $Ta_{30}Si_{18}N_{52}$ (50 nm) / Cu (150 nm) / $Ta_{30}Si_{18}N_{52}$ (50 nm) layer system produced by magnetron sputtering from a Ta_5Si_3 target in nitrogen atmosphere, both on a $LiNbO_3$ substrate, were analysed with respect to depth profiles after annealing for 5 h and 20 h in air (and in vacuum) at different temperatures from 100–400 °C. In system (a) the Ti oxidized in air at $T \geq 300$ °C. A diffusion of Ti into the Cu layer was observed at about 100 °C. The depth profiles show that the TaSiN layer operated as a more effective diffusion barrier on the $LiNbO_3$ substrate for the Cu-SAW technology up to more than 300 °C.[125]

Layered samples of photoresist used for lithography in the production of semiconductor and optical devices have been analyzed directly for trace impurities (Al, Cu, Pt, Au, Th and U) by LA-ICP-MS.[126] The preparation procedure including spiking, mixing, spin coating and baking, plus a matrix matched external calibration in LA-ICP-MS, is discussed by Chi *et al.*[126] The homogeneity of photoresist samples was studied using the radiotracer technique.

Apart from the examples of solid-state mass spectrometric techniques for surface analysis discussed above, ICP-MS can also be employed for the analysis of thick and thin layers in materials research. Possible ICP-MS applications are conceivable in three directions: (a) analysis of matrix composition (stoichiometry) in layers, (b) multi-element trace analysis in layered materials, in both approaches the ICP-MS measurements are carried out after selective dissolution of different layers, and (c) determination of surface contaminants by dissolution of impurities and subsequent mass spectrometric trace analysis. The determination of stoichiometry in semiconducting or non-conducting layers is of relevance for several studies of the growth mechanism and for the control of defects during the manufacture of layered systems. For example, an analytical procedure using different ICP-MS instruments (double-focusing sector field ICP-MS and quadrupole based ICP-MS with and without a collision cell) for the determination of stoichiometry and trace impurities in thin electroceramic $Ba_{0.7}Sr_{0.3}Ti_1O_3$ layers after the dissolution of perovskite layers in a small volume of a HNO_3/H_2O_2 mixture, was developed.[127,128] The stoichiometric composition of thin non-conducting $Ba_{0.7}Sr_{0.3}Ti_1O_3$ perovskite layers determined by ICP-MS was validated by measurements using inductively coupled plasma optical emission spectroscopy (ICP-OES) and X-ray fluorescence analysis (XRF).The results of the stoichiometric determination of a thin $Ba_{0.7}Sr_{0.3}Ti_1O_3$ layer by ICP-MS (using an Elan 6000, PerkinElmer/Sciex) are comparable with those of X-ray fluorescence analysis (XRF) as summarized in Table 9.15.[127] The precision of Ba/Sr and (Ba + Sr) / Ti was found to vary between 0.1 and 0.4 %. Although the accuracy of the (Ba + Sr) / Ti element ratio was in good agreement with the ideal stoichiometry, the Ba/Sr atomic ratio was slightly lower than expected. In comparison to XRF, which has the advantage of measuring the stoichiometry non-destructively, ICP-MS is the only method which is able to determine a multitude of trace impurities in a small volume of thin dissolved layer with low detection limits. Table 9.16 summarizes the results of the determination of trace impurities in a thin $Ba_{0.7}Sr_{0.3}Ti_1O_3$ layer (40 nm) dissolved in 100 µL HNO_3/H_2O_2.[127]

Table 9.15 *Comparison of determinations of stoichiometry in thin* $Ba_{0.7}Sr_{0.3}Ti_1O_3$ *perovskite layer measured by ICP-MS using different instruments and XRF.*

	ICP-SFMS (Element)	ICP-CC-QMS (Platform)	ICP-QMS (Elan 6000)	XRF
Ba/Sr	1.80 ± 0.01	1.78 ± 0.03	2.071 ± 0.016	2.07 ± 0.06
(Ba+Sr)/Ti	0.476 ± 0.003	0.419 ± 0.01	0.486 ± 0.006	0.44 ± 0.03

(S. F. Boulyga, H. J. Dietze and J. S. Becker, J. Anal. At. Spectrom., **16**, 598 (2001). Reproduced by permission of The Royal Society of Chemistry.)

Table 9.16 *Results of trace analysis in a thin dissolved* $Ba_{0.7}Sr_{0.3}Ti_1O_3$ *layer (40 nm) by ICP-QMS; limit of detection (LOD) in aqueous solution, sample volume of layer:* 100 μl

Element	LOD (μg l^{-1})	Amount of impurity (ng)
Ag	0.039	<4
Cd	0.0055	32 ± 1
Ce	0.0013	0.80 ± 0.05
Co	0.0029	<0.3
Cr	0.02	<2
Cu	0.036	42 ± 1
Fe	0.0173	117 ± 16
Mn	0.0017	1.0 ± 0.1
Pb	0.0039	5.2 ± 0.3
Th	0.02	<2
U	0.0027	<0.3
V	0.081	<8
Y	0.0021	0.8 ± 0.1
Zn	0.0037	478 ± 9

(S. F. Boulyga, H. J. Dietze and J. S. Becker, J. Anal. At. Spectrom., **16**, 598 (2001). Reproduced by permission of The Royal Society of Chemistry.)

Several surface analytical techniques have been often employed to characterize thin films for applications in materials science or microelectronics. For example, nanocrystalline anatase TiO_2 thin films with a thickness of several μm have been investigated by X-ray diffraction analysis to study the phase homogeneity, and additionally by atomic force microscopy (AFM) to investigate the surface structure and morphology, by transmission electron microscopy (TEM) to measure the size distribution of the nanocrystalline TiO_2 particles and also by SIMS for depth profiling (Cameca IMS-4f using a Cs^+ primary ion).[129] SIMS was selected as an elemental analytical technique to obtain information about the stoichiometry of matrix elements in thin anatase film and the presence of metallic impurities (K and Na) in the nanoporous thin TiO_2 film deposited by painting nanocrystalline anatase TiO_2 particles in an organic solvent (as a suspension) onto glass substrates. The SIMS measurements demonstrated a very homogeneous distribution of Ti, O and H over the whole film thickness. The film was contaminated with Si and a relatively broad transition region of Si was found between the substrate and film as a result of the annealing procedure at 450 °C.[129]

LA-ICP-MS is a very suitable analytical method for direct trace element analysis on a small area of thin pure foil, because no sample preparation is required. The results of the determination of noble metals in a thin difficult to dissolve rhodium foil measured by LA-ICP-MS are

Table 9.17 *Results of determination of noble metals in a pure rhodium foil by LA-ICP-MS[A].*

Element	Concentration ($\mu g\,g^{-1}$)	Element	Concentration ($\mu g\,g^{-1}$)
Ru	18	Pd	35
Pt	112	Ag	1.0
Os	0.2	Au	3.5
Re	< 0.1	Ir	145

[a] LA-ICP-MS using solution calibration (MCN Aridus, CETAC).

summarized in Table 9.17. Because no standard reference material was available, solution based calibration was applied as a quantification strategy. The noble metal concentrations measured varied between $0.2\,\mu g\,g^{-1}$ for Os and $145\,\mu g\,g^{-1}$ for Ir. Limits of detection were determined for Re at the $0.1\,\mu g\,g^{-1}$ level. Besides the mass spectrometric techniques already discussed, non-mass spectrometric techniques are also utilized in surface analysis. Several mass spectrometric techniques (SIMS, SNMS, LA-ICP-MS, LIMS, dc and rf GDMS) and other non-mass spectrometric techniques (such as AES – Auger electron spectroscopy and including SAM – scanning Auger microscopy; SEM – EDX – scanning electron microscopy with energy dispersive X-ray analysis; XPS – X-ray photoelectron spectroscopy) applied in surface analysis are compared with respect to depth and lateral resolution in Table 9.18. In addition there are several alternative analytical techniques for characterizing the composition and topography of surfaces, interfaces and surface layers such as Rutherford backscattering (RBS), glow discharge optical emission spectrometry (GD-OES), atomic force microscopy (AFM), transmission electron microscopy (TEM), X-ray fluorescence analysis (XRF) or total reflection X-ray fluorescence analysis (TXRF) and others.

9.1.5 Analysis of Surface Contamination and of Process Chemicals Used in Semiconductor Technology

Many important processes such as catalysis, corrosion and adhesion depend on the elemental and molecular composition of the surface or interface region.[130] Therefore, besides the determination of impurities in high purity bulk materials, the surface composition and possible contaminations have to be determined because they influence the physical and chemical properties of, for example, microelectronic devices. For the inspection of silicon surface impurities, TXRF (total reflection X-ray fluorescence analysis) is generally used in a routine mode in the microelectronics industry. *Static* SIMS (SSIMS) using low primary ion beam doses $< 10^{12}$ ions cm^{-2}[130] is the method of choice for characterizing metallic surface contamination on silicon wafers after surface preparation by vapour phase decomposition (VPD).[131] Under static SIMS analysis conditions, only the top few monolayers of the sample surface are analyzed so that contamination is a serious problem using this technique. In addition, the main advantage of SIMS is its high sensitivity resulting in low detection limits for the elements, so that it is also possible to analyze light elements such as H, D, Li, B and C, and to obtain molecular and isotope information and lateral element distribution (images). Disadvantages are the expensive instrumentation, complex spectra due to the high polyatomic ion formation rate, huge matrix effects and consequently quantification problems. Quantification is normally performed using relative sensitivity factors (RSFs) determined on a set of standard reference materials containing the elements of interest in defined concentrations. RSFs are strongly matrix dependent. In general, SSIMS does not tend to be used for the quantification of surface composition because of all the inherent uncertainties as described by Arlingshausen.[132] Quadrupole SIMS

Table 9.18 Comparison of analytical methods for surface analysis.

Method	Process	Particles/ ray measured	Depth resolution	Lateral resolution	Imaging possibility	Detection limits	Strengths of technique	Limitations of technique
SIMS	ions ↘↗ ions	sputtered ions	> 1 nm	50 nm	+++	> ng g^{-1}	high depth resolution good imaging	huge variation of RSCs, intense molecular ions
SNMS	ions ↘↗ neutrals	sputtered neutrals (post ionized by e-beam or laser)	> 3 nm	5 μm	(+)	> μg g^{-1}	high depth resolution elemental information	poor sensitivity
LA-ICP-MS	photons ↘↗ neutrals	ablated particles ionized in ICP	> 100 nm	> 10 μm	++	> ng g^{-1}	high sensitivity	low lateral resolution
LIMS	photons ↘↗ ions	ions	> 100 nm	> 5 μm	+	> ng g^{-1}	insulators	expensive instruments
dc GDMS rf GDMS	Ar* ↘↗ ions	ions	> 10 nm > 150 nm	none none	none none	ng g^{-1} μg g^{-1}	high senstivity insulators	no insulators low depth resolution
AES	e^- ↘↗ e^-	electrons	2–6 mono-layers	20 nm	++	0.2 %	chemical and elemental information (Li – U)	no insulators
SEM-EDX	e^- ↘↗ X-ray	X-ray	none	5 nm	+++	0.1 %	high lateral resolution elements: Na–U	low sensitivity
TXRF	X-ray ↘↗ X-ray	X-ray	1 mono-layer	(none)	(none)	> μg g^{-1}	high depth resolution elemental information	only $Z > 14$
XPS	X-ray ↘↗ e^-	photo electrons	4–8 mono-layers	100 μm	+	< 0.1 %	chemical bonding (Li–U)	low sensitivity

instruments with a transmission of < 1 % have been applied for many years in desorption studies.[133] Because quadrupole based mass spectrometers are restricted in mass resolution and are unsuitable for analyzing large biomolecules on surfaces, the ToF-SIMS instrument (with a maximum mass resolution of $m/\Delta m \approx 10\,000$) developed by Benninghoven's working group[134] is well suited for static SIMS measurements comprising the characterization of the elemental and molecular composition of surfaces. Compared to XPS and AES, the higher surface specificity of SIMS (1–2 monolayers compared with 2–8 monolayers) can be useful for more precise determination of the chemical processes of an outer surface.[132] For instance, SSIMS has been applied in comparison to XPS for studying thin resistant oxide film formation by anodizing aluminium in phosphoric acid.[135] The properties of such very thin films play an important role for the adhesion of aluminium to aluminium in the aircraft industry. SSIMS is also utilized to investigate surface reactions, e.g., the absorption of propylene on ruthenium[136] or the decomposition of ammonia on a silicon surface.[137]

The application of heterogeneous catalysis plays a key role in technological processes. Engineering of the catalytic activities requires the study of the complex chemistry between absorbate and the catalyst at the surface. Static SIMS has been used to determine the surface composition and properties of solid catalysts before and after the catalytic actions by several groups.[138–140] In addition, the dissociation kinetics of NO on Rh (111) surfaces have been studied by temperature programmed static SIMS.[139]

The removal and the control of metallic contaminants on the surface of silicon wafers is extremely important in semiconductor manufacturing processes. For example, metallic cleanness of $< 10^{10}$ atoms cm^{-2} is required on silicon wafer surfaces. The application of TXRF as a surface analytical method after vapour phase decomposition is discussed by Fabry *et al.*[141] The silicon wafer was deposited in a closed box and treated with HF vapour which etched the silicon oxide layer. Furthermore, a small droplet (e.g., 100 μl $H_2O/H_2O_2/HF$ as a scanning solution) was deposited on the wafer surface and scanned using a droplet scanner in order to dissolve surface contaminations. Horn[142] determined Al, Fe, Zn after VPD by ICP-QMS (e.g., Na – 1.16 pg g^{-1} and Fe – 0.29 pg g^{-1}) in a small amount of diluted surface contamination and compared the results with those of AAS. The improvement of the detection limits for vapour phase decomposition ICP-MS for determining trace impurities on silicon wafer is described in reference.[143] Limited sample sizes (200 μl) and different sources of contamination arising from the materials and bottles used to prepare the VPD samples were investigated with respect to Na, Mg, Al, Ca, Ti, V, Cr and Fe.

Surface contamination on silicon surface has been analyzed in our laboratory by O'Brien *et al.*[144] after VPD of different silicon wafers using the direct injection high efficiency nebulizer (DIHEN) for solution introduction in Platform ICP-MS with a collision cell. Difficult to determine elements (Fe, Cr, Co, Cu, Mn and others) were measured in the concentration range from 30 ppt cm^{-2} (Mn) to 600 ppt cm^{-2} (Al). The advantages and disadvantages of VPD and drop etching for recovery of metal impurities on silicon wafer surfaces are compared in reference.[145]

Local contamination on a graphite surface can be measured directly by LA-ICP-QMS as demonstrated by experiments in the author's laboratory. Whereas in Figure 9.12 a the lateral distribution of a thick Si deposit on a graphite surface is analyzed, Figure 9.12 b shows the distribution of $^{27}Al^+$, $^{53}Cr^+$ and $^{57}Fe^+$ on the surface of a graphite electrode. For these measurements, the element distribution on the surface of a graphite electrode used in the semiconductor industry was analyzed by rastering the focused laser beam in a line scan by LA-ICP-QMS. The highest $^{28}Si^+$ intensity (thick deposit) was measured at a distance of 3.5–4.5 cm – near the cone of the electrode, whereas Al, Cr and Fe show a quite different distribution on the graphite surface with local surface impurities at the end and the middle of the graphite electrode (see Figure 9.12 b).[1]

Pritzkow *et al.*[146] described the application of surface and near-surface analytical methods such as ICP-IDMS (after dilution of layer), Rutherford backscattering spectrometry (RBS) and instrumental

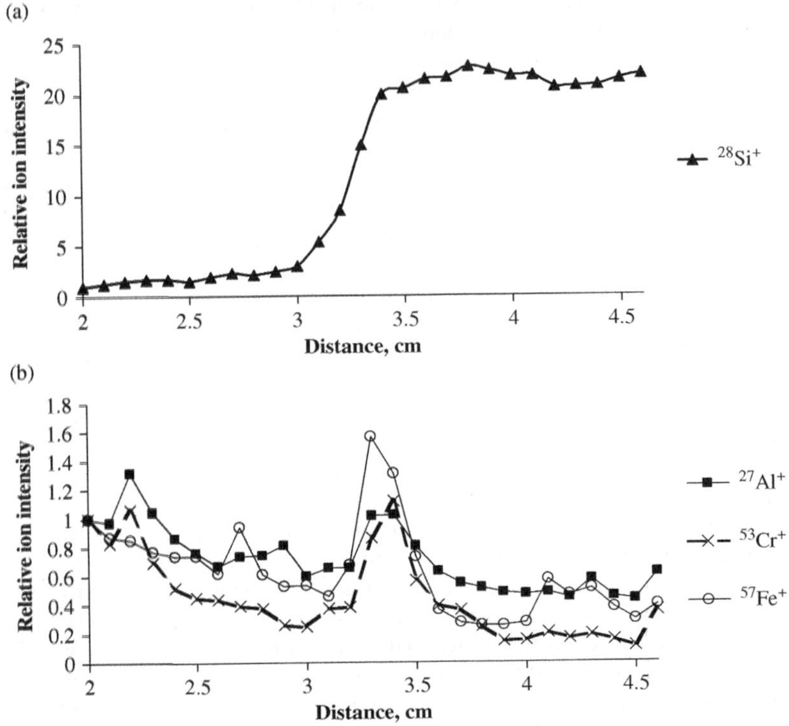

Figure 9.12 Contamination on graphite surface measured by LA-ICP-MS. The measurement started 2 cm from the end of electrode (at this point the relative ion intensity is defined as 1) (a) lateral distribution of Si on graphite surface, (b) distribution of $^{27}Al^+$, $^{48}Ti^+$, $^{53}Cr^+$, $^{56}Fe^+$ and $^{58}Ni^+$ (J. S. Becker, Spectrochim. Acta 57B, 1805 (2002). Reproduced by permission of Elsevier.)

neutron activation analysis (INAA) to characterize a thin layer of reference material (implanted Sb layer on Si/SiO_2), which had been produced and certified. The area density of the implanted Sb layer on Si/SiO_2 was found to be 4.8 10^{16} atoms cm^{-2}.

ICP-MS is used with increasing extent in the microelectronics industry for the determination of ultratrace impurities in process chemicals, especially in ultrapure water, which is the purest chemical available, and in high purity acids, ammonia, hydrogen peroxide and organic solvents. High purity hydrochloric acid (HCl) and nitric acid (HNO_3) are used by semiconductor manufacturers for surface etching, removal of layers, cleaning processes and sample preparation. In general, HCl and HNO_3 are purified using a quartz sub-boiling distillation unit. All chemical procedures for sample preparation are than carried out in clean room laboratories.

A wide application field of ICP-MS is in the trace and ultratrace analysis of high purity solutions[147–152] used in microelectronics, in the analysis of high purity materials, for special procedures in film analysis or cleaning of Si wafers. Selected examples for analysis of high purity process chemicals in the semiconductor industry are summarized in Table 9.19. ICP-MS is the most sensitive trace analytical technique with the lowest detection limits today applied in process control for characterization of impurities in process chemicals used in the semiconductor industry. This feature is advantageous for the analysis of high purity solid samples after digestion and dilution. However, alternative trace analytical techniques, such as graphite furnace and AAS including a high degree of automation in trace matrix/separation and providing reliable and accurate analytical data of trace impurities at ultratrace range are employed in the microelectronics industry.

Table 9.19 *Application of ICP-MS in trace analysis of high purity solutions used in microelectronics.*

Samples	Equipment	Analyzed elements	Limits of detection	References
H_2O	ICP-QMS (HP 4500)	17 elements	0.6 $ng\,l^{-1}$ (Li) 6 $ng\,l^{-1}$ (Ca, Pb)	Hoelzl *et al.*[147]
H_2O	ICP-QMS (Elan 6000)	21 elements	0.3 $ng\,l^{-1}$ (U) 0.15 $\mu g\,l^{-1}$ (As)	Wolf and Grosser[151]
H_2O	ICP-SFMS (PlasmaTrace)	50 elements	sub-$pg\,l^{-1}$ (Ho) 10 $ng\,l^{-1}$ (P, As) 0.1 $\mu g\,l^{-1}$ (S)	Yamasaki and Tsamura[152]
H_2O_2	ICP-QMS (Elan DRC)	25 elements	0.01 $ng\,l^{-1}$ (Rh, In, U) –5 $ng\,l^{-1}$ (Se)	Völlkopf[148]
Isopropyl alcohol (IPA)	ICP-SFMS (Element)	Li, B, Na, Mg, Cr, K, Ca, Cu, Mn, Fe, Ni, Zn, Sn, Pb	0.01 $pg\,l^{-1}$ (Li) 0.2 $pg\,l^{-1}$ (Na, Mg, Cr, K, Cu, Mn, Zn)	Wills *et al.*[150]
H_3PO_4, H_2SO_4, HNO_3 $H_2S_2O_8$, HF, HCl, H_2O_2, H_2O	ICP-SFMS (PlasmaTrace 2)	Si, P, S, Cl, As, Se, Br	0.001 $ng\,g^{-1}$ (As in H_2O) 2 $\mu g\,g^{-1}$ (Cl in H_2SO_4)	Wildner[149]

9.1.6 Microlocal Analysis in Materials Research

Mass spectrometry with spatial resolving power, such as SIMS or LA-ICP-MS, can be used in materials science as a powerful surface analytical technique for the fast survey analysis of a large number of individual micro-objects. For instance, microlocal impurities in solid materials of films have been investigated in order to characterize micrometer size halogenide crystals from the photographic industry.[69] A quantitative surface analysis of silver halide microcrystals using SIMS was performed in comparison to a scanning Auger microprobe (SAM).[153]

An interesting application for LA-ICP-MS is the microlocal analysis of impurities on solid sample surfaces with a spatial resolution of a few μm in ceramics or in non-conducting thick layers. It is known that the spatial resolution depends strongly on the focusing properties of the laser beam, laser power density and properties of the solid sample. The microlocal analysis of impurities in thick inhomogeneous ceramic layers of solid oxide fuel cells was first carried out using LA-ICP-MS with a laser ablation system constructed in our own laboratory.[117] Figure 9.13 shows part of a tested solid oxide fuel cell with local impurities (brown spots) which was analyzed by LA-ICP-QMS (white substrate: yttrium stabilized zirconium oxide (YSZ – electrolyte); black layer: $La_{0.65}Sr_{0.35}MnO_3$ (cathode)). The local impurities were analyzed by LA-ICP-QMS (spot size of laser beam: 100 μm) in these brown spots on the YSZ ceramic surface with a diameter of 75–100 μm. Each measurement consists of ten single measurements. Twenty-seven elements in 19 local impurities were determined with major concentrations of matrix elements of the perovskite layer Mn (24 $mg\,g^{-1}$), Sr (17 $mg\,g^{-1}$) and La (90 $mg\,g^{-1}$) and impurities: B – 6.0 $mg\,g^{-1}$, Ni – 8.5 $mg\,g^{-1}$, Na – 7.0 $mg\,g^{-1}$, Zn – 1.0 $mg\,g^{-1}$, Si – 17.0 $mg\,g^{-1}$, Ce – 2.5 $mg\,g^{-1}$ and P with 5.5 $mg\,g^{-1}$. The stoichiometry of the brown spots (Mn/Sr/La = 1/0.7/3.8)

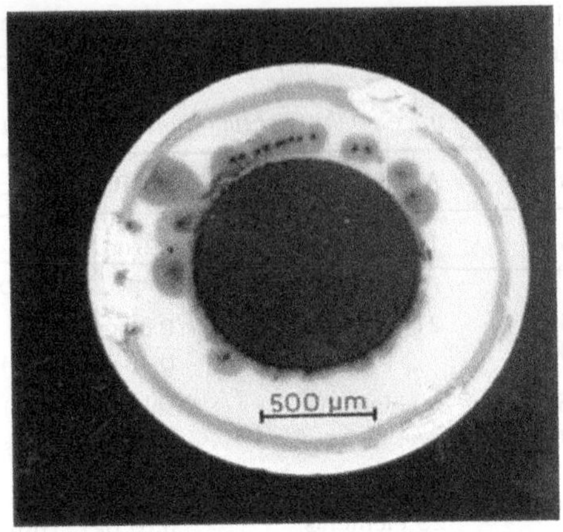

Figure 9.13 Microlocal analysis by LA-ICP-MS on a tested solid oxide fuel cell with local Impurities. The small black spots are the laser craters. (J. S. Becker, Spectrochim. Acta 57B, 1805 (2002). Reproduced by permission of Elsevier.)

is quite different from the stoichiometry of ideal perovskite layers (Mn/Sr/La = 1/0.3/0.65) measured by LA-ICP-QMS (using the Elan 5000, PerkinElmer, Sciex and a laser ablation system constructed in our laboratory). LA-ICP-QMS is very well suited for directly analyzing the concentration of impurities in µm thick ceramic layers as demonstrated in our experiment[81,117] where 26 elements were determined simultaneously in sintered YSZ layers with detection limits in the low $\mu g\, g^{-1}$ range.[1]

9.1.7 Imaging by Inorganic Mass Spectrometry in Materials Science

Imaging mass spectrometric measurements by surface analytical techniques is an important capability provided by modern instruments in order to reflect the abundance of specific mass selected atomic or polyatomic ions and their local concentration on the sample surface. In recent years, interest has strongly increased in the 2D and 3D analysis of material to study phase transformation and microstructure properties. High resolution imaging mass spectrometry by the CAMECA NanoSIMS 50 device with a spatial resolution of 50 nm has been used to investigate the 3D structure of steel[154] and to map the complex oxide structure of intergranular cracks in reactor steels that form during service in an aqueous environment.[155] These grain boundary cracks are filled with multi-layer oxides as demonstrated in FeO^+ and CrO^+ oxide images.[155] Static SIMS and SNMS using low primary ion beam doses ($\leq 1 \times 10^{12}$ ions cm^{-2}) are utilized as the dominant surface analytical techniques for surface characterization. Images are constructed by rastering of the sample surface in a selected area by the focused primary ion beam or in the secondary ion microscope mode with a sensitive ion detector (fluorescence screen). Both 2D imaging analyses of solid samples and 3D elemental imaging analyses with high lateral resolution are performed by SIMS, but quantification is difficult due to the large matrix effects. The 3D distribution characterization of trace element distribution by scanning and imaging SIMS in materials science (e.g., in high-purity chromium) was introduced by Hutter *et al.*[156]

The surface behaviour of a two-phase polymer mixture depends on the chemical structure of the polymer components, the interaction between the two polymers and the processing conditions, which was studied by ToF-SIMS to obtain the molecular surface composition. NanoSIMS as a dynamic SIMS technique provided the elemental distribution of the surface in 2D and 3D images.[157] Thin films of amorphous polymers, polystyrene – PS – and poly(methyl methacrylate) – PMMA – and diblock copolymers spin-coated onto silicon wafers with thicknesses from 5 nm to 50 nm were investigated mass spectrometrically and by atomic force microscopy (AFM) after annealing at 160 °C. NanoSIMS images reveal the formation of nanostructures of annealed surface. The AFM used for topography studies shows these nanostructures to be droplets having a phase shift distinct from the surrounding matrix. The droplet dimensions increase with increasing thickness of the film, but the absolute intensity of the ToF-SIMS spectra for all the annealed films remains almost the same. The surface composition measured by ToF-SIMS was not found to vary when the topography was completely different.[157]

Characterization of (001) n-type diamond[158] by phosphorus doping using microwave plasma enhanced chemical vapour deposition (CVD) was performed by X ray diffraction (XRD) analysis, Raman spectroscopy and Hall effect measurements and also by SIMS. With the imaging mass spectrometry of hydrogen, a high density of hydrogen atoms was observed in the area of non-epitaxial crystallites where in the bulk region hydrogen was determined at the background level to be $\sim 10^{18}\,cm^{-3}$.[158]

The capability of LA-ICP-MS compared to electron probe microanalysis (EPMA) has been studied for quantitative imaging, i.e. the analysis of the lateral elemental distribution of trace elements of an area of $350\,\mu m \times 350\,\mu m$ in multi-phase magnesium based alloy.[159] A lateral resolution of $32\,\mu m$ was achieved with limits of detection in the lower $\mu g\,g^{-1}$ range for Ti, Cr, Mn, Fe, Co, Ni, and Cu in magnesium alloys. Quantification of elements by LA- ICP-MS was carried out using a sum normalization calibration procedure. Multi-element mappings of an area were performed by LA-ICP-MS and the results for the main elements Mg, Al, and Zn were compared to measurements of the same sample area. The agreement for Mg between the two techniques was 2.2 %. The authors investigated the influences of the laser parameters, such as the repetition rate and laser spot size, and optimized the conditions for single spot analysis to achieve high lateral resolution capabilities.

Imaging mass spectrometry by LA-ICP-MS can be employed for studies of homogeneous element distribution in material science (e.g., in semiconductors or ceramics) for future application in the microelectronics industry.

References I

1. Becker, J. S., *Spectrochim. Acta B.* **57**, 1805 (2002).
2. Becker, J. S. and Dietze, H. J., *Int. J. Mass Spectrom.*, **228**, 127 (2003).
3. Becker, J. S. and Dietze, H. J., *Int. J. Mass Spectrom.*, **197**, 1 (2000).
4. DOWA Mining Co., Ltd, www.dowa.co.jp
5. Dietze, H. J. and Becker, J. S., *Fresenius' Z. Anal. Chem.*, **321**, 490 (1985).
6. Heinen, H.-J., Meier, S., Voigt, H. and Wechsung, R., *Int. J. Mass Spectr. Ion Phys.*, **47**, 19 (1983).
7. Matus, L., Seufert, M. and Jochum, K. P., *Int. J. Mass Spectr. Ion Proc.*, **84**, 101 (1988).
8. Svec, H. J., *Anal. Chem. Sympos. Series*, **19**, 89 (1984).
9. Dietze, H. J. and Becker, J. S., in *Laser Ionization Mass Analysis*, Chemical Analysis Series, John Wiley & Sons, Inc., New York, **124**, 453 (1993).
10. Kohler, V. L., Harris, A. and Wallach, E. R., *Microbeam analysis* (San Fransisco, 1989).
11. Panday, V. G., Becker, J. S. and Dietze, H. J., *At. Spectrom.*, **16**, 97 (1995).
12. Panday, V. K., Becker, J. S. and Dietze, H. J., *Anal. Chim. Acta*, **329**, 153 (1996).
13. Panday, V. K., Becker, J. S. and Dietze, H. J., *Fresenius' J. Anal. Chem.*, **352**, 327 (1995).
14. Grasserbauer, M., *Pure and Appl. Chem.*, **60**, 437 (1988).

15. Virag, A., Friedbacher, G., Grasserbauer, M., Ortner, H.M. and Wilhartitz, P., *J. Mater. Res.*, **3**, 694 (1988).
16. Putyera, K. and Sathymurthy, D., *JOM*, **56**, 323 (2004).
17. Kasik, M., Venzago, C. and Dorka, R., *J. Anal At. Spectrom.*, **18**, 603 (2003).
18. Rottmann, L. and Hamester, M., in *Asia Pacific Conference on Plasma Spectrochemistry*. Chiang Mai, Thailand (2005).
19. Matschat, R., Hinrichs, J. and Kipphardt, H. A., *Anal. Bioanal. Chem.*, **386**, 125 (2006).
20. Becker, J. S. and Dietze, H. J., *Mikrochim. Acta*, **118**, 103 (1995).
21. Granfors, M. and Giustavsson, I., *J. Anal. At. Spectrom.*, **16**, 1439 (2001).
22. Coedo, A. G., Dorado, M. T., Padilla, I. and Alguacil, F. J., *J. Anal. At. Spectrom.*, **11**, 1037 (1996).
23. Pattberg, S. and Matschat, R., *Fresenius' J. Anal. Chem.*, **364**, 410 (1999).
24. Becker, J. S., Pickhardt, C. and Dietze, H. J., *J. Anal. At. Spectrom.*, **16**, 603 (2001).
25. Becker, J. S., Pickhardt, C. and Pompe, W., *Int. J. Mass Spectrom.*, **237**, 13 (2004).
26. Van Straaten, M., Swenters, K., Gijbels, R., Verlinden, J. and Adriaenssens, E., *J. Anal. At. Spectrom.*, **9**, 1389 (1994).
27. Leach, A. M. and Hieftje, G. M., *Anal. Chem.*, **73**, 2959 (2001).
28. Leach, A. M. and Hieftje, G. M., *J, Anal. At. Spectrom.*, **17**, 852 (2002).
29. Klingbeil, P., Vogl, J., Pritzkow, W., Riebe, G. and Müller, J., *Anal. Chem.*, **73**, 1881 (2001).
30. Pickhardt, C. and Becker, J. S., *Fresenius' J. Anal. Chem.*, **370**, 534 (2001).
31. Mykytiuk, A. P., Semeniuk, P. and Berman, S., *Spectrochim. Acta Rev.*, **13**, 1 (1990).
32. Venzago, C. and Weigert, M., *Fresenius' J. Anal. Chem.*, **350**, 303 (1994).
33. Beer, B. and Heumann, K. G., *Anal. Chem.*, **65**, 3199 (1993).
34. Beer, B. and Heumann, K. G., *Fresenius' J. Anal. Chem.*, **347**, 351 (1993).
35. Balarama Krishna, M. V., Karunasagar, D. and Arunachalam, J., *Fresenius' J. Anal. Chem.*, **363**, 353 (1999).
36. Boulyga, S. F., Pickhardt, C. and Becker, J. S., *At. Spectr.*, **25**, 53 (2004).
37. Becotte-Haigh, P., Tyson, J. F., Denoyer, E. and Hinds, M. W., *Spectrochim. Acta*, **51B**, 1823 (1996).
38. Grazhulene, S., Karandashew, V., Zadnepruk, I. and Orlova, V., *Fresenius' J. Anal. Chem.*, **355**, 297 (1997).
39. Hattendorf, B., Guenther, D., Schoenbaechler, M. and Halliday, A., *Anal. Chem.*, **73**, 5494 (2001).
40. Kozono, S., Takahashi, S. and Haraguchi, H., *Anal. Bioanal. Chem.*, **372**, 542 (2002).
41. Kozono, S., Takahashi, S. and Haraguchi, H., *Analyst*, **127**, 930 (2002).
42. Pilon, F., Lorthioir, S., Birolleau, J.-C. and Lafontan, S., *J. Anal. At. Spectrom.*, **11**, 759 (1996).
43. Devos, W., Moor, C. and Linemann, P., *J. Anal. At. Spectrom.*, **14**, 621 (1999).
44. Hinds, M. H., Gregoire, D. C. and Ozaki, E. A., *J. Anal. At. Spectrom.*, **12**, 131 (1997).
45. Hoffmann, V., Kasik, M., Robinson, P. K. and Venzago, C., *Anal. Bioanal. Chem.*, **381**, 173 (2005).
46. Becker, J. S., Soman, R. S., Becker, T., Panday, V. K. and Dietze, H. J., *J. Anal. At. Spectrom.*, **13**, 983 (1998).
47. Wiedemann, B., Raedinger, G., Alt, H. C., Heumann, K. G. and Bethge, K., *Fresenius' J. Anal. Chem.*, **364**, 772 (1999).
48. Wiedemann, B., Alt, H. C., Meyer, J. D., Michelmann, R. W. and Bethge, K., *Fresenius' J. Anal. Chem.*, **364**, 768 (1999).
49. Wiedemann, B., Meyer, J. D., Jockel, D., Freyhardt, H. C., Birkmann, B. and Müller, G., *Fresenius' J. Anal. Chem.*, **370**, 541 (2001).
50. Krecar, D., Rosner, M., Draxler, M., Bauer, P. and Hutter, H., *Anal. Bioanal. Chem.*, **384**, 525 (2006).
51. McDaniel, F. D., Datar, S. A., Nigam, M. and Ravi Prasa, G. V., *Nucl. Instrum. Methods B*, **190**, 826 (2002).
52. Becker, J. S., Saprykin, A. I. and Dietze, H. J., *Int. J. Mass Spectr. Ion Proc.*, **164**, 81 (1997).
53. Jäger, R., Saprykin, A. I., Becker, J. S., Dietze, H. J. and Broekaert, J. A. C., *Mikrochim. Acta*, **125**, 41 (1997).
54. Jäger, R., Becker, J. S., Dietze, H. J. and Broekaert, J. A. C., *Int. J. Mass Spectr. Ion Proc.*, **171**, 183 (1997).
55. Saprykin, A. I., Becker, J. S. and Dietze, H. J., *J. Anal At. Spectrom.*, **10**, 897 (1995).

56. Jäger, R., Becker, J. S., Dietze, H. J. and Broekaert, J. A. C., in *Advances in Mass Spectrometry* K. J. Karjalainen, A. E. Hesso, J. E. Jalonen, U. P. Karjalainen (eds.), Elsevier Science Publisher B. V., Amsterdam (1998) MOPO136.
57. Becker, J. S. and Dietze, H. J., *J. Anal At. Spectrom.*, **12**, 881 (1997).
58. Sahayam, A. C., Jiang, S. J. and Wang, C. C., *J. Anal. At. Spectrom.*, **19**, 407 (2004).
59. Ueng, R. L., Jiang, S. J., Wan, C. C. and Sahayam, A. C., *Anal. Chim. Acta*, **536**, 295 (2005).
60. Ueng, R. L., Sahayam, A. C., Jiang, S. J. and Wan, C. C., *J. Anal. At. Spectrom.*, **19**, 681 (2004).
61. Becker, J. S. and Tenzler, D., *Fresenius' J. Anal. Chem.*, **370**, 637 (2001).
62. Becker, J. Su., Pickhardt, C., Hoffmann, N., Hocker, H. and Becker, J. S., *At. Spectr.*, **23**, 1 (2002).
63. Pickhardt, C., Izmer, A., Zoriy, M. V., Schaumlöffel, D. and Becker, J. S., *Int. J. Mass Spectr.*, **248**, 136 (2006).
64. Koch, J., Feldmann, I., Hattendorf, B. *et al.*, *Spectrochim. Acta*, **57B**, 1057 (2002).
65. Fukuda, M. and Sayama, Y., *Fresenius' J. Anal. Chem.*, **357**, 169 (1997).
66. Wende, M. C. and Broekaert, J. A. C., *Fresenius' J. Anal. Chem.*, **370**, 513 (2001).
67. Dobney, A. M., Mank, A. J. G., Grobecker, K. H., Conneely, P. and de Koster, C. G., *Anal. Chim. Acta*, **423**, 9 (2000).
68. Wolf, R. E., Thomas, C. and Bohlke, A., *Appl. Surf. Sci.*, **129**, 299 (1998).
69. Adriaens, A., Van Vaeck, L. and Adams, F., *Mass Spectrom. Rev.*, **18**, 48 (1999).
70. Delcorte, A., Yunus, S., Nieuwjaer, N., Poleunis, C. and Bertrand, P., in *5th European Workshop on Secondary Ion Mass Spectrometry*. Münster (2006). Book of Abstracts, 102.
71. Morea, G., Sabbatini, L., Zambonin, P. G., Swift, A. J., West, R. H. and Vickerman, J. C., *Macromolecules*, **24**, 3630 (1991).
72. Newmann, J. G., Carlson, B. A., Michael, R. S., Mounlder, J. F. and Hohlth, T. A., *Static SIMS Handbook of Polymer Analysis*. Perkin Elmer Corp., Physical electronocs Division, Elden Prairie, Minnesota (1991).
73. Zhang, X., Yi, Y., Liu, Y. *et al.*, *Anal. Chim. Acta*, **555**, 57 (2006).
74. Takaku, Y., Masuda, K., Takahashi, T. and Shimamura, T., *J. Anal. At. Spectrom.*, **8**, 687 (1993).
75. Becker, J. S. and Dietze, H. J., *Fresenius' J. Anal. Chem.*, **344**, 69 (1992).
76. Saprykin, A. I., Becker, J. S. and Dietze, H. J., *Fresenius' J. Anal. Chem.*, **364**, 763 (1999).
77. Harrison, W. W. and Hang, W., *J. Anal. At. Spectrom.*, **11**, 835 (1996).
78. Marcus, R. K. and Broekaert, J. A. C., *Glow discharge plasmas in analytical spectroscopy* (John Wiley & Sons Ltd., Chichester, West Sussex, England, 2003).
79. Schelles, W. and Van Grieeken, R., *J. Anal. At. Spectrom.*, **12**, 49 (1997).
80. Marcus, R. K., *J. Anal. At. Spectrom.*, **11**, 821 (1996).
81. Becker, J. S., Breuer, U., Westheide, J., Saprykin, A. I., Holzbrecher, H., Nickel, H. and Dietze, H. J., *Fresenius' J. Anal. Chem.*, **355**, 626 (1996).
82. Jakubowski, N., Dorka, R., Steers, E., Tempez, A., J. Anal. At. Spectrom. 22, 722 (2007).
83. Jäger, R., Becker, J. S., Dietze, H. J. and Broekaert, J. A. C., *Fresenius' J. Anal. Chem.*, **358**, 214 (1997).
84. Saprykin, A. I., Becker, J. S., Von der Crone, U. and Dietze, H. J., *Fresenius' J. Anal. Chem.*, **358**, 145 (1997).
85. Saprykin, A. I., Becker, J. S. and Dietze, H. J., *Fresenius' J. Anal. Chem.*, **359**, 449 (1997).
86. Beer, K. and Heumann, K. G., *Fresenius' J. Anal. Chem.*, **350**, 284 (1994).
87. Jakubowski, N., Tittes, W., Pollmann, T., Stuewer, D. and Broekaert, J. A. C., *J. Anal. At. Spectrom.*, **11**, 794 (1996).
88. Kohl, F., Jakubowski, N., Brand, R., Pilger, C. and Broekaert, J. A. C., *Fresenius J. Anal. Chem.*, **359**, 317 (1997).
89. Quin, S., Jiang, Z., Hu, B., Qin, Y. and Hu, S., *Fresenius' J. Anal. Chem.*, **367**, 250 (2000).
90. Becker, J. S., Westheide, J., Saprykin, A., Holzbrecher, H., Breuer, U. and Dietze, H. J., *Mikrochim. Acta*, **125**, 153 (1997).
91. Jenett, H., in *Analytikertaschenbuch 16*. Springer Verlag, Berlin, Heidelberg (1997) 43.
92. Benninghoven, A., Janssen, K. T. F., Tümpner, J. and Werner, W. H., *Secondary Ion Mass Spectrometry VIII (Proceedings of the Ninth International Conference on SIMS*, John Wiley & Sons, Ltd, Chichester (1992).

93. Benninghoven, A., Ruedenauer, F. G. and Werner, W. H. in *Secondary Ion Mass Spectrometry* John Wiley & Sons Inc., New York, (1987).
94. Wilson, R. G., Stevie, F. A. and Magee, C. W., *Secondary Ion Mass Spectrometry*. John Wiley & Sons, Inc., New York (1989).
95. Seng, H. L., Yakovlev, N., Zhou, H. L. and Liu, W., in *5th European Workshop on Secondary Ion Mass Spectrometry*. Münster (2006) p6.
96. Sumiya, M., Yoshimura, K., Ohtsuka, K. and Fuke, S., *Appl. Phys. Lett.*, **76**, 2098 (2000).
97. Ehrke, U., Sears, A., Lerch, W. *et al.*, *J. Vac. Sci. Techn.*, **22**, (2004).
98. Ehrke, U. and Maul, H., *Semicond. Proc.*, **8**, 111 (2005).
99. Merkolov, A., de Chambost, E., Schumacher, M. and Peres, P., *Appl. Surf. Sci.*, **231–232**, 640 (2004).
100. Gao, Y., *J. Appl. Phys.*, **64**, 3760 (1988).
101. Niehus, E., Grehl, T., Kollmer, F., Moellers, R. and Rading, D., in *5th European Workshop on Secondary ion Mass Spectrometry*. Münster (2006) 7.
102. Bock, W., Gnaser, H., Wahl, M. and Kopnarski, M., in *5th European Workshop on Secondary Ion Mass Spectrometry*. Münster (2006) 12.
103. De Chambost, E., Monsallut, P., Rasser, B. and Schuhmacher, M., *Semicond. Proc.*, **8**, 111 (2005).
104. Emziane, M., Durose, K., Halliday, D. P., Romeo, N. and Bosio, A., *Thin Sol. Film*, **511–512**, 66 (2006).
105. Lui, R., Wee, A. T. S., Shen, D. H. and Takenaka, H., *Surf. Interface Anal.*, **36**, 172 (2004).
106. Gastel, M., Breuer, U., Holzbrecher, H., Becker, J. S., Dietze, H. J. and Wagner, H., *Fresenius' J. Anal. Chem.*, **358**, 207 (1997).
107. Gastel, M., Breuer, U., Holzbrecher, H. *et al.*, *Fresenius' J. Anal. Chem.*, **353**, 478 (1995).
108. Pidun, M., Lesch, N., Richter, S. *et al.*, *Mikrochim. Acta*, **132**, 429 (2000).
109. Jorzick, J., Loesch, J., Kopnarski, M. and Oechsner, H., *Appl. Phys. A*, **78**, 655 (2004).
110. Oechsner, H. M., Bock, W., Kopnarski, M. and Mueller, M., *Mikrochim. Acta*, **133**, 69 (2000).
111. von Gradowski, M., Jacoby, B., Hilgers, H., Barz, J., Wahl, M. and Kopnarski, M., *Surf. Coatings Techn.*, **200**, 334 (2005).
112. Von Gradowski, M., Wahl, M., Jacoby, B., Hilgers, H. and Kopnarski, M., *Vakuum in Forschung und Praxis*, **17**, 73 (2005).
113. Wahl, M., Gradowski, M. V., Jacoby, B., Hilgers, H. and Kopnarski, M., in *Angewandte Oberflächentagung, AOFA 14*. Kaiserslautern (2006) KV4.
114. Specs GmbH, B., *Application note*.
115. Mitchell, L. J., Ravi Prasa, G. V., Pelicon, P., Smith, E. B. and McDaniel, F. D., *Nucl. Instrum. Methods B*, **119–120**, 455 (2004).
116. Efimov, A., Kasik, M., Putyera, K. and Moreau, O., *Electrochem. Solid-State Lett.*, **3**, 477 (2000).
117. Westheide, J. T., Becker, J. S., Jäeger, R., Dietze, H. J. and Broekaert, J. A. C., *J. Anal. At. Spectr.*, **11**, 661 (1996).
118. Bleiner, D., Lienemann, P., Ulrich, A., Vionmont, H. and Wichser, A., *J. Anal. At. Spectrom.*, **18**, 1146 (2003).
119. Izmer, A., Zoriy, M., Pickhardt, C., Quadakkers, W. J., Shemet, V., Singheiser, L. and Becker, J. S., *J. Anal. At. Spectrom.*, **20**, 918 (2005).
120. Balcean, L. I. L., Lenaerts, J., Moens, L. and Vanhaecke F., *J. Anal. At. Spectrom.*, **20**, 417 (2005).
121. Bleiner, D., Plotnikov, A., Vogt, C., Wetzig, K. and Guenther, D., *Fresenius' J. Anal. Chem.*, **368**, 221 (2000).
122. Plotnikov, A., Vogt, C., Hoffmann, V., Taschner, C. and Wetzig, K., *J. Anal. At. Spectrom.*, **16**, 1290 (2001).
123. Coedo, A. G., Dorado, T., Padilla, I. and Farinas, J. C., *J. Anal. At. Spectrom.*, **20**, 612 (2005).
124. Armelo, L., Bleiner, D., Di Noto, V. *et al.*, *Appl. Surf. Sci.*, **249**, 277 (2005).
125. Baunack, S., Menzel, S., Pekarciková, M. *et al.*, *Anal. Bioanal. Chem.*, **375**, 891 (2003).
126. Chi, P. H., Ko, F. H., Hsu, C. T. *et al.*, *J. Anal. At. Spectrom.*, **17**, 358 (2002).
127. Becker, J. S. and Boulyga, S. F., *Fresenius' J. Anal. Chem.*, **370**, 527 (2001).
128. Boulyga, S. F., Dietze, H. J. and Becker, J. S., *J. Anal At. Spectrom.*, **16**, 598 (2001).
129. Huber, G., Gnaser, H. and Ziegler, C., *Anal. Bioanal. Chem.*, **375**, 917 (2003).

130. Hanton, D. S. and Cornelio Clark, P. A., in *The Encyclopedia of Analytical Chemistry: Instrumentation and Application*. John Wiley & Sons Ltd., Chichester, UK (2000) 11872.
131. Ehrke, H. U., Sears, A. and Schnürer-Patschan, C., in *5th European Workshop on Secondary Ion Mass Spectrometry*. Münster (2006) 25.
132. Arlinghausen, H. F., in *Surface and Thin Film Analysis* H. B. A. H. Jenett (ed.). Wiley VCH, Weinheim, (2002) 86.
133. Krauss, A. R. and Gruen, D. M., *Appl. Phys.*, **14**, 89 (1977).
134. Steffens, P., Niehus, E., Friese, T., Greifendorf, D. and Benninghoven, A., *J. Vac. Sci. Techn. A*, **3**, 1322 (1985).
135. Treverton, J. A., Ball, J., Johnson, D., Vickerman, J. C. and West, R. H., *Surf. Interface Anal.*, **15**, 369 (1990).
136. Sakakini, B. H., Ransley, I. A., Oduoza, C. F., Vickerman, J. C. and Chesters, M. A., *Surf. Sci.*, **271**, 227 (1992).
137. Zhou, X. L., Flores, C. R. and White, J. M., *Surf. Sci. Lett.*, **268**, L267 (1992).
138. Alam, M., Henderson, M. A., Kaviratna, P. D., Herman, G. S. and Oeden, C. H. F., *J. Phys. Chem. B*, **102**, 111 (1998).
139. Borg, H. J., Reijerse, J., van Santen, R. and Niemantsverdriet, J. W., *J. Chem. Phys.*, **101**, 10052 (1994).
140. Henderson, M. A., *J. Chem. Phys. B*, **101**, 221 (1997).
141. Fabry, L., Pahlke, S., Kotz, L., Wobrauschek, P. and Streli, C., *Fresenius' J. Anal. Chem.*, **363**, 98 (1999).
142. Horn, M., *Fresenius' J. Anal. Chem.*, **364**, 385 (1999).
143. Ferrero, E. J. and Posey, D., *J. Anal. At. Spectrom.*, **17**, 1194 (2002).
144. O'Brien, S. E., Acon, B. W., Boulyga, S. F., Becker, J. S., Dietze, H. J. and Montaser, A., *J. Anal At. Spectrom.*, **18**, 230 (2003).
145. Shabani, M. B., Shiina, Y., Kirscht, F. G. and Shimanuki, Y., *Mater. Sci Engin. B-Solid State Mater. Adv. Technol.*, **102**, 238 (2003).
146. Pritzkow, W., Vogl, J., Berger, A. *et al. Fresenius' J. Anal. Chem.*, **371**, 867 (2001).
147. Hoelzl, R., Fabry, L., Kotz, L. and Pahlke, S., *Fresenius' J. Anal. Chem.*, **366**, 64 (2000).
148. Völlkopf, U., Klemm, K. and Pfluger, M. A. S., *At. Spectrosc.*, **29**, 53 (1999).
149. Wildner, H., *J. Anal. At. Spectrom.*, **13**, 573 (1998).
150. Wills, J. and Hamester, M., in *Book of Abstracts of European Winter Conference on Plasma Spectrochemistry, Pau, France*, R. Lobinski (ed.) (1999).
151. Wolf, E. and Grosser, Z. A., *At. Spectroscopy*, **18**, 145 (1997).
152. Yamasaki, S.-I., Tsumura, A. and Takaku, Y., *Mikrochem. J.*, **49**, 305 (1994).
153. Janssens, G., Geuens, J., De Keyzer, R. *et al.*, in *Proceedings of the 10th International Conference on Secondary Ion Mass Spectrometry*, B. H. A. Benninghoven and H. W. Werner (eds.). John Wiley & Sons, Inc., New York (1997) 161.
154. Valle, N., Driller, J., Bouaziz, O. and Migeon, H. N., in *5th Workshop on Secondary ion Mass Spectrometry*. Münster (2006) 97.
155. Kilburn, M. R., Lorzano-Perez., S., English, C. A. and Grovenor, C. R. M., in *5th European Workshop on Secondary Ion Mass Spectrometry*. Münster (2006) 96.
156. Hutter, H., Brunner, C., Piplits, K., Wilhartitz, P. and Grasserbauer, M., in *Proceedings of the 10th International Conference on Secondary Ion Mass Spectrometry*, B. H. A. Benninghoven and H. W. Werner (eds.). John Wiley & Sons, Inc., New York (1997) 141.
157. Kailas, L., Audinot, J. N., Migeon, H. N. and Bertrand, P., *Composite Interfaces*, **13**, 423 (2006).
158. Kato, H., Watanabe, H., Yamasaki, S. and Okushi, H., *Diamond and Related Materials*, **15**, 548 (2006).
159. Latkoczy, C., Müller, Y., Schmutz, P. and Günther, D., *Applied Surface Science*, **252**, 127 (2005).

9.2 Environmental Science and Environmental Control

Inorganic mass spectrometry has found widespread application in the analysis of environmental samples. Pollution of the natural environment (air, soil, plants and animals and also drinking or groundwater, rain, river or sea water) by heavy toxic metals or metalloids from industrial processes, sewage sludge, municipal and industrial waste deposits, by the combustion of fossil fuels, and due to contamination as the result of nuclear fallout leads to serious environmental contamination. Fields of application for mass spectrometry in environmental science and environmental control are summarized in Figure 9.14. The aim of applying analytical chemistry in environmental science is consequently the monitoring of air, water, plants, animals and soils (contaminated soils and solid waste disposal), ensuring the quality and purity of agricultural or manufactured industrial products, mining etc., which cause increasing levels of pollution that threaten the safety and health of the population. In order to study contamination, environmental control requires multi-element, isotope and surface analysis at the trace and ultratrace level and the determination of the individual metal species present in environmental samples.

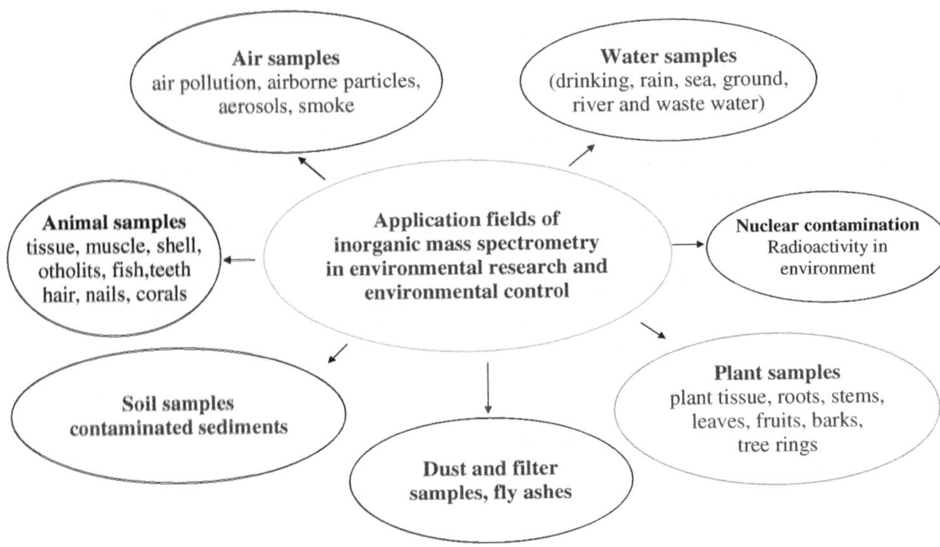

Figure 9.14 *Fields of application in the environmental sciences.*

Different analytical techniques such as ICP-OES (optical emission spectrometry with inductively coupled plasma source), XRF (X-ray fluorescence analysis), AAS (atomic absorption spectrometry) with graphite furnace and flame GF-AAS and FAAS, NAA (neutron activation analysis) and others, are employed for the trace analysis of environmental samples. The main features of selected atomic spectrometric techniques (ICP-MS, ICP-OES and AAS) are summarized in Table 9.20.[1] The detection ranges and LODs of selected analytical techniques for trace analysis on environmental samples are summarized in Figure 9.15.[1]

ICP-MS, in particular, is today the most frequently applied inorganic mass spectrometric technique and has received increasing attention in recent years as demonstrated by a rapidly growing

Table 9.20 *Main features of atomic spectrometric techniques for analysis of liquid samples.*

	ICP-MS	ICP-OES	AAS
Multi-element capapility	+	+	(oligo analytical method)
Matrix and minor elements (stoichiometry)	+	++	+
Trace elements	++	+	++
Ultratrace elements	++	—	+
Limits of detection	pg ml^{-1} to fg ml^{-1}	ng ml^{-1}	pg ml^{-1}
Sample amount of analyte	ng to fg	μg to ng	ng to pg
Isotope analysis (isotope dilution analysis)	++	—	—
Dynamic range	10^9	10^8	10^9

(J. S. Becker, Trends in Anal. Chem., **24**, 243 (2005). Reproduced by permission of Elsevier.)

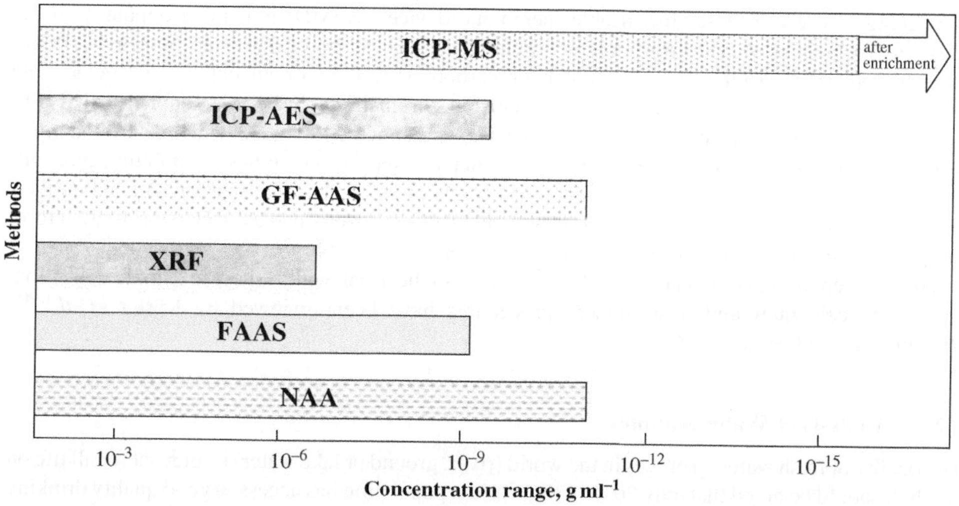

Figure 9.15 *Detection ranges and LODs of selected analytical techniques for trace analysis on environmental samples. (J. S. Becker, Trends in Anal. Chem.,* ***24****, 243 (2005). Reproduced by permission of Elsevier.)*

number of publications on environmental applications.[1–13] On the other hand, the largest application field for ICP-MS is environmental analysis with respect to the environmental control of pollution and also in environmental research. About 40 % of all applications of ICP-MS concern this relevant topic. Whereas in solid mass spectrometry the environmental samples can often be analyzed directly or only need a small number of sample preparation steps, ICP-MS usually requires a more or less time consuming sample preparation which is necessary to obtain valuable data. In environmental science and technology, the familiar sample preparation steps for solid samples are generally applied, such as open vessel digestion on a hot plate and closed vessel digestion with HNO_3 or in a mixture with $HClO_4$, HCl, HF or digestion with aqua regia. Closed vessel digestion procedures are recommended in order to avoid possible contamination and loss of volatile analytes during sample digestion. In addition, selective extraction techniques or leaching of the analyte (e.g., for U isotope analysis in soils) are employed to reduce matrix effects during mass spectrometric measurements. With respect to sample collection and extraction, solid phase extraction (SPE) is one of the leading techniques for the extraction of both organic and inorganic species from aqueous samples. SPE is attractive because it affords easy concentration of the element species of interest, requires minimal amounts of solvent and can be tailored to extract either a broad range of organic and metal containing compounds or provide specific extraction of a pollutant or compound class. New materials for analyte extractions and the development of automated sample processing methods represent trends in SPE.[10] Further developments also concern solid phase microextraction (SPME), hollow fibre and liquid phase microextraction and the introduction of semipermeable membrane devices and others for application in ICP-MS.[10] In SPME, a small polymer coated microextraction fibre is applied to collect analytes of interest. This technique can be utilized as a universal sampling and extraction technique for air, water samples and the headspace above the solids in environmental science and technology, and also to determine time integrated concentrations of pollutants in air and water to estimate the bioaccumulation of toxic pollutants. Development of new materials for SPME to increase sensitivity and selectivity is ongoing.[10] The use of semipermeable membrane devices (SPMD) is gaining popularity as an inexpensive method of collecting organic compounds such as polychlorinated biphenyls (PCB), hexachlorobenzene (HCB), polyaromatic hydrocarbons (PAH) or organochlorine pesticides and others from air and water samples.[10] Size exclusion chromatography (SEC) coupled to ICP-MS promotes a size based separation of macromolecules combined with an elemental determination.[13] SEC-ICP-MS is utilized in the growing area of element speciation studies of mobility, bioavailability and toxicity of different physicochemical forms of elements, which largely depend on their element distribution in the different size fractions of the macromolecules. Detection limits are in the parts per trillion range for most elements of the periodical table.[13]

Trends in environmental analytical chemistry and applications with respect to sample collection, separation techniques and measurement procedures have been reviewed by Köster *et al.*[10,14], Richardson[15] and Butler *et al.*[16]

9.2.1 Analysis of Water Samples

The quality of fresh water resources in the world (river, ground or lake water) is relevant for all life on Earth. It should be noted that only 20 % of the world's population has access to good quality drinking water. The water quality of natural water sources for life is affected by their origin (e.g., mountain aquifers may be contaminated in densely populated regions with industrial and agricultural activities). These numerous sources of pollution have the potential to contaminate the aquifer.

ICP-MS is well suited for the multi-element determination of trace impurities in different types of water samples, such as drinking water, rain, sea, river, groundwater and also waste water samples (and in addition, highest purity water as required for microelectronics) due to its excellent detection limits from the $pg\,ml^{-1}$ to $fg\,ml^{-1}$ concentration range. This powerful and sensitive mass spectrometric technique is applied in routine measurement programs for characterization of water quality and determination of pollution in many laboratories worldwide.

The collection and preparation of water samples requires individual approaches for different analytical tasks. If heavy metals or long-lived radionuclides at the trace and ultratrace concentration range are to be determined in water samples by ICP-MS, especially careful sampling is necessary to avoid possible contamination (using clean bottles and containers washed and cleaned before use, for example, with 2 % nitric acid and high purity water to stabilize traces in the samples), and the loss of analyte by adsorption effects or precipitation should be also considered.

Water samples (drinking water, rain, sea, river or waste water and others) have been characterized by ICP-MS with multi-element capability in respect to metal impurities (such as Ag, Al, As, Ba, Be, Ca, Cd, Cr, Co, Cu, Fe, Hg, K, Na, Sb, Se, Mg, Mn, Mo, Ni, Pb, Tl, Th, U, V and Zn) in many laboratories in routine mode with detection limits at the low $ng\,l^{-1}$ range using ICP-QMS, and below by means of ICP-SFMS. Drinking water samples are controlled in respect of the European legislation (Council Directive 98/83/EC of 3 November 1998 on the quality of water intended for human consumption). For quality control of analytical data, certified standard reference materials e.g. drinking water standard (40CFR 141.51), river water reference material SLRS-4 or CASS-2 certified reference sea-water material and others are employed.

The procedures from water collection and storage via separation of analytes to measurement procedures for different analytical tasks will be described in the following example (for analysis of water samples from Sea of Galilee) in more detail. Samples for measurement of the uranium depth profile in the Sea of Galilee were collected by Halicz *et al.*[90] using a 40 m plastic tube (Tygon R 3603 ID 8.0 mm) connected to a peristaltic pump (MasterFlex by Cole-Parmer) with an online filter (cellulose nitrate filter 0.45 μm Whatman).[90] The sample preparation procedures and trace matrix separation methods are quite different for various analytical studies. In multi-elemental analysis using quadrupole based ICP-MS, major, minor and trace elements of water samples are often measured after dilution alone. As an example, the typical composition of natural water from Israel collected at different locations in the Sea of Galilee measured by quadrupole based ICP-MS is given in Table 9.21. Due to difficulties with the multi-element trace analysis of rare earth elements (REE) in the sub-$\mu g\,l^{-1}$ concentration range in the presence of a relatively high matrix content ($Na = 20–150\,mg\,l^{-1}$, $Mg = 20–50\,mg\,l^{-1}$ and $Ca = 40–100\,mg\,l^{-1}$), lanthanide group separation and purification was proposed before mass spectrometric multi-element analysis using AF-Chelate-650M Toyopearl® resin (TOSOH BIOSCIENCE), as illustrated in Figure 9.16. To accurately determine contents at the ultratrace concentration level of rare earth elements, possible isobaric interferences by polyatomic ions should be investigated and considered. After a careful trace/matrix separation, isobaric interferences with oxide ions e.g. of Ba can be excluded. The results of the trace analysis of REEs from five water samples collected in Israel are summarized in Table 9.22.

In addition, there exist a multitude of different applications in water analysis by ICP-MS for environmental control. For example, Lawrence *et al.*[17] determined rare earth element concentrations in natural waters (these are river, lake, sea or groundwater) by quadrupole ICP-MS using external calibration and employed river water reference material SLRS-4 to validate the analytical data. The speciation of yttrium and lanthanides in water samples by SEC-ICP-MS was studied by Haraguchi *et al.*[18] whereby the detection of La, Ce and Pr corresponded to the occurrence of large organic molecules.

Due to the various health risks of different element species, there are a multitude of applications for natural water samples in this field (e.g., Cr and Sb speciation or Br and I determination).[19] The investigation of heavy metal complexes with humic substances by isotope dilution SEC-ICP-MS has been described, for example, by McSheehy and Mester.[20] Copper, zinc and molybdenum were found to form complexes with similar size fractions of humic substances in seepage water samples from soils. Sturgeon's group proposed the use of solid phase microextraction (SPME)

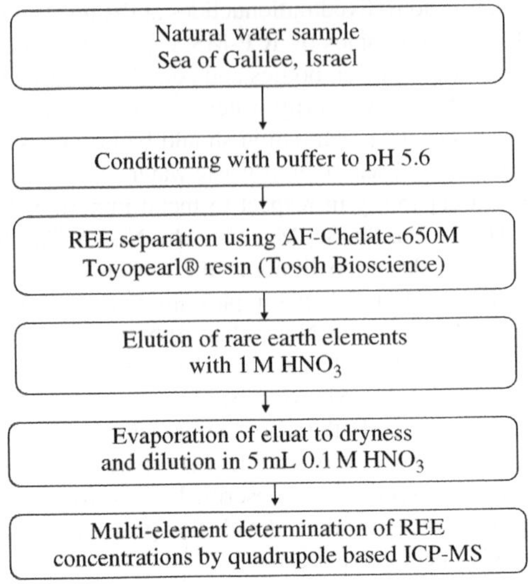

Figure 9.16 *Schematic of lanthanide group separation and purification. (L. Halicz, J. S. Becker et al. Int. J. Mass Spectrom.,* ***249–250****, 296 (2006). Reproduced by permission of Elsevier.)*

Table 9.21 *Composition of natural water from Israel collected at different locations in Sea of Galilee measured by quadrupole based ICP-MS.*

Element	**ICP-MS**
Ca	40–100 mg l^{-1}
Mg	30–40 mg l^{-1}
K	2–8 mg l^{-1}
Na	20–150 mg l^{-1}
B	30–200 μg l^{-1}
Ba	40–200 μg l^{-1}
Cr	0.5–2 μg l^{-1}
Cu	0.5–5 μg l^{-1}
Mn	0.3–70 μg l^{-1}
Ni	0.2–4 μg l^{-1}
Rb	0.5–3 μg l^{-1}
Se	0.3–5 μg l^{-1}
Sr	300–2000 μg l^{-1}
U	0.4–5 μg l^{-1}
Zn	0.2–60 μg l^{-1}

(L. Halicz, J. S. Becker et al. Int. J. Mass Spectrom., **249–250**, 296 (2006). Reproduced by permission of Elsevier.)

Table 9.22 Results of REE determination in five natural water samples (I–V) from Israel (concentration in $ng\,l^{-1}$).

REE	I	II	III	IV	V
La	21.5	12.1	8.4	3.4	0.85
Ce	20.4	6.3	5.3	2.7	1.3
Pr	5.3	3.0	1.9	0.43	0.17
Nd	19.3	13.2	8.6	1.7	0.62
Sm	3.7	2.3	1.6	0.49	0.16
Eu	1.1	0.68	0.39	0.06	0.04
Gd	4.1	2.6	1.9	0.36	0.15
Tb	0.55	0.37	0.29	0.05	0.03
Dy	3.7	2.5	1.9	0.48	0.16
Ho	0.85	0.58	0.42	0.10	0.03
Er	2.4	1.6	1.3	0.33	0.12
Tm	0.37	0.23	0.18	0.07	0.02
Yb	2.1	1.3	1.0	0.53	0.08
Lu	0.34	0.24	0.15	0.15	0.03

(L. Halicz, J. S. Becker et al. Int. J. Mass Spectrom., **249–250**, 296 (2006). Reproduced by permission of Elsevier.)

to extract Cr from sea water after derivatization with β-diketonate ligand 1,1,1-trifluoro-2,4-pentadione (trifluoroacetylacetone – TFA).[21] The chelation reaction of Cr was conducted in a single (aqueous) phase medium. Both liquid phase and headspace extraction were studied by employing a 100 μm polydimethylsiloxane coated SPME fibre. Gas chromatography was coupled to electron impact mass spectrometry (EI-MS) and ICP-MS for the quantitation of Cr. A detection limit for Cr of about 0.01 $ng\,ml^{-1}$ was achieved.[21] Anion exchange chromatography – hydride generation ICP-MS with a dynamic reaction cell yielded instrumental detection limits of 0.15, 0.27 and 0.19 $ng\,l^{-1}$, respectively, for the determination of dissolved inorganic selenium species (selenite, selenate and selenocyanate) in rain and sea water.[22] Methane was employed as the reaction gas to eliminate the interfering $^{40}Ar_2^+$ dimeric ions (to determine $^{80}Se^+$). The oxidation of Se^+ to SeO^+ using O_2 as the reaction gas was also attempted, but yielded incomplete (10–25 %) conversion.[22]

Organotin compounds, which are used in antifouling paints for ships, have been measured widely in coastal waters and sediments. Their toxicity generally follows the order trialkyl > dialkyl > monoalkyl, whereby the dialkyl form is much more neurotoxic, with an effect on brain cells in concentrations as low as 30 $ng\,g^{-1}$.[23] The distribution and concentration of different organotin compounds in marine systems and in freshwater environments and their transformation pathways is discussed by Rosenberg.[24] After extracting organotin compounds from water (e.g., using a poly(dimethylylsiloxane)-coated stir bar) the analysis was performed by thermal desorption GC-ICP-MS.[15]

Unlike other anthropogenic contaminants in waters, the arsenic contamination generally comes from natural sources through the erosion of rocks, minerals and soils.[23] The degree of industrialization is reflected in the amount of As in sewage sludge, as discussed by Prohaska and Stingeder.[25] Waste water from geothermal power plants contains a high amount of arsenate, which provides a further source of pollution.[26]

LA-ICP-MS is the method of choice for directly studying pollution in ice samples. Reinhardt and co-workers[27,28] analyzed ice cores from the polar region by LA-ICP-MS employing a home made cryogenic laser ablation chamber. Selected trace elements (Mg, Al, Fe, Zn, Cd, Pb and some rare earth elements) measured in frozen ice cores of different ages from Greenland can provide information on artificial pollution, but also on climate changes in the past.[27]

9.2.2 Analysis of Air Samples, Particles and Smoke

Air pollution usually means the presence of sufficient concentrations of substances in the atmosphere to interfere with the health and welfare of persons or the environment. This can result in a serious and increasing problem, not only in urbanized and industrialized areas, and it is also a health risk for humans and animals. The investigation of air samples to identify possible anthropogenic pollution plays a particular part in environmental science. A multitude of studies of airborne particulate matter have been concerned with the source and behaviour of atmospheric aerosols, because they can be related to serious health hazards, such as increasing risks of respiratory and other diseases, leading to a higher mortality rate.[29–31] The characterization of aerosol particles from artificial sources (industrial sources such as coal combustion, power plants or especially vehicle exhaust) is required in environmental research, but also in the atmospheric sciences where, besides the particle size distribution, in particular the chemical composition has to be determined to study atmospheric processes and to control environmental pollution. During the last 15 years, aerosol and environmental sciences have been strongly influenced by analytical data from improved measurement techniques for particle characterization.[32] Different analytical possibilities for characterizing individual aerosol particles by direct species, trace and isotope analysis of solids using quite different surface analytical techniques, such as X-ray induced photoelectron spectroscopy – XPS, auger electron spectrometry – AES, Mösbauer spectrometry, micro-Raman spectrometry, SIMS, X-ray absorption techniques: μ-EXAFS and μ-XANES, LA-ICP-MS and others, has been reviewed by Ortner.[33] Air particulates are usually collected on filter surfaces of small pore size using cellulose filters,[34] glass fibre or quartz fibre filters,[35] PTFE filters[36] and others.

Inorganic mass spectrometric techniques for the direct analysis of aerosols without sample preparation, such as LA-ICP-MS and SIMS, have the ability to make sensitive measurements of a multitude of elements and precise isotope analyses to identify the sources of the aerosols. Two pioneering papers on the direct rapid multi-element analysis of atmospheric particulate matter without any additional pretreatment using LA-ICP-MS were contributed by Luedke *et al.* in 1994[35] and by Tanaka *et al.* in 1998.[37] Several years later, LA-ICP-MS was successfully employed to examine large amounts of aerosols in China's urban areas where pollutants are not only emitted from anthropogenic sources but also from natural sources such as soil dust.[30] The chemical composition and the concentration of elements, which varies from day to day, was studied in Beijing by LA-ICP-MS (using a PMS 2000, Yokogawa Electric, Co. coupled with a Nd:YAG laser ablation system).[30] More than 700 filter samples were analyzed in this study over a period of two years. The daily average metal concentrations were determined, e.g., for Al, Cr, As, Cd and Pb as 3.5, 0.019, 0.048, 0.0068 and 0.43 $\mu g\, m^{-3}$, respectively. The As concentrations in aerosol particles were 20-fold higher than those in Tokyo. Soil dust and coal combustion were considered to be the primary source of aerosols in Beijing.[30]

ICP-ToF-MS combined with an inductively heated vaporizer was employed by Luedke *et al.* for trace metal analysis in size separated arctic aerosols (from Spitzbergen).[38] The sampling of aerosol particles was performed by impaction of the particles on cleaned graphite targets. By means of a cascade impactor, eight size classes between 0.35 μm and 16.6 μm aerodynamic diameters were separated. To analyze the metal contents, the targets were rapidly heated up to 2700 °C in an inductively heated vaporizer system. The evaporated material was then transported with Ar into the ion source of an ICP-ToF-MS. The limits of detection for the selected trace elements investigated ranged between 2 and 200 pg. A significant enrichment (up to a factor of 27) of anthropogenic components was found for Li, Mn, Ni, Co, Zn, Pd, Ag, Cd, In, Ba, Tl and Pb.[38]

The trace analysis of environmental samples generally requires special small sample preparation and measurement techniques if the amount of material (e.g. for analysis of aerosols) is limited. Significant improvements in analytical techniques, in order to reduce artefacts and chemical modifications by particle sampling and transportation, are possible by applying the online techniques developed for particle analysis.[32,39–41] The use of special inlet systems that transfer the particles from their natural environment into the mass spectrometer and the development of transportable and reliable instruments with long term stability were the basis for enhanced applications in aerosol science.[32] Two transportable laser mass spectrometers, single particle analysis of sizing system (SPASS) and laser mass analyzer for particles in airborne state (LAMPAS 2), have been developed and applied to determine the size and chemical composition of several thousand single mineral or carbonaceous particles.[32] The comparison of single particle mass spectra and spectral patterns by these two mass spectrometers showed both significant similarities and differences. Five main chemical groups of particles were identified for ambient particle population: mineral particles, salt particles, particles with mixtures of secondary and organic components, particles dominated by secondary components and carbon particles. The reported results are an important basis for a general database of single particle mass spectra, spectral patterns of common and specific particle classes and abundances of these classes for atmospheric aerosols, showing their dependence on particle size, geographic location, meteorological conditions and time of analysis.[32]

A nanoaerosol mass spectrometer (NAMS) is described for the real time characterization of individual airborne nanoparticles (sub-10 nm range) by Wang *et al.*[42] This nanoaerosol mass spectrometer includes an aerodynamic inlet, quadrupole ion guide, quadrupole ion trap, and time-of-flight mass analyzer. Charged particles in the aerosol are drawn through the aerodynamic inlet, focused through the ion guide, and captured in the ion trap. Trapped particles are ablated with a Nd-YAG laser at 532 nm and 150 mJ pulse energy to reach the 'complete ionization limit'[43] where each particle is atomized and ionized to singly and multiply charged ions. The particle numbers are determined with a nanoscanning mobility particle sizer. There is also potential for applying NAMS in atmospheric chemistry for the characterization of nanoparticles, e.g., airborne manufactured nanomaterials. Heavy metals in airborne particles can be determined by the direct introduction of air samples in the ICP source of an ICP-MS, as demonstrated for Zn and Pb.[44,45] An online aerosol laser mass spectrometer with three different and easily interchangeable laser beam ionization techniques using one step laser ionization – LDI, two step laser desorption/photoionization – LDPI and thermal desorption/photoionization – TDPI combined with a ToF-MS for the characterization of inorganic and polyaromatic hydrocarbons (PAH) in particles was presented by Zimmermann *et al.*[46] Single particle analysis of PAH, such as naphthalene ($C_{10}H_8$, $m/z = 128$) or phenanthrene ($C_{14}H_{10}$, $m/z = 178$) is possible by laser desorption resonance enhanced multi-photon ionization mass spectrometry (LD-REMPI-ToF-MS), and inorganic ions of K, Ca, Mg, Mn and Fe have also been detected in wood soot.[46]

Cigarette smoke condensate as the particulate phase has been collected by electrostatic precipitation and introduced as a slurry into the ICP-MS.[47] The direct introduction of cigarette smoke via an automated single port smoking machine in a quadrupole based ICP-MS for multi-element trace metal analysis was proposed by Chang *et al.*[48] The home made smoking machine, the automated smoke collection and the injection apparatus were connected together with an ultrasonic nebulizer (USN) to an ICP-MS Elan 6000 in order to optimize the experimental parameters. Transient signals were measured when a discrete smoke aerosol was introduced into the ICP. The semiquantitative method developed for the determination of metal content in cigarette smoke allowed a rapid screening analysis for test cigarettes. The As, Cd, Hg and Pb concentrations in the mainstream smoke between two reference cigarettes were compared with those measured by a conventional method for sample collection similar to environmental aerosol analysis (measurement of smoke condensate as a slurry), and good agreement of the analytical data was found.[48]

Several working groups have analyzed platinum group elements (PGEs) as possible markers of anthropogenic pollution in aerosols (in airborne particulates from automobile exhaust to check the emission of catalysts), mainly by ICP-MS.[49,50] Airborne particles were also investigated with respect to selected element species (e.g., of Hg and Pb) by GC-ICP-MS.[51,52]

9.2.3 Multi-elemental Analysis of Environmental Samples for Environmental Control

ICP-MS is well established and currently the most widely applied inorganic mass spectrometric technique for the multi-element analysis of environmental samples. 'Common' elements like Fe, Cr, Ni, Zn, Cd, Mn, Pb and Cu are still dealt with in about 60 % of all papers published in environmental ICP-MS.[53] Multi-element analysis of environmental samples with respect to artificial contaminations (from traffic, the mining industry, industrial plants and others) is of increasing significance for environmental control. For example, the metal distribution in samples of road dust[54] collected from different localities (industrial, urban, peripheral) in Italy were analyzed with respect to their chemical composition and to assess the influence of a petrochemical plant and urban traffic in different grain size fractions of street dust. Fifteen trace elements were analyzed by ICP-MS in six fractions of particle sizes ranging from 20 μm to 500 μm in order to study the possible origin of metals in dusts. In the road dust samples, artificial elements from vehicular traffic and a nearby petrochemical plant were also found. High Ba, Cu, Cr, Mo, Pb, Sb and Zn concentrations can be explained by contamination from traffic, whereas high concentrations of Ni, V and, partly, Ba and Cr were associated with emissions from the petrochemical plant. It is interesting to note that Sb was the most highly enriched trace element in the road dust.[54] Multi-element analysis of 52 elements in tree bark sampled from locations close to motorways or industrial sites compared to rural areas was carried out by ICP-QMS after microwave induced digestion to study environmental contamination. The Pt contamination in the tree bark samples ranged from 0.07 to 5.4 ng g^{-1}.[5]

By coupling flow field-flow fractionation (flow FFF) to ICP-MS it is possible to investigate trace metals bound to various size fractions of colloidal and particulate materials.[55] This technique is employed for environmental applications,[55–57] for example to study trace metals associated with sediments. FFF-ICP-MS is an ideal technique for obtaining information on particle size distribution and depth profiles in sediment cores in addition to the metal concentrations (e.g., of Cu, Fe, Mn, Pb, Sr, Ti and Zn with core depths ranging from 0–40 cm).[55] Contaminated river sediments at various depths have been investigated by a combination of selective extraction and FFF-ICP-MS as described by Siripinyanond *et al.*[55]

9.2.4 Environmental Monitoring of Selected Elements, Group Elements and Trace Element Species

Apart from multi-element analysis employed for large scale studies, single element analysis (e.g., especially of toxic elements such as Cd, Hg, Tl or Pb) is performed in environmental science for special applications. For example, Hg and Tl have been determined in environmental samples by slurry sampling using electrothermal vaporization (ETV) ICP-MS. If potassium permanganate is employed as a modifier in ETV at optimized pyrolysis temperatures of 300 °C for Hg and 500 °C for Tl, detection limits of 0.18 $\mu g\,g^{-1}$ (Hg) and 0.07 $\mu g\,g^{-1}$ (Tl) are obtained.[58]

It has been generally accepted for several years that the determination of total element concentrations in environmental samples does not give an accurate estimation of their potential environmental and biological impact. About 1/3 of all ICP-MS applications in environmental analytical chemistry are performed by means of hyphenated techniques such as high performance liquid chromatography (HPLC), size exclusion chromatography (SEC) or capillary electrophoresis (CE) coupled to ICP-MS.[53] As a result, element speciation provides improved information on the fate, uptake, beneficial or toxic aspects, and finally on the translocation of metal between the different compartments of the environment because the uptake, accumulation, transport and interactions of different metals and metalloids in nature is strongly influenced by their specific elemental form.[59] Therefore a complete characterization of a metal or metalloid containing species in the environment (e.g., in a biogeochemical system) is necessary with respect to its oxidation state, associated organic ligands and/or complexed form.[60–62]

Chromium compounds have various industrial applications (e.g., chromium plating, in dyes and pigments, for leather and wood preservation or treatment of cooling tower water) and thus contaminate the environment, while the toxicity depends on the chemical form and concentration. It is well known that traces of Cr (III) are essential for the human and animal body, but Cr (VI) is a poison and carcinogenic for human and animals.[63] These chemical species differ not only in their toxicity, but also their mobility and bioavailability.[64,65] Therefore, special chromium analysis is of increasing relevance. The toxic Cr (VI) compounds are highly soluble, whereas Cr (III) compounds are easily absorbed onto different solid surfaces. Cr (III) is stable over a wide pH range (5–10) in its $Cr(OH)_3$ (s) precipitated form.[66] Chromatographic separation of chromium species is performed by HPLC-ICP-MS with CG5A – CS5A columns obtained from Dionex containing anionic and cationic groups to retain Cr(VI) and Cr (III), respectively, associated with a nitric acid elution. For example, the speciation of chromium has been performed on two real natural water samples collected in two different streams (urban and industrial regions). Cr (III) was found in the industrial effluents in a complex matrix at the $1.65\,\mu g\,l^{-1}$ concentration level by Donard and associates.[66]

Platinum group elements (PGE – mainly Pd, Pt and Rh) are key constituents in the 'three way' catalysts that are the active components in catalytic converters with the role of minimizing emissions from car exhausts, while maximizing the energy release from the fuel.[67] They are finely dispersed on Al_2O_3 and similar inert substrates[68] to create a large surface area. A rising concentration of platinum and palladium in environmental samples can be observed since the introduction of motor vehicle catalytic converters. Therefore, there are growing concerns about increasing levels of PGEs in the environment. The study of environmental contamination with PGEs is of increasing interest from the analytical point of view since an ultratrace analysis is possible, including the characterization of species. Platinum emission from cars equipped with these catalytic converters is well known and has been extensively investigated in environmental samples (and in autocatalysts) by several working groups.[5,67,69–75] Major platinum concentrations have been determined in uncultivated and cultivated soils at $0.14\,ng\,g^{-1}$ and $1.2\,ng\,g^{-1}$, respectively.[76,77] In contaminated soils near motorways, concentrations in the $20\,\mu g\,g^{-1}$ range and higher have been reported.[69] Platinum contents in roadside dust from the $0.0006\,\mu g\,g^{-1}$ to $62.8\,\mu g\,g^{-1}$ level with the highest contamination inside road tunnels have been measured by ICP-MS after microwave assisted acid digestion.[78] ICP-MS has also been employed for the determination of palladium in road dust obtained close to a motorway (at $450\,ng\,g^{-1}$ level) after sample decomposition by aqua regia.[79] For the quantification procedure (via external calibration and mathematical correction of isobaric interferences or isotope dilution technique) after the separation of Pd by extraction to dibutyl sulfide solution. Uncertainties and the accuracy of results were prooved by using a certified reference material CRM TDB – 1 Diabas Rock. The main source of uncertainty with external calibration is represented by the correction of isobaric interferences of atomic and polyatomic ions.[79] Rauch *et al.*[73] analyzed the platinum group elements Pt, Pd and Rh in road dust and sediments from an urban lake by ICP-MS. The samples were prepared by NiS fire assay, which exploits the Pt group elements' affinity for sulfide, followed by cleanup and acidic dissolutions. Pt concentrations in an urban lake (before 1992) were 1 ng g^{-1} and (after 1992) 20 ng g^{-1}. Also an increasing Pd and Rh concentration was found after 1992.[73]

A method for the accurate online ultratrace analysis of palladium in environmental samples (road dust and contaminated soil samples) by ICP-MS after separation of interferent cations by cation exchange chromatography was introduced by Hann and co-workers.[80] Palladium was selectively adsorbed online onto a C_{18} microcolumn which had been reversibly loaded with a complexing agent *N,N*-diethyl-*N'*-benzoylthiourea (DEBT). The palladium complex formed was eluted with methanol and introduced into an ICP-SFMS via microconcentric nebulization with membrane desolvation. Isotope dilution analysis was employed for quantification purposes.[80]

Besides platinum contamination from catalytic converters, platinum from excreted cancerostatic platinum compounds used in anticancer chemotherapy from hospital waste water is also emitted into the sewage in relatively large amounts (e.g. 14.3 kg of platinum was emitted into the public sewage system in Germany in 1996).[81,82] The platinum compound most frequently used in anticancer chemotherapy – cisplatin (*cis*-diaminedichloroplatinum (II) – CPC) is adsorbed in sewage sludge to a considerable degree (about 92 %).[75] Hann *et al.*[83] developed an analytical method for cisplatin, mono- and diaquacisplatin speciation (which were separated in just six min) and their degradation products in aquatic media by HPLC-ICP-MS as a function of the chloride concentrations. At low chloride concentration no equilibrium was established between cisplatin and its degradation products. Monitoring and elimination of cancerostatic platinum compounds present in hospital effluents were examined by Stingeder's working group.[75,84] In order to eliminate the contamination of municipal sewage sludge, a pilot plant for waste water treatment and the optimization of special elimination strategies was set up on the oncological in-patient treatment ward of Vienna University Hospital.[75] CPC speciation is based on HPLC separation on a pentafluorophenylpropyl reverse phase in combination with ICP-MS. Besides cisplatin, also carboplatin and oxaliplatin were separated by HPLC-ICP-MS with detection limits of 0.09, 0.10 and 0.15 $\mu g\, l^{-1}$, respectively. The preliminary evaluation of a pilot membrane bioreactor with complete removal of particles suitable for the treatment of hospital waste water revealed that advanced effluent treatment processes are required for a complete elimination of the cisplatin compound (CPC).[75]

In 2006, the speciation of metals and metalloids (As, Bi, Hg, Pb, Sb, Se and Sn) associated with alkyl groups and biomacromolecules in the environment was critically reviewed by Hirner.[85] More than 60 species of alkylated metals and metalloids have been found in different ecosystems and terrestrial locations all over the world.[85–87] These alkylated metals or metalloids are of interest due to their toxicological properties (e.g. monomethyl mercury, MMHg, which gained worldwide attention during the Minamata tragedy, and are not only known to be produced by microbial methylation within most anaerobic compartments of the environment, but also in the course of enzymatic transformation during human metabolism.[85]

Selected applications for the determination of contaminants in environmental samples are summarized in Table 9.23.

9.2.5 Isotope Ratio Measurements in Environmental Samples

As the most important inorganic mass spectrometric technique, ICP-MS is also employed for the precise and accurate isotope ratio measurements of a multitude of elements (such as Li, B, S, Fe, Sr, Pb, U, Pu) in environmental samples (see Chapter 8).[9,88–90] Isotope ratio measurements of environmental samples require special careful sample preparation techniques including trace/matrix separation and enrichment procedures if the analytes are at the trace and ultratrace level. As an example, the schematic diagrams of the separation and enrichment procedures for the precise isotope analysis of Pu, U and Sr in water samples from the Sea of Galilee using double-focusing

Table 9.23 *Determination of environmental contaminants.*

Samples/matrix	Equipment/sample preparation	Analyte	Detection limits	References
Alpine snow and ice	ICP-SFMS (Element) ice melted, acidified	Pb Pt: $< 2.7\,pg\,g^{-1}$	$3\,pg\,g^{-1}$ $0.008\,pg\,g^{-1}$	Schwikowski *et al.*[62] Brabante[58]
Groundwater	ICP-SFMS (Element), analyte separation	^{226}Ra	$0.02\,pg\,l^{-1}$	Zoriy *et al.*[102]
Sediments	ETV-ICP-MS (slurry sampling) (permanganate as modifier)	Tl, Hg	$0.07\,\mu g\,g^{-1}$ (Tl) $0.18\,\mu g\,g^{-1}$ (Hg)	Da Silva *et al.*[46]
Sediments	ICP-CC-MS (Platform) hot extraction	^{129}I	$0.4\,pg\,g^{-1}$ $^{129}I/^{127}I \sim 5\times10^{-7}$	Izmer *et al.*[107]
Soils (Chernobyl)	ICP-SFMS (Element) and APEX nebulizer ion exchange chromatography	^{236}U	$3\times10^{-14}\,g\,g^{-1}$	Boulyga and Heumann[109]
Road dust Airborne particulate matter	ICP-SFMS (Element) and USN	Pt range: $2\text{–}62\,ng\,g^{-1}$	$1.8\,ng\,g^{-1}$	Caroli *et al.*[59]
Sea water samples (Galilee, Israel)	ICP-SFMS, ICP-QMS trace/matrix separation	REE*, Pu, Sr, Na, K, Ca, Mg, Zn, B, Ba, U, V, Cr, Cu, Mn, Ni, Rb	$10^{-19}\,g\,ml^{-1}$ (^{239}Pu)	Halicz *et al.*[90]

* REE–rare earth elements

sector field ICP-MS with single and multiple ion collection are presented in Fig. 9.17.[90] To avoid matrix effects and clocking effects on the cones, strontium was separated using Sr specific resin supplied by Eichrom with a recovery yield of about 70 %. Detailed descriptions of the separation procedures are given elsewhere.[88,90,91]

With respect to determining its content in environmental studies, e.g., in ice and snow,[74] in contaminated water, soil and plant samples, lead is one of the most extensively investigated elements. It possesses four stable isotopes, and isotope ratios of lead can give additional information on possible pollution in environmental samples. Lead concentration and isotopic composition (^{206}Pb, ^{207}Pb, ^{208}Pb) in melted ice samples from the Canadian High Arctic was determined using double-focusing sector field ICP-MS with detection limits at the $0.06\,pg\,g^{-1}$ level.[92] This detection limit for lead was about two orders of magnitude lower than the lowest concentration of Pb measured in the ice samples (range, $4.3\text{–}1660\,pg\,g^{-1}$; median, $45\,pg\,g^{-1}$). Lead isotope ratios were measured by ICP-SFMS with a precision of $\sim 0.2\,\%$ at low $pg\,g^{-1}$ concentrations of total Pb. It was found that the lead in two ice samples dating from 1974 and 1852 with similar concentrations of $9\,pg\,g^{-1}$ and $6\,pg\,g^{-1}$, respectively, show a different isotopic composition ($^{206}Pb/^{207}Pb$: 1.169 ± 0.002 versus 1.147 ± 0.003). The difference in lead isotope ratios measured by inorganic mass spectrometry (ICP-MS or TIMS) indicates different sources of this heavy metal. Consequently, analytical procedures were used for fingerprinting the predominant sources of anthropogenic Pb in polar snow and ice. Snow from Greenland which receives Pb predominantly from the USA with a lead isotope ratio of $^{206}Pb/^{207}Pb \approx 1.2$ can be distinguished from Devon Island snow with a less radiogenic lead isotope ratio ($^{206}Pb/^{207}Pb \approx 1.15$) measured by ICP-SFMS, as discussed by Shotyk *et al.*[93] In forthcoming studies seasonal variations were found indicating that snow samples containing the greatest

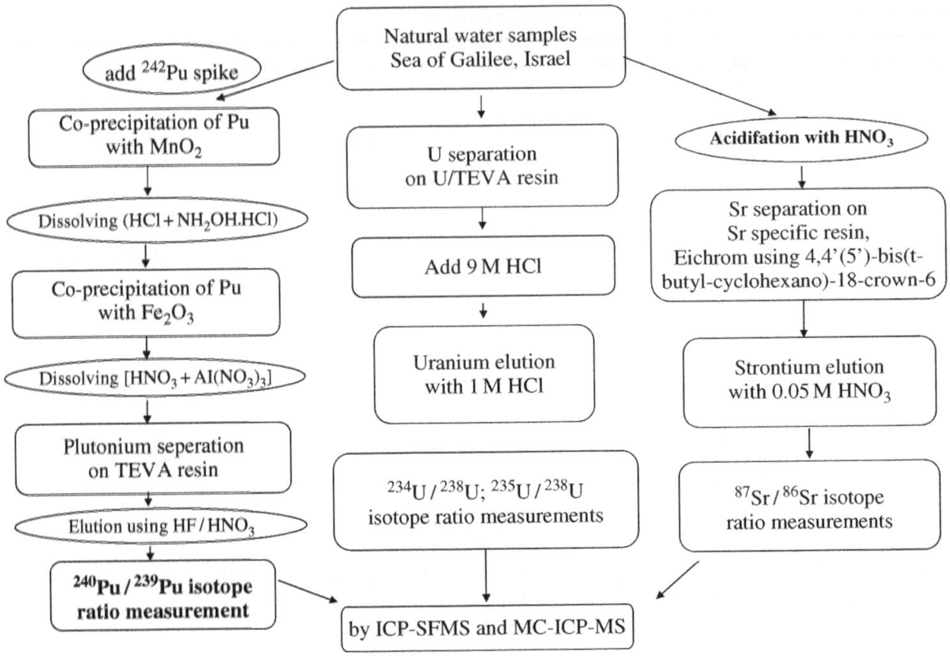

Figure 9.17 Schematic diagram of the separation and enrichment procedures for precise isotope analysis of Pu, U and Sr. (L. Halicz, J. S. Becker et al. Int. J. Mass Spectrom., ***249–250****, 296 (2006). Reproduced by permission of Elsevier.)*

Pb enrichments were from winter when the Arctic is dominated by air masses originating in Eurasia. While the elimination of gasoline lead additives in Europe, North America and Japan has helped to reduce Pb emissions during the past two to three decades, aerosols in the Arctic today are still contaminated by industrial Pb.[93] Lead isotope ratios in lichen have been measured by MC-ICP-MS collected around a copper smelter to detect industrial pollution.[94] In addition, lead isotope analysis in peat cores to reconstruct past environmental changes is challenged by the large variation in Pb concentrations (from $\geq 200\,ng\,g^{-1}$ in pre-anthropogenic peat[95] up to $250\,\mu g\,g^{-1}$ in modern peats[96]). Sample preparation procedures including dry ashing and microwave acid digestion for accurate and precise Pb isotope analysis by MC-ICP-MS (Isoprobe, VG Instruments, Manchester, UK) has been developed by Kylander *et al.*[97] The authors analyzed peat depth profiles (of 160 cm thickness) from Spain to study the Pb accumulation (in the concentration range from 0.4–75 $\mu g\,g^{-1}$) and isotope variation of $^{206}Pb/^{207}Pb$ in peat over the last 2500 years.[97] MC-ICP-MS has also been employed to study such $^{206}Pb/^{207}Pb$ isotope variation in peat bogs for the investigation of possible lead contamination in the environment.[98] Peat deposited in 1880 had a $^{206}Pb/^{207}Pb$ isotope ratio of 1.168–1.178, which already showed the influence of anthropogenic pollution. A higher $^{206}Pb/^{207}Pb$ isotope ratio of 1.193 was detected in older samples (11 000 years old), which were not impacted by human activity.[98]

As an essential element for algal photosynthesis, respiration, and nitrogen fixation in the ocean[99,100] iron also plays a role in climate change,[101] the environment and biogeochemistry.[102,103] However, measuring the Fe content in sea water at the $ng\,l^{-1}$ level remains challenging, mostly because of the ubiquity of iron (causing sample contamination) and the complexity of the sample matrix.[104] Petrov and Quétel studied these problems during the determination of iron concentration in sea water samples by the isotope dilution technique using double-focusing sector field ICP-MS with a single ion collector involving concentrations down to $30\,ng\,l^{-1}$ after matrix separation and Fe enrichment.[104] Matrix separation by means of $Mg(OH)_2$ co-precipitation is only ~96 % to 98.5 % efficient, and the remaining inorganic salt fraction is a source of measurement problems. The authors developed an analytical method to overcome these difficulties, and achieved 2–3 % expanded uncertainty on direct $^{56}Fe/^{57}Fe$ isotope ratio measurements by ICP-SFMS at medium mass resolution for sea water samples with $300–450\,ng\,l^{-1}$ Fe containing 0.1 % residual salinity. The method developed was used to validate iron in CASS-2 certified reference sea water material. Unfortunately, double-focusing sector field ICP-SFMS with single ion collection is not suitable for studying natural isotope fractionation effects of the order of less than 0.1 %, as described in references.[105–107] For these applications MC-ICP-MS instrumentation is more suitable due to the simultaneous multiple ion detection and consequently higher precision of the isotope ratios.[104]

Finally, enriched isotope tracers can be used to monitor small and large scale environmental fluxes of trace elements more extensively.[53]

Compound specific stable isotope analysis using gas chromatography combined with an isotope ratio mass spectrometer – GC-IRMS (see also Chapter 7) – is now a mature analytical technique in environmental science and technology, especially in the area of contaminant source attribution and in assessing the biodegradation of contaminants.[108] Several studies have focused on $^{13}C/^{12}C$, $^{18}O/^{16}O$ and $^{17}O/^{16}O$ isotope ratio measurements for volatile organic and metalorganic compounds to study isotope fractionation effects and to identify contamination in the environment.[109]

The state-of-the art and progress made in isotope analysis by ICP-MS has been summarized in several reviews.[110–112]

9.2.6 Monitoring of Radionuclides in the Environment

Environmental monitoring of nuclear contamination, including the determination of the concentration and isotope ratios of long-lived radionuclides, such as uranium, plutonium isotopes, thorium, ^{237}Np, ^{79}Se, ^{90}Sr, ^{129}I and others, at trace and ultratrace levels, is a fast growing and fascinating application field for inorganic mass spectrometry.[113–120] Among the environmentally important radionuclides, ^{129}I, ^{90}Sr, uranium and transuranium elements are of special importance. For example, the natural ^{129}I inventory in the atmosphere, hydrosphere and biosphere has been estimated to be about 263 kg.[121]

However, the release of ^{129}I from reprocessing plants and the behaviour of ^{129}I in the environment has not been explored in detail due to analytical difficulties. Accelerator mass spectrometry (AMS) has been employed for the ultrasensitive detection of ^{129}I close to nuclear reprocessing facilities (e.g., in Norwegian coastal waters: $3 \times 10^{10}\,atoms\,l^{-1}$). This measured concentration was significantly higher than the Arctic Ocean.[122] The analysis of radionuclides by ICP-MS is of importance in the environmental sciences due to its excellent detection limits, the high sample throughput and the ability to measure precise and accurate isotope ratios.[89,113] It was demonstrated that quadrupole based ICP-MS is useful for isotope ratio measurements of U and Th in environmental materials.[123] Contamination of ^{129}I in environmental materials (sediments from Kazakhstan) were measured by ICP-MS with a detection limit of $0.4\,pg\,g^{-1}$ after the hot extraction of iodine from soil in an oxygen stream and online transfer to the inductively coupled plasma source.[9,124] Using a mixture of oxygen and helium in the hexapole collision cell of quadrupole ICP-CC-MS (Platform), it was possible to neutralize disturbing background $^{129}Xe^+$ ions. This analytical method was applied for isotope ratio measurements of $^{129}I/^{127}I$ down to 10^{-7} in contaminated soil samples.[9,124]

Environmental control in respect of determining concentrations and isotope ratios, e.g. of U, Pu and other actinides, is also required in routine measurements near to nuclear power plants, uranium enrichment facilities or nuclear waste recycling companies. Groundwater samples are analyzed after dilution directly by ICP-MS; for soils a digestion step before mass spectrometric measurement is necessary. If isobaric interferences are observed a trace matrix separation and/or a careful analyte separation (e.g. of U and Pu) is recommended.

The determination of ^{90}Sr and ^{239}Pu and ^{240}Pu isotopes at the ultratrace level was studied on groundwater samples from Kazakhstan. In order to avoid isobaric interferences at $m/z = 90$ for ^{90}Sr determination (from $^{90}Zr^+$, $^{40}Ar^{50}Cr^+$; $^{58}Ni^{16}O_2^{\ +}$; $^{180}Hf^{2+}$ and others) the measurements were performed using a sector field ICP-MS at medium mass resolution under cold plasma conditions. Plutonium traces were separated before mass spectrometric measurements by means of extraction chromatography using Eichrom TEVA resin with a recovery of 83 %. The limits of detection were determined for ^{90}Sr and ^{239}Pu and ^{240}Pu as 11, 0.12 and 0.1 fg ml^{-1}, respectively. The concentrations measured in contaminated groundwater samples for ^{90}Sr ranged from 18 to 32 fg ml^{-1} and for Pu from 28 to 856 fg ml^{-1}. The $^{240}Pu/^{239}Pu$ isotope ratio measured as 0.17 indicated that the most probable source of nuclear contamination in the groundwater samples investigated resulted from nuclear weapons tests at the Semipalatinsk Test Site by the USSR in the 1960s.[125] Extremely low Pu concentrations (3.6×10^{-19} g ml^{-1} = 0.36 ag ml^{-1}) in environmental water samples (Sea of Galilee – Israel) were determined by ICP-SFMS and MC-ICP-MS after preconcentration and matrix separation from 100 litre sea water.[88] The $^{240}Pu/^{239}Pu$ isotope ratios measured by MC-ICP-MS of 0.17 can be explained as contamination due to nuclear global fallout after the nuclear weapons tests in the sixties.[88,90]

While investigating the uranium concentration in the Sea of Galilee, no homogeneous distribution was found as a function of depth of the lake.[90] Fig. 9.18 the shows the depth profile of the uranium concentration from the surface down to 35 m. Whereas the uranium concentration from the lake surface down to 10 m is nearly constant at about 0.6 μg l^{-1}, with increasing depth after a slight enrichment of uranium concentration from 10–20 m, a significant depletion was found at depths greater than 20 m dropping to nearly 0.4 μg l^{-1} uranium. This experimental result obtained by mass spectrometric measurements can be explained by the changing environmental conditions at different

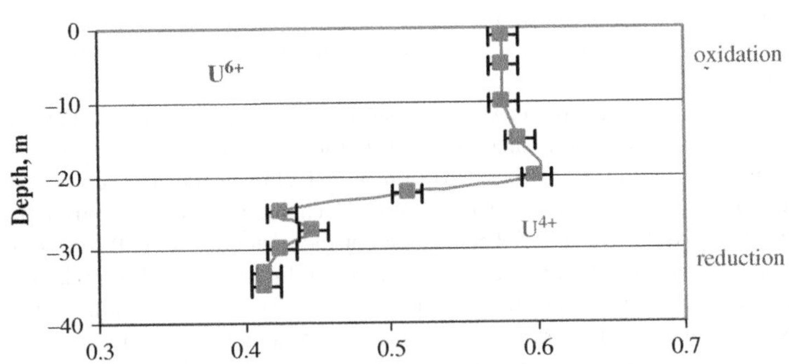

Figure 9.18 Depth profile of uranium concentration in the Sea of Galilee. (L. Halicz, J. S. Becker et al. Int. J. Mass Spectrom., **249–250**, 296 (2006). Reproduced by permission of Elsevier.)

depths. Whereas under oxidation conditions (down to a depth of 20 m in the Sea of Galilee), natural uranium exists as UO_4^{2-} in water, with increasing depth the environmental behaviour of sea water changes to reducing conditions and, consequently, the oxidation state of uranium changes from 6+ to 4+. Consequently, the observed depletion of uranium in the Sea of Galilee at depths > 20 m could be the result of the formation of heavy dissolved uranium compounds which are partly precipitated.[90]

Plutonium at a concentration level of $0.3\,pg\,g^{-1}$ in contaminated soil samples was determined directly by LA-ICP-SFMS using the isotope dilution technique and $^{240}Pu/^{239}Pu$ isotope ratios were measured by Boulyga *et al.*[126]

As a highly selective and very sensitive isotopic analytical technique, accelerator mass spectrometry (AMS) is able to determine radionuclides (^{10}Be, ^{36}Cl, ^{26}Al, ^{99}Tc, ^{129}I, ^{236}U, ^{237}Np, ^{239}Pu or ^{240}Pu) with a very low abundance in the environment, whereby 10^6 ions can be detected by AMS.[117,127] In order to differentiate between nuclear contamination from nuclear weapons tests and releases from nuclear power plants, ^{239}Pu and ^{240}Pu have been determined in environmental samples and bioassays by AMS at extremely low concentration levels.[128] Relatively low $^{240}Pu/^{239}Pu$ isotope ratios were found in Asanov Swamp water and vegetation samples indicate nuclear contamination from early contaminations of weapons grade plutonium.[127] AMS was employed to measure the $^{236}U/^{238}U$ isotope ratio, which is different in nuclear weapons and in civil use.[129] This ultrasensitive mass spectrometric technique allows uranium isotope ratio measurements to be made for ^{236}U with abundance sensitivity in the range from 10^{-10}–10^{-12}.[130,131]

Determination of uranium in soil samples collected in the vicinity of Chernobyl power plant resulted in the detection of ^{236}U, which can be regarded as a marker for environmental contamination.[89,132,133] Using double-focusing sector field ICP-SFMS at a mass resolution ($m/\Delta m$) of 4000 with APEX for highly efficient solution introduction, it was possible to determine ^{236}U determination at concentration ranges down to $3 \times 10^{-14}\,g\,g^{-1}$, and $^{236}U/^{238}U$ isotope ratios in soil samples down to 10^{-7} were determined.[134] ICP-SFMS was employed for ^{226}Ra determination after trace/matrix separation at the low $fg\,ml^{-1}$ concentration level in groundwater.[135]

Radiocarbon (^{14}C) dating is the most common application of AMS, and is also relevant for compound specific measurements. PAHs in sediments from an urban reservoir were ^{14}C-free, but most of the PAHs in these sediments were derived from fossil fuel combustion rather than biomass burning.[136]

A detailed description of the analytical methods and further applications for the monitoring of radionuclides in environmental samples are described in Section 9.9.

References II

1. Becker, J. S., *Trends in Anal. Chem.*, **24**, 243 (2005).
2. Beauchemin, D., *Anal. Chem.*, **76**, 3395 (2004).
3. Beauchemin, D., *Anal. Chem.*, **78**, 4111 (2006).
4. Becker, J. S., *Canad. J. Anal. Sci. Spectr.*, **47**, 98 (2002).
5. Becker, J. Su., Bellis, D., Staton, I., McLeod, C. W., Dombovari, J. and Becker, J. S., *Fresenius' J. Anal. Chem.*, **368**, 490 (2000).
6. Boulyga, S. F. and Becker, J. S., *J. Anal. At. Spectrom.*, **17**, 1143 (2002).
7. Boulyga, S. F., Desideri, D., Meli, M. A., Testa, C. and Becker, J. S., *Int. J. Mass Spectrom.*, **226**, 329 (2003).
8. Dombovari, J., Becker, J. S. and Dietze, H. J., *Fresenius' J. Anal. Chem.*, **367**, 407 (2000).
9. Izmer, A. V., Boulyga, S. F. and Becker, J. S., *J. Anal. At. Spectrom.*, **18**, 1339 (2003).
10. Köster, C. J. and Moulik, A., *Anal. Chem.*, **77**, 3737 (2005).

11. Mironov, V. P., Matusevich, J. L., Kudrjashov, V. P., Boulyga, S. F. and Becker, J. S., *J. Environ. Monit.*, **4**, 997 (2002).
12. Panday, V. K., Hoppstock, K., Becker, J. S. and Dietze, H. J., *At. Spectr.*, **17**, 98 (1996).
13. Sadi, B. B. M., Vonderheide, A. P., Becker, J. S. and Caruso, J. A., SEC-ICP-MS: An Important Analytical Tool for Elemental Speciation in Environmental and Biological Samples in *ACS Symposium Series 893*, A. M. Striegler (ed.), American Chemical Society, Washington, D. C., 168 (2004).
14. Köster, C. J., Simonich, S. I. and Esser, B. K., *Anal. Chem.*, **75**, 2813 (2003).
15. Richardson, S. D., *Anal. Chem.*, **76**, 3337 (2004).
16. Butler, O. T., Cook, J. M., Harrington, C. F., Hill, S. J., Rieuwerts, J. and Milles, D. L., *J. Anal. At. Spectrom.*, **20**, 130 (2005).
17. Lawrence, M. G., Greig, A., Collerson, K. D. and Kamber, B. S., *Appl. Geochem.*, **21**, 839 (2006).
18. Haraguchi, H., Itoh, A., Kimata, C. and Miwa, H., *Analyst*, **123**, 773 (1998).
19. Jain, C. K. and Ali, I., *Water Res.*, **34**, 4304 (2000).
20. McSheehy, D. and Mester, Z., *Trends in Anal. Chem.*, **22**, 311 (2003).
21. Abrankó, L., Yang, L., Sturgeon, R. E., Fodor, P. and Mester, Z., *J. Anal. At. Spectrom.*, **19**, 1098 (2004).
22. Wallschlaeger, D. and London, J., *J. Anal. At. Spectrom.*, **19**, 1119 (2004).
23. Richardson, S. D. and Ternes, T. A., *Anal. Chem.*, **77**, 3807 (2005).
24. Rosenberg, E., in *Handbook of Elemental Speciation II, Species in the Environment, Food, Medicine and Occupational Health*, R. Cornelis, J. Caruso, H. Crews, K. G. Heumann (eds.). John Wiley & Sons Ltd., Chichester (2005).
25. Prohaska, T. and Stingeder, G., in *Handbook of Elemental Speciation II, Species in the Environment, Food, Medicine and Occupational Health*, J. C. R. Cornelis, J. Caruso, H. Crews, K. G. Heumann (eds.). John Wiley & Sons Ltd., Chichester (2005).
26. Takamatsu, T., Aoki, H. and Yoshida, T., *Soil Sci.*, **133**, 239 (1982).
27. Reinhard, H. P., Kriews, M., Miller, H., Luedke, C., Hoffmann, E. and Skole, J., *Anal. Bioanal. Chem.*, **375**, 1265 (2003).
28. Reinhardt, H., Kriews, M., Miller, H. *et al.*, *Fresenius' J. Anal. Chem.*, **370**, 629 (2001).
29. Lee, J. T., Kim, H., Hong, Y. C., Kwon, H. J. and Schwartz, J., *Environ. Res. A*, **84**, 247 (2000).
30. Okuda, T., Kato, J., Mori, J. *et al.*, *Sci. Tot. Envir.*, **330**, 145 (2004).
31. Schwartz, J., *Environ. Res.*, **62**, 7 (1993).
32. Hinz, K. P., Erdmann, N., Grüning, C. and Spengler, B., *Int. J. Mass Spectrom.*, **258**, 151 (2006).
33. Ortner, H. M., in *Handbook of Elemental Speciation: Techniques and Methodology*, Vol. I, R. Cornelis, J. Caruso, H. Crews, K. G. Heumann (eds.). John Wiley & Sons, Ltd, Chichester. (2003) 505.
34. Katoh, T., Akiyama, M., Ohtsuka, H., Nakamura, S., Haraguchi, K. and Akatsuska, K., *J. Anal. At. Spectrom.*, **11**, 69 (1996).
35. Luedke, C., Hoffmann, E. and Skole, J., *Fresenius' J. Anal. Chem.*, **350**, 272 (1994).
36. Wang, C. F., Chen, W. H. and Yang, M. H. and Chiang, P. C., *Analyst*, **120**, 1681 (1995).
37. Tanaka, S., Narita, Y., Sato, N. *et al.*, *J. Anal. At. Spectrom.*, **13**, 135 (1998).
38. Luedke, C., Skole, J., Tauber, K. and Kriews, M., *Spectrochim. Acta B*, **60**, 1412 (2005).
39. Hinz, K. P., Kaufmann, R. and Spengler, B., *Aerosol Sci. Technol.*, **24**, 233 (1996).
40. Kane, D. B., Oktem, B. and Johnston, M. V., *Aerosol Sci. Techn.*, **34**, 520 (2001).
41. Murphy, D. M. and Thomson, D. S., *J. Geophys. Res.*, **102**, 6325 (1997).
42. Wang, S., Zordan, C. A. and Johnston, M. V., *Anal. Chem.*, **78**, 1750 (2006).
43. Reents, W. D. and Schnabel, M. J., *Anal. Chem.*, **73**, 5403 (2001).
44. Kaneco, S., Nomizu, T., Tanaka, K., Mizutani, N. and Kawaguchi, H., *Anal. Sci.*, **11**, 835 (1995).
45. Nomizu, T., Hayashi, H., Hoshino, N. *et al.*, *J. Anal. At. Spectrom.*, **17**, 592 (2002).
46. Bente, M., Adam, T., Ferge, T., Gallavardin, S., Sklorz, M., Streibel, T. and Zimmermann, R., *Int. J. Mass Spectrom.*, **258**, 86 (2006).
47. Torrence, K. M., McSDaniel, R. L., Self, D. A. and Chang, M. J., *Anal. Bioanal. Chem.*, **372**, 723 (2002).
48. Chang, M. J., Naworal, J. D. and Connell, C. T., *J. Anal. At. Spectrom.*, **21**, 574 (2006).
49. Köllensperger, G., Hann, S. and Stingeder, G., *J. Anal. At. Spectrom.*, **15**, 1553 (2000).
50. Rao, C. R. M. and Reddi, G. S., *Trends Anal. Chem.*, **19**, 565 (2000).
51. Leal-Granadillo, I. A., Alonso, A. I. G. and Sanz-Medel, A., *Anal. Chem.*, **423**, 21 (2000).

52. Pecheytan, B., Lalere, B. and Donard, O. F. X., *Environ. Sci. Techn.*, **34**, 27 (2000).
53. Prohaska, T., Köllensperger, G., Hann, S., Stingeder, G., Fitz, W. and Wenzel, W., in *Plasma Source Mass Spectrometry*, G. Holland and S. D. Tanner (eds.). The Royal Society of Chemistry, Cambridge (2003) 93.
54. Manno, E., Varrica, D. and Dongarrà, G., *Atmospheric Environ.*, **40**, 5929 (2006).
55. Siripinyanond, A., Barnes, R. M. and Amarasiriwardena, D. D., *J. Anal. At. Spectrom.*, **17**, 1055 (2002).
56. Chen, B., Shand, C. A. and Beckett, R., *J. Environ. Monit.*, **3**, 7 (2001).
57. Taylor, H. E., Garbarino, J. R., Murphy, D. M. and Beckett, R., *Anal. Chem.*, **64**, 459 (1992).
58. Da Silva, A. F., Dias, L. F., Saint'Pierre, T. D., Curtius, A. J. and Welz, B., *J. Anal. At. Spectrom.*, **18**, 344 (2003).
59. Moldovan, M., Krupp, E. M., Holliday, A. E. and Donard, O. F. X., *J. Anal. At. Spectrom.*, **19**, 815 (2004).
60. Cornelis, R., Caruso, J., Crews, H. and Heumann, K. G. & (eds.), *Handbook of Speciation: Techniques and Methodology*. John Wiley & Sons, Inc., New York (2003).
61. Crews, H., in *Handbook of Elemental Speciation II: Species in the Environment, Food, Medicine & Occupational Health*, Vol. II, R. Cornelis, J. Caruso and K. G. Heumann (eds.). John Wiley & Sons, Ltd, Chichester (2005), 247.
62. Donard, O. F. X., in *Trace Element Speciation for Environment, Food and Health*, L. P. L. Ebdon, R. Cornelis, H. Crews, O. F. X. Donard and P. Quevauviller (eds.). Royal Society of Chemistry, Cambridge (2001) 115.
63. Darrie, G., in *Trace Element Speciation for Environment, Food and Health*, L. P. L. Ebdon, R. Cornelis, H. Crews, O. F. X. Donard and P. Quevauviller (eds). Royal Society of Chemistry, Cambridge (2001) 315.
64. Kotas, J. and Stasicka, Z., *Environ. Poll.*, **107**, 263 (2000).
65. Richard, F. C. and Bourg, A. C. M., *Water Res.*, **25**, 807 (1991).
66. Séby, F., Charles, S., Gagean, M., Garraud, H. and Donard, O. F. X., *J. Anal. At. Spectrom.*, **18**, 1386 (2003).
67. Simpson, L. A., Hearn, R. and Catterick, T., *J. Anal. At. Spectrom.*, **19**, 1244 (2004).
68. Taylor, H. E., Huff, R. A. and Montaser, A., in *Inductively Coupled Plasma Mass Spectrometry*, A. Montaser (ed.). Wiley-VCH, Weinheim (1998).
69. Alt, F., Eschnauer, H. R., Megler, B., Messerschmidt, J. and Toelg, G., *Fresenius' J. Anal. Chem.*, **357**, 1013 (1997).
70. Barbante, C., Cozzi, G., Capodaglio, G. *et al.*, *Anal. Chem.*, **71**, 4125 (1999).
71. Caroli, S., Bocca, B., Petrucci, F. and Alimonzi, A., in *Book of Abstracts of Winter Conference on Plasma Spectrochemistry*, R. Barnes (ed.), ICP Information Newsletters, Fort Lauderdale, Florida (2000) 345.
72. Helmers, E. and Kümmerer, K., *ESPR*, **412**, 99 (1997).
73. Rauch, S., Hemond, H. F. and Peucker-Ehrenbrink, B., *Environ. Sci. Techn.*, **38**, 396 (2004).
74. Schwikowski, M., Barbante, C., Döring, T. *et al.*, *Environ. Sci. Techn.*, **38**, 957 (2004).
75. Stefánka, Z., Hann, S., Lenz, K., Köllensperger, G., Fürhacker, M. and Stingeder, G., in *Proceedings of 9th International Conference on Plasma Source Mass Spectrometry (13–17 September 2004)*, D. Bandura and G. Holland (eds). (The Royal Society, Durham (2005) 235.
76. Angerer, J. and Schaller, K. H. (eds.) *Analysis of Hazardous Substances in Biological Materials*, vol. 4. VCH Verlagsgesellschaft, Weinheim (1994), 1994.
77. Messerschmidt, J., Alt, F., Angerer, J. and Schaller, K. H., *Fresenius' J. Anal. Chem.*, **335**, 813 (1992).
78. Ward, N. I. and Diudding, L. M., *Sci. Tot. Environm.*, **334–335**, 457 (2004).
79. Polák, J., Mestek, O. and Suchánek, M., *Accreditation and Quality Assurance: Journal for Quality, Comparability and Reliability in Chemical Measurement*, **10**, 627 (2006).
80. Rudolph, E., Limbeck, A. and Hann, S., *J. Anal. At. Spectrom.*, **21**, 1287 (2006).
81. Kümmerer, K., *Chemosphere*, **45**, 957 (2001).
82. Kümmerer, K., Helmers, E., Hubner, P. *et al.*, *Sci. Tot. Environ.*, **225**, 155 (1999).
83. Hann, S., Köllensperger, G., Stefánka, Z. *et al.*, *J. Anal. At. Spectrom.*, **18**, 1391 (2003).
84. Lenz, K., Hann, S., Köllensperger, G. *et al.*, *Sci. Tot. Envir.*, **345**, 141 (2005).
85. Hirner, A. V., *Anal. Bioanal. Chem.*, **385**, 555 (2006).
86. Craig, P. J., *Organometallic Compounds in the Environment*. John Wiley & Sons, Chichester, UK (2003).

87. Dopp, E., Hartmann, L. M., Florea, A. M., Rettenmeier, A. W. and Hirner, A. V., *Crit. Rev. Toxicol.*, **34**, 301 (2004).
88. Becker, J. S., Zoriy, M., Halicz, L. *et al.*, *J. Anal. At. Spectrom.*, **19**, 1257 (2004).
89. Boulyga, S. F., Matusevich, J. L., Mironov, V. P. *et al.*, *J. Anal. At. Spectrom.*, **17**, 958 (2002).
90. Halicz, L., Becker, J. S., Pickhardt, C., Gavrieli, I., Burg, A., Nishri, A. and Platzner, I., *Intern. J. Mass Spectr.*, **249–250**, 296 (2006).
91. Segal, I., Halicz, L. and Platzner, I., *Int. J. Mass Spectrom.*, **216**, 177 (2002).
92. Krachler, M., Zheng, J., Fisher, D. and Shotyk, W., *Anal. Chem.*, **76**, 5510 (2004).
93. Shotyk, W., Zheng, J., Krachler, M., Zdanowicz, C., Koerner, R. and Fisher, D., *Geophys. Res. Letters*, **32**, (2005).
94. Spiro, B., Weiss, D. J., Purvis, O. W., Mikhailova, I., Willikamson, B. J., Coles, B. J. and Udachin, V., *Environ. Sci. Techn.*, **38**, 6522 (2004).
95. Brännvall, M. L., Bindler, R., Emteryd, O., Nilsson, M. and Renberg, I., *Water, Air Soil Pollut.*, **100**, 243 (1997).
96. Dunlap, C. E., Steinnes, E. and Flegal, A. R., *Earth Planet. Sci. Lett.*, **167**, 81 (1999).
97. Kylander, M. E., Weiss, D. J., Jeffries, T. and Coles, B. J., *J. Anal. At. Spectrom.*, **19**, 1275 (2004).
98. Novák, M., Emmanuel, S., Vile, M. A. *et al.*, *Environ. Sci. Techn.*, **37**, 437 (2003).
99. Martin, J. and Fitzwater, S., *Nature*, **331**, 341 (1988).
100. Morel, F. M. M. and Price, N. M., *Science*, **300**, 944 (2003).
101. Martin, J. H., *Paleoceanography*, **5**, 1 (1990).
102. Bowie, A. R., Whitworth, D. J., Achterberg, E. P., Fauzi, R., Mantoura, C. and Worsfold, P. J., *Deep-Sea Res. 1*, **49**, 605 (2002).
103. Ussher, S. J., Achterberg, E. P. and Worsfold, P. J., *Environ. Chem.*, **1**, 67 (2004).
104. Petrov, I. and Quétel, C. R., *J. Anal. At. Spectrom.*, **20**, 1095 (2005).
105. Arnold, G. L., Weyer, S. and Anbar, A. D., *Anal. Chem.*, **76**, 322 (2004).
106. Vogl, J., Klingbeil, P., Pritzkow, W. and Riebe, G., *J. Anal. At. Spectrom.*, **18**, 1125 (2003).
107. Weyer, S. and Schwieters, J. B., *Int. J. Mass Spectrom.*, **226**, 355 (2003).
108. Schmidt, T. C., Zwank, L., Elsner, M., Ber, M., Mechdenstock, R. U. and Haderlein, S. B., *Anal. Bioanal. Chem.*, **378**, 283 (2004).
109. Dickhut, R. M., Padma, T. V. and Cincinelli, A., *Environ. Sci. Techn.*, **38**, 3871 (2004).
110. Becker, J. S., *J. Anal. At. Spectrom.*, **17**, 1172 (2002).
111. Becker, J. S., *J. Anal. At. Spectrom.*, **20**, 1173 (2005).
112. Heumann, K. G., *Anal. Bioanal. Chem.*, **378**, 318 (2004).
113. Becker, J. S., *Int. J. Mass Spectrom.*, **242**, 183 (2005).
114. Becker, J. S., *Spectrochimica Acta*, **58B**, 1757 (2003).
115. Desideri, D., Meli, M. A., Roselli, C., Testa, C., Boulyga, S. F. and Becker, J. S., *Anal. Bioanal. Chem.*, **374**, 1091 (2002).
116. Gastel, M., Becker, J. S., Küppers, G. and Dietze, H. J., *Spectrochim. Acta*, **52B**, 2051 (1997).
117. Kutschera, W., *Int. J. Mass Spectrom.*, **242**, 145 (2005).
118. McLean, J. A., Becker, J. S., Boulyga, S. F., Dietze, H. J. and Montaser, A., *Int. J. Mass Spectrom.*, **208**, 193 (2001).
119. Vockenhuber, C., Ahmad, I., Golser, R. *et al.*, *Int. J. Mass Spectrom.*, **223**, 713 (2003).
120. Zoriy, M. V., Halicz, L., Ketterer, M. E., Pickhardt, C., Ostapczuk, P. and Becker, J. S., *J. Anal. At. Spectrom.*, **19**, 362 (2004).
121. Fabryka-Martin, J., Bentley, H., Elmore, D. and Airey, P. L., *Geochim. Cosmochim. Acta.*, **49**, 337 (1989).
122. Alfimov, V., Aldahan, A., Possnert, G. and Winsor, P., *Mar. Poll. Bull.*, **49**, 1097 (2004).
123. Muramatsu, Y., Ruhm, W., Yoshida, S., Tagami, K., Uchida, S. and Wirth, E., *Environ. Sci. Techn.*, **34**, 2913 (2000).
124. Izmer, A., Zoriy, M., Boulyga, S. F. and Becker, J. S., *J. Anal. At. Spectrom.*, **20**, 1278 (2005).
125. Zoriy, M. V., Ostapczuk, P., Halicz, L., Hille, R. and Becker, J. S., *Int. J. Mass Spectrom.* **242**, 203 (2005).
126. Boulyga, S. F., Tibi, M. and Heumann, K. G., *Anal. Bioanal. Chem.*, **342**, 3478 (2004).
127. Skipperud, L. and Oughton, D. H., *Environ. Int.*, **30**, 815 (2004).

128. Marchetti, A. A., Brown T. A., Cox, C. C., Hamilton, T. F. and Martinelli, R. E., *J. Radioanal. Nucl. Chem.*, **263**, 483 (2005).
129. Borretzen, P., Standring, W. J. F., Oughton, D. H., Dowdall, M. and Fifield, L. K., *Environ. Sci. Techn.*, **39**, 92 (2005).
130. Fifield, L. K., *Nucl. Instrum. Methods*, **B172**, 297 (2000).
131. Paul, M., Berkovits, D., Ahmad, I. *et al., Nucl. Instrum. Methods*, **B172**, 688 (2000).
132. Boulyga, S. F. and Becker, J. S., *Fresenius' J. Anal. Chem.*, **370**, 612 (2001).
133. Boulyga, S. F., Becker, J. S., Matusevitch, J. L. and Dietze, H. J., *Int. J. Mass Spectrom.*, **203**, 143 (2000).
134. Boulyga, S. F. and Heumann, K. G., *J. Environ. Radioact.*, **88**, 1 (2006).
135. Zoriy, M., Varga, Z., Pickhardt, C. Ostapczuk, P., Hille, R., Halicz, L., Segal, I. and Becker, J. S., *J. Environ. Monit.*, **7**, 514 (2005).
136. Kanke, H., Uchida, M., Okuda, T. *et al., Nucl. Instrum. Methods Phys. Res. B*, **223–224**, 545 (2004).

9.3 Biology

Mineral elements contained in plant, animals or microorganisms, often at the trace and ultratrace concentration level, play a vital role in biological systems and biological processes. Several elements (e.g., P, S, Mg, Ca, K, Fe, Cu, Mn, Zn, Ni and Mo) are essential for health or are beneficial for plant growth (such as Si, Na, V, Cr, Co and Se) and are thus of vital interest. Several other elements are mostly heavy and potentially toxic elements (e.g., Cd, Hg, Pb, Tl). The life of an organism is critically dependent on the proper regulation of uptake, assimilation, intracellular compartmentation and intercellular translocation of trace metals.[1] Increasingly, knowledge of the chemical composition, chemical form and distribution of species is also required to study the life of an organism or to assess the extent and impact of environmental contamination in biological systems. In addition, metal contents and isotope ratios can vary because of various biological effects. Therefore trace, ultratrace, isotope and distribution (imaging) analysis of biological samples in the life sciences has become one of the most important tasks in analytical chemistry in the past few years.

Sample digestion is of special importance in the analysis of biological samples. In earlier work, biological matrices for bulk analysis were subjected to an oxidative wet digestion procedure (using HNO_3, $HClO_4$, H_2O_2) in an open vessel. Due to problems with incomplete digestion, lost of volatile analytes (such as As, Hg, Sb, Se, Cd, I, Tl and Pb) and possible contamination, it is recommended that wet chemical digestion in a closed vessel should be used. Also the dry ashing of biological samples in air or O_2 atmosphere at temperatures up to 500 °C is no longer recommended due to the loss of volatile elements. For the digestion of biological materials, wet chemical decomposition in closed vessels using a high pressure asher (HPA e.g., from Anton Paar) with HNO_3 or an acid mixture or by microwave induced digestion (the latter digestion technique is preferred in the author's laboratory) is usually recommended. The most successful procedure available today for complete contamination free digestion without loss of analytes seems to be microwave assisted acid digestion of 20–50 mg homogenized biological sample in high pressure PFA® digestion vessels using a mixture of 2 ml HNO_3 (sub-boiled Suprapur), 0.5 ml H_2O_2 (Suprapur) and 0.2 ml HF (Suprapur) in a microwave oven (e.g., from CEM Corporation, Matthews, NC or from Milestone MLS, Shelton, CT) at a pressure between 1 and 2×10^6 Pa and temperatures (upto 250 °C) at increasing rf power from 150 W to 300 W using a special digestion regime.

An issue not to be underestimated in the multi-element determination of trace elements in biological samples is the blank problem caused by different sources of contamination during sampling, sample preparation including crushing, homogenization, sample digestion, trace matrix separation, dilution, acidification (the reagent blanks should be as low as possible) and the whole mass spectrometric measurement procedure (at low instrumental blank). The topic of possible contamination is not only of enormous relevance for the analysis of elements abundant in the environment (Ca, Fe, Si, K, P, Zn, Cu and others), but it could also be problematic if memory effects occur. Therefore all the individual steps in an analytical protocol must be minimized with respect to blank values. Sample preparation methods and trace/species matrix separation procedures for special analytical applications on biological samples will be discussed in the following sections of this chapter.

It should be noted that advances and progress in life sciences, including the study of biological processes would not have been possible without the use of the powerful analytical techniques of mass spectrometry. Research developments in mass spectrometry are increasingly driven by research demands in the life sciences including the huge field of biology.

9.3.1 Analysis of Trace Elements in Biological Samples

Biological samples are potentially excellent records of their environment and are being examined by inorganic and biomolecular mass spectrometry to an increasing extent. Biomaterials such as carbonate shells, fish otoliths and fish scales or tree rings that increase their mass regularly on an annual basis may record changes in their environment via the concentrations of metals or non-metals and the isotopic composition of elements.[2]

Multi-element analysis by inorganic mass spectrometry to determine the total concentration of elements on quite different biological materials is performed by several working groups and in many laboratories in a routine mode mainly by quadrupole based ICP-MS with and without a collision cell or by LA-ICP-MS.[2–5] With its ability to provide very sensitive, accurate and precise multi-elemental analysis of trace and ultratrace elements and of isotope ratios, ICP-MS has developed into one of the most important inorganic mass spectrometric techniques for the characterization of biological materials (botanic and animal samples). The analysis of different certified reference materials is indispensable for verifying multi-element analytical methods involving ICP-MS.[3,5] The preparation and certification of matrix matched standard reference materials of quite different types of biological samples is of significance for the validation of analytical data. Sturgeon *et al.*[6] reported on the production and certification of fish otolith reference material (FEBS-1) prepared from 4.5 kg of dried saggital otolith harvested from the red snapper (*Lutjanus campechanus*). The sample was ball milled and sieved to < 200 mesh, homogenized, irradiated and bottled into 1 g aliquots and certified with respect to major, minor and trace element composition (using ICP-MS determined by the isotope dilution technique or external calibration, NAA, ICP-OES and X-ray fluorescence analysis). Certified values for seven reference elements (Ba, Ca, Li, Mg, Mn, Na and Sr) in fish otoliths are reported. This certified reference material will be of interest to those laboratories requiring quality assurance of measurements of the bulk elemental composition of otoliths and other marine aragonites.[6]

A homogeneity and stability study of the candidate reference materials for trace element analysis (As, Cd, Cu, Cr, Fe, Mn, Ni and Zn) on the Antarctic bivalve *Admussium colbecki* (IRMM 813) is discussed by Caroli and associates.[7] The scallops were collected at Terra Nova Bay (Ross Sea) during the 2000–2001 Italian expedition to Antarctica.

For multi-element analysis of biological materials, quadrupole based ICP-MS is employed with and without a collision cell and sector field instruments. Sometimes the amount of biological tissue is limited, especially if only a few mg of the biological sample is available (e.g., one seed grain, small amount of tissue or small parts of plant or animal, one hair).[3,5,8–11] Therefore, the lowest possible sample weight for performing multi-elemental trace analysis of biological materials has been investigated for ICP-MS measurements in the author's laboratory.[5] The certified reference materials Bovine Liver (NIST SRM 1577b) and Oriental Tobacco Leaves (CTA OTL-1) were applied at sample weights from 50 mg down to 1 mg, which were significantly lower than those recommended. In addition, the contamination problem was considered by minimizing the analytical techniques with microsample handling. The biological samples were analyzed after microwave-induced digestion using small closed vessels in a mixture of sub-boiled superpure 2 ml nitric acid, 0.5 ml hydrogen peroxide and 0.2 ml fluoric acid (see also Figure 6.28). The measurements demonstrated that in the case of homogeneous SRM the use of very low amounts (1–5 mg) result in reliable element concentrations using this analytical method after microwave-induced sample digestion. The detection limits in ICP-QMS (Elan 6000) varied between 0.12 and 0.38 $\mu g\,g^{-1}$ for Li, Na, Cr, Mn, Ni, Cu, Zn, Sr, Cd, Fe and Ni.[5] In order to analyze small amounts of sample, the homogeneity in powdered SRMs (in the μ m range) was studied by SIMS with respect to major and minor elements. In these SIMS experiments, the lateral resolution was about 1 μm. During SIMS measurements on NIST SRM pine needles for Na, K, Al and Mn, a slightly inhomogeneous distribution was observed in the powder grains with a diameter of 100 μm and below. This means that in this small dimension the inhomogeneity is an important limiting factor for the accuracy of trace analysis.

The performance of ICP-QMS with a hexapole collision cell (Platform from Micromass) for trace element determination in a bioassay has been examined systematically by Boulyga *et al.*[9] A significant improvement in the detection limits was achieved for ions affected by interferences with argon ions and argon based polyatomic ions by the removal of Ar^+ and ArX^+ ions ($X = H$, C, N, O, Ar), as well as for heavier elements ($m > 100\,u$) due to the improved ion transmission of the mass spectrometer with collision cell used. The accuracy of the analytical technique was assessed by determining 17 trace elements in certified standard reference material SRM 1566a Oyster Tissue. Good agreement of measured trace element concentrations was found in the concentration range from 28 $ng\,g^{-1}$ (Th) to 1890 $\mu g\,g^{-1}$ (Ca). The benefit of ICP-MS for multi-element analysis compared to instrumental neutron activation analysis (INAA) has been demonstrated on a thyroid tissue sample. Only six trace elements with long-lived radionuclides could be measured by INAA because of lack of irradiation facilities with high speed transportation of samples for the analysis of short lived radionuclides. It should be noted that INAA does not require sample digestion thus overcoming any possible contamination or element losses (as observed by ICP-MS for volatile Sc during sample digestion). By ICP-MS the concentration of 16 elements in a small amount of thyroid sample were determined.[9]

Multi-element analysis on 57 tree bark samples collected at different locations (near power plants, closed to a motorway, in urban or in uncontaminated rural areas) in the UK in order to study environmental contamination was performed by quadrupole based ICP-MS (Elan 6000, Perkin Elmer Sciex) after microwave induced digestions of samples and dilution. The measured concentration ranges of 52 elements varied from the sub-$ng\,g^{-1}$ range (e.g., Hf or Pt) to the low % range (e.g. for Ca) and are summarized in Table 9.24.[4] Oriental Tobacco Leaves SRM (CTA-OTL-1) was employed to validate the analytical ICP-QMS method.

Often only one selected essential or toxic element (such as Cu, Zn, Fe, As, Se, Cd, Pd, Pt or Hg) in biological materials has been analyzed with respect to the total element concentration.[12–14] For example, iron was determined in plant samples using the isotope dilution technique in ICP-QMS.[14] Dry ashing was shown to be well suited for achieving complete equilibrium of the isotopically enriched spike and analyte. To minimize argon based and calcium based polyatomic interferences on $^{56}Fe^+$ and $^{57}Fe^+$, the collision cell was operated under optimized conditions in the He/H_2 cell mode. It was not necessary to separate the calcium matrix from the analyte by ion exchange chromatography. The accuracy of the analytical method was validated by the analysis of three reference materials with a Ca content of 0.216 to 1.85 % (IAEA-359 – Cabbage, NIST SRM 1515 – Apple Leaves and NIST SRM 8412 – Corn Stalk).[14]

Table 9.24 Concentration range of elements in tree bark samples.

Element	Concentration range, μg g^{-1}	Element	Concentration range, μg g^{-1}
Li	0.2–3.2	Nb	0.02–0.3
Be	0.02–0.15	Mo	0.4–4.5
B	65–600	Rh	<0.00005–0.002
Na	40–300	Pd	<0.0016–0.003
Mg	190–1400	Ag	0.02–0.3
Al	50–1400	Cd	0.05–0.73
K	240–2100	In	0.01–0.16
Ca	2400–17500	Sn	0.26–2.1
Sc	0.1–0.6	Sb	0.08–3.7
Ti	2–90	Cs	0.001–0.2
V	0.5–30	Ba	25–320
Cr	0.7–12	La	0.18–6.5
Mn	10–4100	Ce	0.3–13
Fe	100–2700	Pr	0.04–1.8
Co	0.1–2	Nd	0.15–7
Ni	1.0–10	Sm	0.03–1.7
Co	2.5–30	Eu	0.01–0.4
Zn	5.5–92	Gd	0.03–1.5
Ga	1.2–14	Dy	0.02–1.0
Ge	0.07–3	Hf	0.0005–0.001
As	0.2–4.4	W	0.06–3.5
Se	0.4–3.2	Pt	0.0001–0.0054
Rb	0.3–4.6	Tl	0.001–0.009
Sr	18–75	Pb	8–280
Y	0.07–3.5	Th	0.002–0.2
Zr	0.09–2.7	U	0.0014–0.2

(J. Su. Becker et al. *Fresenius J. Anal. Chem.* **368**, 490 (2000). Reproduced by permission of Springer Science and Business Media.

As a relevant element in biology and life sciences, selenium has been determined in different parts of plants and animals, often with respect to its total element concentration but more frequently with respect to several selenium species (see Section 9.3.2) to study the selenium behaviour and its accumulation in plants. Plants differ in their ability to accumulate Se in their tissues and according to the amounts accumulated it is possible to classify them as hyperaccumulators (accumulation of hundreds to several thousands of mg kg^{-1} Se, dry weight).[15] A second group is formed by plants that can accumulate up to 1000 mg kg^{-1} Se, the so-called accumulators (e.g., *Brassia juncea*) and there are also non-accumulators (such as grasses) with a Se content above 100 mg kg^{-1} Se.[15] For sensitive Se determination (total element concentration) in biological samples, a special solution introduction device for the ICP-MS was employed combining pneumatic nebulization with hydride generation in thin liquid films on the walls of a minicyclonic spray chamber.[13] Biological tissue samples and certified reference materials (NIST SRM 1547 Peach Leaves and NIST SRM 1577b Bovine Liver) were analyzed after microwave induced digestion, and a detection limit of 0.03 μg g^{-1} was obtained in the routine mode using ICP-QMS (Elan 6000, Sciex).[13]

Sanz-Medel and co-workers reported on selenium determination in biological samples by isotope dilution analysis in ICP-QMS with an octopole collision cell.[11] The argon based isobaric interference arising during measurements of ^{80}Se was eliminated by using a hydrogen flow of 4 ml min^{-1} in the octopole cell, so that it is possible to determine Se with a detection limit of 14 pg g^{-1}.

Solid sampling electrothermal vaporization (ETV) was employed in ICP-MS for the direct determination of boron traces in biological samples.[16] This analytical method was tested on biological standard reference materials such as NIST SRM 1570 Spinach Leaves, NIST SRM 1573 Tomato Leaves or BCR CRM 281 Rye Grass. Using the isotope dilution technique, it was possible to determine B with a relative standard deviation (RSD) from 7–9 %, and the detection limit was found to be $8\,ng\,g^{-1}$. Sample consumption was 5 mg per determination.[16] Furthermore, ETV-ICP-MS was utilized for the determination of trace Cd and Pb in biological (and environmental) samples after single drop extraction. 8-Hydroxyquinoline was employed as the extractant dissolved in several μl of chloroform and then an organic microdrop was formed at the tip of the microsyringe needle to extract both analytes. The detection limits for Cd and Pb were $4.6\,pg\,ml^{-1}$ and $2.9\,pg\,ml^{-1}$, respectively with a 140-fold enrichment factor for Cd and 190-fold for Pb.[17]

In addition to ICP-MS for the multi-element analysis of aqueous solutions, LA-ICP-MS allows the direct determination of trace elements in biological samples and due to this feature it is a well suited analytical technique for microlocal analysis with spatial resolution. In 1995, Outridge *et al.*[18] reported on the performance of an LA-ICP-MS analysis for studying incremental biological structures as archives of trace element accumulation. The use of LA-ICP-MS for several biological (and environmental) applications is reviewed by Durrant and Ward.[19] Selected examples for determination of trace elements and species in biological samples are summarized in Table 9.25.

LA-ICP-MS is well suited for monitoring element distribution in annual rings of trees. Annual ring profiles in pine and birch from the Norwegian–Russian border and from northwest Russia measured by LA-ICP-MS as a rapid and sensitive method show the pollution history (especially of the heavy metal content) of the area. No correlation was found between the pollution levels of the soils and the observed metal content in the annual rings by Garbe-Schoenberg *et al.*[20]

Table 9.25 *Determination of trace elements and species in biological samples.*

Samples/matrix	Equipment/sample preparation	Analyte	Detection limits/variation range	References
Tree bark	ICP-QMS (Elan 6000) microwave digestion	52 elements	$0.03\,ng\,g^{-1}$ (Pt) Pt: $0.07–5.4\,ng\,g^{-1}$	Becker *et al.*[4]
Animal tissue	ICP-QMS microwave digestion	Pt	$7.5\,ng\,g^{-1}/ng\,g^{-1}$ range	Ardelt *et al.*[12]
Dill	HPLC-ICP-QMS (Agilent with collision cell) chromatographic separation	Se (total) and Se species MeSeCys, MeSeMet	$<0.1\,\mu g\,g^{-1}/0.17\,\mu g\,g^{-1}$ (control)–$31.8\,\mu g\,g^{-1}$ (supplemented)	Cankur *et al.*[32]
Biological reference materials	ICP-MS (Elan 6000) microwave digestion	15 elements	$0.02–0.38\,g\,g^{-1}$	Dombovari *et al.*[5]
Ouster Tissue (SRM NIST 1566a)	ICP-QMS with collision cell (Platform)	17 elements	$0.0014\,ng\,l^{-1}$ (U in aqueous solution) $0.3\,ng\,l^{-1}$ (As)	Boulyga *et al.*[9]
Fish otholith reference material: FEBS-1	ICP-QMS (Elan 6000) isotope dilution	Ba, Ca, Li, Mg, Na, Sr Cd, Cu, Ni, Pb, Zn	low and sub-$ng\,g^{-1}$ range	Sturgeon *et al.*[6]

Trace element levels in the otoliths of chum salmon (*Oncorhynchus keta*) were examined using LA-ICP-MS by several working groups.[21] It was found that the Mg, Zn, Ba and Sr concentrations of the otoliths differed between the fresh water and sea water growth zones. Mg and Zn concentrations in the fresh water growth zone were significantly higher than those in the sea water growth zone, while Sr and Ba concentrations in the former were significantly lower than in the latter. By comparing the Sr/Ca ratios, a linear relationship between the LA-ICP-MS results and those of X-ray analysis with an electron microprobe (EPMA) was observed. Thus it is possible to distinguish the marine and freshwater life phases using the Sr/Ca ratios. In addition, a correlation of Sr concentration and Mg, Zn and Ba concentrations was detected. These results can be interpreted in such a way that trace element compositions may reflect differences between fresh water and sea water conditions.[21] The disadvantage of this analytical procedure for otoliths is that the fish to be analyzed must be killed. An improved strategy in order to avoid killing the fish is to study the migration behaviour of fish to which sensors have been attached by analyzing the fish scales over a number of years. LA-ICP-MS was employed for the microlocal analysis of scales of North Sea houting caught in the IJsselmeer.[22] The fish scale analysis identified different migration patterns for the houting in the IJsselmeer and provided evidence that this fish is sometimes able to pass the migratory barriers to the North Sea, and does not need to migrate to sea to reach maturity. Figure 9.19 illustrates the fish scale analysis of North Sea houting by LA-ICP-MS. Figure 9.19 a shows the photograph of a scale and Figure 9.19 b illustrates part of a scale with laser craters obtained by LA-ICP-MS measurement in the line scan mode. The LA-ICP-QMS measurements (using ICP-QMS Elan 6000 and the laser ablation system LSX 200) were performed by J. Borcherding (Institute for Zoology, General Ecology and Limnology, University of Cologne) in the author's laboratory. Figure 9.19 c shows the $^{88}Sr/^{44}Ca$ ratio for nine North Sea houting caught in the IJsselmeer in 2001/2002. The results of the LA-ICP-MS measurements of $^{88}Sr/^{44}Ca$ ratios on fish scales were divided into three different types. Fish that only lived in freshwater (Type A) were characterized by $^{88}Sr/^{44}Ca$ ratios of around 0.2 from the nucleus to the edge. Of the analyzed houting from Lake IJsselmeer 72 % belonged to this type; most of them were small – the length of the largest houting was 42 cm. Fish with low $^{88}Sr/^{44}Ca$ ratios at the scale nucleus (Type B) followed by increasing ratios above 0.3 (assigned to brackish/marine periods), lived for a longer period in freshwater after hatching and then moved to seawater (four individuals, size 28–44 cm) and returned to fresh water where they were caught. If the $^{88}Sr/^{44}Ca$ ratio at the nucleus was above 0.3 (Type C), the period from hatching in fresh water until the arrival at sea was too short to measure a ratio indicating fresh water. Fish that obviously moved quickly to sea water after hatching (period from hatching in fresh water until arrival at sea was too short to measure the $^{88}Sr/^{44}Ca$ ratio that normally indicates fresh water) and returned to fresh water before they were caught (seven houting, size varied from 26 to 45 cm).[22]

9.3.2 Elemental Speciation in Biological Samples

Speciation analysis, which is a rapidly growing research field in the life sciences, especially due to the use of ICP-MS with its excellent element sensitivity, includes the determination of different oxidation states of elements or the type of chemical bonding with inorganic or organic partners. In speciation analysis it is important to avoid changes in the composition of compounds in samples during sampling and storage. For speciation analysis, different online coupling techniques are used for the separation of different species. The most frequently applied hyphenated technique is ion chromatography (especially HPLC), for example, for the separation of Cr(III) (essential for living organisms) and the toxic and carcinogenic Cr(VI), or for the separation of a multitude of arsenic compounds with great differences in toxicity. Liang *et al.*[23] determined both chromium species

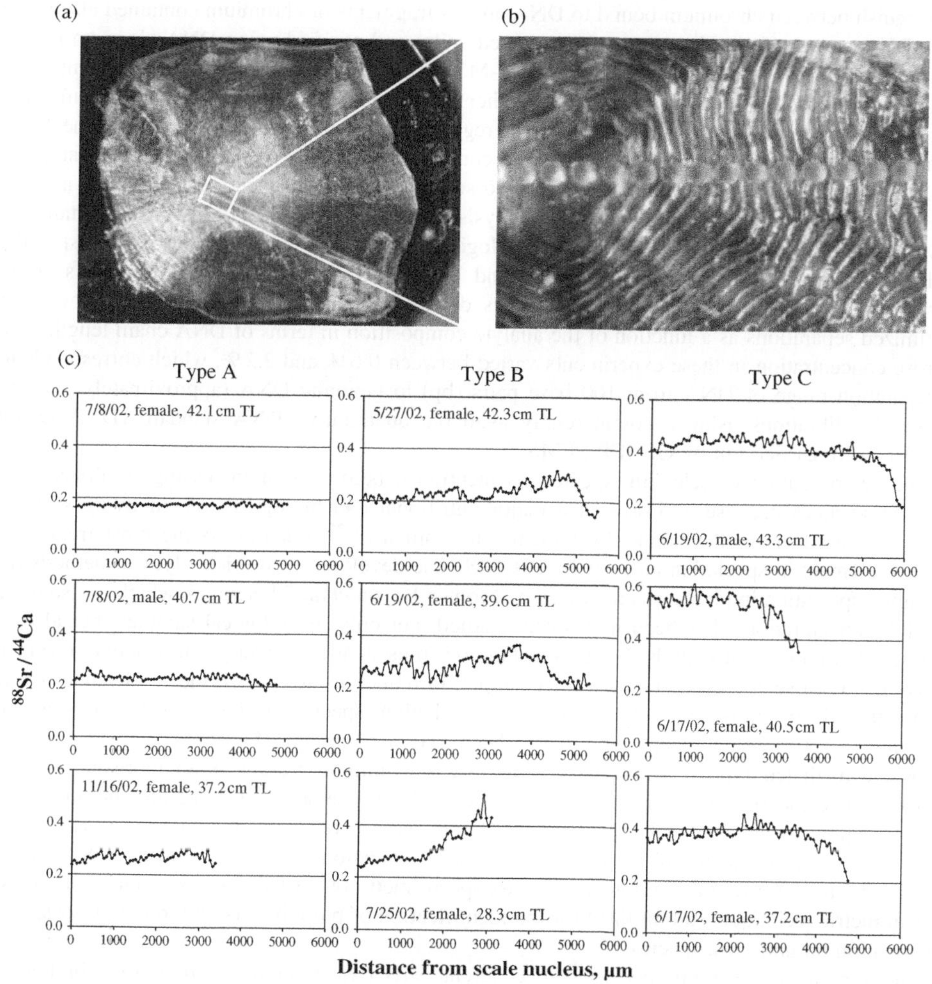

Figure 9.19 *Scale analysis of North Sea houting from Ijsselmeer LA-ICP-MS, a) photograph of a scale, b) part of a scale with laser craters obtained by LA-ICP-MS measurement in line scan modus, c)* $^{88}Sr:^{44}Ca$ *ratio for nine North Sea houting caught in IJsselmeer in 2001/2002. The results were sorted into three different types; A: Fish that only lived in freshwater; B: Fish that lived for a fairly long period in fresh water, then moved to sea water and returned to fresh water where they were caught; C: Fish that obviously moved quickly to sea water after hatching (period from hatching in fresh water until the arrival in the sea was to short to measure a* $^{88}Sr:^{44}Ca$ *ratio that normally indicates fresh water) and returned to fresh water before they were caught.*

in natural water by means of a separation procedure in which Cr(III) was retained on nanometre titanium dioxide at pH 6 and was subsequently eluted with HCl. The limit of detection (LOD) was determined as $0.32\,\mu g\,l^{-1}$ for Cr(III). The potential of chromium for binding to DNA isolated directly from soil was investigated in Caruso's group.[24] An analytical scheme was developed to distinguish between chromium bound to DNA and its fragments or chromium contained elsewhere in a DNA extract (from chromium contaminated soils) and purified using DNA clean up resins. Size exclusion chromatography coupled to ICP-MS with an octopole collision cell was employed to study the binding of chromium to DNA or other components in a DNA extract. Chromium were detected in large and small molecular weight regions of the chromatogram, whereby the DNA samples isolated from the two soils examined contained 0.5–0.7 $\mu g\,l^{-1}$ Cr, indicating that DNA isolated directly from a chromium contaminated soil has chromium bound to the nucleic acids.[24] Valuable quantification procedures for the analysis of nucleic acids are of growing importance in several diagnostic, forensic and molecular biological studies. An analytical procedure using the online coupling of gel electrophoresis (GE) and ICP-SFMS was developed in Bettmer's group for the quantification of DNA via phosphorus determination.[25] GE conditions are chosen for optimized separations as a function of the analyte composition in terms of DNA chain length. The agrose concentration in these experiments varied between 0.6 % and 2.2 %, which corresponds to a separation range of DNA from 100 base pairs (bp) to genomic DNA (approximately 3 Mbp). External calibrations using a commercially available quantitative DNA standard (DNA Quant Ladder) were carried out by GE-ICP-SFMS.[25]

The determination of selenium species, as mentioned before, is of increasing significance in the life sciences because it possesses a major nutritional and biological role in living systems and is required in small amounts by humans and animals.[26,27] Selenium is the most frequently studied element in speciation analysis by ICP-MS. The development of bioanalytical methods for selenium speciation in plants was reviewed by Montes-Bayón *et al.*[15] Selenium, which is also toxic at high concentrations, has been intensively studied, not only in biological samples, but also in biological or human body fluids.[28,29] This metalloid is present as a necessary component to form the active centre, a seleno group (-SeH), for example, in numerous selenoenzymes.[26] Selenocysteine is functionally the most extensively investigated selenium species in biochemical processes and is found in active selenium proteins. It has been classified as the 21st amino acid in terms of ribosome mediated protein synthesis.[30,31] Because selenium is present as an inorganic species (selenite or selenate) or organoselenium compounds (selenocysteine or selenomethionine) in natural water samples, hyphenated techniques such as online coupling of liquid chromatography (LC) with ICP-MS are advantageous. In addition, electrospray ionization mass spectrometry (ESI-MS) or matrix assisted laser desorption ionization mass spectrometry (MALDI-MS), as biomolecular mass spectrometric techniques, are utilized for the identification of biopolymers and to obtain structural information on unknown species.

An example of selenium speciation in different parts (leave, stem and root) of a biological sample (dill – *Anethum graveolens L.*) is illustrated in Figure 9.20.[32] The speciation of selenium in dill, supplemented with sodium selenite during its growth, was performed in Caruso's group using ion pairing reversed phase and cation exchange chromatography by HPLC-ICP-MS with an octopole collision cell. The identifications of species in the sample extracts were based on matching retention time with available standards followed by the standard addition method. The species that eluated at 730 s (see Figure 9.20) was identified as MeSeCys (*Se*-methyl-selenocysteine – (5)), which was the major species in both the root and stem (32 % and 38 %, respectively). Another major Se species identified is MeSeMet (*Se*-methyl-selenomethionine (4)).[32]

Miniaturization of HPLC-ICP-MS is an important issue in bioanalytical chemistry when small amounts of sample (e.g., single cells) need to be investigated.[33] ICP-MS (with an octopole collision cell) in combination with nano-HPLC (75 μm column) was optimized for the detection of selenopeptides in a selenium–yeast protein digest after 100-fold preconcentration on a C_{18} capillary precolumn (300 μm column for salt removal and cleanup).[34] Under identical separation and preconcentration conditions, electrospray MS/MS (using Nanospray qQqToF-MS – QSTAR from Applied

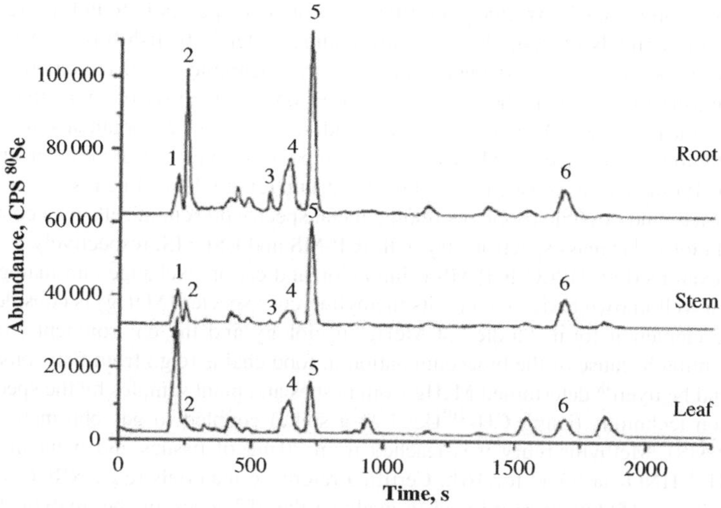

Figure 9.20 *Different selenium species determined by HPLC-ICP-MS in leaves, stems and roots of dill: (1) Se (VI); (2) Se (IV); (3) $SeCys_2$; (4) MeSeMet; (5) MeSeCys; (6) SeMet. (O. Cankur, S.K.V. Yathavakilla, J.A. Caruso, Talanta,* ***70****, 784 (2006). Reproduced by permission of Elsevier.)*

Biosystems) measurements were performed to identify the selenium species. The figures of merit of the nano-HPLC mass spectrometric methods developed using the nanoflow total consumption nebulizer (nDS-200, described elsewhere[35,36]) were demonstrated for a comprehensive online characterization of a mixture of peptides in a tryptic digest of a selenium-containing protein fraction isolated by size exclusion chromatography (SEC) from selenium yeast extract. The performance of the low flow nebulizer used at solution uptake rates of 500 nl min^{-1} for a robust sheathless interface of nano-HPLC and ICP-MS is discussed by Giusti *et al.*[35] The absolute detection limit for selenium was 25 fg (^{80}Se), which allowed the low abundant selenopeptides to be measured by nano-HPLC-ICP-MS. The development of the interface between nano-HPLC and ICP-MS and the application of the on-line isotope dilution technique quantifying selenium-containing peptides in protein tryptic digest is discussed in reference.[37] An overview of quantitative speciation analysis of endogenous trace element compounds in biological systems by the two different approaches of isotope dilution analysis (species unspecific and species specific isotope dilution technique if the biomolecule is unknown or a well defined compound, respectively) is presented by Schaumlöffel and Lobinski.[38]

Simultaneous speciation of inorganic selenium and tellurium in water samples by ICP-MS was performed after selective solid phase extraction (SPE), as discussed by Yu *et al.*[39] Under acidic conditions Se (IV) and Te (IV) complexes with ammonium pyrrolidine dithiocarbamate were formed. The detection limits for Se and Te species in water samples were found to be 7 ng l^{-1} and 3 ng l^{-1}, respectively.

Another intensively studied element in speciation analysis is arsenic. The biological and environmental effects of arsenic species and their transformation pathways have been studied in numerous papers.[40–42] Both arsenite and arsenate accumulate in living tissues because of their affinity for proteins, lipids and other cellular compounds.[43] Arsenic species can undergo transformation via abiotic and biotic processes.[41] Arsenic occurring as an organic species is found in higher concentrations in marine animals compared to the surrounding water.[44] In fish, up to 99 % of arsenic is retained in an organic form. Arsenobetaine and arsenocholine are detected in both marine animals and mushrooms.[45] Gallagher *et al.*[40] reported on the analysis of more than 25 arsenic species in biological tissues. Over the last few decades, more than 20 organoarsenics have been identified in biological matrices.[46] Most of these compounds contain a dimethylarsenoic moiety ($(CH_3)_2As(O)$-R) such as arsenosugars. Feldmann's group reported on labile arsenic compounds in biological matrices and possible problems finding metal species present in cells by a combination of elemental and molecular mass spectrometry with ICP-MS and ESI-MS, respectively.[46] The arsenic species were examined by HPLC-ICP-MS using anion and cation exchange chromatography.

Mercury is a well known toxic element; its methylmercury species (MeHg) is considerably more toxic than the elemental form. Studies of MeHg in biology and the environment have attracted increasing attention because of the bioaccumulation in food chains (e.g., from fish consumption).[47] Hintelmann and Nguyen[48] determined MeHg from tissues and plant samples by the species specific isotope dilution technique (using $CH_3{}^{210}HgCl$ as a spike) coupled to gas chromatography ICP-MS (GC-ICP-MS). Methylmercury was leached from 20 mg of tissues and plant materials with 5 ml of 4 mol l^{-1} HNO_3 at 55 °C for 16 h. Certified reference materials (e.g., NRCC-TORT-2 and NIST Oyster Tissue 1566b) were utilized to evaluate the efficiency of the analytical procedure. The isotope dilution method developed using GC-ICP-MS was applied for MeHg determination in a selection of zooplankton and zoobenthos.[48]

An interesting approach was proposed in Lobinski's group for the analysis of non-covalent Ni species in biological samples.[49] The Ni species in aqueous plant tissue extracts were quantitative determined by SEC-ICP-MS in combination with ESI-ToFMS/MS after purification of Ni species by hydrophilic interaction HPLC (HILIC).[49]

Three different separation techniques – ultrafiltration (UF), high performance size exclusion chromatography (HPSEC) and asymmetric flow field flow fractionation (AsFIFFF) have been employed for the characterization of metal-humic substance complexation in compost leachate.[50] The possible interactions of about 30 elements with different size fractions of humic substances have been investigated by coupling these separation techniques with ICP-MS and UV-vis absorption spectrophotometry. Elements can be classified into roughly three groups according to their association with humic like substances of different size and molecular weight in the leachate. Divalent cations (Co, Ni, Cu or Zn) show an evident tendency to be associated with smaller organic fractions, whereas tri- and tetravalent cations (Al, Fe, Ce or Sn) and rare earth elements tend to be associated to larger fractions.[50]

Szpunar, Lobinski *et al.* have reported on several hyphenated techniques (especially CE-ICP-MS and HPLC-ICP-MS) for elemental speciation in biological systems,[51–54] describing e.g. the quasi-simultaneous determination of more than 30 selenopeptides using HPLC-ICP-MS.

9.3.3 Analysis of Phosphorus, Metals and Metalloids Bonded to Proteins

It is well known that a great variety of biomolecules exist where metals and metalloids are bound to proteins and peptides, coordinated by nucleic acids or complexed by polysaccharides and small organic ligands such as organic acids.[55] Most proteins contain amino acids with covalently bonded heteroelements such as sulphur, selenium, phosphorus or iodine.[51] Several reviews have been published on the development of mass spectrometric techniques for bioanalysis in 'metallomics', which integrate work on metalloproteins, metalloenzymes and other metal containing biomolecules.[1,51,53,54,56–59] The authors consider trace metals, metalloids, P and S (so-called 'hetero-elements') as naturally occurring 'elemental tags' for the detection of biomolecules such as metallo-, seleno- or phosphoproteins.[56]

Phosphorus plays perhaps the most important role in cell biology. Protein phosphorylation (as an essential part of the ribose–deoxyribose phosphorus backbone in the RNA and DNA chain) and de-phosphorylation are key regulatory signalling processes that regulate biochemical pathways.[60] These processes are catalyzed by kinase and phosphatase, respectively.[61] Phosphorus determination in proteins by ICP-MS is difficult due to isobaric interferences of $^{14}N^{16}O^{1}H^{+}$ and $^{15}N^{16}O^{+}$ at $m/z = 31$ (see Figure 6.3) and possible contamination problems during sample preparation. Double-focusing sector field ICP-SFMS at medium mass resolution has been employed for phosphorus determination in small protein samples whereby a small amount of solid protein sample (down to 1 μg) or digest (1–10 μl) protein solution was denatured in nitric acid and hydrogen peroxide by closed microvessel microwave digestion as proposed by Becker *et al.*[62] For quality control of phosphorus determination, a certified reference material (CRM) of single cell proteins (BCR 273) with a high phosphorus content of $26.8 \pm 0.4\,mg\,g^{-1}$ was analyzed. The homogeneity of CRM BCR 273 was investigated for studies on phosphorus determination in proteins while reducing the sample amount as much as possible. Relative standard deviation and measurement accuracy in ICP-QMS) was within 2 %, 3.5 %, 11 % and 12 % when using CRM BCR 273 sample weights of 40 mg, 5 mg, 1 mg and 0.3 mg, respectively. The lowest possible sample weight for an accurate phosphorus analysis in protein samples by ICP-MS was discussed. The analytical method developed was applied for the analysis of homogeneous protein samples in very low amounts (1–100 μg of solid protein sample, e.g., β-casein or down to 1 μl of protein or digest in solution (e.g., tau protein)). A further reduction of the diluted protein solution volume was achieved by the application of flow injection for ICP-SFMS, which was discussed with reference to real protein digests after protein separation using 2D gel electrophoresis. The detection limits for phosphorus in biological samples were determined by ICP-SFMS down to the $ng\,g^{-1}$ level.

Besides phosphorylation oxidation is also one of the most biologically important modifications of proteins and is involved both in normal physiology and in human disease, carcinogenesis and ageing.[63] Boulyga *et al.*[64] determined P and also Fe in proteins (e.g., β-casein, single cell protein, tau protein and phosphorylated peptides after digestion in a HNO_3/H_2O_2 mixture) by sector field ICP-SFMS in comparison to ICP-QMS with a hexapole collision cell, with detection limits in the low $ng\,g^{-1}$ range. The detection limits for P and Fe in high purity water were found by ICP-SFMS to be $20\,pg\,ml^{-1}$ and $3\,pg\,ml^{-1}$, respectively. Sulfur was applied as the internal standard element for evaluating the quantity of sulfur containing proteins. LA-ICP-SFMS was successful applied for the detection and analysis of P and S in proteins separated by 2D gel electrophoresis.[64] The addition of an internal standard element to the gel before sample preparation was found to be advantageous for quantification purposes. In the LA-ICP-MS analysis of P and S as a function of depth in the gel, changing concentrations were observed (see Figure 9.21) with a maximum concentration on the gel surface, but the P/S ratios remained constant as a function of depth within the measurement errors.[64] A possible quantification procedure in LA-ICP-MS to determine the phosphorus content in phosphorylated proteins was proposed via 1D calibration gels as demonstrated for ovalbumin and bovine serum albumin in Figure 9.22a. Transient signals of phosphorus ion intensities of 500, 100, 10, 1 and 0.1 ng of ovalbumin are shown in Figure 9.22b. An unknown amount of P in protein (x ng, see Figure 9.22a) can be determined via the corresponding calibration curve of phosphorus determination in protein bands of 1D gel. The measured calibration curve resulted in a calibration coefficient of only 0.89 due to inhomogeneous protein distribution in the gel analyzed. Becker and co-workers developed a gel screening technique using LA-ICP-SFMS of P, S, Cu, Zn and Fe in proteins and the identification of phosphorylated protein structures by MALDI-FTICR-MS.[65] Yeast mitochondria membrane protein complexes were separated in their native state in the first

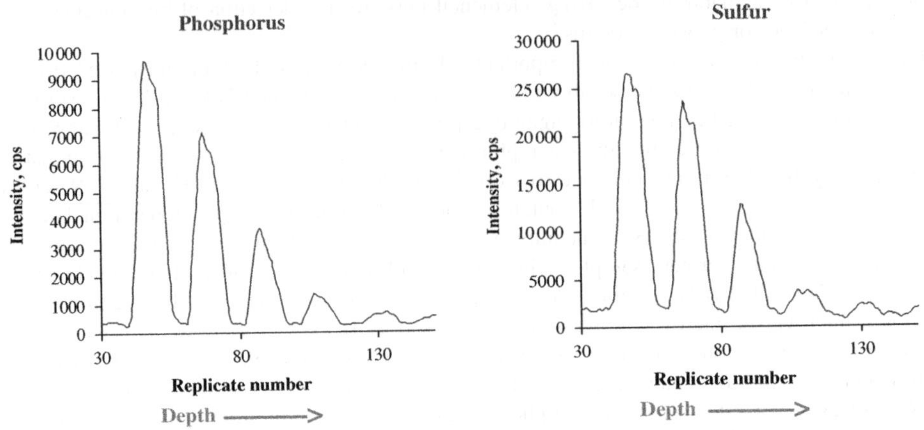

Figure 9.21 Depth profiling of phosphorus and sulfur in protein spots of a 2D gel measured by LA-ICP-MS. Transient signals show decreasing ion intensity with increasing replication number (proportional to measurement time).

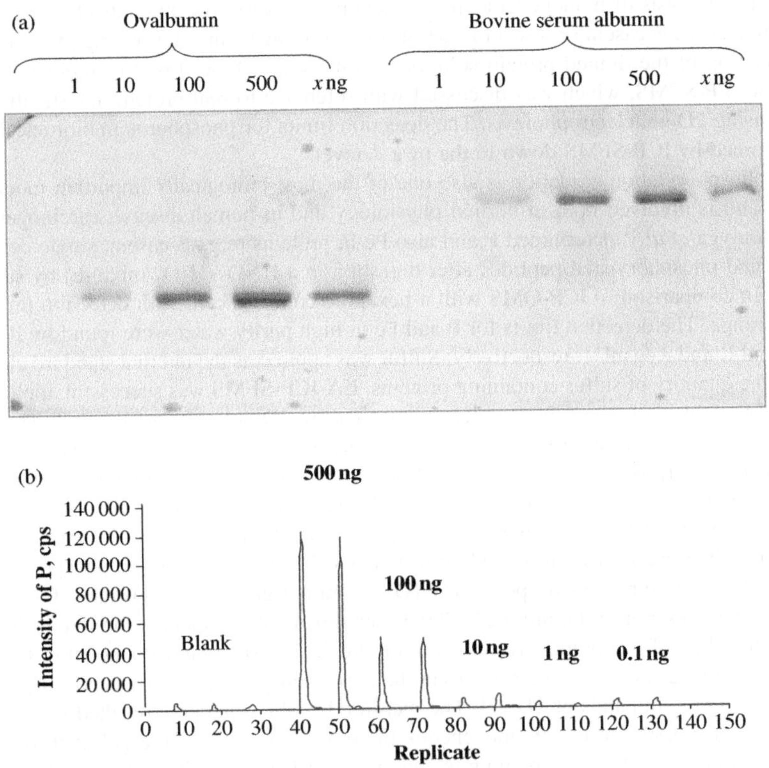

Figure 9.22 Calibration gel for phosphorus determination in proteins (a) 1D gel of ovalbumin and bovine serum albumine for different phosphorus amount and (b) transient signals of P in protein spots (and blank) in 1D gel of ovalbumin measured by LA-ICP-SFMS. (J. S. Becker et al., J. Anal. At. Spectrom., **19**, 1236 (2004). Reproduced by permission of the Royal Society of Chemistry.)

dimension and their sub-unit composition resolved by SDS-PAGE in the second dimension (blue native/SDS 2D-PAGE). The proteins were marked by the silver staining technique (compared to Coomassie staining). P, S and selected essential metals (Cu, Zn and Fe) were detected in a short analysis time with detection limits determined in the gel blank of 0.22 $\mu g\,g^{-1}$ or 8.2 $\mu g\,g^{-1}$ for P and Cu, respectively, for more than 60 mitochondrial protein spots in 2D gels. For quantification purposes, sulfur was used as an internal standard element.[65] Furthermore, the identification of phosphorylated subunits (Atp1p and Atp2p) of yeast mitochondrial ATPase was carried out by a combination of LA-ICP-MS with MALDI-FTICR-MS (on blue native BN/SDS-PAGE). The results were confirmed by Western blot analysis using antibodies directed against phosphorylated amino acids. The combination of LA-ICP-MS and MALDI-FTICR-MS with BN/SDS-PAGE provides a fast and sensitive tool for the structural analysis of phosphorus and metal containing sub-units of membrane protein complexes.[66]

The complementary use of LC-ESI-ToF-MS and hyphenated ICP-MS with dynamic reaction cell (DRC) for the characterization of native and recombinant copper proteins (of molar mass range 10–20 kDa) has been investigated by Hann *et al.*[67] Size exclusion chromatography (SEC) and ion chromatography (IC) were implemented as separation techniques for hyphenated ICP-MS of intact Cu containing proteins.

The contribution of mass spectrometry to the biosciences, especially in proteomics (the determination of the protein constituents of a cell or organism) of individual components and the quantitative definition of cellular processes, was reviewed by Gaskell.[68]

Cloning, sequence analysis and heterologous expression of the gene encoding an (S)-specific alcohol dehydrogenase from *Rhodococcus erythropolis* DSM 43297 is described by Abokitse and Hummel.[69] The nucleotide sequence of 1047 bp (base pairs), coding for 348 amino acids, was cloned in *Escherichia coli* cells and successfully expressed. The sub-unit molecular mass as deduced from the amino acid sequence was determined as 36.0 kDa. The recombinant enzyme exhibited high thermostability, which facilitated its purification by heat treatment, followed by two column chromatography steps. RE-ADH shows great similarity to several zinc containing medium chain alcohol dehydrogenases.[69] The zinc content in RE-ADH enzymes was determined in 1D gel via calibration gels as shown in Figure 9.23. The calibration curve for zinc in RE-ADH, established by measurements of transient signals (see Figure 9.23 a) in 1D gels with different amounts of RE-ADH from 1.5 to 12 μg (Figure 9.23 b), resulted in a correlation coefficient of 0.98. The amount of RE-ADH (native protein) and RE-ADH* (possibly denatured protein) in the 1D gel in Figure 9.23 c was determined, using the calibration curve, as 29 μg and 2 μg, respectively.

Size exclusion chromatography (SEC) together with HPLC-ICP-MS has been employed to examine selenium-containing proteins in chives (*Allium schenoprasum*) grown separately in three different supplementation media (Se (IV), Se (VI) and SeMet).[70] The highest selenium accumulation of up to 700 $\mu g\,g^{-1}$ was observed in the case of the Se (VI)-enriched sample. Reversed phase ion pairing chromatography (RP-IP-HPLC-ICP-MS) was employed for the speciation of selenium containing amino acids.[70] Selenoproteins have been identified in a diverse range of organisms, including bacteria and animals, but their role in the plant kingdom is less well understood. Water soluble selenium containing proteins extracted from selenium accumulating plants (*Brassica juncea*) and non-accumulator species (*Helianthus annuus*) exposed to varying forms and concentrations of selenium have been studied by Caruso's group.[71] Protein extracts were analyzed by SEC-ICP-MS. In addition, the proteolytic digests of plant extracts were analyzed by reverse phase chromatography coupled to ICP-MS in order to investigate the selenoamino acid and selenopeptide content. Selenomethionine was found to be the primary constituent of the proteins of the non-accumulator plant, while the same proportions of selenocysteine and selenomethionine were found in the accumulator extract.[71]

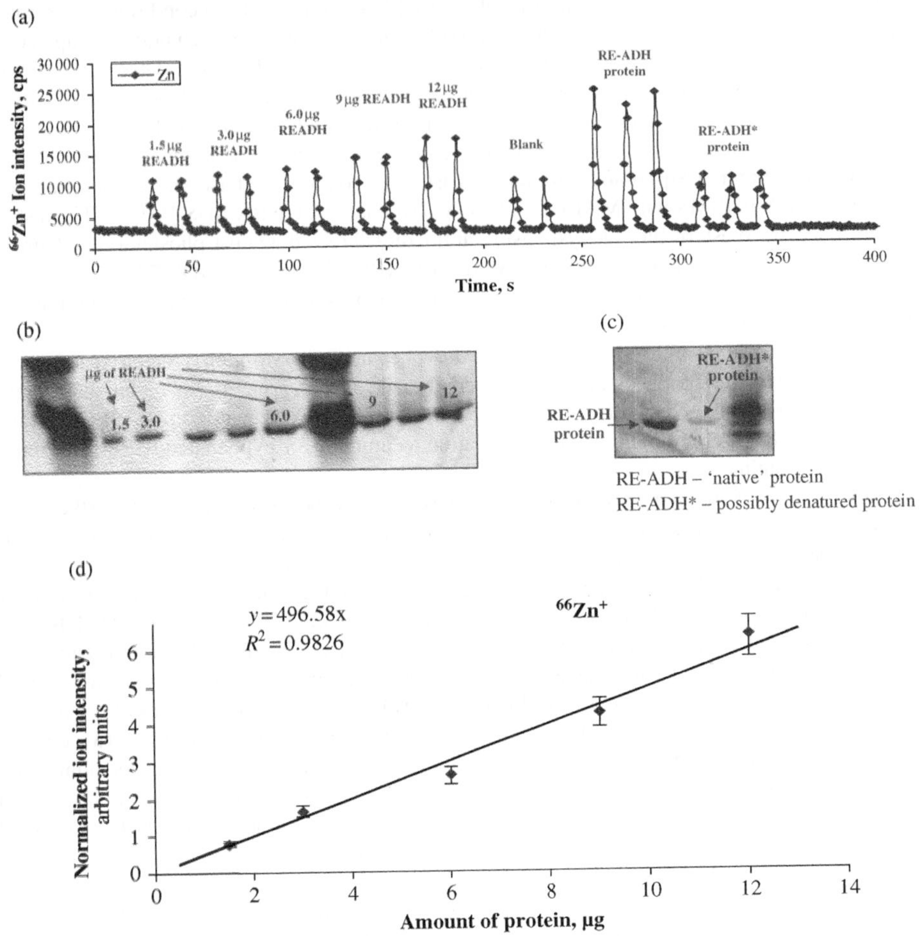

Figure 9.23 *Determination of RE-ADH by LA-ICP-SFMS: (a) transient signals in 1D gels with different amounts of RE-ADH from 1.5 to 12 µg, (b) corresponding 1D calibration gel of RE-ADH, (c) part of 1D gel with unknown amount of RE-ADH – native protein and possibly denaturized RE-ADH* protein, (d) the calibration curve for* ^{66}*Zn in RE-ADH after measurements of transient signals in 1D gel in (b).*

By combining denaturing gel electrophoresis (SDS-PAGE) and LA-ICP-MS an analytical technique was developed for the speciation of selenoproteins in Se contaminated wildlife (waterfowl embryo and fish ovary collected from Se contaminated sites in Wyoming and the San Joaquin Valleys, CA, respectively).[72] Polatajko *et al.* reported on the state of the art of selenium speciation in biological samples.[73]

The applicability of CE-ICP-MS for fast screening of weak metal interactions of Cd^{2+} with several test proteins (such as carbonic anhydrase from bovine erythrocytes, bovine serum albumin, human holotransferrin, ceruloplasmin and superoxide dismutase) has been studied by Cahmoun and Hagège.[74] The sensitivity of element species and their detection limits in CE-ICP-MS have been improved by introducing a preconcentration step involving large volume stacking with polarity switching. The delivery either of solution buffer when performing the preconcentration step or the make up solution in the case of separation of protein mixtures is affected by modifying the interface. As stacking occurs, this method is still limited to standard solutions or interactions with major proteins. Applications to real biological samples require a partial desalting step before the CE-ICP-MS analysis.[74]

9.3.4 Isotope Ratio Measurements of Biological Systems

Carbon isotope fractionation effects of individual compounds were observed in living organisms and also as a result of enzymatic isotope effects and reaction kinetics in biological systems. Such fractionation effects have to be examined by isotope ratio mass spectrometry in order to understand specific processes in life sciences or in environment.[75]

Actinides found as environmental contamination in mosses collected from a bog in the eastern Italian Alps were analyzed after their chemical separation by extraction chromatography (deposited on steel targets) with respect to isotope ratios and their concentration was determined by LA-ICP-MS.[76] Moss samples were contaminated with a variety of actinide isotopes. The detection limits for actinides were determined as $3.6–7.2 \times 10^{-15}\,g\,g^{-1}$ for ^{243}Am and ^{238}U, respectively. The $^{240}Pu/^{239}Pu$ isotope ratio (0.212 ± 0.003) was almost constant within experimental error for all samples investigated. Pu contamination in moss samples was mainly the result of global fallout after nuclear weapons tests. ^{241}Am was found at the $2 \times 10^{14}\,g\,g^{-1}$ level. This example demonstrates that mosses can be used as bioindicators for environmental contamination.[76]

Isotope ratio measurements were employed for the quantification of analytical data using the isotope dilution strategy. For example, isotope dilution analysis was developed by Sanz-Medel's group[77] for the determination of selenomethionine in Se enriched yeast material by HPLC-ICP-MS using a ^{77}Se-enriched selenomethionine spike obtained by growing yeast on a ^{77}Se rich culture medium. For Cr(III)/Cr(VI) determination in yeast, Caruso *et al.*[78] employed the double spike species specific isotope dilution technique measured by HPLC-ICP-MS. The isotope pattern deconvolution approach applied in this work delivers a more intuitive and elegant solution to an otherwise complex data analysis without the need for iterative calculations as widely practised in double spike isotope dilution. The results are in exact agreement with the conventional isotope dilution calculations.[78]

Tracer Experiments for Studying Transport Phenomena

In order to investigate the distribution and transport of mineral elements in plants by tracer experiments, the aqueous nutrient solution was doped with the highly enriched stable isotopes ^{25}Mg, ^{26}Mg, ^{42}Ca, ^{44}Ca and ^{41}K (see Figure 9.24 a). The fine focused primary ion beam can be used for sputtering the solid surface to determine the lateral element distribution of enriched isotope tracers in plant tissues. Whereas the lateral distribution of the highly enriched doped isotopes in a cross section of part of the plant was measured using the SIMS ion microscope mode (isotope ratio measurements were performed as well), the precise isotope ratios and the element concentrations (using the reverse isotope dilution technique) in nutrient solution taken from parts of the plant were determined by ICP-MS. For tracer experiments using ^{26}Mg, ^{44}Ca and ^{41}K spiked plant nutrient solution, magnesium, calcium and potassium isotope ratio measurements were carried out by quadrupole based ICP-MS without a collision cell[79] and also with an octopole collision cell. For calcium and potassium isotope ratio measurements, H_2/He as gas mixtures were employed in an Agilent 7500ce with octopole collision cell to minimize isobaric interferences, such as $^{40}Ar^+$ on $^{40}Ca^+$ or $^{40}Ar^1H^+$ on $^{41}K^+$, and others. The nutrient solutions investigated contained high amounts of calcium and potassium in the 10–30 μg ml^{-1} range.[79] The isotope ratio measurements on certified isotope standard reference materials (NIST SRM 980, and NIST SRM 985 for Mg and K, respectively) yielded an accuracy of 0.1 and 0.3 % (RSD) as demonstrated in routine work in the author's laboratory.

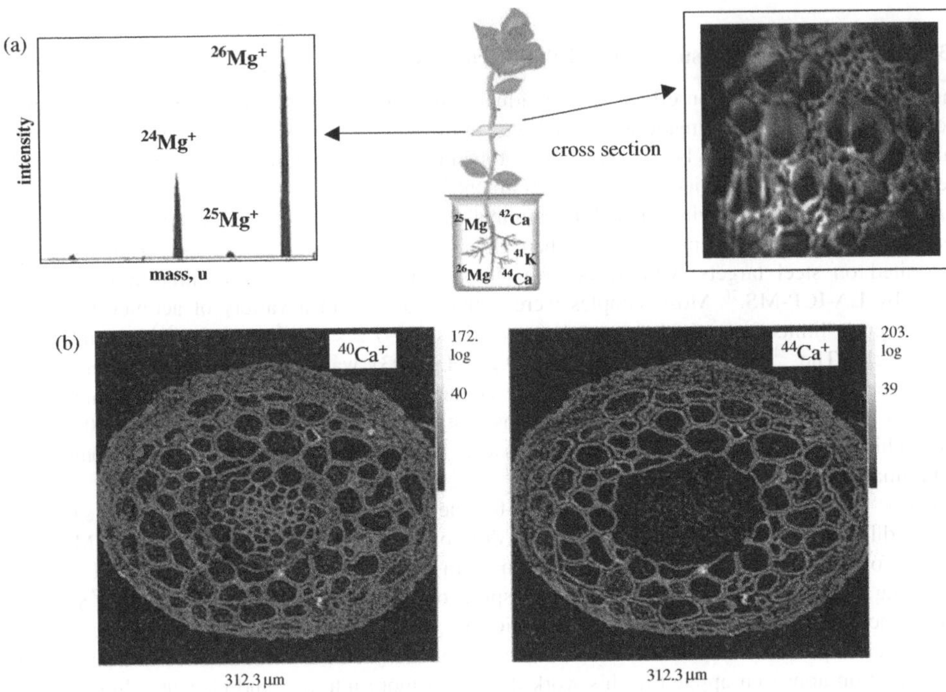

Figure 9.24 *(a) Schematic of tracer experiments on transport phenomena in plants, (b) SIMS images of $^{40}Ca^+$ and $^{44}Ca^+$ of a cross section of a root of a Norway spruce. (J. S. Becker and H. J. Dietze, Int. J. Mass Spectrom. Ion proc.* ***197****, 1 (2000). Reproduced by permission from Elsevier.)*

Figure 9.24 b compares the isotope ion distribution of natural $^{40}Ca^+$ and doped $^{44}Ca^+$ in a cross section of a root of a Norway spruce. These measurements were carried out by SIMS using $^{69}Ga^+$ primary ions (10 keV, 0.1 nA, Ionoptika) with a lateral resolution < 0.1 μm.[80] The investigations can be used to study the mechanism of mineral element uptake and kinetics of transport processes in plants as a function of time. About 20 min after the start of the tracer experiment a barrier was observed for the transport of enriched isotope $^{44}Ca^+$ in the middle of the root.[80]

Tracer Studies on Tau Protein

The study of metal containing proteins is a challenging task in proteomics. Metal containing proteins have been directly determined in separated protein bands in 1D gel by LA-ICP-MS. In order to investigate the binding of Cu and Zn on tau protein isoforms as target proteins in Alzheimer's disease, enriched isotope tracers (^{65}Cu and ^{67}Zn) have been doped to the 1D gel of separated tau protein isoforms after gel electrophoresis. Five bands in the 1D gel were selected for LA-ICP-SFMS measurements (see Figure 9.25 a). Isotope ratio analyses were performed by measuring transient signals of ion intensities of $^{63}Cu^+$, $^{65}Cu^+$, $^{64}Zn^+$ and $^{67}Zn^+$ in the protein bands. The background signals in the blank gel as a result of possible contamination during chemical procedures, especially during silver staining, were considered. The results of $^{65}Cu/^{63}Cu$ isotope ratio measurements in gel bands measured by LA-ICP-MS are summarized in Figure 9.25 b. No Cu was detected in protein band 1. Different $^{65}Cu/^{63}Cu$ isotope ratios were measured in protein bands 2–5 of the tau isoforms. The highest enrichment of ^{65}Cu was found in band 4. This result shows that tau protein isoforms are able to bind Cu. Zn was only detected in band 1 of slice isoform PNS tau and no enrichments with the stable ^{67}Zn isotope were detected. MALDI-FTICR-MS was employed to identify the proteins. Figure 9.24 c shows the MALDI-FTICR mass spectrum of the tryptic digested protein of 1D gel band 5. The protein with a molecular weight of 45.6 kDa was identified as splice isoform tau-F by a database search.

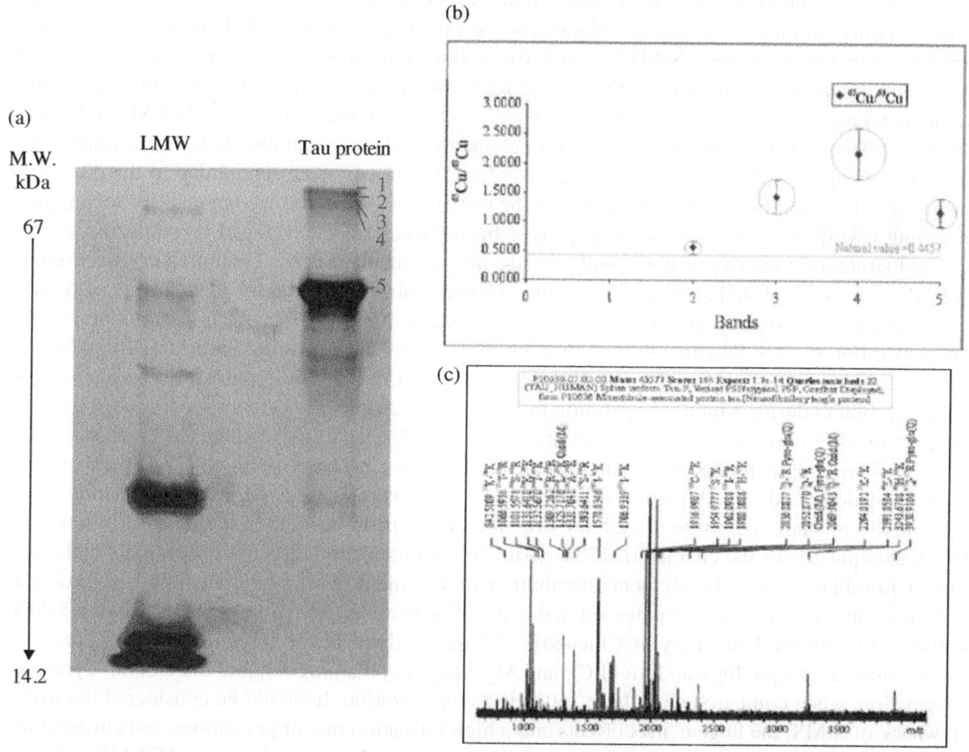

Figure 9.25 *Tracer studies on tau protein: (a) 1D gel, (b)* $^{65}Cu/^{63}Cu$ *isotope ratios in gel bands measured by LA-ICP-MS, (c) MALDI-FTICR mass spectrum of splice isoform tau F (J. Su. Becker, M. Zoriy, M. Przybylski, J. S. Becker, J. Anal. At. Spectrom.,* ***22****, 63 (2007). Reproduced by permission of the Royal Society of Chemistry.)*

9.3.5 Trace and Imaging Analysis on Biological Tissues and Single Cells

An emerging field in bioanalytics is the direct analysis of tissues to study the lateral distribution of elements, element species or biomolecules. Over the past few years, the development and application of imaging mass spectrometric techniques such as matrix assisted laser desorption/ionization mass spectrometry (MALDI-MS),[81–84] secondary ion mass spectrometry (SIMS) and also laser ablation inductively coupled plasma mass spectrometry (LA-ICP-MS)[85,86] has been rapidly growing for the determination of the quantitative distribution of organic compounds, metals, non-metals and metalloids to obtain images of small or large sections of biological tissue.[87,88] Pioneering work on organic ion imaging of biological tissue by MALDI-MS compared to SIMS has been performed by Caprioli's group.[84,89] The quantitative imaging of elements by mass spectrometry in thin tissue slices or on biological surfaces is an especially challenging task in analytical chemistry and is also relevant in different areas of biological research.[87,90,91] One aim of advanced analytical mass spectrometric techniques is to use imaging capability and to perform microlocal analysis on biological tissues, in addition to other existing non-mass-spectrometric techniques (e.g., SEM-EDX), on selected single cells or sub-cellular regions to detect essential and toxic elements of biomolecules. Secondary ion mass spectrometry (SIMS)[86,92,93] is one of the most widely applied surface analytical techniques for the imaging (mapping) of elements on biological surfaces or on tissue sections. SIMS can directly produce ion images of element distribution[93] and of organic compounds[84] in tissue with a lateral resolution in the low μm and sub-μm range using liquid metal primary ion sources or cluster beams to produce secondary ions.[86] ToF-SIMS and laser-SNMS are becoming increasingly important for elemental and molecular distributions in organic and biological samples.[94] Appropriate sample preparation techniques are essential due to the ultrahigh vacuum (UHV) conditions in the ion source of the instrument. Therefore cryopreparation techniques including cryofixation and cryosectioning have been successfully developed to retain the chemical and structural integrity of the sample during measurements. Such a combined cryosectioning ToF-SIMS instrument including an in vacuum cryosectioning technique for the analysis of frozen non-dehydrated biological samples has been developed by Moeller *et al.*[94] This technique allows the preparation of non-dehydrated biological samples via freeze fracturing under UHV conditions without any exposure of the sample to air. However, the absorption of water and residual gas remain a major problem during sample preparation in the vacuum chamber and can be reduced by a slight increase in the sample temperature to around −110 °C. The analysis of quite different samples, including hair, drug capsules, mammalian cells and biofilms, has been reported.[94,95] Dickinson and co-workers described a modification of a dynamic SIMS instrument for imaging of frozen biological samples to study the image of cellular structures with spatial resolution in the sub-μm range.[96] This technique allows the characterization of sub-cellular elemental distributions in plant cells and human fibroblast cells. The element distribution of Ca and Mg in mammalian cells (fibroblast cells) and chromosomes has been determined over cell-size areas (100 μm × 100 μm) using a SIMS instrument (from the University of Chicago).[97] Methanol dried freeze fractured samples preserve the histological morphology and yield Ca and Mg images containing reliable differential dynamic information, when compared with those following lyophilization. It should be considered the main drawback of SIMS: the huge matrix effects and a high formation rate of polyatomic and cluster ions and following the analytical data cannot be easily quantified. Therefore laser LA-ICP-MS is beneficial and has a great potential for future applications in the life sciences, especially in biological research. At present, LA-ICP-MS is the most sensitive technique for imaging mass spectrometric analysis and is employed to an increasing extent for the characterization of biological tissues.[98–100] Due to fewer matrix effects, the quantification of analytical data in LA-ICP-MS is relatively simple compared to SIMS if suitable matrix matched standard reference materials are available because no charging up effects occur. The limits of detection (in the $ng\,g^{-1}$ range) are, in general, lower than in SIMS.[101] For example, Jackson *et al.* reported LA-ICP-MS measurements on microdissected tissues from tail clips of the banded water snake, *Nerodia fasciata*, as a means of assessing contaminant exposure in the frame of ecotoxicological studies.[102] The snakes were split into three exposures

($n = 8$) and were fed three increasing levels (control, medium and high) of fish contaminated with As, Se and Sr for 24 months. The concentrations of tail clips for the three elements measured by LA-ICP-MS correlated significantly with the remaining whole tail concentration determined after homogenization, acid digestion and ICP-MS measurements. Similar concentrations were found for As and Se, which suggests that these elements are homogeneously distributed in the tissue. Statistical analysis showed that the LA-ICP-MS tail clip concentration differed significantly according to dietary treatment.[102] Relatively low arsenic concentrations were measured in the tail ($< 1\,\mu g\,g^{-1}$).

The selenium distribution in thin sections of biological tissues has been quantitatively determined by LA-ICP-MS.[103] The photograph of a cross section of the snail and the selenium LA-ICP-MS image are illustrated in Figure 9.26. Higher concentrations of selenium were found in the skin and gut compared to other parts of the snail. The natural selenium concentration in a 100 μm thin section of snail tissue was observed to be up to $25\,\mu g\,g^{-1}$.[103]

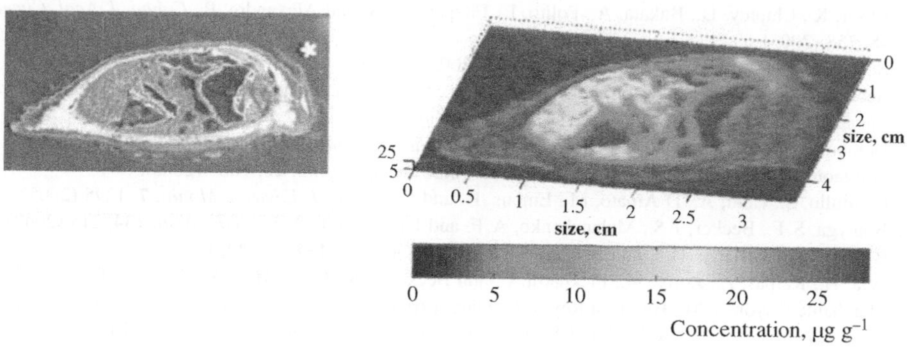

Figure 9.26 Quantitative Se distribution in cross section of snail tissues measured by LA-ICP-SFMS; on the left-hand side, photographs of the investigated sections of snail tissue are shown. (J. S. Becker et al. J. Anal. At. Spectrom. **22**, *736 (2007). Reproduced by permission of the Royal Society of Chemistry.)*

Three dimensional imaging analyses of biological materials have been investigated by ToF-SIMS using a pulsed buckminsterfullerene C_{60}^{+} primary ion gun (20 keV) for depth profile studies.[104] Molecular depth information was acquired using the C_{60}^{+} cluster ion beam to etch through the multi-layer structures of dehydrated organic films onto the chemically etched silicon substrate of the fatty acid dipalmitoyl-phosphatidylcholine (DPPC; it is a phospholipid) to determine the susceptibility of the cell membrane lipid material. The dry, porous organic film deposited onto the silicon substrate was measured by atomic force microscopy (AFM) as being 1.3 μm in thickness and having 100 nm surface roughness. In addition, the DPP–sucrose frozen multi-layer structure was prepared by supplying 5 μl of an aqueous sucrose solution and a saturated lipid chloroform solution (each $15\,mg\,l^{-1}$) was then dropped onto the aqueous sucrose layer. After the evaporation of the chloroform, the multi-layer system was immersed in liquid nitrogen. The frozen sucrose crystalline ice layer was covered by a dry DPPC film 20 nm in thickness. A depth profile analysis of DPPC–sucrose frozen multilayer structures suggests that biomolecule information can be revealed after sputtering removal of the 20 nm overlayer. The experimental results provide evidence of the potential for 3D molecular imaging of biological materials such as single biological cells using cluster ToF-SIMS.[104]

Biomolecular mass spectrometric techniques (MALDI- and ESI-MS) for the analysis of biopolymers are leading representatives of biochemical techniques. The sensitivity of biomolecular mass spectrometric techniques remains modest in the context of the need to define biological processes at the cellular level.[68]

In 2006, Lobinski *et al.*[1] reported on the imaging and speciation analysis of trace elements to study the element distribution, oxidation state, metal site and metal structure in biological environments using mass spectrometric techniques (LA-ICP-MS, SIMS, MALDI-MS) and non-mass-spectrometric techniques such as micro-PIXE (proton induced X-ray emission), XANES (X-ray absorption near edge structure) and EXAFS (extended X-ray absorption fine structure) – the latter two techniques are very sensitive due the use of a more intense synchrotron beam.[1]

References III

1. Lobinski, R., Moulin, C. and Ortega, R., *Biochimie*, **88**, 1591 (2006).
2. Kyser, K., Chipley, D., Bukata, A., Polito, P., Fitzpatrick, A. and Alexandre, P., *Canad. J. Anal. Chem.*, **48**, 258 (2003).
3. Becker, J. S., *Can. J. Anal. Sci. Spectr.*, **47**, 98 (2002).
4. Becker, J. Su., Bellis, D., Staton, I., McLeod, C. W., Dombovari, J. and Becker, J. S., *Fresenius' J. Anal. Chem.*, **368**, 490 (2000).
5. Dombovari, J., Becker, J. S. and Dietze, H. J., *Fresenius' J. Anal. Chem.*, **367**, 407 (2000).
6. Sturgeon, R. E., Willie, S. N., Yang, L. *et al.*, *J. Anal. At. Spectrom.*, **20**, 1067 (2005).
7. Ciardullo, S., Held, A., D'Amato, M., Emons, H. and Caroli, S., *J. Environ. Monit.*, **7**, 1295 (2005).
8. Boulyga, S. F., Becker, J. S., Malenchenko, A. F. and Dietze, H. J., *Mikrochim. Acta*, **134**, 215 (2000).
9. Boulyga, S. F., Dietze, H. J. and Becker, J. S., *Mikrochim. Acta*, **137**, 93 (2001).
10. Sela, H., Karpas, Z., Zoriy, M., Pickhardt, C. and Becker, J. S., *Int. J. Mass Spectrom.*, (2006).
11. Marchante-Gayón, J. M., Hinojosa Reyes, L., Garcia Alonso, J. I. and Sanz-Medel, A., in *Plasma Source Mass Spectrometry*, G. Holland and S. D. Tanner (eds.). The Royal Society of Chemistry, Cambridge (2003) 271.
12. Artelt, S., Creutzenberg, O., Kock *et al.*, *Sci. Total Environ.*, **228**, 219 (1999).
13. Boulyga, S., Dombovari, J., Becker, J. S. and Dietze, H. J., *At. Spectr.*, **21**, 149 (2000).
14. Chu, H. S., Yip, Y. C., Chan, K. C. and Sham, W. C., *J. Anal. At. Spectrom.*, **21**, 1068 (2006).
15. Montes-Bayón, M., Grant, T. D., Meja, J. and Caruso, J. A., *J. Anal. At. Spectrom.*, **17**, 1015 (2002).
16. Resano, M., Aramendia, M. and Vanhaecke, F., *J. Anal. At. Spectrom.*, **21**, 1036 (2006).
17. Li, L., Hu, B., Xia, L. B. and Jiang, Z. C., *Talanta*, **70**, 468 (2006).
18. Outridge, P. M., Veinott, G. and Evans, R. D., *Environ. Rev.*, **3**, 160 (1995).
19. Durrant, S. F. and Ward, N. I., *J. Anal. At. Spectrom.*, **20**, 821 (2005).
20. Garbe-Schoenberg, D., Reimann, C. and Pavlov, V. A., *Environ. Geol.*, **32**, 9 (1997).
21. Arrai, T. and Hirata, T., *Fisheries Sci.*, **72**, 799 (2006).
22. Pickhardt, C., Borcherding, J. and Becker, J. S., in *Book of Abstracts of DGMS 2006*. Mainz (2006).
23. Liang, P., Shi, T. Q., Lu, H. B., Jiang, Z. C. and B in, H., *Spectrochim. Acta B*, **58**, 1709 (2003).
24. Mueller-Spitz, S. R., Vonderheide, A. P., Shann, J. R., Caruso, J. A. and Kinkle, B. K., *Anal. Bioanal. Chem.*, **386**, 142 (2006).
25. Bruechert, W. and Bettmer, J., *J. Anal. At. Spectrom.*, **21**, 1271 (2006).
26. B'Hymer, C. and Caruso, J. A., *J. Chrom. A*, **1114**, 1 (2006).
27. Shah, M., Kannamkumarath, S. S., Wuilloud, C. A., Wuilloud, R. G. and Caruso, J. A., *J. Anal. At. Spectrom.*, **19**, 381 (2004).
28. Caruso, J. A. and Montes-Bayon, M., *Ecotoxicol. Environ. Saf.*, **56**, 148 (2003).
29. Thomson, C. D., *Analyst*, **123**, 827 (2005).
30. Fairweather-Tait, S. J., *Fresenius' J. Anal. Chem.*, **363**, 536 (1999).
31. Stadman, T. C., *Ann. Rev. Biochem.*, **59**, 111 (1990).
32. Cankur, O., Yathavakilla, S. K. V. and Caruso, J. A., *Talanta*, **70**, 784 (2006).
33. Nilsson, S. and Laurell, T., *Anal. Bioanal. Chem.*, **378**, 1676 (2004).

34. Giusti, P., Schaumlöffel, D., Preud'homme, H., Szpunar, J. and Lobinski, R., *J. Anal. At. Spectrom.*, **21**, 26 (2006).
35. Giusti, P., Lobinski, R., Szpunar, J. and Schaumlöffel, D., *Anal. Chem.*, **78**, 965 (2006).
36. Schaumlöffel, D., Giusti, P., Szpunar, J. and Lobinski, R., in *French Patent Application FR 05 05884* (2005).
37. Giusti, P., Schaumlöffel, D., Encinar, J. R. and Szpunar, J., *J. Anal. At. Spectrom.*, **20**, 1101 (2005).
38. Schaumlöffel, D. and Lobinski, R., *Int. J. Mass Spectr.*, **242**, 217 (2005).
39. Yu, C., Cai, Q., Guo, Z.-X., Yang, Z. and Khoo, S. B., *J. Anal. At. Spectrom.*, **19**, 410 (2004).
40. Gallagher, P. A., Wei, X., Shoemaker, J. A., Brockhoff, C. A. and Creed, J. T., *J. Anal. At. Spectrom.*, **14**, 1829 (1999).
41. Prohaska, T. and Stingeder, G., in *Handbook of Elemental Speciation II: Species in the Environment, Food, Medicine and Occupational Health*, H. C. R. Cornelis, J. Caruso, K. G. Heumann (eds.). John Wiley & Sons, Chichester (2005).
42. Soignet, S. L., Maslak, P., Wang, Z. G. *et al.*, *N. Engl. J. Med.*, **339**, 1341 (1998).
43. Ferguson, J. F. and Gavis, J., *Water Res.*, **6**, 1259 (1972).
44. Lunde, G., *Nature*, **224**, 186 (1969).
45. Kuehnelt, D., Goessler, W. and Irgolic, K. J., *Appl. Organomet. Comp.*, **11**, 459 (1997).
46. Raab, A., Hansen, H. R. and Feldmann, J., in *Plasma Source Mass Spectrometry*, G. Holland and D. R. Bandura (eds.). The Royal Society of Chemistry, Cambridge (2005) 72.
47. Luque de Castro, M. D. and Jiménez-Carmona, M. M., *Trends Anal. Chem.*, **17**, 441 (1998).
48. Hintelmann, H. and Nguyen, H. T., *Anal. Bioanal. Chem.*, **381**, 360 (2005).
49. Ouerdane, L., Mari, S., Czernic, P., Lebrun, M. and Lobinski, R., *J. Anal. At. Spectrom.*, **21**, 676 (2006).
50. Bolea, E., Gorriz, M. P., Bouby, M., Laborda, F., Castillo, J. R. and Geckeis, H., *J. Chromatogr. A*, **1129**, 236 (2006).
51. Lobinski, R., Schaumlöffel, D. and Szpunar, J., *Mass Spectrom. Rev.*, **25**, 255 (2006).
52. Schaumlöffel, D., Ruiz Encinar, J. and Lobinski, R., *Anal. Chem.*, **75**, 6873 (2004).
53. Szpunar, J., *Anal. Bioanal. Chem.*, **378**, 54 (2004).
54. Szpunar, J. and Lobinski, R., *Hyphenated Techniques in Speciation Analysis*, R. M. Smith (ed.). RSC Chromatic Monographs, The Royal Society of Chemistry, Cambridge (2003).
55. Fernandez, J. M., Bayon, M. M., Alonso, J. I. G. and Sanz-Medel, A., *J. Mass Spectrom.*, **35**, 639 (2000).
56. Bettmer, J., Jakubowski, N. and Prange, A., *Anal. Bioanal. Chem.*, **386**, 7 (2006).
57. Prange, A. and Schaumlöffel, D., *Anal. Bioanal. Chem.*, **373**, 441 (2002).
58. Sanz-Medel, A., Montes-Bayón, M. and Sanchez, M. L. F., *Anal. Bioanal. Chem.*, **377**, 236 (2003).
59. Szpunar, J., *Analyst*, **125**, 963 (2000).
60. Löffler, G. and Petrides, P. E., *Biochemie and Pathobiochemie*. Springer, Berlin, Heidelberg, New York (1998).
61. Bandura, D. R., Baranov, V. I., Ornatsky, O. I. and Quinn, Z. A., in *Plasma Source Mass Spectrometry*, G. Holland and S. D. Tanner (eds.). The Royal Society of Chemistry, Cambridge (2003) 43.
62. Becker, J. S., Boulyga, S. F., Pickhardt, C., Becker, J. Su., Buddrus, S. and Przybylski, M., *Anal. Bioanal. Chem.*, **375**, 561 (2003).
63. Davies, M. J., Dean, R. T. and Davies, D., *Radical-Mediated Protein Oxidation: From Chemistry to Medicine*. Oxford University Press, Oxford, UK (1998).
64. Boulyga, S. F., Pickhardt, C., Becker, J. Su., Przybylski, M. and Becker, J. S., *Plasma Source Mass Spectrometry: Applications and Emerging Technologies, 8th International Conference on Plasma Source Mass Spectrometry*, G. Holland and S. D. Tanner (eds.) *Durham, United Kingdom*, The Royal Society of Chemistry, Cambridge, 54 (2003).
65. Becker, J. Su., Zoriy, M., Przybylski, M. and Becker, J. S., *J. Anal. At. Spectrom.*, **22**, 63 (2007).
66. Becker, J. S., Zoriy, M., Krause-Buchholz, U. *et al.*, *J. Anal. At. Spectrom.*, **19**, 1236 (2004).
67. Hann, S., Obinger, C., Stingeder, G., Paumann, M., Furtmuller, P. G. and Köllensperger, G., *J. Anal. At. Spectrom.*, **21**, 1224 (2006).
68. Gaskell, S. J., *Appl. Surf. Sci.*, **252**, 6404 (2006).
69. Abokitse, K. and Hummel, W., *Appl. Microbiol. Biotechnol.*, **62**, 380 (2004).
70. Kápolna, E., Shah, M., Caruso, J. A. and Fodor, P., *Food Chem.*. **101**, 1415 (2007).

71. Mounicou, S., Shah, M., Meija, J., Caruso, J. A., Vonderheide, A. P. and Shann, J., *J. Anal. At. Spectrom.*, **21**, 404 (2006).
72. Fan, T. W. M., Pruszkowski, E. and Shuttleworth, S., *J. Anal. At. Spectrom.*, **17**, 1621 (2002).
73. Polatajko, A., Jakubowski, N. and Szpunar, J., *J. Anal. At. Spectrom.*, **21**, 639 (2006).
74. Chamoun, J. and Hagège, A., *J. Anal. At. Spectrom.*, **20**, 1030 (2005).
75. Haynes, J. M., *Geochim. Cosmochim. Acta*, **68**, 1691 (2004).
76. Boulyga, S. F., Desideri, D., Meli, M. A., Testa, C. and Becker, J. S., *Int. J. Mass Spectrom.*, **226**, 329 (2003).
77. Reyes, L. H., Moreno Sanz, F., Epilez, P. H., Marchante-Gayón, J. M., Alonso, J. I. G. and Sanz-Medel, A., *J. Anal. At. Spectrom.*, **19**, 1230 (2004).
78. Meija, J., Yang, L., Caruso, J. A. and Mester, Z., *J. Anal. At. Spectrom.*, **21**, 1294 (2006).
79. Dombovari, J., Becker, J. S. and Dietze, H. J., *Int. J. Mass Spectrom.*, **202**, 231 (2000).
80. Becker, J. S. and Dietze, H. J., *Int. J. Mass Spectrom.*, **197**, 1 (2000).
81. Chaurand, P., Cornett, D. S. and Caprioli, R. M., *Curr. Opinion Biotechn.*, **17**, 431 (2006).
82. Heeren, M. A., Mc.Donnell, L. A., Amstalden, E., Luxambourgh, S. L., Altelaar, A. F. M. and Piersma, S. R., *Appl. Surf. Sci.*, **252**, 6827 (2006).
83. McDonnel, L., Piersma. S. R., Maarten Altelaar, A. F. *et al., J. Mass Spectrom.*, **40**, 160 (2005).
84. Todd, P. J., Schaaf, T. G., Chaurand, P. and Caprioli, R. M., *J. Mass Spectrom.*, **36**, 355 (2001).
85. Brunelle, A., Touboul, D. and Laprévote, O., *J. Mass Spectrom.*, **40**, 985 (2005).
86. Touboul, D., Halgand, F., Brunelle, A. *et al., Anal. Chem.*, **76**, 1550 (2004).
87. Becker, J. S., Zoriy, M., Dehnhardt, M., Pickhardt, C. and Zilles, K., *J. Anal. At. Spectrom.*, **20**, (2005).
88. Becker, J. S., Zoriy, M. V., Pickhardt, C., Palomero-Gallagher, N. and Zilles, K., *J. Anal. At. Spectrom.*, **77**, (2005).
89. Caprioli, R. M., Farmer, T. B. and Gile, J., *Anal. Chem.*, **69**, 4751 (1997).
90. Becker, J. S., Zoriy, M., Becker, J. Su. *et al., Anal. Chem.*, **77**, 5851 (2005).
91. Hutchinson, R. W., Cox, A. G., McLeod, C. W., Marshall, P. S., Harper, A., Dawson, E. L. and Howlett, D. R., *Anal. Biochem.*, **346**, 225 (2005).
92. Chandra, S., *Appl. Surf. Sci.*, **203–204**, 679 (2003).
93. Morrison, G. H., Gay, I. and Cahndra, S., *Scanning Microsc., Suppl.*, **8**, 359 (1994).
94. Moeller, J., Tyler, B. J., Lipinski, D. and Arlinghaus, H. F., in *5th European Workshop on Secondary Ion Mass Spectrometry*. Münster (2006) 111.
95. Tyler, B. J., Rangaranjan, S., Möller, J., Beumer, D. and Arlinghaus, H. F., *Appl. Surf. Sci.*, **252**, 6712 (2006).
96. Dickinson, M., Heard, P. J., Barker, J. H. A., Lewis, A. C., Mallard, D. and Allen, G. C., *Appl. Surf. Sci.*, **252**, 6793 (2006).
97. Levi-Setti, R., Gavrilov, K. L. and Neilly, M. E., *Appl. Surf. Sci.*, **252**, 6765 (2006).
98. Feldmann, J., Kindness, A. and Ek, P., *J. Anal. At. Spectrom.*, **17**, 813 (2002).
99. Ghazi, A. M., Wataha, J. C., O'Dell, N. L., Singh, B. B., Simmons, R. and Shuttleworth, S., *J. Anal. At. Spectrom.*, **17**, 1295 (2002).
100. Kindness, A., Sekaran, N. and Feldmann, J., *Clinical Chemistry*, **49**, 1916 (2003).
101. Becker, J. S., *Spectrochim. Acta B*, **57**, 1805 (2002).
102. Jackson, B. P., Hopkins, W. A. and Baionno, J., *Environ. Sci. Techn.*, **37**, 2511 (2003).
103. Becker, J. S., Zoriy, M., Becker, J. Su., Dobrowolska, J. and Matusch, A. *J. Anal. At. Spectrom.*, **22**, 736 (2007).
104. Kozole, J., Szakal, C., Kurczy, M. and Winograd, N., *Appl. Surf. Sci.*, **252**, 6789 (2006).

9.4 Bioengineering

Bioengineering (or biological engineering), including biomedical engineering and nanobiotechnology, is a new, challenging and rapidly growing high tech research field. Its development, application and bioengineered product control needs advanced, highly sensitive and powerful analytical techniques, often with high sensitivity (low detection limits) for bulk analysis and microlocal analysis at μm and nm lateral resolution. Bioengineering is defined as a basic research oriented activity closely related to biotechnology and genetic engineering focusing on the engineering of biological processes, i.e., the modification of animal or plant cells, or parts of cells, to improve plants or animals or to develop new microorganisms for beneficial purposes.[1]

9.4.1 Activities in Bioengineering and Analytics

Activities in bioengineering which need mature analytics include:

- Tissue engineering in the field of biotechnology, to create or modify biological materials, including the design of artificial organs and devices;
- Molecular and cellular engineering to understand and construct cellular and molecular circuits with useful properties (e.g., protein engineering);
- Design and development of biomaterials (e.g., bioimplated materials);
- Biomicro- and nanoelectronics including design of new devices;
- Development and application of nanobiotechnology and biomedical instrumentation, e.g., of biosensors and bioarrays for detection of biological events and their conversion to electrical signals;
- Medical and biological analyzis to detect and analyze bioelectric signals, analysis of biological systems and related materials and study of biological effects on biological tissue;
- Biological and medical imaging from single proteins to the whole human brain to provide graphic displays of anatomic detail and physiological function. Several applications in imaging mass spectrometry in human brain research are summarized in Section 9.5.7;
- Regenerative medicine to prevent birth defects, to retard damage to diseased tissues and to repair injured tissues: application of stem cell technologies to *in vitro* tissue production and the generation of specific cell types for therapeutic use.[2]

The major field of application for bioengineering is, above all, medicine. Bioengineering focuses mainly on human health and is concerned with understanding complex living systems, combining engineering and life sciences and developing new biomedical technologies. Several applications of inorganic mass spectrometry, especially in the analysis of body fluids (e.g., for monitoring patients' urine and blood with respect to the release of metals from implants) and mass spectrometric imaging in brain research are discussed in Section 9.5.

New generation biomaterials or devices capable of reconstruction or integration within biological tissues are being developed for tissue engineering. Tissue replacement by (if possible) biodegradable biomaterials using living cells is of interest for organ or tissue replacement where no donor organs or natural replacements are available.

Metallic biomaterials (metals such as Ti or its alloys and others) are used for the manufacture of orthopaedic implants due to their excellent biocompatibility with respect to electrical and thermal conductivity and their mechanical properties, e.g., for hard tissue replacement such as total hip and knee joints, for fracture healing aids such as bone plates and screws or dental implants. For example, Co–Cr–Mo alloys are employed for metal-on-metal hip bearings in total joint replacements. Problems with implants occur because of ion release in patients with metal implants. To control this ion release, the ultratrace determination of Co, Cr and Mo in the blood (or serum) and urine of patients with Co–Cr–Mo alloy hip implants is carried out routinely in the author's laboratory. The trace metal determination of Co, Cr and Mo in complex matrices such as urine and blood by ICP-MS is not trivial due to the low concentrations expected in the sub-ng ml^{-1} range, the possible danger of contamination during sample collection, sample preparation and the measurement procedure, and the inherent interference problem. However, double-focusing sector field ICP-SFMS has been successfully utilized at medium mass resolution ($m/\Delta m = 4000$).

Other recently developed biomaterials will be used for quite different purposes in tissue engineering such as artificial pancreas and liver, artificial skin, nerve regeneration, gene therapy vascular grafts, cornea replacement and others.[3]

The fundamentals, applications and progress in tissue engineering are reviewed in the comprehensive handbook by Ikada.[4]

9.4.2 Nanobiotechnology

The development of assemblies of inorganic materials with biomolecules has emerged as a novel approach to the controlled fabrication of functionalized nanostructures and networks.[5] The practice of DNA sequence detection is especially relevant for forensic sciences, food safety, genetics and other fields.[6] The immobilization of single strand DNA probes onto solid materials such as noble metal nanoparticles has proved to be the basis for a multitude of quite different nanobiotechnological and biomedical applications, including the DNA driven assembly of nanoparticles and biosensors.[5–11]

Figure 9.27 shows a schematic of a possible combination of protein, peptide or DNA biotechnology with inorganic micro- and nanoelectronic devices to form nanobiotechnological assemblies. The analytical problems occurring in the production of bioarrays are similar to those of microelectronics. Therefore, process control of nanobiotechnological assemblies is an interesting issue of growing importance in order to optimize the preparation of bioarrays. In addition to characterization of microelectronic parts, the analysis of bioorganic high tech materials is required for advanced structural analysis (utilizing proteomics via the analysis of proteins by biomolecular mass spectrometry), elemental composition and speciation.

Biomolecular templating is a technique that allows fundamental aspects of biological self-assembly to be transferred to the emerging field of nanostructure formation and manipulation. It takes advantage of the well defined chemical, physical, and structural properties of biomolecular templates used for a controlled deposition of inorganic material.[12] For example, the fabrication of regular arrays of metallic nanoparticles (Pt, Pd, Au or Ag nanoclusters) by biomolecular templating is of great interest in order to prepare nm structures for use in nanoelectronic devices (e.g., by deposition of nanoclusters on crystalline bacterial surface layers with well defined cluster lattices or on stretched DNA to prepare nanostructures or highly conductive nanowires), optical devices and chemical devices (e.g., biosensors).[12–16] Pompe and co-workers reported on the fabrication of highly oriented platinum and palladium nanocluster arrays by biomolecular templating.[13,14,17] A crystalline bacterial surface layer with a well defined geometry of *Sporosarcina ureae* (so-called S layer) is used as a protein template (unit cell: 13.2 nm × 13.2 nm) on a cell membrane. Noble metal nanoclusters (Pt or Pd) are deposited chemically to produce nanostructures. The spatial distribution of noble metal nanoclusters is characterized by transmission electron microscopy (TEM).

As is well known, impurities at trace levels due to contamination during assembly production can influence the physical, electrical, magnetic and chemical properties of micro- and nanoelectronic arrays (thin layered systems or bulk);[18] therefore powerful and sensitive trace analytical techniques are required for the characterization of deposited noble metal layers and nanoclusters in bioengineered devices. In order to avoid chemical digestion and contamination during sample preparation, direct solid-state analytical techniques such as LA-ICP-MS (laser ablation inductively coupled plasma mass spectrometry) will attract increasing attention. Trace impurities have been rapidly analyzed in bioengineered devices by LA-ICP-MS with detection limits down to the sub-$ng\,g^{-1}$ range. In addition, LA-ICP-MS allows microlocal analysis and depth profiling with lateral

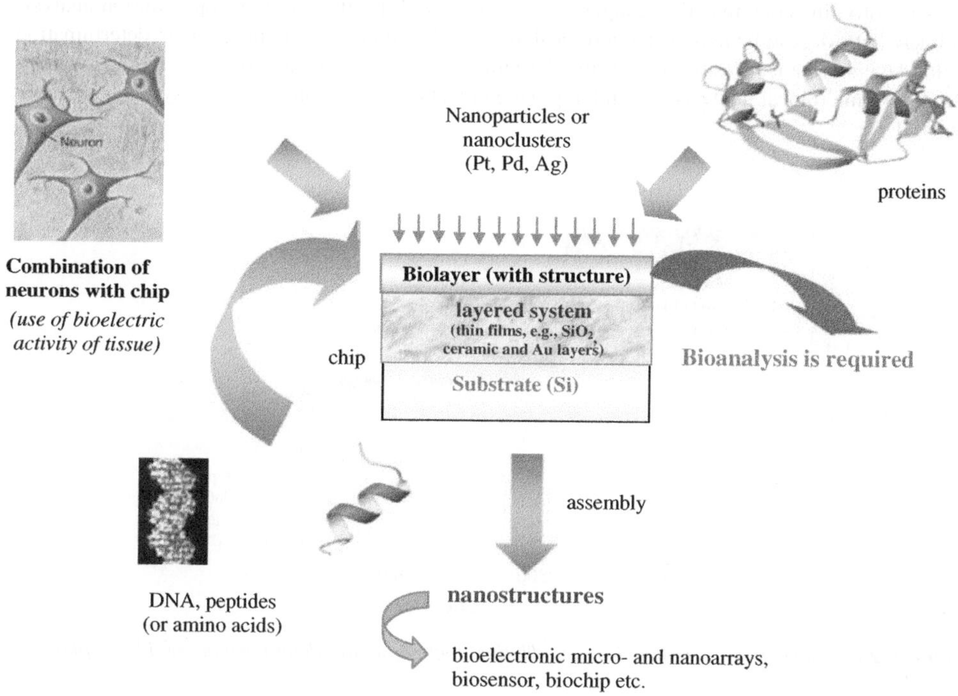

Figure 9.27 *Schematic of combination of protein engineering, biotechnology for micro- and nanoelectronic devices.*

resolution in the low μm range. LA-ICP-MS is a powerful tool for the determination of trace impurities in a few mg of noble metal nanoclusters, as discussed in Section 9.1.[19] The quantification of analytical results for Ag, Tl and Pb has been performed with a modified online isotope dilution analysis in LA-ICP-MS, and could be extended to other analytes of interest.[19]

However, impurities of metals and non-metals in deposited clustered layers also have to be determined. Figure 9.28 gives an example of Pb impurity analysis in high purity nanoclusters by the online isotope dilution technique in LA-ICP-MS. In the first step, the isotope enriched spike solution (^{204}Pb – 52.4 %) is nebulized using a microconcentric nebulizer with desolvator (Aridus, Cetac Technologies) – the experimental arrangement is shown in Figure 16.19. In the second step, the laser ablation is switched on and the ablated platinum is mixed with the nebulized isotope enriched spike solution. Consequently, the ^{204}Pb/^{206}Pb isotope ratio is changed. In the third step, only the platinum is ablated and the lead isotope ratio in nature (with ^{204}Pb – 1.4 %) is measured mass spectrometrically. The lead concentration can be calculated by applying the isotope dilution formula (see Equation 6.3 in Section 6.2.4) taking into consideration the different sensitivities in LA-ICP-MS and ICP-MS measured on platinum SRM NIST 681 (using Pt as the internal standard element). The advantages of the online isotope dilution technique in LA-ICP-MS are due to the fact that because the one point calibration via nebulization using isotope enriched spikes occurs simultaneously to the laser ablation of the sample investigated, nearly ideal matrix matching is achieved and the analysis of even the smallest amount of sample is possible without any sample preparation (no prior mixing with an isotope enriched spike).

Consequently, the contamination danger can be minimized. Furthermore, isotope dilution analysis with LA-ICP-MS can be used as a microlocal analytical technique for multi-element determination in thin layers or nanowires of high purity platinum and palladium nanoclusters on DNA. LA-ICP-MS allows microlocal analysis for studying the elemental distribution in layered systems and bulk material.

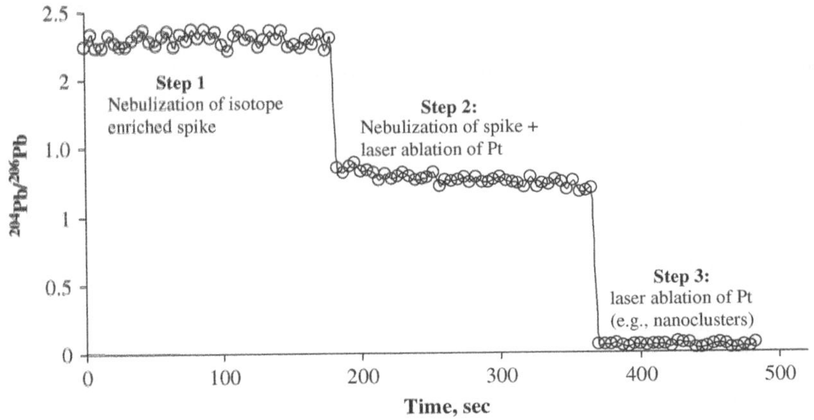

Figure 9.28 *Application of online isotope dilution technique for determination of Pb in platinum nanoclusters.*

Fig 9.29 illustrates Zn determination in a Zn binding protein in a selected protein spot after separation of the proteins of an S layer used as biomolecular template, as described above, by 2D gel measured by LA-ICP-SFMS at medium mass resolution ($m/\Delta m = 4000$).[20] The $^{64}Zn^+$ ion intensity in the protein spot (compared with the background signal in the gel blank) was measured at three different positions in the selected spot. Using LA-ICP-SFMS, the highest $^{64}Zn^+$ intensity was found in the middle of the protein spot investigated. Additional measurements by MALDI-MS on Zn containing protein spot in 2D gel can be useful to identify the structure of the protein.[20]

Of special relevance is the analysis of elemental distribution (bioimaging analysis) in biological hybrid systems and/or an interface analysis between biological layers and microelectronic devices. Kouassi and Irudayaraj have reported on the chemical synthesis and functionalization of magnetic (Fe_3O_4) and gold coated magnetic nanoparticles and the immobilization of single stranded biotinylated oligonucleotides onto these particles.[5] Particle size and oligonucleotide attachment were confirmed by transmission electron microscopy (TEM). The authors demonstrated that single strands of oligonucleotides can be immobilized onto amino and carboxylate functionalized magnetic and also gold coated magnetic nanoparticles by using the streptavidin–biotin interaction after carbodiimidine activation. These studies open up several possibilities for bioconjugate attachment to functionalized iron and iron nanocomposite structures for controlled manipulation and handling using magnetic fields.[5] Nucleation of Pt clusters on biopolymers (DNA and proteins)[13] and the conditions which have to be fulfilled in order to grow chains of nanosized noble metal clusters on DNA templates according to a selectively heterogeneous, template-controlled mechanism have been investigated in Pompe's group.[17] A long incubation of double stranded DNA molecules

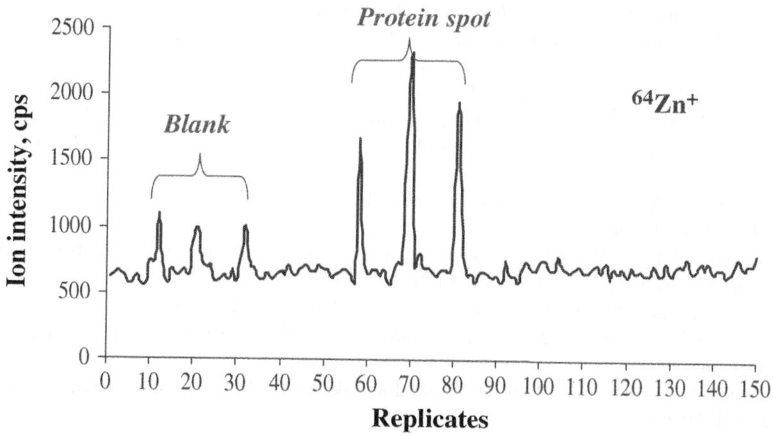

Figure 9.29 Ion intensities of transient $^{64}Zn^+$ signals in one selected protein spot in 2D gel of an S layer compared to the background signal of gel blank.

with Pt (II) complexes is necessary to obtain a template directed formation of thin and uniform cluster chains after chemical reduction of the DNA/salt solution.[17] Due to their electronic properties, Pt, Pd or Au nanowires prepared on DNA molecules (diameter of, for example, the platinum clusters were about 3–5 nm) as some of the thinnest connecting wires ever built by the deposition of noble metal clusters (metallization) on chemically activated DNA templates, can be used in future nanoelectronic devices (e.g., in nanoelectrical circuits). The conductivity of the nanowires can be changed as a function of the amount of clusters deposited.[16,17]

Electrospray ionization (ESI) has been utilized to create an ion beam from a solution containing nanoparticles (gold nanoclusters) or biomolecules (bovine serum albumin) for deposition on graphite substrate surfaces to create organic electronics or biotechnological devices and functional materials (e.g., organic light-emitting diodes).[21] An ESI-ToF-MS combined with a gentle vacuum deposition was applied to monitor the ion beam composition with respect to the ESI fragments and also for process control. Stable ion beams with currents of 20–40 pA were achieved for thin layer deposition.[21]

At present, inorganic mass spectrometry is hardly applied at all in bionanotechnology, but it can be expected that increasing use will be made in future to solve problems that seem intractable.

References IV

1. Bronzino, J., *Handbook of Biomedical Engineering*, CRC Press LLC (2000).
2. http://bioengineering.stanford.edu/research/regenmed.html (2006).
3. http://depts.washington.edu/bioe/research/biomaterials/biomaterials.html (2006).
4. Ikada, Y., *Tissue Engineering*, Vol. 8, A. Hubbard (ed.). Interface Science and Technology, Elsevier, Amsterdam (2006).
5. Kouassi, G. K. and Irudayaraj, J., *Anal. Chem.*, **78**, 3234 (2006).
6. Fritzsche, W. and Taton, T. A., *Nanotechnology*, **14**, 3194 (2003).
7. Kerman, K., Kobayshi, M. and Tamiya, E., *Meas. Sci. Techn.*, **R1–R11**, (2004).

8. Mir, K. U. and Southern, E. M., *Nat. Biotechnol.*, **17**, 788 (1999).
9. Peterson, A. W., Heaton, R. J. and Georgiadis, R. M., *Nucleic Acids Res.*, **29**, 5163 (2001).
10. Petrovikh, D., Kimura-Suda, H. and Whitman, L. J., Tarlov, M. J., *J. Am. Chem. Soc.*, **125**, 5219 (2005).
11. Sekar, M. M., Bloch, W. and St. John, P. M., *Nucleic Acids Res.*, **33**, 366 (2005).
12. Wahl, R., Mertig, M., Raff, J., Selenska-Pobell, S. and Pompe, W., *Adv. Mater.*, **13**, 736 (2001).
13. Ciacchi, L. C., Mertig, M., Seidel, R., Pompe, W. and De Vita, A., *Nanotechnology*, **14**, 840 (2003).
14. Mertig, M., Kirsch, R., Pompe, W. and Engelhardt, H., *Eur. Phys. D*, **9**, 45 (1999).
15. Mertig, M., Kirsch, R., Pompe, W. and Engelhardt, H., *Eur. Phys.*, **D9**, 45 (1999).
16. Richter, J., *Physica E*, **16**, 157 (2003).
17. Seidel, R., Ciacchi, L. C., Weigel, M., Pompe, W. and Mertig, M., *J. Phys. Chem. B*, **108**, 10801 (2004).
18. Becker, J. S., *Spectrochim. Acta B*. **57**, 1805 (2002).
19. Becker, J. S., Pickhardt, C. and Pompe, W., *Int. J. Mass Spectrom.*, **237**, 13 (2004).
20. Becker, J. S., in *Proceedings of 2004 Winter Conference on Plasma Spectrochemistry (Keynote lecture)*. Fort Lauderdale (2004).
21. Rauschenbach, S., Stadler, F. L., Lunedei, E., Malinowski, N., Kolosov, S., Costantini, G. and Kern, K., *Small*, **2**, 540 (2006).

9.5 Medicine

As soon as the importance of trace elements in biological and medical processes became apparent, suitable analytical techniques for survey analysis were sought in order to obtain reliable results for interpreting the influence of element traces on biological and medical systems. In addition to atomic absorption spectroscopy (AAS) and neutron activation analysis (NAA), spark source mass spectrometry (SSMS) and later laser ionization mass spectrometry (LIMS)[1] were employed in the sixties as the oldest inorganic mass spectrometric technique to analyze trace elements in medical samples, e.g., in dried blood plasma,[2] human hair,[3] blood serum,[4] lung tissue[5] and other medical samples. Compared to AAS, which only allows the analysis of selected elements in medical tissues, SSMS was employed for the trace element analysis of medical samples e.g., lung tissue ashes with very low detection limits in the 0.05 and $1\,\mu g\,g^{-1}$ range in the sixties[5] or calcium oxalate uroliths in the seventies[6] due to its multi-element capability (up to 60 elements). The analysis of medical samples was combined with clinical diagnostics, forensic medicine and toxicological investigations.[1,7] Today advanced inorganic and biomolecular mass spectrometric techniques are an indispensable part of medical diagnostics and biomedical studies in research and routine analysis. Biomonitoring of medical samples with respect to essential and toxic elements is gaining importance as a tool for studying the effect of the deficiency or excess of essential and/or toxic trace elements on our health and their contribution to the development of different diseases.[8–10]

9.5.1 Sampling, Sample Handling and Storage of Medical Samples

In order to obtain accurate and precise analytical data, which are reproducible by the analytical procedures applied, the sampling, sample handling (sample preparation) and storage (conservation) of labile compounds or complexes and elemental species are very important steps before mass spectrometric measurements can be made. Systematic errors due to possible contamination using metal containing or contaminated laboratory tools should be avoided. A careful selection of bottle material is required, e.g., pure polyethylene, polypropylene, Teflon or quartz glassware,[11] and clean instruments should be used for sampling and storage and in sample preparation. The use of chemicals (solvents, acids, H_2O_2, EDTA, Milli-Q-water etc.) of the highest purity is necessary if trace metals in medical samples have to be measured at the trace and ultratrace concentration level. To minimize the risk of possible contamination several cleaning procedures consisting of washing procedures for laboratory ware are strongly recommended.[12] An assessment of the potential contamination risk with devices routinely used in hospitals and clinical laboratories for the sampling and storage of whole blood and serum for trace element and species analysis has been made by analyzing leachates from the devices. The devices checked were disposable stainless steel needles, different types of blood collection tubes, serum separation tubes, disposable plastic pipettes and plastic vials used for serum storage.[13] After leaching of tools with $0.05\,mol\,l^{-1}$ HNO_3, 70 trace elements were determined in solutions by double-focusing sector field ICP-SFMS. For elements present at concentrations higher than $10\,ng\,ml^{-1}$ (Na, Ca, Mg, P, Fe, Br, Si, Zn, Cu, Rb, Se and I) the contribution from the devices was, as a rule, negligible, but for the majority of trace and ultratrace elements it may significantly affect or even prevent accurate determination.[13]

Relevant procedures for the collection and storage of blood, urine, tissues and during microdialysis are reviewed in detail by De Cremer.[14] For example, the collection of urine samples is very easy, but the contamination risk from the environment is higher compared to the sampling of blood (using polypropylene or Teflon intravenous catheters). In general, medical samples should be kept at 4 °C in the refrigerator for a short time or frozen at −20 °C. Precipitation of salts and organic compounds from urine results in a co-precipitation of several trace metals, therefore urine is often stored in the presence of an acid or an acid mixture at a well defined pH value (e.g., $0.03\,mol\,l^{-1}$ HNO_3 or HCl, sulfamic acid, Triton X-100).[14] Blood samples have been collected, for example, in lithium heparine monovettes developed for trace metal determination, and analyzed by ICP-MS after dilution (1/10) with 0.1 % Triton-X-100 and 0.5 % ammonia solution.[15] Analytical procedures for sample collection and preparation of medical samples depend on the specific case (e.g., the preparation of thin sections of tissues by cryocutting[16,17]) and are described in detail in a large number of original papers.

Reviews on the stability and storage of individual trace elements have been made by several authors.[14,18–21]

9.5.2 Body Fluids

The investigation of body fluids with respect to nutrient (essential) elements and toxic elements – which are challenging topics for analytical chemistry – include the determination of concentrations at the trace and ultratrace level. However, isotope variation and isotope effects (especially of lighter elements such as hydrogen, carbon, nitrogen, oxygen but also of iron and calcium) have also been studied.[22,23] The most frequently applied mass spectrometric technique for the analysis of body fluids today, which fulfils all requirements and also results in accurate and precise data, is ICP-MS.

Apart from the analysis of blood serum and urine, there are different applications of trace element analysis by ICP-MS, for instance on milk samples.[24–28]

Analysis of Blood and Serum

Body fluids such as blood and serum, which can be readily used for biomonitoring, contain essential (nutritional) trace metals such as Fe, Cu and Zn, but also toxic metals (e.g., Pb, Hg and Hg) at the ultratrace level.[29,30] For example, biomonitoring of 37 trace elements in 130 blood samples from occupationally non-exposed volunteers living in northern Germany (Bremen) was performed by ICP-MS with an octopole collision cell. The studies helped toxicologists and clinical chemists in their research into exposure to metals and health effects caused by this exposure (due to smoking, seafood consumption, dental alloy in teeth etc.).[15] Several working groups have reported on the single and multi-element determination of trace elements in serum or blood samples by double-focusing sector field ICP-MS or a quadrupole based instrument with and without reaction/collision cells.[31–36] Biomonitoring of whole blood with respect to Cd and Pb determination at background levels by ICP-MS compared to well established AAS methods was carried out by Palmer *et al.*[29] 200 μl of blood was analyzed after dilution to 10 ml for analysis by ICP-MS without any additional sample preparation. The analytical method developed was validated using NIST SRM 966 Toxic Metals in Blood. The detection limits obtained are at the sub-$\mu g\,l^{-1}$ level.[29] Parsons and co-workers[37] studied the pyrolysis step and atomization behaviour of Pb in a diluted blood matrix when a tungsten filament electrothermal atomic absorption spectrometer was directly coupled to an ICP-MS, and they proposed a pseudo-simultaneous AAS and ICP-MS measurement of Pb in blood. Rh was added to the sample diluent as a permanent modifier.[37]

An accurate determination of copper and zinc traces in human serum samples from the International Measurement Evaluation Programme-17 launched by IRMM (Geel) has been made by isotope dilution TIMS.[38] An analytical method for the multi-element determination of metals (Ti, V, Cr, Co, Ni and Mo) potentially released from dental implants and prostheses into human body fluids (in blood and urine) by ICP-MS (double-focusing sector field instrument and quadrupole instrument with octopole collision cell) for medical studies was developed in Sanz-Medel's group.[39] The Cr and Co concentrations found in blood samples of patients with chromium–cobalt based alloy varied in the sub-$\mu g\,l^{-1}$ range and were not significantly higher than the basal levels found by other authors.[40]

Biomonitoring of environmental and occupationally relevant trace and ultratrace metals (Al, Co, Cr, Cu, Fe, Mn, Ni, Pt, V and Zn) in human serum and urine was carried out using ICP-SFMS at different mass resolutions by Begerow *et al.*[41] Whereas the elements free from isobaric interferences (Cd, Mn, Pb, Pt and Tl) were measured at low mass resolution ($m/\Delta m = 300$), the determination of Al, Co, Cr, Cu, Fe, Ni, V and Zn was performed in the medium mass resolution mode ($m/\Delta m = 3000$).[41] Trace metal concentrations (Al, Ba, Be, Bi, Cd, Co, Cr, Hg, Li, Mn, Mo, Ni, Pb, Sb, Sn, Sr, Tl, V, W and Zr) in serum and blood samples from patients with Alzheimer's disease and healthy individuals measured by ICP-SFMS were compared by Bocca *et al.*[42] An increment of Hg and Sn in serum, higher levels of Co, Li, Mn and Sn and lower levels of Mo in blood were found in Alzheimer's disease samples.[42]

In clinical studies of trace elements in blood and serum using inorganic mass spectrometry, single element clinical applications dominate in the analytical literature.[43]

As an essential trace metal, iron plays an important role in the human body. Most iron in the adult human body (~ 3 g of 3–5 g total Fe) is bound for oxygen storage and is transported as haemoglobin in red blood cells and as myoglobin in muscle tissue (~ 200 mg). Iron deficiency anaemia develops due to the daily iron loss (0.003 % of total body iron) mainly via intestinal bleeding and if iron losses are not sufficiently balanced by iron supply to the body.[44] It is estimated that 30 % of the world's population are affected by iron deficiency anaemia.[45] Because each iron species in the body has its distinct physiological function, iron speciation is essential for distinguishing between anaemias and disorders of the iron metabolism. Most of the iron bound in haemoproteins can act as oxygen carriers (haemoglobulin and myoglobulin), as oxygen activators (oxidases, peroxidases, oxygenases and oxidoreductases) and electron transporters (cytochromes).[44] Iron speciation is hampered due to problems with the quantification of analytical techniques because human physiology is complex and the iron species concentration can vary significantly during the day. In addition to iron speciation by several analytical techniques, including hyphenated ICP-MS,[44] possible mechanisms of iron metabolism on the physiological level and the iron isotopic composition (possible iron isotope effects) in blood have been studied using multiple collector ICP-MS (Neptune, Thermo Fisher Scientific) and double spike thermal ionization mass spectrometry

(TIMS) by Walczyk and von Blanckenburg.[23] Precise iron isotope ratio measurements of iron from whole blood matrix by MC-ICP-MS (Neptune, Thermo Fisher Scientific) after microwave digestion and ashing decomposition, anion exchange and precipitation procedures are discussed by Stenberg *et al.*[46]

For semiquantitative screening of the distribution of selenium species in cow blood, size exclusion liquid chromatography ICP-QMS (with collision cell) was employed.[47] The blood and milk of cows fed a diet enriched with selenized yeast were compared to control cows. An increase in the selenomethionine and selenized haemoglobin was found as a linear function of the total selenium concentration. No inorganic selenium was detected. The total selenium concentration in the blood correlated well with that in milk. An interesting finding is that the selenium supplementation did not change the protein distribution profiles for other essential elements (Cu, Fe, Mn and Zn).[47]

Aluminium is an environmentally abundant element and has no known biological function in living cells. On the other hand, Al exhibits biological toxicity as demonstrated in experimental models.[48,49] Intake and output of aluminium for the healthy population from various sources, biological monitoring of occupational aluminium exposure in serum and speciation analysis of the amount and composition of Al species with high and low molecular masses, have been reviewed by several groups.[50–52] Besides the use of mass spectrometric and non-mass spectrometric techniques (e.g., ETAAS after separation of human serum proteins[50]) ESI-MS-MS analysis has also been performed to identify the Al species.

Bromate is a regulated disinfection by-product primarily associated with the ozonation of water containing bromide, but it is also a by-product of hypochlorite used to disinfect water. To study the pharmacokinetics of bromate,[53] it is necessary to develop a robust and sensitive analytical method for the identification and quantification of bromate in blood. A critical issue is the extent to which bromate is degraded presystemically and in blood at low (environmentally relevant) doses of ingested bromate as it is delivered to target tissue. A simple isolation procedure was developed using blood plasma spiked with various levels of bromate and bromide. Blood proteins and lipids were precipitated from plasma using acetonitrile. The resulting extracts were analyzed by ion chromatography combined with inductively coupled plasma mass spectrometry (IC-ICP-MS), with a detection limit of $5\,\mathrm{ng\,ml^{-1}}$ plasma for both bromate and bromide. Plasma samples purchased commercially were spiked with bromate and stored for up to seven days. Over the seven day storage period, bromate decay remained under 20 % for two spike doses. Decay studies in plasma samples from spiked blood drawn from live rats showed significant bromate decay within short periods of time preceding sample freezing, although samples which were spiked, centrifuged and frozen immediately after drawing yielded excellent analytical recoveries.

Fitzgerald *et al.*[54] studied the monitoring of calcium metabolism in patients in the final stages of renal disease using serum by AMS after isotopic labelling with a ^{41}Ca radiotracer. The authors hypothesized that bone resorption can be studied directly by serial measurements of the $^{41}Ca/Ca_{total}$ ratio in serum after *in vivo* labelling of the calcium pool with ^{41}Ca.[54]

Accurate determination of the biological important element, selenium, in blood serum by isotope dilution analysis using ICP-QMS with octopole collision cell (Agilent 7500ce, Tokio, Japan) is described by Schaumlöffel and coworkers.[55] A recovery of selenium from human serum reference material was only 78 % when hydrogen was applied as collision gas but 96.7 % using xenon as a non-reactive collision gas to eliminate isobaric interferences. A detection limit of $3.3\,\mu\mathrm{g\,l^{-1}}$ was achieved.[55]

Transition-metal based compounds constitute a discrete class of chemotherapeutics, widely used in medicine as antitumor agents.[56] Several ruthenium complexes enable the body to catalyze oxidation and reduction reactions, depending on the physiological environment, and have attracted much interest as alternative antitumor drugs in the treatment of cancer cells resistant to cisplatin in cancer chemotherapy. Several ruthenium compounds are transported into cells relatively easily and bind to cellular DNA. Brabec and Nováková[56] reported on the DNA binding of ruthenium complexes with antitumor effects. ICP-MS, in addition to GE-AAS and ICP-OES, has been employed for Ru determination in human blood and chicken heart tissue.[57] The binding of certain platinum group metals in drug proteins has been studied by CE-ICP-MS for further insights, especially into the reactivity of a novel antitumor ruthenium(III) complex towards human serum proteins. Using this species selective analytical technique, the interaction of novel ruthenium based anticancer agents with serum plasma proteins was confirmed, whereby the ruthenium was found to be almost all (98–99%) converted into the protein bound from (in the albumin fraction).[58] The total platinum concentration and platinum oxidation states in body fluids, tissue and explants from woman exposed to silicone and saline breast implants has been studied by IC-ICP-MS.[59]

Analysis of Urine

Urine is the most frequently investigated medical matrix[60–64] because samples for evidence of possible contamination with toxic metals or long-lived radionuclides can be easily collected. The difficulties, but also the capabilities, of inorganic mass spectrometric techniques (especially using ICP-MS as an universal and sensitive analytical technique with low detection limits) can be demonstrated with an example of urine analysis with complex matrix composition. Freeze-dried urine reference materials from NIST (SRM 2670) were analyzed with respect to V, Ni and As by flow injection ICP-MS with cryogenic desolvation (in order to minimize interferences with polyatomic ions and to avoid clogging of the sampling orifice) by Houk's working group.[65] The authors observed improved analyte sensitivity by adding a small dose (2 %) of hydrogen to the aerosol gas flow, but the oxide ratios were kept low.

Parsons *et al.*[66] reported on the determination of total mercury as a toxic element in urine at the trace level using ICP-MS. The main sources of exposure in the human body are inhalation of Hg vapour released from dental amalgams and the consumption of fish species that accumulate, for example, methyl mercury (MeHg).

Changes in bone calcium metabolism have been investigated by the isotopic labelling of bone using ^{41}Ca as a long-lived radiotracer and monitoring of urine. A minute amount of ^{41}Ca (100 nCi) was administered orally to women and the kinetics of tracer excretion were assessed by monitoring changes in urinary $^{41}Ca/^{40}Ca$ isotope ratios for up to 700 days measured by accelerator mass spectrometry (AMS) and resonance ionization mass spectrometry (RIMS).[67] Changes in ^{41}Ca radiotracer distribution between different compartments (using a linear three compartment model) were modelled on the basis of kinetic parameters. Changes in bone formation and resorption can be assessed by monitoring urinary ^{41}Ca excretion over the first few weeks post-dosing.[67]

Diluted urine samples (factor 20) of a cancer patient after receiving intravenous chemotherapy with cisplatin were examined by HPLC-ICP-MS with respect to cisplatin (as the parent drug) and intermediately formed mono- and diaquacisplatin products.[68] Analysis of species distribution in diluted human urine from the cancer patient revealed that approximately 40 % of the parent drug was excreted as monoaquacisplatin. The remaining fraction excreted as intact cisplatin was hydrolyzed at a high chloride concentration.[68]

An analytical technique has been developed for the determination of selenium species in urine to identify selenium containing compounds by combining electrospray MS/MS with ICP-MS after preparative solid phase extraction and final reversed phase HPLC separation.[69]

Depleted uranium (DU) is used to reinforce armor shielding and increase penetrability of military munitions. Although the data are conflicting, DU has been invoked as a potential etiological factor in Gulf War syndrome.[70] To control possible contamination of soldiers from the Gulf War or Kosovo conflicts with DU, urine was examined by inorganic mass spectrometry as described in references.[61,62,71,72] TIMS has been used for many years as a standard technique especially for uranium isotope analysis in urine.[73] Today, ICP-MS is the method of choice due to high sample throughput and has been used for the sensitive and precise determination of uranium, thorium and other actinides especially for isotope ratio measurements of uranium at the trace and ultratrace level in urine samples in routine mode in many laboratories.[31–36,71] Because long-lived radionuclides occur at extremely low concentration levels, especially in medical samples, matrix separation and enrichment of the analyte is proposed for their analysis by several authors.[64,74–76] Details of the determination of long-lived radionuclides in urine at trace and ultratrace level are described in Section 9.9.2.

9.5.3 Hair, Nail, Tooth and Bone Analysis

In general, the majority of investigations concerning heavy metals in medical samples deal, not only with blood and urine analysis, but also with hair samples and can be used for the sensitive biomonitoring (for detection of trace elements) of, for example, environmental exposure or ingestion through food or drinking water. The analysis of hair samples has several advantages over the analysis of blood and urine, because hair is stable and robust and its composition does not change over time. Furthermore, sampling is painless, very easy and requires no specific professional skills and no special storage or handling. Third, unlike blood samples, hair has the unique ability to reflect the total body intake over an extended period and it is possible to trace changes over time depending on the length of the hair. Trace elements in hair have been investigated in the criminology field[77,78] where hair has been analyzed for traces of cocaine. An interesting application is in archaeology,[79] where the hair of an ancient iceman was analyzed providing information on his diet, environment[80,81] and occupational exposure[82] to different trace elements. Furthermore, hair analysis is also used in medical research to explore the connection between the concentration of metals in hair and different diseases such as autism[8] and epilepsy.[83]

ToF-SIMS analytical techniques for characterizing human hair surfaces using Ga^+ and C_{60}^+ primary ion sources have been compared.[84] Similar hair fingerprints were obtained with both ion sources. With the C_{60}^+ ion source, on average, secondary ion intensities between two and three orders of magnitude higher than the Ga^+ primary ion source were observed.[84]

An analytical method for determining essential elements (Zn, Fe and Cu) and toxic elements (Cr, Pb and U) on single hair strands by LA-ICP-SFMS has been developed in the author's laboratory.[85] Results for these elements obtained directly by LA-ICP-SFMS are in good agreement with those measured by ICP-QMS on solutions of digested hair samples. In this study, different quantification strategies (such as external calibration, standard addition and isotope dilution) for trace element determination in hair samples were compared. In addition, a solution based calibration strategy for uranium determination in powdered hair was applied by coupling the laser ablation chamber to an ultrasonic nebulizer. The analytical results of the hair analysis of three people measured by ICP-MS[85] compared to the concentrations reported in the literature are summarized in Table 9.26.[86] Whereas in the investigated hair samples from three 'normal', unexposed people, a higher Zn concentration (26–45 $\mu g\,g^{-1}$) compared to Cu (10–26 $\mu g\,g^{-1}$) was measured, chromium and uranium were found at 1 and 0.25 $\mu g\,g^{-1}$ and below, respectively. The concentration of Zn, Fe, Cu, Cr, Pb and U in single hair strands from three volunteers measured by ICP-MS after digestion were fitted to the published concentration range as reported by Rodushkin and Axelsson.[86] The concentration of uranium in 'normal' hair varied between 5 and 1280 $ng\,g^{-1}$, but maximum concentrations of up to 0.3–0.5 $mg\,g^{-1}$ were found for persons who regularly drank natural water sources with a very high uranium content (11 800 $\mu g\,l^{-1}$ in south Finland) in comparison to the recommended value (WHO) of $< 1500\,ng\,l^{-1}$.[85] Single hair analysis collected from one person[85] who had changed their habitat and began ingesting uranium with their drinking water was performed by LA-ICP-MS. Differences in the uranium content observed along the single hair strand correlated with changes in the level of uranium in drinking water. The single hair, including its root, was washed and cut into sections 1 cm in length. The pieces were placed on a carbon pad in a small laser ablation chamber and each one was analyzed by LA-ICP-SFMS in the line scan mode. The experiment was repeated for a few hair strands. Figure 9.30 a presents the amount of uranium in the hair (triangles and solid line) as a function of the distance from the scalp or the root of the hair investigated. The dotted lines show the uranium concentration in water. The uranium concentration in the single hair investigated decreased from 212 to 18 $ng\,g^{-1}$ and correlated with a change in the uranium concentration in drinking water from 2000 to 35 $ng\,l^{-1}$. In addition, measurements of uranium isotope ratios (see Figure 9.30b) showed a natural isotopic composition in the drinking water throughout the whole period, as well as in the single hair sample analyzed by LA-ICP-MS.[85]

Table 9.26 *Concentration of Zn, Fe, Cu, Cr, Pb and U in single hair strands from three volunteers* ($\mu g\,g^{-1}$).

	Person A	**Person B**	**Person C**	**Published range[86]**
Zn	33 ± 4	26 ± 2	45 ± 5	40–327
Fe	53 ± 3	28 ± 3	26 ± 1	3–900
Cu	14 ± 4	10 ± 1	26 ± 1	0.3–293
Cr	1.0 ± 0.1	0.5 ± 0.04	0.7 ± 0.05	0.03–33
Pb	4.7 ± 0.6	1.6 ± 0.07	2.0 ± 0.2	0.22–7.26
U	0.1 ± 0.04	0.06 ± 0.03	0.25 ± 0.01	0.005–1.28

H. Sela, Z. Karpas, M. Zoriy, C. Pickhardt and J. S. Becker, *Int. J. Mass Spectrom.*, **261**, 199 (2007). Reproduced by permission of Elsevier.

Anomalously high Ca concentrations in hair samples (up to 8285 $\mu g\,g^{-1}$) which correlate with high P concentrations (up to 4720 $\mu g\,g^{-1}$) from a group of women from Rio de Janeiro were detected by ICP-MS measurements.[87] These abnormal hair compositions were related to endocrinological pathologies affecting calcium/bone metabolism. Very low Ca concentrations were observed in older women and were related to senile osteoporosis.[87]

Since 1960, it has been demonstrated by various analytical procedures that high concentrations of arsenic were present in Napoleon's hair.[88] Multi-element analysis of two specimens of Napoleon's hair by ICP-MS after mineralization in concentrated nitric acid resulted in arsenic concentrations (42.1 and 37.4 $\mu g\,g^{-1}$) about 40 times higher than normal values, confirming the hypothesis of a significant exposure to arsenic. However, mercury (3.3. and 4.7 $\mu g\,g^{-1}$), antimony (2.1 and 1.8 $\mu g\,g^{-1}$) and lead (229 and 112 $\mu g\,g^{-1}$) were also detected at elevated levels. The elevated concentrations of Sb and Hg are in agreement with the data already known about the therapeutic treatments given to Napoleon (calomel and tartar emetic are compounds of mercury and antimony, respectively).[88]

The longitudinal distribution of thallium in human scalp hair from a person suffering from thallium poisoning compared to control persons has been determined by isotope dilution ETV-ICP-MS.[89] Limits of detection determined by analyzing hair from non-exposed persons were 0.7 pg Tl (5 $ng\,g^{-1}$) in hair when three 10 mm hair segments (150 μg) were analyzed together. Compared to LA-ICP-MS, the analytical technique proposed is more time consuming because the hair strands have to be embedded in pressure hardening glue before cutting them into 10 mm segments and injecting them with solid sample introduction (electrothermal vaporization) in ICP-MS. Several longitudinal concentration gradients for the segments analyzed (length: 10 mm) were obtained

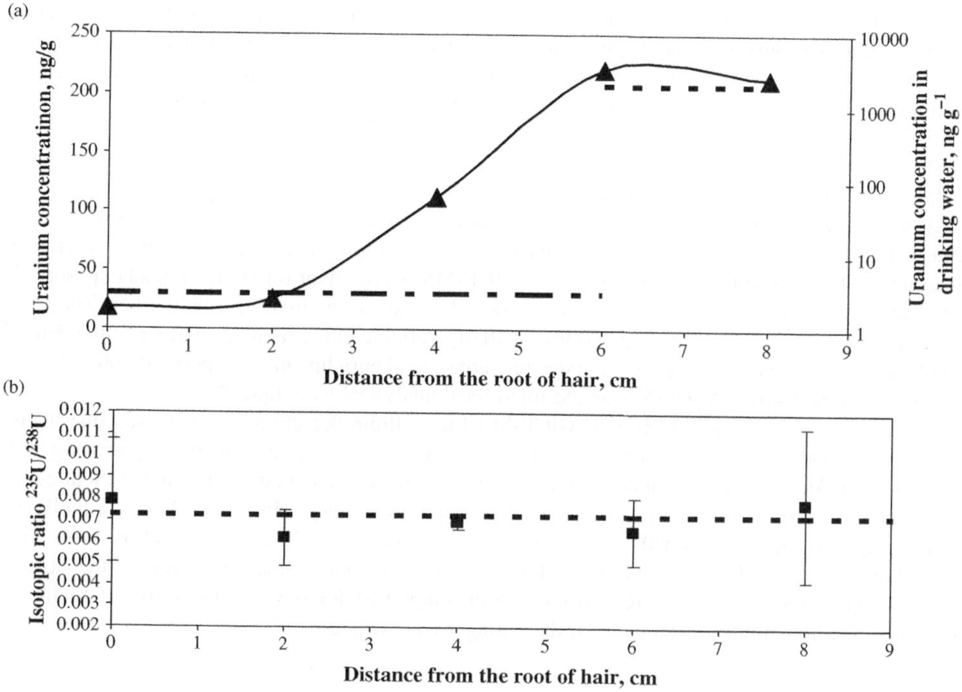

Figure 9.30 *(a) Changes in uranium concentrations along a single hair strand collected from an individual who changed their place of residence in relation to uranium content in drinking water. (b) The* $^{235}U/^{238}U$ *isotope ratio measured along a single hair strand collected from an individual who changed their place of residence; (H. Sela, Z. Karpas, M. Zoriy, C. Pickhardt and J. S. Becker, Int. J. Mass Spectrom.,* **261**, *199 (2007). Reproduced by permission of Elsevier).*

from hair from a person suffering from thallium poisoning.[89] In contrast, LA-ICP-MS allows a significantly improved lateral resolution down to the μm range.

Ingested arsenic is known to be not only excreted by urine, but also to be stored in sulfydryl rich tissue like hair, nails or skin. A method of extracting arsenic species from hair and nail samples has been developed for the analysis of the total arsenic concentration, inorganic As (III) and As (V) and several pentavalent and trivalent methylated arsenic species.[90]

Human tooth enamel provides a nearly permanent and chronological record of an individual's nutritional status and anthropogenic trace metal exposure during development[91] and can be used for biomonitoring. SIMS (compared to X-ray microanalysis) as a microlocal analytical technique has been applied to study the chemical composition in hypermineralized enamel of human teeth (permanent first molars).[92] It was found that hypermineralized enamel had a higher content of carbon and calcium whereas the phosphorus concentration was lower than in normal enamel.[92] In order to determine selected trace elements (Al, Ti, Cr, Mn, Cu, Zn, Sr, Ba, La, and Pb) in ancient tooth samples (about 2000 years old) slurry sampling electrothermal vaporization ICP-MS was employed with polytetrafluoroethylene (PTFE) as the fluorinating reagent.[93] Simple pretreatment (at 500 °C) was applied prior to analysis to remove the organic matrix and separate the dentine and enamel. Microlocal analysis of selected trace elements (with decreasing concentration: Sr>>Mg>Zn>Pb>Fe>Cu) found in sequentially calcifying layers of dental enamel was performed by LA-ICP-MS.[91,94,95] Because deciduous enamel forms from week 13 in the uterus up to 9 months postnatally, the application of LA-ICP-MS allows a retrospective measurement of prenatal and early postnatal trace element uptake during the critical period of child development. Variation in the trace element composition of teeth is relevant for understanding changes in nutrition, pollution and residence.[94]

The exposure pathway of noble metal alloys (platinum and gold) from teeth with noble metal dental alloy restorations has been investigated by ICP-SFMS measurements of urinary gold and platinum.[96] Quantitative lead determination in tooth enamel by LA-ICP-MS (detection limit: 11 $ng\,g^{-1}$) was performed by one point calibration using pelletized bone certified reference material.[97] Measurement of lead in bone by LA-ICP-MS is important for studying lead exposure.[98] Lead accumulates in bone over many years or decades and can be used as a biomarker. The lead content in the bone samples investigated was spatially variable in the range 2–30 $\mu g\,g^{-1}$; the detection limit was estimated as 0.2 $\mu g\,g^{-1}$. In discrete areas, lead enrichment (hot spots) of 100–200 μm width was found using LA-ICP-MS as the microlocal analytical technique.[98]

In recent years, the application of LA-ICP-MS for thallium detection has increased due to its toxicological and environmental impact. In the reconstruction of a case of thallium poisoning (38 years ago in Austria), several human bone samples were investigated by LA-ICP-MS directly and by ICP-SFMS (after microwave induced digestion of samples).[99] In several bones, a thallium concentration in the range of 1.07–2.63 $\mu g\,g^{-1}$ was measured by ICP-SFMS, which is up to 170 times higher compared to concentrations found in control bones. Since nails reflect the health events of previous months, LA-ICP-MS has been employed for the reconstruction of thallium poisoning as a function of distance from the younger edge of the nail.[100]

9.5.4 Microanalysis of Small Amounts of Medical Samples

The greatest health problem after the accident at Chernobyl nuclear power plant (NPP) in 1986 is thyroid pathologies, especially among children. Besides the irradiation of thyroids by radioactive iodine and caesium nuclides, toxic elements from the fallout are a direct risk to health. Studies of small amounts of tissue (1–10 mg) from 90 thyroids obtained during thyroid operations in Minsk hospitals were investigated with respect to trace metal contents of essential and toxic elements by ICP-QMS (Elan 6000, Perkin Elmer Sciex).[101] The accuracy of ICP-MS measurements was checked by analyzing certified reference material SRM 1566a Oyster Tissue (20 mg was digested). In order to validate the ICP-MS results some thyroid tissues were analyzed using instrumental neutron activation analysis (INAA). Inhomogeneous element distribution, which could reduce the accuracy and precision of ICP-MS results, was found by additional SIMS measurements. Almost all essential elements, as well as a number of toxic metals such as Cr, As, Cd, Pb, Hg, U and others, were determined in human thyroid tissue after microwave digestion and dilution of the digest. In general, the thyroid tissues affected by pathology were characterized by higher Ca, Fe, Mn, V, As, Cr, Ni, Pb, U and Ba content in comparison to control tissues. It was found that an increasing calcium concentration of pathological thyroid tissues correlates with increasing uranium, lead, nickel or arsenic content. In comparison to other expensive and time consuming trace analytical techniques, such as instrumental neutron activation analysis (INAA), which requires several sample irradiations to cover such a wide range of elements, ICP-MS is faster and cheaper. Table 9.27 a presents a comparison of results measured by ICP-MS and INAA. The detection limits observed with ICP-MS in the low $ng\,g^{-1}$ range (for As, Bi, Cd, Ce, Co, Mo, Pb, Rb, and V) and in the sub-$ng\,g^{-1}$ range for U and Sb are mostly better compared to INAA. ICP-MS permits a number of relevant trace elements to be determined (e.g., Ni, Cu, Sn, Sr, Cd, Hg, Pb), in addition to INAA.[101]

Table 9.27 Determination of element concentrations in 90 tyroids by ICP-MS Comparison of ICP-MS and INAA (conc. in $\mu g g^{-1}$).

Element	ICP-MS	INAA
Ca	3630 ± 400	3780 ± 390
Cr	2.1 ± 0.2	2.0 ± 0.3
Fe	182 ± 20	191 ± 22
K	250 ± 29	249 ± 29
Na	435 ± 55	438 ± 35
Sb	42 ± 4	40 ± 6
Se	1.8 ± 0.3	1.4 ± 0.2
U	0.065 ± 0.004	0.068 ± 0.010
Zn	323 ± 35	394 ± 39

* S. F. Boulyga, J. S. Becker, A. F. Malenchenko and H. J. Dietze, *Mikrochim. Acta*, **132**, 215 (2000). Reproduced by permission of Springer Science and Business Media.

The application of microanalytical techniques, such as flow injection in ICP-MS (FI-ICP-SFMS), is also of special interest in medical research where very small sample volumes have to be characterized, e.g. for Cr determination in DNA by sector-field ICP-MS.[102] For the separation of isobaric interferences of $^{52}Cr^+$ and $^{40}Ar^{12}C^+$ the measurements were performed at a mass resolution of $m/\Delta m = 3000$. Transient signals of $^{52}Cr^+$ and $^{52}Cr^+$ analyzed by FI-ICP-SFMS of a $10\,\mu g\,l^{-1}$ chromium solution (sample loop: $20\,\mu l$) are illustrated in Figure 9.31a. For quantification by the isotope dilution technique, the small volume of DNA available (diluted 1:10) was injected into a continuous flow of 2 % HNO_3, which is spiked with high-enriched ^{53}Cr solution (the isotope abundance of ^{53}Cr was 83.4%). The application of the isotope dilution technique in flow injection ICP-SFMS is shown in Fig. 9.31 b.[102]

ICP-MS has been combined with flow injection and a selective extraction procedure (which isolates Alzheimer's amyloid plaques) for the selective analysis of plaque cores.[103] The results of multi-element analysis indicate the accumulation of several elements (Al, Fe, Zn) in the plaques.

Microdialysis to monitor the chemistry of extracellular space in human tissue is a challenging technique that has been applied in the neurosciences to measure neurochemical composition *in vivo*.[104] Microdialysis coupled online with in-tube solid phase extraction (SPE) and ICP-MS has been developed by Sun *et al.*[105] for the *in vivo* determination of several trace metals in order to study the transfer kinetics of trace elements in brain extracellular fluid.

9.5.5 P, S, Se and Metal Determination in Proteins

The proteomic area requires versatile tools allowing the detection of a particular protein within a large pool. Of special importance in medical research and the life sciences is the determination of the essential element phosphorus, especially in protein samples.[106] Phosphorylation is thus one of the most important modifications of proteins, and is implicated both in normal physiology and in human disease, carcinogenesis and ageing.[107] Reversible protein phosphorylation is a key determinant in many fundamental cellular functions, such as survival, differentiation, structural organization and stress responses, and is relevant for many physiological processes in carcinogenesis or neurodegenerative diseases, such as Alzheimer's and Parkinson's diseases or Hallervorden–Spatz syndrome.[108,109] The relation of phosphorylation to many neurodegenerative diseases, such as Alzheimer's or Parkinson's, is recognized in a multitude of papers.[110–112] Inductively coupled plasma mass spectrometry with a sector field instrument compared to a quadrupole based ICP-MS

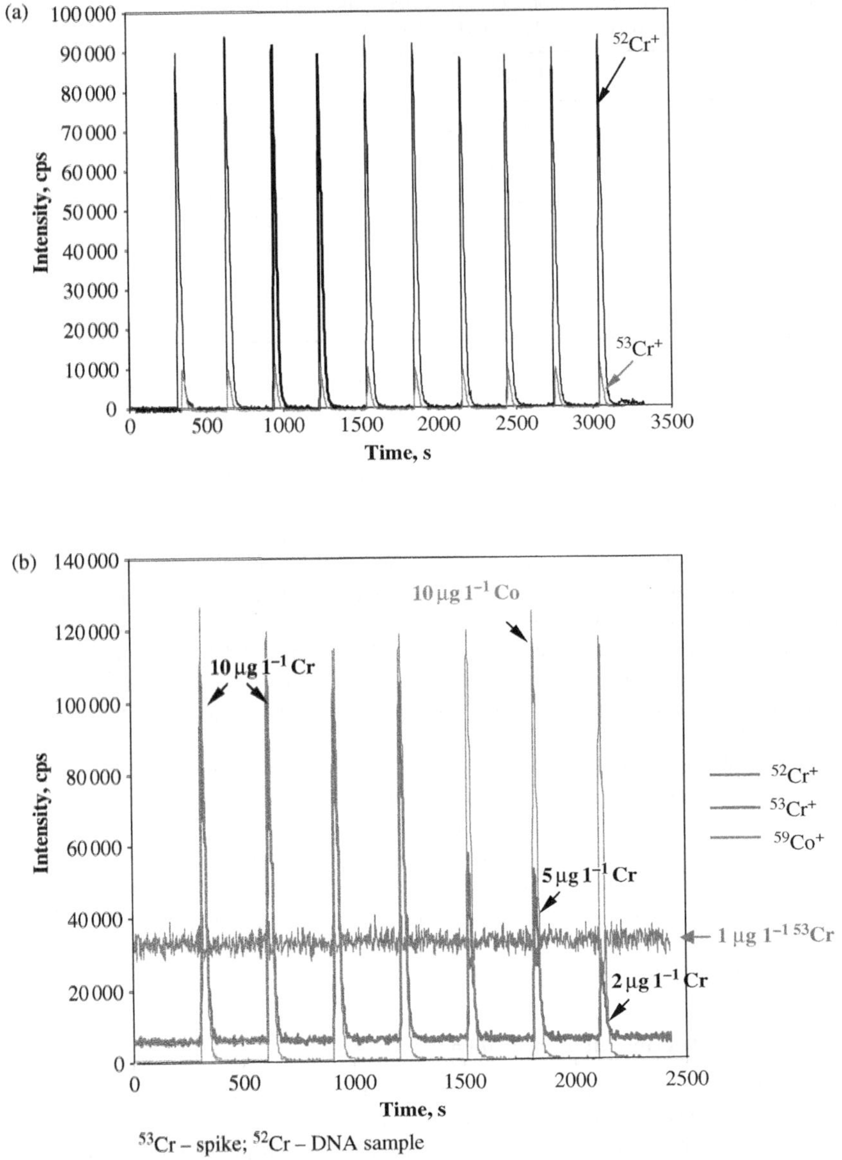

Figure 9.31 *(a) Determination of chromium in DNA using flow injection ICP-SFMS (sample loop: 20 µl); (b) Application of isotope dilution analysis in flow injection ICP-SFMS for chromium determination in DNA. 1 µg l^{-1} ^{53}Cr enriched isotope spike solution (84.3 %) was used for calibration. (J. S. Becker and H. J. Dietze, Int. J. Mass Spectrom. Ion Proc.* ***197****, 1 (2000). Reproduced by permission from Elsevier.)*

was used for phosphorus determination in small protein samples (down to 1 μg or 1–10 μl digest of protein solution) by Becker *et al.*[106] For quality control of phosphorus determination, small amounts (down to 0.3 mg) of a certified reference material (CRM), single cell proteins – BCR 273 – was analyzed. The recommended sample size of BCR 273 due to possible inhomogeneous phosphorus distribution was 500 mg. The determination of phosphorus in 0.3 mg of this CRM by ICP-SFMS was possible with a relative standard deviation of 9.3 %. In addition, the phosphorus concentration in 1 μl tau protein digest (tau is an Alzheimer's relevant target protein) was determined as $2.32 \pm 0.23\,mg\,g^{-1}$. A further reduction of the volume of diluted protein solution was achieved by the application of flow injection in ICP-SFMS. The detection limits of phosphorus determination in biological samples were determined by ICP-SFMS down to the $ng\,g^{-1}$ range.[106] The mass spectrometric approaches described in reference[110] enable the identification of phosphorylated protein structures, and the direct quantification of protein phosphorous contents, using Fourier transform ion cyclotron resonance (FTICR-MS) and laser ablation inductively coupled plasma (LA-ICP-MS). An ultrahigh resolution FTICR-MS method has been developed and applied for the structural identification of phosphorylations in proteins, using direct peptide mapping analysis with high mass accuracy of tryptic phosphorylated fragments. The application of this method to human tau proteins, one of the key proteins for the formation of neurofibrillary tangles in Alzheimer's disease, provided the identification of 18 phosphorylation sites, as described by Becker *et al.*[110]

Elliot *et al.*[113] proposed a strategy for phosphoprotein measurements by a combination of gel electrophoresis and LA-ICP-MS. The potential for developing improved procedures for phosphate measurement through combinations of gel electrophoresis and quadrupole based ICP mass spectrometry utilizing $^{31}P^{16}O^{+}$ is being investigated. Laser ablation of gels offers a rapid and direct quantification possibility, but is subject to high blanks due to P impurities in gels and associated reagents. Nevertheless, the optimization of laser sampling afforded improved method sensitivity (limit of detection $0.09\,\mu g\,g^{-1}$). Implementation of whole gel elution (WGE) with FI-ICP-MS (conventional solution nebulization) following gel electrophoresis permitted quantification at the sub-$\mu g\,l^{-1}$ level, and microcolumn processing (activated alumina) was effective at rejecting phosphate contamination.[113] Edler *et al.*[114] compared ICP-MS with hexapole collision cell and ICP-SFMS for phosphorus determination in pharmaceuticals.

Recent trends in phosphorus speciation utilizing separation techniques considering HLPC, CE and gel electrophoresis in ICP-MS have been examined by Shah and Caruso.[115] For speciation studies, ICP-MS of phosphorus is combined with a biomolecular mass spectrometric technique such as (ESI- or MALDI-MS).[116–118]

The quantitative determination of other essential elements (e.g., Cu, Fe, Zn, Mn, Co, Se and others) in brain or medical tissues is important in brain research and the life sciences and is also relevant for studying many neurodegenerative diseases. A deficiency or excess of these essential elements in proteins and in human tissue has been observed in neurodegenerative diseases (including Alzheimer's and Parkinson's diseases), but metals can also catalyze cytotoxic reactions and are toxic at high concentrations.[119,120] Besides the study of phosphorylated proteins (phosphoproteins), metal containing proteins (so-called 'metalloproteins') play an essential role as cofactors in biological systems (e.g., in single cells or cell organelles) and are gaining increasing attention in proteomics research. Therefore the proteomics area requires versatile and powerful analytical techniques for the detection and characterization of phosphorus, selenium and metal containing proteins within a large pool of proteins, e.g., after electrophoretic separation in 2D gels.[110,113,116,118,121,122] Laser ablation inductively coupled plasma mass spectrometry (LA-ICP-MS) is the method of choice for the detection of metals, phosphorus and selenium in protein bands and spots directly in 1D and 2D gels after electrophoretic separation. On the other hand, mass spectrometers that employ soft ionization techniques (electrospray ionization–ESI and matrix assisted laser desorption ionization – MALDI) permit the identification of large biomolecules, such as proteins.[116,123] While MALDI and ESI mass spectrometry can be applied for the identification of phosphorylation sites in proteins, these techniques cannot provide direct quantitative determinations of phosphorus and metals in biological samples. Therefore, the combination of LA-ICP-MS as an element analytical technique with a biomolecular mass spectrometric technique, such as high resolution MALDI-FTICR-MS, allows the molecular identification and quantification of protein phosphorylation and of metal concentrations. This also enables the investigation of post-translational modifications of proteins, e.g., phosphorylation,[110,116,117,121,122] which are relevant in many fundamental cellular functions, such as survival, differentiation, structural organization, stress responses, for many pathophysiological processes in carcinogenesis or neurodegenerative diseases.[116] In addition, it is well known that metals (e.g., Cu, Fe, Zn, Mn, Ca, Mg and others) and non-metals, such as P and S, are inhomogeneously distributed in biological or medical tissues. Therefore investigating the element distribution in thin tissue slices requires sensitive analytical techniques with high spatial resolution and represents a challenging task in analytical chemistry where LA-ICP-MS is the method of choice. The advantages of LA-ICP-MS compared to other surface analytical techniques, such as scanning electron microscopy with energy disperse X-ray analysis (SEM-EDX),[124] microproton induced X-ray emission (PIXE),[125] autoradiography[126] or secondary ion mass spectrometry (SIMS),[127–129] are its high sensitivities (and consequently very low detection limits) for trace element determination and its ability to quantify analytical data on medical tissues due to the significantly lower matrix effects. The application of LA-ICP-MS in life sciences and in medicine is focused at present on individual tasks, e.g., the mapping of copper and zinc in liver sections of sheep,[130] the extreme ultratrace and isotope analysis of actinides (especially of plutonium) in moss samples[131] or in body fluids (urine).[72] In our recent studies, LA-ICP-MS was applied for the microlocal analysis of selected protein spots in gels after 2D gel electrophoresis and also for the determination of elemental distribution in thin sections of brain tissues (imaging).[16,116,117]

Possible strategies for LA-ICP-MS in brain research and the life sciences are summarized in Figure 9.32. They include the imaging of thin slices of brain tissues in order to obtain images of elements (left-hand side of Figure 9.32) or application in proteome analysis. The combination of both approaches is a future task. LA-ICP-MS has been developed as a microanalytical method for the determination of P, S, Si and metal concentrations (Al, Zn, Cu und Fe) in well separated protein spots after 2D gel electrophoresis in brain samples of patients with Alzheimer's and in control brains[132], where a combination of atomic and biomolecular mass spectrometric methods (LA-ICP-MS and MALDI-FTICR-MS, respectively) was helpful for the characterization and identification of several human Alzheimer brain proteins.[110,117] Of the investigated elements in human brain proteins, essential elements such as P, Cu and Zn are of interest. For these studies, a brain protein mixture was separated by 2D gel electrophoresis (see 2D gel in Figure 9.32, top right) and the protein spots were fast screened by microlocal analysis using LA-ICP-MS with respect to the P, S, Cu, Zn and Fe content. A double-focusing sector field mass spectrometer was applied at mass resolution ($m/\Delta m \sim 4400$) for the analysis of small protein samples or during imaging analysis in brain tissues to solve the interference problem arising in LA-ICP-MS of isobaric interferences on atomic ions of analyte by the formation of polyatomic ions such as $^{15}N^{16}O^+$, $^{14}N^{17}O^+$ and $^{14}N^{16}O^1H^+$ at mass 31 u for phosphorus or $^{32}S_2^+$ during $^{64}Zn^+$ determination, respectively. In Figure 9.33, the transient signals of $^{31}P^+$, $^{63}Cu^+$, $^{56}Fe^+$ and $^{64}Zn^+$ are shown in protein spots 6–9 compared to the background signal of the blank gel. Selected protein spots in 2D gel containing these elements were investigated after tryptic digestion by matrix assisted laser desorption ionization Fourier transform ion cyclotron resonance mass spectrometry (MALDI-FTICR-MS). Protein spot 6 was identified as creatine kinase, β-chain ($M_r = 42.91$ kDa and a phosphorylation state of 1). In Figure 9.34 the MALDI-FTICR mass spectrum of protein spot 8 (ATP synthase) is illustrated with the identified

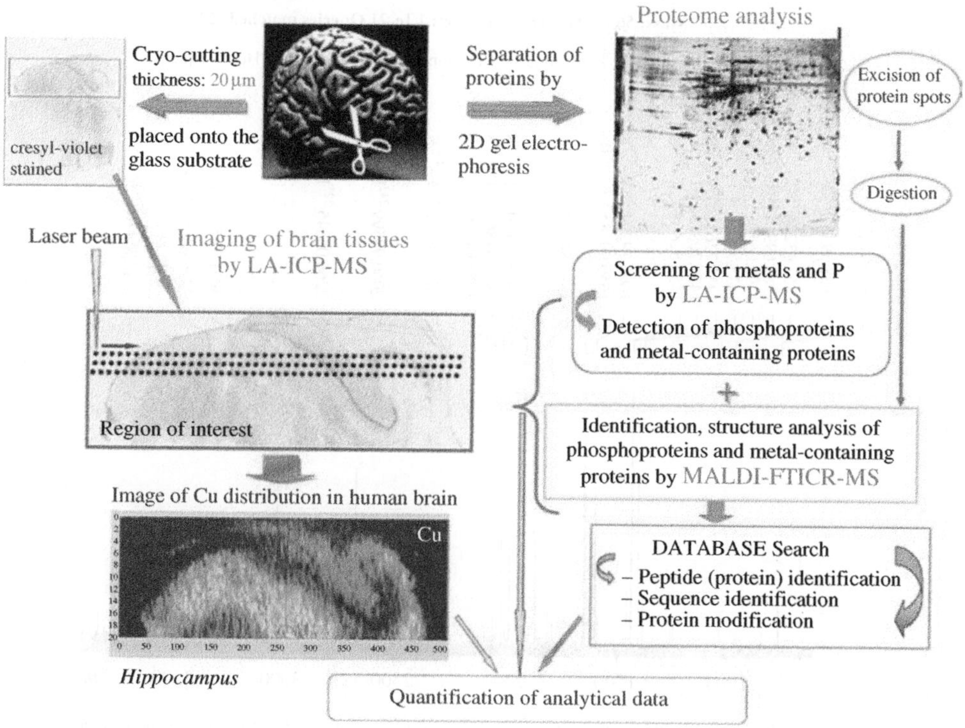

Figure 9.32 *LA-ICP-MS strategies in brain research: imaging of thin sections of brain tissues and screening of 2D gel with respect to phosphoproteins and metal containing proteins.*

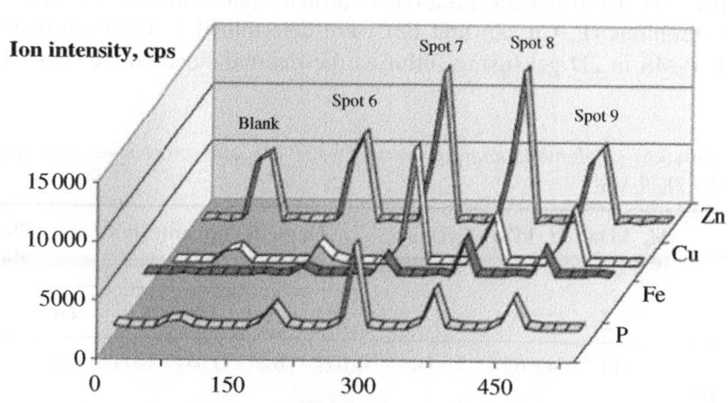

Figure 9.33 *Transient signals of $^{31}P^{+}$, $^{63}Cu^{+}$, $^{56}Fe^{+}$ and $^{64}Zn^{+}$ in protein spots 7–9 measured by LA-ICP-MS.*

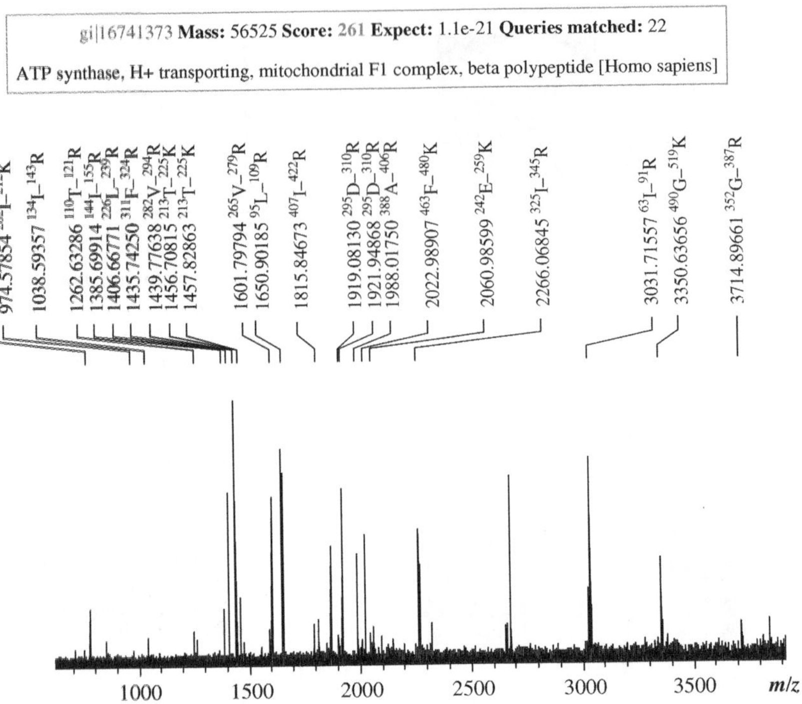

Figure 9.34 *MALDI-FTICR-MS of protein spot 8 (ATP synthase) with identified tryptic peptides and results of database search. (J. Su. Becker, M. Zoriy, M. Przybylski and J. S. Becker. Int. J. Mass Spectr.,* ***261****, 69 (2007). Reproduced by permission of Elsevier.)*

tryptic peptides and results of a database search.[111] Results of the structure analysis of human brain proteins by MALDI-FTICR-MS were combined with those from the direct determination of phosphorus, copper, zinc and iron concentrations in protein spots with LA-ICP-MS. Subsequently, the element concentrations (P, Cu, Zn and Fe) were determined in the identified human brain proteins by LA-ICP-MS in 2D gel (using sulfur as the internal standard element). The results of

Table 9.28 *Combination of element analysis using LA-ICP-MS and protein identification (Alzheimer disease) by MALDI-FTICR-MS.*

Spot	Protein	M_r kDa (exp)	M_r kDa	Cystein residues	Element concentration[a] ($mg\,g^{-1}$)					Phosphorylation State
					P	S	Fe	Cu	Zn	
6	Creatine kinase B-type	~43	42.62	16	0.027	8.7	0.31	1.05	1.61	1
8	ATP synthase	~56	56.54	14	0.041	7.2	0.62	0.64	1.3	3
9	Mutant β-actin	~42	42.14	24	0.041	9.2	0.55	1.46	0.92	4

[a] Relative standard deviation (RSD) of measured element concentration is about 30%.

the combination of element analysis using LA-ICP-MS and protein identification (Alzheimer's disease) by MALDI-FTICR-MS are summarized in Table 9.28. It is interesting is to note that neurodegenerative diseases appear to be directly related to Cu metabolism. Menckes' syndrome and Wilson's disease are hereditary disorders with a functional disturbance of two membrane located ATPases for the transport of Cu ions. In Wilson's disease, Cu excretion via bile ducts is disturbed and consequently Cu is accumulated mainly in the liver. Also Al has been associated with several neurodegenerative diseases, such as Alzheimer's disease, Parkinson's dementia or encephalopathy.[50] The effect of Al on the nervous system and the characteristics of Al neurotoxicity have been studied by Kawahara.[133]

The study of metal containing proteins is a new challenging task in proteomics. Human brain proteins containing P, Cu and Zn have been detected directly in protein spots in 2D gels of a healthy human brain sample (somatomotor cortex) by LA-ICP-MS (see Figure 9.35).[116] Of 176 protein spots in 2D gel from a human brain sample, phosphorus, copper and zinc were detected in 31, 43 and 49 protein spots, respectively. Uranium was found in 20 protein spots. In Figure 9.36, transient signals of $^{31}P^+$, $^{63}Cu^+$, $^{64}Zn^+$ and $^{238}U^+$ are summarized, whereby only in spot 11e were all elements measured simultaneously with relatively high ion intensities. Protein spots 11a and 11d contain P, Cu and also Zn. The detection limits for P, S, Cu, Zn and U were determined in singular protein spots as 0.0013, 1.29, 0.029, 0.063 and 0.000 01 mg g^{-1}, respectively. The identification

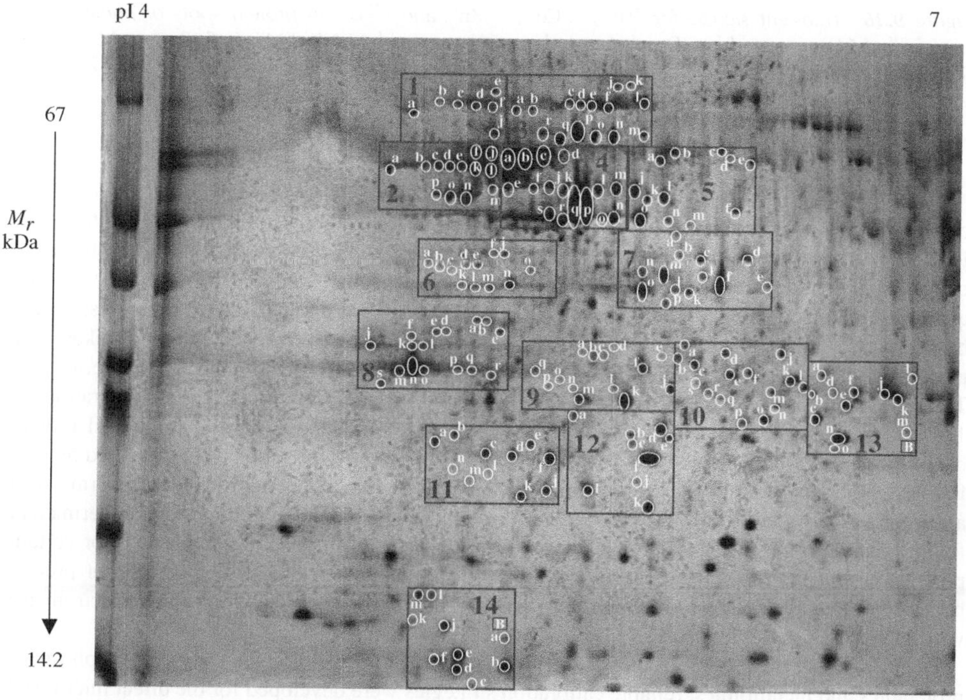

Figure 9.35 *Two-dimensional gel after electrophoretic separation of proteins from human brain sample (somatomotor cortex), 176 separated protein spots were analyzed by LA-ICP-MS. (J. S. Becker, M. Zoriy, J. Su. Becker, C. Pickhardt, E. Damoc, G. Juhacz, M. Palkovits and M. Przybylski, Anal. Chem.,* ***77****, 5851 (2005). Reproduced by permission of American Chemical Society.)*

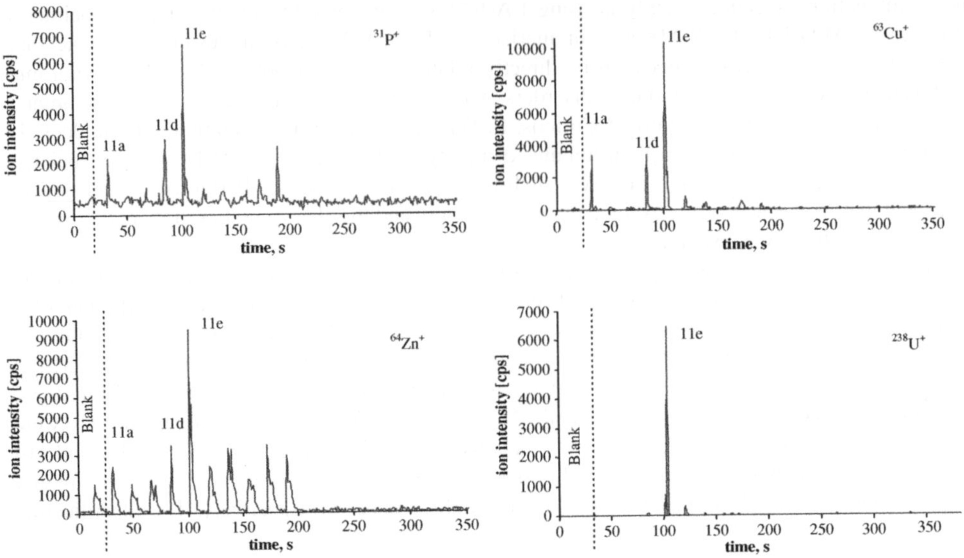

Figure 9.36 *Transient signals for $^{31}P^+$, $^{63}Cu^+$, $^{64}Zn^+$ and $^{238}U^+$ in protein spots of human brain sample (cut 11) separated by 2D gel electrophoresis measured by LA-ICP-MS. (J. S. Becker, M. Zoriy, J. Su. Becker, C. Pickhardt, E. Damoc, G. Juhacz, M. Palkovits and M. Przybylski, Anal. Chem.,* **77**, *5851 (2005). Reproduced by permission of American Chemical Society.)*

of protein spots was performed by MALDI-FTICR-MS. For example, in Figure 9.37 the MALDI-FTICR mass spectrum of protein spot 3n (α-internexin) is presented with the identified peptides and one phosphopeptide K^{448}–K^{461} (E^{449}–K^{462}).[116] In Table 9.29 the results of the identification and quantification of sulfur in proteins using MALDI-FTICR-MS are summarized. In order to study the binding of Cu and Zn in metalloproteins, tracer experiments using isotopically enriched spikes were performed and studied by LA-ICP-MS and MALDI-FTICR-MS by J. Susanne Becker *et al.*[134,135] Tau protein isoforms as target proteins in Alzheimer's disease were studied after separation on 1D gels via tracer experiments using enriched ^{65}Cu and ^{67}Zn isotope tracers. In several protein bands metal ions were detected and $^{65}Cu/^{63}Cu$ and $^{67}Zn/^{64}Zn$ isotope ratios were measured by LA-ICP-MS. The isotope ratio measurements by LA-ICP-MS indicate certain protein isoforms with a changed isotope ratio in comparison to the isotopic composition in nature. This experimental finding demonstrates the formation of new metal containing tau protein complexes. For certain proteins no change in the isotope ratio of Cu and Zn was detected. An identification of protein bands separated after 1D gel electrophoresis was possible by using biopolymer mass spectrometry with MALDI-FTICR-MS.[135]

Due to the lack of suitable standard reference material for quantification purposes in phosphoproteomics and metallomics, reliable calibration strategies were developed for the direct microlocal analysis of phosphorus and metals in protein spots and in thin sections of brain tissue using LA-ICP-MS.[16,17,116] For quantification of analytical data, the application of a solution based calibration strategy was proposed with LA-ICP-MS[6] and the simultaneous determination of P, S, Si, Al, Cu and Zn concentrations in human brain proteins (Alzheimer's disease) or for imaging thin

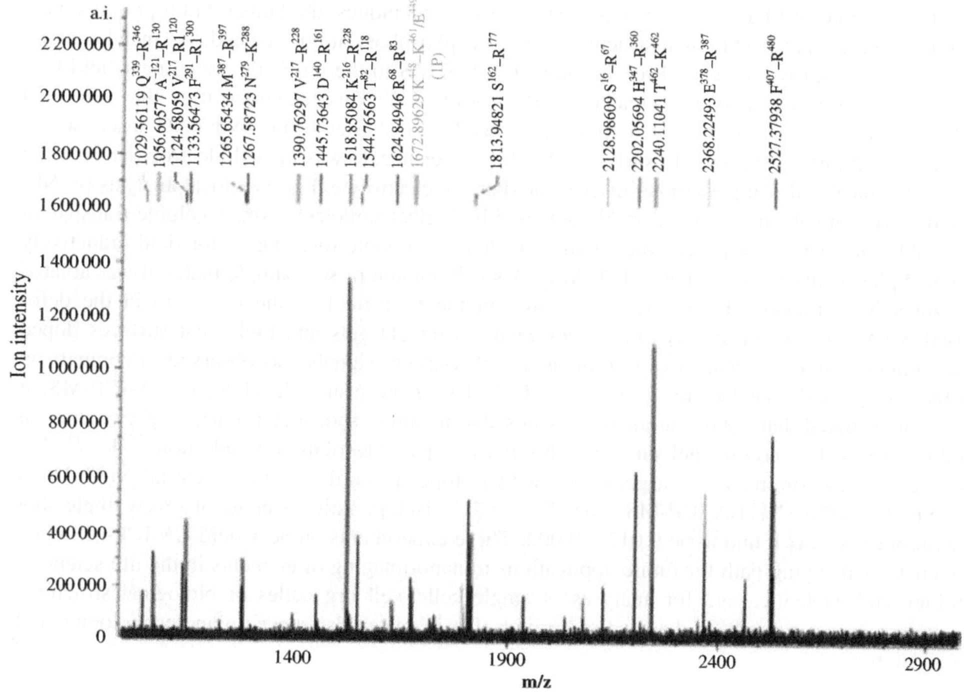

Figure 9.37 MALDI-FTICR mass spectrum of protein spot 3n (α-internexin) with the identified peptides and phosphopeptide K^{448}–K^{461} (E^{449}–K^{462}). (J. S. Becker, M. Zoriy, J. Su. Becker, C. Pickhardt, E. Damoc, G. Juhacz, M. Palkovits and M. Przybylski, Anal. Chem., 77, 5851 (2005). Reproduced by permission of American Chemical Society.)

Table 9.29 Identification and quantification of sulfur in proteins using MALDI-FTICR-MS.

Spot	Protein	M_r, theor (kDa)	M_r, exp (kDa)	Cysteine residues	S, mg g^{-1}
2g	Tubulin α-4 chain	49.9	~50	24	12.6
2l	Tubulin β-4 chain	50.4	~50	27	16.8
3f	Vacuolar ATP synthase catalytic subunit A, ubiquitous isoform	68.3	~69	29	17.2
3n	α-Internexin	55.4	~55	6	3.2
4m	GFAP protein	49.7	~50	12	6.2
4n	Creatine kinase, β chain	42.6	~43	16	7.4
4p	ACTB protein	40.2	~40	22	11.9
4q	ACTB protein	40.2	~40	22	12.4
4o	Creatine kinase, β chain	42.6	~43	16	7.0
5a	Vacuolar ATP synthase subunit B, brain isoform	56.5	~57	22	12.1
7f	Chain B, Human Heart L-Lactate Dehydrogenase H Chain	36.5	~37	14	75.3
12f	Peroxiredoxin 2 isoform b	16.0	~16	2	1.2

sections of brain tissues.[17] In a special arrangement for solution based calibration, a micronebulizer was inserted directly into the laser ablation chamber for quantification purposes.[136] From the results of atomic and biomolecular mass spectrometric techniques, the human brain proteins were characterized with respect to their structure, sequence, phosphorylation state and also metal content.

In order to improve the lateral resolution of LA-ICP-MS to the nm scale range, near field LA-ICP-MS (NF-LA-ICP-MS) was created by Becker *et al.*[137] This technique uses the near field enhancement effect at the tip of a thin silver needle in a laser beam (Nd:YAG laser, wavelength = 532 nm) on the sample surface. The thin silver needle was etched electrolytically in an electrochemical cell using a droplet of citric acid as the electrolyte. For nanolocal analysis by NF-LA-ICP-MS on soft matter (e.g., on 2D gels and biological samples) a small volume transparent laser ablation chamber was constructed and coupled to a double-focusing sector field inductively coupled plasma mass spectrometer (ICP-MS). A small amount of soft sample material was ablated at atmospheric pressure by a single laser shot in the near field of the silver tip in the defocused Nd:YAG laser beam. By single shot analysis on 2D gels and biological surfaces doped with uranium, an enhancement of ion intensities of transient signals was observed in comparison to the background signal of up to a factor of 60. Using the near field effect in LA-ICP-MS, it was demonstrated that nanolocal analysis is possible in single shot measurements of elements on biological samples and on a gel surface with nm scale spatial resolution. In addition, $^{235}U^+/^{238}U^+$ isotope ratio measurements on a gel doped with isotope standard reference material NIST U020 were performed by NF-LA-ICP-MS. The $^{235}U^+/^{238}U^+$ isotope ratio (average of seven single shot measurements) was found to be 0.017 ± 0.002. These experiments on near field LA-ICP-MS open up a new challenging path for future applications in nano-imaging of elements in the life sciences, biology and medicine, e.g., for analyses of single cells, cell organelles or biological structures in the nm range in order to detect disease, but also in materials science, nanotechnologies and nanoelectronics.

9.5.6 Analysis of Tissues

Quantitative determination of essential elements (e.g., P, Cu, Fe, Zn, Mn, Co, Se and others) in brain or medical tissues is relevant in brain research and life sciences, e.g., for studying many neurodegenerative diseases. The deficit or the surplus of these essential elements in proteins and in human tissue has been observed in neurodegenerative diseases (including Alzheimer's and Parkinson's diseases), but metals can also catalyze cytotoxic reactions or are toxic at high concentrations.[119,120] Jackson *et al.*[138] reported on the application of laser ablation quadrupole ICP-MS for the elemental mapping of 100 μm thick sections of rat brain. The laser spot size used was 60 μm, and the laser scan speed was 120 $\mu m\, s^{-1}$. The analysis was relatively rapid, allowing mapping of a whole brain thin section ($\approx 1\, cm^2$) in about 2 h. Furthermore, the method was amenable to multi-element data collection including the physiologically important elements P and S and afforded sub-$\mu g\, g^{-1}$ detection limits for the important trace elements Cu and Zn. Calibrations were performed with pressed pellets of biological certified reference materials, and the elemental distributions and concentrations of Cu, Zn, and Fe were determined in whole rat brain sections. The distributions and concentration ranges for these elements were consistent with previous studies and demonstrate the utility of this technique for rapid mapping of brain thin sections.[138] The multi-element determination of major and trace elements by ICP-MS and ICP-OES in the liver of Wistar rats 1 week to 12 months old is discussed by Takahashi *et al.*[139] Age-related changes in the concentrations of Fe, Mn, Mo, Co and Cd (increasing concentration with increasing age) were found, whereas with increasing age a decreasing concentration was detected for Cu and Zn.[139] The analysis of dissected tissue by LA-ICP-MS of the band water snake was demonstrated by Jackson *et al.*[140]

Table 9.30 *Selected applications for investigations of human tissues and body fluids by ICP-MS and LA-ICP-MS.*

Samples	Equipment	Elements	Limit of detection (LOD)	Results	Ref.
Thyroid tissues of persons living in the Chernobyl area (90 samples)	ICP-QMS (Elan 6000) (after digestion of 1–10 mg sample)	25 trace elements	LOD: As, Bi, Ba, Cd, Ce, Co, Hg, Mn, Mo, Pb, Rb, V – low $ng\,g^{-1}$; Sb, U $< 1\,ng\,g^{-1}$	Higher Ca, Fe, Mn, V, As, Cr, Ni, Pb, U, Ba content in the thyroid tissue affected by pathology	Boulyga *et al.*[101]
Liver, lung, kidney, brain, bone (Autopsy tissue collected from 78 subjects, Tarragona)	ICP-QMS (Elan 6000) (after digestion)	As, Cd, Co, Cr, Cu, Pb, Mn, Hg, Ni, Sn, V, Zn	Recovery rates: 90–95 %	Cu 1.5 (bone) – 6 (liver) $\mu g\,g^{-1}$ Zn* 18 (lung) – 58 (liver) $\mu g\,g^{-1}$ Pb 0.2 (lung) – 1.8 (bone) $\mu g\,g^{-1}$ Mn* 0.2 (bone) – 1.4 (liver) $\mu g\,g^{-1}$ Sn 0.2 (liver) – 0.5 (bone) $\mu g\,g^{-1}$	García *et al.*[143]
Brain, lung, spleen, kidney, heart, liver (aborted fetuses from 21 mothers)	ICP-MS (after digestion)	26 major and trace elements	LOD: Li, Co, Ga, Sr, Mo, Cd, Sn, Sb, Tl, Bi, U $> 5\,ng\,g^{-1}$ RSD: < 10 % (SRM Bovine Liver)	Cu 4 (lung) – 235 (liver) $\mu g\,g^{-1}$ Zn 75 (brain) – 980 (liver) $\mu g\,g^{-1}$ Pb 0.1 (lung) – 0.3 (spleen) $\mu g\,g^{-1}$ Mn 1 (lung) – 7 (liver) $\mu g\,g^{-1}$ Ca 200 (spleen) – 770 (brain) $\mu g\,g^{-1}$	Gélinas *et al.*[144]
Brain healthy and disease Alzheimer's disease (A); 'normal' (N)	ICP-MS (Elan 6000) (after digestion)	Cd, Zn	LOD: Cd – 0.006 $ng\,g^{-1}$ Zn – 10.7 $ng\,g^{-1}$	Cd: (N) 5–145 $ng\,g^{-1}$, (A) 4–104 $ng\,g^{-1}$ Zn: (N) 6–29 $\mu g\,g^{-1}$, (A) 7–19 $\mu g\,g^{-1}$	Panayi *et al.*[145]

Table 9.30 (*Continued*).

Samples	Equipment	Elements	Limit of detection (LOD)	Results	Ref.
Rat brain (18 brain regions of 36 normal Wistar)	ICP-QMS (SPQ-9000) (after digestion of 2–15 mg sample)	Mn, Cu, Zn, Fe	LOD: Mn – 8, Fe – 70, Cu – 27, Zn – 30 $pg\,g^{-1}$ in aqueous solution	Mn, Fe, Cu, Zn concentration increased region-specifically with age. Higher concentrations of Mn, Fe, Zn were observed in the substantia nigra compared with those in neighbouring regions.	Tarohda *et al.*[147]
Rat brain tissue (small size tumor tissue)	LA-ICP-MS (Element)	Cu, Zn, P, S	LOD: 0.8 $\mu g\,g^{-1}$ for Zn, Cu	Inhomogeneous P, S, Cu, Zn distribution, deficits in tumor region	Becker *et al.*[16]
Wistar rats liver (1 week to 12 months old)	ICP-QMS (HP4500) (after digestion of 0.2 mg sample in HNO_3 – HF mixture)	Multi-element analysis	Mo, Cd, Cs – 0.001–0.008 $ng\,ml^{-1}$; Al, Mn, Co, Cu, Zn, Rb, Sr, Pb – 0.01–0.07 $ng\,ml^{-1}$; Ca – 23 $ng\,ml^{-1}$, Na, Mg, P, K, Fe – 230–1150 $ng\,ml^{-1}$ in aqueous solutions	Remaining essentially constant throughout the ages (K, P, Na, Mg, Ca, Rb, Sr, Cs, Pb); increasing with age (Fe, Mn, Mo, Co, Cd); decreasing with age, especially in the early stages of growth, as observed for Cu and Zn	Takahashi *et al.*[139]
Sheep liver	LA-ICP-ToF-MS (Renaissance, Leco)	Cu, Zn	Homogeneous CRM sample: RSD: Cu 2–3 %; Zn 2–5 %	2D element images of thin sections revealed the zonation of Cu in sheep liver with low Cu concentration.	Kindness *et al.*[130]
Bird eggs, faeces, blood, liver, kidney	LA-ICP-QMS (Elan 6000)	Pt, Pd, Rh (PGE)	LOD: Pt $< 0.2\,ng\,g^{-1}$ (blood, eggs) Pd $< 0.5\,ng\,g^{-1}$ (blood) in digest	PGE concentrations were higher in blood compared to both faeces and eggs	Ek *et al.*[142]

Lung, liver, kidney	LA-ICP-SFMS (Element) IDA	Pt	LOD: 20 $pg\,g^{-1}$ for lung and liver tissues and 34 $pg\,g^{-1}$ for kidney tissue in aqueous solution	Pt levels in investigated tissues from five individuals ranged between 0.03–1.42 $ng\,g^{-1}$	Rudolph *et al.*[146]
Brain (hippocampus)	LA-ICP-SFMS (Element)	P, S, Fe, Cu, Zn, Th, U	LOD: 10 $ng\,g^{-1}$ for Th and U	Inhomogeneous distribution of P, S, Cu and Zn in thin section of the hippocampus: Zn: 5 $\mu g\,g^{-1}$–10 $\mu g\,g^{-1}$, Cu < 14 $\mu g\,g^{-1}$	Becker *et al.*[17]
Tooth	LA-ICP-QMS (Elan 6000)	Cu, Fe, Mg, Sr, Pb, Zn		Heterogeneous distribution of Pb, Zn, Sr, Fe in a deciduous tooth: study of anthropogenic trace metal exposure.	Kang *et al.*[95]

PGE - platinum group elements: IDA - isotope dilution analysis; RSD - relative standard deviation.

Brain accumulation of depleted uranium (DU) in rats following treatment with implanted uranium pellets was studied by Fitsanakis *et al.*[70] The authors examined regional brain DU accumulation following surgical implantation of DU metal pellets in male rats for three or six months compared to control rats. Six regions of brain (cerebellum, brainstem, midbrain, hippocampus, striatum and cortex were analyzed after digestion by ICP-SFMS. It was found that DU implanted in peripheral tissues can preferentially accumulate in specific brain regions. After three months postimplantation, DU significantly accumulated in all brain regions except the hippocampus. By six months a significant accumulation was measured only in cortex, midbrain and cerebellum.[70] On multi-element analysis of brain tissues from healthy Wistar rats by ICP-SFMS was reported by Paul *et al.*[141]

Selected applications for the investigation of tissues by ICP-MS and LA-ICP-MS are summarized in Table 9.30.[95,130,142–147]

9.5.7 Imaging Mass Spectrometry of Medical Tissues

In recent years, there has been a growing interest in studying the elemental and biomolecular distribution in biological and especially in medical tissues. Several imaging mass spectrometric techniques such as SIMS, LA-ICP-MS or MALDI-MS enable measurements of element distribution in thin sections of tissues for biomedical research. A breakthrough in imaging mass spectrometry was possible with MALDI-MS[148] with its ability to analyze large biomolecules of thousands of daltons. Imaging mass spectrometry of organic molecules was introduced by Caprioli *et al.*[149,150] using MALDI-MS. A critical evaluation of imaging mass spectrometry is presented by Heeren *et al.*[151] The main drawback of SIMS is the huge matrix effects and high polyatomic ion formation rate, which makes the quantification of analytical data extremely difficult even in imaging mass spectrometry. To improve the spatial resolution and to enhance the yield of secondary ions, cluster ion guns (gold, bismuth and C_{60}^+) as primary ion sources have been applied for imaging SIMS to an increasing extent.[129] For example, the spatial distribution of various specific lipids (including cholesterol, sulfatides, phosphatidychinolins and others) in freeze dried mouse brain sections have been analyzed using Au_3^+ primary ions in negative ToF-SIMS mode (ToF-SIMS IV instrument from ION-TOF GmbH).[152] Compared to LA-ICP-MS, which allows the quantitative imaging of elements in tissues, with SIMS, biomolecules were detected over a wide mass range (up to $m/z \sim 900$). ToF-SIMS together with liquid metal ion guns (cluster ion sources were gold and bismuth and, in addition, fullerenes) to produce elemental and molecular ion images of biological tissues (e.g., rat brain sections) is described by Brunelle *et al.*[153] The authors found that primary bismuth cluster beams (Bi_n^{q+} with $n = 1$–7 and $q = 1, 2$) have much better intensities and efficiencies and improved lateral and mass resolution in comparison to the Au_3^+ cluster ions used in imaging mass spectrometry.[154] An additional possibility of employing imaging mass spectrometry is matrix enhanced SIMS using a ToF-SIMS instrument which has been applied for the sub-cellular imaging of brain tissues.[155] Molecular ion images of phospholipids and peptides in mouse brain slices have also been produced by the bombardment of thin tissues with gold cluster ion beams in ToF-SIMS.[129] The approach using heavy cluster ion beams offers new opportunities, not only in medical research, but also in pharmacological and biological research fields by localizing the compounds of interest, such as drugs or metabolites in tissues.[129] SIMS as an imaging mass spectrometric technique was also employed for nuclear medical applications during radiotracer experiments using ^{99}Tc, ^{89}Sr and ^{76}Br radionuclides. Knowledge of the spatial distribution (with spatial resolution down to 0.5 µm) of these radionuclides in tissues is essential for dosimetric studies in metabolic radiotherapy.[156]

At present the application of LA-ICP-MS in biological research and in medicine focuses on individual tasks, e.g., the determination of element distribution in tree barks,[157] in green leaves,[158] in fish tissues[159] or mapping copper and zinc in liver sections from sheep.[130] An analytical technique using LA-ICP-MS for the quantitative determination of 2D element distribution (mapping) of Cu and Zn in sections of tissue (sheep liver) is described by Feldmann's group.[130] A homogeneous certified reference material (CRM LGC 7112, pig liver) was applied to calibrate the analytical procedure.[130] The authors used a commercial cryogenically cooled laser ablation ICP-MS for the direct analysis of tissues.[160]

The reason for the relatively limited application of LA-ICP-MS in medical and biological research (compared to the geological sciences) can be explained by difficulties in the laser ablation of biological matrices (often with high water content), by the lack of suitable analytical procedures for quantitatively determining element concentration and the relatively high price of modern and powerful laser ablation systems with a good lateral resolution power. At the small laser focus (low μm range) often required for the determination of element distribution in tissues, older commercial laser ablation systems have a low laser power density, mostly $< 10^8\,\mathrm{W\,cm^{-2}}$, and therefore not enough material is ablated for a sensitive element analysis and fractionation effects have also been observed. In order to avoid fractionation effects in laser microlocal analysis at lateral resolution in the low μm range, a laser power density $> 10^9\,\mathrm{W\,cm^{-2}}$ should be applied.

A cooled laser ablation chamber (using two Peltier elements behind the target holder made of aluminium) for the LA-ICP-MS analysis of thin brain sections was developed in author's laboratory in 2004 to analyze medical and biological tissues.[161] The experimental arrangement (see Fig 5.22) resulted in a significant improvement in the stability of the ion currents, thus leading to better precision and higher accuracy of the analytical data as demonstrated for uranium isotope ratio measurements on a droplet of isotope standard solutions from NIST on the surface of biological tissue in previous work.[161]

LA-ICP-MS enables images to be produced of the distribution of essential elements such as Zn, Cu, Fe, S, P, Se and Mn as well as of toxic and also radioactive metals (e.g., Hg, Pb, Cd, Th and U) in thin tissue sections with a spatial resolution in the μm range. This spatial resolution of element distribution analysis is sufficient to distinguish between several layered structures in human brain tissue from the hippocampus, as described by Zilles *et al.*[162]

By two dimensional (2D) imaging of brain tissues, a defined sample area (several $\mathrm{cm^2}$) of a thin section of brain tissue (thickness: 20 μm) was ablated line by line with a focused laser beam in a cooled laser ablation chamber (using two Peltier elements behind the target holder made of aluminium).[161] The spot size of the laser beams was about 50 μm, the laser power density $3 \times 10^9\,\mathrm{W\,cm^{-2}}$. Ion intensities of the analytes, e.g., $^{31}P^+$, $^{32}S^+$, $^{63}Cu^+$ or $^{64}Zn^+$, were measured by LA-ICP-MS within the area of interest in different regions of the human brain tissue (e.g., human hippocampus). Matrix matched homogenized laboratory standards with well defined element concentrations of analytes were prepared and utilized for quantification of LA-ICP-MS data in imaging brain tissues in the routine mode.[16,17,163] Furthermore, online solution based calibration was applied with LA-ICP-MS. The experimental arrangement of the laser ablation chamber with the inserted micronebulizer is shown in Figure 6.23. The capability of online solution based calibration has been demonstrated for the uranium concentration in the prepared synthetic matrix matched laboratory brain standard. The uranium concentration of $100\,\mathrm{ng\,g^{-1}}$ was determined during laser ablation by standard addition mode via solution based calibration at $97.8 \pm 2.5\,\mathrm{ng\,g^{-1}}$ in LA-ICP-MS. For this new microanalytical technique using a direct method of solution based calibration in LA-ICP-MS with cooled laser ablation chamber, the precision and accuracy was measured as 2–3 %. As the result of LA-ICP-MS studies, layered structures for P, S, Cu, Zn and Pb in different section of brain tissues were detected in the $\mathrm{\mu g\,g^{-1}}$ concentration range. The Zn and Cu distribution in human hippocampus is presented in Figure 9.38. The maximal zinc concentration (10 μg g^{-1}) is restricted to a small region of the hippocampus (in the hilus region and lucidum layer). Copper was found in higher overall concentrations (maximum: 14 μg g^{-1}) in the hippocampus. Furthermore, it was not only present in discrete concentrations in the hilus, but also in higher concentrations in the pyramidal and lacunosum molecular layers of the cornu ammonis. In contrast, Th and U in the low ng g^{-1} range were found in brain proteins and also tissues (with a more homogeneous distribution).[17]

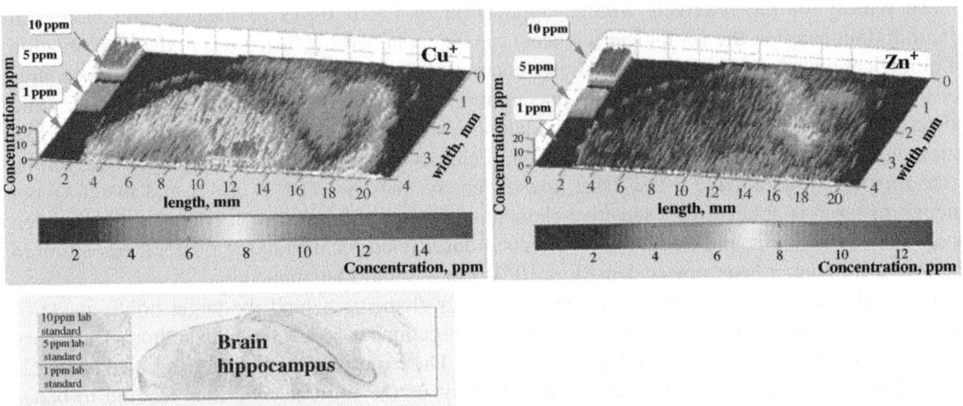

Figure 9.38 *Concentration profile of copper and zinc in the scanned area of interest measured by LA-ICP-MS on brain samples (hippocampus). Calibration was performed via synthetic matrix matched laboratory standards for 1, 5 and 10 μg g^{-1} of analyte (see inserted figures on left). Bottom: histologically processed brain tissue in which cell bodies were stained (cresyl violet staining) in order to demonstrate the layered structure of the analyzed region. (J. S. Becker, M. Zoriy, C. Pickhardt, N. Palomero-Gallagher and K. Zilles, Anal. Chem., **77**, 3208 (2005). Reproduced by permission of American Chemical Society.)*

In addition, the quantitative Zn, Cu and Pb distribution in several sub-regions of the human brain hemisphere were determined by LA-ICP-SFMS.[163] The distribution of Zn and Cu in the insular region of the human brain measured by LA-ICP-MS is illustrated in Figure 9.39. The layered distribution pattern of both elements is clearly visible. In additional studies, copper and zinc distribution within various sub-regions of human brain were quantitatively measured for the first time.[163] A fine band inside the dentate gyrus corresponding to its granular layer contains the highest Cu concentrations. The Cu concentration is high along the pyramidal layer of the cornu ammonis. In the hilus region, only that part of the pyramidal layer in the convexity closest to the dentate gyrus displays Cu accumulation. Zn is most highly concentrated, for example, in the dentate gyrus.

Using a large laser ablation chamber, commercially available from CETAC Technologies, we quantitatively analyzed the Cu distribution in a 20 μm thin section of the whole human hemisphere (110 mm × 65 mm). The quantitative Cu distribution of part of the human hemisphere is presented in Figure 9.40. A layered structure of cortex was found with higher metal concentration in the grey matter compared to the white matter.

Platinum complexes are actually a major constituent in the treatment of testicular, ovarian, bladder, colon and non-small cellular lung cancer. The anticancer activity of cisplatin (cis-diammine-dichloro-platinum (II)) was discovered in 1969.[164] Two additional compounds,

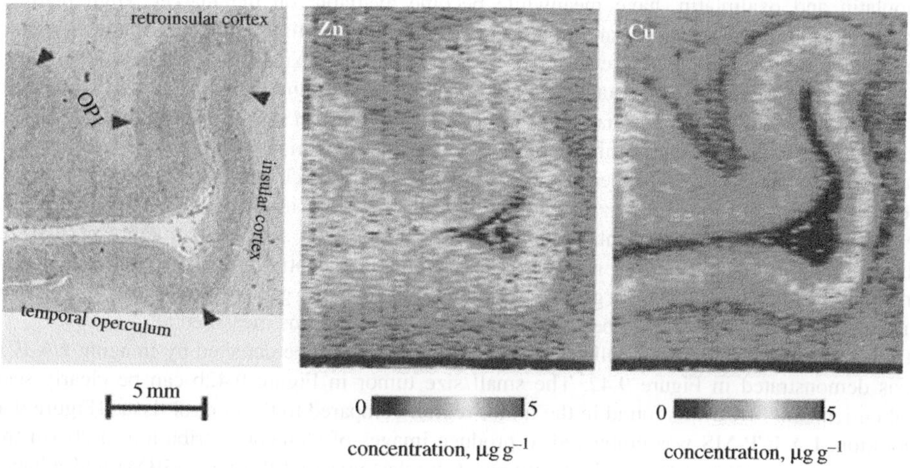

Figure 9.39 *Distribution of Zn and Cu in insular region of human brain measured by LA-ICP-MS compared to cross section of this tissue region stained by cresyl violet.*

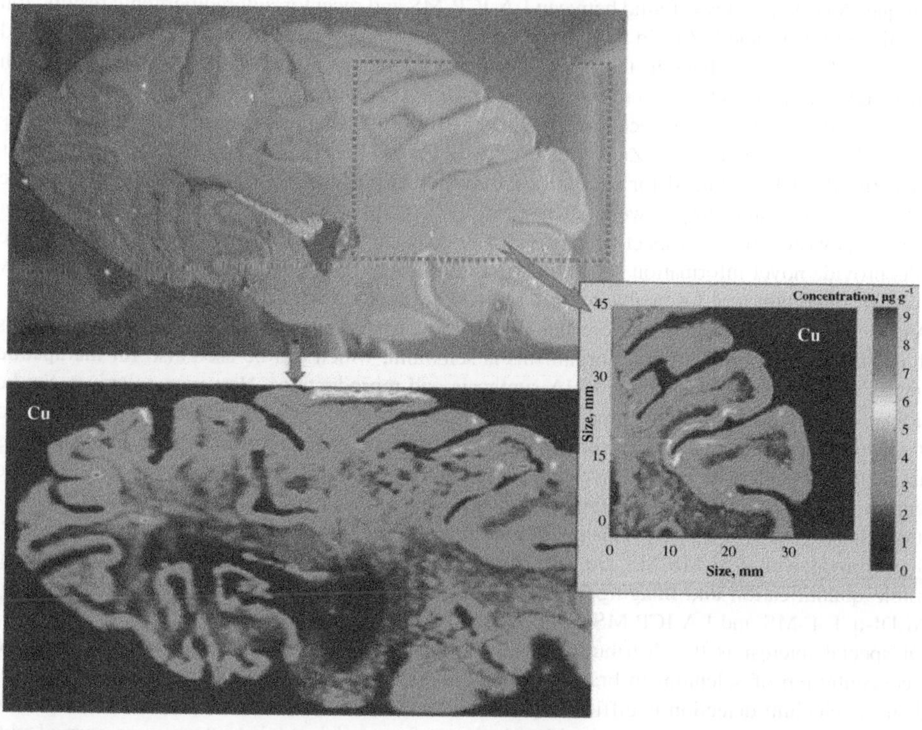

Figure 9.40 *Cu image of human brain hemisphere including the quantitative Cu distribution of part of hemisphere measured by LA-ICP-MS compared to cross section stained by cresyl violet.*

carboplatin and oxaliplatin, have meanwhile become available on the market. High platinum concentrations in the target tissue and low concentrations in dose limiting tissue structures such as renal tubules are desirable to assure selective toxicity. Microlocal analysis of platinum distribution in tissue sections may thus contribute to the optimization of platinum therapy. Scanning laser ablation inductively coupled plasma mass spectrometry (LA-ICP-SFMS) was used to produce images of element distribution in 14 μm thin sections of kidney tissue from a mouse treated with cisplatin 60 min prior to slaughter. The experimental details are described elsewhere.[165] Figure 9.41 shows the quantitative images of copper, zinc and platinum distributions in a cross section of renal tissue determined by LA-ICP-MS. An inhomogeneous distribution with a layered structure of analyzed tissue was found for all elements measured. The average concentrations of Cu, Zn and Pt were found to be in the ranges 3–7 $\mu g\,g^{-1}$, 2–6 $\mu g\,g^{-1}$ and 10–25 $\mu g\,g^{-1}$, respectively. The element distributions correlated very well between the two adjacent sections measured.[165]

It is interesting to note that small size tumors in rat brain can be detected by imaging LA-ICP-MS as demonstrated in Figure 9.42. The small size tumor in Figure 9.42b can be clearly seen. Local enrichment of Cu was found in the tumor region compared to the control tissue (Figure 9.a). In addition, LA-ICP-MS was employed to produce images of element distribution in 20 μm thin tissue sections of primary human brain tumors (glioblastoma multiforme – GBM) and adjacent non-neoplastic brain tissue. The concentration and distribution of selected elements (Cu, Zn, Pb and U) are compared in control tissues and regions affected by GBM.[166] Labelling for A1AR and the pBR (peripheral benzodiazepine receptor) was employed as the receptor-autoradiographic technique. A correlation was found between LA-ICP-MS and receptor-autoradiographic results. Ion intensities of $^{63}Cu^+$ and $^{64}Zn^+$ in four line scans of sections through a human GBM are illustrated in Figure 9.43. The red lines in the inset indicate the position in the scanned tissue from which copper and zinc distributions were obtained. Figure 9.44 summarizes the 2D distribution of (a) the peripheral benzodiazepine receptor (pBR) and (b) the A1 adenosine receptor ligand (A_1A) as well as LA-ICP-MS imaging of Zn, Cu, Pb and U in the glioblastoma sample GBM-III. The limits of detection (LODs) obtained for Cu and Zn were 340 $ng\,g^{-1}$ and 140 $ng\,g^{-1}$, respectively, while LODs of 12 $ng\,g^{-1}$ and 7 $ng\,g^{-1}$ were determined for Pb and U, respectively.[166] High density tumor and tumor invasion zones, respectively, are clearly detected by LA-ICP-MS measurements. These results provide novel information on the distribution of elements in human brain tumors, and may also be an important tool for analysing brain tumor metabolism and pathogenesis.

As discussed in the chapters before one of the most frequently investigated essential elements and as a crucial nutrient for higher organisms is selenium, which can be used also for therapeutic purposes (e.g., antioxidant defence, DNA synthesis and reproduction). However, selenium can be toxic at concentration levels higher than those required for health. Several studies have demonstrated that selenium supplementation can reduce the risk of cancer and other diseases,[167] with selenized yeast being the most extensively investigated selenium containing supplement. For example, the selenium speciation analysis of selenium enriched supplements (in extracts of Se-yeast and Se-methylselenocysteine) by size exclusion, anion exchange and reversed phase ion pair HPLC-ICP-MS combined with ESI-MS-MS are compared in reference[168]. The characterization, including possible quantification and imaging strategies, of selenodrug metabolites by the combination of MALDI-q-ToF-MS and LA-ICP-MS has been reported by McLeod's working group.[169]

Of special interest is the distribution analysis of selenium in certain brain regions. Because the concentration of selenium in brain tissues is relatively low (mostly in the sub-ng g^{-1} range) and since selenium detection is difficult by LA-ICP-MS due to interference problems by isobaric polyatomic ions, tracer experiments with selenium of natural isotopic pattern were carried out. For example, snails were fed 1000 $\mu g\,ml^{-1}$ selenium solution for 60 h. Cross sections 100 μg

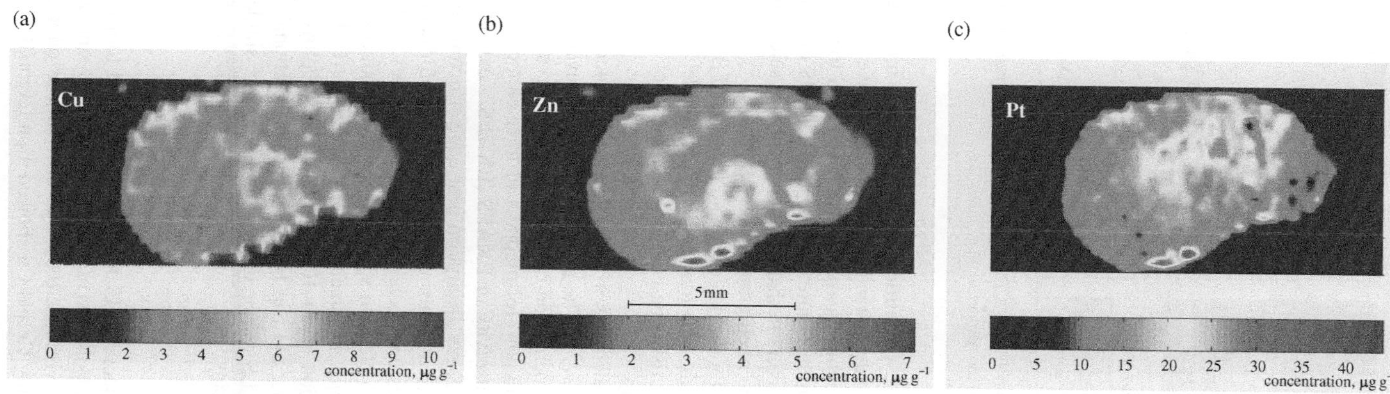

Figure 9.41 *Lateral distributions of (a) Cu, (b) Zn and (c) Pt on thin cross section of mouse kidney measured by LA-ICP-MS. (M. Zoriy, A. Matusch, T. Spruss and J. S. Becker, Int. J. Mass Spectrom.,* **257***, 102 (2006). Reproduced by permission of Elsevier.)*

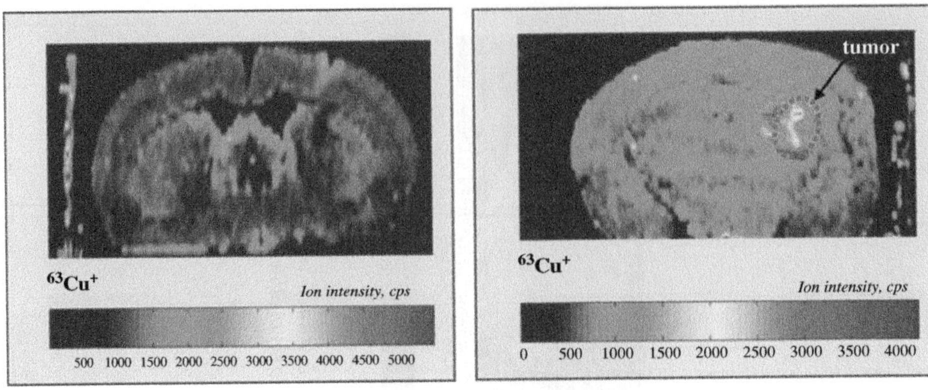

Figure 9.42 *Cu distribution measured by LA-ICP-MS in two cross sections of two rat brains tissues with (a) control tissue compared to (b) tumor region.*

thick were prepared by cryocutting. Quantification of measured ion intensities was performed using prepared matrix matched laboratory standards. The quantitative selenium image on the cross section of a snail labelled with selenium measured by LA-ICP-MS together with a photograph of a cross section of this snail is shown in Figure 9.45 By LA-ICP-MS measurements of selenium in snail tissue, a similar shape and structure to that of layers at higher selenium concentration, e.g., in the skin, was found compared to the photograph.

LA-ICP-MS has been developed for the detection and imaging of β-amyloid protein in immunohistochemical sections from the brain of a transgenic mouse model of Alzheimer's disease. The distribution of β-amyloid deposits in tissue based on measurements of Eu and Ni coupled antibodies has been studied by McLeods group.[170]

Imaging and speciation of trace elements in biological environment, including medical tissues, has been reviewed by Lobinski *et al.*[171] The authors compared several techniques, such as LA-ICP-MS, micro-SXRF (SXRF – synchrotron-radiation X-ray fluorescence analysis), micro-PIXE (PIXE – particle induced X-ray emission) for mapping of trace element distribution in 1D and 2D proteomics gels. The increasing sensitivity of EXAFS (EXAFS – extended X-ray absorption fine structure) and XANES (XANES – X-ray absorption near edge structure) owing to the use of more intense synchrotron beams and efficient focusing optics provides information about oxidation state, fingerprint speciation of metal sites and metal–metal site structures.[171] In addition, transmission electron microscopes fitted with an electron spectrometer that produces an energy loss spectrum for electrons passing through a thin specimen to perform EELS (EELS - electron energy loss spectrometry) with spatial resolution at 50–100 nm range, are employed to study the elemental distribution and the oxidation state and can be employed as additional analytical techniques in life sciences.[171,172] Imaging by EELS can be carried out to analyze single atoms of Ca and Fe in biological structures, such as proteins.[173]

9.5.8 Single Cell Analysis

Of the mass spectrometric techniques, SIMS has become a useful tool for the imaging of single cells.[174] Human liver tumor cells poisoned with Cd, Cu, Cr, Hg and Zn were analyzed using imaging ToF-SIMS (ION-TOF) to visualize the metal distributions on a single cell basis. Among the five metal ions investigated, only Cr and Cu showed a preferential diffusion into the cell after

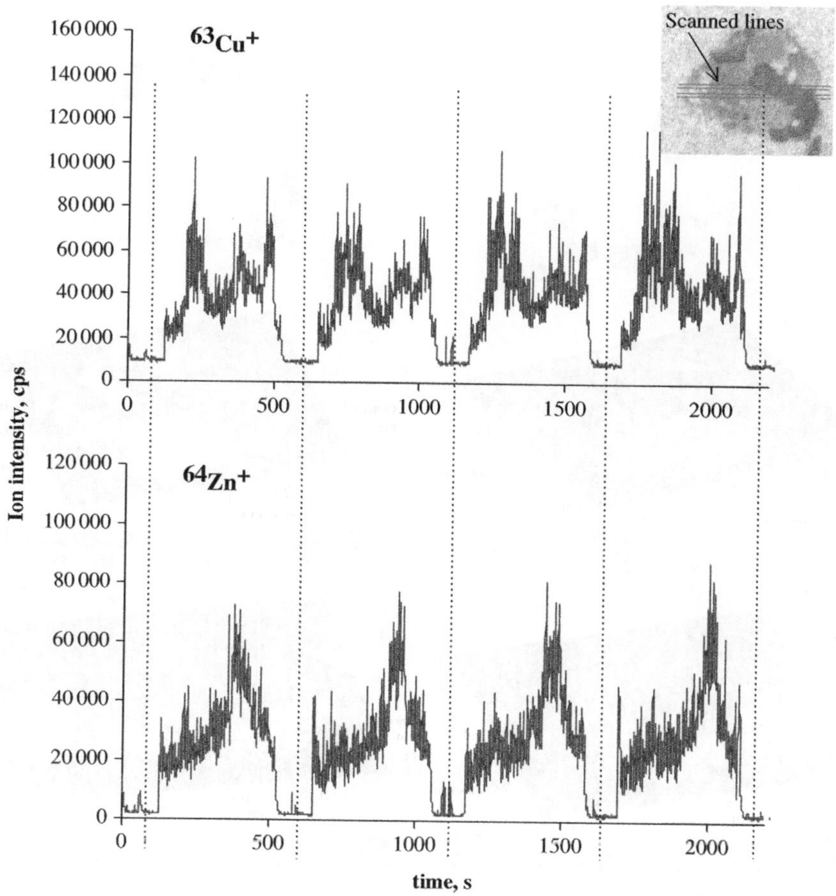

Figure 9.43 *Ion intensities of $^{63}Cu^+$ and $^{64}Zn^+$ in four line scans of sections through a human GBM sample (GBM-III). The red lines in the inset indicate the position in the scanned tissue from which copper and zinc distributions were obtained. (M. Zoriy, M. Dehnhardt, G. Reifenberger, K. Zilles and J. S. Becker, Int. J. Mass Spectrom.,* ***257****, 27 (2006). Reproduced by permission of Elsevier.)*

simulated poisoning. The imaging mass spectrometric measurements were performed on HepG2 cells placed on a silicon wafer and cultured in a special medium described in reference 174 using a focused primary ion beam of 25 keV Ga^+ ions over an analysis area of 50–100 μm × 50–100 μm at a spatial resolution of 3–4 μm (at mass resolution $m/\Delta m \sim 2000$–5000). Images with an improved spatial resolution of 0.5 μm were obtained at $m/\Delta m \sim 200$–1000). Selected ToF-SIMS ion images of Na, K, Cr and Cu from a poisoned single liver tumor cell HepG2 cell are shown in Figure 9.46. Whereas in the upper row of Figure 9.46 images of a HepG2 cell immersed in K_2CrO_4 solution are presented, in the lower rows, the single liver tumor cells are immersed in $CuSO_4$ solution, both solutions having a concentration of 15 μM. Relatively low levels of intracellular Na^+ and relatively high levels of intracellular K^+ were found by imaging mass spectrometry, which indicates that the cells were healthy prior to sample pre-treatment. The results indicate that the cells remain intact after the pre-treatment process.

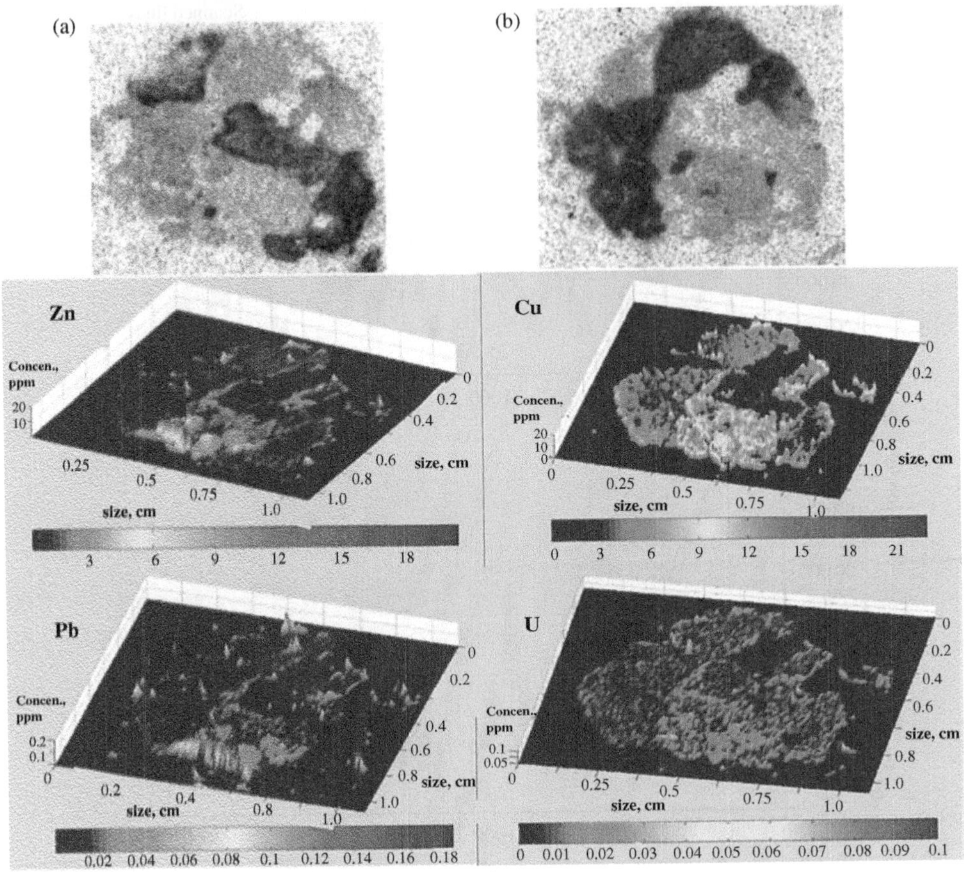

Figure 9.44 *2D distribution of (a) the peripheral benzodiazepine receptor (pBR) and (b) the A1 adenosine receptor ligand (A_1A) as well as LA-ICP-MS imaging of Zn, Cu, Pb and U in the glioblastoma sample GBM-III. (M. Zoriy, M. Dehnhardt, G. Reifenberger, K. Zilles and J. S. Becker, Int. J. Mass Spectrom.,* ***257****, 27 (2006). Reproduced by permission of Elsevier.)*

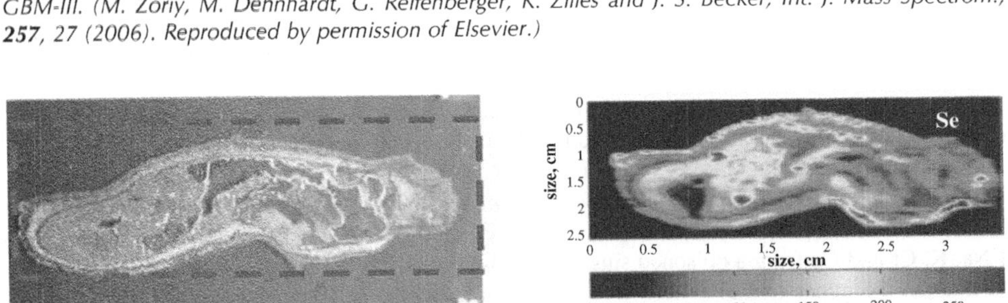

Figure 9.45 *Quantitative selenium image in a cross section of a snail measured by LA-ICP-MS. (J. S. Becker, J. Su. Becker, M. Zoriy, J. Dobrowolska, A. Matusch, Eur. J. Mass Spectrom.,* ***13*** *1 (2007). Reproduced by permission of IM publications.)*

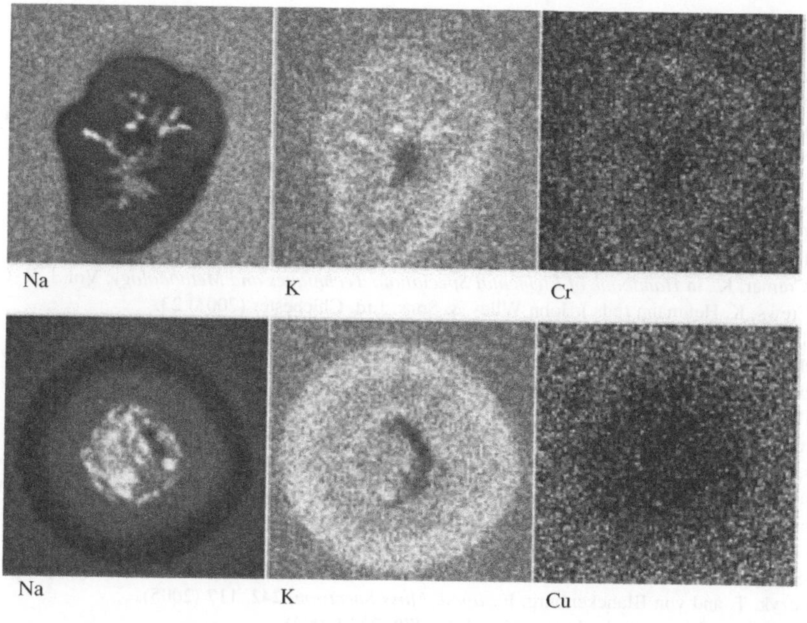

Figure 9.46 *ToF-SIMS images of Na, K and Cu of an poisoned single liver tumor cell – HepG2 cell. In the upper row are images of the HepG2 cell immersed in K_2CrO_4 solution; in the lower ones the cells are in $CuSO_4$ solution – both solutions having a concentration of 15 μM. (F. D. Mai, B. J. Chen, L. C. Wu, F. Y. Li and W. K. Chen, Appl. Surf. Sci.,* ***252****, 6809 (2006). Reproduced by permission of Elsevier.)*

To analyze element distribution, isotope ratios and species in single cells, development of advanced inorganic mass spectrometric techniques in combination with biomolecular mass spectrometry is required for future applications.

9.5.9 Ultrafine Particles and Health

Today the population is becoming increasingly exposed to ultrafine particles (< 20 nm, e.g., Aerosil, TiO_2) in bodycare and household products. Ion microscopy studies revealed that such particles can, for example, penetrate the horny layer of the skin and can result in unexpected interactions. SIMS and Fourier transform laser microprobe mass spectrometry (FT LMMS) have been applied to study TiO_2 stimulated interaction in thin layers of dermatological gels as a result of UV irradiation.[175] For future studies of distribution of ultrafine particles LA-ICP-MS will be employed.

References V

1. Dietze, H. J. and Becker, J. S., *Fresenius Z. Anal. Chem.*, **320**, 490 (1985).
2. Wolstenholme, W. A., *Nature*, **203**, 1284 (1964).
3. Jurachek, J. P., Glemena, G. G. and Harrison, W. W., *Anal. Chem.*, **41**, 1666 (1969).
4. McKernon, M. A. and Harrison, W. W., in *Amer. Chem. Soc. South-Eastern Regional Meeting*. American Chemical Society Richmond, USA (1969).

5. Evans, C. A. and Morrison, G. H., *Anal. Chem.*, **1968**, 869 (1968).
6. Hesse, A., Dietze, H. J., Berg, W. and Hientsch, E., *Eur. Urol.*, **3**, 359 (1977).
7. Williamson, T. G. and Harrison, W. W., *Nat. Bur. Stand. US Spec. Publ.*, **312**, (1969).
8. Fido, A. and Al-Saad, S., *Autism*, **9**, 290 (2005).
9. Klevay, L. M., Christopherson, D. M. and Shuler, T. R., *European J. Clinical Nutrition*, **58**, 1359 (2004).
10. Klys, M., Rojek, S. and Bolechele, F., *J. Chrom. B*, **825**, 38 (2005).
11. Kosta, L., *Talanta*, **29**, 985 (1982).
12. Versieck, J. and Cornelis, R., *Trace Elements in Human Plasma or Serum*. CRC Press, Boca Raton, FL (1989).
13. Rodushkin, I. and Ödman, F., *J. Trace Elem. Medic. Biol.*, **15**, 40 (2001).
14. De Cremer, K., in *Handbook of Elemental Speciation: Techniques and Methodology*, Vol. I, R. Cornelis, H. Crews, K. Heumann (eds.). John Wiley & Sons, Ltd, Chichester (2003) 23.
15. Heiland, P. and Köster, H. D., *J. Trace Elem. Medic. Biol.*, **20**, 253 (2006).
16. Becker, J. S., Zoriy, M., Dehnhardt, M., Pickhardt, C. and Zilles, K., *J. Anal. At. Spectrom.*, **20**, 912 (2005).
17. Becker, J. S., Zoriy, M. V., Pickhardt, C., Palomero-Gallagher, N. and Zilles, K., *Anal. Chem.*, **77**, 5851 (2005).
18. Cornelis, R., Heinzow, B., Herber, R. F. M. *et al.*, *Pure Appl. Chem.*, **67**, 1575 (1995).
19. Das, A. K., Chakraborty, R., Crevera, M. L. and de la Guardia, M., *Mikrochim. Acta*, **122**, 131 (1996).
20. Gomez-Ariza, J. L., Morales, E., Sanchez-Rodas, D. and Giraldez, I., *Trends Anal. Chem.*, **10**, 200 (2000).
21. Quevauviller, P., de la Calle-Guntinas, M. B., Maier, E. A. and Camera, C., *Mikrochim. Acta*, **118**, 131 (1995).
22. MacAvoy, S. E., Arneson, L. S. and Bassett, E., *Oecologia*, **150**, 190 (2006).
23. Walczyk, T. and von Blanckenburg, F., *Int. J. Mass Spectrom*, **242**, 117 (2005).
24. Chen, K. L. and Jiang, S. J., *Anal. Chim. Acta*, **470**, 223 (2002).
25. Frazzoli, C., Cammarone, R. and Caroli, S., *Pure Appl. Chem.*, **78**, 69 (2006).
26. Martino, F. A. R., Sänchez, M. L. and Sanz-Medel, A., *J. Anal. At. Spectrom.*, **15**, 163 (2000).
27. Martino, F. A. R., Sánchez, M. L. F. and Sanz-Medel, A., *Anal. Chim. Acta*, **442**, 191 (2001).
28. Prohaska, T., Köllensperger, G., Krachler, M., De Winne, K., Stingeder, G. and Moens, L., *J. Anal. At. Spectrom.*, **15**, 335 (2000).
29. Palmer, C. D., Lewis, M. E., Geraghty, C. M., Barbosa, F. and Parsons, P. J., *Spectrochim. Acta B*, **61**, 980 (2006).
30. Pruszkowski, E., Neunauer, K. and Thomas, R., *At. Spectrom.*, **19**, 323 (1998).
31. D'Ilio, S., Violante, N., Caimi, N., Di Gregorio, M., Petrucci, F. and Senofonte, O., *Anal. Chim. Acta*, **573–574**, 432 (2006).
32. Krachler, M. and Irgolic, K. J., *J. Trace Elem. Med. Biol.* **13**, 157 (1999).
33. Krachler, M. and Wirnsberger, G. H., *Blood Purification*, **18**, 138 (2000).
34. Riondato, J., Vanhaeke, F., Moens, L. and Dams, R., *J. Anal. At. Spectrom.*, **12**, 933 (1997).
35. Sariego Muniz, C., Marchante-Gayón, J. M., Garcia Alonso, J. I. and Sanz-Medel, A., *J. Anal. At. Spectrom.*, **14**, 193 (1999).
36. Sieniawska, C. E., Mensikov, R. and Delves, H. T., *J. Anal. At. Spectrom.*, **14**, 109 (1999).
37. Parsons, P. J., Zhou, Y., Palmer, C. D., Aldous, K. M. and Brockman, P., *J. Anal. At. Spectrom.*, **18**, 4 (2003).
38. Wang, J. and Zao, M., *Fenxi Huaxue*, **34**, 355 (2006).
39. Sarmiento-Gontález, A., Marchante-Gayòn, J. M., Tejerina-Lobo, J. M., Paz-Jiménez, J. and Sanz-Medel, A., *Anal. Bioanal. Chem.*, **382**, 1001 (2005).
40. Caroli, S., Alimonti, A., Coni, E., Petrucci, F. and Senofonte, O., *Crit. Rev. Anal. Chem.*, **24**, 363 (1994).
41. Begerow, J., Turfeid, M. and Dunemann, L., *J. Anal. At. Spectrom.*, **15**, 347 (2000).
42. Bocca, B., Alimonti, A., Bomboi, G., Giubilei, F. and Fiorte, G., *Trace Elements Electrolytes*, **23**, 270 (2006).
43. Taylor, A., Branch, S., Halls, D., Patriarca, M. and White, M., *J. Anal. At. Spectrom.*, **20**, 323 (2005).
44. Walczyk, T., in *Handbook of Elemental Speciation II – Species in the Environment, Food, Medicine and Occupational Health*, H. C. R. Cornelis, J. Caruso, K. G. Heumann (eds.). John Wiley & Sons, Ltd, Chichester (2005).

45. DeMaeyer, E. and Adiels-Tegma, M., *World Health Stat. Q*, **38**, 302 (1985).
46. Stenberg, A., Malinovsky, D., Rodushkin, I., Andrén, H., Pontér, C., Öhlander, B. and Baxter, D. C., *J. Anal. At. Spectrom.*, **18**, 23 (2003).
47. Palacios, Ò., Enciar, J. R., Bertin, G. and Lobinski, R., *Anal. Bioanal. Chem.*, **383**, 516 (2005).
48. Ward, R. J., Zhang, Y. and Crichton, R. R., *J. Inorg. Biochem.*, **87**, 9 (2001).
49. Zatta, P., Cervellin, D. and Zambenedetti, P., *Toxicol. In Vitro*, **12**, 287 (1998).
50. Milacic, R., in *Handbook of Elemental Speciation II – Species in the Environment, Food, Medicine and Occupational Health*, Vol. II, H. C. R. Cornelis, J. Caruso and K. G. Heumann (eds.). John Wiley & Sons, Ltd, Chichester (2005) 27.
51. Valkonnen, S. and Riihimäki, V., in *Handbook of Elemental Speciation II – Species in the Environment, Food, Medicine and Occupational Health*, Vol. II, H. C. R. Cornelis, J. Caruso and K. G. Heumann (eds.). John Wiley & Sons, Ltd, Chichester (2005) 40.
52. Yokel, R. A. and McNamara, P. L., *Pharmacol. Toxicol.*, **88**, 150 (2001).
53. Quinones, O., Snyder, S. A., Cotruvo, J. A. and Fisher, J. W., *Toxicology*, **221**, 229 (2006).
54. Fitzgerald, R. L., Hillegonds, D. J., Burton, D. W. *et al.*, *Clinical Chemistry*, **51**, 2095 (2005).
55. Schaumlöffel, D., Bierla, K. and Lobinski, R., *J. Anal. At. Spectrom.*, **22**, (2007).
56. Brabec, V. and Nováková, O., *Drug Resistance Updates*, **9**, 111 (2006).
57. Saldivar, L. and Munguia, N., in *Second Asia-Pacific Winter Conference on Plasma Spectrochemistry*. R. M. Barnes (ed.), ICP Information Newsletters Bangkok, Thailand (2006) Poster WP01.
58. Polec-Pawlak, K., Abramski, J. K., Semenova, O., *et al.*, *Electrophoresis*, **27**, 1128 (2006).
59. Lykissa, E. D. and Maharaj, S. V. M., *Anal. Chem.*, **78**, 2925 (2006).
60. Becker, J. S., Burow, M., Boulyga, S. F., Pickhardt, C., Hille, R. and Ostapczuk, P., *Atom. Spectrosc.*, **23**, 177 (2002).
61. Krystek, P., Ritsema, R., *Anal. Bioanal. Chem.*, **374**, 226 (2002).
62. Kurk, D., in *Book of Abstracts of 2004 Winter Conference on Plasma Spectrochemistry, Fort Lauderdale* R. M. Bames (ed.), ICP Information Newsletters (2004).
63. Pappas, R., Ting, G. B. and Pascha, D. C., *J. Anal. At. Spectrom.*, **18**, 1289 (2003).
64. Zoriy, M., Pickhardt, C., Ostapczuk, P., Hille, R. and Becker, J. S., *Int. J. Mass Spectrom.*, (2004).
65. Alves, L. C., Allen, L. A. and Houk, R. S., *Anal. Chem.*, **65**, 2468 (1993).
66. Parsons, P. J., Palmer, C. D., Caldwell, K. L. and Jones, R. L., in *Plasma Source Mass Spectrometry*, D. B. G. Holland (ed.). The Royal Society of Chemistry, Cambridge (2005) 59.
67. Denk, E., Hillegonds, D., Vogel, J. Synal, H.-A., Geppert, C., Wendt, K., Fattinger, K., Hennessy, C., Berglund, M., Hurrell, R. F. and Walczyk, T., *Anal. Bioanal. Chem.*, **386**, 1587 (2006).
68. Hann, S., Köllensperger, G., Stefánka, Z. Stingeder, G., Fürhacker, M., Buchberger, W. and Mader, R. M., *J. Anal. At. Spectrom.*, **18**, 1391 (2003).
69. Huerta, V. D., Szpunar, J., Lobinski, R., Sánchez, F. and Sanz-Medel, A., *J. Anal. At. Spectrom.*, **18**, 1471 (2003).
70. Fitsanakis, V. A., Erikson, K. M., Garcia, S. J., Evje, L., Syversen, T. and Aschner, M., *Biol. Trace Elem. Res.*, **111**, 185 (2006).
71. Becker, J. S., Burow, M., Boulyga, S. F., Pickhardt, C., Hille, R. and Ostapczuk, P., *Atom. Spectrosc.*, **23**, 177 (2002).
72. Becker, J. S., Burow, M., Boulyga, S. F., Pickhardt, C., Hille, R. and Ostapczuk, P., *Atom. Spectrosc.*, **25**, 197 (2004).
73. Horan, P., Dietz, L. and Durakovic, A., *Mil. Med.*, **167**, 620 (2002).
74. Boulyga S. F., Zoriy, M., Ketterer M. E. and Becker J. S., *J. Environ. Monit.*, **5**, 661 (2003).
75. Ketterer, M. E., Hafer, K. M., Link, C. L., Royden, C. S. and Hartsock, W. J., *J. Env. Radioact.*, **67**, 191 (2003).
76. Vonderheide, A. P., Zoriy, M., Izmer, A., Pickhardt, C., Caruso, J., Ostapczuk, P., Hille, R., Becker, J. S., *J. Anal. At. Spectrom.*, **19**, 675 (2004).
77. Pichini, S., Poudevida, S., Pujades, M. Menoyo, E., Pacifici, R., Farre, M. and De La Torre, R., *Therapeutic Drug Monitoring*, **28**, 106 (2006).
78. Srogi, K., *Anal. Lett.*, **39**, 231 (2006).
79. Macko, S. A., Lubec, G., Teschler-Nicola, M., Adrusevich, V. and Engel, M. H., *FASEB J.*, **13**, 559 (1999).

80. Karpas, Z., Lorber, A., Sela, H., Paz-Tal, O., Hagag, Y., Kurttio, P. and Salonen, L., *Health Phys.*, **89**, 315 (2005).
81. Karpas, Z., Paz-Tal, O., Lorber, A., Salonen, L., Komulainen, H., Auvinen, A., Saha, H. and Kuttio, P., *Health Phys.*, **88**, 229 (2005).
82. Nakao, T., Aozasa, O., Ohta, S. and Miyata, H., *Archivus Environm. Contam. Toxic.*, **49**, 124 (2005).
83. Lawthom, C., Williams, J., Dunstan, F. D. *et al.*, *Epilepsia*, **46**, 49 (2005).
84. Delcorte, A., Yunus, S., Nieuwjaer, N., Poleunis, C. and Bertrand, P., in *5th European Workshop on Secondary Ion Mass Spectrometry*. Münster (2006) 102.
85. Sela, H., Karpas, Z., Zoriy, M., Pickhardt, C. and Becker, J. S., *Int. J. Mass Spectrom.*, **261**, 199 (2007).
86. Rodushkin, I. and Axelsson, M. D., *Sci. Tot. Envir.*, **250**, 83 (2000).
87. Miekeley, N., de Carvalho Fortes, L. M., Porto da Silveira, C. L. and Lima, M. B., *J. Trace Elem. Med. Biol.*, **15**, 46 (2001).
88. Kintz, P., Ginet, M. and Cirimele, V., *J. Anal. Toxicol.*, **30**, 621 (2006).
89. Maurice, J. F., Wibetoe, G. and Sjastad, K. E., *J. Anal. At. Spectrom.*, **17**, 485 (2002).
90. Raab, A. and Feldmann, J., *Anal. Bioanal. Chem.*, **381**, 332 (2005).
91. Kang, D., Amarasiriwardena, D. D. and Goodman, A. H., *Anal. Bioanal. Chem.*, **378**, 1608 (2004).
92. Jaelevik, B., Odelius, H., Dietz, W. and Norén, J. G., *Archives Oral Biol.*, **46**, 239 (2001).
93. Xiang, G., Jiang, Z., He, M. and Hu, B., *Mikrochim. Acta*, **154**, 247 (2006).
94. Dolphin, A. E., Goodman, A. H. and Amarasiriwardena, D. D., *Amer. J. Phys. Anthropol.*, **128**, 878 (2005).
95. Kang, D., Awarasiriwardena, D. and Goodman, A. H., *Anal. Bioanal. Chem.*, **378**, 1608 (2004).
96. Benemann, J., Lehmann, N., Bromen, K. Marr, A., Seiwert, M., Schulz, C. and Jöckel, K. H., *Intern. J. Hygiene Environm. Health*, **208**, 499 (2005).
97. Uryu, T., Yoshinaga, J., Yanagisawa, Y., Endo, M. and Takahashi, J., *Anal. Sci.*, **19**, 1413 (2003).
98. Bellis, D. J., Hetter, K. M., Jones, J., Amarasiriwardena, D. D. and Parson, P. J., *J. Anal. At. Spectrom.*, **21**, 948 (2006).
99. Hann, S., Latkoczy, C., Bereuter, T. L., Prohaska, T., Stingeder, G. and Reiter, C., *Intern. J. Legal Medicine*, **119**, 35 (2005).
100. Kleinhanns, L. C., Kreissig, K., Kamber, B. S., Meisel, T., Nägler, T. F. and Kramers, J. S., *Anal. Chem.*, **74**, 67
101. Boulyga, S. F., Becker, J. S., Malenchenko, A. F. and Dietze, H. J., *Mikrochim. Acta*, **134**, 215 (2000).
102. Becker, J. S. and Dietze, H. J., *Int. J. Mass Spectrom.*, **197**, 1 (2000).
103. Beauchemin, D. and Kisilevsky, R., *Anal. Chem.*, **70**, 1026 (1998).
104. Johansen, M. J., Newman, R. A. and Madden, T., *Pharmacotherapy*, **17**, 464 (1997).
105. Sun, Y., Lu, Y. and Chung, Y., *J. Anal. At. Spectrom.*, **22**, (2007).
106. Becker, J. S., Boulyga, S. F., Pickhardt, C., Becker, J. Su., Buddrus, S. and Przybylski, M., *Anal. Bioanal. Chem.*, **375**, 561 (2003).
107. Sefton, B. M. and Hunter, T. (eds.), *Protein Phosphorylation*. Academic Press, San Diego, CA (1998).
108. Becker, J. S., Zoriy, M., Krause-Buchholz, U. Becker, J. Su., Pickhardt, C., Przybylski, M., Pompe, W. and Rödel, G., *J. Anal. At. Spectrom.*, **19**, 1236 (2004).
109. Hasegawa, M., Fijiwara, H., Nonaka, T. *et al.*, *J. Biol. Chem.*, **277**, 49071 (2002).
110. Becker, J. S., Boulyga, S. F., Becker, J. Su., Pickhardt, C., Damoc, E. and Przybylski, M., *Int. J. Mass Spectrom.*, **228**, 985 (2003).
111. Becker, J. Su., Zoriy, M., Przybylski, M. and Becker, J. S., *Int. J. Mass Spectrom.*, **261**, 69 (2007).
112. Boulyga, S. F., Pickhardt, C., Becker, J. Su., Przybylski, M. and Becker, J. S., *Plasma Source Mass Spectrometry: Applications and Emerging Technologies*, (*8th International Conference on Plasma Source Mass Spectrometry*), *Durham, United Kingdom*, 54 (2003).
113. Elliott, V. L., McLeod, C. W. and Marshall, P. S., *Anal. Bioanal. Chem.* **383**, 416 (2005)
114. Edler, M., Busker, E., Feldmann, I., Jakubowski, N. and Venzago, C., in *Plasma Source Mass Spectrometry*, D. Bandura, G. Holland (eds.). The Royal Society of Chemistry, Cambridge (2005) 99.
115. Shah, M. and Caruso, J. A., *J. Sep. Sci.*, **28**, 1969 (2005).
116. Becker, J. S., Zoriy, M., Becker, J. Su. C., Damoc, E., Juhacz, G., Palkovits, M. and Przybylski, M., *Anal. Chem.*, **77**, 5851 (2005).

117. Becker, J. S., Zoriy, M., Becker, J. Su., Pickhardt, C. and Przybylski, M., *J. Anal. At. Spectrom.*, **19**, 149 (2004).
118. Wind, M., Gosenca, D., Kubler, D. and Lehmann, W. D., *J. Anal. At. Spectrom.*, **17**, 21 (2002).
119. Finney, L. A., Halloran T. V., *Science*, **300**, 931 (2003).
120. Szpunar, J., *Anal. Bioanal. Chem.*, **378**, 54 (2004).
121. Lobinski, R., Schaumlöffel, D. and Szpunar, J., *Mass Spectrom. Rev.*, **25**, 255 (2006).
122. Szpunar, J. and Lobinski, R., *Hyphenated Techniques in Speciation Analsis*, R. M. Smith (ed.). RSC Chromatic Monographs, The Royal Society of Chemistry, Cambridge (2003).
123. Fenn, J. B., Mann, M., Meng, C. K., Wong, S. F. and Whithouse, C. M., *Science*, **246**, 64 (1989).
124. Mizuriha, V., Hasegawa, H. and Notoya, M., *Acta Histochem. Cytochem.*, **30**, 3125 (1997).
125. Mesjasz-Przybylowicz, J. and Przybylowicz, W. J., *Nucl. Instr. Meth. B*, **189**, 470 (2002).
126. Takeda, A., Tamano, H., Enomoto, S. and Oku, N., *Cancer Res*, **61**, 5065 (2001).
127. Chandra, S., *Appl. Surf. Sci.*, **203–204**, 679 (2003).
128. Morrison, G. H., Gay, I. and Cahndra, S., *Scanning Microsc., Suppl.*, **8**, 359 (1994).
129. Touboul, D., Halgand, F., Brunelle, A. *et al.*, *Anal. Chem.*, **76**, 1550 (2004).
130. Kindness, A., Sekaran, N. and Feldmann, J., *Clinical Chemistry*, **49**, 1916 (2003).
131. Boulyga, S. F., Desideri, D., Meli, M. A., Testa, C. and Becker, J. S., *Int. J. Mass Spectrom.*, **226**, 329 (2003).
132. Becker, J. S., Zoriy, M., Becker, J. Su., Pickhardt, C. and Przybylsky, M., *J. Anal. At. Spectrom.*, **19**, 149 (2004).
133. Kawahara, M., *J. Alzheimer's Disease*, **8**, 171 (2005).
134. Becker, J. Su., Zoriy, M., Pickhardt, C., Przybylski, M. and Becker, J. S., *Int. J. Mass Spectrom.*, **242**, 135 (2005).
135. Becker, J. Su., Zoriy, M., Przybylski, M. and Becker, J. S., *J. Anal. At. Spectrom.*, **22**, 63 (2007).
136. Pickhardt, C., Izmer, A., Zoriy, M. V., Schaumlöffel, D. and Becker, J. S., *Int. J. Mass Spectrom.*, **248**, 136 (2006).
137. Becker, J. S., Gorbunoff, A., Zoriy, M. V., Izmer, A. V. and Kayser, M., *J. Anal. At. Spectrom.*, **21**, 19 (2006).
138. Jackson, G. S., Harper, S., Smith, L. and Flinn, J., *Anal. Bioanal. Chem.*, **384**, 251 (2006).
139. Takahashi, S., Takahashi, I., Sato, H., Kubota, Y., Yoshida, S. and Muramatsu, Y., *Laboratory Animals*, **34**, 97 (2000).
140. Jackson, B. P., Hopkins, W. A. and Baionno, J., *Environ. Sci. Techn.*, **37**, 2511 (2003).
141. Paul, M. C., Parsons, C. H., Calford, M. B. and von Nagy-Felsobuki, E. I., *Spectrochim Acta B*, **59**, 1485 (2004).
142. Ek, K. H., Rauch, S., Morrison, G. M. and Lindberg, P., *Sci. Tot. Envir.*, **334–335**, 149 (2004).
143. Garcia, F., Ortega, A., Domingo, J. L. and Corbella, J., *J. Environ. Sci. Health A*, **36**, 1767 (2001).
144. Gélinas, Y., Lafond, J. and Schmit, J. P., *Biol. Trace Elem. Res.*, **59**, 63 (1997).
145. Panayi, A. E., Spyou, N. M., Iversen, B. S., White, M. A. and Part, P., *Neurol. Sci.*, **195**, 1 (2002).
146. Rudolph, R., Hann, S., Stingeder, G. and Reiter, C., *Anal. Bioanal. Chem.*, **382**, 1500 (2005).
147. Tarohda, T., Yamamoto, M. and Amamo, R., *Anal. Bioanal. Chem.*, **380**, 240 (2004).
148. Karas, M., Bachmann, D., Bahr, U. and Hillenkamp, F., *Int. J. Mass Spectr. Ion Proc.*, **78**, 53 (1987).
149. Caprioli, R. M., Farmer, T. B. and Gile, J., *Anal. Chem.*, **69**, 4751 (1997).
150. Stoeckli, M., Chaurand, P., Hallahan, D. E. and Caprioli, R. M., *Nat. Med.*, **7**, 493 (2001).
151. Heeren, M. A., Mc.Donnell, L. A., Amstalden, E., Luxambourgh, S. L., Altelaar, A. F. M. and Piersma, S. R., *Appl. Surf. Sci.*, **252**, 6827 (2006).
152. Sjövall, P., Lausmaa, J. and Johansson, B., *Anal. Chem.*, **76**, 4271 (2004).
153. Brunelle, A., Touboul, D. and Laprévote, O., *J. Mass Spectrom.*, **40**, 985 (2005).
154. Touboul, D., Kollmer, F., Niehus, E., Brunelle, A. and Laprévote, O., *J. Am. Soc. Mass Spectrom.*, **16**, 1608 (2005).
155. McDonnel, L., Piersma. S. R., Maarten Altelaar, A. F. *et al.*, *J. Mass Spectrom.*, **40**, 160 (2005).
156. Fourre, C., Clerc, J. and Fragu, P., *J. Anal. At. Spectrom.*, **12**, 1105 (1997).
157. Hoffmann, E., Lüdke, C. and Stephanowitz, H., *Fresenius' J. Anal. Chem.*, **355**, 900 (1996).

158. Hoffmann, E., Lüdke, C., Scole, J., Stephanowitz, H., Ullrich, E. and Colditz, D., *Fresenius' J. Anal. Chem.*, **367**, 579 (2000).
159. Ek, P., *1999 European Winter Conference on Plasma Spectrochemistry*, Book of Abstracts, p. 111 Pau, France.
160. Feldmann, J., Kindness, A. and Ek, P., *J. Anal. At. Spectrom.*, **17**, 813 (2002).
161. Zoriy, M., Kayser, M., Izmer, A., Pickhardt, C. and Becker, J. S., *Int. J. Mass Spectrom.*, **242**, 297 (2005).
162. Zilles, K., Palomero-Gallagher, N., Grefkes, C., Scheperjans, F., Boy, C., Amunts, K. and Schleicher, A., *European Neuropsychopharm.*, **12**, 587 (2002).
163. Dobrowolska, J., Dehnhardt, M., Matusch, A. Zoriy, M., Palomero-Gallagher, N., Koscielniak, P., Zilles, K. and Becker, J. S., Talanta 2007 (published on-line, 25 July 2007).
164. Rosenberg, B., Van Camp, L., Trosko, J. E. and Mansour, V. H., *Nature*, **222**, 385 (1969).
165. Zoriy, M., Matusch, A., Spruss, T. and Becker, J. S., *Int. J. Mass Spectrom.*, **257**, 102 (2006).
166. Zoriy, M., Dehnhardt, M., Reifenberger, G., Zilles, K. and Becker, J. S., *Int. J. Mass Spectrom.*, **257**, 27 (2006).
167. Rayman, M. P., *Proc. Nutr. Soc.*, **61**, 203 (2002).
168. Infante, H. G., O'Connor, G., Rayman, M. *et al.*, *J. Anal. At. Spectrom.*, **19**, 1524 (2004).
169. Dickson, H., Bunch, J., Stokes, S., Mead, R. and McLeod, C. W., in *Book of Abstracts of 17th International Mass Spectrometry Conference*, ThP Vychodeska Tiskama, Prague 143 (2006).
170. Hutchinson, R. W., Cox, A. G., McLeod, C. W. Marshall, P. S., Harper, A., Dawson, E. L. and Howlett, D. R., *Anal. Biochem.*, **346**, 225 (2005).
171. Lobinski, R., Moulin, C. and Ortega, R., *Biochimie*, **88**, 1591 (2006).
172. Leapman, R. D. and Newbury, D. E., *Anal. Chem.*, **65**, 2409 (1993).
173. Leapman, R. D., *J. Microsc.*, **210**, 5 (2003).
174. Mai, F. D., Chen, B. J., Wu, L. C., Li, F. Y. and Chen, W. K., *Appl. Surf. Sci.*, **252**, 6809 (2006).
175. Van Royen, P., Motsio, H., Van Vaeck, L. and Butz, T., in *5th European Workshop on Secondary Ion Mass Spectrometry* Münster 92 (2006).

9.6 Food Analysis

The content of contaminants and toxins (including toxic elements) in food has aroused increasing public interest due to the danger of serious health problems as a result of the ingestion of contaminated food. Ensuring the production and provision of healthy and high quality foodstuffs is necessary and is strictly regulated in the majority of countries. To fulfil the requirements and legislation concerning food quality, fast and reliable analytical techniques with multi-element capability, low detection limits and a high sample throughput should be provided. In order to study traces of pollutants or healthy components, food science in the broadest sense can be extended to include soil and groundwater analysis (including environmental analysis – see Section 9.2), plant uptake and at the end of the food chain, the metabolite fate of elemental species when certain foods are consumed by humans and animals.[1] Compared to older and slower analytical techniques (such as GF-AAS), due to its high sensitivity, multi-element capability and the possibility of analyzing metal species at the trace and ultratrace level, ICP-MS, in particular, has become the method of choice for fast routine analysis of toxic and essential (nutrient) trace elements in foodstuffs. International food legislation 'protecting the health of consumers and ensuring fair practices in the food trade' is laid down under the auspices of the Codex Alimentarius Commission (CAC), a United Nations organization established by the FAO and WHO.[2] Since 1962 the CAC has developed specific standards for individual groups of food.

9.6.1 Determination of Trace Elements and Species in Foodstuffs

Multi-element trace analysis is an important prerequisite for the quality assurance of foodstuffs with respect to the characterization of non-essential, toxic and essential (nutrient) elements as pollutions or as mineral elements relevant to health. Contamination with heavy metals such as Cd, Pb or Hg has become a serious problem with increasing environmental (artificial) contamination e.g., due to industrial pollution. The increasing use of inorganic mass spectrometric techniques (especially of ICP-MS) in the analysis of foodstuffs for multi-element analysis of trace elements or the detection of selected elements and species at a low concentration level has resulted from advances in very sensitive and quantitative measurements of metals, metalloids and several non-metals, including their speciation.

Reliable matrix matched certified reference materials (CRM) have to be created and characterized in order to evaluate analytical data measured on foodstuffs by inorganic mass spectrometry. For example, the measurement of As, Se, Fe, Mn, Rb, Cu and Zn concentrations in a natural matrix candidate fish tissue Certified Reference Material from the National Institute of Standards and Technology (NIST) – SRM 1947 – has been performed using a multi-element quantification strategy based on the method of standard addition incorporating internal standards in ICP-MS with a collision cell.[3] Concentrations for multiple trace elements in SRM 1946 Lake Superior Fish Tissue from 0.073 $mg\,kg^{-1}$ (Mn) to 8.04 $mg\,kg^{-1}$ (Rb) were successfully predicted by this calibration strategy. Myors and co-workers[4] analyzed ten trace elements (Cd, Cu, Cr, Hg, Mo, Ni, Pb, Se, Sn and Zn) in a wheat flour candidate sample for the provision of reference values. The authors applied a double isotope dilution technique for measurements of selected trace elements by quadrupole ICP-MS with an octopole reaction cell (Agilent 7500c), and compared the obtained data with reference values and the results of a proficiency test with 11 participant laboratories. Expanded relative uncertainties at the 95 % confidence level for the ten trace elements in the wheat flour varied from 3.1–14 %.[4] Furthermore, an Agilent 7500ce ICP-MS with an octopole collision cell (in order to solve the interference problem) was successfully employed for the measurement of 12 trace elements (Mg, P, Ca, Cr, Mn, Fe, Ni, Cu, Zn, As, Cd and Pb) in the concentration range from 0.0059 $mg\,kg^{-1}$ (Pb) to 3378 $mg\,kg^{-1}$ (P) in SRM NIST 2387 Peanut Butter after digestion in an open vessel using a mixture of nitric acid and hydrogen peroxide.[5]

Of all the toxic elements, mercury is widely considered to be among the environmental pollutants with the highest priority and is regarded as the most highly bioconcentrated trace metal in the human food chain.[6] The toxicity of mercury depends on its chemical species, whereby alkylmercury compounds have the highest toxicity. For fish a guideline level of 0.5 $mg\,kg^{-1}$ for methylmercury (most Hg in fish will be present as $HgCH_3$ – MeHg) has been recommended. Methylmercury in prepared tuna material (with a relatively high mercury content of about 4.3 $mg\,kg^{-1}$ Hg) was quantitatively measured in 20 specialized laboratories worldwide and the different analytical techniques employed and results obtained by GC-ICP-MS were compared.[7] Alkaline digestion at room temperature (with manual shaking) or high temperature (under sonication, oven or hot plate conditions) with hydrochloric acid leaching was the second most popular choice for most participants together with the isotope dilution approach or external calibration. As a result it was found that 80 % of study participants are potentially able to supply accurate results for MeHg in fish type matrix (low $mg\,kg^{-1}$ Hg level) within $\pm$ 10 % uncertainty.[7]

Clough and co-workers reported on the uncertainty contribution using single and double isotopically enriched spikes in the isotope dilution approach employing a multiple ion collector ICP-MS combined with an HPLC and with cold vapour generation (CV) to form HPLC-CV-MC-ICP-MS for the determination of MeHg in fish tissue.[8] Two fish tissue reference materials DORM-2 (NRC, Canada) and BCR 464 were employed in these studies. The amount of each certified reference material (CRM) and of ^{199}Hg enriched methylmercury chloride was chosen to give a $^{199}Hg/^{200}Hg$ isotopic ratio of close to unity at complete equilibration. Both IDMS procedures gave comparable accuracy of analytical data. The application of double isotopic enriched spikes resulted in more precise data.[8]

Specific recommendations on contaminants (e.g., maximum levels of lead, cadmium or mercury) in foodstuffs are summarized in the European Union Regulations.[9] The specifications concerning trace elements in foods and food additives are As $< 3\,mg\,kg^{-1}$, Pb $< 10\,mg\,kg^{-1}$, and Hg and Cd $< 1\,mg\,kg^{-1}$.

For arsenic the toxic species are inorganic arsenite and arsenate, whereas other species found in fish are the less toxic monomethylarsonic acid, dimethylarsinic acid and arsenosugars as well as largely non-toxic arsenicals such as arsenobetaine, arsenocholine and the tetramethylarsonium ion.[2] The recommended WHO guideline level for inorganic arsenic in drinking water is $10\,\mu g\,l^{-1}$. High arsenic concentrations in drinking water above $0.2\,mg\,l^{-1}$ have been reported.[10] In the marine environment the total arsenic concentration in animals and plants varies between 0.5 and $50\,mg\,kg^{-1}$, whereby, in general, freshwater fish contain lower arsenic concentrations compared to ocean fish. Arsenic mostly occurs in plants in the low concentration range, below $0.02\,mg\,kg^{-1}$, except rice and certain mushrooms.[2] The intake of arsenic and consequently exposure of humans is mainly due to the consumption of contaminated fish in countries with a large consumption of seafood and seaweed products and possibly via the drinking water. The average adult intake from food has been estimated to be $30–120\,\mu g\,day^{-1}$. However, commercially available peanut butter also contains measurable amounts of inorganic and organic species of As. The total arsenic concentration ranges from $6.5–21.4\,ng\,g^{-1}$. Four arsenic species As(V), As(III) including two organoarsenic species (DMA and MMA) have been detected in the low $ng\,g^{-1}$ range for NIST SRM 2387 Peanut Butter by online coupling of ion chromatography and hydride generation to form ICP-MS (IC-HG-ICP-MS).[11] Data for arsenic species in market foods are rarely available in the literature.[2]

One of the most widely investigated elements in foods is selenium as an essential element which is also permitted as a food additive in food supplements. Selenium is a powerful antioxidant and is employed for heart disease and the prevention of cancer. There is a need for selenium speciation because the bioavailability and the toxicity depend on its binding form.[12] Depending on the concentration and species present, it is known for selenium that the 'biological window' between essentiality and toxicity is very narrow.[13] The main essential role identified for selenium is as a component of the enzyme glutathione peroxidase (GSH-PX) – one of the body's antioxidant defence systems.[14] A selenium deficit in food can lead, for example, to Keshan disease. The increasing use of selenium supplements where selenium occurs as different selenium species (e.g., Se-cysteine as selenoamino acid) has been studied by several working groups.[15,16] Selenomethionine (SeMet), a major species of selenium found in selenium rich foods, is usually incorporated non-selectively into proteins as a substitute for sulfur-containing methionine.[17] Selenium speciation in dill supplemented with sodium selenite during its growth has been performed using ion pairing reversed phase and cation exchange chromatography by HPLC-ICP-MS employing an Agilent 7500ce with octopole collision cell in Caruso's group.[18] *Se*-methyl-selenocysteine (MeSeCys) was found to be the major species in all parts of the plant (see Figure 9.19). Selenium supplemented dill plants can be used as a source of MeSeCys. The presence of MeSeMet suggests that the plant could volatilize Se, since it is a volatile (less toxic) selenium species.[18] Selenium containing proteins in chives (*Allium schenoprasum*), grown separately in three different Se supplementation media (Se(IV), Se(VI) and SeMet) were analyzed by size exclusion chromatography (SEC) together with HPLC-ICP-MS.[19] Reversed phase ion pairing chromatography (RP-IP-HPLC-ICP-MS) was applied for the speciation of selenium containing amino acids in chives.[19] Since selenoamino acids are soluble in water, leaching with hot water has been applied to recover selenium species.[20] The sample is homogenized with water, sonicated or heated, and ultracentrifuged. An evaluation of yeast based selenium food supplements containing organic selenium species was studied by HPLC-ICP-MS.[15] Lobinski's working group[21] developed a method for the identification of selenium containing peptides in Brazil nuts, which have been classified among the foodstuffs that contain the highest level of unadulterated selenium (up to 500 mg kg^{-1}). Fifteen seleno-containing peptides were characterized after size exclusion chromatography (SEC) of the soluble (tryptic digested) protein fraction by HPLC-ES-QToF-MS/MS and ICP-MS with a detection limit of 1.3 fmol.[21] Selected volatile selenoamino acids in food supplements were analyzed by GC-ICP-MS.[22,23]

For high throughput determination of selected toxic trace elements (such as Sb, As, Cd, Cr, Pb, Hg and Sn) in a wide range of food samples (rice, oyster meat, spinach, pepper, sausage and others), a simple and fast analytical method using ICP-MS after closed vessel digestion of food samples in a microwave oven with concentrated nitric acid, followed by microwave assisted evaporation to concentrate the analytes in the sample solution, is recommended.[24] The instrumental detection limits obtained for seven toxic elements investigated varied between 0.01 μg l^{-1} (Cd) and 0.08 μg l^{-1} (Sn) and are mostly one to two orders of magnitude better than measured by AAS and ICP-OES. The analytical method was validated and the recoveries were evaluated on CRMs, which were spiked with seven toxic elements of interest and analyzed by ICP-MS.[24] Recoveries of the seven elements from spiked samples ranged from 93.1 to 103.6 %, whereas the precision of the determined concentrations varied between 3.1 % and 4.3 %.

Food analysis was performed by flow injection ICP-MS in order to study the bio-accessibility of elements (Zn and Pb in SRM NIST 8433 corn bran) after online leaching with artificial gastrointestinal fluids,[25] which offers a quick and easy method to assess the maximum bio-accessibility of elements from foodstuffs. The method is based on a single line flow injection manifold to repeatedly inject 100 μl aliquots of a given reagent that is then pumped through a microcolumn of food connected to an ICP-MS. This online leaching approach allows the continuous monitoring of the progressive release of elements from food by a given reagent (e.g., gastric juice).[25]

In food analysis, multi-element speciation of a tea infusion was investigated by size exclusion chromatography (SEC) coupled to ICP-MS.[26] Of the 14 elements investigated, only Fe, Ni, Cu, Sr, Ba, Pb and Al were found to be associated with organic complexes. Further experiments with cation exchange chromatography led to the conclusion that the metal binding organic ligands were large polyphenolic compounds.[26] Spzunar *et al.*[27] studied the speciation analysis of biomolecular complexes of lead in wine by SEC-ICP-MS.

Size exclusion chromatography coupled to an ICP-QMS with an octopole collision cell was employed for the multi-elemental speciation of essential elements (P, S, Cr, Mn, Fe, Co, Cu, Zn, Br, Se and I) and Al as a toxic element in premature human milk.[28] Comparison of speciation results demonstrates that premature human mothers' milk differs significantly from formula milks in terms of the element binding pattern to the biomolecules. It was found that premature human milk is very rich in high molecular weight species associated with metals. The authors concluded that more attention must be paid to the chemical form in which essential elements are added to the formulas, particularly those used for the nutrition of premature babies.[28]

Chromium, iron and selenium in foodstuffs from animal sources have been analyzed after closed vessel microwave digestion by collision cell ICP-MS.[29] The limits of quantification (LOQ) of the analytical procedure were estimated under optimized experimental conditions with 0.025, 0.086 and 0.041 mg kg^{-1} for Cr, Fe and Se, respectively. The results obtained for the three elements in nine different certified reference materials were, in all cases, in good agreement with the certified values.[29]

Of special interest is the occurrence, the concentration of Pb and of its species in food and drink. Lead can be present as a contaminant in foodstuffs or drinking water with a remarkably high content, especially if Pb plumbing or Pb lined storage tanks and pipes have been used. Crews has reviewed the speciation of lead in food and wine including speciation analysis.[30]

Pesticides, which are generally persistent compounds, display a widespread distribution in the environment due to their extensive usage in agriculture. Their lipophilic character causes bioaccumulation and therefore a presence in the food chain. Prange and co-workers[31] developed analytical techniques for the sensitive determination of a wide range of compounds. A multi-compound pesticide mixture was separated by gas chromatography which was coupled in a heated transfer line to an ICP-QMS with an octopole collision cell. Sample injection using an autosampler allowed a fast screening of samples with respect to S, P or halogen containing pesticides. Several P, S Cl, Br and I containing pesticides (e.g., chlorpyrifos, dildrin, mirex, fensulfothion and ethoprop) were detected in fruit and vegetable extracts, such as tomato extract, at the $ng\,g^{-1}$ and sub-$ng\,g^{-1}$ concentration levels.[31] The determination of perchlorate contamination in food and environmental samples was possible by employing a selective method using ion chromatography-ion association electrospray ionization mass spectrometry (IC/IA-ESI-MS).[32] To improve the selectivity, perchlorate is detected as $DClO_4^+$ after the (postcolumn) addition of D^{2+}, which is associated with the perchlorate in the gas phase.

The oral aluminium bioavailability of a representative food containing food additive acidic sodium aluminium phosphate (SALP), a leavening agent in baked goods, has been studied by accelerator mass spectrometry.[33] In the experiments, rats were fed a special diet (biscuit containing 1 % or 2 % acidic SALP, synthesized to contain ^{26}Al). ^{26}Al, which was used as a biological tracer, can be ultrasensitively analyzed by AMS. The rats received concurrent ^{27}Al infusion for comparison with oral ingestion.[33]

Human exposure to platinum group elements (PGE) has been causing more and more concern due to increasing use of car catalytic converters based on PGE. The uptake of PGE such as Pd, Pt and Rh as the result of PGE contamination in the diet (through bread and cow milk) has been studied by sector field ICP-MS.[34] The Pd concentration, which was the PGE with the highest concentration detected in all the investigated food samples, varied between $27.5\,ng\,g^{-1}$ (in white bread), $12.4\,ng\,g^{-1}$ (in skim milk) and $3.8\,ng\,g^{-1}$ (in full cream milk).[34] The platinum concentrations measured ranged from $0.083\,ng\,g^{-1}$ in milk to $0.257\,ng\,g^{-1}$ in white bread. The largest variations were observed for Rh ($0.14\,ng\,g^{-1}$ in wholemeal bread up to $2.2\,ng\,g^{-1}$ in white bread). The Rh in skim and full cream milk was found at the $1.1–1.7\,ng\,g^{-1}$ level, respectively.[34]

A rapid and accurate method using double-focusing ICP-SFMS (Element, Finnigan) with an ultrasonic nebulizer and desolvation unit (Cetac USN 6000) and ion chromatography has been applied for the determination of plutonium at the ultratrace level in food.[35] The samples were prepared by HNO_3 closed vessel microwave digestion, evaporated to dryness and diluted into a mobile phase (1.5 M HNO_3 and 0.1 mM 2,6-pyridinedicarboxylic acid). By online separation using a polystyrene–divinylbenzene ion chromatography column, ^{239}Pu and ^{238}U were separated in order to reduce the $^{238}U^1H^+$ interference. A further reduction of $^{238}U^1H^+$ interference was achieved by the application of an ultrasonic nebulizer (USN). The detection level for Pu of $0.020\,pg\,g^{-1}$ ($4.6 \times 10^{-2}\,Bq\,kg^{-1}$) is significantly below 1/10 of the European Union legislation for baby food ($1\,Bq\,kg^{-1} - 0.436\,pg\,g^{-1}$).[35]

Amselfelder red wine from Kosovo has been investigated with respect to uranium and plutonium content using ICP-SFMS (Element). The microconcentric nebulizer Aridus (Cetac Technologies) was applied for solution introduction into the ICP source. In different red wine samples, uranium with a natural isotopic composition was found at a concentration of $0.145 \pm 0.011\,\mu g\,l^{-1}$, which corresponds to the 'normal' uranium background concentration. Plutonium was not detected at a detection limit of $10^{-5}\,\mu g\,l^{-1}$.[36]

9.6.2 Analysis of Mineral and Bottled Water

For the trace analysis of bromate, chlorate, iodate and perchlorate (at sub-$\mu g\,g^{-1}$ quantities) in bottled, drinking (and natural) waters, a simple and rapid method has been developed after the removal of sulfate and carbonate (using barium and hydronium cartridges, respectively) and measurement by LC-MS/MS (liquid chromatography – tandem triple quadrupole mass spectrometer). Ionization is accomplished by using electrospray ionization in negative mode. By means of liquid chromatography, these four species were well separated with limits of detection from $0.021\,\mu g\,l^{-1}$ for perchlorate to $0.070\,\mu g\,l^{-1}$ for iodate. Drinking water treatment plants were evaluated using this analytical method. Due to this highly sensitive method, bottled waters were found to contain detectable levels of perchlorate.[37]

The detection of antimony species in drinking water stored in (polyethylene terephthalate) PET containers has been reported by several groups.[38,39] As a result most commercially available PET materials typically contain $190–300\,\mu g\,g^{-1}$ Sb,[40] whereby antimony trioxide is a suspected carcinogen. Sb that is leached from the PET containers in drinking waters was observed to be 100 times elevated compared with uncontaminated ground or drinking water ($2.2–3.8\,ng\,l^{-1}$).[38]

As a naturally occurring long-lived radionuclide ($t_{1/2} = 1600$ a), ^{226}Ra is widely distributed in natural samples and also occurs in foodstuffs and beverages at the ultratrace concentration level. Radium has been recognized as one of the most toxic natural radioelements. Furthermore, because of its similarity to the alkaline earth metals, radium follows the calcium pathway in biological organisms, so that it is strongly absorbed into bones, cell and tissues. Mineral water samples of different origins were investigated with respect to ^{226}Ra concentration at the low femtogram per ml concentration level after trace matrix separation by ICP-SFMS.[41] ^{226}Ra concentrations in mineral water samples were found to be between 0.7 and $14\,fg\,ml^{-1}$. Relatively high ^{226}Ra concentrations (of 10.3 and $14.2\,fg\,ml^{-1}$) in mineral water correlate with a high uranium content of 17.3 and $19.2\,ng\,ml^{-1}$ in these samples. The detection limit for ^{226}Ra determination was found to be $0.02\,fg\,ml^{-1}$.[41]

9.6.3 Fingerprinting of Foods by Trace Analysis and Isotope Ratio Measurements

The application of trace metal analysis for food authenticity has been known for many decades. Certain geographical origins attract a premium price so there may be a financial incentive to misdescribe food or adulterate it with cheaper varieties.[42] In addition, isotope ratio measurements extend the field of studies to include the possibility of tracing the origins of contamination, the origins of plants and fruit, and applications for the quality control of agricultural products.[43] That means foodstuffs can be easily investigated with respect to their geographical origin by means of trace and isotope analyses. Fingerprinting of foods, including isotope ratio measurements of light elements (C, N, O, H and S), is very useful in the area of the declaration and origin testing of food products and agricultural raw materials. The determination, e.g., of the stable nitrogen and carbon isotopic compositions ($\delta^{15}N$ and $\delta^{13}C$ values, respectively) of different organisms in a food chain provides information about the origin and 'history' of a foodstuff. By means of isotope ratio measurements on carbon and nitrogen by stable isotope ratio mass spectrometry it can be demonstrated that the selective metabolism of the lighter isotopes of these elements (^{14}N and ^{12}C) during food assimilation and waste excretion causes animals to become enriched in the heavier isotopes (^{15}N and ^{13}C) relative to their diets. Different isotopic compositions (e.g., for H/D, C, O, N, S) are found worldwide in water and in the organic substance of plants and animals. Typical isotope δ-values can be identified for the different continents and each single country of the continents. On the basis of the isotope values, e.g., for hydrogen or/and oxygen, products can be easily assigned to a continent or a certain country.[44] So far, isotope ratio measurements have been used in the area of food and animal feed by industry, trade associations, customs and public authorities for quality control and these measurements can be performed in specialized routine isotope analytical laboratories (e.g., www.agroisolab.de). The application of fingerprinting has been extended to analyze the vintage of wine by a combination of trace metal and $\delta^{18}O$, $\delta^{2}H$ and $\delta^{13}C$ determination.[45,46] Stable isotope variation of carbon and nitrogen in milk (and urine) of cattle has been examined by isotope ratio mass spectrometry (IRMS) under different feeding regimes (over six months). The $\delta^{13}C$ values were more negative for milk compared to urine in the case of feeding grass than if maize was fed. The difference between the $^{15}N/^{14}N$ isotope ratios for the two feeding regimes was less pronounced than the $^{13}C/^{12}C$ ratios.[47] Stable isotope variations of H, C, N, O and S as a tool to trace the authenticity of beef from Germany have been examined by Boner and Förstel.[48]

The trace element composition of Florida and Brazilian orange juices was studied using plasma spectroscopy by Winefordner *et al.* in 1979.[49] Nowadays, ICP-MS has been applied to differentiate and classify orange juice from six or more different countries.[50,51] The Se, Cd, Sr and Pb concentration and, in addition, $\delta^{13}C$ and $\delta^{15}N$ values determined in wheat using ICP-MS and gas source mass spectrometry, respectively, have been applied for differentiating USA, Canada and European samples on the basis of their fingerprints due to underlying geogenic and anthropogenic differences.[42]

$^{87}Sr/^{86}Sr$ isotope ratios can be used to characterize the authenticity of food. The original isotopic composition of Sr in the biomass should reflect the composition of the environment where the food is produced. The Sr isotope ratio measurement may support the information given by stable isotope ratios of the bioelements (C, O, H, N, S) by adding an independent parameter. In difficult legal cases, the data may be helpful and should be available for food control. $^{87}Sr/^{86}Sr$ isotope ratios of milk samples from different origins measured after careful trace/matrix separation by ICP-SFMS are close to the IUPAC table value (0.7099[52]) and vary as a function of origin.[53]

References VI

1. Olivas, R. M. and Cámera, C., in *Handbook of Elemental Speciation: Techniques and Methodology*, Vol I, R. Cornelis, J. Caruso, H. Crews and K. Heumann (eds.). John Wiley & Sons, Chichester (2003) 73.
2. Berg, T., in *Handbook of Elemental Speciation: Techniques and Methodology*, Vol I, R. Cornelis, J. Caruso, H. Crews and K. Heumann (eds.) John Wiley & Sons, Ltd, Chichester (2003) 629
3. Christopher, S. J., Day, R. D., Bryan, C. E. and Turk, G. C., *J, Anal. At. Spectrom.*, **20**, 1035 (2005).
4. Myors, R. B., Nolan, A. L., Askew, S., Saxby, D. L., Hearn, R. and Mackay, L. G., *J. Anal. At. Spectrom.*, **20**, 1051 (2005).
5. www.agilent.com/chem/icpms. (2005).
6. Horvat, M. and Gibicar, D., in *Handbook of Elemental Speciation II, Species in the Environment, Food, Medicine and Occupational Health*, Vol. II, H. C. R. Cornelis, J. Caruso and K. G. Heumann (eds.). John Wiley & Sons, Ltd, Chichester (2005) 281.
7. Quétel, C. R., Snell, J. P., Aregbe, Y., *J. Anal. At. Spectrom.*, **20**, 1058 (2005).
8. Clough, R., Belt, S. T., Fairman, B., Catterick, T. and Evans, E. H., *J. Anal. At. Spectrom.*, **20**, 1072 (2005).
9. European Commission Regulation No. 466/2001 of 8 March 2001, (2001) 1.
10. Larsen, E. H. and Berg, T., in *Trace Element Speciation for Environment, Food and Health*, L. P. L. Ebdon, R. Cornelis, H. Crews, O. F. X. Donard and P. Quevauviller (eds.). Royal Society of Chemistry, Cambridge (2001) 251.
11. Hovanec, B. M., *J. Anal. At. Spectrom.*, **19**, 1141 (2004).

12. Uden, P. C., in *Handbook of Elemental Speciation II, Species in the Environment, Food, Medicine and Occupational Health*, Vol. II, H. C. R. Cornelis, J. Caruso and K. G. Heumann (eds.). (John Wiley & Sons, Chichester, 2005), vol. II, p. 346.
13. Garcia Alonso, J. I. and Encinar, J. R., in *Handbook of Element Speciation, Techniques and Methodologies*, Vol. I, R. Cornelis, J. Caruso, H. Crews and K. Heumann (eds.). John Wiley & Sons, Ltd, Chichester (2003) 163.
14. Moens, L., de Smaele, T., Dams, R., van den Broeck, P. and Sandra, P., *Anal. Chem.*, **69**, 1604 (1997).
15. B'Hymer, C. and Caruso, J. A., *J. Anal. At. Spectrom.*, **15**, 1531 (2000).
16. Salih, B., *Spectrochim. Acta B*, **55**, 1115 (2000).
17. Patching, S. G. and Gardiner, P. H. E., *J. Trace Elements Med. Biol.*, **13**, 193 (1999).
18. Cankur, O., Yathavakilla, S. K. V. and Caruso, J. A., *Talanta*, **70**, 784 (2006).
19. Kápolna, E., Shah, M., Caruso, J. A. and Fodor, P., *Food Chemistry*, **101**, 1415 (2007).
20. Bird, S. M., Ge, H., Uden, P. C., Tyson, J. F., Block, E. and Denoyer, E., *J. Anal. At. Spectrom.*, **12**, 785 (1997).
21. Dernovics, M., Giusti, P. and Lobinski, R., *J. Anal. At. Spectrom.*, **21**, 41 (2006).
22. Péréz Méndez, S., Blanco González, E. and Sanz-Medel, A., *J. Anal. At. Spectrom.*, **15**, 1109 (2000).
23. Péréz Méndez, S., Montes Bayón, M., Blanco González, E. and Sanz-Medel, A., *J. Anal. At. Spectrom.*, **14**, 1333 (1999).
24. Chan, K. C., Yip, Y. C., Chu, H. S. and Sham, W. C., *J. AOAC. Int.*, **89**, 469 (2006).
25. Chu, M. and Beauchemin, D., *J, Anal. At. Spectrom.*, **19**, 1213 (2004).
26. Oedegard, K. E. and Lund, W., *J. Anal. At. Spectrom.*, **12**, 403 (1997).
27. Szpunar, J., Pellerin, P., Makarov, A., Doco, T., Williams, P., Medina, B. and Lobinski, R., *J. Anal. At. Spectrom.*, **13**, 749 (1998).
28. de la Flor St Remy, R. R., Fernández Sánchez, M. L., Lopez-Sastre, J. B. and Sanz-Medel, A., *J. Anal. At. Spectrom.*, **19**, 1104 (2004).
29. Dufailly, V., Noël, L. and Guérin, T., *Anal. Chim. Acta*, **565** 214 (2006).
30. Crews, H., in *Handbook of Elemental Speciation II: Species in Environment, Food, Medicine & Occupational Health*, Vol. II, H. C. R. Cornelis, J. Caruso and K. G. Heumann (eds.). John Wiley & Sons, Ltd, Chichester (2005) 247.
31. Proefrock, D., Leonard, P., Wilbur, S. and Prange, A., *J. Anal. At. Spectrom.*, **19**, 623 (2004).
32. Martinelango, P. K., Anderson, J. L., Dasgupta, P. K., Armstrong, D. W., Al-Horr, R. S. and Slingsby, W., *Anal. Chem.*, **77**, 4829 (2005).
33. Yokel, R. A. and Florence, R. L., *Toxicology*, **227**, 86 (2006).
34. Frazzoli, C., Cammarone, R. and Caroli, S., *Pure and Appl. Chem.*, **78**, 69 (2006).
35. Evans, P., Elahi, S., Lee, K. and Fairman, B., *J. Environ. Monit.*, **5**, 175 (2003).
36. Becker, J. S., *J. Anal. At. Spectrom.*, **17**, 1172 (2002).
37. Snyder, S. A., Vanderford, B. J. and Rexing, D. J., *Environ. Sci. Techn.*, **39**, 4586 (2005).
38. Shotyk, W., Krachler, M. and Chen, B., *J. Environ. Monit.*, **8**, 288 (2006).
39. Hansen, H. R. and Pergantis, S. A., *J. Anal. At. Spectrom.*, **21**, 731 (2006).
40. Duh, B., *Polymer*, **43**, 3147 (2002).
41. Zoriy, M., Varga, Z., Pickhardt, C., Ostapczuk, P., Hille, R., Halicz, L., Segal, I. and Becker, J. S., *J. Environ. Monit.*, **7**, 514 (2005).
42. Branch, S., Burke, S., Evans, P., Fairman, B. and Wolff Briche, C. S. J., *J. Anal. At. Spectrom.*, **18**, 17 (2003).
43. Prohaska, T., Köllensperger, G., Hann, S., Stingeder, G., Fitz, W. and Wenzel, W., in *Plasma Source Mass Spectrometry*, G. Holland and S. D. Tanner (eds.). The Royal Society of Chemistry, Cambridge (2003) 93.
44. www.agroisolab.de.
45. Day, M., Zhang, B. L., Martin, G. J., Asselin, C. and Morlat, R., *J. Int. Sci. Vigne Vin*, **29**, 75 (1995).
46. Greenough, J. D., Longerich, H. P. and Jackson, B., *Can. J. Appl. Spectrosc.*, **41**, 76 (1996).
47. Knobbe, N., Vogl, J., Pritzkow, W. *et al.*, *Anal. Bioanal. Chem.*, **386**, 104 (2006).
48. Boner, M. and Förstel, H., *Anal. Bioanal. Chem.* **378**, 301 (2004).
49. McHard, J. A., Foulk, S. J. and Winefordner, J. D., *J. Agricult. Food Chem.*, **27**, 1326 (1979).
50. Dettmar, H. P., Barbour, G. S., Blackwell, K. T. *et al.*, *Comput. Chem.*, **20**, 261 (1996).

51. Martin, G. J., Founier, J. B., Allain, P. and Mauras, Y., *Analusis*, **25**, 7 (1997).
52. *IUPAC Isotopic Composition of the Elements 1997, J. Anal. At. Spectrom.*, **14**, 5N (1999).
53. Zoriy, M. V., Rashad, A., Pickhardt, C., Mohsen, H. J., Förstel, H., Helal, A. I., Zahran, N. F. and Becker, J. S., *At. Spectr.*, **24**, 195 (2003).

9.7 Geology and Geochemistry

The topics of geological and geochemical research comprise studies of the chemical and isotopic composition of elements and their distribution in geological materials in order to understand the coherence of the physical and chemical evolution of the Earth over about 4.5 billion years, to explain the origin of the solar system and also geochemical processes, e.g., insight into melting and fractionation processes during the formation of rocks or to explain weathering and changes in climate.

9.7.1 Sample Preparation Techniques for Geological Samples

As a function of the analytical task, different sample preparation steps have to be employed for geological material before mass spectrometric measurements. For the direct analysis of geological samples (e.g., of a small piece of mineral, rock or gemstone), for bulk analysis, for microlocal or in depth analysis, e.g., for the characterization of inhomogeneities, microinclusions (solid, fluid or gaseous inclusions) in geological samples by solid-state inorganic mass spectrometric techniques, such as SIMS, LA-ICP-MS and LIMS, only a few sample preparation steps are necessary and sometimes none at all. Due to possible surface contamination, it is recommended for some applications that the surface should be cleaned prior to analysis e.g., by dilute HNO_3, by presputtering in SIMS or pre-ablation in LA-ICP-MS. Practically no chemical sample preparation is performed for the analysis of powdered geological samples by LA-ICP-MS. Fine powdered geological material (conducting or non-conducting) can be measured directly by LA-ICP-MS as a thin layer fixed on double sided sticky tape.

Single particles are analyzed directly by being fixed on such tape or embedded in epoxy, and mass spectrometric measurements are carried out on cut and polished cross sections.

Chemical elements in compact geological samples, such as rocks, minerals, sediments, stones and soils etc., are mostly very inhomogeneously distributed so that for a reliable bulk analysis to determine matrix, minor and trace elements, the preparation of homogeneous samples by careful milling and homogenizing and pressing to targets or electrodes is recommended before measurements are made by solid-state mass spectrometry. In addition, for SSMS and direct current (dc) GDMS as solid-state mass spectrometric techniques, the non-conducting sample should be converted by pressing the mixed powdered sample with a carrier (binder) to form electrically conducting electrodes or targets. A less abundant metal in the geological sample in the high mass range is selected, with only one or at most two stable isotopes (such as high purity gold or silver powder), as the carrier material for producing electrically conducting electrodes or targets from non-conducting geological samples. On the other hand, high purity carbon powder as the binder is much cheaper and therefore recommended for SSMS and GDMS because carbon is rarely analyzed by inorganic mass spectrometry and very stable targets or electrodes for mass spectrometric measurements can be produced. Any possible contamination during electrode and target preparation is avoided by applying graphite of the highest purity as binder. In addition, it is advantageous to mix carbon with the sample because many carbon cluster ions of the type C_n^+

aid orientation on the mass scale in solid-state mass spectrometry if the mass resolution is sufficient to resolve isobaric interferences of analyte ions from carbon cluster ions. The typical carbon cluster ion distribution (C_n^+/C^+ as a function of cluster size, *n* observed in solid mass spectrometry (SSMS, GDMS and LA-ICP-MS) is compared in Figure 9.55. However, carbide cluster ion formation (MC_n^+) with matrix elements M (such as Si, B, W or REE) also occurs in plasmas, which results in a multitude of isobaric interferences (e.g. for ^{53}Cr and ^{54}Fe determination see Figure 6.9). These mixed and pressed graphite targets are also suitable for multi-element trace and isotope analysis using LA-ICP-MS. For LA-ICP-MS measurements, pressed powder pellets have been utilized in a similar way to those made for XRF analysis.[1] A simple and rapid method of stable target preparation for the bulk analysis of powdered samples by LA-ICP-MS, is the so-called 'glue technique'. The glue solution (methyl methacrylate resin, 10 g) is dissolved in acetone (100 ml). This solution (1 ml) is mixed with powdered sample (500 mg) until the mixture becomes a paste. The paste is shaped into a target which can be analyzed after being stored for 2 h at room temperature.[2] Another possibility is the preparation of homogeneous samples before measurement by direct fusion of geological samples to form prepared glassy targets[3–5] or by fusion with a lithium borate/metaborate mixture[6–9] as described in Chapter 6. The preparation of very homogeneous targets of geological sample with a lithium borate mixture (90 % $Li_2B_4O_7$ and 10 % $LiBO_2$) in a muffle furnace at 1050 °C from inhomogeneous geological materials is discussed briefly in Chapter 6.3 (see Figure 6.27).[6,10] The homogeneity of fused targets can be then verified by SIMS measurements or LA-ICP-MS itself. This sample preparation procedure has been applied, in LA-ICP-MS for example, for the determination of 20 or more trace elements in geological materials, such as fused basalt glasses[3] or zeolites.[9] Zeolites, which are microporous aluminosilicates, were utilized due to their catalytic and ion exchange performance in toxic and radioactive waste treatment. Detection limits determined in LA-ICP-MS on geological samples including zeolites via lithium borate fused targets, depending on the purity of the chemicals used, are in the low $\mu g\, g^{-1}$ and $ng\, g^{-1}$ concentration level as found by measurements with quadrupole based ICP-MS. These detection limits can be improved significantly by applying sector field ICP-MS due to better sensitivity and lower background signal (signal-to-noise ratio).[3] Odegard *et al.*[7] observed detection limits at the low $ng\, g^{-1}$ concentration level in double-focusing sector field LA-ICP-MS for most elements using lithium borate of the highest purity for preparing homogeneous fused targets. A rapid analytical method for the determination of trace elements in geological samples by LA-ICP-MS after direct fusion at high temperature to prepare homogeneous glasses was developed by Federowich *et al.*[11] Using a similar protocol, homogeneous silicate glasses (basalt, andesite, komatiite, peridotite, rhyolite and quartz-diorite) were prepared by directly fusing and stirring at 1400–1600 °C as proposed by Dingwell and co-workers (MPI-DING reference glasses).[3,4,12]

A thin metallic layer (e.g., Au) deposited on geological sample is helpful before direct mass spectrometric measurements because in SIMS charging effects occur due to the sputtering process of non-conducting samples.

The application of ICP-MS in geological research and for routine measurements requires a (time consuming) chemical digestion procedure of often difficult to dissolve samples followed by the dilution of digested solution. Geological samples are usually digested by conventional digestion procedures using concentrated HF-HNO_3 mixtures (10:1) in Teflon beakers. The Teflon beakers are placed in an ultrasonic bath in order to disaggregate granular material and render it more susceptible to acid attack.[13] To avoid possible contamination, it is recommended that sample material be digested in acid in a closed vessel in a microwave oven at controlled higher pressure and temperature (compared to an open vessel digestion on a hot plate). Leaching of soil or dust samples with acids or an acid mixture in an open vessel on a hot plate does not result, in general, in a quantitative transfer of all analytes. For example, the leaching of uranium from dust and soils with nitric acid is not complete and therefore this procedure is not recommended for determining the uranium content. However, leaching of geological samples can be applied in cases where no quantitative element concentration data are necessary (e.g., for uranium isotope ratio measurements).

Another analytical procedure for sample preparation including analyte separation and enrichment is the coprecipitation of the trace elements to be determined. The co-precipitation behaviour of Ti, Mo, Sn and Sb under two different fluoride forming conditions (at $< 70\,°C$ in an ultrasonic bath and at $245\,°C$ using a Teflon bomb) has been studied to improve the accuracy of the trace analysis of these elements in Ca–Al–Mg fluorides, by ICP-MS.[14] The applicability of this analytical method (including isotope dilution technique) was demonstrated for four carbonaceous chondrites and silicate reference materials of basalt or andesite.[14]

An analytical procedure for the determination of about 20 trace elements in coal samples after microwave digestion is described by Wang *et al.*[15] It was found that high temperature digestion with HNO_3 alone (250 °C, 7.5 MPa) led to an extensive decomposition of the organic matrix and clay in coal, whereby solid carbon remained in the final solution after evaporation. Most of the trace elements were quantitatively dissolved in three standard coals. A lost of Hg was observed during the evaporation step.

Matrix separation steps are required, in particular, for various applications of mass spectrometry for geochronological studies. To study the fine isotope variation of geological materials and to avoid isobaric interferences and matrix effects, a careful and selective separation of analytes at trace concentrations from bulk material, often including its enrichment, is proposed. Possible analytes for isotope ratio measurements are, for example, 'daughter products' of radioactive decay, such as Sr, Nd, Hf, Pb and Os, and their respective 'parents': Rb, Sm, Lu, U and Re. For example, in the Rb–Sr method for the age dating of geological samples, the isobaric interference of $^{87}Rb^+$ and $^{87}Sr^+$ is a serious problem. Mathematic correction of isobaric interference at $m/z = 87$ is often unsatisfactory. The interference problem can be solved only by careful chemical separation of both analytes, e.g., by using ion exchange chromatography. A matrix separation step follows after a complete microwave induced digestion of a defined amount of powdered geological sample (100–200 mg) in an acid mixture of HNO_3/HF (or on a hot plate at a temperature of about 200–250 °C). Removal of rubidium from the sample solution has been performed in the author's laboratory by means of extraction chromatography (using crown ether–Eichrom's Sr resin).[16]

In addition, selected rare elements, such as precious metals, were extracted from matrix elements and preconcentrated to avoid isobaric interferences with polyatomic ions as demonstrated for platinum determination in the presence of a relatively high concentration of Hf, resulting in isobaric interferences at $m/z = 194$, 195 and 196 due to HfO^+ diatomic ion formation.[17] An improved method for extracting marine sediment fractions and its application for Sr and Nd isotopic analysis is described by Bayon *et al.*[18]

Dowall *et al.*[19] have reported on the role of chemical pre-concentration in isotope geochemistry.

9.7.2 Fractionation Effects in LA-ICP-MS

A serious problem in LA-ICP-MS described in the literature on many occasions is the time-dependent elemental fraction (so-called 'ablation fractionation') occurring during laser ablation and the transport process of ablated material, or during atomization and ionization processes in the inductively coupled plasma.[20–22] Numerous papers focus on the study of fraction effects in LA-ICP-MS as a function of experimental parameters applied during laser ablation (such as laser energy, laser power density, laser pulse duration, wave length of laser beam, ablation spot size, repetition rate, beam profile, the physical, chemical and optical properties of the sample surface and others).[23,24] Elemental fractionation during the laser ablation of solid materials has been observed as time dependent ion intensity variation (e.g., of Pb/U, Pb/Th and U/Th ratios in geochronological studies), which results in analytical errors in the LA-ICP-MS measurements. It was found that fractionation effects during laser ablation in LA-ICP-MS strongly depend on the wavelength of the laser used, but ultimately on the laser energy E and following on laser power density Φ applied. By reducing the wavelength from 1064 nm (Nd:YAG) to 532 nm (second harmonic of Nd:YAG), 266 nm (fourth harmonic) to 213 nm (fifth harmonic) and 193 nm (ArF excimer laser) a decrease in fractionation effects was observed. This correlates with a decrease in mean particle sizes ablated from the solid surface.[23] Fractionation effects are also described as a function of diameter-to-depth ratios of the ablated laser crater.[25,26]

An important parameter in order to avoid elemental fractionation in laser ablation ICP-MS is the laser power density (Φ) which is a function of laser energy, of laser pulse duration and focusing conditions, as described in the following equation:

$$\Phi = E/t \times a \ [\mathrm{W\,cm^{-2}}] \tag{9.1}$$

where E = energy of laser beam in J, t = the duration of laser pulse in s and a = the area of laser crater in $\mathrm{cm^2}$. Elemental fractionation effects – i.e., the preferred ablation of an element with low evaporation point compared to an element with a higher one – are serious problems, especially if small crater sizes in single point analysis (single laser shot) are applied and/or if the laser power density is smaller than $10^9\,\mathrm{W\,cm^{-2}}$. The laser power density Φ affects the relative sensitivity coefficients of elements (RSC= measured value/certified element concentration or recommend value) as illustrated in Figure 2.20. The optimum range for laser power density in laser mass spectrometry is observed from 10^9–$10^{10}\,\mathrm{W\,cm^{-2}}$. In this range of laser power density the RSCs of most elements are about one. With decreasing Φ, the fractionation effects and consequently the RSCs of elements (independent of ionization potential) change. Fractionation of trace elements during laser ablation of glass and copper was found at a laser power density of $0.62 \times 10^4\,\mathrm{W\,cm^{-2}}$ to $1.35 \times 10^4\,\mathrm{W\,cm^{-2}}$, as described by Outridge *et al.*[21] or by laser ablation in a single shot regime. Elemental fractionation can thus be observed during depth profiling in LA-ICP-MS and LIMS (in single point mode) due to possible defocusing of the laser beam as a function of depth. During the measurements a decrease in the ion intensity of the analyte with increasing depth of laser crater is observed due to decreasing laser power density during the measurements. Minimum element fractionations are achieved if the solid sample is ablated with the laser beam under constant focusing conditions and if the laser power density is higher than $10^9\,\mathrm{W\,cm^{-2}}$ as demonstrated in several papers.[8,9,27–29] Practical recommendations for routine analytics by LA-ICP-MS are: (a) depending on the matrix composition and analytes to be determined, each analytical task requires an optimization of the analytical procedure with respect to experimental parameters (to obtain maximum analyte ion intensity and minimum isobaric interferences), (b) possible fractionation and matrix effects should be minimized under optimal operating conditions, and the mass spectrometric measurements should be carried out at exactly constant experimental parameters (laser ablation at laser power density $\Phi > 10^9\,\mathrm{W\,cm^{-2}}$) to avoid fractionation effects and (c) an internal standard element should be employed for the correction of plasma instabilities of instrumental drifts.

9.7.3 Multi-element Analysis of Geological Samples

For many decades, spark source mass spectrometry (SSMS) was the method of choice for survey analyses of quite different types of geological materials, especially for the sensitive multi-element trace element analysis of geological samples with detection limits in the $\mu g\,g^{-1}$ and $ng\,g^{-1}$ concentration range.[30–35] SSMS has been employed for the characterization of a multitude of geological samples with respect to chemical composition (quantitative determination of matrix, minor and trace elements, especially for REE analysis). Table 9.31 summarizes the results of the determination of rare earth elements (REEs) in zircon and monazite from the Precambrian Baltic Shield by SSMS. In monazite, high concentrations of light REEs were observed and the element concentrations decreased with increasing mass. The typical alternating element distribution with higher concentrations for REEs with even atomic numbers, compared to those with odd atomic numbers is clearly demonstrated by the monazite sample. For zircon from the Precambrian Baltic Shield relatively low concentrations for lighter REEs were found. The measurements were performed using a double-focusing SSMS with Mattauch–Herzog geometry (see Figure 5.24) and an ion sensitive photoplate for ion detection, which allows the whole mass range to be detected from 7–238 u in one mass spectrum.[36,37] For geological exploration purposes, trace element measurements of precious metals or rare earth elements (REEs) in rocks and minerals are of special interest.[17,38] REE determination in geological materials can provide insights into melting and fractionation processes during mineral formation.[39] The composition of REEs in zircons where an increasing concentration for heavy elements was found, has been studied by different analytical techniques.[31,40]

Table 9.31 *Determination of rare earth elements in Precambrian zircon and monazite from Baltic Shield by SSMS (concentration in $\mu g\,g^{-1}$)*[36].

	Zircon	Monazite		Zircon	Monazite
La	22	39 600	Tb	19	1 015
Ce	90	79 700	Yb	910	320
Pr	7.2	13 600	Dy	150	2 900
Nd	45	54 000	Er	380	900
Sm	18	8 860	Tm	85	80
Eu	50	265	Ho	80	490
Gd	50	8 800	Lu	120	23

For the analysis of geological samples, Jochum *et al.*[41] equipped a SSMS instrument (AEI.MS 702R) with a multiple ion counting system (MIC-SSMS) in order to improve the SSMS capability. The authors studied geological reference materials from the United States Geological Survey (USGS), MPI-DING reference glasses and Hawaiian basalt samples. Compared to conventional SSMS measurements with photoplate ion detection, the detection limits were improved by using MIC-SSMS down to the $ng\,g^{-1}$ range. It was thus possible to investigate such geochemically interesting rare elements as Nb, Ta, Zr, Hf and Y. Using MIC-SSMS, eight geological standard glasses were characterized by Stoll and Jochum.[42] The characterization of eight geological MPI-DING reference glasses for *in situ* microanalysis by a variety of bulk and microanalytical methods (such as SSMS, LIMS, SIMS, INAA, LA-ICP-MS, ICP-MS, TIMS, XRF and PIXE) in a number of laboratories is presented by Jochum *et al.*[3,4] Reference values with RSDs between 1 % and 10 % were calculated from the analytical data of 60 elements. Today, these homogeneous silicate glasses are investigated intensively by LA-ICP-MS as a sensitive multi-element trace analytical technique for isotope ratio measurements in different laboratories.[4,8,12,23,43,44] Certified standard reference materials (such as NIST glass SRM series or BCR-2G basaltic glass prepared from BCR-2 reference material[13]) are utilized for the multi-element analysis of geological samples, in general, and of the MPI-DING glasses, in particular, using LA-ICP-MS. NIST glass SRMs are usually employed for the calibration of geological samples. Figure 9.47 compares the matrix composition of glass NIST SRM 612 with that of the analyzed geological glasses described in reference.[43] It can be clearly seen that the matrix composition of the fused BCR 2G geological glass standard agrees better with those of the geological glasses investigated.[43] Therefore, BCR 2G geological glass standard was used for quantification in the author's laboratory.[3,43] In addition, solution based calibration in LA-ICP-MS (see Section 6.2.6.) was applied for quantification purposes on homogeneous geological glass samples, and the concentrations for 26 trace elements of direct bulk analysis on geological glass samples were compared with those on fused lithium borate targets and the reference values.[8] LA-ICP-MS with solution based calibration provided analytical results with a deviation of 1–13 % from the reference value for most of the elements.[8] The results of LA-ICP-MS on MPI-DING Geological Glass Standard Basalt ML3B-G (Mauna Loa tholeiitic basalt glass) in comparison to reference values are summarized in Table 9.32. Good agreement with the reference value was measured for most trace elements.[3,43]

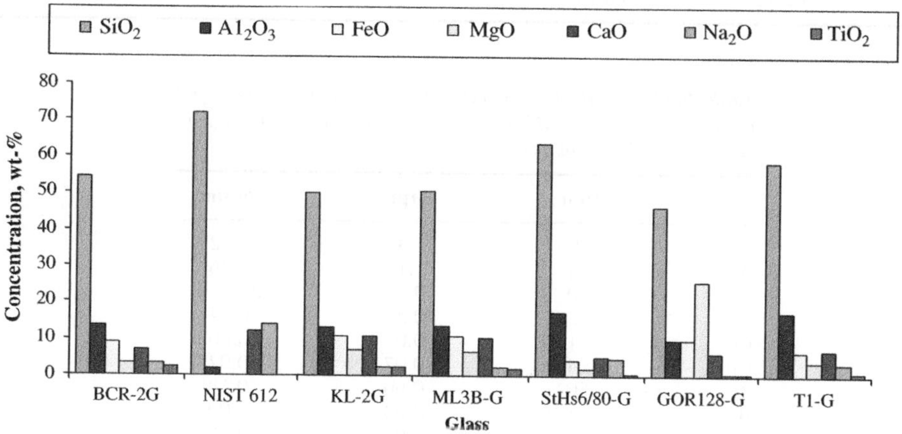

Figure 9.47 *Comparison of matrix composition of geological glasses (BCR- 2G – basaltic glass prepared from BCR-2 reference material,[127,13] theoleiitic basalt samples KL2-G and ML3B-G from the Kilauea and Mauna Loa volcanoes of Hawaii, respectively,[128] St Helens andensite ash StHs6-80-G, Gorgona Island komatiite glass GOR 128-G,[129] Italian Alps quartz diorite glass Ti-G[130] and glass NIST SRM 612.[43] (J. S. Becker, C. Pickhardt and H. J. Dietze, Mikrochim. Acta,* ***135****, 71 (2000). Reproduced by permission of Springer Science and Business Media.)*

In general, for the bulk analysis of geological samples by LA-ICP-MS, measurements are carried out with a relatively large laser beam diameter (100 μm–400 μm).[9,43,45] Since possible small heterogeneities in the μm range do not influence the analytical results, the precision and accuracy of the analytical data is improved compared to LA-ICP-MS measurements at a small laser beam diameter and the limits of detection (LODs) for elements are optimized. The LODs of elements depend on the laser beam parameters, laser power density, wavelength of laser beam and the matrix composition. The LODs of selected elements in LA-ICP-MS for different spot (laser crater) diameters of 4 μm, 16 μm, 30 μm and 60 μm in analyzing MPI-DING reference glass T1-G (Italian Alps quartz diorite glass) are compared in Table 9.33.[23] The LODs of elements measured by LA-ICP-MS via the direct multi-element trace analysis of homogeneous geological (glass) sample

Table 9.32 Results of LA-ICP-MS on MPI-DING Geological Glass Standard Basalt ML3B-G in comparison to reference value (concentration in $\mu g\, g^{-1}$)[8].

Basalt ML3B-G					
	Measured value	**Reference value**		**Measured value**	**Reference value**
Sc	30.8 ± 1.5	31.3 ± 0.8	Sm	4.91 ± 0.13	4.79 ± 0.05
Cr	143 ± 6	170 ± 10	Eu	1.68 ± 0.06	1.68 ± 0.01
Co	38.2 ± 0.7	39 ± 3	Gd	5.29 ± 0.20	5.23 ± 0.08
Cu	108 ± 2	115 ± 4	Dy	4.85 ± 0.03	4.81 ± 0.06
Ni	108 ± 5	105 ± 2	Ho	0.93 ± 0.04	0.91 ± 0.01
Y	24.0 ± 0.6	24.3 ± 0.7	Er	2.44 ± 0.09	2.46 ± 0.05
Zr	126 ± 2	126 ± 2	Tm	0.33 ± 0.01	0.326 ± 0.004
Cs	0.12 ± 0.01	0.14 ± 0.1	Yb	2.00 ± 0.09	2.05 ± 0.02
La	8.8 ± 0.1	8.96 ± 0.07	Lu	0.26 ± 0.01	0.286 ± 0.005
Pr	3.48 ± 0.05	3.47 ± 0.04	Ta	0.53 ± 0.05	0.55 ± 0.01
Nd	170 ± 0.2	16.8 ± 0.1			

Table 9.33 Limits of detection ($\mu g\, g^{-1}$) for different laser spot diameters analyzing MPI-DING reference glass T1-G (Italian Alps quartz diorite glass)[23].

	4 μm	30 μm	60 μm
Na	79	1.3	0.22
Al	84	1.1	0.36
Si	5800	70	29
Ti	2.3	0.25	0.14
Co	1.8	0.02	0.005
Cu	4.2	0.07	0.022
Zn	50	0.66	0.18
Ba	1.0	0.03	0.005
Eu	0.65	0.002	0.0007
W	0.51	0.004	0.003
Pb	2.2	0.02	0.007

varied from the $\mu g\, g^{-1}$ down to the sub-$ng\, g^{-1}$ range. With increasing laser spot diameter, more material is ablated and consequently the LODs for the elements studied decreased. The lowest limit of detection in MPI-DING reference glass T1-G was observed for the rare earth element lutetium (at spot diameter – 60 μm) at 0.2 $ng\, g^{-1}$.[23] Interesting approaches for enhancing elemental sensitivity in LA-ICP-MS include using helium as the carrier gas[46] and modifying the interface to obtain a higher ion extraction efficiency (smaller sample cone orifice and lower background pressure).[47]

The quantitative analysis of silicate reference materials (BCR-2G, ATHO and NIST glass 610) using solution based calibration in LA-ICP-MS is described by Halicz and Günther.[48] The calibration solution in an argon stream was introduced into the ICP by an Aridus micronebulizer after mixing with the laser ablated material using He as the carrier gas in a glass coaxial bulb mixer. An ArF 193 nm excimer laser system (GeoLaserC, Göttingen, Germany) was employed for the laser ablation of silicates. In order to quantify the analytical data it was assumed that the sum of all of the determined analyte oxides was 100 % (altogether 42 elements including the rare earth elements were analyzed). Unknown amounts of water, Cl or CO_2 and any other elements or molecules not measurable by LA-ICP-MS can contribute to the uncertainty.[48] This calibration technique requires an optimization of two gas flow rates (the He carrier gas flow rate from the laser ablation chamber and the Ar gas flow rate from the Aridus nebulizer) whereby the sum of the two (He and Ar) gas flows of $1.31\,min^{-1}$ is higher than the optimized flow rates for each sample introduction tool. Therefore, the online solution based calibration procedure in LA-ICP-MS as developed in the author's laboratory is advantageous.[8] For this calibration strategy an ultrasonic nebulizer (USN) or micronebulizer with desolvator (Aridus) was coupled directly to the laser ablation chamber of an LA-ICP-MS system. In the proposed arrangement (see Figure 6.19) the nebulizer gas flow coming from the USN or Aridus nebulizer is also used as the carrier gas flow for the transport of laser ablated material from the solid sample (with a Nd-YAG laser at a wavelength of 266 nm) into the ICP. The optimized gas flow rate of the coupling technique is the same for both sample introduction systems (laser ablation of solid sample and nebulization of standard solution) and consequently optimal signal-to-noise ratios were obtained. The nebulized analytes in the calibration solution transported by argon into the laser ablation chamber are mixed with the ablated material by a focused laser beam in this laser ablation chamber. Two different calibration procedures via calibration curves and by the application of relative sensitivity coefficients (RSCs) are compared, whereby in solution-based calibration using LA-ICP-MS, mostly performed via calibration curves, more accurate element concentrations are obtained compared to one-point calibration using RSCs. LA-ICP-MS with solution based calibration can provide analytical data for 26 trace elements (or even more) with a relative standard deviation of 2–10 % compared to the reference value for most elements investigated in the basalt sample from the Hawaiian Mauna Loa volcano (ML3B) and the andesitic ash from the St Helens (USA) eruption (StHs6/80).[8]

Inorganic mass spectrometric techniques provide suitable analytical procedures for fast qualitative and semiquantitative characterization (fingerprinting) of trace elements in unknown geological samples. In particular, laser ablation as a sample introduction technique for solid materials in ICP-MS offers many advantages including application to the multi-element and isotope analysis of electrically conducting and also of non-conducting materials involving no or hardly any sample preparation steps. LA-ICP-MS allows a rapid survey analysis of an unknown material (much faster and less expensive than by SSMS or GDMS) with respect to major, minor and trace elements in geological samples and has become the most versatile analytical technique for sensitive multi-element analysis in geological samples.[3,4,6,8,9,12,45,49]

A minimally destructive fingerprinting of gemstones (sapphires) by LA-ICP-MS to distinguish sources from different locations is discussed by Guillong and Günther.[50] A homogenized 193 excimer laser with flat-top beam profile (pulse energy – 70 mJ, laser crater diameter – 120 μm, depth – 1.2 μm) was used for the determination of 12 minor and trace elements. Al was used as the internal standard element, and NIST glass SRM 612 for calibration as an external standard. The reproducibility of the trace element determination was limited by the heterogeneity of the gemstones. The limits of detection for most light elements were in the $\mu g\,g^{-1}$ range, and for heavier elements $< 1\,\mu g\,g^{-1}$ (down to $0.12\,\mu g\,g^{-1}$ for Pr). The detection limits of about 40 elements in LA-ICP-MS are significantly improved in comparison to those determined by XRF (seven elements were analyzed only). The mean amount of sample ablated was 55 ng. The authors discussed that LA-ICP-MS is significantly less expensive than SIMS and more sensitive than laser induced breakdown spectroscopy (LIBS) for the rapid and direct multi-element analysis – even of ultratrace elements – of gemstones with the possibility of microlocal analysis with spatial resolution.[51]

One of the most stable minerals investigated in geosciences is zircon ($ZrSiO_4$). The geochemically extremely robust naturally occurring zircon crystals[32,52–54] are often investigated because they include information about complex geological history. The, for geologists, very famous zircon single crystal ('91500') with an original mass of 238 g provided by the Harvard Mineralogical Museum in Cambridge, Massachusetts, was utilized for geochemical studies. The original collection site of the sample is recorded as Kuehl Lake in Ontario, Canada. An interlaboratory characterization of the single zircon crystal ('91500', also used as a zircon standard with known U–Pb isotope ratio for *in situ* age dating) with respect to content of REEs, Y, Ba, Pb, Sr, Nb, Hf, Nb, Th and U, using LA-ICP-MS, ICP-MS and SIMS compared to EMPA (electron microprobe analysis) is discussed by Wiedenbeck *et al.*[54] In comparing SIMS trace element determination for REEs, a calibration was made of Y and Hf against NIST SRM 610 glass using the recommended values of Pearce *et al.*[55] Schmidt *et al.*[56] measured the trace element compositions of the coexisting gabbroic and granitic melts by LA-ICP-MS and demonstrated that the effect of melt structure contributes about one order of magnitude to crystal/melt partition coefficients. The quantitative determination of some major and transition trace elements in several silicate reference materials by LA-ICP-MS was studied by Jarvis and Williams in 1993.[1] The accuracy of major element determination was generally better than ±5 % with a precision of 10 % RSD. The limits of detection were in the range of 0.05 $\mu g\,g^{-1}$ (for U)–13 $\mu g\,g^{-1}$ (for Ca).[1] A cordiente sample (42/IA) from Kiranur, South India, and an ilvaite sample (ISX1) from Serifos, Greece, were characterized by different high precision analytical methods and prepared for use as potential electron microprobe reference samples.[57] By combining LA-ICP-MS, ICP-OES and EMPA (electron microprobe analysis), the major, minor and trace element compositions of these geological samples were determined and their near stoichiometric composition and high homogeneity at the μm level demonstrated.[57]

For multi-element analysis, ICP-MS is usually applied after digestion of geological samples as a rapid analytical method with a large dynamic range. After digestion of geological material, a possible inhomogeneous element distribution is no longer of significance if a representative amount of homogenized sample material is digested and diluted. A simple method for the precise determination of more than 40 trace elements in geological samples by ICP-MS is described by Eggins *et al.*[13] The combination of enriched isotopes along with conventional internal standard elements permits (by correction of mass dependent element sensitivities) the precise determination of trace elements over the whole mass range in a broad spectrum of geological matrices such as magmatic rocks. A number of element pairs possess virtually identical ionic radii for a given coordination number and are well known for their similar geochemical behaviours, in particular Zr and Hf, Nb and Ta, and Y and Ho. ICP-MS has been applied to demonstrate subtle geochemical differences by the analysis of Nb, Ta, Zr and Hf in magmas from ocean islands and subduction zones.[13] The application of the multi-element isotope dilution technique in sector field ICP-MS for quantitative trace analysis on geological reference materials after digestion, has been investigated by Willbold and Jochum.[58]

9.7.4 Trace Analysis of Selected Elements in Geological Materials

Special analytical tasks for the inorganic sciences in geochemistry require the trace element analysis of selected elements in geological samples. For the direct determination of chlorine, bromine and iodine in powdered geological and environmental samples, Boulyga and Heumann applied isotope dilution LA-ICP-MS with a special powerful laser ablation system (so-called 'LINA-Spark™- Atomizer').[59] Under optimized experimental conditions, concentrations in the range of 30 $\mu g\,g^{-1}$–16 $mg\,g^{-1}$ for chlorine, 2–140 $\mu g\,g^{-1}$ for bromine and 0.1–31 $\mu g\,g^{-1}$ for iodine in two sediment reference materials (Estuarine Sediment – SRM 1646; Buffalo River Sediment – SRM 2704) and three rock reference samples (granite – GS-N; bauxite – BX-N and disthene – DT-N) were determined by isotope dilution LA-ICP-MS. The detection limits of LA-ICP-MS are 8 $\mu g\,g^{-1}$

for chlorine, 1.7 μg g^{-1} for bromine and 0.1 μg g^{-1} for iodine. Quite a good correlation of measured values with the indicative values was found (see Figure 9.48).[59]

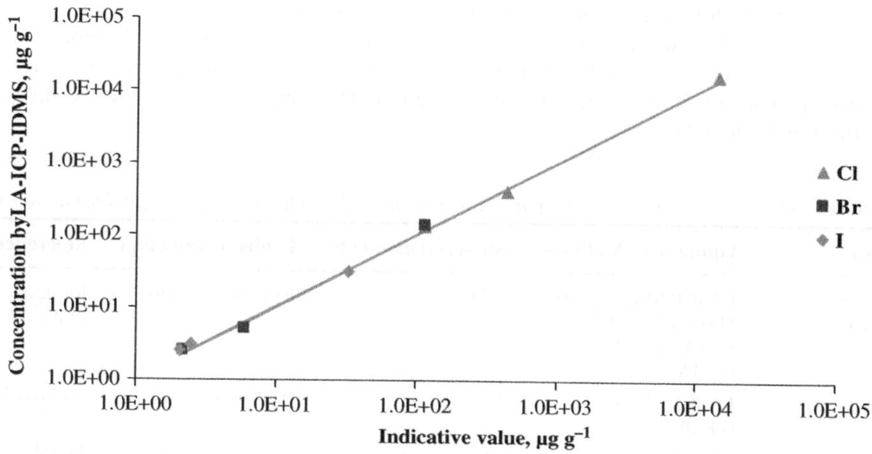

Figure 9.48 *Correlation of halogen concentration measured in sediments and rock samples by isotope dilution LA-ICP-MS using a high ablation system LINA-Spark™-Atomizer and double-focusing sector-field ICP-MS 'Element 2' with the indicative values;[59] Samples: SRM 2704 (Buffalo River Sediment), SRM 1646 (Estuarine Sediment), Granite GS-N, Bauxite BX-N and Disthene DT-N. (S. Boulyga and K. G. Heumann, Int. J. Mass Spectrom.,* ***242****, 291 (2005). Reproduced by permission of Elsevier.)*

As a petrogenic tracer in glassy rock samples from (Köfels, Ötz Valley, Tyrol), iridium was analyzed after microwave induced digestion of the silicate matrix (using a mixture of HNO_3-HCl-HF) with respect to its content, by double-focusing sector field ICP-MS.[60] The procedure for iridium matrix separation by anion exchange was investigated and optimized. The limits of detection for iridium determination by ICP-SFMS (at low mass resolution) using, in all approaches, an ultrasonic nebulizer with membrane desolvation, was 6 pg g^{-1}, compared to 15 pg g^{-1} at a mass resolution of 9500 (the USN was used without membrane desolvation) and 59 pg g^{-1} for the ion exchange procedure. The iridium concentration in two köfelsite stones was found to be 150 pg g^{-1} and 260 pg g^{-1}.[60]

The analysis of platinum group elements (PGEs at trace concentration level such as Os, Ir, Ru, Pt and Re) in kimberlite and other geological materials (komatiite and periodite) after chemical separation by double-focusing sector field ICP-SFMS was carried out by Pretorius *et al.*[61] The detection limits were found to be in the low pg g^{-1} range and below (e.g., 0.9 pg g^{-1}) in aqueous solution. The concentrations of all PGEs were measured in the low ng g^{-1} range, in general, in the following sequence: Pt > Ru > Ir. Re was found in the sub-ng g^{-1} range.[61] The heterogeneity of selected PGEs in geological materials, which often results in irreproducible results, was investigated on the Re–Os system using ICP-QMS, MC-ICP-MS and TIMS.[62] It was demonstrated, by the determination of Re, Os and Ir for three pyroxenite and three komatiite samples, that sample inhomogeneity is the major contributor to the high variability of analytical results.[62] An analytical procedure for palladium determination in rock samples (certified rock reference materials GBW 07291 and GWB 07293 from the Institute of Geophysical and Geochemical Exploration, Langfang,

China) after digestion with aqua regia and using flow injection online solid phase extraction ICP-MS in order to minimize the isobaric interferences is discussed by Fang *et al.*[63]

Precise and accurate measurements of ultralow Nb, Ta, Zr and Hf concentrations and the chondritic Zr/Hf and Nb/Ta ratios by MC-ICP-MS (Isoprobe, Micromass) were examined by Weyer *et al.*[64] For the development of the isotope dilution technique, enriched ^{180}Ta, ^{94}Zr and ^{180}Hf isotopes were applied for quantitative Zr, Ta and Hf determination, and the monoisotopic element Nb was measured relative to Zr after quantitative separation from the matrix by ion exchange.

Selected applications of inorganic mass spectrometry in trace analysis of geological samples are summarized in Table 9.34.[3,8,9,52]

Table 9.34 *Selected applications of inorganic mass spectrometry in trace analysis of geological samples.*

Samples	Equipment/ Methods	Analyzed elements	Limits of detection	References
MPI-DING reference glasses	LA-ICP-MS, ICP-MS SIMS, SSMS,TIMS, INAA, XRF, PIXE, EPMA	~70	low ng g^{-1} range	Jochum *et al.*[3,4]
Zeolites	LA-ICP-QMS (Elan 6000)	38	0.03 μg g^{-1} (U)	Pickhardt *et al.*[9]
Zircons (Egypt)	LA-ICP-MS, JMS-Plasmax2 SEM-EDX, RBS	13	sub-μg g^{-1}	Helal *et al.*[52]
Basalt, andesite	LA-ICP-MS, Elan 6000 (solution based calibration)	27	0.0006 μg g^{-1} (U)	Pickhardt *et al.*[8]
Kimberlite, komatiite Periodotite	ICP-SFMS (ELEMENT), extraction	PGE – Os, Ir, Ru, Pt, Re	0.002 8 pg g^{-1} (Os) 0.00016 pg g^{-1} (Rh)	Pretorius[61]

9.7.5 Isotope Analysis Including Age Determination of Minerals and Rocks by Mass Spectrometry

Study of Isotope Fine Variation in Nature

Improvements to the precision and accuracy of isotope ratio measurements, especially by the introduction of mass spectrometers with multiple ion collectors, in recent years, has led to increasing interest in isotope ratio measurements in many fields of geosciences to study fine variation of stable isotopes, especially of light elements (such as Li, C, O, S, N and H), or the age dating of minerals by geochronology. Geological interest in lithium isotope analysis has arisen from its use as a powerful tracer of recycled crustal materials (e.g., oceanic crust and sediments) that find their way back into the source regions of mantle-derived melts.[65] The lithium isotopic composition of glass standard reference materials from the United States Geological Survey, USGS (the natural glass samples: BHVO-2G, BCR-2G and BIR-2G, synthetic glasses of basaltic composition: GSA-1G, GSC-1G, GSD-1G and GSE-1G) and from the National Institute of Standards and Technology, NIST (NIST SRM 610, SRM 612 and SRM 614 with the nominal concentration of 61 trace elements of 500, 50 and 5 μg g^{-1}, respectively) were determined by multi-collector ICP-MS (Neptune,

Thermo Fisher Scientific), TIMS (Triton, Thermo Fisher Scientific) and SIMS (Cameca 4f), as discussed by Kasemann *et al.*[65] USGS glasses of basaltic composition provide a relatively new set of microanalytical calibration materials in addition to the NIST glasses SRM commonly used as reference materials in geological research and for routine measurements. The Li isotope data of all glass samples measured by MC-ICP-MS and MC-TIMS agree within the analytical error of 2‰, independently of the matrix composition of the geological sample or Li concentration. For SIMS analysis, the authors proposed the use of the USGS basaltic glasses GSD-1G (δ^7Li 31.14 ± 0.8‰, 2σ) and BCR-2G (δ^7Li 4.08 ± 1.0‰, 2σ) as suitable standards that cover a wide range of Li isotope compositions.[65] Lithium isotope analysis of the silica rich NIST 600 glass series, measured by MC-ICP-MS and MC-TIMS, agrees with a precision of 0.8‰ (RSD).

Tanimizu and Hirata studied natural isotopic variation of terrestrial nickel using MC-ICP-MS (Nu Plasma 500, Nu Instruments, Wrexham, UK) equipped with a fixed Faraday detector array with a variable zoom lens system.[66] An external precision was routinely obtained by MC-ICP-MS at 0.02 % and 0.01 % ($2\ \sigma$) for ^{58}Ni/^{62}Ni and ^{60}Ni/^{62}Ni isotope ratios on Ni isotopic standard reference material (NIST SRM 986) by correcting the mass discrimination effect with a ^{65}Cu/^{63}Cu isotope ratio of 0.4456. Very small isotope variations in mafic and ultramafic rocks, in iron meteorite (Campo del Cielo) and man-made Ni products were detected. Tungsten isotope ratios in six geochemical reference materials, using the rhenium external correction approach, were analyzed by the same working group.[67] The reproducibility in tungsten isotope ratio measurements was better than 0.05‰ for all isotope ratios. These measurements clearly demonstrate that tungsten isotopes can be used as a new tool for stable isotope geochemistry to study fine variation in nature, e.g., the radiogenic growth of the ^{182}W isotope due to the radioactive decay of ^{182}Hf ($t_{1/2} = 9 \times 10^6$ a) in the framework of Hf–W chronometry. An analytical method using an MC-ICP-MS (Neptune, Thermo Fisher Scientific) in medium mass resolution mode has been developed to study very small isotope variations of iron in inorganic matrices (including geological samples).[68] Small isotope variations of Eu in two pieces of Allende meteorite (in the ‰ range), measured by MC-ICP-MS, were detected by Moynier *et al.*[69] (see Section 9.8).

Isotope analysis by SIMS can exhibit matrix dependent mass discrimination effects. For SIMS, due to the known significant matrix effects, a systematic error was found in isotope ratio measurements, especially for light elements such as lithium. *In situ* studies using lithium and boron isotopes as geochemical markers were carried out for the characterization of well prepared homogeneous reference glassy samples (basaltic rock JB-2 from the Geological Survey of Japan) by SIMS investigations.[5] The main sources of instrumental mass discrimination (mass bias) occurring in geological materials are inherent in secondary ion sputtering and ionization, transmission of secondary ions (e.g., in ToF mass spectrometers), and ion detection.[70] Consequently, a careful calibration of isotope ratio measurements using oxygen isotope standard reference materials is required[71]

The zircon reference material '91500' discussed before seems to be very suitable for calibrating *in situ* oxygen isotope analysis. For example, SIMS was employed to study the δ^{18}O value on the zircon single crystal '91500'. No systematic differences in SIMS δ^{18}O data were observed between individual fragments of zircon crystal.[54] Peck *et al.*[72] studied oxygen isotope ratios by ion microprobe analysis for evidence of oxygen isotope variation in continental crust and oceans in 3.3 to 4.4×10^9 year old zircons. In addition, the authors also analyzed rare earth elements (REEs).[72]

Age Dating Methods in Geosciences

Geochronology for age dating of minerals is a major field in earth sciences, includes measurements of isotope ratios and requires advanced sensitive, precise and accurate mass spectrometric techniques. The determination of the age of the Earth has been of significant interest for hundreds of years, but it was not until radioactivity was discovered at close of the nineteenth century that the possibility of a physical estimate became feasible.[73] As is well known, the most exact way of determining the ages of rocks depends upon the regularity of radioactive processes, as Paneth *et al.*[74] recognized in 1928.

The physical methods of geochronology are based on the radioactive decay of unstable (parent) radionuclides into stable (daughter) isotopes such as ^{87}Rb to ^{87}Sr; ^{238}U to ^{206}Pb; ^{232}Th to ^{208}Pb; ^{235}U to ^{207}Pb; ^{147}Sm to ^{143}Nd; ^{187}Re to ^{187}Os; ^{176}Lu to ^{176}Hf and ^{40}K to ^{40}Ar. There are a huge number of papers on this field in the geological sciences including element determination, isotope analysis and age dating by inorganic mass spectrometry, so that only a few relevant works will be discussed to characterize the state-of-the-art in this topic. Selected age dating techniques in terrestrial geological samples and meteorites based on natural radioactivity, the abundance of mother and daughter isotopes, the half-life, the radioactive decay and preferable application fields in geochronology are summarized in Table 9.35. The geochronological techniques in this table are ordered according to the mass of their parent nuclides, whereas in the following sub-sections the order is related to their significance in geosciences.

U–Pb, Th–Pb and Pb–Pb Methods for Age Dating

Age dating using the U, Th–Pb methods is based on the radioactive decay of ^{238}U, ^{235}U and ^{232}Th isotopes (as parent radioisotopes of a chain of radioactive daughters ending with stable isotopes of radiogenic lead) to ^{206}Pb, ^{207}Pb and ^{208}Pb, respectively, via the well known uranium and thorium decay lines. These dating techniques are the best known, most versatile and valuable geochronological techniques.[75] For example, the decay of ^{238}U with a half-life of $t_{1/2} \sim 4.5 \times 10^9$ a gives rise to the uranium series (uranium–radium decay line) via eight α (^4He) decays and six β^- decays, which include ^{234}U as an intermediate daughter isotope and end in stable ^{206}Pb:

$$^{238}U \rightarrow {}^{206}Pb + 8\,^4He + 6\beta^- + Q \tag{9.2}$$

where Q represents the sum of the decay energies ($Q = 47.4\,MeV\,atom^{-1}$).[76]

The decay of radioisotope ^{238}U to stable ^{206}Pb in a closed system is described by the following equation:

$$^{206}Pb/^{204}Pb = (^{206}Pb/^{204}Pb)_0 + (^{238}U/^{204}Pb)(e^{\lambda t} - 1) \tag{9.3}$$

with $^{206}Pb/^{204}Pb$ = ratio of lead isotopes at the present time t; $(^{206}Pb/^{204}Pb)_0$ = the initial lead isotope ratio incorporated into the system at the time of its formation t years ago, i.e. $(t=0)$; $(^{238}U/^{204}Pb)$ = ratio of mother radioisotope ^{238}U and stable daughter isotope ^{204}Pb at the present time t; λ = decay constant of ^{238}U ($1.55 \times 10^{-10}\,a^{-1}$); and t = time elapsed since the system became closed to U and Pb.

Similar equations describe the decays of radionuclide ^{235}U to radiogenic ^{207}Pb and of ^{232}Th to ^{208}Pb. The decay of radionuclide ^{235}U ($t_{1/2} \sim 0.72 \times 10^9$ a) gives rise to the actinium series, which ends in the stable ^{207}Pb after the emission of seven α (^4He) decays and four β^- decays:

$$^{235}U \rightarrow {}^{207}Pb + 7\,^4He + 4\beta^- + Q \tag{9.4}$$

where $Q = 45.2\,MeV\,atom$.

The decay of radionuclide ^{232}Th ($t_{1/2} \sim 1.41 \times 10^{10}$ a) via the thorium decay line results in the emission of six alpha and four beta particles leading to the formation of the stable and most abundant lead isotope ^{208}Pb:

$$^{232}Th \rightarrow {}^{208}Pb + 6\,^4He + 4\beta^- + Q \tag{9.5}$$

where $Q = 39.8\,MeV\,atom^{-1}$.

Table 9.35 Natural radioactivity used for age dating in terrestrial geological samples and meterorites.

Mother isotope	Abundance of mother isotope [a], %	Half life $t_{1/2}$, a	Radioactive decay	Stable radiogene daughter nuclide	Abundance of daughter isotope [a], %	Preferable application fields in geochronology
^{14}C	$^{14}C/^{12}C = 1.2 \cdot 10^{-12}$	5730 ± 40	β^-	^{14}N	99.634	Carbonaceous material up to ~ 60 000 a
^{40}K	0.0117	1.26×10^9	β^- (88.8 %) K capture (11.2 %)	^{40}Ca ^{40}Ar	96.93 99.6	Rarely used in K rich minerals, K bearing minerals and rocks (biotite, muskovite, feldspar) Precambrian age dating
^{87}Rb	27.83	4.88×10^{10}	β^-	^{87}Sr	7.00	Rb bearing minerals (granite, biotite, felspar, mica, sediments) $\sim 0.1–0.5 \times 10^9$ a
^{147}Sm	14.99	1.06×10^{11}	α	^{143}Nd	12.2	Igneous rocks, meteorites Precambrian age dating
^{176}Lu ^{187}Re	2.59 62.6	3.57×10^{10} 4.23×10^{10}	β^- β^-	^{176}Hf ^{187}Os	5.26 1.96	Precambrian zircons, Re rich rocks, molybdenite $\sim 0.1–2.7 \times 10^9$ a
^{232}Th	100	1.41×10^{10}	Thorium decay line	$^{208}Pb + 6\,^4He$	52.4	Th-bearing minerals and rocks
^{238}U	99.27	4.51×10^9	Uranium–radium decay line	$^{206}Pb + 8\,^4He$	24.1	U-bearing minerals and rocks
^{235}U	0.72	7.13×10^8	Uranium–actinium decay line	$^{207}Pb + 7\,^4He$	22.1	e.g., zircons, monazite, sphene, apatite, titanite, perovskite, rutile up to 4.5×10^9 a

[a] in Nature: IUPAC table value.

It is well known that each decay chain of ^{238}U, ^{235}U and ^{232}Th always leads to the formation of a specific stable lead isotope with different mass, (^{206}Pb, ^{207}Pb and ^{208}Pb, respectively). These three different decay chains result in three independent geochronometers. Not only have the resulting $^{207}Pb/^{235}U$, $^{206}Pb/^{238}U$ and $^{208}Pb/^{232}Th$ ages been determined on old U rich and Th rich minerals, but the $^{207}Pb/^{206}Pb$ age could also be determined because this isotope ratio changes systematically with time due to different decay rates of the two parent isotopes of uranium. The pioneering work on $^{207}Pb/^{206}Pb$ age dating was done by Nier and co-workers in 1941.[77] The authors determined the $^{207}Pb/^{206}Pb$ age as $2.57 \pm 0.07 \times 10^9$ a from monazite rich samples (from the Huron district in Canada). It was found that the $^{207}Pb/^{206}Pb$ age is more reliable than the other ages because the $^{207}Pb/^{206}Pb$ system is less severely affected by recent radiogenic Pb loss.[75] The Pb–Pb age of a rock is then estimated from the isochrone plot (compare the schematic with isochrone in Figure 8.9 in Section 8.8. for Rb–Sr age dating) constructed as the diagram of the isotope composition of $^{207}Pb/^{204}Pb$ versus $^{206}Pb/^{204}Pb$ for different minerals of the same age.

Thermal ionization mass spectrometry (TIMS) was for many decades the standard analytical method for most of the geochronological techniques. In addition to TIMS, today SIMS and LA-ICP-MS are being employed to an increasing extent, e.g., for *in situ* geochronological investigations on very old zircon and monazite crystals, since both minerals are rich in uranium.[20,78] Imaging mass spectrometric measurements of radiogenic isotopes and elements by SIMS or LA-ICP-MS and particularly by cathodoluminescence, respectively, on polished zircons, allows growth zones to be characterized and they can therefore be used to guide subsequent *in situ* analysis (including zircon geochronology in the 20–30 μm range). Jeffries *et al.*[79] collected zircons from tephra (volcanic rock fragments) from two separate geographic localities, but from the same stratigraphic horizon. $^{207}Pb/^{206}Pb$ isotope ratios were plotted in the Tera–Wasserburg concordia diagram against $^{238}U/^{206}Pb$ isotope ratios from two samples (tephra 1 and tephra 2) measured by LA-ICP-MS, which resulted in U–Pb ages of 14.35 ± 0.27 Ma and 14.15 ± 0.14 Ma (Ma – million years), respectively.[79] Generally, zircon does not incorporate excessive amounts of common Pb into its lattice during its crystallization. Therefore many laboratories do not perform a common Pb correction to their U–Pb data due to problems with isobaric interferences (e.g., of stable ^{204}Pb isotope with the ^{204}Hg isotope). Such an isobaric interference of lead and mercury isotopes at $m/z = 204$ can be avoided by separating the two elements. Jeffries *et al.*[79] reduced this source of contamination by incorporating a gold trap in the carrier gas line and absorbed Hg traces from the carrier gas. The $^{238}U/^{206}Pb$ age determined by means of this approach in LA-ICP-MS can be shown to be accurate.

Tiepolo[80] developed *in situ* lead geochronology by LA-ICP-MS using a double-focusing sector field ICP-SFMS (Element, Thermo Fisher Scientific) with a Nd:YAG laser, at a wavelength of 213 nm for three zircon crystals of different ages (150, 294 and 577 Ma) and thus different radiogenic lead contents of 0.7, 10 and 40 $\mu g\,g^{-1}$, respectively. With a 40 μm spot size of the laser beam, an *in situ* Pb geochronology with an internal precision of 1.1 % was possible (Pb concentration was about 40 $\mu g\,g^{-1}$). A spatial resolution of 20 μm can be adopted for relatively Pb rich zircons even if the internal precision is about 1.5 times lower than with a 40 nm laser spot size.[80]

Besides studies of zircon composition[32] and age dating,[53] microlocal analysis is the preferred method of deciphering the geological record.[75] For instance, the $^{206}Pb/^{238}U$ age of single zircon grains (1846 ± 0.072 Ma) measured by quadrupole based LA-ICP-MS agrees with the TIMS value (1884 ± 0.005 Ma), where TIMS requires more time consuming sample preparation steps before mass spectrometric measurements. On a single zircon grain, with a diameter of less than 100 μm, the age of the mineral can be determined by U–Pb and/or Pb–Pb age dating techniques, e.g., using TIMS after chemical separation of analytes, or LA-ICP-MS directly if possible isobaric interferences are considered carefully, as demonstrated in several papers.[81,82] The microstructure and the U–Pb ages (measured by SHRIMP – sensitive high resolution ion probe mass spectrometry: *in situ* U–Pb SIMS) of several zones in a zircon crystal (with a diameter of 200 μm) from a leucocratic gneiss from the Narryer Terrane in Western Australia, record several high grade thermal events between 3.94 and 4.19 Ma, as found by Nelson *et al.*[83] Several applications of SHRIMP for geochronological studies, especially for age dating on zircons (including lunar zircons), monazite, apatite, perovskite and other minerals have been described by different working groups.[84–88]

SIMS and LA-ICP-MS not only enable U rich minerals to be characterized, but also Th rich minerals *in situ* in order to obtain accurate U/Th–Pb ages.[25,53,75,89–92] An important requirement for producing an accurate age is the careful mass spectrometric determination of the $^{207}Pb/^{206}Pb$ and $^{206}Pb/^{238}U$ ratios. The capability for U–Pb dating of the mineral perovskite with elevated uranium and thorium content, making it a potentially useful geochronometer, has been studied by Cox and Wilton.[93] Due to the lack of a suitable perovskite standard for age dating, zircon standard reference materials were utilized for calibrating the isotope ratio measurements. The authors[93] found that the weighted mean $^{206}Pb/^{238}U$ age from the Oka carbonatite, Quebec, Canada, of 131 ± 7 Ma is consistent with published geochronology data from the region. Cox and Wilton suggested that LA-ICP-MS dating of perovskite could be a useful analytical technique. The low cost and relative speed with which this type of age determination can be performed should make the LA-ICP-MS an attractive alternative to ID-TIMS and SHRIMP.[93]

Rb–Sr Method for Age Dating

The rubidium–strontium geochronometer used in the Rb–Sr geochronological method is based on the radioactive β^--decay of ^{87}Rb to ^{87}Sr. The growth of radiogenic ^{87}Sr in a Rb rich mineral can be described by the following Equation (9.6). In the rubidium–strontium age dating method, the radioactive ^{87}Rb isotope with a natural isotope abundance of 27.85 % and a half-life of 4.88×10^{10} years is fundamental to the β^- decay to the isobar ^{87}Sr. The equation for the Rb–Sr method can be derived from the general equation of radioactive decay (Equation 8.8 in Section 8.8):

$$^{87}Sr/^{86}Sr = (^{87}Sr/^{86}Sr)_0 + (^{87}Rb/^{86}Sr)\ (e^{\lambda t} - 1) \tag{9.6}$$

where $^{87}Sr/^{86}Sr$ = ratio of these strontium isotopes at the present time t; $(^{87}Sr/^{86}Sr)_0$ = initial isotope ratio of these strontium isotopes at the time ($t = 0$) the system became closed to Rb and Sr; $(^{87}Rb/^{86}Sr)$ = ratio of these isotopes at the present time t; λ = decay constant of ^{87}Rb ($1.42 \times 10^{-11}\ a^{-1}$); and t = time elapsed since the system became closed to Rb and Sr.

Whereas the abundance of ^{87}Sr in rubidium rich rocks changes over time due to the radioactive β^- decay of ^{87}Rb as a function of the primordial rubidium concentration and the age of the mineral, the abundance of the stable ^{86}Sr isotope and consequently the $^{86}Sr/^{88}Sr$ is constant in nature. The constant $^{86}Sr/^{88}Sr$ isotope ratio is often used for internal standardization (mass bias correction) during strontium isotope ratio measurements of $^{87}Sr/^{88}Sr$. In the rubidium–strontium age dating method, the isotope ratios $^{87}Sr/^{86}Sr$ and $^{87}Rb/^{86}Sr$ are measured mass spectrometrically (mainly by TIMS or nowadays by ICP-MS) and the primordial strontium ratio $(^{87}Sr/^{86}Sr)_0$ at $t = 0$ and the age t of the rock can be derived from the isochrone (graph of measured $^{87}Sr/^{86}Sr$ isotope ratios (represented on the ordinate) as a function of the $^{87}Rb/^{86}Sr$ ratio (on the abscissa) in several minerals with different primordial Rb concentrations). The age of the minerals will be determined from the slope of the isochrone $(e^{\lambda t} - 1)$, and the primordial isotope ratio $(^{87}Sr/^{86}Sr)_0$ from the point of intersection with the ordinate (see Figure 8.9). Rb–Sr age dating is today an established geochronological technique using mass spectrometry (TIMS and ICP-MS after analyte separation) for Rb bearing rocks and minerals (such as granite, biotite, feldspar, mica, sediments and others.

Nebel and Mezger[94] reported on the reassessment of standard K feldspar NBS SRM 607, which is widely used as a reference material for high precision Rb/Sr ratio and Sr isotope ratio measurements by MC-ICP-MS and TIMS. The Rb/Sr ratio of the standard was obtained by the isotope dilution technique. The rubidium measurements were performed on a MC-ICP-MS (Isoprobe, Micromass), and the strontium isotope ratios were determined with a MC-TIMS (Triton, Thermo Fisher Scientific).[94] Double-focusing sector field ICP-MS (Element, Thermo Fisher Scientific) was applied for the age dating of geological samples from Egypt via strontium isotope ratio measurements after digestion and Rb and Sr separation by extraction chromatography with crown ether. The Rb–Sr age of geological samples from different archaeological sites in the eastern desert of Egypt was determined via a Rb/Sr isochrone as 455 ± 34 Ma.[16]

Biotite is usually a primary magmatic mineral of granitoids and is widely used for Rb–Sr (but also for K–Ar) age dating. Weathering biotite releases inorganic nutrients essential for plant growth and Sr isotopes useful in tracing regional and global hydrological cycles.[95] During the transformation to oxidized biotite, ^{87}Sr and ^{40}Ar are preferentially released relative to Rb and K, respectively, via solid-state diffusion through the biotite lattice, resulting in a drastic reduction of original isotopic age. Sr isotope ratios during complex weathering processes have been studied, for example, by MC-TIMS (VG 54-30, equipped with nine Faraday cups).[96] Regional and local variations in isotopic systems affected by particular weathering processes should be considered when dating biotite or biotite bearing rocks in weathering environments.[96]

Sm–Nd Methods for Age Dating

Samarium and neodymium are REEs whereby the stable ^{147}Nd isotope is formed by α decay from the mother radioisotope ^{147}Sm ($t_{1/2} = 1.06 \times 10^{11}$ a). The long half-life of ^{147}Sm allows it to be used for the age dating of extremely old geological samples via Sm–Nd age dating. Both elements are widely distributed in minerals and rocks with concentrations in the low $\mu g\,g^{-1}$ range and below. The original application of the Sm–Nd method focused on cosmochemical work for characterizing meteorites and lunar samples.[75] The geochronological technique allows the dating of igneous rocks, achondrite and chondrite meteorites back to the Precambrian era. The isotopic evolution of Nd in the earth is described by the decay of ^{147}Sm in a 'chondritic uniform reservoir' (so-called CHUR).[76]

Lu–Hf Methods for Age Dating

It has been recognized that the Lu–Hf isotopic system in zircon is a powerful tool for deciphering the evolution of the earth's crust and mantle.[97–99] Zircon normally contains 0.5–2 wt % Hf, which results in an extremely low Lu/Hf ratio ($^{176}Lu/^{177}Hf < 0.002$) and consequently a negligible radiogenic growth of ^{176}Hf due to the β^- decay of ^{176}Lu. Therefore, the $^{176}Hf/^{177}Hf$ ratio of zircon can be regarded as the initial value at the time when it crystallized. LA-ICP-MS with a multiple ion collector system has also been employed to study the hafnium isotopic composition of zircon and baddeleyite standards in U–Pb geochronology.[100]

The Re–Os Method for Age Dating

The Re-Os method is of special interest for the dating of very old Re rich ores, minerals and meteorites, where ^{187}Os is formed by the β^- decay of the long-lived ^{187}Re isotope with a half-life of 4.23×10^{10}. In 1937, Nier performed an isotopic analysis of osmium using OsO_4.[101] The first evidence of highly enriched ^{187}Os (~ 99.5 %) in molybdenite was found by Hintenberger and co-workers in 1954.[102] Since this time ^{187}Os has been used as a powerful geochemical tracer measured via the sensitive and precise isotope analysis of Os.[62,103] Highly enriched ^{187}Os (formed due to the radioactive decay of ^{187}Re) can be prepared from Precambrian Re rich ores with a low initial osmium content. Herr *et al.*[104,105] demonstrated that the Re–Os method could be used to date iron meteorites and terrestrial samples such as molybdenite.

The age of a mineral can be calculated via the isochrone technique using the equation:

$$^{187}Os/^{186}Os = (^{187}Os/^{186}Os)_0 + (^{187}Re/^{186}Os)\ (e^{\lambda t} - 1) \tag{9.7}$$

where $^{187}Os/^{186}Os$ = ratio of these osmium isotopes at the present time; $(^{187}Os/^{186}Os)_0$ = initial ratio of these osmium isotopes at the time the system became closed to Re and Os (t = 0); $(^{187}Re/^{186}Os)$ = ratio of these isotopes at the present time; λ = decay constant of ^{187}Re ($1.5 \times 10^{-11}\ a^{-1}$); and t = time elapsed since the system became closed to Re and Os.

The β^- decay of ^{187}Re to the stable isobar ^{187}Os leads to a significant isotopic variation in old minerals. A highly enriched natural ^{187}Os sample with an isotope abundance of 99.44 %[106] (IUPAC table value: 1.96 %[107]) was characterized in several European laboratories by ICP-QMS in comparison to SIMS, SNMS and GDMS. Two different ^{187}Os enriched natural Os samples were analyzed by double-focusing sector field ICP-MS with single ion and multiple ion collector systems.[108] Using different mass spectrometers, the ^{187}Os abundance of 98.93 % was found to be in good agreement (see Figure 9.49 b). This enrichment of ^{187}Os in both samples is the result of the β–decay of ^{187}Re in old minerals as compared with the isotopic composition on metallic osmium from Johnson Matthey using different ICP-MS set-ups (ICP-SFMS, ICP-CC-QMS and MC-ICP-MS) in Figure 9.49 a. A 17 % enrichment of ^{187}Os in comparison to the IUPAC table value of 1.96 % was determined. The precision of $^{188}Os/^{192}Os$ isotope ratio measurement on the metallic osmium sample was 0.09 % for double-focusing sector field ICP-MS with a single ion collector, 0.08 % for quadrupole ICP-MS with a hexapole collision cell (ICP-CC-QMS) and 0.003 % using the MC-ICP-MS from Nu Instruments.[108] *In situ* Os isotope ratio analysis of iridosmines with a

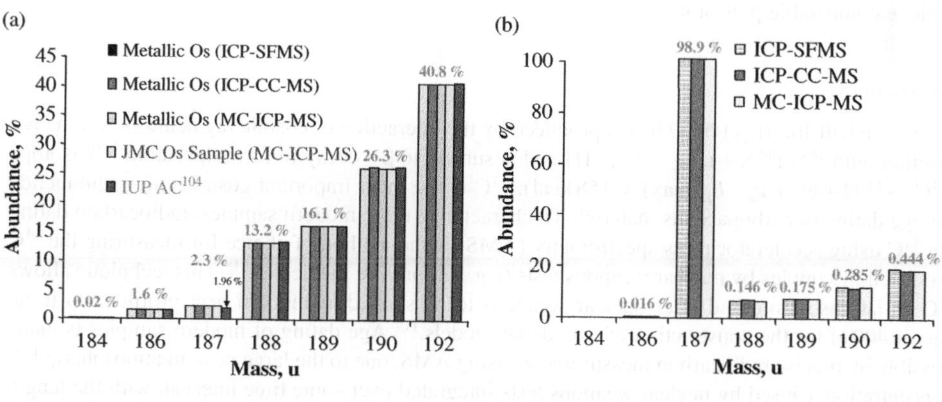

Figure 9.49 *Comparison of isotope abundances of metallic Os and natural highly enriched* ^{187}Os *samples: (a) results of isotope analysis on metallic osmium (Johnson Matthey Chemicals); (b) isotope analysis on natural* ^{187}Os *enriched samples due to* β *decay of* ^{187}Re.[108]

precision for microlocal isotope ratio measurement of 0.05 % was performed by Hirata and co-workers with LA-MC-ICP-MS using the VG Plasma 54.[109] Pearson *et al.* reported on *in situ* measurements of Re–Os isotopes in mantle sulfides by LA-MC-ICP-MS (Nu Instruments).[110]

The K–Ar/Ca System for Age Dating

^{40}K is a radioactive nuclide with a half-life of $t_{1/2} = 1.26 \times 10^9$ a, 11.2 % of which decays to ^{40}K by electron capture and 88.8 % to ^{40}Ca. Both decays can be used for geochronological studies.

The potassium–argon age dating technique is one of the earliest geochronological methods using gas source mass spectrometry to determine radiogenic ^{40}Ar in potassium rich minerals and was created by Aldrich and Nier in 1948.[111] A static gas source mass spectrometer was combined with online argon extraction.[112] Obviously, the mother radionuclide ^{40}K is less abundant (0.017 %), but the abundance of K in most minerals is high and argon is a rare noble gas in minerals so that the K–Ar geochronometer is a very useful technique still used today on K bearing minerals and rocks that quantitatively retain radiogenic argon. K–Ar geochronology is applied for dating biotites, muskovites and hornblende from plutonic igneous metamorphic rocks and feldspar from volcanic rocks of younger materials of up to Precambrian age.[76] Difficulties in K–Ar age dating occur due to Ar loss from different mineral phases, such as feldspar or biotite, which is higher than in hornblende resulting in the incorrect dating of rocks.[113]

Because ^{40}Ca is the most abundant naturally occurring Ca isotope, with an isotope abundance of 96.93 %, and Ca is an abundant element in the Earth's crust, whereas the mother nuclide ^{40}K (0.017 %) is a low abundance isotope of potassium, the enrichment of ^{40}Ca due to radioactive decay of ^{40}K in geological samples is very small and thus the detection of radiogenic ^{40}Ca in the presence of 'common' Ca is extremely difficult. In addition, extensive fraction effects of Ca isotopes on the earth and other solar system materials[114] and fraction effects during sample preparation and mass spectrometric measurements (instrumental isotope fractionation) are possible, and therefore the K–Ca system is less robust and more difficult to use for geochronology except for special applications.[75] The problems of a K–Ca geochronometer for petrogenic studies and Archaean micas are similar to those of the Rb–Sr system using MC-TIMS, and have been examined by Nelson and McCulloch[115] and Fletcher *et al.*[116,117] The K–Ca method is unlikely to replace the K–Ar and Rb–Sr techniques for dating crystalline igneous rocks because it is more difficult to apply and to achieve comparable precision.[76]

^{14}C Dating

^{14}C with a half-life ($t_{1/2}$) of 5730 a is produced by the interaction of cosmic ray neutrons via the n/p reaction with ^{14}N ($^{14}N + n \rightarrow {}^{14}C + {}^{1}H$) and is subject to β^- decay to form the stable ^{14}N isotope ($^{14}C \rightarrow {}^{14}N + e^- + \nu_e$, $E_\beta(\max) = 158$ kcal). ^{14}C is the most important cosmogenic radionuclide for age dating of carbonaceous material. To characterize more modern samples, radiocarbon dating via ^{14}C using accelerator mass spectrometry (AMS) is the method of choice for measuring the ^{14}C produced in samples by nuclear weapons tests (e.g., in forensic medicine).[118] This technique allows ^{14}C variations before and after nuclear weapons to be studied in modern peat profiles (up to an age of 400 a) for the construction of age–depth models.[119] Age dating of modern samples is made possible by precise radiocarbon measurements using AMS, due to the large peak in atmospheric ^{14}C concentration, caused by nuclear weapons tests integrated over some time interval, with the length specific to the peat section. In the peat layers covered by the weapons testing peak, calendar ages of individual peat samples can be determined almost with an accuracy of 2–3 years.[119] Radiocarbon dating has been employed to determine, for instance, the age of the famous Iceman 'Ötzi' by analyzing small pieces of tissue and bone. By using $^{14}C/^{12}C$ measurements with AMS it was found that Ötzi had lived between 5100 and 5350 years ago.[118]

9.7.6 Mass Spectrometric Microlocal and Imaging Analysis of Geological Samples

In situ microanalytical techniques for trace element determination, such as LA-ICP-MS, LIMS, SIMS and also non-mass spectrometric techniques (e.g., synchrotron radiation induced X-ray fluorescence analysis – SR-XRF[120]) have become increasingly employed in the geological and also in the cosmological sciences (see Section 9.8) for studying small solid samples (μm- and sub-μm sized particles, e.g., presolar dust grains), liquid or gaseous inclusions or heterogeneities in minerals and rocks. In addition, element distribution and images of isotopes are of increasing interest in the geosciences. Groener and Hoppe[121] developed an automated ion imaging technique using the Cameca NanoSIMS for analyzing μm sized particles, e.g., for refractory presolar minerals (SiC) separated from meteorites. The reproducibility of the NanoSIMS technique was examined on an ion image of $^{16}O^+$, $^{17}O^+$, $^{18}O^+$, $^{28}Si^+$ and $^{27}Al^{16}O^+$ of $\sim 1\,\mu m$ Al_2O_3 grains dispersed on a gold foil. In 27 h 440 grains were measured (image raster size: $30\,\mu m \times 30\,\mu m$). This technique can also be employed for geochemical studies, for instance of very small mineral inclusions in geological samples.

The geological MPI-DING references glasses prepared and characterized by Jochum *et al.* employing several analytical techniques as described in Chapter 6 are recommended to improve possible quantification procedures for *in situ* microlocal analysis.[4,12]

As a microanalytical technique, LA-ICP-MS is well suited for analyzing fluid inclusions in minerals.[122–125] Audétat *et al.* reported on the formation of a magmatic hydrothermal ore deposit using LA-ICP-MS measurement of element concentrations in a series of fluid inclusions representing the fluid before, during and after the deposition of cassiterite (SnO_2).[126] The authors found that a thin precipitation was driven by the mixing of hot magmatic brine with cooler meteoric water. At the same time, a separate magmatic vapour phase selectively transported copper and boron into the liquid mixture.[126]

For quantitative imaging of elements on cross section of geological samples LA-ICP-MS can be employed.

References VII

1. Jarvis, K. E. and Williams, J. G., *Chem. Geol.*, **106**, 251 (1993).
2. Klemm, K. and Bombach, G., *Fresenius' J. Anal. Chem.*, **370**, 641 (2001).
3. Jochum, K. P., Dingwell, D. B., Rocholl, A. *et al., Geostand. Newslett.*, **24**, 87 (2000).
4. Jochum, K. P., Stoll, B., Herwig, K. *et al., Geochem. Geophys. Geosyst.*, **7**, 1 (2006).
5. Le Fevre, B. and Ottolini, L., *Mikrochim. Acta*, **155**, 189 (2006).
6. Becker, J. S. and Dietze, H. J., *Fresenius' J. Anal. Chem.*, **365**, 429 (1999).
7. Odegard, M., Dundas, S. H., Flem, B. and Grimstedt, A., *Fresenius' J. Anal. Chem.*, **362**, 477 (1998).
8. Pickhardt, C., Becker, J. S. and Dietze, H. J., *Fresenius' J. Anal. Chem.*, **368**, 173 (2000).
9. Pickhardt, C., Brenner, I. B., Becker, J. S. and Dietze, H. J., *Fresenius' J. Anal. Chem.*, **368**, 79 (2000).
10. Jarvis, K. E., *Chem. Geol.*, **68**, 31 (1988).
11. Fedorowich, J. S., Richards, J. P., Jain, J. C., Kerrich, R. and Fan, J., *Chem. Geol.*, **106**, 229 (1993).
12. Jochum, K. P., Pfänder, J., Woodhead, J. D. *et al., Geochem. Geophys. Geosyst.*, **6**, 1 (2005).
13. Eggins, S. M., Woodhead, J. D., Kinsley, L. P. J. *et al., Chem. Geol.*, **134**, 311 (1997).
14. Lu, Y., Makishima, A. and Nakamura, E., *Chem. Geol.*, **236**, 13 (2007).
15. Wang, J., Nakazato, T., Sakanishi, K., Yamada, O., Tao, H. and Saito, I., *Talanta*, **68**, 1584 (2006).

16. Zoriy, M. V., Rashad, A., Pickhardt, C. Mohsen, H. T., Förstel, H., Helal, A. I., Zahran, N. F. and Becker, J. S. *At. Spectr.*, **24**, 195 (2003).
17. Barefoot, R. R., *J. Anal. At. Spectrom.*, **13**, 1077 (1998).
18. Bayon, G., German, C. R., Boella, R. M., Milton, J. A., Taylor, R. N. and Nesbitt, R. W., *Chem. Geol.*, **187**, 179 (2002).
19. Nowell, G. M., Pearson, D. G., Ottley, C. J., Schwieters, J. and Dowall, D. P., in *Plasma Source Mass Spectrometry*, G. Holland and S. D. Tanner (eds.). The Royal Society of Chemistry, Cambridge (2003) 307.
20. Hirata, T. and Nesbitt, R. W., *Geochim. Cosmochim. Acta*, **59**, 2491 (1995).
21. Outridge, P. M., Doherty, W. and Gregoire, D. C., *Spectrochim. Acta*, **52B**, 278 (1997).
22. Tunheng, A. and Hirata, T., in *Plasma Source Mass Spectrometry*, D. Bandura and G. Holland (eds.) The Royal Society of Chemistry, Proceedings of Durham Conference 2004) **301**, 177 (2005).
23. Hattendorf, B., Latkoczy, C. and Günther, D., *Anal. Chem.*, **75**, 341A (2003).
24. Kroslakova, I. and Günther, D., *J. Anal. At. Spectrom.*, **22**, 51 (2007).
25. Horn, I., Rudnick, R. L. and McDonough, W. F., *Chem. Geol.*, **167**, 405 (2000).
26. Mank, A. J. G. and Mason, P. R. D., *J. Anal. At. Spectrom.*, **14**, 1143 (1999).
27. Becker, J. S. and Dietze, H. J., *Fresenius' J. Anal. Chem.*, **344**, 69 (1992).
28. Cromwell, E. F. and Arrowsmith, P., *Appl. Spectrosc.*, **49**, 1652 (1995).
29. Figg, D. J., Cross, J. B. and Brink, C., *Appl. Surf. Sci.*, **129**, 287 (1998).
30. Bacon, R. and Ure, A. M., *Analyst*, **109**, 1229 (1984).
31. Becker, J. S. and Dietze, H. J., *Z. Angew. Geol.*, **29**, 136 (1983).
32. Becker, J. S. and Dietze, H. J., *Z. Angew. Geol.*, **32**, 299 (1986).
33. Dietze, H. J., *Massenspektroskopische Spurenanalyse*, Akademischer Verlagsgesellschaft Geest & Portig K.-G., Leipzig (1975).
34. Dietze, H. J., Becker, J. S. and Gerstenberger, H., *Int. J. Mass Spectrom. Ion Proc.*, **54**, 243 (1984).
35. Ramendik, G. in *Chemical Analysis*, F. Adams, R. Gijbels and R. Van Grieken (eds.). New York (1988), 95.
36. Becker, J. S. and Dietze, H. J., *Zeitschrift Angew. Geol.*, **29**, 599 (1983).
37. Dietze, H. J., *Analytikertaschenbuch*, **10** 249 (1991).
38. Olesik, J. W., in *Inorganic Mass Spectrometry*, C. M. Barshick, D. C. Duckworth and D. H. Smith (eds.). Marcel Dekker, Inc., New York, Basel (2000).
39. Feigenson, M. D., Patino, L. C. and Carr, M. J., *Geophys. Res. [Solid Earth]*, **101**, 118515 (1996).
40. Hoskin, P. W. O., *J. Trace Microprobe Techn.*, **16**, 301 (1998).
41. Jochum, K. P., Stoll B., Pfänder, J. *et al., Fresenius' J. Anal. Chem.*, **370**, 647 (2001).
42. Stoll, B. and Jochum, K. P., *Fresenius' J. Anal. Chem.*, **364**, 380 (1999).
43. Becker, J. S., Pickhardt, C. and Dietze, H. J., *Mikrochim. Acta*, **135**, 71 (2000).
44. Jochum, K. P., Stoll, B., Herwig, K. and M., W., *J. Anal. At. Spectrom.*, **21**, 666 (2006).
45. Becker, J. S., Pickhardt, C., Hoffmann, N., Hocker, H. and Becker, J. S., *At. Spectr.*, **23**, 1 (2002).
46. Horn, I. and Günther, D., *Appl. Surf. Sci.*, **207**, 144 (2003).
47. Günther, D., Longerich, H. P., Jackson, S. E. and Forsythe, L., *Fresenius' J. Anal. Chem.*, **355**, 771 (1996).
48. Halicz, L. and Günther, D., *J. Anal. At. Spectrom.*, **19**, 1539 (2004).
49. Jochum, K. P., Nohl, U., Herwig, K., Lammel, E., Stoll, B. and Hofmann, A. W., *Geostand. Geoanal. Res.*, **29**, 333 (2005).
50. Guillong, M. and Günther, D., *Spectrochim. Acta*, **56B**, 1219 (2001).
51. Abuduryim, A. and Kitawaki, H., *Gems and Gemology*, **42**, 98 (2006).
52. Helal, A. I., Zahran, N. F., Mohamed, R. A. Mohsen, H. T., Becker, J. S., Kolsev, A. P. and Hashad, A. H., *Int. J. Mass Spectrom.*, **221**, 139 (2002).
53. Li, X., Liang, M., Sun, M., Guan, H. and Malpas, J. G., *Chem. Geol.*, **175**, 209 (2001).
54. Wiedenbeck, M., Hanchar, J. M., Peck, W. H. *et al., Geostand. Geoanal. Res.*, **28**, 9 (2004).
55. Pearce, N. J. G., Perkins, W. T., Westgate, J. A. *et al., Geostand. Newslett.*, **21**, 115 (1997).
56. Schmidt, M. W. D., Connolly, J. A., Günther, D. and Bogaerts, M., *Science*, **312** 1646 (2006).
57. Bertoldi, C., Hauzenberger, C., Rahmoun, N. S. *et al., Amer. Mineralogist*, **91**, 1184 (2006).
58. Willbold, M. and Jochum, K. P., *Geostand. Geoanal. Res.*, **29**, 63 (2005).

59. Boulyga, S. and Heumann, K. G., *Int. J. Mass Spectrom.*, **242**, 291 (2005).
60. Köllensperger, G., Hann, S., Prinz, G., Stingeder, G. and Bujatti-Narbeshuber, M., *Fresenius' J. Anal. Chem.*, **370**, 559 (2001).
61. Pretorius, W., Chipley, D., Kyser, K. and Helmstaedt, H., *J. Anal. At. Spectrom.*, **18**, 302 (2003).
62. Meisel, T., Moser, J. and Wegscheider, W., *Fresenius' J. Anal. Chem.*, **370**, 566 (2001).
63. Weyer, S., Munker, C., Rehkaemper, M. and Mezger, K., *Chem. Geol.*, **187**, 295 (2002).
64. Fang, J., Lui, L. W. and Yan, X. P., *Spectrochim. Acta*, **61B**, 864 (2006).
65. Kasemann, S. A., Jeffcoate, A. B. and Elliott, T., *Anal. Chem.*, **77**, 5251 (2005).
66. Tanimizu, M. and Hirata, T., *J. Anal. At. Spectrom.*, **21**, 1423 (2006).
67. Irisawa, K. and Hirata, T., *J. Anal. At. Spectrom.*, **21**, 1387 (2006).
68. Schoenberg, R. and von Blanckenburg, F., *Int. J. Mass Spectrom.*, **242**, 257 (2005).
69. Moynier, F., Bouvier, A., Blichert-Toft, J., Telouk, P., Gasperini, D. and Albarède, F., *Geochim. Cosmochim. Acta*, **70**, 4287 (2006).
70. Eiler, J. M., Graham, C. and Valley, J. W., *Chem. Geol.*, **138**, 221 (1997).
71. Tamborini, G., Phinney, D., Bildstein, O. and Betti, M., *Anal. Chem.*, **74**, 6098 (2002).
72. Peck, W. H., Valley, J. W., Wilde, S. A. and Graham, C. M., *Geochim. Cosmochim. Acta*, **65**, 4215 (2001).
73. De Laeter, J. R., in *The History of Meteoritics and Key Meteoritic Collections: Fireballs, Falls and Finds*, G. J. H. McCall, A. J. Bowden and R. J. Howarth (eds.). Geology Society, Special Publications, London, **256**, 363 (2006).
74. Paneth, F. A., *Z. Elektrochem.*, **34**, 645 (1928).
75. De Laeter, J. R., *Application of Inorganic Mass Spectrometry*, Wiley-Interscience Series on Mass Spectrometry, John Wiley & Sons, Inc., New York, (2001).
76. Faure, G., *Principles of Isotope Geology*, John Wiley & Sons, Inc., New York (1986).
77. Nier, A. O., Thompson, R. W. and Murphy, B. F., *Phys. Rev.*, **60**, 112 (1941).
78. Fryer, B. J., Jackson, S. E. and Longerich, H. P., *Chem. Geol.*, **109**, 1 (1993).
79. Jeffries, T., Storey, C. and Fernandez-Suarez, J., *Spectroscopy EUROPE*, **18**, 8 (2006).
80. Tiepolo, M., *Chem. Geolog.*, **199**, 159 (2003).
81. Sela, H., Karpas, Z. and Becker, J. S., *Int. J. Mass Spectrom.*, **261**, 199 (2006).
82. Wetzel, K., Becker, J. S. and Dietze, H. J., *Isotopenpraxis*, **19**, 221 (1983).
83. Nelson, D. R., Robinson, B. W. and Myers, J. S., *Earth Planet. Sci. Lett.*, **181**, 89 (2000).
84. Cocheri, A., Baudin, T., Autran, A., Guerrot, C., Fanning, C. M. and Laumonier, B., *Bull. Soc.Geol. France*, **176**, 269 (2005).
85. Compston, W., *J. Roy. Soc. W. Aust.*, **79**, 109 (1996).
86. Compston, W., Williams, I. S. and Meyer, C., *Research School of Earth Sciences Annual Report, Australian National University*, 116 (1983).
87. Ireland, T. R. and Wlotzka, F., *Earth Planet. Sci. Lett.*, **109**, 1 (1992).
88. Zeitler, P. K., Sutter, J. F., Williams, I. S., Zartmann, R. and Tahirkheli, R. A., *Geol. Soc. Am.*, **23**, 1 (1989).
89. Cox, A., Wilton, D. H. C. and Kosler, J., *Can. Mineral.*, **41**, 273 (2003).
90. De Laeter, J. R. and Kennedy, A. K., *Int. J. Mass Spectrom.*, **178**, 43 (1998).
91. Kosler, J., Fonneland, H., Sylvester, P., Tubrett, M. and Pedersen, R. B., *Chem. Geol.*, **182**, 605 (2002).
92. Platzner, I. T., *Modern Isotope Ratio Mass Spectrometry*, John Wiley & Sons, Inc., New York (1997).
93. Cox, R. A. and Wilton, D. H. C., *Chem. Geol.*, **235**, 21 (2006).
94. Nebel, O. and Mezger, K., *Chem. Geol.*, **233**, 337 (2006).
95. Erel, J. D., Blum, E., Roueff, E. and Ganor, J., *Geochim. Cosmochim. Acta*, **68**, 4549 (2004).
96. Joeng, G. Y., Cheong, C. S. and Kim, J., *Geochim. Cosmochim. Acta*, **70**, 4734 (2006).
97. Hakesworth, C. and Kemp, T., *Chem. Geol.*, **226**, 144 (2006).
98. Harrison, T. M., Blichert-Toft, J., Muller, W., Albarede, F., Holden, P. and Mojzsis, S. J., *Science*, **310**, 1497 (2005).
99. Kinny, P. D. and Maas, R. *Rev. Mineral. Geochem.* **53**, 327 (2003)
100. Wu, F. Y., Yang, Y. H., Xie, L. W., Yang, J. H. and Xu, P., *Chem. Geol.*, **234**, 105 (2006).
101. Nier, A. O., *Phys. Rev.*, **52**, 885 (1937).
102. Hintenberger, H., Herr, H. and Voshage, H., *Phys. Rev.*, **95**, 1960 (1954).

103. Völkening, J., Walczyk, T. and Heumann, K. G., *Int. J. Mass Spectrom. Ion Proc.*, **105**, 147 (1991).
104. Herr, W., Hoffmeister, W. and Hirt, B., Houtemans, *Z. Naturforsch.*, **16a**, 1053 (1961).
105. Herr, W. and Merz, E., *Naturforschung*, **10a**, 613 (1955).
106. Becker, J. S. and Dietze, H. J., *Mikrochim. Acta*, **118**, 103 (1995).
107. Boehlke, J. K., De Laeter, J. R., De Bièvre, P. *et al.*, *IUPAC Isotopic Composition of the Elements 2001, J. Phys. Chem. Ref. Data*, **34**, 57 (2005).
108. Boulyga, S. F., Segal. I., Platzner, I., Halicz, L. and Becker, J. S., *Int. J. Mass Spectrom.*, **218**, 245 (2002).
109. Hirata, T., Hattori, M. and Tanaka, T., *Chem. Geol.*, **1444**, 269 (1998).
110. Pearson, N. J., Alard, O., Griffin, W. L., Jackson, S. E. and O'Reilly, Y., *Geochim. Cosmochim. Acta*, **66**, 1037 (2002).
111. Aldrich, L. T. and Nier, A. O., *Phys. Rev.*, **74**, 876 (1948).
112. Dalrymple, G. B. and Lanphere, M. A., *Potassium Argon Dating*, Freeman, New York (1969).
113. Hart, S. R., *J. Geol.*, **72**, 493 (1964).
114. Russell, W. A., Papanastassiou, D. A. and Tombrello, T. A., *Geochim. Cosmochim. Acta*, **42**, 1075 (1978).
115. Nelson, D. R. and McCulloch, M. T., *Chem. Geol.*, **79**, 275 (1989).
116. Fletcher, J. S., Maggi, A. L., Riosman, K. J. R. and McNaughton, M. J., *Int. J. Mass Spectrom. Ion Proc.*, **163**, 1 (1997).
117. Fletcher, J. S., McNaughton, M. J., Pidgeon, R. T. and Rosman, K. J. R., *Chem. Geol.*, **138**, 289 (1997).
118. Kutschera, W., *Int. J. Mass Spectrom.*, **242**, 145 (2005).
119. Goslar, T., van der Knaap, W. O., Hicks, S. *et al.*, *Radiocarbon*, **47**, 115 (2005).
120. Gomez-Morilla, I., Simon, A., Simon, R., Williams, C. T., Kiss, A. Z. and Grime, G. W., *Nucl. Instr. Meth. B*, **249**, 897 (2006).
121. Groener, M. and Hoppe, P., *Appl. Surf. Sci.*, **252**, 7148 (2006).
122. Audétat, A., Günther, D. and Heinrich, C. A., *Geochim. Cosmochim. Acta*, **64**, 3373 (2000).
123. McCandless, T. E., Lajack, D. J., Ruiz, J. and Ghazi, A. M., *Geostand. Newslett.*, **21**, 279 (1997).
124. Moissette, A., Shepherd, T. J. and Chenery, S. P., *J. Anal. At. Spectrom.*, **11**, 177 (1996).
125. Sherpherd, T. J. and Chenery, S. R., *Geochim. Cosmochim. Acta*, **59**, 3997 (1995).
126. Audétat, A., Günther, D. and Heinrich, C. A., *Science*, **279**, 2091 (1988).
127. Jochum, K. P., *Spectroscopy Europa*, **9**, 22 (1997).
128. Newsom, H. E., White, W. M., Jochum, K. P. and Hofmann, A. W., *Earth Planet. Sci. Lett.*, **80**, 299 (1986).
129. Echeverria, L. M., *Contr. Min. Petrol.*, **73**, 253 (1980).
130. Klein, M., Stosch, H. G. and Seck, H. A., *Chem. Geol.*, **138**, 257 (1997).

9.8 Cosmochemistry, Planetary and Space Science

This chapter concerns the fields that use inorganic mass spectrometry to investigate the composition and evolution of matter in the universe and in the solar system. Cosmochemistry is related to nuclear astrophysics, because almost all the chemical elements were synthesized by nuclear reactions in the interior of stars.[1] Mass spectrometric analyses of elemental composition, the distribution and variation of isotope abundances are very helpful, especially for cosmochronological studies, in order to explain the formation, history and evolution of stars in our universe and to understand the chemical and nuclear processes.

9.8.1 Cosmochemical Trace Analysis

Fundamental geochemical research – also relevant for a comparison between the elemental distribution of extraterrestrial, meteoritic, solar and stellar material and of interest for the large field of cosmochemistry – was performed by Clarke at the US Geological Survey as long ago as 1889. Clarke's pioneering work, based on the analysis of materials from the Earth's crust, concerned the geochemical study of the 'relative abundances' of the chemical elements as a function of their atomic weight in nature.[2] The Earth's crust is not well suited to study the average abundance of chemical elements in nature because of terrestrial fractionation processes over the past 4.55×10^9 years. It is interesting to note that the Earth's moon is even more fractionated than the Earth, as it is depleted in volatile elements and enriched in refractory elements with respect to the elemental composition of the Earth's crust.[1] About 50 years later, Goldschmidt provided the first adequate table of relative abundances of the elements in nature. In his classic paper, Goldschmidt included analytical data from different types of phases in meteorites (chondrites such as stone, iron and troilite – sulfide) and discussed the geochemical distribution law of elements in nature.[3] 'Cosmic' abundances of chemical elements provide the basic data for theories of cosmochemistry and nuclear astrophysics.[1] A curve of elemental abundances for the Sun and similar main sequence stars as a function of their mass number with maximum abundance for H and He, followed by Si and a local maximum for the Fe group was presented by Suess and Urey in 1956.[4] This 'cosmic abundance curve' is representative of main sequence stars originating in the solar system.[1] A comprehensive review on abundances of chemical elements in meteoritic and solar samples was published in 1989 by Anders and Grevesse.[5] The authors utilized thermal ionization mass spectrometry in multi-element analysis via the isotope dilution technique for cosmic studies. The quantitative measurement of low concentration rare earth elements (REEs) and Ba in carbonaceous chondrites was described.[5] Nowadays, ICP-MS and LA-ICP-MS are employed for the multi-element analysis of extraterrestrial materials. In this way, precise REEs, thorium and uranium concentrations in chondritic meteorites have been measured by ICP-MS in comparison with radiochemical neutron activation analysis (NAA).[6] SIMS, especially, as a microanalytical technique (preferentially Cameca NanoSIMS with a spatial resolution down to 50 nm) has been employed to an increasing extent in cosmochemistry to study the elemental and isotopic composition (including ion imaging of presolar single grains) of micrometeorites or μm sized presolar SiC and spinel grains.[7–9] SIMS was applied for studies of zoning patterns of light lithophile elements (Li, Be and B) in pyroxenes of some basaltic meteorites.[10] Wasserburg and co-workers developed an analytical technique for measuring Os, Ir, Pt and Au concentrations *in situ* in five iron meteorites by SIMS, using a Cs^+ primary ion beam in the microscale range.[11] At a spatial resolution of 10–20 μm, detection limits of less than $1\,\mu g\,g^{-1}$ for Os, $0.1\,\mu g\,g^{-1}$ for Ir and $10–20\,ng\,g^{-1}$ for Pt and Au were achieved.

Sample materials often investigated in cosmochemistry, planetary and space sciences are cometary and interstellar dust to study the elemental and isotopic composition of extraterrestrial material. For example, cometary and interstellar dust was analyzed intensively in the NASA Stardust mission (collected in January 2004 during the passage through the coma of comet 81P/Wild 2). After recovery of the Stardust return capsule in the Utah desert, the mass spectrometric analyzes were performed by ToF-SIMS as a key technique with respect to elemental, mineralogical and isotopic composition.[12] Selected particles removed from the aerogel were embedded in epoxy and sliced with an ultramicrotome to study the element distribution. These investigations are useful to explain the formation of the solar system. NASA's Stardust mission will provide answers to fundamental questions about comets, the origin of the solar system and possibly even the origin of life itself.

In 2006 Beauchamp reported on the expanding role of mass spectrometry in developing new instruments for laboratory simulations and *in situ* exploration of the space environment.[13] The cometary and interstellar dust analyzer (CIDA), which was designed for the direct analysis of the space environment and mounted on the spacecraft, is a time-of-flight mass spectrometer for measuring, *in situ*, the chemical composition and the original mass of the individual dust grains.

For quantitative interstellar dust analysis in the sub-μm range, a new secondary neutral mass spectrometer on the basis of a ToF-SIMS has been developed by Henkel *et al.*[14] The authors successfully combined a ToF-SIMS based upon the BIOTOF design[15] with a laser SNMS to analyze presolar stardust particles. This instrument allows SIMS mapping with a mass resolution ($m/\Delta m$) of 3700 and high spatial resolution of 250 nm using a field emission Ga^+ primary ion source (IOG25, Ionoptika, Southampton, UK) with an ion pulse length of ~ 2 ns. In order to improve the ionization efficiency and to avoid inherent SIMS matrix effects for more precise quantification procedures compared to SIMS, non-resonant laser post-ionization (using a GAM F_2 excimer laser) of sputtered neutrals was employed. Efficient post-ionization was achieved in SNMS by using this excimer laser at a wavelength of 157 nm of the laser beam (4.5 mJ per pulse at 750 Hz) of sputtered neutrals. With a photon energy of 7.9 eV, around 60 % of the elements of the periodic table may be non-resonantly photo-ionized by a single photon ionization.[14] The secondary ions were analyzed using a ToF mass spectrometer (R500, Kore Technology; Ely, UK) consisted of a two stage reflectron type flight tube with a microchannel plate detector (MCP). The high spatial resolution achieved enabled structures to be resolved with micron sized interstellar dust grains. To study isotope anomalies on presolar grains or cometary dust, isotope ratio measurements were performed with an accuracy of a few %. A 3D image could be obtained by depth profiling of a single grain.[14]

Diamonds, accounting for $400\,\mu g\,g^{-1}$ of the bulk meteorite, were found to have the highest abundance in presolar grains of primitive meteorites[16], but, due to their small average size of only 2 nm, mass spectrometric microlocal and imaging (and also isotope) analysis was extremely difficult. The diamonds may have been produced by a supernova shock wave passing through molecular clouds. Individual presolar grains discovered in meteorites contain measurable concentrations of trace elements and show isotope anomalies (C, N, Si, Mg, Ti, Ba, Nd and Sm).[1] Several types of presolar grains with large isotopic anomalies have been identified in primitive meteorites in recent years.[17] These grains predate the formation of the solar system materials into which they were incorporated, without completely losing their isotopic and mineralogical identities.[18] These investigations of the preserved memory of their stellar origins allow the stellar processes to be identified that led to their formation and also the physical and chemical conditions during their condensation.[19,20] Laboratory studies of presolar grains have contributed to explaining nucleosynthesis, stellar evolution and dust formation in stellar environments.[17,21]

9.8.2 Isotope Analysis in Cosmochemistry

The discovery of isotopic anomalies of light and also of heavier elements in meteorites has been challenging and was only possible by inorganic mass spectrometry. Imaging mass spectrometry using SIMS, especially, has enabled the isotopic distribution and anomalies to be studied in presolar dust grains. For these studies, grains were deposited on gold foil, and the SIMS measurements were performed with the aid of an O^- primary ion beam for positively charged secondary ions or by means of a Cs^+ primary ion beam for analyzing negatively charged secondary ions to produce ion images of individual isotopes. In addition, $^{18}O/^{16}O$, $^{13}C/^{12}C$ and $^{30}Si/^{28}Si$ isotope ratios in presolar dust grains from primitive meteorites have been measured.[16] Various presolar dust grains discovered in carbonaceous chondrites (corundum, silicon nitride and silicon carbides) preserve the isotopic pattern of presolar matter.

Carbon, oxygen, nitrogen and titanium isotopic compositions in presolar graphite spherules (supernova graphite) and their TiC sub-components ranging in size from 15 to 500 nm have been studied directly on TEM (transmission electron microscope) ultramicrotome sections of the spherule using a multiple ion collector SIMS instrument with lateral resolution in the nm range (NanoSIMS from Cameca).[17] The typical NanoSIMS secondary ion images on the TEM ultramicrotome sections acquired with a primary Cs^+ ion beam have sufficient spatial resolution to discern many of the internal features of this presolar graphite grain. Some of the layers of turbostratic graphite (observed in TEM – transmission electron microscopy) appear compact in the centre and less cohesive towards the outside. The entire 40 layer SIMS measurements took about seven hours. Isotopic ratio measurements of the 'bulk' particle using the CAMECA IMS 3f SIMS indicate a supernova origin for this graphite spherule. Isotopic gradients in $^{12}C/^{13}C$ and $^{16}O/^{18}O$ from the core of the graphite spherule to its perimeter were detected, with the most anomalous composition being present in the centre. These gradients may be the result of isotopic exchange with isotopically 'normal' material either in the laboratory or during the particle's history.[17] No isotope gradient for $^{16}O/^{17}O$ and $^{14}N/^{15}N$ was found. TiC grains located by the NanoSIMS in supernova graphite show significant variations in their oxygen isotope ratios, with $^{16}O/^{18}O$ ratios ranging from 14 to 150 (compared to a terrestrial value of 499). No variation in Ti isotope ratios in three TiC grains was observed.[17]

In 1999, Thiemens[22] reported in *Science* on mass independent isotope effects in planetary atmospheres and the early solar system. Europium isotopic variations were detected in the Allende meteorite and the nature of mass dependent fractionation in the solar nebula is being investigated by Moynier *et al.*[23] The famous Allende meteorite (carbonaceous chondrite – a stone meteorite) entered the atmosphere on February 8, 1969 and exploded. Fragments fell, widely distributed, near the village of Pueblito de Allende, in the Mexican state of Chihuahua. Whereas by measuring the $^{153}Eu/^{151}Eu$ isotope ratio by MC-ICP-MS for a terrestrial basalt, two terrestrial soils and four meteorites (whole rocks and/or chondrules from Bjurböle, Forrest City, Murchison, and Allende) no isotopic variations were found; two different pieces of Allende show a ^{153}Eu deficit of up to 1 ‰.[23] Precise Ni isotope ratio measurements were performed using the large geometry Nu Plasma 1700 multi-collector ICP-MS (see Figure 5.12 e) on Ni after quantitative extraction of analyte to study nickel isotope fine variation in a large number of terrestrial and extraterrestrial samples such as basaltic meteorites, iron meteorites, refractory inclusions from chondrites and individual chondrules.[24,25]

Isotope ratios of Ca, Ti and Si on 37 presolar silicon carbide grains from supernova were examined using the NanoSIMS (Cameca).[26] In all grains of type X, a large enrichment in ^{28}Si (up to five times Solar) was detected. This mass spectrometric finding implies that most X grains were formed from well mixed regions in supernova ejects.

9.8.3 Cosmogenic Radionuclides and Age Dating

One important analytical task in cosmochemistry is to study the formation and evolution of the solar system including the nucleosynthesis of chemical elements and to determine their age. Many radionuclides are produced in the atmosphere of the Earth by nuclear reaction caused by cosmic rays.[27] The cosmic rays (mainly protons and α particles with a very large energy range from 0.01–10^2 GeV) originating from the Sun and the Milky Way Galaxy interact with the atoms of the atmosphere resulting in cosmogenic and anthropogenic radionuclides such as ^{3}H, ^{10}Be, ^{14}C, ^{26}Al, ^{32}Si, ^{36}Cl, ^{39}Ar and ^{81}Kr. These radionuclides have been detected at the ultratrace level in the atmosphere, biosphere, hydrosphere, cryosphere, lithosphere and cosmosphere (in meteorites and lunar material) by AMS and partly by RIMS.[28]

The most famous cosmogenic radionuclide is ^{14}C ($t_{1/2} = 5730\,a$), which is produced by the interaction of cosmic ray neutrons via an (n,p) reaction with nitrogen [$^{14}N(n,p)^{14}C$], whereas the radioactive decay of ^{14}C takes place by β^- decay to form the stable ^{14}N isotope. ^{14}C is the most important cosmogenic radionuclide for dating (see Section 9.7.5) in archaeology and can be analyzed using isotope sensitive accelerator mass spectrometry. Extremely small isotope ratios $^{14}C/^{12}C = 10^{-12}$ in nature can be measured by means of AMS.[28]

A number of extinct radionuclides (^{7}B, ^{10}Be, ^{14}C, ^{22}Na, ^{26}Al, ^{36}Cl, ^{35}S, ^{37}Ar, ^{41}Ca) have been identified in meteorites or lunar materials.[1] These radionuclides can be also used in isotope geoscience and for age dating.[27] For example, in 1983 Nishiizumi *et al.* reported on measurements of ^{26}Al ($t_{1/2} = 7.1 \times 10^5$ a) in drill cores of lunar samples from NASA's Apollo 15 manned mission, using accelerator mass spectrometry.[29] High precision SIMS was applied to analyze ^{26}Al–^{26}Mg ratios for early solar system chronology and for evidence of ^{26}Al in ferromagnesian chondrules.[30,31]

Other applications of inorganic mass spectrometry in planetary sciences were reviewed by de Laeter, including the analysis of the planets Mars (Viking spacecraft mission), Venus (Pioneer mission) and Jupiter (Galileo mission), where small molecules were discovered on the planets, and extraterrestrial materials from comets, Moon and meteorites.[32]

References VIII

1. De Laeter, J. R., *Application of Inorganic Mass Spectrometry*, Wiley-Interscience Series on Mass Spectrometry, John Wiley & Sons, Inc. New York (2001).
2. Clarke, F. W., *Bull. Phil. Soc.*, **11**, 131 (1889).
3. Goldschmidt, V. M., *Nor. Vidensk. Akad. Oslo Mat. Natl. Kl.*, **4**, 148 (1938).
4. Suess, H. E. and Urey, H. C., *Rev. Mod. Phys.*, **28**, 53 (1956).
5. Anders, E. and Grevesse, N., *Geochim. Cosmochim. Acta*, **53**, 197 (1989).
6. Shinotsuka, K. and Ebihara, M., *Anal. Chim. Acta*, **338**, 237 (1997).
7. Gröner, E. and Hoppe, P., *Appl. Surf. Sci.*, **252**, 7148 (2006).
8. Hoppe, P., *Appl. Surf. Sci.*, **252**, 7102 (2006).
9. McPhail, D. S., *Appl. Surf. Sci.*, **252**, 7107 (2006).
10. Treiman, A. H., Musselwhite, D. S., Herd, C. D. K. and Shearer Jr., C. K., *Geochim. Cosmochim. Acta*, **70**, 2919 (2006).
11. Hsu, W., Huss, G. R. and Wasserburg, G. J., *Geochim. Cosmochim. Acta*, **64**, 1133 (2000).
12. Stephan, T., in *Conference Proceedings of 5th European Workshop on Secondary Ion Mass Spectrometry*. Münster (2006).
13. Beauchamp, J., in *Conference Proceedings of 17th International Mass Spectrometry Conference*. Prague (2006) L4.
14. Henkel, T., Tizard, J., Blagburn, D. and Lyon, I., *Appl. Surf. Sci.*, **252**, 7117 (2006).
15. Braun, R. M., *Rapid Commun. Mass Spectrom.*, **12**, 1246 (1998).
16. Zinner, E. K., *Meteorit. Planet. Sci.*, **33**, 549 (1998).
17. Stadermann, F. J., Croat, T. K., Bernatowicz, T. J. *et al.*, *Geochim. Cosmochim. Acta*, **69**, 177 (2005).
18. Bernatowicz, T. J. and Zinner, E., *AIP Conf. Proc.*, **402**, 750 (1997).
19. Huss, G. R. and Lewis, R. S., *Geochim. Cosmochim. Acta*, **59**, 115 (1995).
20. Nittler, L. R., *Earth Planet. Sci. Lett.*, **209**, 259 (2003).
21. Zinner, E., *Ann. Rev. Earth Planet. Sci.*, **26**, 147 (1998).
22. Thiemens, M. H., *Science*, **283**, 341 (1999).
23. Moynier, F., Bouvier, A., Blichert-Toft, J., Telouk, P., Gasperini, D. and Albarède, F., *Geochim. Cosmochim. Acta*, **70**, 4287 (2006).
24. Quitté, G., Halliday, A., Meyer, B. S., Markowski, C., Latkoczy, C. and Günther, D., *Astrophys. J.* (in press) (2007).
25. Quitté, G. and Oberli, F., *J. Anal. At. Spectrom.*, **21**, 1249 (2006).
26. Besmehn, A. and Hoppe, P., *Geochim. Cosmochim. Acta*, **67**, 4693 (2003).
27. Faure, G., *Principles of Isotope Geology*, John Wiley & Sons, Inc., New York (1986).
28. Kutschera, W., *Int. J. Mass Spectrom.*, **242**, 145 (2005).
29. Nishiizumi, K., Arnold, J. R., Klein, J. and Middleton, R., *Lunar Planet.Sci.*, **14**, 560 (1983).
30. Kita, N., Mostefaoui, S., Lui, Y. Z., Togashi, S. and Morishita, Y., *Appl. Surf. Sci.*, **203–204**, 806 (2003).
31. Kita, N., Nagahra, H., Togashi, S. and Morishita, Y., *Geochim. Cosmochim. Acta*, **64**, 3913 (2000).
32. Wieser, M. E. and De Laeter, J. R., *Phys. Rev.*, **C 64**, 024307 (2001).

9.9 Determination of Long-lived Radionuclides

The characterization of highly radioactive nuclear materials, the control of nuclear waste of high radiological toxicity, and the management of radioactive waste for final storage and disposal with regard to long-lived radionuclides are important fields of application for analytical chemistry.[1–5] Experts have estimated that worldwide more than 160 000 tons of long-lived radioactive waste (LLRW) have been created and some tens of thousands of tons are added every year. Furthermore, the mobilization of radionuclides in the environment has been studied in order to trace the routes from the soil via plants into the food chain. Therefore, long-lived radionuclide analysis at the trace and ultratrace level is of increasing relevance in areas including radiobioassay, environmental monitoring, decontamination and environmental remediation, health safety and food control and certain nuclear forensic studies.[1–3,5–8] For example, isotope ratio measurements of uranium (e.g., of depleted uranium or the evidence of ^{236}U) and plutonium in the environment or in body fluids (see Section 8.8) can help distinguish different sources of nuclear contamination in the environment such as, for example, global fallout from nuclear weapons testing, nuclear power plants or nuclear accidents.[5,9–12] In addition, the determination of ultratrace concentrations and precise isotope ratio measurements of uranium, plutonium, neptunium and americium are required for the characterization of microparticles in swipe samples to reveal ongoing undeclared nuclear activities.[1–3,5,10,13–15] The determination of the isotopic composition of single particles, especially, has been recognized to be of the utmost importance in environmental monitoring for non-proliferation in nuclear safeguards and forensic investigations.[5] The fields of application for inorganic mass spectrometry in determining long-lived radionuclides are expected to increase with improvements in sensitivity and precision and decreasing detection limits.

Mass spectrometric techniques play a dominant role for the determination of transuranium elements in bulk samples as well as in microparticles.[1,5] The radioactive element most frequently investigated by inorganic mass spectrometry is uranium.[15,16] The determination of the concentrations and the precise isotopic analysis of naturally occurring radioactive elements (e.g. ^{238}U, ^{235}U, ^{232}Th and the decay nuclides) by inorganic mass spectrometry as terrestrial sources of radioactivity is of interest in environmental science, geology, health and food control as well as in solid state research and material control (e.g., for the characterization of high purity metals, alloys, semiconductors, insulators and chemicals applied for micro- and nanoelectronics) because impurities of radioactive elements can significantly disturb the micro- and nanoelectronic properties.[17] An analytical method for uranium and thorium determination at the ppt and sub-ppt levels in lead of ultrapurity, by ICP-MS has been proposed by Grinberg *et al.*[18] after extraction chromatography. U and Th were analyzed after acid digestion and trace/matrix separation using extraction chromatography (UTEVA resin). Pickhardt *et al.* used LA-ICP-MS to analyze uranium and thorium concentrations in natural zeolites, which can be used to clean up radioactive materials, with detection limits at the $ng\,g^{-1}$ level.[19]

The characterization of radioactive waste is especially required with respect to the long-lived transuranium elements ^{237}Np ($t_{1/2} = 2.14 \times 10^6$ a) ^{239}Pu ($t_{1/2} = 2.41 \times 10^4$ a), ^{240}Pu ($t_{1/2} = 6.5 \times 10^3$ a), ^{242}Pu ($t_{1/2} = 3.8 \times 10^5$ a), ^{243}Am ($t_{1/2} = 7.4 \times 10^3$ a), fission fragments and activated products ^{79}Se ($t_{1/2} = 2.9 \times 10^5$ a), ^{93}Zr ($t_{1/2} = 1.5 \times 10^6$ a), ^{99}Tc ($t_{1/2} = 2.13 \times 10^5$ a), ^{107}Pd ($t_{1/2} = 6.5 \times 10^6$ a), ^{126}Sn ($t_{1/2} = 1.0 \times 10^4$ a), ^{129}I ($t_{1/2} = 1.57 \times 10^7$ a) and ^{135}Cs ($t_{1/2} = 2.3 \times 10^6$ a). But also ^{241}Am with a half-life of $t_{1/2} = 432$ a has to be determined in waste and environmental samples by mass spectrometry. The determination of ^{238}Pu ($t_{1/2} \sim 88$ a) at the trace level is difficult in the presence of uranium due to isobaric interference with $^{238}U^+$ in mass spectra measured by ICP-MS or LA-ICP-MS. Therefore the application of a sensitive and selective technique such as RIMS or AMS is advantageous. The determination of ^{238}Pu for plutonium isotope analysis in irradiated nuclear fuel samples after careful online separation of plutonium from uranium has been proposed for measurements of plutonium isotope ratios using HPLC-MC-ICP-MS by Günther-Leopold and co-workers.[20] Chromatographic separation by HPLC was successfully applied to avoid the isobaric interference of a ^{238}Pu trace with ^{238}U present as the main component in the fuel sample.

Besides the analysis of nuclear fuel and of radioactive waste materials, the determination of contamination and enrichment of selected radioactive nuclides, e.g., ^{129}I, which is one of the most important environmental indicators of nuclear accidents,[21,22] ^{79}Se, ^{99}Tc, ^{237}Np, ^{239}Pu, ^{240}Pu and ^{241}Am at ultratrace concentration levels, is useful for environmental monitoring of fallout from nuclear weapons testing, nuclear power plants or nuclear accidents.[23,24]. Selected application fields for the determination of natural and artificial long-lived radionuclides (LLR) and radionuclides investigated by mass spectrometric techniques are summarized in Tables 9.36 and 9.37, respectively.

Table 9.36 *Application fields for determination of natural and artificial long-lived radionuclides (LLR).*

Application field	Purpose/Materials	Radionuclides
Environmental monitoring	*Study of natural radioactivity* – in sea, river and rain water, plants, animals, soils,	U and Th (natural)
	Contamination of environment (artificial radionuclides) – nuclear power plants and accidents – nuclear weapons and weapon tests (e.g., nuclear fallout)	Tc, ^{90}Sr, ^{129}I, Np, Pu, Th, U, Am, Cm, Ra $^{236}U/^{238}U$, $^{240}Pu/^{239}Pu$
Solid-state research	*Material control of natural radioactivity* in high purity metals, semiconductors, ceramics	U and Th impurities
Geology	*Dating of geological samples* – rocks and minerals	^{87}Rb–^{87}Sr, ^{147}Sm–^{143}Nd, ^{187}Re–^{187}Os, ^{238}U–^{206}Pb
Health control	*Control of exposed and contaminated persons* working in nuclear power plants or other nuclear facilities (nuclear laboratories in research and medicine) and of persons living in contaminated areas – in urine, blood, faeces, – in tissue, hair, nails	e.g., Th, U, Pu, ^{90}Sr, Am, Np, Cm
Food control	drinking, mineral waters, foods	U, Th, Pu, Ra
Isotope dilution techniques	*Quantification of analytical data*	e.g., iodine determination using ^{129}I
Tracer experiments	*Process control* – any material	
Radioactive waste control	*Characterization of waste* – nuclear facilities (e.g., nuclear reactor and power plants), – nuclear weapons, medicine **for recycling and final storage of radioactive waste**	e.g., ^{129}I, ^{79}Se, ^{236}U, Pu, transuranics

Table 9.37 *Most important radionuclides investigated by mass spectrometry.*

Radionuclide	Mass (u)	Half-life ($t_{1/2}$, a)	Mode of decay
^{79}Se	78.918 50	2.9×10^5	β^- to ^{79}Br
^{90}Sr	89.907 74	29.1	β^- to ^{90}Y
^{99}Tc	98.907 00	2.13×10^5	β^- to ^{99}Ru
^{129}I	128.904 99	1.57×10^7	β^- to ^{129}Xe
^{210}Pb	209.984 17	22.6	β^- to ^{210}Bi; α to ^{206}Hg
^{226}Ra	226.025 40	1.6×10^3	α to ^{222}Ra
^{228}Ra	228.031 06	5.76	β^- to ^{228}Ac
^{230}Th	230.033 13	7.54×10^4	α to ^{226}Ra, SF
^{232}Th	232.038 05	1.4×10^{10}	α to ^{228}Ra, SF
^{233}U	233.039 63	1.59×10^5	α to ^{229}Th, SF
^{234}U	234.040 95	2.45×10^5	α to ^{230}Th, SF
^{235}U	235.043 94	7.04×10^8	α to ^{231}Th, SF
^{236}U	236.045 56	2.34×10^7	α to ^{232}Th, SF
^{237}Np	237.048 17	2.14×10^6	α to ^{233}Pa, SF
^{238}U	238.050 78	4.47×10^9	α to ^{234}Th, SF
^{238}Pu	238.049 55	87.74	α to ^{234}U, SF
^{239}Pu	239.052 16	2.41×10^4	α to ^{235}U, SF
^{240}Pu	240.053 81	6.54×10^3	α to ^{236}U, SF
^{241}Pu	241.056 84	14.29	α to ^{237}U; SF, β^- to ^{241}Am
^{242}Pu	242.058 74	3.76×10^5	α to ^{238}U, SF
^{244}Pu	244.064 20	8.2×10^7	α to ^{240}U, SF
^{241}Am	241.056 82	432.7	α to ^{237}Np, SF

SF – spontaneous fisson.

Inorganic mass spectrometry – especially ICP-MS – has been established in the past decade as a method able to compete with the well established classical radioanalytical techniques due to easier sample preparation steps, excellent detection limits and the ability to carry out precise isotope ratio measurements. ICP-MS is, at present, the most widely used analytical technique for determining the concentration and isotopic composition of long-lived radionuclides and represents a valuable alternative to radioanalytical techniques for the determination of actinides owing to its high sensitivity and the possibility of isotope ratio measurements at the ultratrace level. In comparison to α-spectrometry, smaller sized samples can be analyzed by ICP-MS and the time consuming and labour intensive chemical separation and enrichment procedures can be largely reduced or eliminated. The plutonium isotopes ^{239}Pu and ^{240}Pu, especially, are very difficult to analyze by α-spectrometry due to the similar α energies of ^{239}Pu and ^{240}Pu (5.24 and 5.25 MeV, respectively). However, these Pu isotopes can be easily determined by ICP-MS in a separated plutonium fraction or when uranium hydride formation (possible interference of $^{238}U^1H^+$ at $^{239}Pu^+$) is minimized.[25–27] The uranium hydride formation rate $^{238}U^1H^+/^{238}U^+$ in ICP-MS, which is relevant for ^{239}Pu and ^{236}U determination at the ultratrace level, was studied using different nebulizers and varied between 1.2×10^{-4} (using DIHEN[27]), 1.9×10^{-5} and 2.1×10^{-5} for the cross flow and MicroMist nebulizer, respectively,[28] and 1.2×10^{-5} and 1.0×10^{-5} for the ultrasonic nebulizer and Aridus, respectively.[29] A low hydride formation rate of 1.1×10^{-6} was observed in ICP-MS with the APEX-Q (with desolvator).[30] To minimize the hydride formation in ICP-MS, heavy water (D_2O) was applied as a solvent whereby the $^{238}U^1H^+/^{238}U^+$ ratio was found to be 9×10^{-7} by applying the Aridus with desolvator in sector field ICP-MS.[29]

Conventional radiochemical methods for the determination of long-lived radionuclides at low concentration levels require a careful chemical separation of the analyte, e.g., by liquid–liquid, solid phase extraction or ion chromatography. The chemical separation of the interferents from the long-lived radionuclide at the ultratrace level and its enrichment in order to achieve low detection limits is often very time consuming. Inorganic mass spectrometry is especially advantageous in comparison to radioanalytical techniques for the characterization of radionuclides with long half-lives ($> 10^4$ a) at the ultratrace level and very low radioactive environmental or waste samples.

In addition to the analysis of artificial radionuclides, mass spectrometry is useful for characterizing naturally occurring radioactive materials (NORMs). NORMs consist of ^{238}U, ^{235}U, ^{232}Th and series daughters and are released into the environment from many anthropogenic sources. Ketterer *et al.*[31] applied quadrupole based ICP-MS after selective analyte extraction for envirogeochemical exploration of NORM wastes with respect to uranium ($^{234}U/^{238}U$) and thorium ($^{232}Th/^{230}Th$) isotope ratios. The determination of natural uranium and thorium in environmental samples by ETV-ICP-MS after matrix removal by online solid phase extraction is described by Truscott and co-workers.[32]

Table 9.38 *Detection limits of mass spectrometry for determination of long-lived radionuclides.*[2]

Analytical method	Detection limits
Solid-state mass spectrometry	(μg g^{-1})
SSMS	1–0.001
GDMS	0.1–0.000 1
SIMS	10–0.002
LA-ICP-MS	0.1–0.000 01
ICP-MS	(ng l^{-1})
ICP-QMS	0.01–0.6
ICP-SFMS ($m/\Delta m = 300$)	0.000 04–0.005
ICP-CC-QMS (with hexapole collision cell)	0.003–0.01
ICP-TOFMS	0.1–1
MC-ICP-MS (ion counting)	0.0001–0.0002

Henry *et al.*[33] reported that improvements in quadrupole ICP-MS resulted in ag mass detection capability. Consequently the analysis of radionuclides with shorter half-lives is also possible. In Table 9.38 the detection limits of several mass spectrometric techniques for the determination of long-lived radionuclides are compared.

9.9.1 Determination of Half-life of Long-lived Radionuclides

A major topic in isotope mass spectrometry is the determination of the half-lives of long-lived radionuclides. De Bièvre and Verbruggen[34] determined the half-life of ^{241}Pu for β-decay in the isobaric radionuclide ^{241}Am on material from Oak Ridge that had initially been about 93 % isotopically enriched. Due to the isobaric interference of ^{241}Pu and ^{241}Am radionuclides during mass spectrometric measurements by TIMS, Am had to be removed by chemical separation immediately (less than 48 h) prior to measurements as described in reference [34]. On the basis of all the measurements performed over an extended period of more than 20 years and after considering the possible effects of systematic errors during these measurements, a half-life for the β^- decay of ^{241}Pu of ($t_{1/2} = 14.290 \pm 0.006$ a) was reported.[34]

The alpha decay half-life of ^{244}Pu on two samples of enriched ^{244}Pu, using ^{240}Pu and ^{242}Pu isotopes as reference nuclides, was redetermined using TIMS and α-spectrometry and found to be $(8.12 \pm 0.03) \times 10^7$ a.[35]

9.9.2 Methodological Developments and Applications of ICP-MS for Determination of Long-lived Radionuclides Including Trace/Matrix Separation

Long-lived radionuclides occur at extremely low concentrations, especially in environmental samples, therefore several authors have proposed matrix separation and enrichment of the analytes before analysis.[21,24,26,36–39] Radiochemical methods often require very careful and time consuming separation and enrichment processes and measurement procedures of α-, β- and γ-emitting radioactive species at the trace and ultratrace level using conventional radioanalytical techniques.[40–43] Trace/matrix separation, which is performed offline or online in order to avoid possible isobaric interferences, matrix effects and to reduce the detection limits for the determination of long-lived radionuclides, is also advantageous before ICP-MS measurements as the most widely applied mass spectrometric technique.

Analytical procedures have been developed and applied for plutonium and uranium determination in natural water samples from Israel by Halicz *et al.*[44] For uranium, a natural variation of the ^{234}U/^{238}U isotope ratio by a factor of up to two compared to the IUPAC table value[45] was found by applying double-focusing sector field ICP-MS with single and multiple ion collectors. Pu contamination at the extreme ultratrace level ($< 10^{-19}$ g ml^{-1}) was detected in the Sea of Galilee. Pu isotope ratio measurements using MC-ICP-MS demonstrated that the Pu contamination was the result of global nuclear fallout after nuclear weapons tests in the 1960s.[44]

Karpas *et al.* reported on the determination of the ^{234}U/^{238}U isotope ratio (54.9×10^{-6}) by MC-ICP-MS and ICP-QMS compared to α-spectrometry in hair, nail and water samples.[46] A correlation of 0.99 was found between the two ICP-MS methods and of 0.98 with α-spectrometry for the water sample. The measurement time of the water sample after dilution for ICP-MS was ~ 1 min and 1000 min (including uranium separation and counting time) for α-spectrometry.[46]

For the determination of ^{99}Tc in environmental and radioactive waste samples by ICP-MS, which is disturbed by isobaric interference with $^{99}Ru^+$ (and $^{98}MoH^+$) ions, Yamamoto *et al.*[47] suggested a separation of ^{99}Tc using different solvent extraction and purification techniques with anion exchange. In this article, the authors describe the determination of ^{99}Tc in sediments from the Irish Sea with an absolute detection limit of 0.25 pg, (0.16 mBq), using a double-focusing sector field ICP-MS (PlasmaTrace, VG Elemental Ltd.). Eroglu *et al.*[24] studied the separation and enrichment of ^{99}Tc from sea water by anion exchange with a detection limit of 0.03 ng l^{-1} using a quadrupole based ICP-MS (HP 4500, Hewlett-Packard). For the separation of ^{99}Ru and ^{99}Tc from environmental samples, online IC-ICP-MS was employed by Betti.[48] In a highly radioactive evaporator concentrate from a nuclear power plant, ^{99}Tc was measured in our laboratory after chemical separation procedures using ICP-MS. The detection limit for ^{99}Tc determination in separated radioactive waste solution was determined by double-focusing ICP-SFMS to be 5 pg l^{-1}, corresponding to an activity of 3 μBq ml^{-1}. In comparison to ICP-SFMS using quadrupole-based ICP-MS (ELAN 6000, PerkinElmer Sciex), a detection limit of 0.1 ng l^{-1} was determined.[28]

Barrero Moreno *et al.*[49] determined neptunium and plutonium in the presence of high concentrations of uranium by ion chromatography coupled to ICP-MS.

^{226}Ra determination in environmental samples (highly saline thermal waters) by ICP-QMS with ultrasonic nebulization has been described by several working groups.[39,50,51] Radium was preconcentrated and isolated from the matrix elements by selective extraction using a radium specific solid phase extraction membrane disk designed for radioactive counting methods as proposed by

Joannon and Pin.[50] A low detection limit was achieved in quadrupole ICP-MS when the pressure in the interface was reduced from approx. 2 to 0.85 mbar. Larivière *et al.*[51] developed a selective extraction procedure for the preconcentration of ^{226}Ra from uranium ores and biological samples. The measurements were performed by ICP-QMS with a hexapole collision cell in order to reduce possible interferences. An absolute detection limit of 0.02 fg (0.75 mBq) was obtained using less than 4 mg of solid sample or 25 ml of liquid sample.

ICP-MS was used for many years in the author's laboratory for the sensitive and precise determination of thorium and uranium and for isotope ratio measurements of uranium at the trace and ultratrace level in urine samples in a routine mode. An analytical method for the rapid determination of both long-lived radionuclides in urine samples was proposed, whereby different sample preparation and quantification procedures were studied. In order to avoid matrix effects during the mass spectrometric measurements and clogging effects on the nickel cones and torch, the urine samples were digested in a mixture of nitric acid, hydrogen peroxide and hydrogen fluoride by closed vessel microwave digestion. Recovery studies for trace elements spiked in a small volume of urine matrix resulted in good agreement of analytical data with the expected results even for the 0.5 $\mu g\,l^{-1}$ U and Th spiked samples. The detection limits for the determination of uranium and thorium in urine samples by ICP-MS using quadrupole and double-focusing sector field ICP-MS were determined down to the sub-ng g^{-1} level. The main factors affecting the accurate and precise determination of ^{236}U using ICP-MS are the instrumental background, the isobaric interference of $^{235}UH^{+}$ polyatomic ions on $^{236}U^{+}$ analyte ions, and the presence of $^{238}U^{+}$ and $^{235}U^{+}$ peak tails. In order to reduce $^{235}UH^{+}$ formation, D_2O (heavy water) is used as a solvent for the dissolution and dilution of uranium samples. Abundance sensitivity was improved by the use of medium mass resolution ($m/\Delta m = 4450$) in comparison to low mass resolution in double-focusing sector field ICP-MS (ICP-SFMS). The performance of several different sample introduction systems was compared for solution introduction. The performance of several nebulizers (Meinhard, Aridus and ultrasonic nebulizer) for $^{236}U/^{238}U$ isotope ratio measurements in ICP-SFMS was studied. It has been shown that for all nebulization systems, a diminution in UH^{+}/U^{+} is observed in D_2O as compared to H_2O as the solvent.[29] Selected results of the determination of uranium concentration and isotope ratios in urine samples are summarized in Table 9.39. In comparison to a table value for $^{235}U/^{238}U$ of 0.007 25, a contamination with depleted uranium (DU) was detected in urine samples 2–5. The analytical technique developed was applied for $^{235}U/^{238}U$ isotopic ratio measurements in urine samples, whereby possible contamination with depleted or enriched ^{235}U was found.

Table 9.39 *Results of determination of uranium concentration and isotope ratio measurements by LA-ICP-SFMS on several urine samples (n = 3).*

Sample	U conc. ($\mu g\,l^{-1}$)	$^{235}U/^{238}U$
1	0.128 ± 0.030	0.0073 ± 0.002
2	0.066 ± 0.005	0.0026 ± 0.0002
3	0.130 ± 0.060	0.0036 ± 0.010
4	0.170 ± 0.040	0.0061 ± 0.001
5	0.073 ± 0.002	0.0029 ± 0.0001

Since it is easy to sample, urine is frequently used to check for possible contamination of the human body with toxic elements or long-lived radionuclides. As naturally occurring α emitting long-lived radionuclides, ^{238}U, ^{235}U ^{234}U and ^{232}Th are taken up daily at a low concentration level with drinking water and food. Possible contamination with non-natural uranium is detected rapidly by mass spectrometric isotope ratio measurements when the $^{235}U/^{238}U$ and the $^{234}U/^{238}U$ ratio deviate from the IUPAC table values[52] (e.g., $^{235}U/^{238}U = 0.00725$) and/or ^{236}U is found in urine sample with an isotope ratio $^{236}U/^{238}U > 10^{10}$ higher than that observed in nature. Urine samples are easy to collect so that rapid monitoring techniques for routine control is available, especially for measurements of long-lived radionuclides at ultratrace concentration level.

Because long-lived radionuclides occur at extremely low concentrations, especially in medical samples, several authors have proposed matrix separation and enrichment of the analyte before beginning analysis.[12,26,53,54] Trace/matrix separation, which is performed offline or online, is also advantageous in order to avoid possible isobaric interferences, matrix effects and to reduce the detection limits for the determination of long-lived radionuclides.

For example, for the determination of Pu in urine at the low ag ml^{-1} concentration level we developed an analytical method using ICP-SFMS, whereby 1 L of urine doped with pg ^{242}Pu was analyzed after co-precipitation with $Ca_3(PO_4)_2$ followed by extraction chromatography on TEVA resin in order to enrich the Pu and remove uranium and matrix elements.[26] The capability of double-focusing ICP-SFMS for the determination of Pu in aqueous solution and urine was studied using two nebulizers, PFA-100 and DIHEN (direct injection high efficiency nebulizer), for solution introduction with uptake rates of 0.06 and 0.58 ml min^{-1}, respectively. The sensitivity for Pu in ICP-SFMS was determined as 2000 and 1380 MHz ppm^{-1} for the PFA-100 and DIHEN nebulizers, respectively. Due to the low solution uptake rate of the DIHEN, the absolute sensitivity was about seven times better and yielded 1380 counts fg^{-1} in comparison to 207 counts fg^{-1} measured with the PFA-100 nebulizer. Recovery using a ^{242}Pu tracer was about 70 %. The limits of detection (LOD) for ^{239}Pu in 1 litre of urine, based on an enrichment factor of 100 for the PFA-100 nebulizer and 1000 for the DIHEN, were 9×10^{-18} and 1.02×10^{-18} g ml^{-1}, respectively. Measurements of the $^{240}Pu/^{239}Pu$ isotopic ratio in synthetically prepared urine standard solution yielded a precision of 1.8 % and 1.9 % and an accuracy of 1.5 % for the PFA-100 and 1.8 % for the DIHEN nebulizers.[26] Flow injection and online preconcentration using an anion exchange resin and matrix separation after microwave digestion was proposed for the determination of Pu isotopes in leaves by ICP-DRC-MS with a dynamic reaction cell, by Epov *et al.*[55] Detection limits for several Pu isotopes (except ^{238}Pu) were obtained by analyzing 10 g apple leaves (NIST SRM 1515) in the low fg g^{-1} and sub-fg g^{-1} ranges. To study plutonium and neptunium speciation in groundwater containing natural humic acid at different pH values, Trautmann's group coupled capillary electrophoresis (CE) to ICP-MS.[56]

Of similar interest is the determination of ^{90}Sr ($t_{1/2} = 29.1$ a) at the ultratrace level, since it appears as a radionuclide in the decay series of nuclear fission generated in the nuclear industry and can therefore be found in nuclear waste or released by nuclear accidents. Current methods for the detection of this radionuclide are time consuming and may be prone to a large variety of interferents. ICP-SFMS has been explored for the determination of ^{90}Sr in the presence of zirconium in urine.[54] Specific techniques have been investigated for removing zirconium as well as other contributions to the background at $m/z = 90$. A quadrupole ICP-MS equipped with a hexapole collision cell was first explored (final LOD = 2 ng l^{-1} for water samples), however, the desired limit of detection for ^{90}Sr in urine was quite low (0.02 pg l^{-1}). The performance of a double-focusing sector field ICP mass spectrometer (ICP-SFMS) was further investigated (optimization curve is shown in Figure 2.9), which allowed ^{90}Sr to be measured under cold plasma conditions (optimum rf power: 750 W) at the ultratrace level. Other potential interferences (such as possible polyatomic ion formation) were investigated, minimized and instrumental detection limits calculated as 3 pg l^{-1} for water samples. Final parameters included the use of a cold plasma and medium mass resolution in ICP-SFMS. The method was applied for the analysis of ^{90}Sr extracted from urine using a crown ether extraction resin.

The analyte was concentrated (enrichment factor: 200) but high levels of natural strontium in the separated fraction (of about 1 $\mu g\,ml^{-1}$) meant higher detection limits (80 $pg\,l^{-1}$) due to 'peak tailing' of $^{88}Sr^+$ at $m/z = 90$ and the relatively low abundance sensitivity of ICP-SFMS at a medium mass resolution of 6×10^{-7}. This detection limit in the separated fraction corresponded to a detection limit of 0.4 $pg\,l^{-1}$ in the original urine sample. The recovery of ^{90}Sr, determined by the described analytical method in spiked urine samples, was in the range 82–86 %. Decreasing the detection limit for ^{90}Sr determination is recommended by the application of a multiple ion collector ICP-MS due to improved abundance sensitivity. The analytical methods described can also be applied for the analysis of other body fluids, such as blood or human milk or for the determination of ^{90}Sr in bones.

TIMS has been used for many years as the benchmark technique especially for uranium isotope analysis.[57] Instrumental improvements have enabled ICP-MS to approach the accuracy and precision obtained by TIMS in measuring data. In addition, due to time consuming sample preparation steps and the need for a large volume of urine, the method has been replaced by the more powerful ICP-MS in many laboratories.[58] An interlaboratory analytical exercise on the determination of natural and depleted uranium in urine was carried out by different ICP-MS instruments, by thermal ionization mass spectrometry (TIMS) and instrumental neutron activation analysis.[59] TIMS has also been employed to determine fg quantities of ^{239}Pu and ^{240}Pu in bioassay samples (such as human urine and artificial urine),[60] in an interlaboratory comparison for the analysis of the ^{239}Pu and $^{239}Pu/^{240}Pu$ atomic ratios in synthetic urine by TIMS and AMS as reported in reference.[61]

Betti reported on the civil use of depleted uranium and various relevant scenarios (e.g., the case of an aircraft accident which occurred in Amsterdam in 1992 involving a fire) in terms of radiological exposure to bystanders.[62]

In past years, online ion chromatography ICP-MS, in particular, has been increasingly applied for the characterization of radioactive materials and environmental samples using ICP-MS.[37,63–65] For example, analytical procedures for the determination of spallation nuclides in the tantalum target of a spallation neutron source irradiated with 800 MeV protons were developed in the author's laboratory.[63,64,66] In Figure 6.6 a and b, the mass spectra of rare earth elements (REEs) of natural isotopic pattern are compared with those in an irradiated tantalum target. In order to separate isobars of rare earth elements, such as long-lived ^{173}Lu, from stable ^{173}Yb, online high performance liquid chromatography (HPLC – for a chromatographic separation of a lanthanide mixture into the individual elements) was coupled to the mass spectrometer. The concentrations of different spallation nuclides (see Table 9.40) in the first tantalum plate of the spallation neutron source (sample 1 was collected from the centre of the plate, sample 3 is from the edge and sample 2 was taken from area between samples 1 and 2) were measured by HPLC-ICP-MS using the reverse isotope dilution technique.[67] With increasing distance from the centre of the plate, that means with decreasing irradiation density of the 800 MeV proton beam on the tantalum target, the concentration of the spallation nuclides decreased. We also coupled capillary electrophoresis (CE) to a double-focusing sector field ICP-MS for the same analytical task.[64] This approach reduces the sample volume, from the 100 μl range using HPLC-ICP-MS, to the nl range, which is extremely important for the analyzis of highly radioactive solutions. The spallation nuclides in the irradiated tantalum were analyzed by both mass spectrometric coupling techniques after the dissolution of highly radioactive tantalum in a HNO_3/HF mixture and after (offline) matrix separation by liquid–liquid extraction of the tantalum matrix (in order to reduce the high ^{182}Ta activity). The theoretical results of the spallation yields of tantalum were verified by mass spectrometric measurements of the concentration of spallation nuclides of the irradiated tantalum target by both HPLC-ICP-MS and CE-ICP-MS.[63,64,66,67] One example is shown in Table 9.41 where the nuclide abundances of gadolinium produced via spallation reactions in an irradiated tantalum target and measured by

Table 9.40 *Concentration of different spallation nuclides of the lanthanides* ($\mu g\,g^{-1}$) *in an irradiated tantalum target.*

Nuclide	Half-life	Sample 1	Sample 2	Sample 3
^{141}Pr	stable	9.6 ± 1.7	2.1 ± 0.6	0.7 ± 0.2
^{145}Pm	17.7 a	21.2 ± 3.6	4.4 ± 1.1	0.9 ± 0.3
^{150}Gd	1.8×10^6 a	28.7 ± 4	6.3 ± 1.2	1.1 ± 0.3
^{163}Ho	33 a	80.3 ± 10.4	21 ± 4	3 ± 0.6
^{173}Yb	stable	111.1 ± 13.3	29.6 ± 5.3	4.6 ± 0.8
^{173}Lu	1.4 a	16.5 ± 2.8	5 ± 1	0.8 ± 0.2

Table 9.41 *Nuclide abundances of gadolinium (%) produced via spallation reactions in an irradiated tantalum target via spallation reactions.*

Nuclide	Nature	Theory	CE-ICP-MS	HPLC-ICP-MS
^{148}Gd	—	15.8	19.4	20.7
^{150}Gd	—	18.3	18.8	18.4
^{152}Gd	0.2	27.2	22.1	22.7
^{154}Gd	2.15	2.6	< 3.4	< 1.1
^{155}Gd	14.7	34.4	33.4	33.9
^{156}Gd	20.5	0.57	—	< 1.1
^{157}Gd	15.7	0.40	—	< 1.9
^{158}Gd	24.9	—	—	—
^{160}Gd	21.9	—	—	—

HPLC-ICP-MS and CE-ICP-MS, are compared with the table values and theoretical results. ^{155}Gd is the most highly abundant nuclide. Compared to the isotope pattern in natural samples more light gadolinium nuclides with a typical isotopic pattern were formed by the irradiation of Ta with high energy protons.

The quasi-simultaneous separation and determination of lanthanides and actinides by ion chromatography inductively coupled plasma mass spectrometry combined with the isotope dilution technique and the further use of ion chromatography for the determination of fission products and actinides in nuclear applications are described by Betti *et al.*[10,48,68]

$^{90}Sr/^{238}U$ ratios in irradiated fuels were measured by sector field ICP-MS with multiple ion collection, equipped with a hexapole collision cell (Isoprobe, GV Instruments Ltd, Manchester, UK).[69] The interference problem of isobars ^{90}Sr and ^{90}Zr is solved by the addition of oxygen as a reactant gas to the collision cell. Whereas Zr^+ is oxidized to ZrO^+, ions Sr^+ ions are not reactive under the experimental conditions applied. The $^{90}Sr/^{238}U$ atomic ratio has been determined mass spectrometrically in different spent fuel sample solutions (UOX – uranium oxide and MOX – mixed oxide spent fuel samples) by the double spike isotope dilution method (using enriched ^{84}Sr and ^{235}U spikes with isotopic abundances of 99.9 % and 93.3 %, respectively). Primary U/Pu fission products separation with anion exchange resin was performed to eliminate the U/Pu matrix before measuring the $^{90}Sr/^{238}U$ ratio, as described by Isnard *et al.*[69] The double spike isotope dilution technique allows the $^{90}Sr/^{238}U$ ratio in fuel samples to be determined directly and in a very accurate way with an external precision (RSD, 1σ) better than 0.4 %.[69]

Different applications of mass spectrometry in the trace and ultratrace determination of long lived radionuclides in nuclear fuel, solid radioactive waste samples, radioactive solutions and

Table 9.42 *Determination of long-lived radionuclides in nuclear fuel and radioactive waste samples.*

Samples	Method	Radionuclides	Detection limits	References
Radioactive waste solution	ICP-SFMS (Element, USN)	^{226}Ra, ^{230}Th, ^{233}U, ^{237}Np, ^{239}Pu, ^{241}Am	0.05 pg l^{-1} (^{241}Am) 0.04 pg l^{-1} (^{239}Pu)	Becker and Dietze[28]
Radioactive waste solution	ICP-SFMS (Element, DIHEN)	^{226}Ra, ^{230}Th, ^{230}Th, ^{233}U, ^{237}Np, ^{238}U, ^{241}Am	0.1 pg l^{-1} (^{241}Am) 0.1 pg l^{-1} (^{237}Np)	McLean *et al.*[8]
Spent uranium	ICP-SFMS (Element, Aridus) ion exchange	^{236}U, ^{239}Pu	0.2 pg l^{-1} (solution) 0.04 pg g^{-1} (soil)	Boulyga and Becker[62]
Radioactive concrete	LA-ICP-SFMS Element	^{99}Tc, ^{232}Th, ^{233}U, ^{235}U, ^{236}U, ^{238}U, ^{237}Np	0.02 ng g^{-1} (^{236}U) 1.4 ng g^{-1} (^{99}Tc)	Becker *et al.*[6]
UO_2 fuel	ICP-QMS, (Elan 5000), ion chromatography, isotope dilution	^{238}U, ^{237}Np, ^{239}Pu	0.06 μg l^{-1} (Np)	Barrero Moreno *et al.*[36]
Fission products	ETV-ICP-MS	^{79}Se	13 ng l^{-1}	Compte *et al.*[72]
Spent nuclear fuel	ICP-QMS, (Elan 5000), online ion chromatograpy	^{238}U, ^{237}Np, ^{244}Pu, ^{243}Am, ^{248}Cm,	0.45 ng ml^{-1}	Perna *et al.*[71]
Nuclear fuel samples	HPLC-MC-ICP-MS (Neptune)	^{238}Pu, ^{239}Pu, ^{240}Pu, ^{241}Pu ^{240}Pu/^{239}Pu $\sim$ 0.56	^{238}Pu/^{239}Pu $\sim$ 0.067	Günther-Leopold *et al.*[20]

environmental, biological, medical samples and food are summarized in Tables 9.42[8,20,28,48,68,70–72] and 9.43,[21,32,39,50,73,74] respectively. Further applications for the isotope ratio measurements of long-lived radionuclides are discussed in Section 8.9.

9.9.3 Ultratrace Analysis of Long-lived Radionuclides in Very Small Sample Volumes

Micro- and nanovolume analytical techniques are required for the analysis of long-lived radionuclides at the ultratrace concentration level in highly radioactive samples, to reduce the radioactivity of the sample analyzed, the waste, contamination of instruments and tools, and the dose to the operator. Furthermore, for a multitude of applications, especially for biomonitoring (for monitoring the contamination arising from radioactive waste in the environment and for evidence of nuclear fallout), the concentrations of long-lived radionuclides are extremely low and the amount of sample is often restricted. In order to analyze small sample volumes, several micronebulizers (e.g., MCN Aridus, Cetac Technologies, USA and MicroMist, Glass Expansion, Australia see Figure 5.14 a)[19,28,75,76] and DIHEN[75] (DIHEN, J. E. Meinhard Associates, USA) have been used for determining long-lived radionuclides by solution introduction into the ICP-MS instead of ultrasonic nebulizers, which consume high volumes of solution. Using the DIHEN[75] the sample solution introduction into the inductively coupled plasma is achieved with an analyte transport efficiency of 100 %, but a high formation rate of polyatomic ions is also observed.[8,75]

The application of flow injection, as mentioned above, is extremely helpful for the introduction of small volumes of solution into the ICP in order to minimize radioactive contamination in the

Table 9.43 *Determination of long-lived radionuclides in biological, environmental and medical samples and food.*

Samples	Equipment	Radionuclides	Detection limits	References
Thermal water	ICP-QMS + USN (PlasmaQuad PQ2) extraction, ion exchange	^{226}Ra	$2\,fg\,ml^{-1}$	Joannon and Pin[49]
Mineral and ground water	ICP-SFMS (Element), analyte separation	^{226}Ra	$0.02\,pg\,l^{-1}$	Zoriy *et al.*[38]
Food	ICP-SFMS + USN (Element), ion extraction	Pu	$0.03\,pg\,g^{-1}$	Evans *et al.*[65]
Urine	ICP-SFMS (Element) ICP-CC-MS (Platform)	^{90}Sr	$0.4\,fg\,ml^{-1}$ (Element) $2\,pg\,ml^{-1}$ (Platform)	Vonderheide *et al.*[72]
Sediments	ICP-CC-MS (Platform) hot extraction	^{129}I	$0.4\,pg\,g^{-1}$ $^{129}I/^{127}I \sim 5 \times 10^{-7}$	Izmer *et al.*[21]
Biological SRMs, Sea and river water	ICP-QMS, Elan 5000 PlasmaQuad PQ2 ETV, on-line solid phase extraction	^{238}U, ^{232}Th	ETV-ICP-QMS $0.9\,pg\,ml^{-1}$ (^{238}U) $0.3\,pg\,ml^{-1}$ (^{232}Th)	Truscott *et al.*[36]
Human Brain (imaging)	LA-ICP-SFMS (Element)	U, Th	$10\,ng\,g^{-1}$ (U)	Becker *et al.*[64]

instrument.[70,76] Microlitre volumes of an aqueous solution can be handled by a commercial HPLC injection valve coupled, for example, to a microconcentric nebulizer (Micromist, Glass Expansion) for small droplet formation and a minicyclonic spray chamber. It is thus possible to analyze small sample volumes (sample loop: $> 1\,\mu l$) of radioactive waste solution introduced by this HPLC injection valve into a continuous flow of 2 % nitric acid. In Figure 9.50, an example is given of the application of flow injection ICP-MS (using a quadrupole ICP-MS, Elan 6000) for the determination of ^{232}Th in small volumes of aqueous solution. In the left part of Figure 9.50, transient signals for 1, 2 and $4\,ng\,l^{-1}$ (at a sample loop of $20\,\mu l$) solutions are demonstrated. The flow injection isotope dilution technique (right part of Figure 9.50) was developed for the accurate determination of radionuclide concentration. In this experiment, a ^{232}Th solution (continuous flow) was spiked with $20\,\mu l$ of $5\,\mu g\,l^{-1}$ highly enriched ^{230}Th (99.85 %) for quantitative Th determination in radioactive solutions.[76] Lower analyte concentrations in radioactive waste solution can be measured by more sensitive ICP-SFMS. For example, transient signals of a ^{237}Np standard solution (sample loop: $20\,\mu l$; Np concentrations: $10\,pg\,ml^{-1}$ and $100\,pg\,ml^{-1}$) were measured with a precision of 2.0 and 1.6 % (RSD, $N = 5$), respectively.[70] Another possibility for analyzing small quantities of sample is the application of electrothermal vaporization for sample introduction in ICP-MS for the determination of long-lived radioisotopes, which is described in reference[77].

A low flow micronebulizer (e.g., MicroMist or Aridus) together with ICP-MS was successfully applied in the author's laboratory in routine measurements for research, e.g., for the environmental monitoring of spent reactor uranium, radioactive waste solution,[28,76] evaluating origins of contamination with nuclear fuel in Chernobyl samples, studying radionuclide behaviour in the environment or for the analysis of depleted uranium in samples from Kosovo.[40,78] In Table 9.44 the figures of merit of several nebulizers are compared, microflow nebulizer DS[79] and Aridus[29] (both from Cetac Technologies), Meinhard,[8] MicroMist,[8] PFA[29] nebulizers and quartz DIHEN,[8] for ^{238}U determination by ICP-SFMS. Considering the different solution uptake rates of several nebulizers, the highest

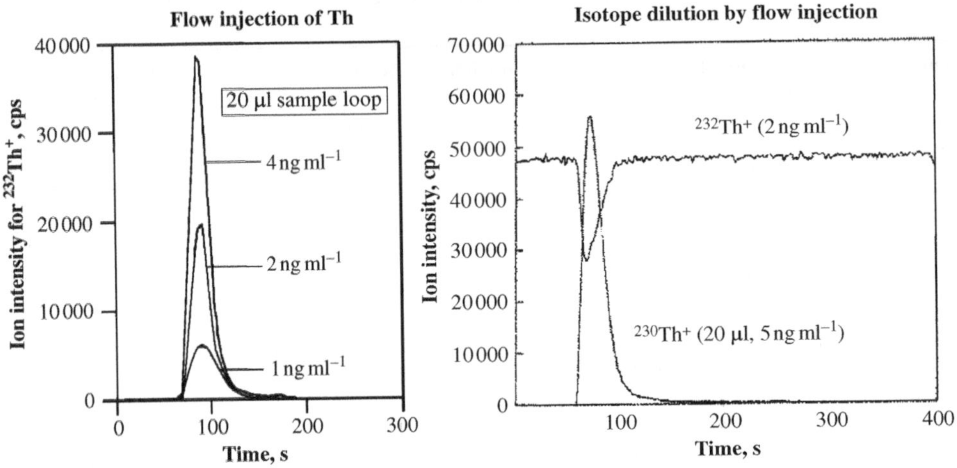

Figure 9.50 *Determination of ^{232}Th in small volumes of aqueous solution by flow injection ICP-MS (using Micro Mist). (J. S. Becker, R. S. Soman, K. L. Sutlon, J. A. Caruso and H. J. Dietze, J. Anal. At. Spectrom.* ***14****, 933(1999). Reproduced by permission of the Royal Society of Chemistry.)*

absolute sensitivity for the determination of ^{238}U (2418 cps fg^{-1}) was measured using the DS 5 microflow nebulizer in sector field ICP-MS in the flow injection mode at a low solution uptake rate of 0.007 ml min^{-1}. The transient signal of the injected 10 ng l^{-1} uranium solution (sample loop of flow injection system = 54 nl) in a continuous flow of carrier liquid at 7 µl min^{-1} measured by ICP-SFMS is illustrated in Figure 9.51.[79] The absolute detection limits were determined for ^{242}Pu as 1.5×10^{-17}g (6×10^{-20} mol; ~ 38 000 ^{242}Pu atoms).[79]

Table 9.44 *Comparison of the absolute sensitivity of different nebulizer types for $^{238}U^+$ measured with ICP-SFMS (Element).*

Nebulizer	Solution uptake rate, ml min^{-1}	Sample volume, ml	Absolute sensitivity, counts fg^{-1}	Reference
DS-5	0.007	0.000 05[a]	2418	Schaumlöffel *et al.*[70]
q-DIHEN	0.06	0.12	1680	McLean et *al.*[8]
Aridus	0.1	0.2	2340	Zoriy et *al.*[29]
PFA	0.2	0.4	580	Zoriy et *al.*[29]
MicroMist	0.2	0.4	820	McLean *et al.*[8]
Meinhardt	1	2	97	McLean *et al.*[8]

[a] using flow injection.

Online trace enrichment by flow injection using a microcolumn of activated alumina and mass spectrometric determination of uranium in mineral, river and sea water is relevant for analyzing small sample volumes.[80] Flow injection with online preconcentration using solid

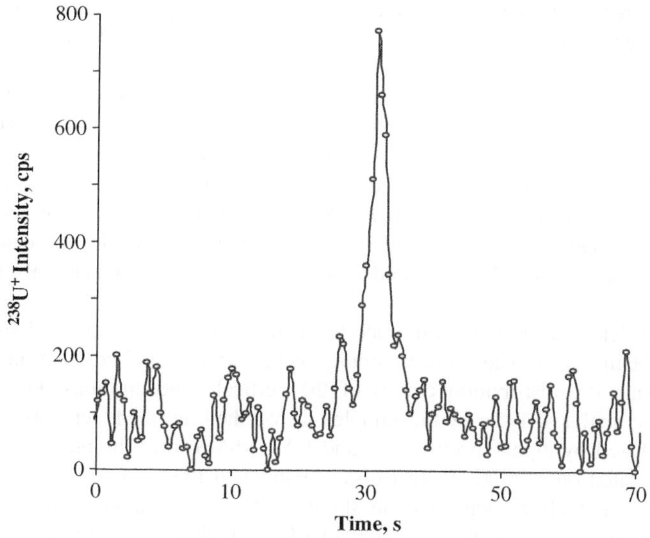

Figure 9.51 *Transient signal of $^{238}U^+$ using nanovolume flow injection of a 10 ng l^{-1} uranium solution (sample loop = 54 nm). (D. Schaumlöffel, P. Giusti, M. Zoriy, C. Pickhardt, J. Szpunar, R. Lobinski and J. S. Becker, J. Anal. At. Spectrom.,* **20**, *17(2005). Reproduced by permission of the Royal Society of chemistry.)*

phase adsorption on a minicolumn of ^{99}Tc, ^{230}Th and ^{234}U, can be achieved at the ultratrace level in soils with detection limits in the soil for ^{99}Tc, ^{230}Th and ^{234}U of 11 mBq g^{-1} (0.02 ng g^{-1}), 3.7 mBq g^{-1} (0.005 ng g^{-1}) and 0.74 mBq g^{-1} (0.003 ng g^{-1}), respectively.[81]

9.9.4 Determination of Long-lived Radionuclides by LA-ICP-MS and ETV-ICP-MS

In order to reduce possible contamination problems during sample preparation, LA-ICP-MS is applied for the precise and accurate determination of ^{235}U/^{238}U isotope ratios at the trace and ultratrace level in medical samples, such as urine, after deposition on a quartz substrate. LA-ICP-MS is advantageous for providing fast evidence of possible contamination of urine samples (screening) of exposed persons with depleted or enriched ^{235}U or any other contaminations.[82] An analytical procedure has been developed in the author's laboratory for the determination of U and Th concentration and the ^{235}U/^{238}U isotope ratio in urine at the trace and ultratrace level by LA-ICP-MS. After careful homogenization, the dried urine samples were analyzed directly by LA-ICP-MS without any time consuming digestion procedure. Matrix matched synthetic laboratory standards doped with ^{230}Th (IRMM 60), uranium with natural isotope composition (^{235}U/^{238}U = 0.007 25) and uranium isotope standard reference material (NIST U-930) at the low pg ml^{-1} level were prepared in order to study the figures of merit of the analytical methods developed. The recovery rate for thorium and uranium concentration measured on a synthetic urine laboratory standard by LA-ICP-MS varied between 91 and 104 %. The precision and accuracy of the analytical methods was found to be 7 % and <1 %, respectively, for uranium concentration, by using urine laboratory standards (at uranium concentration 0.1 ng ml^{-1}.

To an increasing extent, LA-ICP-MS is the method of choice for the direct analysis of solid samples with respect to the analysis of long-lived radionuclides. Most applications of LA-ICP-MS are described with respect to the analysis of the naturally occurring radioactive elements U and Th in geological and environmental samples[19,83,84] Furthermore, LA-ICP-MS was employed for the characterization of radionuclides in solid radioactive waste materials.[14,70] For the determination of long-lived radionuclides in non-conducting materials, a synthetic laboratory standard with a concrete matrix was doped with low levels of long-lived radionuclides (e.g. ^{99}Tc, ^{129}I, ^{232}Th, ^{233}U, ^{237}Np and ^{238}U). The detection limits determined for Tc, U and Np in a blank concrete sample were in the low $pg\,g^{-1}$ concentration range. The detection limits are lower by more than one order of magnitude if double-focusing sector field ICP-MS (Element) is used in comparison to quadrupole LA-ICP-MS (Elan 6000).[14,70] The capability of laser ablation ICP-MS for determining long-lived radionuclides for trace, ultratrace and isotope analysis in solid materials is discussed in reference 85.

Actinides were determined at the ultratrace level in moss samples collected from the eastern Italian Alps (1500 m a.s.l.). The frozen samples were cut into 1–2 cm sections and analyzed separately to obtain the distribution curves of the vertical concentrations. For plutonium and americium isotope analysis, 1–2 g of the samples were ashed, leached, separated with respect to analytes and analyzed by alpha spectrometry and LA-ICP-MS after the plutonium or americium had been electroplated on a stainless steel disk.[23] Estimated limits of quantification of LA-ICP-MS for actinide radionuclides deposited on stainless steel plates after chemical separation are summarized in Table 9.45. For most of the long-lived radionuclides in moss samples, lower limits of determination were found at the $10^{-15}\,g\,g^{-1}$ concentration level compared to those of α-spectrometry.[23]

Table 9.45 *Detection limits of LA-ICP-MS for radionuclides in mosses deposited on stainless steel plates after chemical separation.*

Radionuclide	half life (a)	LA-ICP-SFMS ($g\,g^{-1}$)	α-spectrometry ($g\,g^{-1}$)
^{230}Th	7.5×10^{4}	3.8×10^{-15}	2.6×10^{-14}
^{232}Th	1.4×10^{10}	6.8×10^{-15}	1.2×10^{-8}
^{234}U	2.4×10^{5}	3.8×10^{-15}	8.7×10^{-14}
^{235}U	7.0×10^{8}	4.9×10^{-15}	2.5×10^{-10}
^{236}U	2.3×10^{7}	4.0×10^{-14}	1.2×10^{-11}
^{238}U	4.5×10^{9}	7.1×10^{-15}	4.0×10^{-9}
^{239}Pu	2.4×10^{4}	3.9×10^{-15}	2.2×10^{-14}
^{240}Pu	6.6×10^{3}	3.7×10^{-15}	5.9×10^{-15}
^{241}Am	432	3.7×10^{-15}	6.3×10^{-16}

Online isotope dilution in LA-ICP-MS using the microflow total consumption nebulizer DS-5 inserted into the laser ablation chamber was introduced to determine the long-lived radionuclides U and Th in biological tissues in Becker's group.[73,86] The low solution uptake rate of $7\,\mu l\,min^{-1}$ for DS-5 reduces the amount of isotope enriched tracer solution needed for online isotope dilution in LA-ICP-MS. A uranium concentration of $11.19 \pm 1.11\,\mu g\,g^{-1}$ was determined in NIST apple leaves SRM 1515 doped with a uranium concentration of $10\,\mu g\,g^{-1}$ by online isotope dilution with a nebulizing isotope enriched uranium tracer solution (NIST U350) during the ablation of biological samples. Online isotope dilution is used for single shot analysis (microlocal analysis) whereby the isotope enriched tracer solution is nebulized and the sample is ablated with a single laser shot. This procedure also allows quantitative analysis in the case of element imaging, for example, in materials science or in biological and biomedical applications.[86]

Uranium (and thorium) was also detected by LA-ICP-MS in human brain samples (thin sections of hippocampus[73]) and protein spots of human brain separated by 2D gel electrophoresis.[87] The detection limits for uranium and thorium determination in thin sections of brain tissues were determined as $10\,ng\,g^{-1}$. Figure 9.52 shows transient signals of $^{238}U^+$ in protein spots 9a and 9d from a human brain sample (somatomotor cortex) measured by LA-ICP-MS. Protein spots were separated by 2D gel electrophoresis are illustrated.[87]

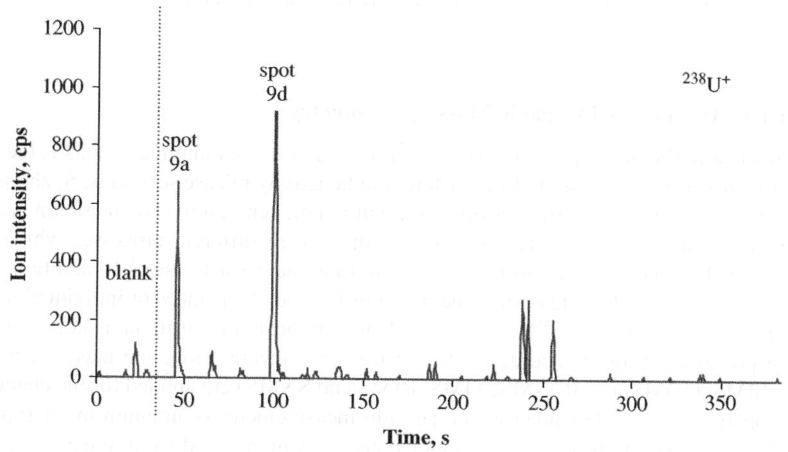

Figure 9.52 *Transient signals of* $^{238}U^+$ *in protein spots 9a and 9d from a human brain sample measured by LA-ICP-MS, protein spots were separated by 2D gel electrophoresis.*

The advantages and disadvantages of several sample introduction systems for the determination of long-lived radionuclides in liquid samples using quadrupole based ICP-MS (ELAN DRC II ICP-MS, Perkin Elmer Sciex) coupled to a laser ablation system (UP 266MACRO, New Wave Research, California) for the analysis of dried droplets, using electrothermal vaporization (ETV HGA-600MS from PerkinElmer) and solution nebulization (SN) have been compared by Sturgeon's working group.[41] The authors determined several radionuclides in digested biological tissue, nearshore sea water and river water. Samples were subjected to $Ca_3(PO_4)_2$ coprecipitation preconcentration prior to analysis. ETV- and LA-ICP-MS achieved similar sensitivities, and the detection limits of ETV-ICP-MS ranged from 0.017 to $0.029\,pg\,ml^{-1}$. The best precision was observed for SN (2 % RSD versus 7 and 8 % for ETV- and LA-ICP-MS, respectively). ETV accommodates samples with a higher dissolved solid content. The accuracy of the analytical methods was validated by determining, for example, U and Th in NIST SRM 1566b Oyster Tissue.[41] For the determination of ^{79}Se in fission product solutions, Compte *et al.* employed ETV-ICP-MS after a single chemical separation step to extract selenium from a highly radioactive solution.[72] ETV-ICP-MS after analyte separation eliminated all interferences normally associated with the determination of ^{79}Se. The concentration of ^{79}Se measured in fission product solution was $0.43\,mg\,l^{-1}$.[72] Sturgeon's group have reported on the application of ETV as a sample introduction procedure with ICP-MS for the determination of U, Th and Pu in natural water, biological materials and urine.[88] These elements suffer strong interactions with the graphite substrate at high temperatures, leading to carbide formation, severe to memory effects and poor sensitivity.[41] This problem was avoided by the use of sample vaporization from a tantalum surface and the addition of Freon gas as a modifier, and consequently under dry plasma conditions the formation of disturbing polyatomic interferences (especially hydride, oxide or hydroxide formation) was reduced.[88] A further drawback of ETV-ICP-MS is the requirement for a modifier (e.g., sodium chlorate and nitric acid) as described for ^{99}Tc determination by Song and Probst,[89] therefore for most applications alternative analytical techniques were selected.

Further applications of laser ablation inductively coupled plasma mass spectrometry for the trace and ultratrace and isotope analysis of long-lived radionuclides are discussed in references[1–3,90]. An excellent review of speciation for actinide elements is presented by Geipel.[91]

9.9.5 Particle Analysis by Inorganic Mass Spectrometry

Of special interest in the mass spectrometric determination of transuranium elements is the characterization of microparticles stemming from different radioactivity release scenarios. Such microparticles bearing radionuclides, in particular uranium, plutonium, neptunium and americium, can enter the environment and therefore the human food chain through different processes which can be related to the nuclear fuel cycle as well as to clandestine nuclear activities.[5] In addition, nuclear safeguards programmes seek to determine the uranium isotope abundances of individual μm sized particles. Anomalous amounts of ^{233}U, ^{235}U or ^{236}U may indicate that artificial isotope enrichment or depletion processes were the source of these particles.[5] Several inorganic mass spectrometric techniques (SIMS, GDMS, LA-ICP-MS, TIMS, RIMS and SNMS) are applied for the characterization of microparticles.[5,92–95] For precise isotope ratio measurements of uranium in microparticles, multiple collector mass spectrometry is advantageous, as demonstrated by Bouman *et al.*[93] in an experiment where five multiple ion counters in MC-TIMS (Triton, Thermo Fisher Scientific) were used to simultaneously measure ^{233}U, ^{234}U, ^{235}U, ^{236}U and ^{238}U.

Danesi *et al.*[96] applied SIMS, in addition to X-ray fluorescence imaging, by using a microbeam (μ-XRF) and scanning electron microscope equipped with an energy dispersive X-ray fluorescence analyzer (SEM-EDXRF) to characterize soil samples and to identify small DU particles collected in Kosovo locations where depleted uranium (DU) ammunition was employed during the 1999 Balkan conflict. Knowledge of DU particles is needed as a basis for the assessment of the potential environmental and health impacts of military use of DU, since it provides information on possible resuspension and inhalation. The measurements indicated 'spots' where hundreds of thousands of particles may be present in a few mg of contaminated soil. The particle size distribution showed that most of the DU particles were $< 5\,\mu m$ in diameter and more than 50 % of the particles had a diameter of $< 1.5\,\mu m$.[96]

Six plutonium containing particles stemming from soil (Marshall Island) were characterized by SIMS, SEM-EDX-WDX and sychrotron radiation by Jernström *et al.*[97] All the particles were identified as nuclear fuel fragments of exploded weapons components. Since they contained plutonium with a low ^{240}Pu/^{239}Pu atomic ratio (less than 0.065), which corresponds to weapons grade plutonium or a detonation with low fission yield, the particles were identified as originating from the safety test and low yield tests conducted on Runit Island.[97]

Resonant and non-resonant laser post-ionization of sputtered uranium atoms using SIRIS (sputtered initited resonance ionization spectroscopy) and SNMS (secondary neutral mass spectrometry) in one instrument for the characterization of sub-μm sized single microparticles was suggested by Erdmann *et al.*[94] Resonant ionization mass spectrometry allows a selective and sensitive isotope analysis without isobaric interferences as demonstrated for the ultratrace analysis of plutonium from bulk samples.[94] Unfortunately, no instrumental equipment combining both techniques is commercially available.

Comparative studies of neodymium measurements in UO_2 fuel for the future application of local burn up calculations have been carried out by the SIMS analysis of a ^{143}Nd implanted UO_2 single crystal.[98] Samples for a round robin test were produced by the implantation of ^{143}Nd (5×10^{10} at cm^{-2}, 400 keV) in a UO_2 single crystal. Different depth profiles of the $^{143}Nd^+/^{238}U^+$ ratio and especially the NdO^+/Nd^+, UO^+/U^+ and $UO_2{}^+/U^+$ ratios obtained by SIMS are not constant over time and indicate evidence of some fluctuations during the analysis.[98] SIMS plays a dominant role for particle analysis, but also LA-ICP-MS with lower spatial resolution can be applied to identify anomalies in the isotopic composition of radionuclides.

References IX

1. Becker, J. S., *Int. J. Mass Spectrom.*, **242**, 183 (2005).
2. Becker, J. S., *Spectrochim. Acta*, **58B**, 1757 (2003).
3. Becker, J. S., *Spectroscopy Europe*, **14**, 8 (2002).
4. Becker, J. S. and Dietze H. J., *Encyclopedia Analytical Chemistry*, A. Meyers (ed.). John Wiley & Sons, Ltd, Chichester, 12947 (2000).
5. Betti, M., Aldava de las Heras, L. and Tamborini, G., *Appl. Spectroscopy Rev.*, **41**, 491 (2006).
6. Becker, J. S. and Dietze, H. J., *Advances in Mass Spectrometry*, **14**, 687 (1998).
7. Henry, R., Koller, D., Liezers, M. *et al., J. Radioanal. Nucl. Chem.*, **249**, 103 (2001).
8. McLean, J. A., Becker, J. S., Boulyga, S. F., Dietze, H. J. and Montaser, A., *Int. J. Mass Spectrom.*, **208**, 193 (2001).
9. Agarande, M., Benzoubir, S., Neiva-Marques, A. M. and Bouisset, P., *J. Environ. Radioactivity*, **72**, 169 (2004).
10. Betti, M., *Microchem. J.*, **67**, 363 (2000).
11. Desideri, D., Meli, M. A., Roselli, C., Testa, C., Boulyga, S. F. and Becker, J. S., *Anal. Bioanal. Chem.*, **374**, 1091 (2002).
12. Ketterer, M. E. Hafer, K. M., Link, C. L., Royden, C. S. and Hartsock, W. J., *J. Environ. Radioact.*, **67**, 191 (2003).
13. Betti, M., *Int. J. Mass Spectrom.*, **242**, 169 (2005).
14. Gastel, M., Becker, J. S., Küppers, G. and Dietze, H. J., *Spectrochim. Acta*, **52B**, 2051 (1997).
15. Larivière, D., Taylor, V. F., Evans, R. D. and Cornett, R. J., *Spectrochim. Acta*, **61B** 877 (2006).
16. Becker, J. S., *J. Anal. At. Spectrom.*, **17**, 1172 (2002).
17. Becker, J. S., *Spectrochim. Acta*, **57B**, 1805 (2002).
18. Grinberg, P., Scott, W. and Sturgeon, R. E., *Anal. Chem.*, **77**, 2432 (2005).
19. Pickhardt, C., Brenner, I. B., Becker, J. S. and Dietze, H. J., *Fresenius' J. Anal. Chem.*, **368**, 79 (2000).
20. Günther-Leopold, I., Kobler Waldis, J., Wernli, B. and Kopajtic, Z., *Int. J. Mass Spectrom.*, **242**, 197 (2005).
21. Izmer, A., Zoriy, M., Boulyga, S. F. and Becker, J. S., *J. Anal. At. Spectrom.*, **20**, 1278 (2005).
22. Izmer, A. V., Boulyga, S. F. and Becker, J. S., *J. Anal. At. Spectrom.*, **18**, 1339 (2003).
23. Boulyga, S. F., Desideri, D., Meli, M. A., Testa, C. and Becker, J. S., *Int. J. Mass Spectrom.*, **226**, 329 (2003).
24. Eroglu, A. E., McLeod, C. W., Leonard, K. S. and McCubbin, D., *J. Anal. At. Spectrom.*, **13**, 875 (1998).
25. Taylor, R. N., Warnecke, T., Milton, J. A., Croudace, I. W., Warwick P. E. and Nesbitt, R., *J. Anal. At. Spectrom.*, **16**, 279 (2001).
26. Zoriy, M., Ostapczuk, P., Halicz, L., Hille, R. and Becker, J. S., *Int. J. Mass Spectrom.*, **242**, 203 (2005).
27. Zoriy, M., Pickhardt, C., Ostapczuk, P., Hille, R. and Becker, J. S., *Int. J. Mass Spectrom.*, **232**, 217 (2004).
28. Becker, J. S. and Dietze, H. J., *J. Anal. At. Spectrom.*, **14**, 1493 (1999).
29. Zoriy, M. V., Halicz, L., Ketterer, M. E., Pickhardt, C., Ostapczuk, P. and Becker, J. S., *J. Anal. At. Spectrom.*, **19**, 362 (2004).
30. Boulyga, S. F. and Heumann, K. G., *J. Environ. Radioact.*, **88**, 1 (2006).

31. Ketterer, M. E., Jordan, J. A., Szechenyi, S. H., Hudson, D. D. and Layman, R. R., *J. Anal. At. Spectrom.*, **15**, 1569 (2000).
32. Truscott, J. B., Bromley, L., Jones, P., Evans, E. H., Turner, J. and Fairman, B., *J. Anal. At. Spectrom.*, **14**, 627 (1999).
33. Henry, R., Koller, D., Liezers, M. *et al., J. Radioanal. Nucl. Chem.*, **249**, 103 (2001).
34. De Bievre, P. and Verbueken, A., *Metrologia*, **36**, 25 (1999).
35. Aggarwal, S. K., *Radiochim. Acta*, **94**, 397 (2006).
36. Barrero Moreno, J. M., Alonso, J. I. G., Arbore, P., Nicolaou, G. and Koch., L., *J. Anal. At. Spectrom.*, **11**, 929 (1996).
37. Truscott, J. B., Jones, P., Fairman, B. E. and Evans, E. H., *J. Chrom. A*, **928**, 91 (2001).
38. Vonderheide, A. P., Zoriy, M. V., Izmer, A. V. Pickhardt, C., Caruso, J. A., Ostapczuk, P., Hille, R. and Becker, J. S., J. Anal. Atom. Spectrom., **19**, 675 (2004).
39. Zoriy, M., Varga, Z., Pickhardt, C. *et al., J. Environ. Monit.*, **7**, 514 (2005).
40. Desideri, D., Meli, M. A., Roselli, C., Testa, C., Boulyga, S. F. and Becker, J. S., *Anal. Bioanalyt. Chem.*, **374**, 1091 (2002).
41. Grinberg, P., Yang, L., Mester, Z., Willie, S. and Sturgeon, R. E., *J. Anal. At. Spectrom.*, **21**, 1201 (2006).
42. Hrnecek, E., Heras, L. A. D. and Betti, M., *Radiochim. Acta*, **90**, 721 (2002).
43. Kwong, L. L. W., Gastaud, J., La Rosa, J., Lee, S. D., Povinec, P. P. and Wyse, E., *J. Radioanal. Nucl. Chem.*, **261**, 283 (2004).
44. Halicz, L., Becker, J. S., Pickhardt, C. Gavrieli, I., Burg, A., Nishri, A., and Platzner, I. *Int. J. Mass Spectrom.*, **249–250**, 296 (2006).
45. Boehlke, J. K., De Leater, J. R., De Bievre, P. *et al., J. Phys., Chem. Ref. Data*, **34**, 57 (2005).
46. Karpas, Z., Lorber, A., Sela, H. *et al., Radiation Protection Dosimetry*, **118**, 106 (2006).
47. Yamamoto, M., Sayarbaini, K., Kofuji, K. *et al., J. Radioanal. Nucl. Chem.*, **197**, 185 (1995).
48. Betti, M., *Microchem. J.*, **67**, 363 (2000).
49. Barrero Moreno, J. M., Betti, M. and Alonso, J. I. G., *J. Anal. At. Spectrom.*, **12**, 355 (1997).
50. Joannon, S. and Pin, C., *J. Anal. At. Spectrom.*, **16**, 32 (2001).
51. Larivière, D., Epov, V. N., Evans, R. D. and Cornett, R. J., *J. Anal. At. Spectrom.*, **18**, 338 (2003).
52. Bohlke, J. K., De Laeter, J. R., DeBievre, P., Hidaka, H., Pieser, H. S., Rosman, K. J. R. and Taylor, P. D. P., *J. Phys. Chem. Ref. Data.*, **34**, 57 (2005).
53. Boulyga S. F., Zoriy, M., Ketterer M. E. and Becker J. S., *J. Environ. Monit.*, **5**, 661 (2003).
54. Vonderheide, A. P., Zoriy, M., Izmer, A. *et al., J. Anal. At. Spectrom.*, **19**, (2004).
55. Epov, V. N., Benkhedda, K. and Evans, R. D., *J. Anal. At. Spectrom.*, **20**, 990 (2005).
56. Kuczewski, B., Marquart, C. M., Seibert, A., Geckheis, H., Kratz, J. V. and Trautmann, N., *Anal. Chem.*, **75**, 6769 (2003).
57. Richter, S., Alonso, A., De Bolle, W. *et al., J. Anal. At. Spectrom.*, **20**, 1381 (2005).
58. Pappas, R. S., Ting, B. G. and Paschal, D. C., *J. Anal. At. Spectrom.*, **18**, 1289 (2003).
59. Ough, E. A., Lewis, B. J., Andrews, W. S., Bennett, L. G. I., Hancock, R. G. V. and D'Agistino, P. A., *Health Phys.*, **90**, 494 (2006).
60. Elliot, N. L., Bickel, G. A., Linauskas, S. H. and Paterson, L. M., *J. Radioanal. Nucl. Chem.*, **267**, 637 (2006).
61. McCurdy, D., Lin, Z., Inn, K. G. W. *et al., J. Radioanal. Nucl. Chem.*, **263**, 447 (2005).
62. Betti, M., *J. Environ. Radioact.*, **64**, 113 (2003).
63. Becker, J. S., Kerl, W. and Dietze, H. J., *Anal. Chim. Acta*, **387**, 145 (1999).
64. Day, J. A., Caruso, J. A., Becker, J. S. and Dietze, H. J., *J. Anal. At. Spectrom.*, **15**, 1343 (2000).
65. Marumatsu, Y., Uchida, S., Tagami, K., Yoshida, S. and Fujikawa, T., *J. Anal. At. Spectrom.*, **14**, 859 (1999).
66. Kerl, W., Becker, J. S., Dannecker, W. and Dietze, H. J., *Fresenius' J. Anal. Chem.*, **362**, 433 (1998).
67. Kerl, W., *Ultrapuren- und Isotopenanalyse langlebiger Radionuklide mittels doppel-fokussierender Sektorfeld-ICP-Massenspectrometrie* PhD Thesis, Research Centre Juelich, Juelich (1998).
68. Perna, L., Bocci, P., Aldava de las Heras, L., De Pablo, J. and Betti, M., *J. Anal. At. Spectrom.*, **17**, 1166 (2002).

69. Isnard, H., Aubert, M., Blanchet, P., Brennetot, R., Chartier, F., Geersten, V. and Manuguerra, F., *Spectrochim. Acta B*, **61**, 150 (2006).
70. Becker, J. S., Gastel, M., Tenzler, D. and Dietze, H. J., *Adv. Mass Spectrom.*, **14**, 141 (1998).
71. Boulyga, S. F., Becker, J. S., Matusevitch, J. L. and Dietze, H. J., *Int. J. Mass Spectrom.*, **203**, 143 (2000).
72. Compte, J., Bienvenu, P., Brochard, E., Fernandez, J. M. and Andreoletti, G., *J. Anal. At. Spectrom.*, **18**, 702 (2003).
73. Becker, J. S., Zoriy, M. V., Pickhardt, C., Palomero-Gallagher, N. and Zilles, K., *Anal. Chem.*, **77**, 5851 (2005).
74. Evans, P., Elahi, S., Lee, K. and Fairman, B., *J. Environ. Monit.*, **5**, 175 (2003).
75. Becker, J. S., Dietze, H. J., McLean, J. A. and Montaser, A., *Anal. Chem.*, **71**, 3077 (1999).
76. Becker, J. S., Soman, R. S., Sutton, K. L., Caruso, J. A. and Dietze, H. J., *J. Anal. At. Spectrom.*, **14**, 933 (1999).
77. Alvarado, J. S. and Erickson, M. D., *J. Anal. At. Spectrom.*, **11**, 923 (1996).
78. Boulyga, S. F., Testa, C., Desideri, D. and Becker, J. S., *J. Anal. At. Spectrom.*, **16**, 1283 (2001).
79. Schaumlöffel, D., Giusti, P., Zoriy, M., Pickhardt, C., Szpunar, J., Lobinski, R. and Becker, J. S., *J. Anal. At. Spectrom.*, **20**, 17 (2005).
80. Dadfarnia, S. and McLeod, C. W., *Appl. Spectrosc.*, **48**, 1331 (1994).
81. Hollenbach, M., Grohs, J., Mamich, S., Kroft, M. and Denoyer, E. R., *J. Anal. At. Spectrom.*, **9**, 927 (1994).
82. Becker, J. S., Burow, M., Zoriy, M., Pickhardt, C., Ostapczuk, P., Hille, R., *At. Spectrom.*, **25**, 197 (2004).
83. Becker, J. S. and Dietze, H. J., *Fresenius' J. Anal. Chem.*, **365**, 429 (1999).
84. Jochum, K. P., Dingwell, D. B., Rocholl, A. *et al.*, *Geostand. Newlett.*, **24**, 87 (2000).
85. Becker, J. S., Pickhardt, C. and Dietze, H. J., *Int. J. Mass Spectrom.*, **203**, 283 (2000).
86. Pickhardt, C., Izmer, A., Zoriy, M. V., Schaumlöffel, D. and Becker, J. S., *Int. J. Mass Spectr.*, **248**, 136 (2006).
87. Becker, J. S., Zoriy, M., Becker, J. Su. Pickhardt, C., Damoc, E., Juhacz, G., Palkovits, M. and Przybylski, M., *Anal. Chem.*, **77**, 5851 (2005).
88. Grinberg, P., Willie, S. N. and Sturgeon, R. E., *J. Anal. At. Spectrom.*, **20**, 717 (2005).
89. Song, M. and Probst, T. U., *Anal. Chim. Acta*, **413**, 207 (2000).
90. Pickhardt, C., Dietze, H. J. and Becker, J. S., *Int. J. Mass Spectr.*, **242**, 273 (2005).
91. Geipel, G., in *Handbook of Elemental Species II – Species in the Environment, Food, Medicine and Occupational Health*, Vol. II, H. C. R. Cornelis, J. Caruso and K. G. Heumann (eds.). John Wiley & Sons, Ltd, Chichester (2005) 509.
92. Betti, M., Tamborini, G. and Koch, L., *Anal. Chem.*, **71**, 2616 (1999).
93. Bouman, C., Cocheri, A., Robert, M., Schwieters, J. B. and Wieser, M. E., *Geochem. Cosmochim. Acta (Suppl. 1)*, **67**, A 44 (2003).
94. Erdmann, N., Betti, M., Kollmer, F. *et al.*, *Anal. Chem.*, **75**, 3175 (2003).
95. Wieser, M. E. and Schwieters, J. B., *Int. J. Mass Spectrom.*, **242**, 97 (2005).
96. Danesi, P. R., Markowicz, A., Chinea-Cano, E. *et al.*, *J. Environ. Radioact.*, **64**, 143 (2003).
97. Jernström, J., Eriksson, M., Simon, R. *et al.*, *Spectrochim. Acta*, **61B**, 971 (2006).
98. Portier, S., Brémier, S., Hasnaoui, R. *et al.*, in *5th European Workshop on Secondary Ion Mass Spectrometry*. Münster (2006) 95.

9.10 Forensic Analysis

The goal of investigating forensic samples is to determine an association between a sample in a crime scene and a possible source of known origin. Information resulting from the forensic analysis of a wide range of samples such as projectiles, bullets, glass shards, metal fragments, textile and synthetic fibres, plastics, paints, adhesives, soils and ceramics, and also weapons (including nuclear weapons), explosives, ignitable liquids or illicit drugs can help to solve crimes. Analytical techniques such as X-ray fluorescence analysis (XRF), scanning electron microscopy with energy-disperse X-ray analysis (SEM-EDX),[1,2] atomic absorption spectrometry (AAS),[3,4] neutron activation analysis (NAA),[5,6] ICP optical emission spectrometry (ICP-OES)[7] or laser induced breakdown spectroscopy (LIBS)[8] have been used for forensic purposes for many years. Most of these analytical techniques have drawbacks for the analysis of small sample sizes and in distinguishing between materials with physical, chemical and visual similarities. Furthermore, isotope analysis could be relevant in forensic science.

Nowadays, inorganic mass spectrometric techniques (such as ICP-MS,[9] LA-ICP-MS,[10] SIMS, SSMS,[11] TIMS, GDMS, AMS and RIMS), mostly with multi-element capability and very high sensitivity, offer superb capability for crime applications. Especially new ultrasensitive mass spectrometers developed for multi-element trace and ultratrace analysis and also isotope ratio measurements have found increasing application in forensic science. Therefore, trace, ultra-trace and especially isotope analysis, are of increasing importance for studying the element and isotopic distribution in forensic samples and can be helpful to characterize small sized fragments (e.g., a splinter of glass, a single hair, one synthetic fibre, one drop of blood, a splash or a trace from a gun) found on the suspect in comparison to the source material. Stable isotope mass spectrometry is an additional technique that can be utilized to test a given hypothesis.[12] However, mass spectrometric techniques of organic compounds or biomolecules are also indispensable in routine forensic work. Imaging mass spectrometry, in particular, is a relatively new analytical technology that takes advantage of the methodology and instrumentation of MALDI or ESI mass spectrometry, which can be used to locate specific molecules such as drugs, lipids, peptides and proteins directly from the surface of fresh frozen tissue sections.[13]

9.10.1 Fingerprinting in Forensic Studies

Quick fingerprinting of microscopic specimens (such as glass shards or fibres of clothing) by LA-ICP-MS or SIMS helps forensic investigators analyze crime scenes to determine the origin of the sample. The main advantages of LA-ICP-MS in forensic analysis concern the accurate and fast fingerprinting of unknown samples compared to the original material by identifying trace elements and the isotopic composition, which is specific for each individual sample. Furthermore, it has a very low sample consumption (< 1 mg enables very small fragments of forensic samples to be analyzed) without sample destruction and consequently preservation of the original sample for further crime investigations. Different applications of LA-ICP-MS in forensic science and also archaeology are described by Watling.[14] The most useful application of fingerprinting by LA-ICP-MS is, apart from sourcing the provenance of glass samples, the analysis of steel and lead from projectiles found at the crime scene.[14] LA-ICP-MS has been employed for quick fingerprinting of trace element composition in gold nuggets and to determine the source of a particular artefact[15] or to characterize diamonds and indicator minerals for diamond exploration.[16] The fundamental principle of gold fingerprinting is based on the concept that there is an individual mineralizing event which results in different trace metal compositions (the so-called 'trace element association pattern') depending on the environment of the ore during formation (mineralization) and/or in the manufacturing processes of the gold nugget or gold bar to produce gold jewellery. Each manufactured piece of gold possesses a specific trace element pattern that depends on the source and manufacturing processes. Fingerprinting can thus provide information on the origin and history of the gold sample under investigation.

ToF-SIMS has been utilized to obtain characteristic mass spectra and images from different smokeless gunpowders and black powder samples.[17] In the mass spectra obtained, peaks indicative of both the organic and inorganic additive constituents in the gunpowder samples were observed.

Consequently, ToF-SIMS enables a differentiation to be made between different gunpowder samples with the aid of principal component analysis (PCA).[17]

In spite of the high cost of secondary ion mass spectrometers and the technical expertise required, this sensitive microlocal and surface analytical technique can be applied successfully in forensic studies, e.g., for measurement of trace elements and inks on counterfeit banknotes.[18]

9.10.2 Multi-element Analysis for Forensic Studies

Of the inorganic mass spectrometric techniques, ICP-MS and LA-ICP-MS are the methods most frequently applied in forensic science due to their low detection limits, multi-element capability and the possibility to measure isotope ratios. Forensic toxicologists typically work with body fluids such as urine, blood or visceral tissues.[19,20] The most commonly investigated sample materials are glass samples and glass fragments.[21–24] Glass is a fragile material that is often found at the scenes of crimes such as hit-and-run accidents, burglaries and violent crimes. If a small fragment of glass is the only evidence that links a suspect to a crime scene, sensitive analytical techniques are required. The conventional technique for distinguishing between glass samples of different origins was for a long time the measurement of the refractive index (RI). This is possible with sufficient precision and accuracy.[25] The capability of LA-ICP-MS and SIMS to detect trace elements and analyze them as a function of depth has proved to be of assistance in the discrimination of glass fragments which have an identical RI. For example, in 1991, glass fragments from 81 automobile side windows were collected and analyzed by the FBI Laboratory using ICP-OES.[26] By additional measurements using ICP-MS after digestion in an acid mixture (HF: HNO_3: HCl, 2: 1: 1) of a small amount of the glass samples (5–10 mg) a large number of trace elements (45) especially of selected metals (including different isotopes) were determined, which significantly improves the discrimination of glass samples in comparison to refraction index measurements alone.[26] NIST glass SRM 612 was employed in the development of the analytical procedure. The main drawback of ICP-MS for the analysis of difficult to dissolve forensic samples is the time consuming sample preparation, especially for hazardous materials and the increasing risk of contamination during sample preparation. For the direct analysis of glass fragments LA-ICP-MS is therefore the method of choice. With its possibility of direct analysis and its multi-element capability, LA-ICP-MS allows considerably better fingerprinting and fast characterization of very small pieces of glass samples for forensic purposes and by comparing them the origin of sample can be discovered at a crime scene and the possible source of the glass.

An analytical method and its validation for the forensic analysis of float glass samples using LA-ICP-MS is described in reference[27] whereby ten elements (K, Ti, Mn, Rb, Sr, Zr, Ba, La, Ce and Pb) enable glass fragments with sizes down to 1 mm^2 to be characterized directly. However, inhomogeneity studies of different glass panes from various sources are also helpful for forensic investigations of float glass samples.[27] Small fragments are embedded in epoxy stubs before mass spectrometric measurement. A multitude of certified glass standards (e.g., from NIST) are available for calibration procedures. The concentrations are determined with limits of detection in the low and sub-$\mu g\ g^{-1}$ ranges (e.g., Pb – $1.5\,\mu g\,g^{-1}$, Ti – $17\,\mu g\,g^{-1}$ and Sr – $0.2\,\mu g\,g^{-1}$), and accuracy and precision are $< 10\,\%$ for most elements, measured by LA-ICP-MS. When defining appropriate matching criteria for the comparison of two glass samples for forensic purposes, a balance must always be struck between defining criteria that are too small and thus increasing the chance of false exclusion. On the other hand, if the criteria are too large the chance of a false inclusion increases. Based on these results, matching criteria for the overlapping of intervals defined by the concentration values $\pm$ twice the standard deviation (2σ) have been set for the forensic comparison of float glass samples.[27]

Forensic analysis of glass samples was performed in different laboratories within the NITE-CRIME (Natural Isotopes and Trace Elements in Criminalistics and Environmental Forensics) European Network using a variety of LA-ICP-MS systems.[10] This network aims at providing a forum for the development of analytical protocols involving trace element and isotope analysis in forensic science whereby several glass standard reference materials (e.g., NIST SRM 610 and NIST SRM 612) and matrix matched glass reference standards (FGS 1 and FGS 2, produced by the German glass manufacturer SCHOTT Glass in 2002) are characterized with respect to matrix and trace elements by interlaboratory tests. Small pieces of glass samples are cut using a diamond wire saw. Pieces of float samples are embedded using a Heraeus Kulzer Technovit 2000® resin or fixed with a double-sided tape on a target before LA-ICP-MS analysis.[10] As a result of various interlaboratory tests an improved precision and accuracy of concentration values has been observed. This indicates that LA-ICP-MS is to become a routine technique in forensic studies. The capabilities of LA-ICP-MS are compared with those of LIBS, whereby LA-ICP-MS provides the best analytical data and the best discrimination between different glass samples.[8] The same set of 21 float glass samples was analyzed by LA-ICP-MS resulting in only two indistinguishable pairs of samples out of 210 possible pairs, which represents a 99 % discrimination. One of those indistinguishable pairs was formed by fragments of glass from each of the two layers of the same windscreen.[8]

9.10.3 Trace Element Analysis of Selected Elements and Speciation

One of the most famous applications in forensic science is the analysis of Napoleon's hair by ICP-MS after mineralization in concentrated nitric acid whereby an arsenic concentration about 40 times higher than normal (about $40\,\mu g\,g^{-1}$) was measured (see Section 9.5). Ingested arsenic is known to be stored in sulfydryl rich tissue, like hair, nails or skin. ETV-ICP-MS combined with isotope dilution has been employed to measure thallium in human scalp hair from a person poisoned by thallium compared to control subjects, whereby several longitudinal concentration gradients for the analyzed segments (length: 10 mm) were obtained.[28]

In Stingeder's group the unique capabilities of sector field ICP-MS and LA-ICP-MS were employed to reconstruct details of a homicide by thallium poisoning, which took place about 40 years ago in Austria.[29] Thallium was determined in several human bone samples after acid digestion in a microwave oven. The thallium concentration measured by ICP-MS and GF-AAS varied from $1.07–2.63\,\mu g\,g^{-1}$, which was up to 170 times higher compared to the concentration found in persons who died due of natural causes. LA-ICP-MS was employed to analyze a thumbnail from the poisoned person compared to a control person. Thallium peaks were detected in the nail of the victim at a distance of 2.5 mm from the younger edge of nail.[29]

The separation and analysis of three organophosphorus chemical warfare degradation products is described by Caruso's group.[30] Ethyl methylphosphonic acid is the major hydrolysis product of VX, produced as a human-made chemical warfare agent classified as a nerve agent. Nerve agents are the most toxic and rapidly acting of all known chemical warfare agents. As an agent of chemical warfare, VX ($C_{11}H_{26}NO_2PS$) is an organophosphate similar to pesticides but much more potent. In addition, isopropyl methylphosphonic acid, as the major hydrolysis product of sarin, and methylphosphonic acid, as the final hydrolysis product of both VX and sarin, have been separated as analytes by reversed phase ion pairing HPLC with the use of myristyl trimethylammonium bromide as the ion pairing reagent and an ammonium acetate–acetic acid buffer system at pH 4.85. An ICP-QMS Agilent 7500ce with an octopole collision cell (to reduced polyatomic interferences) was employed for the $^{31}P^+$ and $^{31}P^{16}O^+$ ion detection of organophosphorus degradation products of chemical warfare agents. Detection limits for organophosphorus species were found between 139 and 263 pg ml^{-1}. The analytical technique developed was successfully applied to the analysis of spiked environmental water and soil samples.[30] The rapid screening and analysis of degradation compounds of chemical warfare agents using liquid chromatography and mass spectrometry with electrospray ionization or ion mobility mass spectrometry (this technique is discussed elsewhere) has been reported in several publications.[31–33]

9.10.4 Nuclear Forensic Studies

Millions of radioactive sources exist around the world, usually distributed not only at nuclear power plants, but also medical radiotherapy facilities and industrial irradiators. Unfortunately, the radioactive materials housed in these places are often not under adequate control and are therefore susceptible to theft by terrorists.[34] The appalling events of September 11, 2001, spawned a major international initiative to strengthen security for such materials and facilities worldwide. Highly toxic radionuclides (plutonium radionuclides, ^{210}Po or ^{137}Cs) at trace level are increasingly being used as 'modern weapons' to kill undesirable persons.

For several decades the IAEA (International Atomic Energy Agency) has identified radioactive sources used in nuclear power plants, and also for the illegal production of nuclear weapons, in industrial radiography, radiotherapy, in industrial irradiators and thermo-electric generators, as the sources that are most susceptible to acts of terrorism.[34] These sources contain significant amounts of radioactive material – such as ^{60}Co, ^{90}Sr, ^{137}Cs and ^{192}Ir.[35] The resulting new field of nuclear forensics employs many analytical spectrometric techniques – including inorganic mass spectrometry (SIMS, TIMS, AMS, GDMS, ICP-MS, LA-ICP-MS and RIMS) – to ensure that nuclear sources are safe, and also to track them down in the event of attempted murder or sabotage. Nuclear forensic studies concern the determination of the concentration and isotope ratios of long-lived radionuclides, such as uranium, plutonium isotopes, thorium, ^{237}Np, ^{79}Se, ^{90}Sr, ^{129}I and others.[36–43] For the analysis of body fluids with respect to long-lived radionuclides, ICP-MS is the method of choice for fast and low cost routine measurements. The best known investigations focus on the evidence of depleted uranium in urine from soldiers serving during the Gulf War or in the Kosovo conflict (see Section 9.9). The analytical techniques developed were applied for routine measurements in many nuclear laboratories, but were also employed for safety and control of persons working with nuclear materials in nuclear science, medicine, nuclear power stations or other nuclear facilities. Hair or nail samples can also be used for the biomonitoring of nuclear contamination. The uranium distribution in one single hair (see Figure 9.30) measured by double-focusing sector field LA-ICP-MS as a function of uranium in the drinking water consumed was studied in the author's laboratory.[44] RIMS and AMS are advantageous for ultrasensitive and selective isotope analysis of low abundance long-lived radionuclides.

SIMS is mostly employed for microlocal analytical investigations. For example, isotope ratio measurements by SIMS have been performed in nuclear forensic studies to determine the age of Pu particles.[45] For the age determination of Pu particles, relative sensitivity coefficients (RSC) were determined as correction factors for the different ionization efficiency of Pu compared to U. The age of a sample of known origin calculated from $^{238}Pu/^{234}U$ and $^{240}Pu/^{236}U$ ratios agreed well with the reported age of 2.3 years.[45] SIMS was employed for oxygen isotope ratio measurements in three different uranium oxide microparticles of nuclear forensic interest by Betti's working group.[46] The magnitude of instrumental mass discrimination and reproducibility for isotope standards in SIMS has been evaluated in order to exploit oxygen isotope ratio measurements by SIMS (using Cameca 6f and 3f at the Institute for Transuranium Elements, Karlsruhe) as a diagnostic tool in nuclear forensics.[46] The $^{18}O/^{16}O$ results for 12 uranium oxide samples measured by TIMS in the same Institute have been reported.[47] The TIMS $^{18}O/^{16}O$ results on three uranium oxide microparticles are in good agreement with the SIMS data.[46] Besides the oxygen isotope ratios, uranium isotopic ratios (and/or plutonium isotopic ratios if detectable), especially, can provide information on the origin of particles.

Ultratraces of ^{210}Po are distributed worldwide as nuclear contamination in the environment, but ^{210}Po also occurs as a natural radioactive decay product in the decay chain of ^{238}U and can be analyzed by α-spectrometry. In spite of the short ^{210}Po half life ($t_{1/2} = 138$ days), traces can also be detected after careful separation (due to isobaric interference with ^{210}Pb in nuclear samples and environmental materials) using advanced mass spectrometric techniques such as CE-ICP-MS or AMS as demonstrated in references[48,49]

9.10.5 Forensic Investigations by Isotope Ratio Measurements

In the possible event of a (biological, chemical or nuclear) weapons intercept or attack there is a need for more diagnostic forensic analyses to explain the agent, the source and its production method. Besides the chemical (elemental) composition, in particular precise isotope ratio measurements of light elements (H, C, N and O) or heavier elements (such as Sr, Nd, Pb, U and Pu) may provide insights into the production processes of weapon, explosives, ignitable liquids or illicit drugs. A contamination of polycyclic aromatic hydrocarbons (PAHs) in sediments near the site of a former gas manufacturing plant has been investigated by environmental forensics via carbon isotope ratios measured with an isotope ratio mass spectrometer coupled to a gas chromatograph (GC-IRMS).[50] It was found that the compound-specific isotope ratios of carbon in the PAHs obtained by GC-IRMS were not the same as those of soil samples from the site of the former gas manufacturing plant. The biogeochemical natural isotope and trace element (NITE) signatures consist of elemental and isotopic profiles related to regional climate (H and O isotopes), bio-environment (C and N isotopes) and geology (elements and S, Sr, Nd, Pb and other systems). A worldwide geographical analytical profiling platform which enables the geographical profiling of food commodities and can be used as a general forensic profiling tool is to be established. For measurements of isotope ratios of Sr, Nd and Pb in forensic samples, ICP-MS and TIMS are employed, whereas for high precision isotope analysis multiple collector instruments are recommended.

Ghazi and Millette have reported on the *in situ* environmental forensic application of the lead isotope ratio determination of six layers of a paint chip sample by LA-ICP-MS.[51] The layers were analyzed for isotopes of lead (^{204}Pb, ^{206}Pb, ^{207}Pb and ^{208}Pb) and two mercury isotopes (^{202}Hg and ^{204}Hg) to correct possible interferences. Four of the six paint layers contained significant amounts of lead. An appreciable amount of mercury was detected only in one layer. The lead isotope ratios for the layer with the highest Pb concentration had a radiogenic signature similar to that of the Pb–Zn–Ag mines in eastern Missouri. In another layer, less radiogenic Pb was detected similar to that of Pb–Zn–Ag mines of Sierra Madre in east central Mexico. The lead isotope ratio of a third layer had no radiogenic signature.

Further applications of analytical chemistry in forensic sciences are discussed in a special issue on Forensic Analysis in *Analytical Bioanalytical Chemistry*.[52]

References X

1. Andrasko, J. and Maehly, A., *J. Forensic Sci.*, **22**, 250 (1978).
2. Reeve, B., Mathiesen, J. and Fong, W., *J. Forensic Sci.*, **21**, 291 (1976).
3. Hickman, D. A., Harbottle, G. and Sayre, E. V., *Forensic Sci. Int.*, **23**, 189 (1983).
4. Hughes, J. C., Catterick, T., and Southeard, G., *Forensic Sci. Int.*, **8**, 217 (1976).
5. Coleman, R. F. and Goode, G. C., *J. Radioanal. Chem.*, **15**, 367 (1973).
6. Pitts, S. J. and Kratochvil, B., *J. Forensic Sci.*, **36**, 122 (1991).
7. Buscaglia, J., *Anal. Chim. Acta*, **288**, 17 (1994).
8. Almirali, J. R., Umpierrez, S., Castro, W., Gornushkin, I. and Winefordner, J. D., in *Proceedings of SPIE*, E. M. Carapezza (ed). Bellingham, WA (2005) 5778.
9. Mank, A. J. G. and Mason, P. R. D., *J. Anal. At. Spectrom.*, **14**, 1143 (1999).
10. Latkoczy, C., Duecking, M., Becker, S. *et al.*, *Forensic Sci. Int.*, **50**, 1327 (2005).
11. Haney, M., *J. Forensic Sci.*, **22**, 534 (1977).
12. Benson, S., Lennard, C., Maynard, P. and Roux, C., *Forensic Sci. Int.*, **157**, 1 (2006).
13. Chaurand, P., Cornett, D. S. and Caprioli, R. M., *Curr. Opinion Biotechn.*, **17**, 431 (2006).
14. Watling, R. J., *Spectroscopy*, **14**, 16 (1999).
15. Watling, R. J., Herbert, H. K., Detlev, D. and Abell, I. D., *Spectrochim. Acta B*, **198**, 205 (1994).
16. Watling, R. J., Herbert, H. K., Barrow, I. S. and Thomas, A. G., *Analyst*, **120**, 1357 (1995).
17. Mahoney, C. M., Gillen, G. and Fahey, A. J., *Forensic Sci. Int.*, **158**, 39 (2006).
18. Grant, P., Chambers, D., Grace, L., Phinney, D. and Hutcheon, I. D., *Phys. Today*, **Oct 1998**, 32 (1998).
19. Becker, J. S., *Canad. J. Anal. Sci. Spectr.*, **47**, 98 (2002).
20. Watterson, J., *Analyst*, **9**, 961 (2006).
21. Hickman, D. A., *Anal. Chem.*, **56**, 844 (1984).
22. Hickman, D. A., *Forensic Sci. Int.*, **17**, 265 (1981).
23. Koons, R. D., Fiedler, C. and Rawalt, R. C., *J. Forensic Sci.*, **33**, 49 (1988).
24. Ryland, S., *J. Forensic Sci.*, **31**, 1314 (1986).
25. Montero, S., Hobbs, A. L., French, T. A and Admirall, J. R., *J. Forens. Sci*, **48**, 1101 (2003).
26. Duckworth, D. C., Morton, S. J., Bayne, C. K., Koons, R. D., Montero, S. and Almirall, J. R., *J. Anal. At. Spectrom.*, **17**, 662 (2002).
27. Berends-Montero, S., Wiarda, W., de Joode, P. and van der Peijl, G., *J. Anal. At. Spectrom.*, **2006**, 1185 (2006).
28. Maurice, J. F., Wibetoe, G. and Sjastad, K. E., *J. Anal. At. Spectrom.*, **17**, 485 (2002).
29. Hann, S., Latkoczy, C., Bereuter, T. L., Prohaska, T., Stingeder, G. and Reiter, C., *Intern. J. Legal Medicine*, **119**, 35 (2005).
30. Richardson, S. D., Sadi, B. B. M. and Caruso, J. A., *J. Anal. At. Spectrom.*, **21**, 396 (2006).
31. Lui, Q., Hu, X. and Xie, J., *Anal. Chim. Acta*, **512**, 93 (2004).
32. Smith, J. R. and Shih, M. L., *J. Appl. Toxicology*, **21**, 27 (2001).
33. Steiner, W. E., Clowers, B. H., Matz, L. M., Siems, W. F. and Hill Jr, H. H., *Anal. Chem.*, **74**, 4343 (2002).
34. Hou, X., Chen, W., He, Y. and Jones, B. T., *Appl. Spectr. Rev.*, **40**, 245 (2005).
35. *International Atomic Energy Agency (IAEA)*, Press Release 2002/09 (2002).
36. Becker, J. S., *Int. J. Mass Spectrom.*, **242**, 183 (2005).
37. Becker, J. S., *Spectrochim. Acta* **58B**, 1757 (2003).
38. Desideri, D., Meli, M. A., Roselli, C., Testa, C., Boulyga, S. F. and Becker, J. S., *Anal. Bioanal. Chem.*, **374**, 1091 (2002).
39. Gastel, M., Becker, J. S., Kuppers, G. and Dietze, H. J., *Spectrochim. Acta*, **52B**, 2051 (1997).
40. Kutschera, W., *Int. J. Mass Spectrom.*, **242**, 145 (2005).
41. McLean, J. A., Becker, J. S., Boulyga, S. F., Dietze, H. J. and Montaser, A., *Int. J. Mass Spectrom.*, **208**, 193 (2001).
42. Vockenhuber, C., Ahmad, I., Golser, R. *et al.*, *Int. J. Mass Spectrom.*, **223**, 713 (2003).
43. Zoriy, M. V., Halicz, L., Ketterer, M. E., Pickhardt, C., Ostapczuk, P. and Becker, J. S., *J. Anal. At. Spectrom.*, **19**, 362 (2004).
44. Sela, H., Karpas, Z., Zoriy, M., Pickhardt, C. and Becker, J. S., *Int. J. Mass Spectrom.*, **261**, 199 (2007)

45. Tamborini, G., Wallenius, M., Bildstein, O., Pajo, L. and Betti, M., *Mikrochim. Acta*, **139**, 185 (2002).
46. Tamborini, G., Phinney, D., Bildstein, O. and Betti, M., *Anal. Chem.*, **74**, 6098 (2002).
47. Pajo, L., Tamborini, G., Rasmussen, G., Mayer, K. and Koch, L., *Spectrochim. Acta*, **56B**, 541 (2001).
48. Casey, N. and Olesik, J. W., in *Book of Abstracts of 33 FACSS*. FACSS Lake Buena Vista, FL (2006).
49. Peck, G. A., Szymcak, R. and Jeffree, R. A., in *IAPSO-IAEA Workshop WS01: Isotopes as Tracers in Marine Environmental Studies*. Monaco (2006).
50. Saber, D., Mauro, D. and Sirivedhin, T., *Environ. Forensics*, **7**, 65 (2006).
51. Ghazi, A. M. and Millette, J. R., *Environ. Forensics*, **5**, 97 (2006).
52. Ballou, S., Goddpaster, J. V., MacCrehan, W. and Reeder, D. (eds.), *Anal. Bioanal. Chem.* **376**, 1149–1297 (2003).

9.11 Study of Cluster and Polyatomic Ion Formation by Mass Spectrometry

In all the mass spectrometric techniques discussed, polyatomic and cluster ions are formed, sometimes with high intensities, in plasmas, during evaporation or thermal surface ionization or during the sputter process, and they appear in the mass spectra. Therefore, polyatomic ions, also called 'molecular ions' in the literature, are of special interest in analytical chemistry. The polyatomic ions are small (diatomic, triatomic or more) 'molecular ions'. The word 'cluster' has come to mean a group of atoms or molecules formed by interactions ranging from very weak van der Waals contracts to strong ionic bonds.[1] As discussed in the previous chapters, the quantification of analytical results and the determination of isotope ratios is often difficult in all inorganic mass spectrometric techniques due to incomplete knowledge of ion formation processes, owing to various degrees of ionization of elements and dissociation and ionization of molecules, and, in part, because of the high proportion of polyatomic and cluster ions formed in the plasma or by sputtering, which complicates the analysis of trace impurities through isobaric interferences with atomic ions of analyte.[2–6] Whereas in solid-state mass spectrometry, different degrees of ionization of the analytes and also matrix effects can be corrected by the application of relative sensitivity coefficients (RSCs), systematic investigations of cluster and polyatomic ion intensities at defined experimental parameters are useful for estimating and correcting mass spectral interferences. In addition, inorganic mass spectrometric techniques – such as SSMS,[7] GDMS,[8] LIMS,[9,10] ICP-MS,[2,11] LA-ICP-MS[12] and SIMS[13,14] – are not only sensitive and powerful for the multi-element determination of trace contaminants or isotope ratios in inorganic materials or bioorganic and environmental samples, but these techniques can also be utilized for fundamental studies of cluster and polyatomic ion formation and dissociation of species.[2,4]

Investigations of cluster formation serve to explain the evaporation and atomization of sample material and ion formation processes. A further aim of cluster research is to find out under what conditions cluster or polyatomic ion formation can be influenced in order to avoid disturbing interferences and decrease the detection limits of elements. On the other hand, polyatomic ions have also been used as analyte ions for analysis, e.g. the application of MCs^+ and MCs_2^+ dimeric and trimeric ions as analyte[11] or of cluster primary ion beams (e.g., of bismuth and gold primary clusters)[15,16] by the bombardment and sputtering of a solid surface in SIMS.[17–21] Especially in SIMS, a multitude of cluster ions with high ion formation rates are observed.[18,22,23]

Mass spectrometric studies of the formation of clusters and their ionization in plasmas have been of interest for more than four decades. Increasing research has been performed together with activities in cluster physics and chemistry because clusters may also be viewed as a link in the chain from atoms or molecules through small crystallites to a complete solid-state structure (important for materials research, nanotechnology or catalytic processes).[24–26] Consequently, clusters can also be considered as a bridge between the disciplines of physics and chemistry.[27] Therefore, the aim of cluster research is the synthesis of new clusters in order to produce solid-state (nm size) structures with new and interesting properties.[24,26,28] Clusters of non-magnetic elements become magnetic, semiconducting materials exhibit metallic properties, metallic systems become semiconducting or the colour of the particles changes with size.[27] Whereas cluster physics, which has extensive research potential, studies the growth of small crystallites in the initial phase, in cluster chemistry, cluster formation processes are investigated (e.g., by plasma chemical reactions) and the stability of reaction products. A classification of different cluster types is possible as a function of interatomic bonding: weak van der Waals bond clusters, molecular clusters, H-bonding clusters, ionic bond clusters, valence clusters with covalent chemical bonding energy of several eV and metallic clusters (see Figure 9.53). The ways of generating such clusters including thermal (in a Knudsen cell[29]) or laser evaporation of a solid sample,[30] adiabatic expansion of a condensed gas through a nozzle,[31] inert gas condensation,[32] ion bombardment of a solid surface with highly energetic ions[33] or plasma chemical reactions[2] are summarized in Figure 9.54. All types of clusters were studied by mass spectroscopy amongst other techniques (e.g., photoelectron spectroscopy) often in combination with a theoretical calculation of their electronically stability. The cluster abundances observed in the mass spectra are proportional to the stability of the cluster ions from the ion formation process to their ion detection in the mass spectrometer.

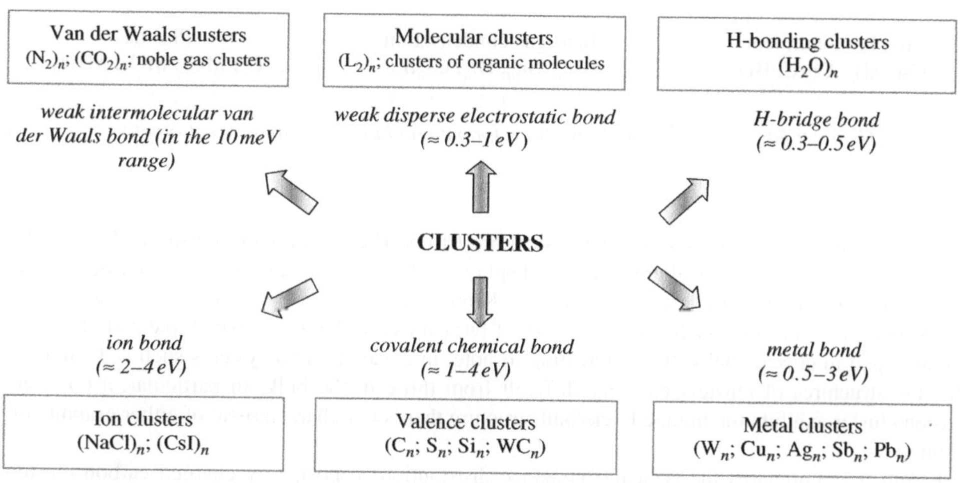

Figure 9.53 *Classification of clusters.*

Pioneering work on the mass spectra of Na clusters by Knight *et al.*[34] provided the first insight that clusters and nuclear physics have something in common. They observed that Na clusters consisting of 2, 8, 10 and 40 atoms were unusually stable and coincided with the magic number in nuclear physics where nuclei with the same numbers of protons and/or neutrons were known to be very stable.[34]

9.11.1 Carbon and Boron Nitride Cluster Ion Formation

An extensively investigated material in the study of cluster formation in different plasmas is graphite, also from the perspective that it is possible to detect very stable configurations with

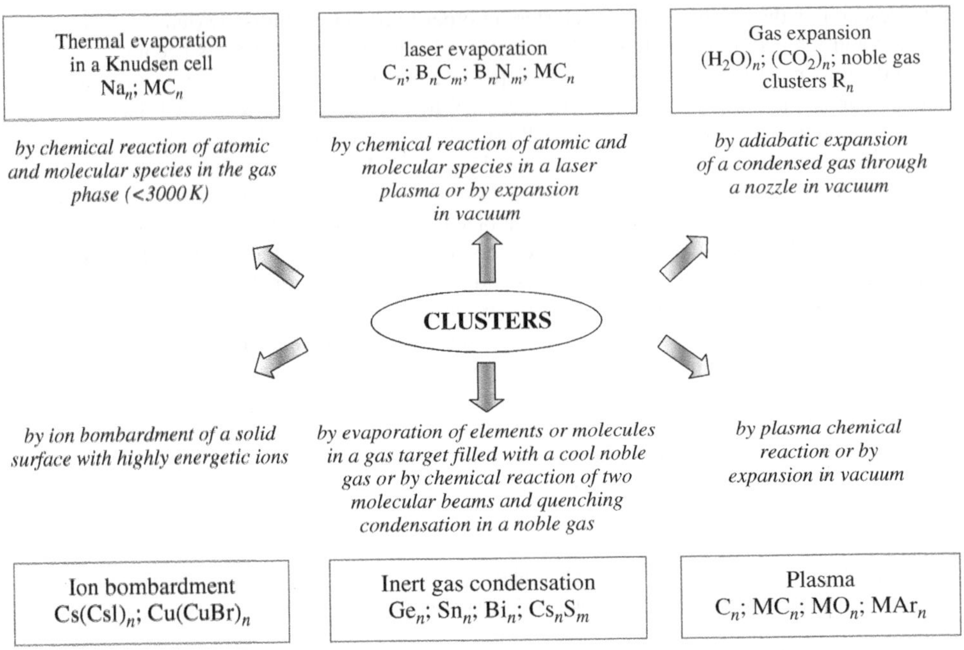

Figure 9.54 *Schematic of cluster (ion) formation in inorganic mass spectrometry.*

highly symmetrical structures and new physical properties. The most famous cluster is the remarkably stable C_{60} buckminsterfullerene which displays high symmetry and was detected during the vaporization of graphite by laser irradiation by Kroto *et al.*[30,35] In 1996, Kroto was honoured by the Nobel prize in chemistry for his discovery of buckminsterfullerene, in which the carbon atoms are arranged in closed shells. It is interesting to note, in covalent bond systems such as C and Si, that the structures of clusters are very difficult from those in the bulk. In particular, a 60 atom carbon cluster exhibits the unique buckyball structure that is not characteristic of either graphite or diamond.

Figure 9.55 compares the typical abundance distribution of positively charged carbon cluster ions in SSMS,[36] LIMS,[2,36] rf GDMS[37] and LA-ICP-MS.[2] In all experiments a similar distribution with a general decrease of cluster intensities with increasing cluster size was found, with local maxima for cluster ions with odd numbered atoms. This typically alternating abundance distribution of carbon cluster ions with alternating abundance distribution, local maxima for odd numbered cluster size and four periodicity for carbon clusters of up to nine atoms was described using SSMS by Franzen and Hintenberger in 1961[38] and in a laser plasma using a LAMMA by Fürstenau *et al.*[39] The change of periodicity for positively charged cluster ions was interpreted by theoretical investigations using the LCAO-Xα method[40] as a change of cluster structure from chains for small clusters to rings for clusters greater than nine carbon atoms. In addition, the cluster formation rate can be influenced in SSMS by changing the spark voltage, in rf GDMS by varying the rf power and/or Ar pressure and in LIMS by changing the laser power density.[40] For example, whereas at laser power densities $\Phi > 5 \times 10^9\ \mathrm{W\,cm^{-2}}$ in LIMS the atomic ions C^+ were observed with higher intensities in comparison to the C_n^+ cluster ions (with $n \geq 2$), at a lower laser power density of

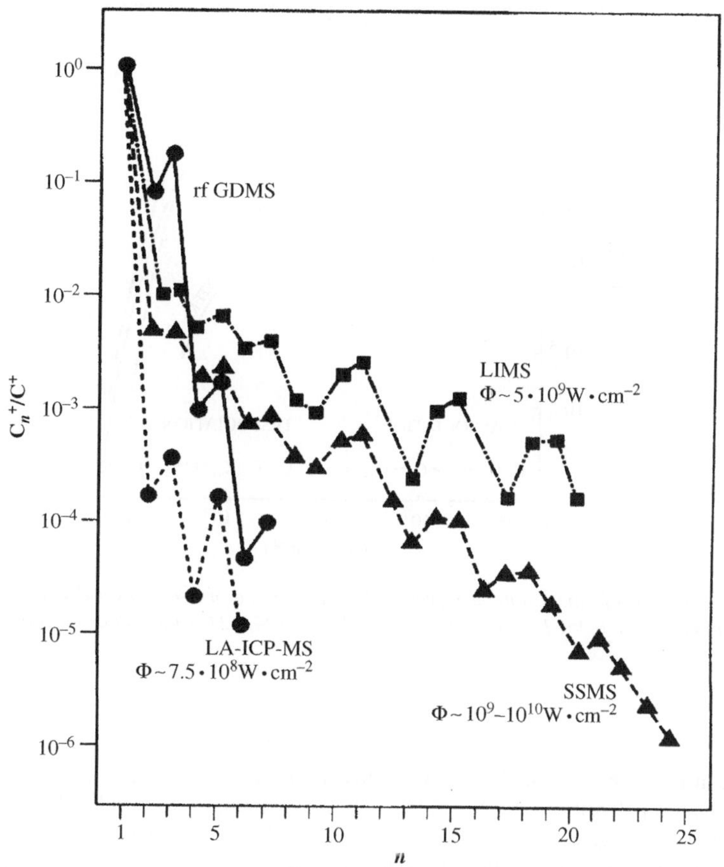

Figure 9.55 *Carbon cluster ion distribution measured in SSMS, LIMS, rf GDMS and LA-ICP-MS (Φ – power density). (J. S. Becker and H. J. Dietze, Fresenius' J. Anal. Chem.,* ***359****, 338(1997). Reproduced by permission of Springer Science and Business Media.)*

$\Phi \sim 10^8\,W\,cm^{-2}$ the C_{11}^+ species were the most stable cluster ions with the highest intensities in LIMS.[3] The influence of laser power density is illustrated in Figure 9.56 At low laser power density small clusters (C_3) are observed to have the highest intensity. The abundance distribution in laser evaporation is similar to that observed from thermal vaporization in Knudsen effusion cells with the following abundance sequence ($C_3 > C_2 > C_1 >> C_5 > C_4$).[40] By increasing the laser power density up to $\Phi \sim 10^8\,W\,cm^{-2}$ the formation of bigger clusters becomes evident whereby C_{11}^+ clusters were observed with the highest ion intensity. A further increase of laser power density shifts the maximum of the abundance distribution of the clusters from C_{11}^+ through C_7^+, C_5^+, C_3^+ to atomic ions C^+ due to the dissociation of clusters in the plasma with increasing electron density. These fragmentation processes may be attributed to unimolecular decays of electronically excited clusters, photodissociation or collision induced processes.[40] In LA-ICP-MS, the cluster formation rate can be changed by varying the rf power and carrier gas flow rate in the inductively coupled plasma.

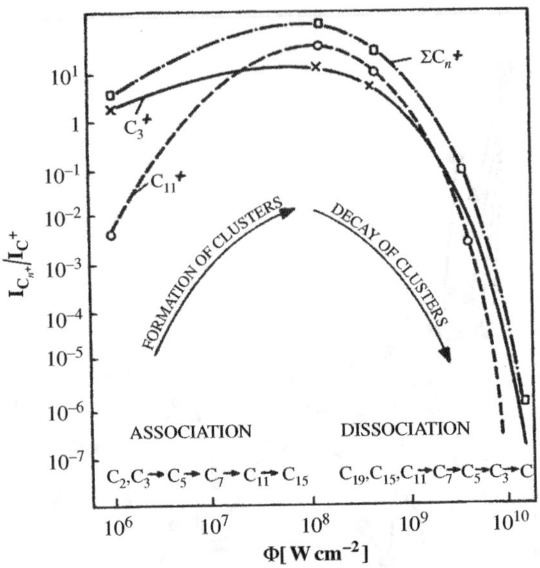

Figure 9.56 Cluster ion formation in laser plasma in dependence of laser power density Φ. (G. Seifert, J. S. Becker and H. J. Dietze, Int. J. Mass Spectrom. Ion Proc., **184**, 121(1988). Reproduced by permission of Elsevier)

The abundance distribution for C_n^+ and C_n^- cluster ions, observed in laser plasmas of a graphite target at laser power density of $10^8\,\mathrm{W\,cm^{-2}}$, is shown in Figure 9.57. Under these experimental conditions, the small negatively charged cluster ions are observed in comparison to the positively charged cluster ions with higher ion intensities. For larger cluster sizes ($n > 10$), the intensity of the negatively charged cluster ions decreases rapidly. For positively charged carbon cluster ions C_n^+ the typical abundance distribution with maximum intensity for the C_{11}^+ cluster ions in laser mass spectra (at a laser power density of $10^8\,\mathrm{W\,cm^{-2}}$) is found with local maxima for odd numbered cluster sizes and four periodicity for carbon clusters of up to nine atoms for high purity graphite (used for the preparation of electrodes in SSMS), pyrolytic and polycrystalline graphite and amorphous carbon.[41]

Similar to graphite, the hexagonal α-boron nitride with graphite structure forms a variety of clusters with alternating abundance distribution for odd and even numbered cluster ions (see Figure 9.58). The details of cluster formation by irradiation of a boron nitride target with a focused laser beam and the theoretical interpretation of possible ways of cluster formation are described in references[40,42]. For both types of cluster ions investigated (C_n^+ and $B_nN_m^+$), the theoretical calculation yields a higher stability of species with odd numbers of atoms in comparison to neighbouring cluster ions with even numbers of atoms by reason of paired electrons. Therefore in a laser mass spectrum (at a laser power density of $\Phi \sim 10^9\,\mathrm{W\,cm^{-2}}$) of a boron nitride target, an alternating cluster distribution is found (see Figure 9.58). The $B_nN_{n-1}^+$ and $B_nN_{n-2}^+$ cluster ions appear with local maximum and minimum intensities, respectively. The higher intensity of $B_nN_{n-1}^+$ cluster ions compared to the $B_nN_{n-2}^+$ correlates with the higher stability of these species. The typical distribution of boron nitride clusters shown in Figure 9.58 is very similar for three

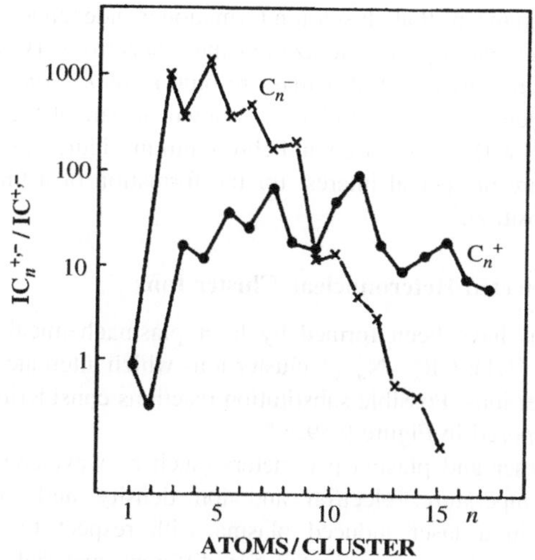

Figure 9.57 Ion intensities of positively and negatively charge carbon cluster ions in laser plasma. (G. Seifert, J. S. Becker and H. J. Dietze, Int. J. Mass Spectrom. Ion Proc., **184**, 121(1988). Reproduced by permission of Elsevier)

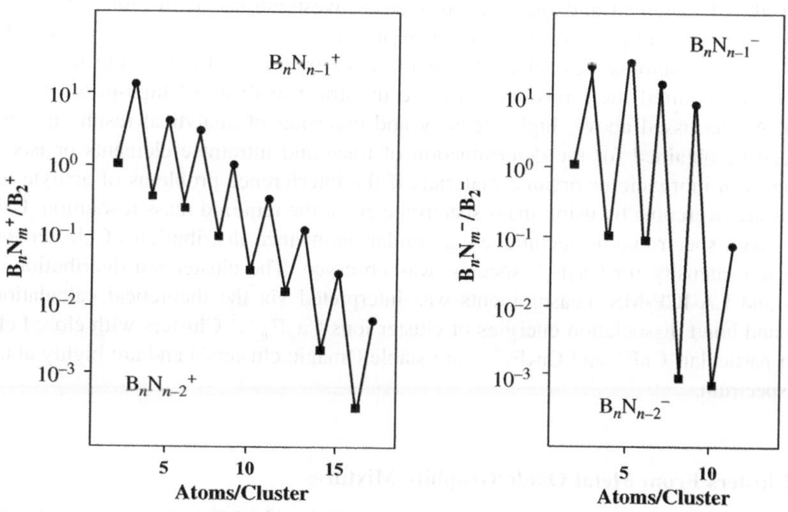

Figure 9.58 Cluster ion distribution ($B_nN_{n-1}^+$ and $B_nN_{n-1}^-$) for boron nitride in a laser plasma.

different boron nitride modifications (hexagonal α-boron nitride with graphite structure, cubic β-boron nitride with diamond structure and hexagonal γ-boron nitride resembling the wurtzite type) as measured by LIMS,[43] which means that cluster ion formation is independent of the structure of solid boron nitride. All measurements by laser ionization mass spectrometry using a double-focusing sector field instrument were performed at a mass resolution of $m/\Delta m \sim 10\,000$. Consequently, most of the isobaric interferences were resolved as shown in part of the laser mass spectrum of boron nitride contaminated with carbon, iron and chromium in Figure 6.9. In addition, the clusters formed in laser plasmas are of special interest for the formation of a thin layer composition by laser induced plasma deposition.[43]

9.11.2 Formation of Selected Heteronuclear Cluster Ions

New types of cluster ions have been formed by laser plasmachemical reactions in a graphite and boron nitride mixture (1:1): $CB_{n-1}N_{n-1}^+$ cluster ions which alternate with $CB_{n-1}N_{n-2}^+$ ionic species and BC_{n-1}^+ cluster ions. Possible substitution reactions considering the cluster ions of the initial compounds are proposed in Figure 9.59.[2,44]

The influence of the laser and plasma parameters (such as wavelength, laser power density, pulse length, plasma temperature, electron and ion density and others) on the physical and chemical processes in a laser induced plasma with respect to the formation of polyatomic and cluster ions has been investigated for different materials (e.g. graphite,[36] boron nitride,[44] boron nitride/graphite mixture,[45] boron carbide,[46] tungsten oxide/graphite mixture[47] and superconductors[48]).

The result of intensity alternation for the boron nitride cluster ions is in accordance with the generation of high cluster ion beams $M_nX_{n-1}^+$ for several systems (e.g. NaCl, CuBr or CsI[26,49]) by other methods of cluster formation such as by quenching condensation in a cold rare gas or by ion bombardment of halides.

A similar distribution of cluster ions $Ca_nF_m^+$ was observed by SSMS and LA-ICP-MS measurements in CaF_2. Theoretical and mass spectrometric investigations of the formation of calcium fluoride cluster ions were performed in the frame of ultratrace analysis on high purity CaF_2 single crystals. Table 9.46 summarizes selected possible interferences of $Ca_nF_m^+$ cluster ions in mass spectra and the required mass resolution by the ultratrace analysis of high-purity CaF_2 (purity 99.996 %). As discussed above, high accuracy and precision of analytical results in mass spectrometry can be obtained for the determination of trace and ultratrace elements or isotope ratio measurements in inorganic or organic materials if the interference problems of analyte ions with cluster ions are overcome by using mass spectrometers at the required mass resolution. Using both solid-state mass spectrometric techniques, a similar abundance distribution of cluster ions, with maximum ion intensity for $Ca_2F_3^+$ species, was observed. The cluster ion distribution obtained by SSMS and LA-ICP-MS measurements was interpreted via the theoretical calculation of the structures and bond dissociation energies of cluster ions $Ca_nF_m^+$.[5] Clusters with closed electronic shells – in particular, CaF^+ and $Ca_2F_3^+$ – are stable ('magic clusters') and are highly abundant in the mass spectrum.

9.11.3 Clusters From Metal Oxide/Graphite Mixtures

Besides the carbon cluster ions, carbide cluster ions are formed in different plasmas with remarkably high ion intensities. The abundance distribution of these carbide cluster ions as a function of cluster size has been observed to be very similar using the different solid-state mass spectrometric methods, with higher intensities for cluster ions with even numbered carbon atoms than for neighbouring

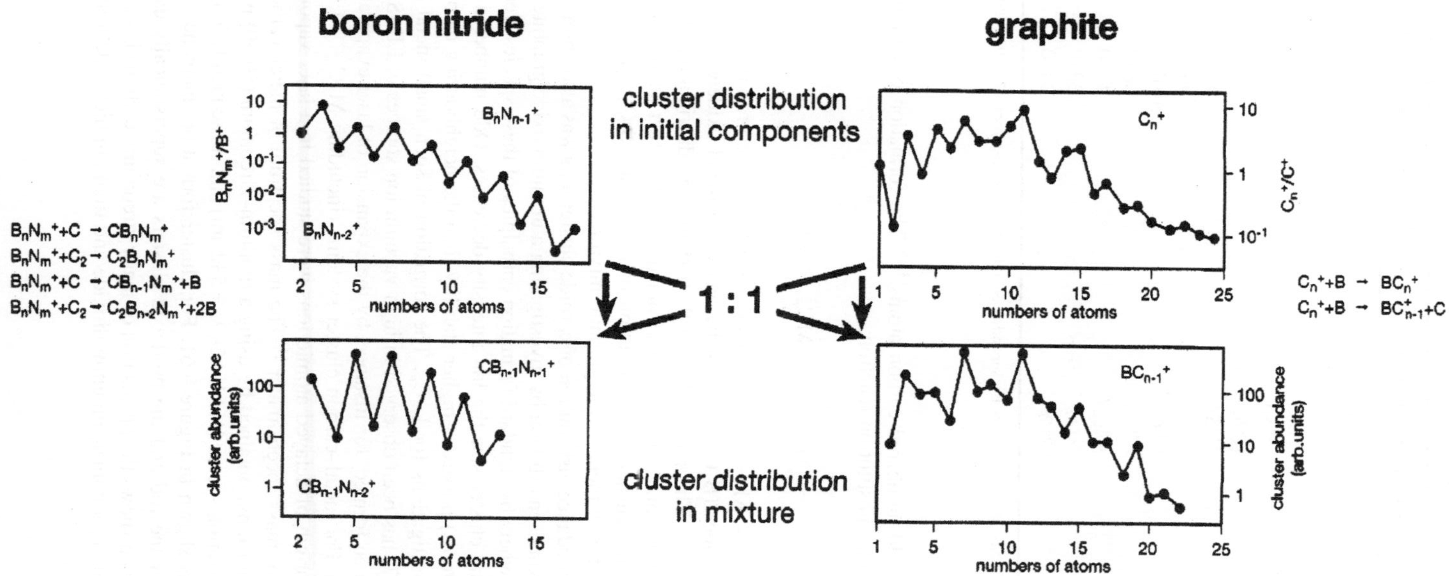

Figure 9.59 Cluster ion formation of a boron nitride/graphite mixture in a laser plasma. (J. S. Becker and H. J. Dietze, Fresenius' J. Anal. Chem., **359**, 338(1997). Reproduced by permission of Springer Science and Business Media.)

Table 9.46 *Possible interferences of cluster ions in mass spectra and the required mass resolution by the ultratrace analysis of high purity CaF_2.*

Cluster ion	Atomic ion of analyte	Required mass resolution ($m/\Delta m$)
$^{40}Ca^{19}F^+$ ($^{40}Ar^{19}F^+$)[a]	$^{59}Co^+$	2120
$^{40}Ca^{19}F_2^+$ ($^{40}Ar^{19}F_2^+$)[a]	$^{78}Se^+$	1850
$^{40}Ca^{19}F_3^+$	$^{97}Mo^+$	1850
$^{40}Ca^{19}F_4^+$	$^{116}Sn^+$	2130
$^{40}Ca_2^{19}F^+$	($^{99}Tc^+$)	
$^{40}Ca_2^{19}F_2^+$	$^{118}Sn^+$	5790
$^{40}Ca_2^{19}F_3^+$	$^{137}Ba^+$	9400
$^{40}Ca_2^{19}F_4^+$	$^{156}Gd^+$	46 850
$^{40}Ca_3^{19}F^+$	$^{139}La^+$	6840
$^{40}Ca_3^{19}F_2^+$	$^{158}Gd^+$	4000
$^{40}Ca_3^{19}F_3^+$	$^{177}Hf^+$	2940
$^{40}Ca_3^{19}F_4^+$	$^{196}Pt^+$	2350

[a] observed interference in LA-ICP-MS. (R. Kaschner, J. S. Becker and G. Seifert, *Int. J. Mass Spectrom*, **176**, 103 (1998). Reproduced by permission of Elsevier.)

carbide cluster ions with odd numbered carbon atoms.[3,46] The formation of carbide cluster ions can be described by a simple substitution reaction:

$$M^+ + C_n \rightarrow MC_{n-1}^+ + C \tag{9.8}$$

Carbide cluster ions (MC_n^+ – M = matrix element) have been measured by investigating them directly from the solid carbides (B_4C,[46] SiC) or by analyzing metal oxide/graphite mixtures (for M = rare earth element,[3] Si,[46] Th or U[36]). Figure 9.60 shows the distribution of silicon carbide cluster ions (SiC_n^+) in laser ionization mass spectrometry by the direct analysis of compact SiC in comparison to the carbide cluster ion distribution of LaC_n^+ and SrC_n^+ in spark source mass spectrometry, by investigating a metal oxide/graphite mixture.

A similar alternating abundance distribution of carbide cluster ions was also observed in e^--beam mass spectrometry with Knudsen effusion by investigating a uranium oxide/graphite and a thorium oxide/graphite system,[50] where by standard formation enthalpies of these species were determined. In general, with increasing cluster size the ion intensity decreases. Odd numbered clusters (with even numbered carbon atoms) possess a higher stability than the neighbouring carbide clusters as shown by theoretical investigations for LaC_n. The formation of such small metal carbon cluster ions (MC_n^+ with $n = 1$–7) has been observed in high vacuum ion sources of LIMS and SSMS.

Bigger clusters have been formed, for instance, by the expansion of laser evaporated material in a gas still under vacuum. For metal–carbon cluster systems (including $M_nC_m^+$ of Ti, Zr and V), their formation and the origin of delayed atomic ions were studied in a laser vaporization source coupled to a time-of-flight mass spectrometer.[51] The mass spectrum of metal–carbon cluster ions (TiC_2^+ and $Zr_nC_m^+$ cluster ions) obtained by using a titanium–zirconium (50:50) mixed alloy rod produced in a laser vaporization source (Nd:YAG, $\lambda = 532$ nm) and subsequently ionized by a XeCl excimer laser (308 nm) is shown in Figure 9.61. For cluster formation, methane (~ 15 % seeded in helium) is pulsed over the rod and the produced clusters are supersonically expanded in the vacuum. The mass spectrum shows the production of many zirconium–carbon clusters. Under these conditions only the titanium monomer, titanium dioxide and titanium dicarbide ions are formed.

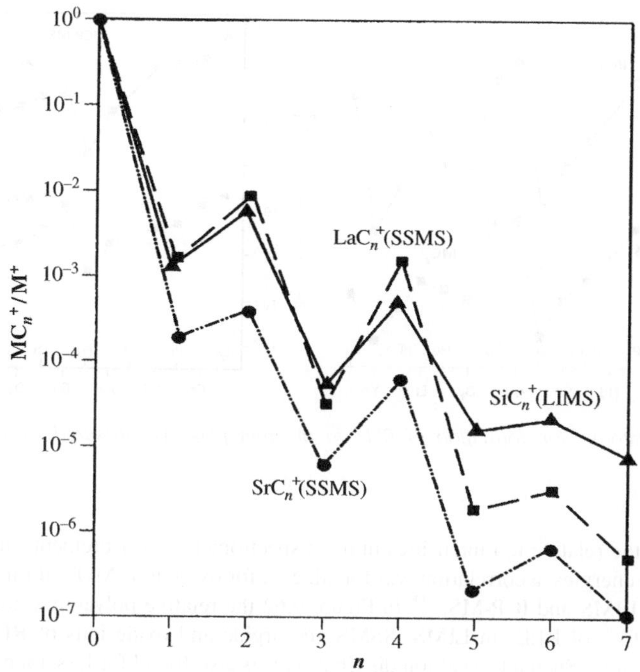

Figure 9.60 *Metal carbide formation in SSMS and LIMS. (J. S. Becker and H. J. Dietze, Fresenius' J. Anal. Chem.,* ***359****, 338(1997). Reproduced by permission of Springer Science and Business Media.)*

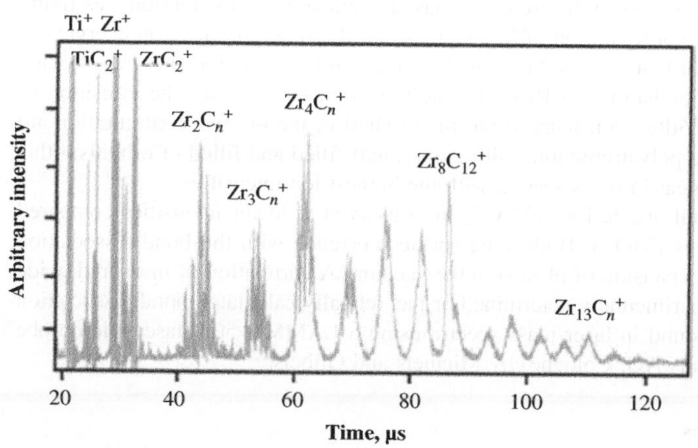

Figure 9.61 *ToF mass spectrum of metal–carbon cluster ions (TiC_2^+ and $Zr_nC_m^+$ cluster ions) using a titanium–zirconium (50:50) mixed alloy rod produced in a laser vaporization source (Nd:YAG, λ = 532 nm) and ionization by a XeCl excimer laser (308 nm). (K. M. Davis, S. J. Peppernick and A. W Castleman, J. Chem. Phys.,* ***124****, 164304(2006). Reproduced by permission of American Institute of Physics.)*

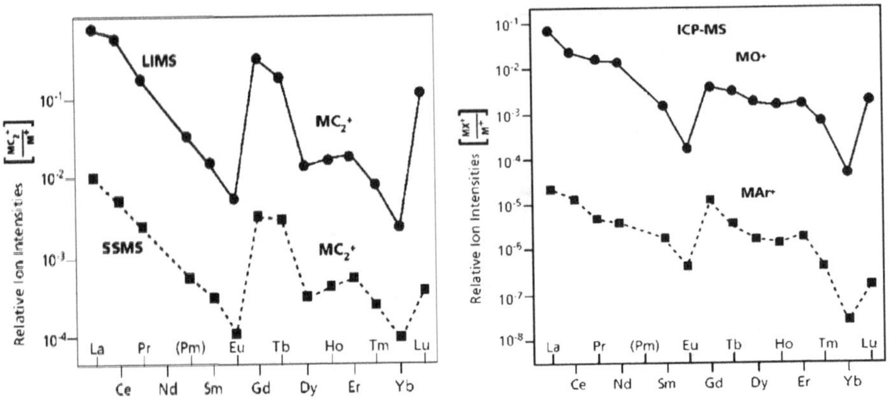

Figure 9.62 *Polyatomic ion formation of REEs in different plasmas measured by LIMS, SSMS and ICP-MS.*

By comparing the relative ion intensities in mass spectrometric measurements and experimental bond dissociation energies, a correlation was found, e.g., for oxide ions MO^+ of rare earth elements (REE) in SSMS, LIMS and ICP-MS.[2,11] In Figure 9.62 the relative polyatomic ion intensities (of dicarbide ions MC_2^+ of REE) in LIMS, SSMS and argide and oxide ions of REE measured by ICP-MS are compared. Such a typical ion distribution was also found for less abundant MAr^+ ions by using ICP-MS (with intensities about three to four orders of magnitude lower in comparison to MO^+)[2] as shown at the bottom of Figure 9.62. By the examination of abundances of cluster ions, e.g. monocarbide and dicarbide cluster ions MC^+ and MC_2^+ or oxide ions MO^+ of lanthanides, as a function of their atomic numbers in different plasmas, a characteristic distribution was found correlating with the electronic configuration of lanthanides and the bond dissociation energies of oxides, carbides or dicarbides of lanthanides. In Figure 9.63 the bond dissociation energies of oxide ions of REE, the sublimation enthalpies of REE and the transition energies for the transition of the 5d electron from the $4f^{n-1}5d6s^2$ configuration to the 4f shell of the $4f^n$-$6s^2$ configuration are compared. Energetically stable polyatomic ions with empty, half filled and filled 4f orbitals – this is true of La, Gd and Lu – appear in mass spectra with the highest ion intensities.[3]

In general, in ICP-MS, metal argide ions (MAr^+) are observed at lower intensities compared to the dimeric metal oxide ions (MO^+). Both ionic species correlate with the bond dissociation energies in the ICP or by the expansion of plasma in the vacuum. A correlation of measured oxide ion intensities (MO^+) and experimentally determined or theoretically calculated bond dissociation energies of oxides has been found in laser mass spectra using a LAMMA 500 (laser microprobe mass analyzer, Leybold Hereaus AG, Cologne) by Michiels and Gijbels.[52]

9.11.4 Argon Diatomic Ions

In Figure 9.64, a correlation is observed for the metal argide ion intensities in ICP-MS and binding dissociation energies. In general, binding energies and therefore intensities of metal argide ions containing transition metal atoms (e.g. $CoAr^+$, $NiAr^+$) are higher than for those with a main group element with low ionization potential (e.g. $MgAr^+$, $NaAr^+$). In these metal argide ions the positive charge is located at the metal atom (M^+Ar^o). The large ionization potential of the transition metals induce M^+Ar^o, a component of M^0Ar^+, which causes an additional stabilization of argide

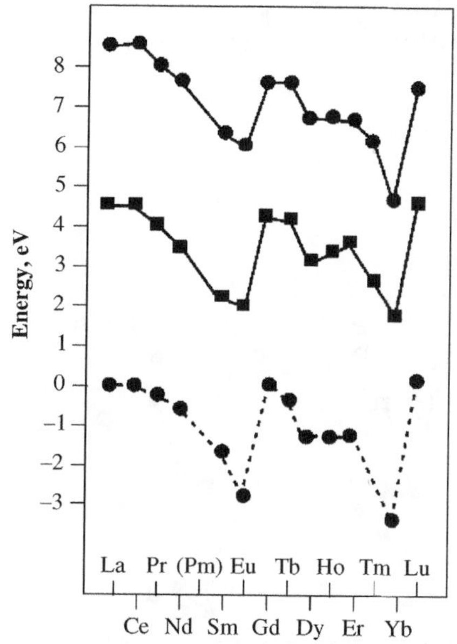

Figure 9.63 Properties of REEs (from above: dissociation energies of lanthanide oxides (•), sublimation enthalpies of REEs (■) and transition energies for the transition of the 5d electron from $4f^{n-1}5d6s^2$ configuration to the 4f shell of the $4f^n6s^2$ configuration (– – –). (J. S. Becker and H. J. Dietze, J. Anal. At. Spectrom., **10**, 637 (1995). Reproduced by permission of the Royal Society of Chemistry.)

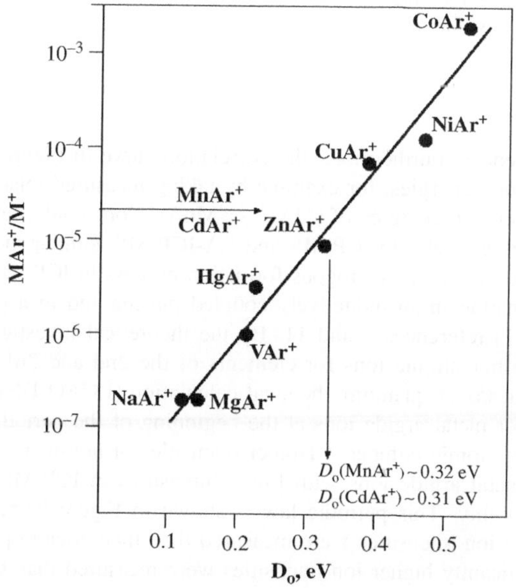

Figure 9.64 Relative metal argide ion intensities vs dissociation energies. (J. S. Becker and H. J. Dietze, Fresenius' J. Anal. Chem., **359**, 338(1997). Reproduced by permission of Springer Science and Business Media.)

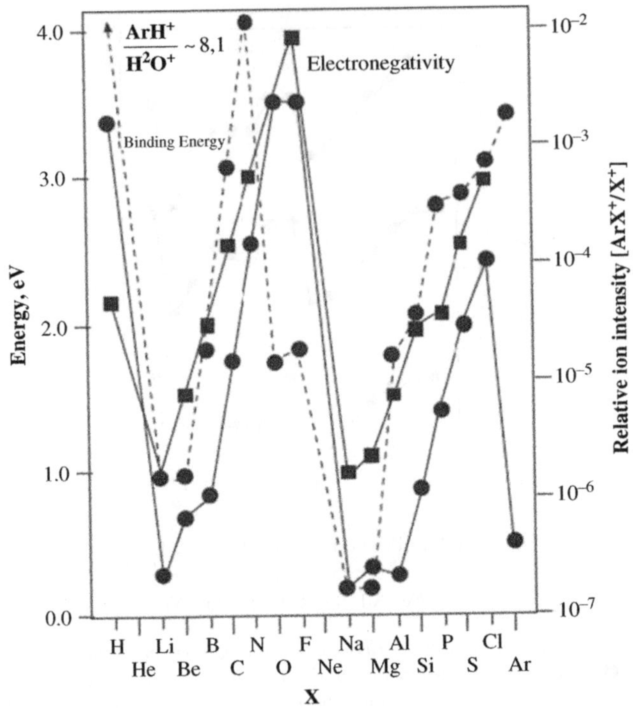

Figure 9.65 Binding energy of argon polyatomic ions (ArX^+) and relative ion intensity [ArX^+/X^+]. (J. S. Becker and H. J. Dietze, Fresenius' J. Anal. Chem., **359**, 338(1997). Reproduced by permission of Springer Science and Business Media.)

ions for transition elements.[53] Furthermore, the correlation curve in Figure 9.64 can help to estimate unknown dissociation energies, for example by taking measured relative ion intensities into consideration, the dissociation energies of $CdAr^+$ or $MnAr^+$ ions could be estimated to be about 0.31 eV and 0.32 eV, respectively. In ICP-MS and LA-ICP-MS, similar metal argide ion intensities have been measured, due to an analogous formation process in ICP. Details of investigations on polyatomic ion formation in an inductively coupled plasma and in a glow discharge plasma are discussed in detail in references 2 and 11. By the theoretical investigation of bond dissociation energies of argon polyatomic ions for elements of the 2nd and 3rd period in the periodic table of elements, by means of quantum chemical calculation (LCAO-DFT-LDA), small binding energies are obtained for metal argide ions at the beginning of the period and the binding energies rise with increasing atomic number.[11] Higher intensities of non-metal argon polyatomic ions in comparison to the metal argide ions with lower intensities in ICP-MS can be interpreted as being due to higher stability. This periodic law is shown in Figure 9.65, except for ArO^+ and ArF^+ (in ICP-MS lower ion intensities were measured than theoretically predicted) and $Ar_2{}^+$ (in ICP mass spectra significantly higher ion intensities were measured than theoretically expected). A further explanation of different ion intensities and the stability of non-metal argon polyatomic ions and metal argide ions is the different chemical bonding, which is strictly located at the metal in the metal argide ions, which possess bond dissociation energies of some 10 meV, whereas non-metal argon polyatomic ions are strongly bound with binding energies in the order of the usual chemical bond, i.e. a few eV.[11] Although the theoretical calculation predicted that the non-metal argon polyatomic ions would be more stable in comparison to metal argide ions in GDMS, significantly higher concentrations of metal argide ions (sometimes in the percent range) were observed. Although in GDMS a correlation of measured ion intensities with bond dissociation energies for different metal argide ions was found similar to that in ICP-MS, in GDMS some metal argide ions (e.g. $ZnAr^+$) were observed with significantly higher intensities than expected.[54] From the investigation of metal ions and metal argide ions as a function of argon pressure in rf glow discharge by studying a GaAs sample,[11] the associative reaction was assumed to be the dominating process: $X^+ + Ar \rightarrow ArX^+$.

9.11.5 Oxide Ion Formation of Long-lived Radionuclides in ICP-MS

The oxide ion formation of long-lived radionuclides ^{226}Ra, ^{230}Th, ^{237}Np, ^{238}U, ^{239}Pu and ^{241}Am has been studied in mixed aqueous solution by double-focusing sector field ICP-MS using different nebulizers. For the oxide ion formation rate, isotope analysis via oxide ions MO^+

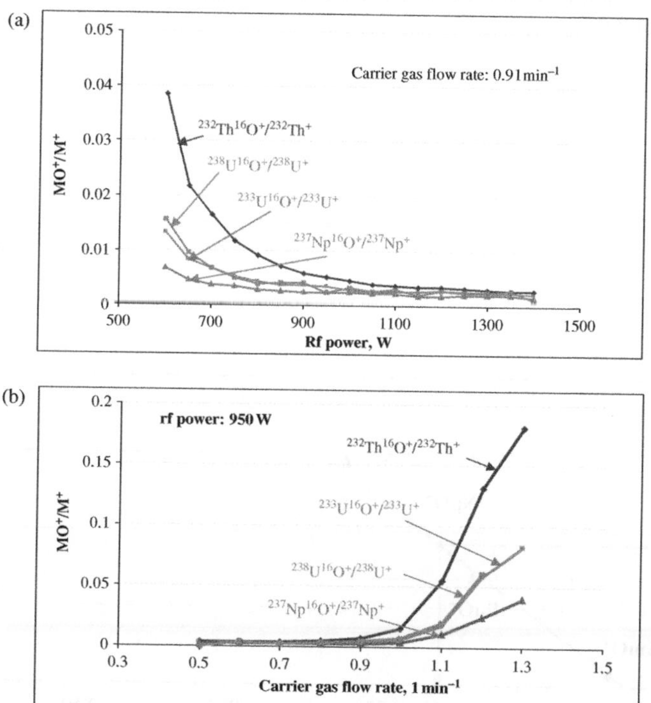

Figure 9.66 *Oxide formation rate of long-lived radionuclides as a function of rf power (a) and carrier gas flow rate (b) measured in a concrete matrix by LA-ICP-MS. (J. S. Becker, C. Pickhardt and H. J. Dietze, Int. J. Mass Spectrom.,* ***203****, 283(2000). Reproduced by permission of Elsevier.)*

was performed for uranium because of the high oxide formation rate, especially using the direct injection high efficiency nebulizer (DIHEN) in ICP-SFMS.[55] These measurements of the $^{235}U^{16}O^+/^{238}U^{16}O^+$ isotope ratio can be helpful if interferences are observed on the atomic ions or it can be used to validate the analytical data of isotope ratios via atomic ions of uranium ($^{235}U^+/^{238}U^+$).

The oxide formation rate (MO^+/M^+) can be minimized by optimizing the experimental parameters. For example, the relative oxide ion intensities as a function of the rf power of ICP (at a constant carrier gas flow rate of 0.91 l min^{-1}) and carrier gas flow rate (at a constant rf power of 950 W) obtained using LA-ICP-MS on a concrete matrix are presented in Figure 9.66 a and b. The decreasing oxide ion formation with increasing rf power of the ICP (i.e., an increase of plasma temperature, kinetic energy of electrons and electron density) could be explained as an increased dissociation of oxide compound in the plasma. Similar behaviour has also been found in ICP-MS in the analysis of aqueous solutions. Increasing the carrier gas flow rate in LA-ICP-MS, which corresponds to increasing nebulizer gas flow rate in ICP-MS, results in an increasing oxide formation rate, because at a higher gas flow rate the residence time of the aerosol in the plasma region is lower and the oxides are not completely dissociated. Relative oxide ion intensities of selected actinides in high purity water are plotted as a function of dissociation energies of species in Figure 9.67. The observed correlation between relative oxide ion intensities and known bond dissociation energies allows estimates to be made of unknown bond dissociation energies (e.g., for AmO ~ 670 kJ mol^{-1}).

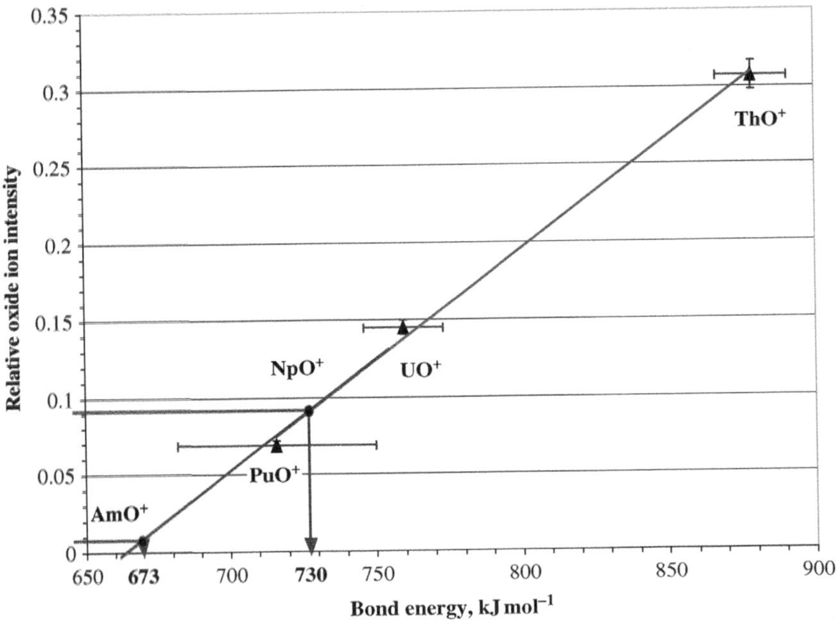

Figure 9.67 *Oxide formation rate of long-lived radionuclides (in concrete matrix) measured by LA-ICP-MS in dependence of bond energies. (J. S. Becker, C. Pickhardt and H. J. Dietze, Int. J. Mass Spectrom.,* ***203****, 283(2000). Reproduced by permission of Elsevier.)*

The formation of polyatomic and cluster ions as studied by mass spectrometry has been reviewed by several groups.[1,2,4,27,56,57] For example, studies of the chemistry of organometallic complexes consisting of metal atoms, metal clusters and metal surfaces with organic molecules have been carried out for a long time using conventional chemical synthesis procedures.[27] Laser vaporization techniques together with mass spectrometric analysis have been employed to study the reactivity of mass selected species and dissociation energies combined with a theoretical calculation of their electronic stability providing information on the atomic and electronic structure of clusters of relevance, for example, in catalysis and surface sciences.[58,59]

References XI

1. Castleman, A. W. and Jena, P., *Proc. Nat. Academy Sci. (PNAS)*, **103**, 10552 (2006).
2. Becker, J. S. and Dietze, H. J., *Fresenius' J. Anal. Chem.*, **359**, 338 (1997).
3. Becker, J. S. and Dietze, H. J., *J. Anal. At. Spectrom.*, **10**, 637 (1995).
4. Houk, R. S. and Praphairaksit, N., *Spectrochim. Acta*, **56B**, 1069 (2001).
5. Kaschner, R., Becker, J. S. and Seifert, G., *Int. J. Mass Spectr.*, **176**, 103 (1998).
6. Niu, H. and Houk, R. S., *Spectrochim. Acta B*, **1996**, 779 (1996).
7. Dietze, H. J., *Massenspektroskopische Spurenanalyse*, Akademischer Verlagsgesellschaft Geest & Portig K.-G., Leipzig, (1975).
8. Marcus, R. K. and Broekaert, J. A. C., *Glow Discharge Plasmas in Analytical Spectroscopy*. John Wiley & Sons Ltd., Chichester, West Sussex, England (2003).
9. Dietze, H. J. and Becker, J. S., *Fresenius Z. Anal. Chem.*, **321** 490 (1985).
10. Dietze, H. J. and Becker, J. S., in *Laser Ionization Mass Analysis*, A. Vertes, R. Gijbels and F. Adams (eds.) Chemical Analysis Series, John Wiley & Sons, Inc., New York, **124**, 453 (1993).
11. Becker, J. S., Seifert, G., Saprykin, A. and Dietze, H. J., *J. Anal. At. Spectrom.*, **11**, 643 (1996).
12. Leach, J. J., Allen, L. A., Aeschliman, D. B. and Houk, R. S., *Anal. Chem.*, **71**, 440 (1999).
13. Becker, J. S. and Dietze, H. J., *Int. J. Mass Spectr.*, **228**, 127 (2003).
14. Benninghoven, A., Ruedenauer, F. G. and Werner, W. H. (eds.), *Secondary Ion Mass Spectrometry*. John Wiley & Sons, Inc., New York (1987) Chapter I.
15. Brunelle, A., Touboul, D. and Laprévote, O., *J. Mass. Spectrom.*, **40**, 985 (2005).
16. Touboul, D., Kollmer, F., Niehus, E., Brunelle, A. and Laprévote, O., *J. Am. Soc. Mass Spectrom.*, **16**, 1608 (2005).
17. Marie, Y., Gao, Y. and Migneon, H. N., *Surface and Interface Analysis*, **23**, 38 (1995).
18. Gao, Y., Erickson, J. W. and Hockett, R. A., *SIMS X, (Proceedings of the 10th Int. Conf. Second. Ion Mass Spectrom., 1995, Münster)*, A. Benninghoven, B. Hagenhoff and H. W. Werner (eds.). John Wiley & Sons, Ltd, Chichester (1997) 339.
19. Gillen, G. and Fahey, A., *Appl. Surf. Sci.*, **203–204**, 209 (2003).
20. Gnaser, H., *J. Vac. Sci. Techn.*, **12**, 452 (1994).
21. Jones, E. A., Fletcher, J. S., Thompson, C. E., Jackson, D. A., Kockyer, N. P. and Vickerman, J. C., *Appl. Surf. Sci.*, **252**, 6844 (2006).
22. Benninghoven, A., Janssen, K. T. F., Tümpner, J. and Werner, W. H., *Secondary Ion Mass Spectrometry VIII (Proceedings of the Ninth International Conference on SIMS*, John Wiley & Sons, Ltd, Chichester (1992).
23. Betti, M., *Int. J. Mass Spectr.*, **242**, 169 (2005).
24. Märk, T. D., *Adv. Mass Spectrom.*, **13**, 71 (1995).
25. Märk, T. D. and Castleman, A. W., *Adv. At. Mol. Phys.*, **28**, 65 (1985).
26. Martin, T. P., *Angew. Chem.*, **98**, 197 (1986).
27. Jena, P. and Castleman, A. W., *PNAS*, **103**, 10560 (2006).
28. Becker, J. S., Pickhardt, C. and Pompe, W., *Int. J. Mass Spectrom.*, **237**, 13 (2004).
29. Hilpert, K. and Miller, M., *J. Electrochem. Soc.*, **141**, 2769 (1994).
30. Rohlfing, E. A., Cox, D. M. and Kaldor, A., *J. Chem. Phys.*, **81**, 3322 (1984).
31. Echt, O., Kreisle, D., Knapp, M. and Recknagel, E., *Chem. Phys. Lett.*, **91**, 413 (1982).

32. Sattler, K., Muehlbach, J. and Recknagel, E., *Phys Rev*, **45**, 821 (1980).
33. Betz, G., Kirchner, R., Nicolussi, A. and Husindky, W., *Secondary Ion Mass Spectrometry IX (Proceedings of the Ninth International Conference on SIMS, Yokohama,)*. John Wiley & Sons, Ltd, Chichester (1994) 57.
34. Knigth, W. D., Clemenger, K., de Heer, W. A., Saunders, W. A., Chou, M. Y. and Cohen, M. L., *Phys. Rev. Lett.*, **52**, 2141 (1984).
35. Kroto, H. W., Heath, J. R., O'Brien, S. C., Curl, R. F. and Smalley, R. G., *Nature*, **318**, 162 (1985).
36. Dietze, H. J., Becker, J. S., Opauszky, I., Matus, L., Nyary, I. and Frecska, J., *Mikrochim. Acta (Wien)*, **81** 263 (1983).
37. Jäger, R., Saprykin, A. I., Becker, J. S., Dietze, H. J. and Broekaert, J. A. C., *Mikrochim. Acta*, **125**, 41 (1997).
38. Franzen, J. and Hintenberger, H., *Z. Naturforsch.*, **16A**, 535 (1961).
39. Fürstenau, N., Hillenkamp, F. and Nitsche, R., *Int. J. Mass Spectrom. Ion Proc.*, **31**, 85 (1979).
40. Seifert, G., Becker, J. S. and Dietze, H. J., *Int. J. Mass Spectrom. Ion Proc.*, **184**, 121 (1988).
41. Becker, J. S. and Dietze, H. J., *Z. Chem.*, **26**, 833 (1986).
42. Becker, J. S. and Dietze, H. J., *Int. J. Mass Spectrom. Ion Proc.*, **73**, 157 (1986).
43. Becker, J. S., Dietze, H. J., Kessler, G., Bauer, H.-D. and Pompe, W., *Z. Phys. B – Condensed Matter*, **81**, 47 (1990).
44. Becker, J. S. and Dietze, H. J., *Int. J. Mass Spectrom.*, **73**, 157 (1986).
45. Becker, J. S. and Dietze, H. J., *Int. J. Mass Spectrom. Ion Proc.*, **67**, 57 (1985).
46. Becker, J. S. and Dietze, H. J., *Int. J. Mass Spectrom. Ion Proc.*, **82**, 287 (1988).
47. Becker, J. S. and Dietze, H. J., *Int. J. Mass Spectrom. Ion Proc.*, **82**, 47 (1988).
48. Becker, J. S. and Dietze, H. J., *Int. J. Mass Spectrom. Ion Proc.*, **82**, R1 (1986).
49. Martin, T. P., *Ber. Bunsenges. Phys. Chem.*, **88**, 300 (1984).
50. Gupta, S. K. and Gingerich, K. A., *J. Chem. Phys.*, **72**, 2795 (1980).
51. Davis, K. M., Peppernick, S. J. and Castleman, A. W., *J. Chem. Phys.*, **124**, 164304 (2006).
52. Michels, E. and Gijbels, R., *Anal. Chem.*, **56**, 1115 (1984).
53. Bauschlicher, C., Partridge, H. and Langhoff, S. R., *Chem. Phys. Lett.*, **165**, 272 (1990).
54. Barshick, C. M., Smith, D. H., Johnson, E., Kling, F. L., Bastug, T. and Fricke, B., *Appl. Spectrosc.*, **49**, 885 (1995).
55. Becker, J. S., Dietze, H. J., McLean, J. A. and Montaser, A., *Anal. Chem.*, **71**, 3077 (1999).
56. Armentrout, P. B., *Int. J. Mass Spectrom.*, **227**, 289 (2003).
57. O'Hair, R. A. J. and Khairallah, G. N., *J. Cluster Sci.*, **15**, 331 (2004).
58. Meyer, F., Khan, F. A. and Armentrout, P. B., *J. Am. Chem. Soc.*, **117**, 9740 (1995).
59. Willey, K. F., Cheng, P. Y., Bishop, M. B. and Duncan, M. A., *J. Am. Chem. Soc.*, **113**, 4721 (1991).

9.12 Further Applications

Inorganic mass spectrometric techniques (especially ICP-MS) are increasingly used in the pharmaceutical industry.[1] Atomic spectrometric techniques (such as AAS) are being replaced for product and process control and for providing 'impurity profiles' of inorganic impurities (e.g., heavy metals or residual noble metals from catalyzers) from starting materials, intermediate or final products in the pharmaceutical industry, by ICP-MS due to its high sensitivity and multi-element capability. A semiquantitative analysis with information about the concentration limits is often sufficient for this analytical task. A multitude of analytical biochemistry laboratories provide development services to the pharmaceutical and chemical industries and the analytical techniques developed in these labs are employed in routine measurements. In addition to several applications described in Section 9.5, only selected examples will be discussed in the following section.

9.12.1 Pharmaceutical Applications and Analysis of Drugs

Two main types of analysis are required: (a) qualitative determination of the presence of elements and (b) quantitative determination of the amount of elements or species of interest contained in pharmaceutical products. Most analyses for pharmaceutical applications involve separation steps combined with ICP-MS, such as HPLC-ICP-MS or gel electrophoresis and the analysis of gel blots by LA-ICP-MS. Phosphorylated proteins (e.g., β-casein) have been measured by LA-ICP-MS with a detection limit of 16 pmol. HPLC-ICP-MS has been employed for the identification and quantification of metabolites of bradykinin in human and rat plasma.[1]

In recent years, inorganic mass spectrometric techniques have been employed to an increasing extent for analysis of drugs. Krystek and Ritsema[2] described the application of HPLC-ICP-MS for the study of medicine containing organogermanium, which has been used for special treatments of, for instance, cancer and AIDS. Germanium is not essential for humans and the toxicity of the investigated organogermanium compounds (β-carboxyethylgermanium sesquioxide, spirogermanium or germanium-lactade-citrate or unspecific forms) is low. Acute and chronic toxic effects of inorganic GeO_2 have been demonstrated, so that a reliable analytical species analysis is indispensable. Sector field ICP-SFMS has been employed for multi-element screening (56 elements) in germanium containing drugs.[2]

The therapeutic utility of gallium compounds has been the subject of much research.[3] Gallium maltolate is an orally bioavailable form of gallium being developed for the treatment of cancer and bone disease.[4] Oral Ga utilizes the iron uptake pathway in the gastrointestinal tract for tissue distribution via the iron transport protein, transferrin. Treatment with an intravenous infusion of gallium nitrate for hypercalcemia of malignancy resulted in nephrotoxity due to the precipitation of unbound gallate in the blood. The serum protein binding behaviour of oral gallium maltolate compared to intravenous gallium nitrate has been examined by size exclusion chromatography (SEC) coupled to ICP-MS and Western blot analysis.[4]

ICP-MS has been employed, as discussed in Section 9.5, for the determination of platinum originating from cisplatin, carboplatin and oxaliplatin in human plasma ultrafiltrate. The method developed was successfully used to support pharmacokinetic studies in cancer patients treated with cisplatin, carboplatin or oxaliplatin.[5] Counterfeit products on the drug market, which have important implications for pharmaceutical companies and human health, can be clarified by mass spectrometric isotope ratio measurements. For example, precise and accurate sulfur isotope measurements ($\sigma^{34}S$) by MC-ICP-MS, were employed to study the isotope variation of pharmaceuticals and to detect to the origin of counterfeits by Clough *et al.*[6]

9.12.2 Archaeology

Inorganic mass spectrometry provides analytical techniques well suited for studying the provenance of archaeological objects. The materials science of man made and natural 'ancient' materials is generally more complex than twenty-first century materials. McPhail studied the conservation of glass artefacts from glass vessels by secondary ion mass spectrometry (SIMS). Small, micron size inclusions in ancient glass samples were characterized with respect to matrix composition in order to study the glass manufacturing technology and to explain the provenance of glass artefacts by revealing differences in matrix composition.[7]

In another study, ToF-SIMS was employed to study Hannibal's actual route over the Alps taken during the invasion of Italy in the Second Punic War, by examining the burnt outcrop from one of the possible passes (the Col du Clapier). ToF-SIMS investigations of a 100 μm thick burnt crust in the hornblende schist interlaced with veins of quartz-feldspatic minerals yielded various elements, C, Mg, Na, Ca, Si, Ti, P, Al and Fe, and hydrocarbon fragments as combustion products whereas Ca and Na along with Mg–Fe silicates were derived from the local rock. The presence of K has been attributed directly to combustion, resulting from the ash residue that would remain after burning timber.[8]

An examination of the potential for provenance studies of variation in mineralogical and chemical composition on very old ceramic assemblages (of ancient Japanese pottery and of Mesoamerican plumbate pottery surfaces) has been performed by LA-ICP-MS. Results suggested that the region's geological complexity may inhibit successful provenance studies of ceramics.[9–11]

About 30 elements were investigated by LA-ICP-MS for the characterization of antique glazed ceramics manufactured in the Aragon area during the fourteenth to eighteen centuries. These pottery samples were covered by a vitreous Pb enriched layer ($\sim 100\,\mu m$ in thickness), to which Co was added as a colouring pigment. It was found that the differences in the concentration levels of Cu, As and Mn allow antique pottery to be classified. In addition to the more established techniques routinely applied for archaeological analysis (SEM, XRF, NAA and ICP-MS/ICP-OES after digestion) an increasing use of LA-ICP-MS can be expected for the analysis of archaeological and art objects.[12]

This is mostly used for archaeological studies based on the geochronological system of the uranium/thorium–lead decay whereby the isotope variation of lead is investigated as a result of the radioactive decay of ^{238}U, ^{235}U and ^{232}Th into the stable lead isotopes ^{206}Pb, ^{207}Pb and ^{208}Pb, respectively[13] (see Section 9.7.5).

References XII

1. Marshall, P. S., Heudi, O., Ramirez-Molina, C. *et al.*, in *Plasma Source Mass Spectrometry*, G. Holland and S. D. Tanner (eds.). The Royal Society of Chemistry, Cambridge (2003) 28.
2. Krystek, P. and Ritsema, R., *J. Trace Elem. Medicine and Biology*, **18**, 9 (2004).
3. Bernstein, L. R., *Pharmacol. Rev.*, **50**, 665 (1998).
4. Gray, D. J., Burns, R., Brunswick, P. *et al.*, in *Plasma Source Mass Spectrometry: Current Trends and Future Developments*. The Royal Society of Chemistry, Cambridge (2005).
5. Brouwers, E. E. M., Tibben, M. M., Rosing, H. *et al.*, *J. Mass Spectrom.*, **41**, 1186 (2006).
6. Clough, R., Evans, P., Catterick, T. and Evans, H., *Anal. Chem.*, **78**, 6126 (2006).
7. McPhail, D. S., *Appl. Surf. Sci.*, **252**, 7107 (2006).
8. Sodhi, R. N. S., Mahaney, W. C. and Milner, M. W., *Appl. Surf. Sci.*, **252**, 7140 (2006).
9. Cochrane, E. E. and Neff, H., *J. Archaeol. Sci.*, **33**, 378 (2006).
10. Fitzpatrick, S. M., Takamiya, H., Neff, H. and Dickinson, W. R., *Geoarchaeology*, **21**, 803 (2006).
11. Neff, H., *J. Archaeol. Sci.*, **30**, 21 (2003).
12. Resano, M., Peréz-Arantegui, J., Garcia-Ruiz, E., Vanhaecke, F. and Wills, J., *J. Anal. At. Spectrom.*, **20**, 508 (2005).
13. Tunstall, S. and Amarasiriwardena, D., *Microchem. J.*, **73**, 335 (2002).

10

Future Developments and Trends in Inorganic Mass Spectrometry

At present, advanced mass spectrometric techniques have been successfully established among a multitude of quite different analytical methods as very powerful tools which are increasingly being employed for high tech research topics and for daily routine analyses in many laboratories worldwide. Numerous different applications in various fields of use, as demonstrated in the several chapters of this book, illustrate the excellent current capability of inorganic mass spectrometry in the multi-element determination of elements in a wide dynamic range (from % range for determination of stoichiometries, e.g., in layered materials, down to the extreme ultratrace level, e.g., in environmental research, speciation analysis and isotope ratio measurements).

What now are the future trends in inorganic mass spectrometry? If we look for an answer we should ask: what are the limits to inorganic mass spectrometry or where is further development required? The main drawback of inorganic mass spectrometry is the inherent interference problem, which has been solved for accelerator mass spectrometry (AMS) and resonance ionization mass spectrometry (RIMS). However, these techniques do not possess multi-element capability and have been employed for quite special applications in ultrasensitive isotope analysis (the most famous application is ^{14}C age dating by AMS). To improve mass spectrometry the collision/reaction cell technology in ICP-MS has been a breakthrough towards obtaining mass spectra less disturbed by isobaric interferences on analytes. This technique should be improved so that it can be employed universally in daily routine work. However, not all interference problems can be solved and new interferences may occur. With respect to the interference problem in mass spectrometry, an improvement of mass resolution (resolving power) by further instrumental development of ion trap or Fourier transform ion cyclotron resonance mass spectrometry in inorganic mass spectrometry with maximum mass resolution of up to 1 000 000, as demonstrated in organic mass spectrometry, would be very challenging for future instrumental developments. This technique would allow a truly interference free analysis. There are several interesting new approaches, for example, in Koppenaal's group (see Ref. 33 in Section 5.1.2) who interfaced an ICP ion source to a linear ion trap (LIT) mass spectrometer which will ultimately be developed into a custom ICP-LIT Orbitrap mass spectrometric system. This system includes a collision/reaction cell to be used in conjunction with the high resolution capability of the Orbitrap (Thermo Fisher Scientific).

In the future development of inorganic mass spectrometers, the following major general trends can be recognized. On the one hand, the improvement of mass spectrometric instrumentation and techniques focuses on providing higher instrumental sensitivity together with lower detection limits of elements, that means increasing the signal-to-noise ratio. On the other hand, the sample amount will be minimized further to avoid contamination and to reduce waste, and an increasing throughput of samples in the routine mode is also required. Many areas in science are dominated by developments towards miniaturization. The reduction in the amount of sample to be analyzed is not only important for characterizing radioactive samples, but also for microelectronic, biological and biomedical applications for studying, for instance, single cells and looking inside one single cell by mass spectrometry.

Activities for miniaturizing mass spectrometers (e.g., microplasma on chip or insertion of diode lasers in RIMS), for constructing cheaper and more compact instrumentation with the same performance or improved properties compared to existing instruments are required as the next generation mass spectrometers. The introduction of microwave induced plasmas or of μ-torches to reduce Ar gas consumption involves developments in this future direction.

In order to reduce the instrumental background in mass spectrometers, new materials are required or an improvement to the vacuum system. New ion detectors with very low noise should be developed for the measurement of very low isotope ratios. Multi-stage MS-MS instruments could be applied to improve abundance sensitivity.

In addition, a maximum degree of automation of user friendly and easy-to-handle mass spectrometric equipment is necessary, especially in routine laboratory work in industry, but also in research laboratories for daily routine tasks. Besides instrumental developments, there is a need for improvements to the software for steering the apparatus, for optimization and control of experimental parameters and evaluation of measured data, including the graphic presentation of mass spectrometric results (such as 3D profiles for imaging mass spectrometry). For practical applications of commercial mass spectrometers, improved diagnostic tools for maintenance of the complete instrumentation during operation are required to reduce the time for troubleshooting should any technical fault occur.

With respect to sample preparation, it is necessary to develop effective and fast procedures involving only a few steps in order to avoid contamination, reduce analysis time and to improve the quality of analytical work. Microsampling and the use of smaller sample sizes is required and also the further development of analytical techniques. In particular, there is a need for the development of online and/or hyphenated techniques in ICP-MS. Microsampling combined with the separation of small amounts of analytes will be relevant for several chromatographic techniques (such as the development of micro- and nano-HPLC). There is a demand for further development of the combination of LA-ICP-MS as an element analytical technique with a biomolecular mass spectrometric technique such as MALDI- or ESI-MS for molecular identification and quantification of protein phosphorylation as well as of metal concentrations, this also enables the study of post-translational modifications of proteins, e.g. phosphorylation.

Inorganic mass spectrometry requires the development of suitable reference materials, such as matrix matched standard reference materials for trace, surface (including depth profiling and microlocal) analysis and/or the creation of matrix independent calibration procedures. The development of species specific standards will be intensified for speciation studies in the future.

Furthermore, isotope standard reference materials are required in different chemical forms for a multitude of elements in order to calibrate and evaluate isotope ratio measurements by mass spectrometry. In addition, reliable analytical data required in science and routine work will only be obtained by improving the quality assurance of the measurement procedures and also using interlaboratory comparisons of analytical results in round robin tests.

The significance of inorganic mass spectrometry in precise isotope ratio measurements will be further increased for sector field mass spectrometers with multiple ion collector systems and also for mass spectrometers with single ion detection or quadrupole based instruments with and without a collision cell.

The development of surface analytical techniques such as LA-ICP-MS, GDMS and SIMS focuses on improvements to sensitivity and detection limits in order to obtain precise and accurate analytical data. With respect to surface analytical investigations, an improvement of spatial and depth resolution is required, e.g., by the establishment of a near field effect or the application of fs lasers in LA-ICP-MS. There is a need for the improvement of analytical techniques in the μm and nm range, in depth profiling analysis and especially in imaging mass spectrometry techniques to perform surface analyses faster and provide more accurate data on different materials to produce quantitative 3D elemental, isotopic and molecular distribution patterns of increased areas of interest with high spatial and depth resolution over an acceptable analysis time.

Appendix I

Table of Isotopic Abundances,[1] Atomic Masses[2] and Ionization Energies[3] of Elements

Atomic number, Z	Element	Symbol	Mass number, A	Isotopic abundance, χ(%)	Atomic mass, m_a (u)	Ionization energy E_i (eV)	
						First E_i	Second E_i
1	Hydrogen	H	1	99.9885	1.007 825	13.598	
	Deuterium	D	2	0.0115	2.014 102		
2	Helium	He	3	0.000134	3.016 029	24.587	54.416
			4	99.999866	4.002 603		
3	Lithium	Li	6	7.59	6.015 122	5.392	75.638
			7	92.41	7.016 004		
4	Beryllium	Be	9	100	9.012 182	9.322	18.211
5	Boron	B	10	19.9	10.012 937	8.298	25.154
			11	80.1	11.009 305		
6	Carbon	C	12	98.93	12.0 (by definition)	11.260	24.383
			13	1.07	13.003 355		
7	Nitrogen	N	14	99.636	14.003 074	14.534	29.601
			15	0.364	15.000 109		
8	Oxygen	O	16	99.757	15.994 915	13.618	35.116
			17	0.038	16.999 132		
			18	0.205	17.999 160		
9	Fluorine	F	19	100	18.998 403	17.422	34.970
10	Neon	Ne	20	90.48	19.992 440	21.564	40.962
			21	0.27	20.993 847		
			22	9.25	21.991 386		
11	Sodium	Na	23	100	22.989 770	5.139	47.286

Inorganic Mass Spectrometry: Principles and Applications J. S. Becker

(Continued)

Atomic number, Z	Element	Symbol	Mass number, A	Isotopic abundance, χ(%)	Atomic mass, m_a (u)	Ionization energy E_i (eV)	
						First E_i	Second E_i
12	Magnesium	Mg	24	78.99	23.985 042	7.646	15.035
			25	10.00	24.985 837		
			26	11.01	25.982 593		
13	Aluminum	Al	27	100	26.981 538	5.986	18.828
14	Silicon	Si	28	92.223	27.976 926	8.151	16.345
			29	4.685	28.976 495		
			30	3.092	29.973 770		
15	Phosphorus	P	31	100	30.973 761	10.486	19.725
16	Sulfur	S	32	94.99	31.972 071	10.360	23.33
			33	0.75	32.971 459		
			34	4.25	33.967 867		
			36	0.01	35.967 081		
17	Chlorine	Cl	35	75.76	34.968 853	12.967	23.81
			37	24.24	36.965 903		
18	Argon	Ar	36	0.3365	35.967 546	15.759	27.629
			38	0.0632	37.962 732		
			40	99.6003	39.962 383		
19	Potassium	K	39	93.2581	38.963 707	4.341	31.625
			40	0.0117	39.963 999		
			41	6.7302	40.961 826		
20	Calcium	Ca	40	96.941	39.962 591	6.113	11.871
			42	0.647	41.958 618		
			43	0.135	42.958 767		
			44	2.086	43.955 481		
			46	0.004	45.953 693		
			48	0.187	47.952 533		
21	Scandium	Sc	45	100	44.955 910	6.54	12.80
22	Titanium	Ti	46	8.25	45.952 630	6.82	13.58
			47	7.44	46.951 764		
			48	73.72	47.947 947		
			49	5.401	48.947 871		
			50	5.185	49.944 792		
23	Vanadium	V	50	0.250	49.947 163	6.74	14.65
			51	99.750	50.943 964		
24	Chromium	Cr	50	4.345	49.946 050	6.766	16.60
			52	83.789	51.940 512		
			53	9.501	52.940 653		
			54	2.365	53.938 885		
25	Manganese	Mn	55	100	54.938 049	7.435	15.640
26	Iron	Fe	54	5.845	53.939 615	7.870	16.18
			56	91.754	55.934 942		
			57	2.119	56.935 398		
			58	0.282	57.933 280		
27	Cobalt	Co	59	100	58.933 200	7.86	17.06
28	Nickel	Ni	58	68.0769	57.935 348	7.653	18.168
			60	26.2231	59.930 790		

			61	1.1399	60.931 060		
			62	3.6345	61.928 348		
			64	0.9256	63.927 969		
29	Copper	Cu	63	69.15	62.929 601	7.726	20.292
			65	30.85	64.927 794		
30	Zinc	Zn	64	48.268	63.929 146	9.394	17.964
			66	27.975	65.926 036		
			67	4.102	66.927 130		
			68	19.024	67.924 847		
			70	0.631	69.925 325		
31	Gallium	Ga	69	60.108	68.925 581	5.999	20.51
			71	39.892	70.924 707		
32	Germanium	Ge	70	20.38	69.924 250	7.899	15.934
			72	27.31	71.922 076		
			73	7.76	72.923 460		
			74	36.72	73.921 178		
			76	7.83	75.921 403		
33	Arsenic	As	75	100	74.921 597	9.81	18.633
34	Selenium	Se	74	0.89	73.922 477	9.752	21.19
			76	9.37	75.919 214		
			77	7.63	76.919 915		
			78	23.77	77.917 310		
			80	49.61	79.916 522		
			82	8.73	81.916 700		
35	Bromine	Br	79	50.69	78.918 338	11.814	21.8
			81	49.31	80.916 291		
36	Krypton	Kr	78	0.355	77.920 388	13.999	24.359
			80	2.286	79.916 379		
			82	11.593	81.913 485		
			83	11.500	82.914 137		
			84	56.987	83.911 508		
			86	17.279	85.910 615		
37	Rubidium	Rb	85	72.17	84.911 792	4.177	27.28
			87	27.83	86.909 186		
38	Strontium	Sr	84	0.56	83.913 426	5.695	11.030
			86	9.86	85.909 265		
			87	7.00	86.908 882		
			88	82.58	87.905 617		
39	Yttrium	Y	89	100	88.905 848	6.38	12.24
40	Zirconium	Zr	90	51.45	89.904 702	6.84	13.13
			91	11.22	90.905 643		
			92	17.15	91.905 039		
			94	17.38	93.906 314		
			96	2.80	95.908 275		
41	Niobium	Nb	93	100	92.906 376	6.88	14.32
42	Molybdenum	Mo	92	14.77	91.906 810	7.099	16.15
			94	9.23	93.905 087		
			95	15.90	94.905 841		
			96	16.68	95.904 678		
			97	9.56	96.906 020		
			98	24.19	97.905 407		
			100	9.67	99.907 476		
43	Technetium	Tc				7.28	15.26

(Continued)

Atomic number, Z	Element	Symbol	Mass number, A	Isotopic abundance, χ(%)	Atomic mass, m_a (u)	Ionization energy E_i (eV)	
						First E_i	Second E_i
44	Ruthenium	Ru	96	5.54	95.907 604	7.37	16.76
			98	1.87	97.905 287		
			99	12.76	98.905 938		
			100	12.60	99.904 219		
			101	17.06	100.905 582		
			102	31.55	101.904 349		
			104	18.62	103.905 430		
45	Rhodium	Rh	103	100	102.905 504	7.46	18.08
46	Palladium	Pd	102	1.02	101.905 607	8.38	19.43
			104	11.14	103.904 034		
			105	22.33	104.905 083		
			106	27.33	105.903 484		
			108	26.46	107.903 895		
			110	11.72	109.905 153		
47	Silver	Ag	107	51.839	106.905 093	7.576	21.49
			109	48.161	108.904 756		
48	Cadmium	Cd	106	1.25	105.906 458	8.993	16.908
			108	0.89	107.904 183		
			110	12.49	109.903 006		
			111	12.80	110.904 158		
			112	24.13	111.902 761		
			113	12.22	112.904 401		
			114	28.73	113.903 359		
			116	7.49	115.904 756		
49	Indium	In	113	4.29	112.904 062	5.786	18.869
			115	95.71	114.903 879		
50	Tin	Sn	112	0.97	111.904 822	7.344	14.632
			114	0.66	113.902 783		
			115	0.34	114.903 347		
			116	14.54	115.901 745		
			117	7.68	116.902 955		
			118	24.22	117.901 608		
			119	8.59	118.903 311		
			120	32.58	119.902 198		
			122	4.63	121.903 441		
			124	5.79	123.905 274		
51	Antimony	Sb	121	57.21	120.903 822	8.641	16.53
			123	42.79	122.904 216		
52	Tellurium	Te	120	0.09	119.904 026	9.009	18.6
			122	2.55	121.903 056		
			123	0.89	122.904 271		
			124	4.74	123.902 819		
			125	7.07	124.904 424		
			126	18.84	125.903 305		
			128	31.74	127.904 462		
			130	34.08	129.906 223		
53	Iodine	I	127	100	126.904 468	10.451	19.131
54	Xenon	Xe	124	0.0952	123.905895	12.130	21.21
			126	0.0890	125.904 268		

			128	1.9102	127.903 531		
			129	26.4006	128.904 780		
			130	4.0710	129.903 509		
			131	21.2324	130.905 083		
			132	26.9086	131.904 155		
			134	10.4357	133.905 395		
			136	8.8573	135.905 220		
55	Cesium	Cs	133	100	132.905 447	3.894	25.1
56	Barium	Ba	130	0.106	129.906 311	5.212	10.004
			132	0.101	131.905 056		
			134	2.417	133.904 504		
			135	6.592	134.905 684		
			136	7.854	135.904 571		
			137	11.232	136.905 822		
			138	71.698	137.905 242		
57	Lanthanum	La	138	0.090	137.907 108	5.577	11.06
			139	99.910	138.906 349		
58	Cerium	Ce	136	0.185	135.907 140	5.47	10.85
			138	0.251	137.905 986		
			140	88.450	139.905 435		
			142	11.114	141.909 241		
59	Praseodymium	Pr	141	100	140.907 648	5.42	10.55
60	Neodymium	Nd	142	27.2	141.907 719	5.49	10.72
			143	12.2	142.909 810		
			144	23.8	143.910 083		
			145	8.3	144.912 596		
			146	17.2	145.913 113		
			148	5.7	147.916 889		
			150	5.6	149.920 887		
61	Promethium	Pm				5.55	10.90
62	Samarium	Sm	144	3.07	143.911 996	5.63	11.07
			147	14.99	146.914 894		
			148	11.24	147.914 818		
			149	13.82	148.917 180		
			150	7.38	149.917 272		
			152	26.75	151.919 729		
			154	22.75	153.922 206		
63	Europium	Eu	151	47.81	150.919 846	5.67	11.25
			153	52.19	152.921 227		
64	Gadolinium	Gd	152	0.20	151.919 789	6.14	12.1
			154	2.18	153.920 862		
			155	14.80	154.922 619		
			156	20.47	155.922 120		
			157	15.65	156.923 957		
			158	24.84	157.924 101		
			160	21.86	159.927 051		
65	Terbium	Tb	159	100	158.925 343	5.85	11.52
66	Dysprosium	Dy	156	0.056	155.924 278	5.93	11.67
			158	0.095	157.924 405		
			160	2.329	159.925 194		
			161	18.889	160.926 930		
			162	25.475	161.926 795		
			163	24.896	162.928 728		
			164	28.260	163.929 171		

(Continued)

Atomic number, Z	Element	Symbol	Mass number, A	Isotopic abundance, χ(%)	Atomic mass, m_a (u)	Ionization energy E_i (eV)	
						First E_i	Second E_i
67	Holmium	Ho	165	100	164.930 319	6.02	11.80
68	Erbium	Er	162	0.139	161.928 775	6.10	11.93
			164	1.601	163.929 197		
			166	33.503	165.930 290		
			167	22.869	166.932 046		
			168	26.978	167.932 368		
			170	14.910	169.935 461		
69	Thulium	Tm	169	100	168.934 211	6.18	12.05
70	Ytterbium	Yb	168	0.13	167.933 895	6.254	12.17
			170	3.04	169.934 759		
			171	14.28	170.936 323		
			172	21.83	171.936 378		
			173	16.13	172.938 207		
			174	31.83	173.938 858		
			176	12.76	175.942 569		
71	Lutetium	Lu	175	97.41	174.940 768	5.426	13.9
			176	2.59	175.942 683		
72	Hafnium	Hf	174	0.16	173.940 042	7.0	14.9
			176	5.26	175.941 403		
			177	18.60	176.943 220		
			178	27.28	177.943 698		
			179	13.62	178.945 815		
			180	35.08	179.946 549		
73	Tantalum	Ta	180	0.012	179.947 466	7.89	
			181	99.988	180.947966		
74	Tungsten	W	180	0.12	179.946 706	7.98	
			182	26.50	181.948 205		
			183	14.31	182.950 224		
			184	30.64	183.950 932		
			186	28.43	185.954 362		
75	Rhenium	Re	185	37.40	184.952 955	7.88	
			187	62.60	186.955 750		
76	Osmium	Os	184	0.02	183.952 491	8.7	
			186	1.59	185.953 838		
			187	1.96	186.955 748		
			188	13.24	187.955 836		
			189	16.15	188.958 145		
			190	26.26	189.958 445		
			192	40.78	191.961 479		
77	Iridium	Ir	191	37.30	190.960 591	9.1	
			193	62.70	192.962 923		
78	Platinum	Pt	190	0.014	189.959 930	9.0	18.563
			192	0.782	191.961 035		
			194	32.967	193.962 663		
			195	33.832	194.964 774		
			196	25.242	195.964 934		
			198	7.163	197.967 875		

79	Gold	Au	197	100	196.966 551	9.225	20.5
80	Mercury	Hg	196	0.150	195.965 814	10.437	18.756
			198	9.97	197.966 752		
			199	16.87	198.968 262		
			200	23.10	199.968 309		
			201	13.18	200.970 285		
			202	29.86	201.970 625		
			204	6.87	203.973 475		
81	Thallium	Tl	203	29.52	202.972 329	6.108	20.428
			205	70.48	204.974 412		
82	Lead	Pb	204	1.4	203.973 028	7.416	15.028
			206	24.1	205.974 449		
			207	22.1	206.975 880		
			208	52.4	207.976 636		
83	Bismuth	Bi	209	100	208.980 384	7.289	16.69
84	Polonium	Po				8.42	
86	Radon	Rn				10.748	
88	Radium	Ra				5.279	10.147
89	Actinium	Ac				6.9	12.1
90	Thorium	Th	232	100	232.038 050	6.95	11.2
91		Pa		100			
92	Uranium	U	234	0.0054	234.040 945	6.08	
			235	0.7204	235.043 922		
			238	99.2742	238.050 784		

[1] J. K. Böhlke, J. R. De Laeter, P. DeBievre, H. Hidaka, H. S. Peiser, K. J. R. Rosman, and P. D. P. Taylor, "Isotopic Compositions of the Elements 2001", *J. Phys. Chem. Ref. Data.*, **34**, 57 (2005).

[2] J. R. De Laeter, J. K. Böhlke, P. DeBievre, H. Hidaka, H. S. Peiser, K. J. R. Rosman, P. D. P. Taylor, *Pure Appl. Chem.* **75**, 683 (2003).

[3] The ionization energies are from the *Handbook of Chemistry and Physics*, 70th edn., 1989–90, CRC Press Inc., Boca Raton, FL, Tables E 80–81. Columns [1] and [2] are the first and the second ionization potentials respectively.

Appendix II

Table of Atomic Weights of Elements

Atomic Number	Element	Symbol	Atomic weights, $A_r(E)$	Notes
1	Hydrogen	H	1.007 94(7)	1, 2, 3
2	Helium	He	4.002 602(2)	1, 2
3	Lithium	Li	[6.941 (2)]	1, 2, 3, 4
4	Beryllium	Be	9.012 182(3)	
5	Boron	B	10.811 (7)	1, 2, 3
6	Carbon	C	12.010 7(8)	1, 2
7	Nitrogen	N	14.006 7(2)	1, 2
8	Oxygen	O	15.999 4(3)	1, 2
9	Fluorine	F	18.998 403 2(5)	
10	Neon	Ne	20.179 7(6)	1, 3
11	Sodium	Na	22.989 769 28(2)	
12	Magnesium	Mg	24.305 0(6)	
13	Aluminium	Al	26.981 538 6(8)	
14	Silicon	Si	28.085 5(3)	2
15	Phosphorus	P	30.973 762 (2)	
16	Sulfur	S	32.065 (5)	1, 2
17	Chlorine	Cl	35.453 (2)	3
18	Argon	Ar	39.948 (1)	1, 2
19	Potassium	K	39.098 3(1)	1
20	Calcium	Ca	40.078 (4)	1
21	Scandium	Sc	44.955 912 (6)	
22	Titanium	Ti	47.867 (1)	
23	Vanadium	V	50.941 5(1)	
24	Chromium	Cr	51.996 1(6)	
25	Manganese	Mn	54.938 045 (5)	
26	Iron	Fe	55.845 (2)	
27	Cobalt	Co	58.933 195 (5)	
28	Nickel	Ni	58.693 4(2)	

29	Copper	Cu	63.546(3)	2
30	Zinc	Zn	65.409(4)	
31	Gallium	Ga	69.723(1)	
32	Germanium	Ge	72.64(1)	
33	Arsenic	As	74.921 60(2)	
34	Selenium	Se	78.96(3)	
35	Bromine	Br	79.904(1)	
36	Krypton	Kr	83.798(2)	1, 3
37	Rubidium	Rb	85.467 8(3)	1
38	Strontium	Sr	87.62(1)	1, 2
39	Yttrium	Y	88.905 85(2)	
40	Zirconium	Zr	91.224(2)	1
41	Niobium	Nb	92.906 38(2)	
42	Molybdenum	Mo	95.94(2)	1
43	Technetium	Tc	[98]	5
44	Ruthenium	Ru	101.07(2)	1
45	Rhodium	Rh	102.905 50(2)	
46	Palladium	Pd	106.42(1)	1
47	Silver	Ag	107.868 2(2)	1
48	Cadmium	Cd	112.411(8)	1
49	Indium	In	114.818(3)	
50	Tin	Sn	118.710(7)	1
51	Antimony	Sb	121.760(1)	1
52	Tellurium	Te	127.60(3)	1
53	Iodine	I	126.904 47(3)	
54	Xenon	Xe	131.293(6)	1, 3
55	Caesium	Cs	132.905 451 9(2)	
56	Barium	Ba	137.327(7)	
57	Lanthanum	La	138.905 47(7)	1
58	Cerium	Ce	140.116(1)	1
59	Praseodymium	Pr	140.907 65(2)	
60	Neodymium	Nd	144.242(3)	1
61	Promethium	Pm	[145]	5
62	Samarium	Sm	150.36(2)	1
63	Europium	Eu	151.964(1)	1
64	Gadolinium	Gd	157.25(3)	1
65	Terbium	Tb	158.925 35(2)	
66	Dysprosium	Dy	162.500(1)	1
67	Holmium	Ho	164.930 32(2)	
68	Erbium	Er	167.259(3)	1
69	Thulium	Tm	168.934 21(2)	
70	Ytterbium	Yb	173.04(3)	1
71	Lutetium	Lu	174.967(1)	1
72	Hafnium	Hf	178.49(2)	
73	Tantalum	Ta	180.947 88(2)	
74	Tungsten	W	183.84(1)	
75	Rhenium	Re	186.207(1)	
76	Osmium	Os	190.23(3)	1
77	Iridium	Ir	192.217(3)	
78	Platinum	Pt	195.084(9)	
79	Gold	Au	196.966 569(4)	
80	Mercury	Hg	200.59(2)	
81	Thallium	Tl	204.383 3(2)	
82	Lead	Pb	207.2(1)	1, 2
83	Bismuth	Bi	208.980 40(1)	
84	Polonium*	Po	[209]	5
85	Astatine*	At	[210]	5
86	Radon*	Rn	[222]	5

(Continued)

Atomic Number	Element	Symbol	Atomic weights, $A_r(E)$	Notes
87	Francium*	Fr	[223]	5
88	Radium*	Ra	[226]	5
89	Actinium*	Ac	[227]	5
90	Thorium*	Th	232.038 06(2)	1, 5
91	Protactinium*	Pa	231.035 88(2)	5
92	Uranium*	U	238.028 91(3)	1, 3, 5
93	Neptunium*	Np	[237]	5
94	Plutonium*	Pu	[244]	5
95	Americium*	Am	[243]	5
96	Curium*	Cm	[247]	5
97	Berkelium*	Bk	[247]	5
98	Californium*	Cf	[251]	5
99	Einsteinium*	Es	[252]	5
100	Fermium*	Fm	[257]	5
101	Mendelevium*	Md	[258]	5
102	Nobelium*	No	[259]	5
103	Lawrencium*	Lr	[262]	5
104	Rutherfordium*	Rf	[261]	5, 6
105	Dubnium*	Db	[262]	5, 6
106	Seaborgium*	Sg	[266]	5, 6
107	Bohrium*	Bh	[264]	5, 6
108	Hassium*	Hs	[277]	5, 6
109	Meitnerium*	Mt	[268]	5, 6
110	Darmstadtium*	Ds	[281]	5, 6
111	Roentgenium*	Rg	[272]	5, 6
112	Ununbium*	Uub	[285]	5, 6
113	Ununtrium*	Uut	[284]	5, 7
114	Ununquadium*	Uuq	[289]	5, 6
115	Ununpentium*	Uup	[288]	5, 7
116	Ununhexium*	Uuh	[291]	5, 8, see Note above
118	Ununoctium*	Uuo	[294]	5, 8, see Note above

1 Geological specimens are known in which the element has an isotopic composition outside the limits for normal material. The difference between the atomic weight of the element in such specimens and that given in the Table may exceed the stated uncertainty.

2 Range in isotopic composition of normal terrestrial material prevents a more precise value being given; the tabulated value should be applicable to any normal material.

3 Modified isotopic compositions may be found in commercially available material because it has been subject to an undisclosed or inadvertant isotopic fractionation. Substantial deviations in atomic weight of the element from that given in the table can occur.

4 Commercially available Li materials have atomic weights that range between 6.939 and 6.996; if a more accurate value is required, it must be determined for the specific material [range quoted for 1995 Tables 6.94 and 6.99].

5 Element has no stable nuclides. The value enclosed in brackets, e.g. [209], indicates the mass number of the longest lived isotope of the element. However, three such elements (Th, Pa, and U) do have a characteristic terrestrial isotopic composition, and for these an atomic weight is tabulated.

6 The names and symbols for elements 112–118 are under review. The temporary system recommended by J. Chatt, *Pure Appl. Chem.*, **51**, 381–384 (1979) is used above. The names of elements 101–109 were agreed in 1997 (See *Pure Appl. Chem.*, **69**, 2471–2473 (1997)), for element 110 in 2003 (see *Pure Appl. Chem.*, **75**, 1613–1615 (2003)) and for element 111 in 2004 (see *Pure Appl. Chem.*, **76**, 2101–2103 (2004)).

7 A report in *Phys. Rev. C*, 69, 021601 (R) (2004) claims the detection of elements 113 and 115.

8 A report in *Pure Appl. Chem.*, **78**, 889–904 (2006) claims the detection of elements 116 and 118.

* Element has no stable nuclides.

Source: M. Wieser, IUPAC Technical Report, *Pure Appl. Chem.*, **78**, 2051 (2006).

Appendix III

Definitions

Abundance sensitivity	Ratio of the maximum ion current recorded at a mass ***m*** to the ion current arising from the same species recorded at an adjacent mass ($\boldsymbol{m \pm 1}$).
Acceleration voltage	Electrical potential used to impact translational energy to ions in a mass spectrometer.[1]
Accuracy	Closeness of agreement between measured quantity values and a true quantity of the measurand.[2]
Atomic mass, m (u)	Mass of an individual isotope (or radionuclide).
Atomic number (Z)	of element = number of protons; determines the place of element in the periodical table.
Atomic weight (A_r(E))	of an element is the average of atomic masses of all naturally occurring isotopes of the chemical element as found in a particular environment (list of A_r(E), see Appendix II).
Bias	Difference between the expected value of the measurement and the true (reference) value.[2]
Double-focusing mass spectrometer	Combination of magnetic sector field, B and electric sector fields, E which uses both direction and velocity focusing of ions. The ion beam of a given mass/charge ratio is brought to a focus on imaging curve when the ion beam with ions of different kinetic energies is initially diverging.
Faraday cup	(or cylinder collector) is a hollow collector, open at end and closed at the other, used to collect beams of ions.

Inductively coupled plasma ionization	Ionization of sample (and plasma components) with high efficiency in a plasma (e.g., from argon) at atmospheric pressure and plasma gas temperature of about 5000–8000 K. The inductively coupled plasma is formed in a stream of plasma gas (Ar) flowing through the plasma torch encircled at the top by an induction coil mostly from copper (load coil) coupled to an rf generator. The rf energy is provide to the electrons. The plasma is generated if the Ar gas is seeded with energetic electrons produced by a Tesla discharge. The main ionization processes of analyte in an inductively plasma are electron ionization, charge transfer ionization and Penning ionization.
Isotope abundance (χ)	refers to the abundance of isotope in nature (see Appendix I).
Ion collector	is a device for detection (collection) of separated ion beams such as Faraday cup (with amplifier) or an electron multiplier.
Ion source	is the part of the mass spectrometer that ionizes the material under analysis. In inorganic mass spectrometry inductively coupled plasma, glow discharge, thermal surface, laser, spark and sputtered ion sources are applied.
Isotope	is defined by its atomic number (Z) and mass (m). Isotopes of a chemical element possess the same atomic number (number of protons), but the different number of neutrons and hence a different mass. The word isotope (meaning at the same place) indicates that all isotopes of an element are located at the same place on the periodical table.
Isotope dilution mass spectrometry	A quantitative mass spectrometry technique in which an isotopically enriched compound is used as an internal standard.[1]
Laser ablation	is a process of removing material from a sample surface by the interaction of photons of the laser beam with the atoms and compounds of the sample surface. The ablated sample is transported with a carrier gas (e.g., Ar) into the inductively coupled plasma of an ICP-MS.
Laser plasma ionization	produced ions by the interaction of photons of the laser beam with atoms and compounds of sample surface (in a vacuum ion source). The generated ions were utilized for analysis.
Limit of detection (LOD)	Lowest amount of the analyte in the sample which can be detected but not necessarily quantified as an exact value. LOD is expressed as a concentration or a quantity derived from the smallest signal $x_{L,}$ which can be detected with a reasonable certainty for an analytical procedure. The LOD can be determined using the equation: $x_L = x_{bl} + ks_{bl}$ with x_{bl} and s_{bl} as the mean value and the standard deviation of blank, respectively. LOD of an element or element species is often used as $3s_{bl}$ (3σ) of blank.[3]

Limit of determination	(or limit of quantification) is not officially defined. The limit of determination (often used $6s_{bl}$ of blank) is the lowest amount of an analyte in the sample which can be determined as an exact value using an analytical procedure.[3]
Magnetic deflection	Deflection of an ion beam as a result of the motion of the ions in a magnetic sector field (B). Generally the direction of motion of the ions is at right angles to the direction of the magnetic field, and the motion is uniform.[4]
Mass analyzer	separates the ion beams according to their mass per charge (*m/z*). There are many types of mass analyzers which are categorized based on the principles of operation (e.g., single and double focusing sector field, quadrupole based, time of flight, Fourier transform ion cyclotron resonance mass analyzers).
Mass defect	of a nucleus is equivalent to the binding energy of the nucleons in the nucleus and corresponds to: $\Delta m = \Delta E/c^2$.
Mass range	Range of mass numbers which can be characterized by a mass spectrometer with sufficient mass resolution to differentiate adjacent peaks.
Mass resolution ($m/\Delta m$) (resolving power)	is commonly acceptable defined in terms of the overlap (or 'valley') between two peaks with the mass m_1 and m_2. (10% of peak high is recommended when there is an overlap between two peaks.)
Mass spectrograph	Instrument that separates a beam of ions according to their m/z values in which the ions are directed onto a focal plane detector such as a photographic plate.[1]
Mass spectrometer	instrument that measures the m/z values and relative abundances of ions.
Mass spectrometry	Branch of science dealing with all aspects of mass spectrometers and the results obtained with these instruments.
Mass spectrum	is a 2D graph the relative abundances of ions as a function of mass of charge (m/z) ratio.
Mass-to-charge ratio (*m/z*)	is dimensionless by definition; it is formed by dividing the mass of an ion by its charge number.
Mass unit (m_u)	is defined as 1/12 of the mass of a ^{12}C atom $m_u = 1\,u = 1/12$ of mass of ^{12}C (inorganic mass spectrometry) or $m_u = 1\,u = 1\,Da$ (dalton) $= 1.6605402 \times 10^{-27}$ kg.
Matrix-assisted laser desorption/ionization (MALDI)	Formation of gas-phase ions from molecules that are present in a solid or liquid matrix that is irradiated by a laser beam.[1]
Mattauch-Herzog mass spectrometer	Double-focusing mass spectrometer (EB) in which a deflection in a 31.8° electric sector is followed by a deflection in a 90° magnetic sector field.
Microchannel plate	A thin plate that contains a closely spaced array of channels that each act as a continuous dynode particle multiplier. A charged particle, fast neutral particle, or photon striking the plate causes

	a cascade of secondary electrons that ultimately exits the opposite side of the plate.[1]
Molar mass	Mass of one mole (6.022 1415(10) x 10^{23} atoms or molecules) of a compound.[1]
Monoisotopic element	is an element considered by CIAAW (Commission on Isotopic Abundances and Atomic Weights of the IUPAC)[3] if it has only one stable isotope or a half-life greater than 1×10^{10} a, e.g., Th is considered to be monoisotopic, because ^{232}Th has a half-life of 1.4×10^{10} a.
Precision	Closeness of agreement between the independent results by replicate measurements carried out under defined specific conditions.
Nier-Johnson mass spectrometer	Double-focusing mass spectrometer with EB configuration in which a deflection of 90° electrostatic analyser (E) is followed by a 90° magnetic analyser (B) (or reverse geometry – BE).
Quadrupole mass spectrometer	Mass spectrometer that consists of four parallel rods whose opposing poles are connected. The voltage applied to the rods is a superposition of a static potential and a sinusoidal radio frequency potential. The motion of an ion in the x and y dimensions is described by the Mathieu equation whose solutions show that ions in a particular m/z range can be transmitted along the z-axis.[1]
Reference material	Material sufficiently homogeneous and stable regarding one or more properties, used in calibration, in assignment of a value to another material, or in quality assurance.[5]
Reproducibility	Closeness of agreement between the results of the measurements of the same measurand carried out under changed conditions of measurements.[2]
Secondary electron multiplier (SEM)	A device to multiply ion current in an electron beam (by first conversion to electrons). These electrons are then accelerated to another electrode, which in turn emits further secondary electrons so that the process can be repeated.[4]
Secondary ionization	Process in which secondary ions are ejected from a sample surface during the bombardment by a primary beam of atoms or ions.[1]
Sensitivity of mass spectrometer	is, for an analyte investigated, the measured ion intensity in cps (or peak height) relative to the concentration (in $\mu g\, g^{-1}$ – ppm) or required sample amount in ng. The sensitivity of an analytical procedure results from the slope of the calibration curve of the analyte measured.
Single-focusing mass spectrometer	Instrument in which an ion beam with a given value of mass to charge is brought to a focus although the initial direction of ions diverge.[4]
Spark ionization	Formation of ions from a solid material by an intermittent electrical discharge.[1]

Static field	Electric or magnetic field that does not change in time.[1]
Tandem mass spectrometer	Mass spectrometer designed for mass spectrometry/mass spectrometry.[1]
Thermal surface ionization	Formation of ions from neutral species (atoms or molecules) through contact with a high temperature surface.[1]
Time-of-flight mass spectrometer	Instrument that separates ions by m/z in a field-free region after acceleration to a fixed acceleration energy.[1]

[1] K. K. Murray *et al.* Standard Definition of Terms Relating to Mass Spectrometry (IUPAC Recommendations 2006) prepared for publication.

[2] International Standards Organization (1993) International Vocabulary of Basic and General Term in Metrology, 2nd edn. ISO, Geneva, ISBN 92-67-01075-1.

[3] Fleming, J., Neidhardt, B. Tausch, C. and Wegscheider, W. *Accred. Qual. Assur.* **2**, 51, 160, 348 (1997).

[4] J. F. J. Todd, *Pure Appl. Chem.* 63, 1541 (1991).

[5] I. Mills *et al.* Quantities, Units and Symbols in Physical Chemistry, second edition, Blackwell Science, in preparation.

Appendix IV

Abbreviations and Acronyms

A	mass number, nucleon number
AAS	atomic absorption spectrometry
AES/SAM	auger electron spectrometry/scanning auger microscopy
AFM	atomic force microscopy
AMS	accelerator mass spectrometry
amu	atomic mass unit (symbol: u)
$A_r(E)$	atomic weight of a chemical element is the average of all masses of all naturally occurring stable isotopes
α	degree of ionization: $\alpha =$ number of single charge atomic ions/number of uncharged atoms
B	magnetic field
BAM	Federal Institute of Materials Research and Testing, Germany (Bundesanstalt für Materialforschung und Prüfung, Berlin)
BCR	Community Bureau of Reference
c	speed of light in a vacuum with $2.99792458 \times 10^8\ \mathrm{m\,s^{-1}}$
CIAAW	Commission on Isotopic Abundances and Atomic Weights of the IUPAC
CE	capillary electrophoresis
CRM	certified reference material
CVD	chemical vapour deposition
CW	continuous wave
dc GDMS	direct current glow discharge mass spectrometry
DIHEN	direct injection high efficiency nebulizer
e	elementary charge: $e = 1.60217733(49) \times 10^{-19}\ \mathrm{C}$
E	electrical field
EDX	energy dispersed X-ray analysis
E_d, *D*	dissociation energy (D_0 from the ground state)
E_i	ionization energy

E_{ea}	electron affinity
EMPA	electron microprobe analysis
ESA	electrostatic analyzer
ESI	electrospray ionization
ETV	electrothermal vaporization
EXAFS	extended X-ray absorption fine structure
FTICR	Fourier transform ion cyclotron resonance
FTICR-MS	Fourier transform ion cyclotron resonance mass spectrometry
GC	gas chromatography
GDMS	glow discharge mass spectrometry
GD-OES	glow discharge optical emission spectrometry
GSMS	gas source mass spectrometry
χ	isotopic abundance is given as mole fraction of the corresponding atoms in %
HPLC	high performance liquid chromatography
I	ion current
IAEA	International Atomic Energy Agency
ICP-OES	inductively coupled plasma optical emission spectrometry
ICP-MS	inductively coupled plasma mass spectrometry
ICP-QMS	quadrupole based inductively coupled plasma mass spectrometry
ICP-SFMS	double-focusing sector field inductively coupled plasma mass spectrometry
IDA	isotope dilution analysis
IRM	isotopic reference material
IRMM	Institute for Reference Materials and Measurement, Belgium, formerly CBNM
IRMS	isotope ratio mass spectrometry
IUPAC	International Union of Pure and Applied Chemistry
LA-ICP-MS	laser ablation inductively coupled plasma mass spectrometry
LIBS	laser induced break down spectroscopy
LOD	limit of detection
LLRW	long-lived radioactive waste
MALDI	matrix assisted laser desorption/ionization
MC-ICP-MS	multiple ion collectors inductively coupled plasma mass spectrometry
MCN	microconcentric nebulizer
m_a, m	mass of an atom, atomic mass
m_e	electron rest mass: $m_e = 9.109\,389\,7(54) \times 10^{-31}$ kg
m_n	mass of neutron: $m_n = 1.674\,93 \times 10^{-27}$ kg
m_p	mass of proton: $m_p = 1.672\,62 \times 10^{-27}$ kg
m_r	molecular weight or relative molecular mass (ratio of the mass of a molecule to the unified atomic mass unit)
MS	mass spectrometry (or mass spectrometer)
MS/MS	tandem mass spectrometry
m_u	unified atomic mass unit: $1\,u = m_a(^{12}C)/12 \approx 1.660\,54 \times 10^{-27}$ kg
m/z	mass-to-charge ratio
n	number of charges of the ion (for a single charged ion M^+ is $n = 1$; for a double charged ion M^{2+} is $n = 2$ etc.)
N	neutron number: $N = A - Z$

N_A	Avogadro's constant: 6.02205×10^{23} mol^{-1}
NAA	neutron activation analysis
NEXAFS	near edge X-ray absorption fine structure spectrometry
NIST	National Institute of Standards and Technology, USA formely NBS
OES	optical emission spectrometry
PIXE	proton induced X-ray emission
RBS	Rutherford backscattering spectrometry
rf GDMS	radio frequency glow discharge mass spectrometry
REE	rare earth elements
RIMS	resonance ionization mass spectrometry
RSC	relative sensitivity coefficient
SDS-PAGE	sodium dodecyl sulfate polyacrylamide gel electrophoresis
SEC	size exclusion chromatography
SEM	scanning (reflection) electron microscopy
SHRIMP	sensitive high resolution ion probe mass spectrometry
SI	International System of Units
SIMS	secondary ion mass spectrometry
SNMS	secondary neutral mass spectrometry
SRM	standard reference material
STEM	scanning transmission (electron) microscopy
STM	scanning tunnelling (electron) microscopy
SXRF	synchrotron radiation X-ray fluorescence
T	temperature (in degrees Kelvin)
TEM	transmission electron microscopy
t	time
$t_{1/2}$	half-life of radionuclide
TIMS	thermal ionization mass spectrometry
TXRF	total reflection X-ray fluorescence analysis
ToF	time-of-flight analyzer
ToF-SIMS	time of flight secondary ion mass spectrometry
UHV	ultrahigh vacuum
USN	ultrasonic nebulizer
V	accelerating voltage
XANES	X-ray absorption near edge structure
XPS	X-ray induced photoelectron spectroscopy
XRF	X-ray fluorescence analysis
z	charge of an ion ($z = n \cdot e$)
Z	atomic number, proton number

Appendix V

List of Standard Reference Materials for Isotope Ratio Measurements

Element	Identification	Year of Issue	Chemical Form
Boron	IRMM 011	1969	H_3BO_3
	NIST SRM 951	1970	H_3BO_3
Boron-10	NIST SRM 952	1969	H_3BO_3
Bromine	NIST SRM 977	1964	NaBr
*Cadmium	IRMM 621–623	1997	Sol. in HNO_3
Chlorine	NIST SRM 975	1962	NaCl
*Chlorine	IRMM 641	2000	Aq. sol.
*Chlorine	IRMM 642	2000	Aq. sol.
Chromium	NIST SRM 979	1966	$Cr(NO_3)_3$
Chromium	IRMM 012	1999	Sol. in HCl
*Chromium-50	IRMM 624	1999	Sol. in HCl
*Chromium	IRMM 625	1999	Sol. in HCl
Copper	NIST SRM 994	1964	Cu
*Copper-65	IRMM 632	1999	Sol. in HNO_3
*Copper	IRMM 633	1999	Sol. in HNO_3
Gallium	NIST SRM 994	1986	Ga
Iron	IRMM 014	1992	Fe
*Iron-57	IRMM 620	1996	Sol. in HCL
*Iron	IRMM 634	1999	Sol. in HCL
Lead normal	NIST SRM 981	1968	Pb
Lead radiogen	NIST SRM 983	1968	Pb
*Lead-206	NIST SRM 991	1968	Pb
Lithium	IRMM 016	1984	Li_2CO_3
Lithium-6	IRMM 015	1986	Li_2CO_3
Magnesium	NIST SRM 980	1966	Mg
*Magnesium	IRMM 009	1999	Sol. in HNO_3
*Magnesium	IRMM 637	1999	Sol. in HNO_3
*Magnesium	IRMM 638	1999	Sol. in HNO_3

(Continued)

Element	Identification	Year of Issue	Chemical Form
*Mercury	IRMM 639	1999	Sol. in HCL
*Mercury	IRMM 640	2000	Sol. in HCL
Nickel	NIST SRM 986	1989	Ni
*Nitrogen-15 (nitrate)	IRMM 627	1998	Aq. Sol.
*Nitrogen-15 (nitrate)	IRMM 628	1998	Aq. Sol.
Platinum	IRMM 010	1999	Metal
*Platinum-194	IRMM 630a	1998	Sol. in HCl
*Platinum	IRMM 631	1998	Sol. in HCl
Plutonium	IRMM 290 A-G	1993	$Pu(NO_3)_4$
Potassium	NIST SRM 985	1975	KCl
Rhenium	NIST SRM 989	1973	Re
Rubidium	NIST SRM 984	1969	RbCl
Silicon	NIST SRM 990	1975	Si
Silicon	IRMM 017	1989	Si
Silicon	IRMM 018	1989	SiO_2
Silver	NIST SRM 978a	1962	$AgNO_3$
Strontium	NIST SRM 987	1982	$SrCO_3$
*Strontium	IRMM 635	1997	Sol. in HCL
Thallium	NIST SRM 997	1980	Th
Thorium	IRMM 035		Sol. in HCL
Thorium	IRMM 036		Sol. in HCL
Uranium	NIST SRM U002	1966	U_3O_8
Uranium	NIST SRM U970	1987	Sol. in HCL
Uranium	IRMM 183–187	1988/1993	UF

CRMs issued by NIST (previously NBS) and IRMM (previously CBNM) based on calibrated isotope ratio mass spectrometry. The * indicates the sample is a 'spike' (i.e. can be used for isotope dilution). Further information are presented in www.irmm.jrc.be and www.nist.gov.

P. D. P. Taylor, P. DeBievre and S. Valkiers, *Handbook of Stable Isotope Analytical Techniques*, Vol. 1, 909 (2004) P. A. Groot (ed.), Elsevier.

Index

Printed and bound by CPI Group (UK) Ltd, Croydon, CR0 4YY

16/09/2026

14963387-0007